D0874153

# Theory of Applied Robotics

Kinematics, Dynamics, and Control

# Theory of Applied Robotics
## Kinematics, Dynamics, and Control

**Reza N. Jazar**
*Department of Mechanical Engineering*
*Manhattan College*
*Riverdale, NY*

 Springer

Reza N. Jazar
Department of Mechanical Engineering
Manhattan College
Riverdale, NY 10471

Theory of Applied Robotics: Kinematics, Dynamics, and Control

Library of Congress Control Number: 2006939285

ISBN-10: 0-387-32475-5          e-ISBN-10: 0-387-68964-8
ISBN-13: 978-0-387-32475-3      e-ISBN-13: 978-0-387-68964-7

Printed on acid-free paper.

9 8 7 6 5 4 3 2 1

springer.com

Dedicated to my wife,
*Mojgan*
and our children,
*Vazan*
and
*Kavosh*
and to my supportive father.

Everything depends on
"What you like to do, what you have to do, and what you do."
Happiness means "what you do is what you like to do."
Success means "what you do is what you have to do."

# Preface

This book is designed to serve as a text for engineering students. It introduces the fundamental knowledge used in robotics. This knowledge can be utilized to develop computer programs for analyzing the kinematics, dynamics, and control of robotic systems.

The subject of robotics may appear overdosed by the number of available texts because the field has been growing rapidly since 1970. However, the topic remains alive with modern developments, which are closely related to the classical material. It is evident that no single text can cover the vast scope of classical and modern materials in robotics. Thus the demand for new books arises because the field continues to progress. Another factor is the trend toward analytical unification of kinematics, dynamics, and control.

Classical kinematics and dynamics of robots has its roots in the work of great scientists of the past four centuries who established the methodology and understanding of the behavior of dynamic systems. The development of dynamic science, since the beginning of the twentieth century, has moved toward analysis of controllable man-made systems. Therefore, merging the kinematics and dynamics with control theory is the expected development for robotic analysis.

The other important development is the fast growing capability of accurate and rapid numerical calculations, along with intelligent computer programming.

### Level of the Book

This book has evolved from nearly a decade of research in nonlinear dynamic systems, and teaching undergraduate-graduate level courses in robotics. It is addressed primarily to the last year of undergraduate study and the first year graduate student in engineering. Hence, it is an intermediate textbook. This book can even be the first exposure to topics in spatial kinematics and dynamics of mechanical systems. Therefore, it provides both fundamental and advanced topics on the kinematics and dynamics of robots. The whole book can be covered in two successive courses however, it is possible to jump over some sections and cover the book in one course. The students are required to know the fundamentals of kinematics and dynamics, as well as a basic knowledge of numerical methods.

The contents of the book have been kept at a fairly theoretical-practical level. Many concepts are deeply explained and their use emphasized, and most of the related theory and formal proofs have been explained. Throughout the book, a strong emphasis is put on the physical meaning of the concepts introduced. Topics that have been selected are of high interest in the field. An attempt has been made to expose the students to a broad range of topics and approaches.

### Organization of the Book

The text is organized so it can be used for teaching or for self-study. Chapter 1 "Introduction," contains general preliminaries with a brief review of the historical development and classification of robots.

Part I "Kinematics," presents the forward and inverse kinematics of robots. Kinematics analysis refers to position, velocity, and acceleration analysis of robots in both joint and base coordinate spaces. It establishes kinematic relations among the end-effecter and the joint variables. The method of Denavit-Hartenberg for representing body coordinate frames is introduced and utilized for forward kinematics analysis. The concept of modular treatment of robots is well covered to show how we may combine simple links to make the forward kinematics of a complex robot. For inverse kinematics analysis, the idea of decoupling, the inverse matrix method, and the iterative technique are introduced. It is shown that the presence of a spherical wrist is what we need to apply analytic methods in inverse kinematics.

Part II "Dynamics," presents a detailed discussion of robot dynamics. An attempt is made to review the basic approaches and demonstrate how these can be adapted for the active displacement framework utilized for robot kinematics in the earlier chapters. The concepts of the recursive Newton-Euler dynamics, Lagrangian function, manipulator inertia matrix, and generalized forces are introduced and applied for derivation of dynamic equations of motion.

Part III "Control," presents the floating time technique for time-optimal control of robots. The outcome of the technique is applied for an open-loop control algorithm. Then, a computed-torque method is introduced, in which a combination of feedforward and feedback signals are utilized to render the system error dynamics.

### Method of Presentation

The structure of presentation is in a "*fact-reason-application*" fashion. The "fact" is the main subject we introduce in each section. Then the reason is given as a "proof." Finally the application of the fact is examined in some "examples." The "examples" are a very important part of the book because they show how to implement the knowledge introduced in "facts." They also cover some other facts that are needed to expand the subject.

### Prerequisites

Since the book is written for senior undergraduate and first-year graduate level students of engineering, the assumption is that users are familiar with matrix algebra as well as basic feedback control. Prerequisites for readers of this book consist of the fundamentals of kinematics, dynamics, vector analysis, and matrix theory. These basics are usually taught in the first three undergraduate years.

### Unit System

The system of units adopted in this book is, unless otherwise stated, the international system of units (SI). The units of degree (deg) or radian (rad) are utilized for variables representing angular quantities.

### Symbols

- Lowercase bold letters indicate a vector. Vectors may be expressed in an $n$ dimensional Euclidian space. Example:

$$\mathbf{r} \quad , \quad \mathbf{s} \quad , \quad \mathbf{d} \quad , \quad \mathbf{a} \quad , \quad \mathbf{b} \quad , \quad \mathbf{c}$$
$$\mathbf{p} \quad , \quad \mathbf{q} \quad , \quad \mathbf{v} \quad , \quad \mathbf{w} \quad , \quad \mathbf{y} \quad , \quad \mathbf{z}$$
$$\boldsymbol{\omega} \quad , \quad \boldsymbol{\alpha} \quad , \quad \boldsymbol{\epsilon} \quad , \quad \boldsymbol{\theta} \quad , \quad \boldsymbol{\delta} \quad , \quad \boldsymbol{\phi}$$

- Uppercase bold letters indicate a dynamic vector or a dynamic matrix, such as and Jacobian. Example:

$$\mathbf{F} \quad , \quad \mathbf{M} \quad , \quad \mathbf{J}$$

- Lowercase letters with a hat indicate a unit vector. Unit vectors are not bolded. Example:

$$\hat{\imath} \quad , \quad \hat{\jmath} \quad , \quad \hat{k} \quad , \quad \hat{e} \quad , \quad \hat{u} \quad , \quad \hat{n}$$
$$\hat{I} \quad , \quad \hat{J} \quad , \quad \hat{K} \quad , \quad \hat{e}_\theta \quad , \quad \hat{e}_\varphi \quad , \quad \hat{e}_\psi$$

- Lowercase letters with a tilde indicate a $3 \times 3$ skew symmetric matrix associated to a vector. Example:

$$\tilde{a} = \begin{bmatrix} 0 & -a_3 & a_2 \\ a_3 & 0 & -a_1 \\ -a_2 & a_1 & 0 \end{bmatrix} \quad , \quad \mathbf{a} = \begin{bmatrix} a_1 \\ a_2 \\ a_3 \end{bmatrix}$$

- An arrow above two uppercase letters indicates the start and end points of a position vector. Example:

$$\overrightarrow{ON} = \text{a position vector from point } O \text{ to point } N$$

- A double arrow above a lowercase letter indicates a $4 \times 4$ matrix associated to a quaternion. Example:

$$\overleftrightarrow{q} = \begin{bmatrix} q_0 & -q_1 & -q_2 & -q_3 \\ q_1 & q_0 & -q_3 & q_2 \\ q_2 & q_3 & q_0 & -q_1 \\ q_3 & -q_2 & q_1 & q_0 \end{bmatrix}$$

$$q = q_0 + q_1 i + q_2 j + q_3 k$$

- The length of a vector is indicated by a non-bold lowercase letter. Example:

$$r = |\mathbf{r}| \quad , \quad a = |\mathbf{a}| \quad , \quad b = |\mathbf{b}| \quad , \quad s = |\mathbf{s}|$$

- Capital letters $A$, $Q$, $R$, and $T$ indicate rotation or transformation matrices. Example:

$$Q_{Z,\alpha} = \begin{bmatrix} \cos\alpha & -\sin\alpha & 0 \\ \sin\alpha & \cos\alpha & 0 \\ 0 & 0 & 1 \end{bmatrix} \quad , \quad {}^{G}T_B = \begin{bmatrix} c\alpha & 0 & -s\alpha & -1 \\ 0 & 1 & 0 & 0.5 \\ s\alpha & 0 & c\alpha & 0.2 \\ 0 & 0 & 0 & 1 \end{bmatrix}$$

- Capital letter $B$ is utilized to denote a body coordinate frame. Example:

$$B(oxyz) \quad , \quad B(Oxyz) \quad , \quad B_1(o_1 x_1 y_1 z_1)$$

- Capital letter $G$ is utilized to denote a global, inertial, or fixed coordinate frame. Example:

$$G \quad , \quad G(XYZ) \quad , \quad G(OXYZ)$$

- Right subscript on a transformation matrix indicates the *departure* frames. Example:

$$T_B = \text{transformation matrix from frame } B(oxyz)$$

- Left superscript on a transformation matrix indicates the *destination* frame. Example:

$${}^{G}T_B = \text{transformation matrix from frame } B(oxyz)$$
$$\text{to frame } G(OXYZ)$$

- Whenever there is no sub or superscript, the matrices are shown in a bracket. Example:

$$[T] = \begin{bmatrix} c\alpha & 0 & -s\alpha & -1 \\ 0 & 1 & 0 & 0.5 \\ s\alpha & 0 & c\alpha & 0.2 \\ 0 & 0 & 0 & 1 \end{bmatrix}$$

- Left superscript on a vector denotes the frame in which the vector is expressed. That superscript indicates the frame that the vector belongs to; so the vector is expressed using the unit vectors of that frame. Example:

$$^G\mathbf{r} = \text{position vector expressed in frame } G(OXYZ)$$

- Right subscript on a vector denotes the tip point that the vector is referred to. Example:

$$^G\mathbf{r}_P = \text{position vector of point } P$$
$$\text{expressed in coordinate frame } G(OXYZ)$$

- Right subscript on an angular velocity vector indicates the frame that the angular vector is referred to. Example:

$$\boldsymbol{\omega}_B = \text{angular velocity of the body coordinate frame } B(oxyz)$$

- Left subscript on an angular velocity vector indicates the frame that the angular vector is measured with respect to. Example:

$$_G\boldsymbol{\omega}_B = \text{angular velocity of the body coordinate frame } B(oxyz)$$
$$\text{with respect to the global coordinate frame } G(OXYZ)$$

- Left superscript on an angular velocity vector denotes the frame in which the angular velocity is expressed. Example:

$$^{B_2}_G\boldsymbol{\omega}_{B_1} = \text{angular velocity of the body coordinate frame } B_1$$
$$\text{with respect to the global coordinate frame } G,$$
$$\text{and expressed in body coordinate frame } B_2$$

Whenever the subscript and superscript of an angular velocity are the same, we usually drop the left superscript. Example:

$$_G\boldsymbol{\omega}_B \equiv {}^G_G\boldsymbol{\omega}_B$$

Also for position, velocity, and acceleration vectors, we drop the left subscripts if it is the same as the left superscript. Example:

$$^B_B\mathbf{v}_P \equiv {}^B\mathbf{v}_P$$

- If the right subscript on a force vector is a number, it indicates the number of coordinate frames in a serial robot. Coordinate frame $B_i$ is set up at joint $i + 1$. Example:

$$\mathbf{F}_i = \text{force vector at joint } i + 1 \text{ measured at the origin of } B_i(oxyz)$$

At joint $i$ there is always an action force $\mathbf{F}_i$, that link $(i)$ applies on link $(i+1)$, and a reaction force $-\mathbf{F}_i$, that link $(i+1)$ applies on link $(i)$. On link $(i)$ there is always an action force $\mathbf{F}_{i-1}$ coming from link $(i-1)$, and a reaction force $-\mathbf{F}_i$ coming from link $(i+1)$. Action force is called *driving force*, and reaction force is called *driven force*.

- If the right subscript on a moment vector is a number, it indicates the number of coordinate frames in a serial robot. Coordinate frame $B_i$ is set up at joint $i+1$. Example:

$\mathbf{M}_i =$ moment vector at joint $i+1$ measured at the origin of $B_i(oxyz)$

At joint $i$ there is always an action moment $\mathbf{M}_i$, that link $(i)$ applies on link $(i+1)$, and a reaction moment $-\mathbf{M}_i$, that link $(i+1)$ applies on link $(i)$. On link $(i)$ there is always an action moment $\mathbf{M}_{i-1}$ coming from link $(i-1)$, and a reaction moment $-\mathbf{M}_i$ coming from link $(i+1)$. Action moment is called *driving moment*, and reaction moment is called *driven moment*.

- Left superscript on derivative operators indicates the frame in which the derivative of a variable is taken. Example:

$$\frac{^Gd}{dt}x \quad , \quad \frac{^Gd}{dt}{}^B\mathbf{r}_P \quad , \quad \frac{^Bd}{dt}{}^G_B\mathbf{r}_P$$

If the variable is a vector function, and also the frame in which the vector is defined is the same as the frame in which a time derivative is taken, we may use the following short notation,

$$\frac{^Gd}{dt}{}^G\mathbf{r}_P = {}^G\dot{\mathbf{r}}_P \quad , \quad \frac{^Bd}{dt}{}^B_o\mathbf{r}_P = {}^B_o\dot{\mathbf{r}}_P$$

and write equations simpler. Example:

$$^G\mathbf{v} = \frac{^Gd}{dt}{}^G\mathbf{r}(t) = {}^G\dot{\mathbf{r}}$$

- If followed by angles, lowercase $c$ and $s$ denote $cos$ and $sin$ functions in mathematical equations. Example:

$$c\alpha = \cos\alpha \quad , \quad s\varphi = \sin\varphi$$

- Capital bold letter $\mathbf{I}$ indicates a unit matrix, which, depending on the dimension of the matrix equation, could be a $3 \times 3$ or a $4 \times 4$ unit matrix. $\mathbf{I}_3$ or $\mathbf{I}_4$ are also being used to clarify the dimension of $\mathbf{I}$. Example:

$$\mathbf{I} = \mathbf{I}_3 = \begin{bmatrix} 1 & 0 & 0 \\ 0 & 1 & 0 \\ 0 & 0 & 1 \end{bmatrix}$$

- An asterisk ★ indicates a more advanced subject or example that is not designed for undergraduate teaching and can be dropped in the first reading.

- Two parallel joint axes are indicated by a parallel sign, ($\parallel$).

- Two orthogonal joint axes are indicated by an orthogonal sign, ($\vdash$). Two orthogonal joint axes are intersecting at a right angle.

- Two perpendicular joint axes are indicated by a perpendicular sign, ($\perp$). Two perpendicular joint axes are at a right angle with respect to their common normal.

# Contents

# II  Dynamics                                                     417

# III  Control                                                     565

# 1

# Introduction

Law Zero: A robot may not injure humanity, or, through inaction, allow humanity to come to harm.

Law One: A robot may not injure a human being, or, through inaction, allow a human being to come to harm, unless this would violate a higher order law.

Law Two: A robot must obey orders given it by human beings, except where such orders would conflict with a higher order law.

Law Three: A robot must protect its own existence as long as such protection does not conflict with a higher order law.

Isaac Asimov proposed these four refined laws of "robotics" to protect us from intelligent generations of robots. Although we are not too far from that time when we really do need to apply Asimov's rules, there is no immediate need however, it is good to have a plan.

The term *robotics* refers to the study and use of *robots*. The term was first adopted by Asimov in 1941 through his short story, Runaround.

Based on the Robotics Institute of America (RIA) definition: "A robot is a reprogrammable multifunctional manipulator designed to move material, parts, tools, or specialized devices through variable programmed motions for the performance of a variety of tasks."

From the engineering point of view, robots are complex, versatile devices that contain a mechanical structure, a sensory system, and an automatic control system. Theoretical fundamentals of robotics rely on the results of research in mechanics, electric, electronics, automatic control, mathematics, and computer sciences.

## 1.1 Historical Development

The first position controlling apparatus was invented around 1938 for spray painting. However, the first industrial modern robots were the Unimates, made by J. Engelberger in the early 60s. Unimation was the first to market robots. Therefore, Engelberger has been called the father of robotics. In the 80s the robot industry grew very fast primarily because of the huge investments by the automotive industry.

In the research community the first automata were probably Grey Walter's machina (1940s) and the John's Hopkins beast. The first programmable robot was designed by George Devol in 1954. Devol funded Unimation.

The theory of Denavit and Hartenberg developed in 1955 and unified the forward kinematics of robotic manipulators. In 1959 the first commercially available robot appeared on the market. Robotic manipulators were used in industries after 1960, and saw sky rocketing growth in the 80s.

Robots appeared as a result of combination two technologies: teleoperators, and computer numerical control (CNC) of milling machines. Teleoperators were developed during World War II to handle radioactive materials, and CNC was developed to increase the precision required in machining of new technologic parts. Therefore, the first robots were nothing but numerical control of mechanical linkages that were basically designed to transfer material from point A to B.

Today, more complicated applications, such as welding, painting, and assembling, require much more motion capability and sensing. Hence, a robot is a multi-disciplinary engineering device. Mechanical engineering deals with the design of mechanical components, arms, end-effectors, and also is responsible for kinematics, dynamics and control analyses of robots. Electrical engineering works on robot actuators, sensors, power, and control systems. System design engineering deals with perception, sensing, and control methods of robots. Programming, or software engineering, is responsible for logic, intelligence, communication, and networking.

Today we have more than 1000 robotics-related organizations, associations, and clubs; more than 500 robotics-related magazines, journals, and newsletters; more than 100 robotics-related conferences, and competitions each year; and more than 50 robotics-related courses in colleges. Robots find a vast amount industrial applications and are used for various technological operations. Robots enhance labor productivity in industry and deliver relief from tiresome, monotonous, or hazardous works. Moreover, robots perform many operations better than people do, and they provide higher accuracy and repeatability. In many fields, high technological standards are hardly attainable without robots. Apart from industry, robots are used in extreme environments. They can work at low and high temperatures; they don't even need lights, rest, fresh air, a salary, or promotions. Robots are prospective machines whose application area is widening.

It is claimed that robots appeared to perform in 4A for 4D, or 3D3H environments. 4A performances are automation, augmentation, assistance, and autonomous; and 4D environments are dangerous, dirty, dull, and difficult. 3D3H means dull, dirty, dangerous, hot, heavy, and hazardous.

## 1.2   Components and Mechanisms of a Robotic System

Robotic manipulators are kinematically composed of links connected by joints to form a kinematic chain. However, a robot as a system, consists

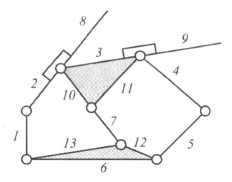

FIGURE 1.1. A two-loop planar linkage with 7 links and 8 revolute joints.

of a *manipulator* or *rover*, a *wrist*, an *end-effector*, *actuators*, *sensors*, *controllers*, *processors*, and *software*.

## 1.2.1 Link

The individual rigid bodies that make up a robot are called *links*. In robotics we sometimes use *arm* to mean link. A robot arm or a robot link is a rigid member that may have relative motion with respect to all other links. From the kinematic point of view, two or more members connected together such that no relative motion can occur among them are considered a single link.

**Example 1** *Number of links.*

*Figure 1.1 shows a mechanism with 7 links. There can not be any relative motion among bars 3, 10, and 11. Hence, they are counted as one link, say link 3. Bars 6, 12, and 13 have the same situation and are counted as one link, say link 6. Bars 2 and 8 are rigidly attached, making one link only, say link 2. Bars 3 and 9 have the same relationship as bars 2 and 8, and they are also one link, say link 3.*

## 1.2.2 Joint

Two links are connected by contact at a *joint* where their relative motion can be expressed by a single coordinate. Joints are typically *revolute* (rotary) or *prismatic* (linear). Figure 1.2 depicts the geometric form of a revolute and a prismatic joint. A *revolute joint* (R), is like a hinge and allows relative rotation between two links. A *prismatic joint* (P), allows a translation of relative motion between two links.

Relative rotation of connected links by a revolute joint occurs about a line called *axis of joint*. Also, translation of two connected links by a prismatic joint occurs along a line also called *axis of joint*. The value of

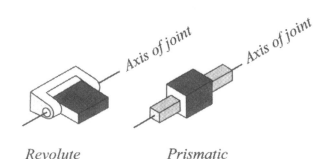

FIGURE 1.2. Illustration of revolute and prismatic joints.

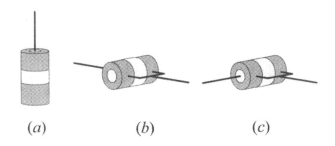

FIGURE 1.3. Symbolic illustration of revolute joints in robotic modeles.

the single coordinate describing the relative position of two connected links at a joint is called *joint coordinate* or *joint variable*. It is an *angle* for a revolute joint, and a *distance* for a prismatic joint.

A symbolic illustration of revolute and prismatic joints in robotics are shown in Figure 1.3 (a)-(c), and 1.4 (a)-(c) respectively.

The coordinate of an *active joint* is controlled by an actuator. A *passive joint* does not have any actuators and its coordinate is a function of the coordinates of active joints and the geometry of the robot arms. Passive joints are also called *inactive* or *free joints*.

Active joints are usually prismatic or revolute, however, passive joints may be any of the *lower pair joints* that provide surface contact. There are six different lower pair joints: *revolute, prismatic, cylindrical, screw, spherical,* and *planar*.

Revolute and prismatic joints are the most common joints that are utilized in serial robotic manipulators. The other joint types are merely implementations to achieve the same function or provide additional degrees of freedom. Prismatic and revolute joints provide one degree of freedom. Therefore, the number of joints of a manipulator is the *degrees-of-freedom* (DOF) of the manipulator. Typically the manipulator should possess at least six DOF: three for positioning and three for orientation. A manipula-

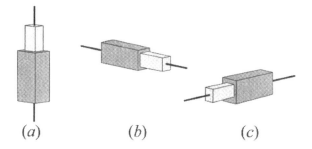

$(a)$    $(b)$    $(c)$

FIGURE 1.4. Symbolic illustration of prismatic joints in robotic models.

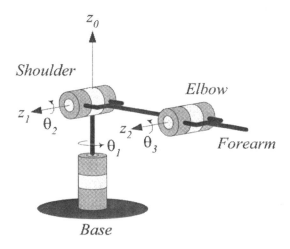

FIGURE 1.5. Illustation of a 3R manipulator.

tor having more than six DOF is referred to as a kinematically *redundant* manipulator.

### 1.2.3 Manipulator

The main body of a robot consisting of the links, joints, and other structural elements, is called the *manipulator*. A manipulator becomes a robot when the wrist and gripper are attached, and the control system is implemented. However, in literature robots and manipulators are utilized equivalently and both refer to robots. Figure 1.5 schematically illustrates a 3R manipulator.

### 1.2.4 Wrist

The joints in the kinematic chain of a robot between the forebeam and end-effector are referred to as the *wrist*. It is common to design manipulators

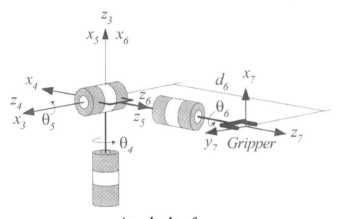

*Attached to forearm*

FIGURE 1.6. Illustration of a spherical wrist kinematics.

with spherical wrists, by which it means three revolute joint axes intersect at a common point called the *wrist point*. Figure 1.6 shows a schematic illustration of a spherical wrist, which is a R⊢R⊢R mechanism.

The spherical wrist greatly simplifies the kinematic analysis effectively, allowing us to decouple the positioning and orienting of the end effector. Therefore, the manipulator will possess three degrees-of-freedom for position, which are produced by three joints in the arm. The number of DOF for orientation will then depend on the wrist. We may design a wrist having one, two, or three DOF depending on the application.

## 1.2.5  End-effector

The *end-effector* is the part mounted on the last link to do the required job of the robot. The simplest end-effector is a gripper, which is usually capable of only two actions: opening and closing. The arm and wrist assemblies of a robot are used primarily for positioning the end-effector and any tool it may carry. It is the end-effector or tool that actually performs the work. A great deal of research is devoted to the design of special purpose end-effectors and tools. There is also extensive research on the development of anthropomorphic hands. Such hands have been developed for prosthetic use in manufacturing. Hence, a robot is composed of a manipulator or *mainframe* and a wrist plus a tool. The wrist and end-effector assembly is also called a *hand*.

## 1.2.6  Actuators

*Actuators* are drivers acting as the muscles of robots to change their configuration. The actuators provide power to act on the mechanical structure against gravity, inertia, and other external forces to modify the geometric location of the robot's hand. The actuators can be of electric, hydraulic, or pneumatic type and have to be controllable.

## 1.2.7  Sensors

The elements used for detecting and collecting information about internal and environmental states are *sensors*. According to the scope of this book, joint position, velocity, acceleration, and force are the most important information to be sensed. Sensors, integrated into the robot, send information about each link and joint to the control unit, and the control unit determines the configuration of the robot.

## 1.2.8  Controller

The *controller* or *control unit* has three roles.

1-*Information role*, which consists of collecting and processing the information provided by the robot's sensors.

2-*Decision role*, which consists of planning the geometric motion of the robot structure.

3-*Communication role*, which consists of organizing the information between the robot and its environment. The control unit includes the processor and software.

# 1.3   Robot Classifications

The Robotics Institute of America (RIA) considers classes 3-6 of the following classification to be robots, and the Association Francaise de Robotique (AFR) combines classes 2, 3, and 4 as the same type and divides robots in 4 types. However, the Japanese Industrial Robot Association divides robots in 6 different classes:

Class 1: *Manual handling devices*: A device with multi degrees of freedom that is actuated by an operator.

Class 2: *Fixed sequence robot*: A device that performs the successive stages of a task according to a predetermined and fixed program.

Class 3: *Variable sequence robot*: A device that performs the successive stages of a task according to a predetermined but programmable method.

Class 4: *Playback robot*: A human operator performs the task manually by leading the robot, which records the motions for later playback. The robot repeats the same motions according to the recorded information.

Class 5: *Numerical control robot*: The operator supplies the robot with a motion program rather than teaching it the task manually.

Class 6: *Intelligent robot*: A robot with the ability to understand its environment and the ability to successfully complete a task despite changes in the surrounding conditions under which it is to be performed.

Other than these official classifications, robots can be classified by other criteria such as geometry, workspace, actuation, control, and application.

## 1.3.1  Geometry

A robot is called a *serial* or *open-loop* manipulator if its kinematic structure does not make a loop chain. It is called a *parallel* or *closed-loop* manipulator if its structure makes a loop chain. A robot is a *hybrid* manipulator if its structure consists of both open and closed-loop chains.

As a mechanical system, we may think of a robot as a set of rigid bodies connected together at some joints. The joints can be either *revolute* (R) or *prismatic* (P), because any other kind of joint can be modeled as a combination of these two simple joints.

Most industrial manipulators have six DOF. The open-loop manipulators can be classified based on their first three joints starting from the grounded joint. From the two types of joints there are mathematically 72 different manipulator configurations, simply because each joint can be P or R, and the axes of two adjacent joints can be *parallel* ($\parallel$), *orthogonal* ($\vdash$), or *perpendicular* ($\perp$). Two orthogonal joint axes intersect at a right angle, however two perpendicular joint axes are in right-angle with respect to their common normal. Two perpendicular joint axes become parallel if one axis turns 90 deg about the common normal. Two perpendicular joint axes become orthogonal if the length of their common normal tends to zero.

Out of the 72 possible manipulators, the important ones are: $R\parallel R\parallel P$ (SCARA), $R\vdash R\perp R$ (articulated), $R\vdash R\perp P$ (spherical), $R\parallel P\vdash P$ (cylindrical), and $P\vdash P\vdash P$ (Cartesian).

1. $R\parallel R\parallel P$

   The SCARA arm (Selective Compliant Articulated Robot for Assembly) shown in Figure 1.7 is a popular manipulator, which, as its name suggests, is made for assembly operations.

2. $R\vdash R\perp R$

   The $R\vdash R\perp R$ configuration, illustrated in Figure 1.5, is called *elbow*, *revolute*, *articulated*, or *anthropomorphic*. It is a suitable configuration for industrial robots. Almost 25% of industrial robots, PUMA for instance, are made of this kind. Because of its importance, a better illustration of an articulated robot is shown in Figure 1.8 to indicate the name of different components.

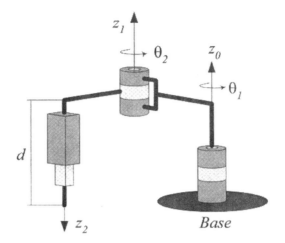

FIGURE 1.7. Illustration of an R‖R‖P manipulator.

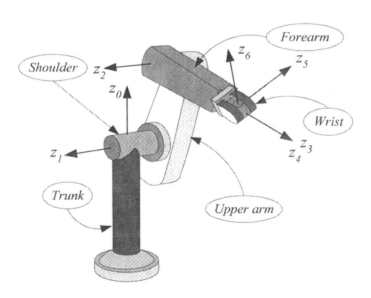

FIGURE 1.8. Structure and terminology of a R⊢R⊥R elbow manipulator equipped with a spherical wrist.

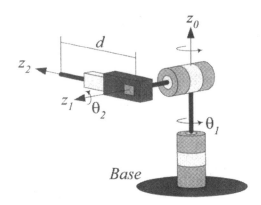

FIGURE 1.9. The R⊢R⊥P spherical configuration of robotic manipulators.

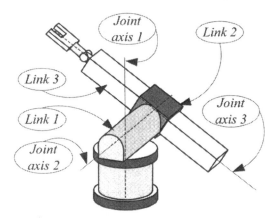

FIGURE 1.10. Illustration of Stanford arm; an R⊢R⊥P spherical manipulator.

3. R⊢R⊥P

The spherical configuration is a suitable configuration for small robots. Almost 15% of industrial robots, Stanford arm for instance, are made of this configuration. The R⊢R⊥P configuration is illustrated in Figure 1.9.

By replacing the third joint of an articulate manipulator with a prismatic joint, we obtain the spherical manipulator. The term spherical manipulator derives from the fact that the spherical coordinates define the position of the end-effector with respect to its base frame. Figure 1.10 schematically illustrates the Stanford arm, one of the most well-known spherical robots.

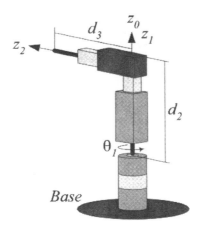

FIGURE 1.11. The R‖P⊥P configuration of robotic manipulators.

4. R‖P⊢P

The cylindrical configuration is a suitable configuration for medium load capacity robots. Almost 45% of industrial robots are made of this kind. The R‖P⊢P configuration is illustrated in Figure 1.11. The first joint of a cylindrical manipulator is revolute and produces a rotation about the base, while the second and third joints are prismatic. As the name suggests, the joint variables are the cylindrical coordinates of the end-effector with respect to the base.

5. P⊢P⊢P

The Cartesian configuration is a suitable configuration for heavy load capacity and large robots. Almost 15% of industrial robots are made of this configuration. The P⊢P⊢P configuration is illustrated in Figure 1.12.

For a Cartesian manipulator, the joint variables are the Cartesian coordinates of the end-effector with respect to the base. As might be expected, the kinematic description of this manipulator is the simplest of all manipulators. Cartesian manipulators are useful for table-top assembly applications and, as gantry robots, for transfer of cargo.

## 1.3.2 Workspace

The *workspace* of a manipulator is the total volume of space the end-effector can reach. The workspace is constrained by the geometry of the manipulator as well as the mechanical constraints on the joints. The workspace is broken into a *reachable* workspace and a *dexterous* workspace. The reachable workspace is the volume of space within which every point is reachable

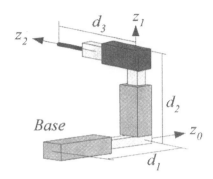

FIGURE 1.12. The P⊢P⊢P Cartesian configuration of robotic manipulators.

by the end-effector in at least one orientation. The dexterous workspace is the volume of space within which every point can be reached by the end-effector in all possible orientations. The dexterous workspace is a subset of the reachable workspace.

Most of the open-loop chain manipulators are designed with a wrist subassembly attached to the main three links assembly. Therefore, the first three links are long and are utilized for positioning while the wrist is utilized for control and orientation of the end-effector. This is why the subassembly made by the first three links is called the *arm*, and the subassembly made by the other links is called the *wrist*.

## 1.3.3  Actuation

Actuators translate power into motion. Robots are typically actuated electrically, hydraulically, or pneumatically. Other types of actuation might be considered as piezoelectric, magnetostriction, shape memory alloy, and polymeric.

Electrically actuated robots are powered by AC or DC motors and are considered the most acceptable robots. They are cleaner, quieter, and more precise compared to the hydraulic and pneumatic actuated. Electric motors are efficient at high speeds so a high ratio gearbox is needed to reduce the high RPM. Non-backdriveability and self-braking is an advantage of high ratio gearboxes in case of power loss. However, when high speed or high load-carrying capabilities are needed, electric drivers are unable to compete with hydraulic drivers.

Hydraulic actuators are satisfactory because of high speed and high torque/mass or power/mass ratios. Therefore, hydraulic driven robots are used primarily for lifting heavy loads. Negative aspects of hydraulics, besides their noisiness and tendency to leak, include a necessary pump and other hardware.

Pneumatic actuated robots are inexpensive and simple but cannot be controlled precisely. Besides the lower precise motion, they have almost the same advantages and disadvantages as hydraulic actuated robots.

## 1.3.4 Control

Robots can be classified by control method into *servo* (closed loop control) and *non-servo* (open loop control) robots. Servo robots use closed-loop computer control to determine their motion and are thus capable of being truly multifunctional reprogrammable devices. Servo controlled robots are further classified according to the method that the controller uses to guide the end-effector.

The simplest type of a servo robot is the *point-to-point* robot. A point-to-point robot can be taught a discrete set of points, called *control points*, but there is no control on the path of the end-effector in between the points. On the other hand, in *continuous path* robots, the entire path of the end-effector can be controlled. For example, the robot end-effector can be taught to follow a straight line between two points or even to follow a contour such as a welding seam. In addition, the velocity and/or acceleration of the end-effector can often be controlled. These are the most advanced robots and require the most sophisticated computer controllers and software development.

Non-servo robots are essentially open-loop devices whose movement is limited to predetermined mechanical stops, and they are primarily used for materials transfer.

## 1.3.5 Application

Regardless of size, robots can mainly be classified according to their application into *assembly* and *non-assembly* robots. However, in the industry they are classified by the category of application such as *machine loading, pick and place, welding, painting, assembling, inspecting, sampling, manufacturing, biomedical, assisting, remote controlled mobile,* and *telerobot*.

According to design characteristics, most industrial robot arms are anthropomorphic, in the sense that they have a "shoulder," (first two joints) an "elbow," (third joint) and a "wrist" (last three joints). Therefore, in total, they usually have six degrees of freedom needed to put an object in any position and orientation.

Most commercial serial manipulators have only revolute joints. Compared to prismatic joints, revolute joints cost less and provide a larger dextrous workspace for the same robot volume. Serial robots are very heavy, compared to the maximum load they can move without loosing their accuracy. Their useful load-to-weight ratio is less than 1/10. The robots are so heavy because the links must be stiff in order to work rigidly. Simplicity of the forward and inverse position and velocity kinematics has always been one of the major design criteria for industrial manipulators. Hence, almost all of them have a special kinematic structure.

## 1.4 Introduction to Robot's Kinematics, Dynamics, and Control

The forward kinematics problem is when the kinematical data are known for the joint coordinates and are utilized to find the data in the base coordinate frame. The inverse kinematics problem is when the kinematics data are known for the end-effecter in Cartesian space. Inverse kinematics is highly nonlinear and usually a much more difficult problem than the forward kinematics problem. The inverse velocity and acceleration problems are linear, and much simpler, once the inverse position problem has been solved. An inverse position solution is said to have a closed form if it is not iterative.

*Kinematics*, which is the English version of the French word *cinématique* from the Greek κίνημα (movement), is a branch of science that analyzes motion with no attention to what causes the motion. By *motion* we mean any type of displacement, which includes changes in position and orientation. Therefore, *displacement*, and the successive derivatives with respect to time, velocity, acceleration, and jerk, all combine into kinematics.

*Positioning* is to bring the end-effector to an arbitrary point within dextrose, while *orientation* is to move the end-effector to the required orientation at the position. The positioning is the job of the arm, and orientation is the job of the wrist. To simplify the kinematic analysis, we may decouple the positioning and orientation of the end-effector.

In terms of the kinematic formation, a 6 DOF robot comprises six sequential moveable links and six joints with at least the last two links having zero length.

Generally speaking, almost all problems of kinematics can be interpreted as a vector addition. However, every vector in a vectorial equation must be transformed and expressed in a common reference frame.

*Dynamics* is the study of systems that undergo changes of state as time evolves. In mechanical systems such as robots, the change of states involves motion. Derivation of the equations of motion for the system is the main step in dynamic analysis of the system, since equations of motion are essential in the design, analysis, and control of the system.

The dynamic equations of motion describe dynamic behavior. They can be used for computer simulation of the robot's motion, design of suitable control equations, and evaluation of the dynamic performance of the design.

Similar to kinematics, the problem of robot dynamics may be considered as *direct* and *inverse dynamics* problems. In direct dynamics, we should predict the motion of the robot for a given set of initial conditions and torques at active joints. In the inverse dynamics problem, we should compute the forces and torques necessary to generate the prescribed trajectory for a given set of positions, velocities, and accelerations.

The robot control problem may be characterized as the desired motion of the end-effector. Such a desired motion is specified as a trajectory in Cartesian coordinates while the control system requires input in joint coordinates.

Sensors generate data to find the actual state of the robot at joint space. This implies a requirement for expressing the kinematic variables in Cartesian space to be transformed into their equivalent joint coordinate space. These transformations are highly dependent on the kinematic geometry of the manipulator. Hence, the robot control comprises three computational problems:

1- Determination of the trajectory in Cartesian coordinate space,

2- Transformation of the Cartesian trajectory into equivalent joint coordinate space, and

3- Generation of the motor torque commands to realize the trajectory.

### 1.4.1   Triad

Take any four non-coplanar points $O$, $A$, $B$, $C$. The *triad OABC* is defined as consisting of the three lines $OA$, $OB$, $OC$ forming a rigid body. The position of $A$ on $OA$ is immaterial provided it is maintained on the same side of $O$, and similarly $B$ and $C$. Rotate $OB$ about $O$ in the plane $OAB$ so that the angle $AOB$ becomes 90 deg, the direction of rotation of $OB$ being such that $OB$ moves through an angle less than 90 deg. Next, rotate $OC$ about the line in $AOB$ to which it is perpendicular, until it becomes perpendicular to the plane $AOB$, in such a way that $OC$ moves through an angle less than 90 deg. Calling now the new position of $OABC$ a triad, we say it is an *orthogonal triad* derived by continuous deformation. Any orthogonal triad can be superposed on the $OABC$.

Given an orthogonal triad $OABC$, another triad $OA'BC$ may be derived by moving $A$ to the other side of $O$ to make the *opposite triad OA'BC*.

All orthogonal triads can be superposed either on a given orthogonal triad $OABC$ or on its opposite $OA'BC$. One of the two triads $OABC$ and $OA'BC$ is defined as being a *positive triad* and used as a standard. The other is then defined as *negative triad*. It is immaterial which one is chosen as positive, however, usually the *right-handed convention* is chosen as positive, the one for which the direction of rotation from $OA$ to $OB$ propels a *right-handed screw* in the direction $OC$. A right-handed (positive) orthogonal triad cannot be superposed to a left-handed (negative) triad. Thus there are just two essentially distinct types of triad. This is an essential property of three-dimensional space.

### 1.4.2  Unit Vectors

An orthogonal triad made of *unit vectors* $\hat{\imath}$, $\hat{\jmath}$, $\hat{k}$ is a set of three unit vectors whose directions form a positive orthogonal triad. From this definition,

$$\hat{\imath}^2 = 1 \quad , \quad \hat{\jmath}^2 = 1 \quad , \quad \hat{k}^2 = 1. \tag{1.1}$$

Moreover, since $j \times k$ is parallel to and in the same sense as $i$, by definition of the vector product we have

$$\hat{\jmath} \times \hat{k} = \hat{\imath} \tag{1.2}$$

and similarly

$$\hat{k} \times \hat{\imath} = \hat{\jmath} \tag{1.3}$$
$$\hat{\imath} \times \hat{\jmath} = \hat{k}. \tag{1.4}$$

Further

$$\hat{\imath} \cdot \hat{\jmath} = 0 \tag{1.5}$$
$$\hat{\jmath} \cdot \hat{k} = 0 \tag{1.6}$$
$$\hat{k} \cdot \hat{\imath} = 0 \tag{1.7}$$

and

$$(\hat{\jmath} \times \hat{k}) \cdot \hat{\imath} = \hat{\imath} \cdot \hat{\imath} = +1. \tag{1.8}$$

Any vector $\mathbf{r}$ may be put in the form

$$\mathbf{r} = (\mathbf{r} \cdot \hat{\imath})\hat{\imath} + (\mathbf{r} \cdot \hat{\jmath})\hat{\jmath} + (\mathbf{r} \cdot \hat{k})\hat{k}. \tag{1.9}$$

Vector addition is the key operation in kinematics. However, special attention must be taken since vectors can be added only when they are expressed in the same frame. Thus, a vector equation such as

$$\mathbf{a} = \mathbf{b} + \mathbf{c} \tag{1.10}$$

is meaningless without indicating the frame they are expressed in, such that

$$^B\mathbf{a} = {}^B\mathbf{b} + {}^B\mathbf{c}. \tag{1.11}$$

### 1.4.3  Reference Frame and Coordinate System

In robotics, we assign one or more coordinate frames to each link of the robot and each object of the robot's environment. Thus, communication among the coordinate frames, which is called *transformation of frames*, is a fundamental concept in the modeling and programming of a robot.

The angular motion of a rigid body can be described in one of several ways, the most popular being:

1. A set of rotations about a right-handed globally fixed Cartesian axis,

2. A set of rotations about a right-handed moving Cartesian axis, and

3. Angular rotation about a fixed axis in space.

Reference frames are a particular perspective employed by the analyst to describe the motion of links. A *fixed frame* is a reference frame that is motionless and attached to the ground. The motion of a robot takes place in a fixed frame called the *global reference frame*. A *moving frame* is a reference frame that moves with a link. Every moving link has an attached reference frame that sticks to the link and accepts every motion of the link. The moving reference frame is called the *local reference frame*. The position and orientation of a link with respect to the ground is explained by the position and orientation of its local reference frame in the global reference frame. In robotic analysis, we fix a global reference frame to the ground and attach a local reference frame to every single link.

A *coordinate system* is slightly different from reference frames. The coordinate system determines the way we describe the motion in each reference frame. A *Cartesian system* is the most popular coordinate system used in robotics, but cylindrical, spherical and other systems may be used as well.

Hereafter, we use "reference frame," "coordinate frame," and "coordinate system" equivalently, because a Cartesian system is the only system we use.

The position of a point $P$ of a rigid body $B$ is indicated by a vector $\mathbf{r}$. As shown in Figure 1.13, the position vector of $P$ can be decomposed in global coordinate frame

$$^G\mathbf{r} = X\hat{I} + Y\hat{J} + Z\hat{K} \tag{1.12}$$

or in body coordinate frame

$$^B\mathbf{r} = x\hat{i} + y\hat{j} + z\hat{k}. \tag{1.13}$$

The coefficients $(X, Y, Z)$ and $(x, y, z)$ are called *coordinates* or *components* of the point $P$ in global and local coordinate frames respectively. It is efficient for mathematical calculations to show vectors $^G\mathbf{r}$ and $^B\mathbf{r}$ by a vertical array made by its components

$$^G\mathbf{r} = \begin{bmatrix} X \\ Y \\ Z \end{bmatrix} \tag{1.14}$$

$$^B\mathbf{r} = \begin{bmatrix} x \\ y \\ z \end{bmatrix}. \tag{1.15}$$

Kinematics can be called the study of positions, velocities, and accelerations, without regards to the forces that cause these motions. Vectors and

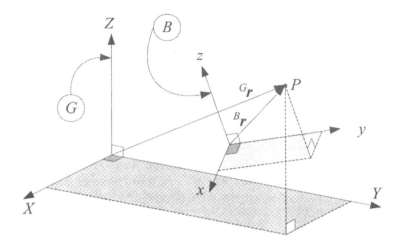

FIGURE 1.13. Position vector of a point $P$ may be decomposed in body frame $B$, or global frame $G$.

reference frames are essential tools for analyzing motions of complex systems, especially when the motion is three dimensional and involves many parts.

A coordinate frame is defined by a set of basis vectors, such as unit vectors along the three coordinate axes. So, a rotation matrix, as a coordinate transformation, can also be viewed as defining a change of basis from one frame to another.

A rotation matrix can be interpreted in three distinct ways:

1. *Mapping.* It represents a coordinate transformation, mapping and relating the coordinates of a point $P$ in two different frames.

2. *Description of a frame.* It gives the orientation of a transformed coordinate frame with respect to a fixed coordinate frame.

3. *Operator.* It is an operator taking a vector and rotating it to a new vector.

Rotation of a rigid body can be described by *rotation matrix R, Euler angles, angle-axis convention,* and *quaternion,* each with advantages and disadvantages.

The advantage of $R$ is direct interpretation in change of basis while its disadvantage is that nine dependent parameters must be stored. The physical role of individual parameters is lost, and only the matrix as a whole has meaning.

Euler angles are roughly defined by three successive rotations about three axes of local (and sometimes global) coordinate frames. The advantage of using Euler angles is that the rotation is described by three independent parameters with plain physical interpretations. Their disadvantage is that their representation is not unique and leads to a problem with singularities. There is also no simple way to compute multiple rotations except expansion into a matrix.

Angle-axis convention is the most intuitive representation of rotation. However, it requires four parameters to store a single rotation, computation of combined rotations is not simple, and it is ill-conditioned for small rotations.

Quaternions are good in preserving most of the intuition of the angle-axis representation while overcoming the ill conditioning for small rotations and admitting a group structure that allows computation of combined rotations. The disadvantage of quaternion is that four parameters are needed to express a rotation. The parameterization is more complicated than angle-axis and sometimes loses physical meaning. Quaternion multiplication is not as plain as matrix multiplication.

### 1.4.4   Vector Function

Vectors serve as the basis of our study of kinematics and dynamics. Positions, velocities, accelerations, momenta, forces, and moments all are vectors. Vectors locate a point according to a known reference. As such, a vector consists of a magnitude, a direction, and an origin of a reference point. We must explicitly denote these elements of the vector.

If either the magnitude of a vector $\mathbf{r}$ and/or the direction of $\mathbf{r}$ in a reference frame $B$ depends on a scalar variable, say $\theta$, then $\mathbf{r}$ is called a *vector function* of $\theta$ in $B$. A vector $\mathbf{r}$ may be a function of a variable in one reference frame, but be independent of this variable in another reference frame.

**Example 2** *Reference frame and parameter dependency.*

*In Figure 1.14, P represents a point that is free to move on and in a circle, made by three revolute jointed links. $\theta$, $\varphi$, and $\psi$ are the angles shown, then $\mathbf{r}$ is a vector function of $\theta$, $\varphi$, and $\psi$ in the reference frame $G(X, Y)$. The length and direction of $\mathbf{r}$ depend on $\theta$, $\varphi$, and $\psi$.*

*If $G(X, Y)$, and $B(x, y)$ designate reference frames attached to the ground and link 2, and P is the tip point of link 3 as shown in Figure 1.15, then the position vector $\mathbf{r}$ of point P in reference frame B is a function of $\varphi$ and $\psi$, but is independent of $\theta$.*

## 1.5   Problems of Robot Dynamics

There are three basic and systematic methods to represent the relative position and orientation of a manipulator link. The first and most popular

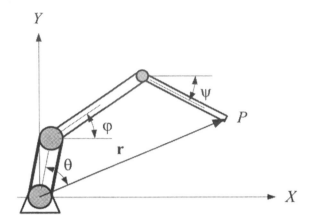

FIGURE 1.14. A planar 3R manipulator and position vector of the tip point P in global coordinate G(X, Y).

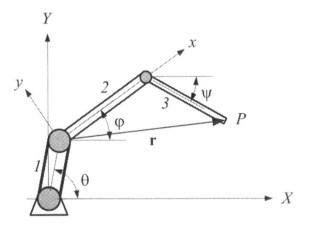

FIGURE 1.15. A planar 3R manipulator and position vector of the tip point P in second link local coordinate B(x, y).

method used in robot kinematics is based on the Denavit-Hartenberg notation for definition of spatial mechanisms and on the homogeneous transformation of points. The $4 \times 4$ matrix or the homogeneous transformation is utilized to represent spatial transformations of point vectors. In robotics, this matrix is used to describe one coordinate system with respect to another. The transformation matrix method is the most popular technique for describing robot motions.

Researchers in robot kinematics tried alternative methods to represent rigid body transformations based on concepts introduced by mathematicians and physicists such as the *screw theory*, *Lie algebra*, and *Epsilon algebra*. The transformation of a rigid body or a coordinate frame with respect to a reference coordinate frame can be expressed by a *screw displacement*, which is a translation along an axis with a rotation by an angle about the same axis. Although screw theory and Lie algebra can successfully be utilized for robot analysis, their result should finally be expressed in matrices.

## 1.6   Preview of Covered Topics

The book is arranged in three parts: I-Kinematics, II-Dynamics, and III-Control. Part I is important because it defines and describes the fundamental tools for robot analysis.

Rotational analysis of rigid bodies is a main subject in relative kinematic analysis of coordinate frames. It is about how we describe the orientation of a coordinate frame with respect to the others. In Chapters 2 and 3, we define and describe the rotational kinematics for the coordinate frames having a common origin. So, Chapters 2 and 3 are about the motion of two directly connected links via a revolute joint. The origin of coordinate frames may move with respect to each other, so, Chapter 4 is about the motion of two indirectly connected links.

In Chapter 5, the position and orientation kinematics of rigid links are utilized to systematically describe the configuration of the final link of a robot in a global Cartesian coordinate frame. Such an analysis is called forward kinematics, in which we are interested to find the end-effector configuration based on measured joint coordinates. The Denavit-Hartenberg convention is the main tool in forward kinematics. In this Chapter, we have shown how we may kinematically disassemble a robot to basic mechanisms with 1 or 2 DOF, and how we may kinematically assemble the basic mechanisms to make an arbitrary robot.

Chapter 6 deals with kinematics of robots from a Cartesian to joint space viewpoint that is called inverse kinematics. We start with a known position and orientation of the end-effector and search for a proper set of joint coordinates.

Velocity relationships between rigid links of a robot is the subject of Chapters 7 and 8. The definitions of angular velocity vector and angular velocity matrix are introduced in Chapter 7. The velocity relationship between robot links, as well as differential motion in joint and Cartesian spaces, are covered in Chapter 8. The Jacobian matrix is the main concept of this Chapter.

Part I is concluded by describing the applied numerical methods in robot kinematics. In Chapter 9 we introduce efficient and applied methods that can be used to ease computerized calculations in robotics.

In part II, the techniques needed to develop the equations of robot motion are explained. This part starts with acceleration analysis of relative links in Chapter 10. The methods for deriving the robots' equations of motion are described in Chapter 11. The Lagrange method is the main subject of dynamics development. The Newton-Euler method is described alternatively as tool to find the equations of motion. The Euler-Lagrange method has a simpler concept, however it provides the unneeded internal joint forces. On the other hand, the Lagrange method is more systematic and provides a basis for computer calculation.

In part III, we start with a brief description of path analysis. Then, the optimal control of robots is described using the floating time method. The floating time technique provides the required torques to make a robot follow a prescribed path of motion in an open loop control. To compensate a possible error between the desired and the actual kinematics, we explain the computed torque control method and the concept of the closed loop control algorithm.

## 1.7   Robots as Multi-disciplinary Machines

Let us note that the mechanical structure of a robot is only the visible part of the robot. Robotics is an essentially multidisciplinary field in which engineers from various branches such as mechanical, systems, electrical, electronics, and computer sciences play equally important roles. Therefore, it is fundamental for a *robotical engineer* to attain a sufficient level of understanding of the main concepts of the involved disciplines and communicate with engineers in these disciplines.

## 1.8   Summary

There are two kinds of robots: serial and parallel. A serial robot is made from a series of rigid links, where each pair of links is connected by a revolute (R) or prismatic (P) joint. An R or P joint provides only one degree of freedom, which is rotational or translational respectively. The final link of a robot, also called the end-effector, is the operating member of a robot that interacts with the environment.

To reach any point in a desired orientation, within a robot's workspace, a robot needs at least 6 DOF. Hence, it must have at least 6 links and 6 joints. Most robots use 3 DOF to position the wrist point, and use the other 3 DOF to orient the end-effector about the wrist point.

We attach a Cartesian coordinate frame to each link of a robot and determine the position and orientation of each frame with respect to the others. Therefore, to determine the position and orientation of the end-effector, we need to find the end-effector frame in the base frame.

# Exercises

1. Meaning of indexes for a position vector.

   Explain the meaning of ${}^G\mathbf{r}_P$, ${}^G_B\mathbf{r}_P$, and ${}^G_G\mathbf{r}_P$, if $\mathbf{r}$ is a position vector.

2. Meaning of indexes for a velocity vector.

   Explain the meaning of ${}^G\mathbf{v}_P$, ${}^G_B\mathbf{v}_P$, and ${}^B_B\mathbf{v}_P$, if $\mathbf{v}$ is a velocity vector.

3. Meaning of indexes for an angular velocity vector.

   Explain the meaning of ${}^{B_2}\boldsymbol{\omega}_{B_1}$, ${}^G_{B_2}\boldsymbol{\omega}_{B_1}$, ${}^B_G\boldsymbol{\omega}_B$, and ${}^{B_2}_{B_3}\boldsymbol{\omega}_{B_1}$, if $\boldsymbol{\omega}$ is an angular velocity vector.

4. Meaning of indexes for a transformation matrix.

   Explain the meaning of ${}^{B_2}T_{B_1}$, and ${}^G T_B$, if $T$ is a transformation matrix.

5. Laws of robotics.

   What is the difference between law zero and law one of robotics?

6. New law of robotics.

   What do you think about adding a fourth law to robotics, such as: A robot must protect the other robots as long as such protection does not conflict with a higher order law.

7. The word "robot."

   Find the origin and meaning of the word "robot."

8. Robot classification.

   Do you consider a crane as a robot?

9. Robot market.

   Most small robot manufacturers went out of the market around 1990, and only a few large companies remained. Why do you think this happened?

10. Humanoid robots.

    The mobile robot industry is trying to make robots as similar to humans as possible. What do you think is the reason for this: humans have the best structural design, humans have the simplest design, or these kinds of robots can be sold in the market better?

11. Robotic person.

    Why do you think we call somebody who works or behaves mechanically, showing no emotion and often responding to orders without question, a robotic person?

12. Robotic journals.

    Find the name of 10 technical journals related to robotics.

13. Number of robots in industrial poles.

    Search the robotic literatures and find out, approximately, how many industrial robots are currently in operation in the United States, Europe, and Japan.

14. Robotic countries.

    Search the robotic literatures and find out what countries are ranked first, second, and third according to the number of industrial robots in use?

15. Advantages and disadvantages of robots.

    Search the robotic literatures and name 10 advantages and 2 disadvantages of robots.

16. Mechanisms and robots.

    Why do we not replace every mechanism, in an assembly line for example, with robots?

17. Higher pairs and lower pairs.

    Joints can be classified as *lower pairs* and *higher pairs*. Find the meaning of "lower pairs" and "higher pairs" in mechanics of machinery.

18. Degrees of freedom (DOF) elimination.

    Joints can be classified by the number of degrees of freedom they provide, or by the number of degrees of freedom they eliminate. There are, therefore, 5 classes for joints. Find the name and the number of joints in each class.

19. Human wrist DOF.

    Examine and count the number of DOF of your wrist.

20. Human hand is a redundant manipulator.

    An arm (including shoulder, elbow, and wrist) has 7 DOF. What is the advantage of having one extra DOF (with respect to 6 DOF) in our hand?

21. Usefulness of redundant manipulators.

    Discuss possible applications in which the redundant manipulators would be useful.

22. ★ Independent Cartesian coordinate systems in Euclidean spaces.

In 3D Euclidean space, we need a triad to locate a point. There are two independent and non-superposable triads. How many different non-superposable Cartesian coordinate systems can be imagined in 4D Euclidean space? How many Cartesian coordinate systems do we have in an $n$D Euclidean space?

23. ★ Disadvantages of a non-orthogonal triad.

Why do we use an orthogonal triad to define a Cartesian space? Can we define a 3D space with non-orthogonal triads?

24. ★ Usefulness of an orthogonal triad.

Orthogonality is the common property of all useful coordinate systems such as Cartesian, cylindrical, spherical, parabolical, and ellipsoidal coordinate systems. Why do we only define and use orthogonal coordinate systems? Do you think ability to define a vector, based on inner product and unit vectors of the coordinate system, such as,

$$\mathbf{r} = (\mathbf{r} \cdot \hat{\imath})\hat{\imath} + (\mathbf{r} \cdot \hat{\jmath})\hat{\jmath} + (\mathbf{r} \cdot \hat{k})\hat{k}$$

is the main reason for defining the orthogonal coordinate systems?

25. ★ Three coplanar vectors.

Show that if $\mathbf{a} \times \mathbf{b} \cdot \mathbf{c} = 0$, then $\mathbf{a}$, $\mathbf{b}$, $\mathbf{c}$ are coplanar.

26. ★ Vector function, vector variable.

A vector function is defined as a dependent vectorial variable that relates to a scalar independent variable.

$$\mathbf{r} = \mathbf{r}(t)$$

Describe the meaning and define an example for a vector function of a vector variable,

$$\mathbf{a} = \mathbf{a}(\mathbf{b})$$

and a scalar function of a vector variable

$$f = f(\mathbf{b}).$$

27. ★ Frame-dependent and frame-independent.

A vector function of scalar variables is a frame-dependent quantity. Is a vector function of vector variables frame-dependent? What about a scalar function of vector variables?

28. ★ Coordinate frame and vector function.

Explain the meaning of $^B\mathbf{v}_P(^G\mathbf{r}_P)$, if $\mathbf{r}$ is a position vector, $\mathbf{v}$ is a velocity vector, and $\mathbf{v}(\mathbf{r})$ means $\mathbf{v}$ is a function of $\mathbf{r}$.

# Part I

# Kinematics

Kinematics is the science of *geometry in motion*. It is restricted to a pure geometrical description of motion by means of position, orientation, and their time derivatives. In robotics, the kinematic descriptions of manipulators and their assigned tasks are utilized to set up the fundamental equations for dynamics and control.

Because the links and arms in a robotic system are modeled as rigid bodies, the properties of rigid body displacement takes a central place in robotics. Vector and matrix algebra are utilized to develop a systematic and generalized approach to describe and represent the location of the arms of a robot with respect to a global fixed reference frame $G$. Since the arms of a robot may rotate or translate with respect to each other, body-attached coordinate frames $A, B, C, \cdots$ or $B_1, B_2, B_3, \cdots$ will be established along with the joint axis for each link to find their relative configurations, and within the reference frame $G$. The position of one link $B$ relative to another link $A$ is defined kinematically by a coordinate transformation $^AT_B$ between reference frames attached to the link.

The direct kinematics problem is reduced to finding a transformation matrix $^GT_B$ that relates the body attached local coordinate frame $B$ to the global reference coordinate frame $G$. A $3 \times 3$ rotation matrix is utilized to describe the rotational operations of the local frame with respect to the global frame. The homogeneous coordinates are then introduced to represent position vectors and directional vectors in a three dimensional space. The rotation matrices are expanded to $4 \times 4$ homogeneous transformation matrices to include both the rotational and translational motions. Homogeneous matrices that express the relative rigid links of a robot are made by a special set of rules and are called *Denavit-Hartenberg matrices* after Denavit and Hartenberg (1955). The advantage of using the Denavit-Hartenberg matrix is its algorithmic universality in deriving the kinematic equation of a robot link.

The analytical description of displacement of a rigid body is based on the notion that all points in a rigid body must retain their original relative positions regardless of the new position and orientation of the body. The total rigid body displacement can always be reduced to the sum of its two basic components: the translation displacement of an arbitrary reference point fixed in the rigid body plus the unique rotation of the body about a line through that point.

Study of displacement motion of rigid bodies leads to the relation between the time rate of change of a vector in a global frame and the time rate of change of the same vector in a local frame.

*Transformation* from a local coordinate frame $B$ to a global coordinate frame $G$ is expressed by

$$^G\mathbf{r} = {}^GR_B \, {}^B\mathbf{r} + {}^G\mathbf{d}_B$$

where $^B\mathbf{r}$ is the position vector of a point in $B$, $^G\mathbf{r}$ is the position vector of the same point expressed in $G$, and $\mathbf{d}$ is the position vector of the origin

$o$ of the body coordinate frame $B(oxyz)$ with respect to the origin $O$ of the global coordinate frame $G(OXYZ)$. Therefore, a transformation has two parts: a *translation* $\mathbf{d}$ that brings the origin $o$ on the origin $O$, plus a *rotation*, $^{G}R_B$ that brings the axes of $oxyz$ on the corresponding axes of $OXYZ$.

The transformation formula $^{G}\mathbf{r} = {}^{G}R_B\,{}^{B}\mathbf{r} + {}^{G}\mathbf{d}_B$, can be expanded to connect more than two coordinate frames. The combination formula for a transformation from a local coordinate frame $B_1$ to another coordinate frame $B_2$ followed by a transformation from $B_2$ to the global coordinate $G$ is

$$
\begin{aligned}
^{G}\mathbf{r} &= \left( {}^{G}R_{B_2}\,{}^{B_2}\mathbf{r} + {}^{G}\mathbf{d}_{B_2} \right) + \left( {}^{B_2}R_{B_1}\,{}^{B_1}\mathbf{r} + {}^{B_2}\mathbf{d}_{B_1} \right) \\
&= {}^{G}R_{B_2}\,{}^{B_2}R_{B_1}\,{}^{B_1}\mathbf{r} + {}^{G}\mathbf{d}_{B_2} + {}^{B_2}\mathbf{d}_{B_1} \\
&= {}^{G}R_{B_1}\,{}^{B_1}\mathbf{r} + {}^{G}\mathbf{d}_{B_1}.
\end{aligned}
$$

A robot consists of $n$ rigid links with relative motions. The link attached to the ground is link $(0)$ and the link attached to the final moving link, the end-effector, is link $(n)$. There are two important problems in kinematic analysis of robots: the *forward kinematics problem* and the *inverse kinematics problem*.

In forward kinematics, the problem is that the position vector of a point $P$ is in the coordinate frame $B_n$ attached to the end-effector as $^{n}\mathbf{r}_p$, and we are looking for the position of $P$ in the base frame $B_0$ shown by $^{0}\mathbf{r}_p$. The forward kinematics problem is equivalent to having the values of joints variable and asking for the position of the end-effector.

In inverse kinematics, the problem is that we have the position vector of a point $P$ in the base coordinate frame $B_0$ as $^{0}\mathbf{r}_p$, and we are looking for $^{n}\mathbf{r}_p$, the position of $P$ in the base frame $B_0$. This problem is equivalent to having the position of the end-effector and asking for a set of joint variables that make the robot reach the point $P$.

In this part, we develop the transformation formula to move the kinematic information back and forth from a coordinate frame to another coordinate frame.

# 2

# Rotation Kinematics

## 2.1  Rotation About Global Cartesian Axes

Consider a rigid body $B$ with a local coordinate frame $Oxyz$ that is originally coincident with a global coordinate frame $OXYZ$. Point $O$ of the body $B$ is fixed to the ground $G$ and is the origin of both coordinate frames. If the rigid body $B$ rotates $\alpha$ degrees about the $Z$-axis of the global coordinate frame, then coordinates of any point $P$ of the rigid body in the local and global coordinate frames are related by the following equation

$$^G\mathbf{r} = Q_{Z,\alpha}\,^B\mathbf{r} \tag{2.1}$$

where,

$$^G\mathbf{r} = \begin{bmatrix} X \\ Y \\ Z \end{bmatrix} \tag{2.2}$$

$$^B\mathbf{r} = \begin{bmatrix} x \\ y \\ z \end{bmatrix} \tag{2.3}$$

and

$$Q_{Z,\alpha} = \begin{bmatrix} \cos\alpha & -\sin\alpha & 0 \\ \sin\alpha & \cos\alpha & 0 \\ 0 & 0 & 1 \end{bmatrix}. \tag{2.4}$$

Similarly, rotation $\beta$ degrees about the $Y$-axis, and $\gamma$ degrees about the $X$-axis of the global frame relate the local and global coordinates of point $P$ by the following equations

$$^G\mathbf{r} = Q_{Y,\beta}\,^B\mathbf{r} \tag{2.5}$$

$$^G\mathbf{r} = Q_{X,\gamma}\,^B\mathbf{r} \tag{2.6}$$

where,

$$Q_{Y,\beta} = \begin{bmatrix} \cos\beta & 0 & \sin\beta \\ 0 & 1 & 0 \\ -\sin\beta & 0 & \cos\beta \end{bmatrix} \tag{2.7}$$

$$Q_{X,\gamma} = \begin{bmatrix} 1 & 0 & 0 \\ 0 & \cos\gamma & -\sin\gamma \\ 0 & \sin\gamma & \cos\gamma \end{bmatrix}. \tag{2.8}$$

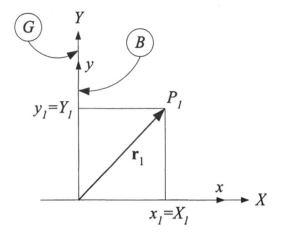

FIGURE 2.1. Position vector of point P when local and global frames are coincident.

**Proof.** Let $(\hat{\imath}, \hat{\jmath}, \hat{k})$ and $(\hat{I}, \hat{J}, \hat{K})$ be the unit vectors along the coordinate axes of $Oxyz$ and $OXYZ$ respectively. The rigid body has a space fixed point at $O$, that is the common origin of $Oxyz$ and $OXYZ$. Figure 2.1 illustrates the top view of the system.

The initial position of a point $P$ is indicated by $P_1$. The position vector $r_1$ of $P_1$ can be expressed in body and global coordinate frames by

$$
\begin{aligned}
{}^{B}\mathbf{r}_1 &= x_1\hat{\imath} + y_1\hat{\jmath} + z_1\hat{k} & (2.9) \\
{}^{G}\mathbf{r}_1 &= X_1\hat{I} + Y_1\hat{J} + Z_1\hat{K} & (2.10)
\end{aligned}
$$

$$
\begin{aligned}
x_1 &= X_1 \\
y_1 &= Y_1 & (2.11) \\
z_1 &= Z_1
\end{aligned}
$$

where ${}^{B}\mathbf{r}_1$ refers to the position vector $\mathbf{r}_1$ expressed in the body coordinate frame $B$, and ${}^{G}\mathbf{r}_1$ refers to the position vector $\mathbf{r}_1$ expressed in the global coordinate frame $G$.

If the rigid body undergoes a rotation $\alpha$ about the $Z$-axis, then the local frame $Oxyz$, point $P$, and the position vector $\mathbf{r}$ will be seen in a second position, as shown in Figure 2.2. Now the position vector $\mathbf{r}_2$ of $P_2$ is expressed in both coordinate frames by

$$
\begin{aligned}
{}^{B}\mathbf{r}_2 &= x_2\hat{\imath} + y_2\hat{\jmath} + z_2\hat{k} & (2.12) \\
{}^{G}\mathbf{r}_2 &= X_2\hat{I} + Y_2\hat{J} + Z_2\hat{K}. & (2.13)
\end{aligned}
$$

Using the definition of the inner product and Equation (2.12) we may

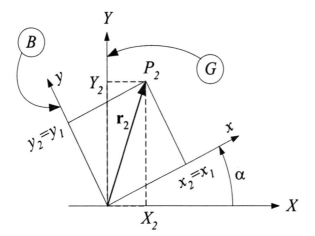

FIGURE 2.2. Position vector of point P when local frames are rotated about the Z-axis.

write

$$X_2 = \hat{I} \cdot \mathbf{r}_2 = \hat{I} \cdot x_2\hat{\imath} + \hat{I} \cdot y_2\hat{\jmath} + \hat{I} \cdot z_2\hat{k} \qquad (2.14)$$

$$Y_2 = \hat{J} \cdot \mathbf{r}_2 = \hat{J} \cdot x_2\hat{\imath} + \hat{J} \cdot y_2\hat{\jmath} + \hat{J} \cdot z_2\hat{k} \qquad (2.15)$$

$$Z_2 = \hat{K} \cdot \mathbf{r}_2 = \hat{K} \cdot x_2\hat{\imath} + \hat{K} \cdot y_2\hat{\jmath} + \hat{K} \cdot z_2\hat{k} \qquad (2.16)$$

or equivalently

$$\begin{bmatrix} X_2 \\ Y_2 \\ Z_2 \end{bmatrix} = \begin{bmatrix} \hat{I} \cdot \hat{\imath} & \hat{I} \cdot \hat{\jmath} & \hat{I} \cdot \hat{k} \\ \hat{J} \cdot \hat{\imath} & \hat{J} \cdot \hat{\jmath} & \hat{J} \cdot \hat{k} \\ \hat{K} \cdot \hat{\imath} & \hat{K} \cdot \hat{\jmath} & \hat{K} \cdot \hat{k} \end{bmatrix} \begin{bmatrix} x_2 \\ y_2 \\ z_2 \end{bmatrix}. \qquad (2.17)$$

The elements of the *Z-rotation matrix*, $Q_{Z,\alpha}$, are called the *direction cosines* of $^B\mathbf{r}_2$ with respect to $OXYZ$. Figure 2.3 shows the top view of the initial and final configurations of $\mathbf{r}$ in both coordinate systems $Oxyz$ and $OXYZ$. Analyzing Figure 2.3 indicates that

$$\begin{aligned} \hat{I} \cdot \hat{\imath} = \cos\alpha, \quad &\hat{I} \cdot \hat{\jmath} = -\sin\alpha, \quad \hat{I} \cdot \hat{k} = 0 \\ \hat{J} \cdot \hat{\imath} = \sin\alpha, \quad &\hat{J} \cdot \hat{\jmath} = \cos\alpha, \quad \hat{J} \cdot \hat{k} = 0 \\ \hat{K} \cdot \hat{\imath} = 0, \quad &\hat{K} \cdot \hat{\jmath} = 0, \quad \hat{K} \cdot \hat{k} = 1. \end{aligned} \qquad (2.18)$$

Combining Equations (2.17) and (2.18) shows that we can find the components of $^G\mathbf{r}_2$ by multiplying the Z-rotation matrix $Q_{Z,\alpha}$ and the vector $^B\mathbf{r}_2$.

$$\begin{bmatrix} X_2 \\ Y_2 \\ Z_2 \end{bmatrix} = \begin{bmatrix} \cos\alpha & -\sin\alpha & 0 \\ \sin\alpha & \cos\alpha & 0 \\ 0 & 0 & 1 \end{bmatrix} \begin{bmatrix} x_2 \\ y_2 \\ z_2 \end{bmatrix}. \qquad (2.19)$$

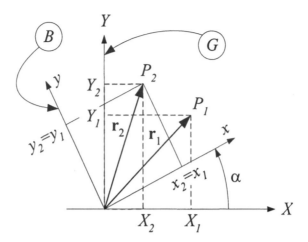

FIGURE 2.3. Position vectors of point P before and after the rotation of the local frame about the Z-axis of the global frame.

It can also be shown in the following short notation

$$^{G}\mathbf{r}_2 = Q_{Z,\alpha}\,^{B}\mathbf{r}_2 \tag{2.20}$$

where

$$Q_{Z,\alpha} = \begin{bmatrix} \cos\alpha & -\sin\alpha & 0 \\ \sin\alpha & \cos\alpha & 0 \\ 0 & 0 & 1 \end{bmatrix}. \tag{2.21}$$

Equation (2.20) says that the vector $\mathbf{r}$ at the second position in the global coordinate frame is equal to $Q_Z$ times the position vector in the local coordinate frame. Hence, we are able to find the global coordinates of a point of a rigid body after rotation about the $Z$-axis, if we have its local coordinates.

Similarly, rotation $\beta$ about the $Y$-axis and rotation $\gamma$ about the $X$-axis are described by the $Y$-rotation matrix $Q_{Y,\beta}$ and the $X$-rotation matrix $Q_{X,\gamma}$ respectively.

$$Q_{Y,\beta} = \begin{bmatrix} \cos\beta & 0 & \sin\beta \\ 0 & 1 & 0 \\ -\sin\beta & 0 & \cos\beta \end{bmatrix} \tag{2.22}$$

$$Q_{X,\gamma} = \begin{bmatrix} 1 & 0 & 0 \\ 0 & \cos\gamma & -\sin\gamma \\ 0 & \sin\gamma & \cos\gamma \end{bmatrix} \tag{2.23}$$

The rotation matrices $Q_{Z,\alpha}$, $Q_{Y,\beta}$, and $Q_{X,\gamma}$ are called *basic global rotation matrices*. We usually refer to the first, second and third rotations about the axes of the global coordinate frame by $\alpha$, $\beta$, and $\gamma$ respectively.

■

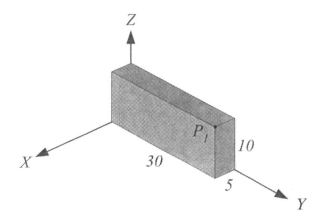

FIGURE 2.4. Corner $P$ of the slab at first position.

**Example 3** *Successive rotation about global axes.*

The final position of the corner $P(5, 30, 10)$ of the slab shown in Figure 2.4 after 30 deg rotation about the Z-axis, followed by 30 deg about the X-axis, and then 90 deg about the Y-axis can be found by first multiplying $Q_{Z,30}$ by $[5, 30, 10]^T$ to get the new global position after first rotation

$$\begin{bmatrix} X_2 \\ Y_2 \\ Z_2 \end{bmatrix} = \begin{bmatrix} \cos 30 & -\sin 30 & 0 \\ \sin 30 & \cos 30 & 0 \\ 0 & 0 & 1 \end{bmatrix} \begin{bmatrix} 5 \\ 30 \\ 10 \end{bmatrix} = \begin{bmatrix} -10.68 \\ 28.48 \\ 10.0 \end{bmatrix},$$

and then multiplying $Q_{X,30}$ and $[-10.68, 28.48, 10.0]^T$ to get the position of $P$ after the second rotation

$$\begin{bmatrix} X_3 \\ Y_3 \\ Z_3 \end{bmatrix} = \begin{bmatrix} 1 & 0 & 0 \\ 0 & \cos 30 & -\sin 30 \\ 0 & \sin 30 & \cos 30 \end{bmatrix} \begin{bmatrix} -10.68 \\ 28.48 \\ 10.0 \end{bmatrix} = \begin{bmatrix} -10.68 \\ 19.66 \\ 22.9 \end{bmatrix},$$

and finally multiplying $Q_{Y,90}$ and $[-10.68, 19.66, 22.9]^T$ to get the final position of $P$ after the third rotation. The slab and the point $P$ in first, second, third, and fourth positions are shown in Figure 2.5.

$$\begin{bmatrix} X_4 \\ Y_4 \\ Z_4 \end{bmatrix} = \begin{bmatrix} \cos 90 & 0 & \sin 90 \\ 0 & 1 & 0 \\ -\sin 90 & 0 & \cos 90 \end{bmatrix} \begin{bmatrix} -10.68 \\ 19.66 \\ 22.9 \end{bmatrix} = \begin{bmatrix} 22.90 \\ 19.66 \\ 10.68 \end{bmatrix}$$

**Example 4** *Global rotation, local position.*

If a point $P$ is moved to $^G\mathbf{r}_2 = [4, 3, 2]^T$ after a 60 deg rotation about the

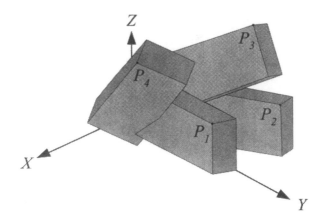

FIGURE 2.5. Corner $P$ and the slab at first, second, third, and final positions.

*Z-axis, its position in the local coordinate is*

$$^{B}\mathbf{r}_2 = Q_{Z,60}^{-1}\,{}^{G}\mathbf{r}_2$$

$$\begin{bmatrix} x_2 \\ y_2 \\ z_2 \end{bmatrix} = \begin{bmatrix} \cos 60 & -\sin 60 & 0 \\ \sin 60 & \cos 60 & 0 \\ 0 & 0 & 1 \end{bmatrix}^{-1} \begin{bmatrix} 4 \\ 3 \\ 2 \end{bmatrix} = \begin{bmatrix} 4.60 \\ -1.95 \\ 2.0 \end{bmatrix}.$$

*The local coordinate frame was coincident with the global coordinate frame before rotation, thus the global coordinates of $P$ before rotation was also $^{G}\mathbf{r}_1 = [4.60, -1.95, 2.0]^T$. Positions of $P$ before and after rotation are shown in Figure 2.6.*

## 2.2 Successive Rotation About Global Cartesian Axes

The final global position of a point $P$ in a rigid body $B$ with position vector $\mathbf{r}$, after a sequence of rotations $Q_1$, $Q_2$, $Q_3$, ..., $Q_n$ about the global axes can be found by

$$^{G}\mathbf{r} = {}^{G}Q_B\,{}^{B}\mathbf{r} \tag{2.24}$$

where,

$$^{G}Q_B = Q_n \cdots Q_3 Q_2 Q_1 \tag{2.25}$$

and, $^{G}\mathbf{r}$ and $^{B}\mathbf{r}$ indicate the position vector $\mathbf{r}$ in the global and local coordinate frames. $^{G}Q_B$ is called the *global rotation matrix*. It maps the local coordinates to their corresponding global coordinates.

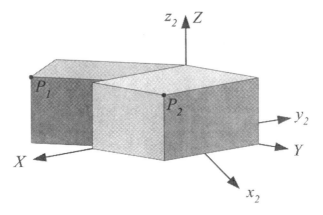

FIGURE 2.6. Positions of point $P$ in Example 4 before and after rotation.

Since matrix multiplications do not commute, the sequence of performing rotations is important. A rotation matrix is *orthogonal;* i.e., its transpose $Q^T$ is equal to its inverse $Q^{-1}$.

$$Q^T = Q^{-1} \qquad (2.26)$$

Rotation about the global coordinate axes is conceptually simple because the axes of rotations are fixed in space. Assume we have the coordinates of every point of a rigid body in the global frame that is equal to the local coordinates initially. The rigid body rotates about a global axis, then the proper global rotation matrix gives us the new global coordinate of the points. When we get the coordinates of points of the rigid body after the first rotation, our situation before the second rotation is similar to what we had before the first rotation.

**Example 5** *Successive global rotation matrix.*
*The global rotation matrix after a rotation $Q_{Z,\alpha}$ followed by $Q_{Y,\beta}$ and then $Q_{X,\gamma}$ is*

$$
\begin{aligned}
{}^G Q_B &= Q_{X,\gamma} Q_{Y,\beta} Q_{Z,\alpha} \\
&= \begin{bmatrix} c\alpha c\beta & -c\beta s\alpha & s\beta \\ c\gamma s\alpha + c\alpha s\beta s\gamma & c\alpha c\gamma - s\alpha s\beta s\gamma & -c\beta s\gamma \\ s\alpha s\gamma - c\alpha c\gamma s\beta & c\alpha s\gamma + c\gamma s\alpha s\beta & c\beta c\gamma \end{bmatrix}. \quad (2.27)
\end{aligned}
$$

**Example 6** *Successive global rotations, global position.*
*The end point $P$ of the arm shown in Figure 2.7 is located at*

$$
\begin{bmatrix} X_1 \\ Y_1 \\ Z_1 \end{bmatrix} = \begin{bmatrix} 0 \\ l\cos\theta \\ l\sin\theta \end{bmatrix} = \begin{bmatrix} 0 \\ 1\cos 75 \\ 1\sin 75 \end{bmatrix} = \begin{bmatrix} 0.0 \\ 0.26 \\ 0.97 \end{bmatrix}.
$$

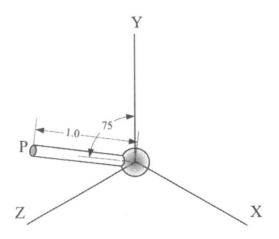

FIGURE 2.7. The arm of Example 6.

*The rotation matrix to find the new position of the end point after $-29$ deg rotation about the $X$-axis, followed by $30$ deg about the $Z$-axis, and again $132$ deg about the $X$-axis is*

$$^G Q_B = Q_{X,132} Q_{Z,30} Q_{X,-29} = \begin{bmatrix} 0.87 & -0.44 & -0.24 \\ -0.33 & -0.15 & -0.93 \\ 0.37 & 0.89 & -0.27 \end{bmatrix}$$

*and its new position is at*

$$\begin{bmatrix} X_2 \\ Y_2 \\ Z_2 \end{bmatrix} = \begin{bmatrix} 0.87 & -0.44 & -0.24 \\ -0.33 & -0.15 & -0.93 \\ 0.37 & 0.89 & -0.27 \end{bmatrix} \begin{bmatrix} 0.0 \\ 0.26 \\ 0.97 \end{bmatrix} = \begin{bmatrix} -0.35 \\ -0.94 \\ -0.031 \end{bmatrix}.$$

**Example 7** *Twelve independent triple global rotations.*

*In general, there are 12 different independent combinations of triple rotations about the global axes. They are:*

$$\begin{aligned} 1 &- Q_{X,\gamma} Q_{Y,\beta} Q_{Z,\alpha} \\ 2 &- Q_{Y,\gamma} Q_{Z,\beta} Q_{X,\alpha} \\ 3 &- Q_{Z,\gamma} Q_{X,\beta} Q_{Y,\alpha} \\ 4 &- Q_{Z,\gamma} Q_{Y,\beta} Q_{X,\alpha} \\ 5 &- Q_{Y,\gamma} Q_{X,\beta} Q_{Z,\alpha} \\ 6 &- Q_{X,\gamma} Q_{Z,\beta} Q_{Y,\alpha} \\ 7 &- Q_{X,\gamma} Q_{Y,\beta} Q_{X,\alpha} \\ 8 &- Q_{Y,\gamma} Q_{Z,\beta} Q_{Y,\alpha} \\ 9 &- Q_{Z,\gamma} Q_{X,\beta} Q_{Z,\alpha} \\ 10 &- Q_{X,\gamma} Q_{Z,\beta} Q_{X,\alpha} \\ 11 &- Q_{Y,\gamma} Q_{X,\beta} Q_{Y,\alpha} \\ 12 &- Q_{Z,\gamma} Q_{Y,\beta} Q_{Z,\alpha} \end{aligned} \tag{2.28}$$

*The expanded form of the 12 global axes triple rotations are presented in Appendix A.*

**Example 8** *Order of rotation, and order of matrix multiplication.*

*Changing the order of global rotation matrices is equivalent to changing the order of rotations. The position of a point $P$ of a rigid body $B$ is located at ${}^B\mathbf{r}_P = \begin{bmatrix} 1 & 2 & 3 \end{bmatrix}^T$. Its global position after rotation 30 deg about $X$-axis and then 45 deg about $Y$-axis is at*

$$
\begin{aligned}
\left({}^G\mathbf{r}_P\right)_1 &= Q_{Y,45}\, Q_{X,30}\, {}^B\mathbf{r}_P \\
&= \begin{bmatrix} 0.53 & -0.84 & 0.13 \\ 0.0 & 0.15 & 0.99 \\ -0.85 & -0.52 & 0.081 \end{bmatrix} \begin{bmatrix} 1 \\ 2 \\ 3 \end{bmatrix} = \begin{bmatrix} -0.76 \\ 3.27 \\ -1.64 \end{bmatrix}
\end{aligned}
$$

*and if we change the order of rotations then its position would be at*

$$
\begin{aligned}
\left({}^G\mathbf{r}_P\right)_2 &= Q_{X,30}\, Q_{Y,45}\, {}^B\mathbf{r}_P \\
&= \begin{bmatrix} 0.53 & 0.0 & 0.85 \\ -0.84 & 0.15 & 0.52 \\ -0.13 & -0.99 & 0.081 \end{bmatrix} \begin{bmatrix} 1 \\ 2 \\ 3 \end{bmatrix} = \begin{bmatrix} 3.08 \\ 1.02 \\ -1.86 \end{bmatrix} .
\end{aligned}
$$

*These two final positions of $P$ are $d = \left|\left({}^G\mathbf{r}_P\right)_1 - \left({}^G\mathbf{r}_P\right)_2\right| = 4.456$ apart.*

## 2.3   Global Roll-Pitch-Yaw Angles

The rotation about the $X$-axis of the global coordinate frame is called *roll*, the rotation about the $Y$-axis of the global coordinate frame is called *pitch*, and the rotation about the $Z$-axis of the global coordinate frame is called *yaw*. The global *roll-pitch-yaw rotation matrix* is

$$
\begin{aligned}
{}^G Q_B &= Q_{Z,\gamma} Q_{Y,\beta} Q_{X,\alpha} \\
&= \begin{bmatrix} c\beta c\gamma & -c\alpha s\gamma + c\gamma s\alpha s\beta & s\alpha s\gamma + c\alpha c\gamma s\beta \\ c\beta s\gamma & c\alpha c\gamma + s\alpha s\beta s\gamma & -c\gamma s\alpha + c\alpha s\beta s\gamma \\ -s\beta & c\beta s\alpha & c\alpha c\beta \end{bmatrix} . \quad (2.29)
\end{aligned}
$$

Given the roll, pitch, and yaw angles, we can compute the overall rotation matrix using Equation (2.29). Also we are able to compute the equivalent roll, pitch, and yaw angles when a rotation matrix is given. Suppose that $r_{ij}$ indicates the element of row $i$ and column $j$ of the roll-pitch-yaw rotation matrix (2.29), then the roll angle is

$$
\alpha = \tan^{-1}\left(\frac{r_{32}}{r_{33}}\right) \tag{2.30}
$$

and the pitch angle is

$$\beta = -\sin^{-1}(r_{31}) \qquad (2.31)$$

and the yaw angle is

$$\gamma = \tan^{-1}\left(\frac{r_{21}}{r_{11}}\right) \qquad (2.32)$$

provided that $\cos\beta \neq 0$.

**Example 9** *Global roll, pitch, and yaw rotations.*

*Figures 2.8, 2.9, and 2.10 illustrate 45 deg roll, pitch and yaw about the axes of global coordinate frame.*

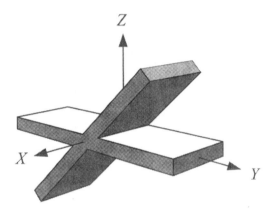

FIGURE 2.8. Global roll.

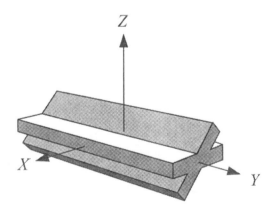

FIGURE 2.9. Global pitch.

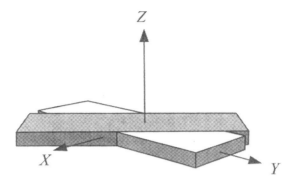

FIGURE 2.10. Global yaw.

## 2.4   Rotation About Local Cartesian Axes

Consider a rigid body $B$ with a space fixed at point $O$. The local body coordinate frame $B(Oxyz)$ is coincident with a global coordinate frame $G(OXYZ)$, where the origin of both frames are on the fixed point $O$. If the body undergoes a rotation $\varphi$ about the $z$ axis of its local coordinate frame, as can be seen in the top view shown in Figure 2.11, then coordinates of any point of the rigid body in local and global frames are related by the following equation

$$^B\mathbf{r} = A_{z,\varphi}\,{}^G\mathbf{r}. \tag{2.33}$$

The vectors $^G\mathbf{r}$ and $^B\mathbf{r}$ are the position vectors of the point in global and local frames respectively

$$^G\mathbf{r} = \begin{bmatrix} X & Y & Z \end{bmatrix}^T \tag{2.34}$$

$$^B\mathbf{r} = \begin{bmatrix} x & y & z \end{bmatrix}^T \tag{2.35}$$

and $A_{z,\varphi}$ is the *z-rotation matrix*

$$A_{z,\varphi} = \begin{bmatrix} \cos\varphi & \sin\varphi & 0 \\ -\sin\varphi & \cos\varphi & 0 \\ 0 & 0 & 1 \end{bmatrix}. \tag{2.36}$$

Similarly, rotation $\theta$ about the $y$-axis and rotation $\psi$ about the $x$-axis are described by the *y-rotation matrix* $A_{y,\theta}$ and the *x-rotation matrix* $A_{x,\psi}$ respectively.

$$A_{y,\theta} = \begin{bmatrix} \cos\theta & 0 & -\sin\theta \\ 0 & 1 & 0 \\ \sin\theta & 0 & \cos\theta \end{bmatrix} \tag{2.37}$$

$$A_{x,\psi} = \begin{bmatrix} 1 & 0 & 0 \\ 0 & \cos\psi & \sin\psi \\ 0 & -\sin\psi & \cos\psi \end{bmatrix} \tag{2.38}$$

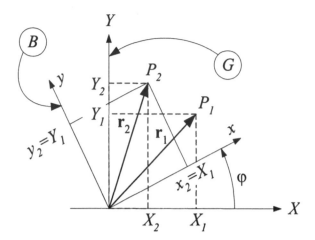

FIGURE 2.11. Position vectors of point P before and after rotation of the local frame about the z-axis of the local frame.

**Proof.** Vector $\mathbf{r}_1$ indicates the position of a point $P$ of the rigid body $B$ where it is initially at $P_1$. Using the unit vectors $(\hat{i}, \hat{j}, \hat{k})$ along the axes of local coordinate frame $B(Oxyz)$, and $(\hat{I}, \hat{J}, \hat{K})$ along the axes of global coordinate frame $B(OXYZ)$, the initial and final position vectors $\mathbf{r}_1$ and $\mathbf{r}_2$ in both coordinate frames can be expressed by

$$
\begin{aligned}
{}^{B}\mathbf{r}_1 &= x_1\hat{i} + y_1\hat{j} + z_1\hat{k} & (2.39)\\
{}^{G}\mathbf{r}_1 &= X_1\hat{I} + Y_1\hat{J} + Z_1\hat{K} & (2.40)
\end{aligned}
$$

$$
\begin{aligned}
{}^{B}\mathbf{r}_2 &= x_2\hat{i} + y_2\hat{j} + z_2\hat{k} & (2.41)\\
{}^{G}\mathbf{r}_2 &= X_2\hat{I} + Y_2\hat{J} + Z_2\hat{K}. & (2.42)
\end{aligned}
$$

The vectors ${}^{B}\mathbf{r}_1$ and ${}^{B}\mathbf{r}_2$ are the initial and final positions of the vector $\mathbf{r}$ expressed in body coordinate frame $Oxyz$, and ${}^{G}\mathbf{r}_1$ and ${}^{G}\mathbf{r}_2$ are the initial and final positions of the vector $\mathbf{r}$ expressed in the global coordinate frame $OXYZ$.

The components of ${}^{B}\mathbf{r}_2$ can be found if we have the components of ${}^{G}\mathbf{r}_2$. Using Equation (2.42) and the definition of the inner product, we may write

$$
\begin{aligned}
x_2 &= \hat{i}\cdot\mathbf{r}_2 = \hat{i}\cdot X_2\hat{I} + \hat{i}\cdot Y_2\hat{J} + \hat{i}\cdot Z_2\hat{K} & (2.43)\\
y_2 &= \hat{j}\cdot\mathbf{r}_2 = \hat{j}\cdot X_2\hat{I} + \hat{j}\cdot Y_2\hat{J} + \hat{j}\cdot Z_2\hat{K} & (2.44)\\
z_2 &= \hat{k}\cdot\mathbf{r}_2 = \hat{k}\cdot X_2\hat{I} + \hat{k}\cdot Y_2\hat{J} + \hat{k}\cdot Z_2\hat{K} & (2.45)
\end{aligned}
$$

or equivalently

$$
\begin{bmatrix} x_2 \\ y_2 \\ z_2 \end{bmatrix} = \begin{bmatrix} \hat{i}\cdot\hat{I} & \hat{i}\cdot\hat{J} & \hat{i}\cdot\hat{K} \\ \hat{j}\cdot\hat{I} & \hat{j}\cdot\hat{J} & \hat{j}\cdot\hat{K} \\ \hat{k}\cdot\hat{I} & \hat{k}\cdot\hat{J} & \hat{k}\cdot\hat{K} \end{bmatrix} \begin{bmatrix} X_2 \\ Y_2 \\ Z_2 \end{bmatrix}. \qquad (2.46)
$$

The elements of the $z$-rotation matrix $A_{z,\varphi}$ are the *direction cosines* of $^G\mathbf{r}_2$ with respect to $Oxyz$. So, the elements of the matrix in Equation (2.46) are

$$
\begin{array}{lll}
\hat{\imath} \cdot \hat{I} = \cos \varphi, & \hat{\imath} \cdot \hat{J} = \sin \varphi, & \hat{\imath} \cdot \hat{K} = 0 \\
\hat{\jmath} \cdot \hat{I} = -\sin \varphi, & \hat{\jmath} \cdot \hat{J} = \cos \varphi, & \hat{\jmath} \cdot \hat{K} = 0 \\
\hat{k} \cdot \hat{I} = 0, & \hat{k} \cdot \hat{J} = 0, & \hat{k} \cdot \hat{K} = 1
\end{array}
\tag{2.47}
$$

Combining Equations (2.46) and (2.47), we can find the components of $^B\mathbf{r}_2$ by multiplying $z$-rotation matrix $A_{z,\varphi}$ and vector $^G\mathbf{r}_2$.

$$
\begin{bmatrix} x_2 \\ y_2 \\ z_2 \end{bmatrix} = \begin{bmatrix} \cos \varphi & \sin \varphi & 0 \\ -\sin \varphi & \cos \varphi & 0 \\ 0 & 0 & 1 \end{bmatrix} \begin{bmatrix} X_2 \\ Y_2 \\ Z_2 \end{bmatrix}.
\tag{2.48}
$$

It can also be shown in the following short form

$$
^B\mathbf{r}_2 = A_{z,\varphi} \, ^G\mathbf{r}_2
\tag{2.49}
$$

where

$$
A_{z,\varphi} = \begin{bmatrix} \cos \varphi & \sin \varphi & 0 \\ -\sin \varphi & \cos \varphi & 0 \\ 0 & 0 & 1 \end{bmatrix}.
\tag{2.50}
$$

Equation (2.49) says that after rotation about the $z$-axis of the local coordinate frame, the position vector in the local frame is equal to $A_{z,\varphi}$ times the position vector in the global frame. Hence, after rotation about the $z$-axis, we are able to find the coordinates of any point of a rigid body in local coordinate frame, if we have its coordinates in the global frame.

Similarly, rotation $\theta$ about the $y$-axis and rotation $\psi$ about the $x$-axis are described by the $y$-rotation matrix $A_{y,\theta}$ and the $x$-rotation matrix $A_{x,\psi}$ respectively.

$$
A_{y,\theta} = \begin{bmatrix} \cos \theta & 0 & -\sin \theta \\ 0 & 1 & 0 \\ \sin \theta & 0 & \cos \theta \end{bmatrix}
\tag{2.51}
$$

$$
A_{x,\psi} = \begin{bmatrix} 1 & 0 & 0 \\ 0 & \cos \psi & \sin \psi \\ 0 & -\sin \psi & \cos \psi \end{bmatrix}
\tag{2.52}
$$

We indicate the first, second, and third rotations about the local axes by $\varphi$, $\theta$, and $\psi$ respectively. ∎

**Example 10** *Local rotation, local position.*

*If a local coordinate frame $Oxyz$ has been rotated 60 deg about the $z$-axis and a point $P$ in the global coordinate frame $OXYZ$ is at $(4, 3, 2)$, its coordinates in the local coordinate frame $Oxyz$ are*

$$
\begin{bmatrix} x \\ y \\ z \end{bmatrix} = \begin{bmatrix} \cos 60 & \sin 60 & 0 \\ -\sin 60 & \cos 60 & 0 \\ 0 & 0 & 1 \end{bmatrix} \begin{bmatrix} 4 \\ 3 \\ 2 \end{bmatrix} = \begin{bmatrix} 4.60 \\ -1.97 \\ 2.0 \end{bmatrix}.
$$

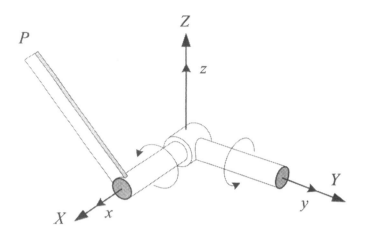

FIGURE 2.12. Arm of Example 12.

**Example 11** *Local rotation, global position.*

*If a local coordinate frame $Oxyz$ has been rotated $60 \deg$ about the z-axis and a point $P$ in the local coordinate frame $Oxyz$ is at $(4, 3, 2)$, its position in the global coordinate frame $OXYZ$ is at*

$$
\begin{bmatrix} X \\ Y \\ Z \end{bmatrix} = \begin{bmatrix} \cos 60 & \sin 60 & 0 \\ -\sin 60 & \cos 60 & 0 \\ 0 & 0 & 1 \end{bmatrix}^T \begin{bmatrix} 4 \\ 3 \\ 2 \end{bmatrix} = \begin{bmatrix} -0.60 \\ 4.96 \\ 2.0 \end{bmatrix}.
$$

**Example 12** *Successive local rotation, global position.*

*The arm shown in Figure 2.12 has two actuators. The first actuator rotates the arm $-90 \deg$ about y-axis and then the second actuator rotates the arm $90 \deg$ about x-axis. If the end point $P$ is at $^B\mathbf{r}_P = \begin{bmatrix} 9.5 & -10.1 & 10.1 \end{bmatrix}^T$, then its position in the global coordinate frame is at*

$$
\begin{aligned}
^G\mathbf{r}_2 &= [A_{x,90}\, A_{y,-90}]^{-1}\, {}^B\mathbf{r}_P \\
&= A_{y,-90}^{-1}\, A_{x,90}^{-1}\, {}^B\mathbf{r}_P \\
&= A_{y,-90}^{T}\, A_{x,90}^{T}\, {}^B\mathbf{r}_P \\
&= \begin{bmatrix} 10.1 \\ -10.1 \\ 9.5 \end{bmatrix}.
\end{aligned}
$$

# 2.5  Successive Rotation About Local Cartesian Axes

The final global position of a point $P$ in a rigid body $B$ with position vector $\mathbf{r}$, after some rotations $A_1$, $A_2$, $A_3$, ..., $A_n$ about the local axes, can be found by

$$^B\mathbf{r} = {}^B A_G\,{}^G\mathbf{r} \tag{2.53}$$

where,

$$^B A_G = A_n \cdots A_3 A_2 A_1. \tag{2.54}$$

$^B A_G$ is called the *local rotation matrix* and it maps the global coordinates to their corresponding local coordinates.

Rotation about the local coordinate axis is conceptually interesting because in a sequence of rotations, each rotation is about one of the axes of the local coordinate frame, which has been moved to its new global position during the last rotation.

Assume that we have the coordinates of every point of a rigid body in a global coordinate frame. The rigid body and its local coordinate frame rotate about a local axis, then the proper local rotation matrix relates the new global coordinates of the points to the corresponding local coordinates. If we introduce an intermediate space-fixed frame coincident with the new position of the body coordinate frame, we may give the rigid body a second rotation about a local coordinate axis. Now another proper local rotation matrix relates the coordinates in the intermediate fixed frame to the corresponding local coordinates. Hence, the final global coordinates of the points must first be transformed to the intermediate fixed frame and second transformed to the original global axes.

**Example 13** *Successive local rotation, local position.*

*A local coordinate frame $B(Oxyz)$ that initially is coincident with a global coordinate frame $G(OXYZ)$ undergoes a rotation $\varphi = 30\deg$ about the z-axis, then $\theta = 30\deg$ about the x-axis, and then $\psi = 30\deg$ about the y-axis. The local coordinates of a point $P$ located at $X = 5$, $Y = 30$, $Z = 10$ can be found by $\begin{bmatrix} x & y & z \end{bmatrix}^T = A_{y,\psi} A_{x,\theta} A_{z,\varphi} \begin{bmatrix} 5 & 30 & 10 \end{bmatrix}^T$. The local rotation matrix is*

$$^B A_G = A_{y,30} A_{x,30} A_{z,30} = \begin{bmatrix} 0.63 & 0.65 & -0.43 \\ -0.43 & 0.75 & 0.50 \\ 0.65 & -0.125 & 0.75 \end{bmatrix}$$

*and coordinates of $P$ in the local frame is*

$$\begin{bmatrix} x \\ y \\ z \end{bmatrix} = \begin{bmatrix} 0.63 & 0.65 & -0.43 \\ -0.43 & 0.75 & 0.50 \\ 0.65 & -0.125 & 0.75 \end{bmatrix} \begin{bmatrix} 5 \\ 30 \\ 10 \end{bmatrix} = \begin{bmatrix} 18.35 \\ 25.35 \\ 7.0 \end{bmatrix}.$$

**Example 14** *Successive local rotation.*

The rotation matrix for a body point $P(x, y, z)$ after rotation $A_{z,\varphi}$ followed by $A_{x,\theta}$ and $A_{y,\psi}$ is

$$
\begin{aligned}
{}^{B}A_G &= A_{y,\psi} A_{x,\theta} A_{z,\varphi} \\
&= \begin{bmatrix} c\varphi c\psi - s\theta s\varphi s\psi & c\psi s\varphi + c\varphi s\theta s\psi & -c\theta s\psi \\ -c\theta s\varphi & c\theta c\varphi & s\theta \\ c\varphi s\psi + s\theta c\psi s\varphi & s\varphi s\psi - c\varphi s\theta c\psi & c\theta c\psi \end{bmatrix} . \quad (2.55)
\end{aligned}
$$

**Example 15** *Twelve independent triple local rotations.*

Euler proved a theorem that stated: *Any two independent orthogonal coordinate frames can be related by a sequence of rotations (not more than three) about coordinate axes, where no two successive rotations may be about the same axis.* In general, there are 12 different independent combinations of triple rotation about local axes. They are:

$$
\begin{aligned}
1 &- A_{x,\psi} A_{y,\theta} A_{z,\varphi} \\
2 &- A_{y,\psi} A_{z,\theta} A_{x,\varphi} \\
3 &- A_{z,\psi} A_{x,\theta} A_{y,\varphi} \\
4 &- A_{z,\psi} A_{y,\theta} A_{x,\varphi} \\
5 &- A_{y,\psi} A_{x,\theta} A_{z,\varphi} \\
6 &- A_{x,\psi} A_{z,\theta} A_{y,\varphi} \\
7 &- A_{x,\psi} A_{y,\theta} A_{x,\varphi} \\
8 &- A_{y,\psi} A_{z,\theta} A_{y,\varphi} \\
9 &- A_{z,\psi} A_{x,\theta} A_{z,\varphi} \\
10 &- A_{x,\psi} A_{z,\theta} A_{x,\varphi} \\
11 &- A_{y,\psi} A_{x,\theta} A_{y,\varphi} \\
12 &- A_{z,\psi} A_{y,\theta} A_{z,\varphi}
\end{aligned}
\quad (2.56)
$$

The expanded form of the 12 local axes' triple rotation are presented in Appendix B.

## 2.6   Euler Angles

The rotation about the $Z$-axis of the global coordinate is called *precession*, the rotation about the $x$-axis of the local coordinate is called *nutation,* and the rotation about the $z$-axis of the local coordinate is called *spin*. The *precession-nutation-spin rotation* angles are also called *Euler angles.* Euler angles rotation matrix has a lot of application in rigid body kinematics. To find the Euler angles rotation matrix to go from the global frame $G(OXYZ)$ to the final body frame $B(Oxyz)$, we employ a body frame $B'(Ox'y'z')$ as shown in Figure 2.13 that before the first rotation coincides with the global frame. Let there be at first a rotation $\varphi$ about the $z'$-axis. Because $Z$-axis

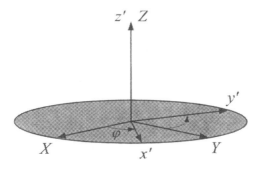

FIGURE 2.13. First Euler angle.

and $z'$-axis are coincident, by our theory

$$^{B'}\mathbf{r} = {}^{B'}A_G\,{}^G\mathbf{r} \tag{2.57}$$

$$^{B'}A_G = A_{z,\varphi} = \begin{bmatrix} \cos\varphi & \sin\varphi & 0 \\ -\sin\varphi & \cos\varphi & 0 \\ 0 & 0 & 1 \end{bmatrix}.$$

Next we consider the $B'(Ox'y'z')$ frame as a new fixed global frame and introduce a new body frame $B''(Ox''y''z'')$. Before the second rotation, the two frames coincide. Then, we execute a $\theta$ rotation about $x''$-axis as shown in Figure 2.14. The transformation between $B'(Ox'y'z')$ and $B''(Ox''y''z'')$ is

$$^{B''}\mathbf{r} = {}^{B''}A_{B'}\,{}^{B'}\mathbf{r} \tag{2.58}$$

$$^{B''}A_{B'} = A_{x,\theta} = \begin{bmatrix} 1 & 0 & 0 \\ 0 & \cos\theta & \sin\theta \\ 0 & -\sin\theta & \cos\theta \end{bmatrix}. \tag{2.59}$$

Finally we consider the $B''(Ox''y''z'')$ frame as a new fixed global frame and consider the final body frame $B(Oxyz)$ to coincide with $B''$ before the third rotation. We now execute a $\psi$ rotation about the $z''$-axis as shown in Figure 2.15. The transformation between $B''(Ox''y''z'')$ and $B(Oxyz)$ is

$$^{B}\mathbf{r} = {}^{B}A_{B''}\,{}^{B''}\mathbf{r} \tag{2.60}$$

$$^{B}A_{B''} = A_{z,\psi} = \begin{bmatrix} \cos\psi & \sin\psi & 0 \\ -\sin\psi & \cos\psi & 0 \\ 0 & 0 & 1 \end{bmatrix}. \tag{2.61}$$

By the rule of composition of rotations, the transformation from $G(OXYZ)$ to $B(Oxyz)$ is

$$^{B}\mathbf{r} = {}^{B}A_G\,{}^G\mathbf{r} \tag{2.62}$$

where

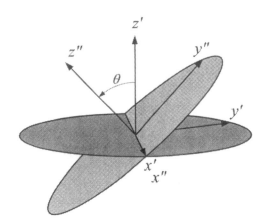

FIGURE 2.14. Second Euler angle.

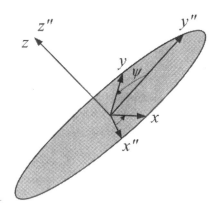

FIGURE 2.15. Third Euler angle.

$$
\begin{aligned}
{}^B A_G &= A_{z,\psi} A_{x,\theta} A_{z,\varphi} \\
&= \begin{bmatrix}
c\varphi c\psi - c\theta s\varphi s\psi & c\psi s\varphi + c\theta c\varphi s\psi & s\theta s\psi \\
-c\varphi s\psi - c\theta c\psi s\varphi & -s\varphi s\psi + c\theta c\varphi c\psi & s\theta c\psi \\
s\theta s\varphi & -c\varphi s\theta & c\theta
\end{bmatrix} \quad (2.63)
\end{aligned}
$$

and therefore,

$$
\begin{aligned}
{}^B A_G^{-1} &= {}^B A_G^T = {}^G Q_B = [A_{z,\psi} A_{x,\theta} A_{z,\varphi}]^T \\
&= \begin{bmatrix}
c\varphi c\psi - c\theta s\varphi s\psi & -c\varphi s\psi - c\theta c\psi s\varphi & s\theta s\varphi \\
c\psi s\varphi + c\theta c\varphi s\psi & -s\varphi s\psi + c\theta c\varphi c\psi & -c\varphi s\theta \\
s\theta s\psi & s\theta c\psi & c\theta
\end{bmatrix} . (2.64)
\end{aligned}
$$

Given the angles of precession $\varphi$, nutation $\theta$, and spin $\psi$, we can compute the overall rotation matrix using Equation (2.63). Also we are able to compute the equivalent precession, nutation, and spin angles when a rotation matrix is given.

If $r_{ij}$ indicates the element of row $i$ and column $j$ of the precession-nutation-spin rotation matrix, then,

$$
\theta = \cos^{-1}(r_{33}) \tag{2.65}
$$

$$
\varphi = -\tan^{-1}\left(\frac{r_{31}}{r_{32}}\right) \tag{2.66}
$$

$$
\psi = \tan^{-1}\left(\frac{r_{13}}{r_{23}}\right) \tag{2.67}
$$

provided that $\sin\theta \neq 0$.

**Example 16** *Euler angle rotation matrix.*

*The Euler or precession-nutation-spin rotation matrix for $\varphi = 79.15$ deg, $\theta = 41.41$ deg, and $\psi = -40.7$ deg would be found by substituting $\varphi$, $\theta$, and $\psi$ in Equation (2.63).*

$$
\begin{aligned}
{}^B A_G &= A_{z,-40.7} A_{x,41.41} A_{z,79.15} \\
&= \begin{bmatrix}
0.63 & 0.65 & -0.43 \\
-0.43 & 0.75 & 0.50 \\
0.65 & -0.125 & 0.75
\end{bmatrix}
\end{aligned}
$$

**Example 17** *Euler angles of a local rotation matrix.*

*The local rotation matrix after rotation 30 deg about the z-axis, then rotation 30 deg about the x-axis, and then 30 deg about the y-axis is*

$$
\begin{aligned}
{}^B A_G &= A_{y,30} A_{x,30} A_{z,30} \\
&= \begin{bmatrix}
0.63 & 0.65 & -0.43 \\
-0.43 & 0.75 & 0.50 \\
0.65 & -0.125 & 0.75
\end{bmatrix}
\end{aligned}
$$

*and therefore, the local coordinates of a sample point at $X = 5$, $Y = 30$,
$Z = 10$ are*

$$
\begin{bmatrix} x \\ y \\ z \end{bmatrix} = \begin{bmatrix} 0.63 & 0.65 & -0.43 \\ -0.43 & 0.75 & 0.50 \\ 0.65 & -0.125 & 0.75 \end{bmatrix} \begin{bmatrix} 5 \\ 30 \\ 10 \end{bmatrix} = \begin{bmatrix} 18.35 \\ 25.35 \\ 7.0 \end{bmatrix}.
$$

*The Euler angles of the corresponding precession-nutation-spin rotation
matrix are*

$$
\theta = \cos^{-1}(0.75) = 41.41 \deg
$$

$$
\varphi = -\tan^{-1}\left(\frac{0.65}{-0.125}\right) = 79.15 \deg
$$

$$
\psi = \tan^{-1}\left(\frac{-0.43}{0.50}\right) = -40.7 \deg.
$$

*Hence, $A_{y,30}A_{x,30}A_{z,30} = A_{z,\psi}A_{x,\theta}A_{z,\varphi}$ when $\varphi = 79.15$ deg, $\theta = 41.41$ deg,
and $\psi = -40.7$ deg. In other words, the rigid body attached to the local
frame moves to the final configuration by undergoing either three consecutive rotations $\varphi = 79.15$ deg, $\theta = 41.41$ deg, and $\psi = -40.7$ deg about z, x,
and z axes respectively, or three consecutive rotations 30 deg, 30 deg, and
30 deg about z, x, and y axes.*

**Example 18** *Relative rotation matrix of two bodies.*

*Consider a rigid body $B_1$ with an orientation matrix $^{B_1}A_G$ made by Euler
angles $\varphi = 30$ deg, $\theta = -45$ deg, $\psi = 60$ deg, and another rigid body $B_2$
having $\varphi = 10$ deg, $\theta = 25$ deg, $\psi = -15$ deg, with respect to the global
frame. To find the relative rotation matrix $^{B_1}A_{B_2}$ to map the coordinates
of second body frame $B_2$ to the first body frame $B_1$, we need to find the
individual rotation matrices first.*

$$
\begin{aligned}
^{B_1}A_G &= A_{z,60}A_{x,-45}A_{z,30} \\
&= \begin{bmatrix} 0.127 & 0.78 & -0.612 \\ -0.927 & -0.127 & -0.354 \\ -0.354 & 0.612 & 0.707 \end{bmatrix}
\end{aligned}
$$

$$
\begin{aligned}
^{B_2}A_G &= A_{z,10}A_{x,25}A_{z,-15} \\
&= \begin{bmatrix} 0.992 & -6.33 \times 10^{-2} & -0.109 \\ 0.103 & 0.907 & 0.408 \\ 7.34 \times 10^{-2} & -0.416 & 0.906 \end{bmatrix}
\end{aligned}
$$

*The desired rotation matrix $^{B_1}A_{B_2}$ may be found by*

$$
^{B_1}A_{B_2} = {}^{B_1}A_G \, {}^{G}A_{B_2} \tag{2.68}
$$

2. Rotation Kinematics 53

*which is equal to*

$$^{B_1}A_{B_2} = {}^{B_1}A_G {}^{B_2}A_G^T$$

$$= \begin{bmatrix} 0.992 & 0.103 & 7.34 \times 10^{-2} \\ -6.33 \times 10^{-2} & 0.907 & -0.416 \\ -0.109 & 0.408 & 0.906 \end{bmatrix}.$$

**Example 19** *Euler angles rotation matrix for small angles.*

The Euler rotation matrix ${}^B A_G = A_{z,\psi} A_{x,\theta} A_{z,\varphi}$ for very small Euler angles $\varphi, \theta$, and $\psi$ is approximated by

$$^B A_G = \begin{bmatrix} 1 & \gamma & 0 \\ -\gamma & 1 & \theta \\ 0 & -\theta & 1 \end{bmatrix} \tag{2.69}$$

*where,*

$$\gamma = \varphi + \psi. \tag{2.70}$$

*Therefore, in case of small angles of rotation, the angles $\varphi$ and $\psi$ are indistinguishable.*

**Example 20** *Small second Euler angle.*

If $\theta \to 0$ then the Euler rotation matrix ${}^B A_G = A_{z,\psi} A_{x,\theta} A_{z,\varphi}$ approaches to

$$^B A_C = \begin{bmatrix} c\varphi c\psi - s\varphi s\psi & c\psi s\varphi + c\varphi s\psi & 0 \\ -c\varphi s\psi - c\psi s\varphi & -s\varphi s\psi + c\varphi c\psi & 0 \\ 0 & 0 & 1 \end{bmatrix}$$

$$= \begin{bmatrix} \cos(\varphi + \psi) & \sin(\varphi + \psi) & 0 \\ -\sin(\varphi + \psi) & \cos(\varphi + \psi) & 0 \\ 0 & 0 & 1 \end{bmatrix} \tag{2.71}$$

*and therefore, the angles $\varphi$ and $\psi$ are indistinguishable even if the value of $\varphi$ and $\psi$ are finite. Hence, the Euler set of angles in rotation matrix (2.63) is not unique when $\theta = 0$.*

**Example 21** *Euler angles application in motion of rigid bodies.*

The $zxz$ Euler angles are good parameters to describe the configuration of a rigid body with a fixed point. The Euler angles to show the configuration of a top are shown in Figure 2.16 as an example.

**Example 22** ★ *Angular velocity vector in terms of Euler frequencies.*

A Eulerian local frame $E\left(o, \hat{e}_\varphi, \hat{e}_\theta, \hat{e}_\psi\right)$ can be introduced by defining unit vectors $\hat{e}_\varphi$, $\hat{e}_\theta$, and $\hat{e}_\psi$ as shown in Figure 2.17. Although the Eulerian frame is not necessarily orthogonal, it is very useful in rigid body kinematic analysis.

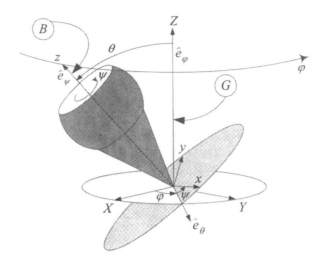

FIGURE 2.16. Application of Euler angles in describing the configuration of a top.

The angular velocity vector $_G\omega_B$ of the body frame $B(Oxyz)$ with respect to the global frame $G(OXYZ)$ can be written in Euler angles frame $E$ as the sum of three Euler angle rate vectors

$$_G^E\omega_B = \dot{\varphi}\hat{e}_\varphi + \dot{\theta}\hat{e}_\theta + \dot{\psi}\hat{e}_\psi. \tag{2.72}$$

where, the rate of Euler angles, $\dot{\varphi}$, $\dot{\theta}$, and $\dot{\psi}$ are called Euler frequencies.

To find $_G\omega_B$ in body frame we must express the unit vectors $\hat{e}_\varphi$, $\hat{e}_\theta$, and $\hat{e}_\psi$ shown in Figure 2.17, in the body frame. The unit vector $\hat{e}_\varphi = \begin{bmatrix} 0 & 0 & 1 \end{bmatrix}^T = \hat{K}$ is in the global frame and can be transformed to the body frame after three rotations

$$
\begin{aligned}
^B\hat{e}_\varphi &= {}^BA_G\,\hat{K} \\
&= A_{z,\psi}A_{x,\theta}A_{z,\varphi}\hat{K} \\
&= \begin{bmatrix} \sin\theta\sin\psi \\ \sin\theta\cos\psi \\ \cos\theta \end{bmatrix}.
\end{aligned} \tag{2.73}
$$

The unit vector $\hat{e}_\theta = \begin{bmatrix} 1 & 0 & 0 \end{bmatrix}^T = \hat{\imath}'$ is in the intermediate frame $Ox'y'z'$ and needs to get two rotations $A_{x,\theta}$ and $A_{z,\psi}$ to be transformed

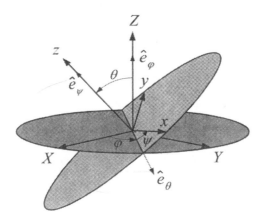

FIGURE 2.17. Euler angles frame $\hat{e}_\varphi, \hat{e}_\theta, \hat{e}_\psi$.

*to the body frame*

$$
\begin{aligned}
{}^B\hat{e}_\theta &= {}^B A_{Ox'y'z'}\,\hat{\imath}' \\
&= A_{z,\psi}\,A_{x,\theta}\,\hat{\imath}' \\
&= \begin{bmatrix} \cos\psi \\ -\sin\psi \\ 0 \end{bmatrix}.
\end{aligned}
\tag{2.74}
$$

*The unit vector $\hat{e}_\psi$ is already in the body frame, $\hat{e}_\psi = \begin{bmatrix} 0 & 0 & 1 \end{bmatrix}^T = \hat{k}$. Therefore, ${}_G\boldsymbol{\omega}_B$ is expressed in body coordinate frame as*

$$
\begin{aligned}
{}^B_G\boldsymbol{\omega}_B &= \dot\varphi \begin{bmatrix} \sin\theta\sin\psi \\ \sin\theta\cos\psi \\ \cos\theta \end{bmatrix} + \dot\theta \begin{bmatrix} \cos\psi \\ -\sin\psi \\ 0 \end{bmatrix} + \dot\psi \begin{bmatrix} 0 \\ 0 \\ 1 \end{bmatrix} \\
&= \left( \dot\varphi\sin\theta\sin\psi + \dot\theta\cos\psi \right)\hat{\imath} \\
&\quad + \left( \dot\varphi\sin\theta\cos\psi - \dot\theta\sin\psi \right)\hat{\jmath} \\
&\quad + \left( \dot\varphi\cos\theta + \dot\psi \right)\hat{k}
\end{aligned}
\tag{2.75}
$$

*and therefore, components of ${}_G\boldsymbol{\omega}_B$ in body frame $Oxyz$ are related to the Euler angle frame $O\varphi\theta\psi$ by the following relationship:*

$$
{}^B_G\boldsymbol{\omega}_B = {}^B A_E \, {}^E_G\boldsymbol{\omega}_B
\tag{2.76}
$$

$$
\begin{bmatrix} \omega_x \\ \omega_y \\ \omega_z \end{bmatrix} = \begin{bmatrix} \sin\theta\sin\psi & \cos\psi & 0 \\ \sin\theta\cos\psi & -\sin\psi & 0 \\ \cos\theta & 0 & 1 \end{bmatrix} \begin{bmatrix} \dot\varphi \\ \dot\theta \\ \dot\psi \end{bmatrix}.
\tag{2.77}
$$

*Then, ${}_G\boldsymbol{\omega}_B$ can be expressed in the global frame using an inverse transfor-*

*mation of Euler rotation matrix (2.63)*

$$
\begin{aligned}
{}^G_G\omega_B &= {}^BA_G^{-1}\,{}^B_G\omega_B \\[4pt]
&= {}^BA_G^{-1}
\begin{bmatrix}
\dot\varphi\sin\theta\sin\psi+\dot\theta\cos\psi \\
\dot\varphi\sin\theta\cos\psi-\dot\theta\sin\psi \\
\dot\varphi\cos\theta+\dot\psi
\end{bmatrix}
\end{aligned}
\tag{2.78}
$$

$$
\begin{aligned}
&= \left(\dot\theta\cos\varphi+\dot\psi\sin\theta\sin\varphi\right)\hat I \\
&\quad+\left(\dot\theta\sin\varphi-\dot\psi\cos\varphi\sin\theta\right)\hat J \\
&\quad+\left(\dot\varphi+\dot\psi\cos\theta\right)\hat K
\end{aligned}
\tag{2.79}
$$

*and hence, components of $_G\omega_B$ in global coordinate frame $OXYZ$ are related to the Euler angle coordinate frame $O\varphi\theta\psi$ by the following relationship:*

$$
{}^G_G\omega_B = {}^GQ_E\,{}^E_G\omega_B
\tag{2.80}
$$

$$
\begin{bmatrix}
\omega_X \\ \omega_Y \\ \omega_Z
\end{bmatrix}
=
\begin{bmatrix}
0 & \cos\varphi & \sin\theta\sin\varphi \\
0 & \sin\varphi & -\cos\varphi\sin\theta \\
1 & 0 & \cos\theta
\end{bmatrix}
\begin{bmatrix}
\dot\varphi \\ \dot\theta \\ \dot\psi
\end{bmatrix}.
\tag{2.81}
$$

**Example 23 ★** *Euler frequencies based on a Cartesian angular velocity vector.*

The vector $^B_G\omega_B$, that indicates the angular velocity of a rigid body $B$ with respect to the global frame $G$ written in frame $B$, is related to the Euler frequencies by

$$
{}^B_G\omega_B = {}^BA_E\,{}^E_G\omega_B
$$

$$
{}^B_G\omega_B =
\begin{bmatrix}
\omega_x \\ \omega_y \\ \omega_z
\end{bmatrix}
=
\begin{bmatrix}
\sin\theta\sin\psi & \cos\psi & 0 \\
\sin\theta\cos\psi & -\sin\psi & 0 \\
\cos\theta & 0 & 1
\end{bmatrix}
\begin{bmatrix}
\dot\varphi \\ \dot\theta \\ \dot\psi
\end{bmatrix}.
\tag{2.82}
$$

*The matrix of coefficients is not an orthogonal matrix because,*

$$
{}^BA_E^T \neq {}^BA_E^{-1}
$$

$$
{}^BA_E^T =
\begin{bmatrix}
\sin\theta\sin\psi & \sin\theta\cos\psi & \cos\theta \\
\cos\psi & -\sin\psi & 0 \\
0 & 0 & 1
\end{bmatrix}
\tag{2.83}
$$

$$
{}^BA_E^{-1} = \frac{1}{\sin\theta}
\begin{bmatrix}
\sin\psi & \cos\psi & 0 \\
\sin\theta\cos\psi & -\sin\theta\sin\psi & 0 \\
-\cos\theta\sin\psi & -\cos\theta\cos\psi & 1
\end{bmatrix}.
\tag{2.84}
$$

*It is because the Euler angles coordinate frame $O\varphi\theta\psi$ is not an orthogonal frame. For the same reason, the matrix of coefficients that relates the Euler* *frequencies and the components of* ${}^G_G\omega_B$

$$
{}^G_G\omega_B = {}^GQ_E\,{}^E_G\omega_B
$$

$$
{}^G_G\omega_B = \begin{bmatrix} \omega_X \\ \omega_Y \\ \omega_Z \end{bmatrix} = \begin{bmatrix} 0 & \cos\varphi & \sin\theta\sin\varphi \\ 0 & \sin\varphi & -\cos\varphi\sin\theta \\ 1 & 0 & \cos\theta \end{bmatrix} \begin{bmatrix} \dot\varphi \\ \dot\theta \\ \dot\psi \end{bmatrix}
$$

*is not an orthogonal matrix. Therefore, the Euler frequencies based on local
and global decomposition of the angular velocity vector* ${}_G\omega_B$ *must solely be
found by the inverse of coefficient matrices*

$$
{}^E_G\omega_B = {}^BA_E^{-1}\,{}^B_G\omega_B \tag{2.85}
$$

$$
\begin{bmatrix} \dot\varphi \\ \dot\theta \\ \dot\psi \end{bmatrix} = \frac{1}{\sin\theta} \begin{bmatrix} \sin\psi & \cos\psi & 0 \\ \sin\theta\cos\psi & -\sin\theta\sin\psi & 0 \\ -\cos\theta\sin\psi & -\cos\theta\cos\psi & 1 \end{bmatrix} \begin{bmatrix} \omega_x \\ \omega_y \\ \omega_z \end{bmatrix} \tag{2.86}
$$

*and*

$$
{}^E_G\omega_B = {}^GQ_E^{-1}\,{}^G_G\omega_B \tag{2.87}
$$

$$
\begin{bmatrix} \dot\varphi \\ \dot\theta \\ \dot\psi \end{bmatrix} = \frac{1}{\sin\theta} \begin{bmatrix} -\cos\theta\sin\varphi & \cos\theta\cos\varphi & 1 \\ \sin\theta\cos\varphi & \sin\theta\sin\varphi & 0 \\ \sin\varphi & -\cos\varphi & 0 \end{bmatrix} \begin{bmatrix} \omega_X \\ \omega_Y \\ \omega_Z \end{bmatrix}. \tag{2.88}
$$

*Using (2.85) and (2.87), it can be verified that the transformation ma-
trix* ${}^BA_G = {}^BA_E\,{}^GQ_E^{-1}$ *would be the same as Euler transformation matrix
(2.63).*

*The angular velocity vector can thus be expressed as*

$$
{}_G\omega_B = \begin{bmatrix} \hat\imath & \hat\jmath & \hat k \end{bmatrix} \begin{bmatrix} \omega_x \\ \omega_y \\ \omega_z \end{bmatrix}
$$

$$
= \begin{bmatrix} \hat I & \hat J & \hat K \end{bmatrix} \begin{bmatrix} \omega_X \\ \omega_Y \\ \omega_Z \end{bmatrix}
$$

$$
= \begin{bmatrix} \hat K & \hat e_\theta & \hat k \end{bmatrix} \begin{bmatrix} \dot\varphi \\ \dot\theta \\ \dot\psi \end{bmatrix}.
$$

**Example 24** ★ *Integrability of the angular velocity components.*
   *The integrability condition for an arbitrary total differential*

$$
dz = P\,dx + Q\,dy \tag{2.89}
$$

*is*

$$
\frac{\partial P}{\partial y} = \frac{\partial Q}{\partial x}. \tag{2.90}
$$

The angular velocity components $\omega_x$, $\omega_y$, and $\omega_z$ along the body coordinate axes $x$, $y$, and $z$ can not be integrated to obtain the associated angles because

$$\omega_x dt = \sin\theta \sin\psi \, d\varphi + \cos\psi \, d\theta \qquad (2.91)$$

and

$$\frac{\partial (\sin\theta \sin\psi)}{\partial \theta} \neq \frac{\partial \cos\psi}{\partial \varphi}.$$

However, the integrability condition (2.90) is satisfied by the Euler frequencies because, for example,

$$d\theta = \cos\psi \, (\omega_x \, dt) - \sin\psi \, (\omega_y \, dt) \qquad (2.92)$$

and

$$\frac{\partial (\cos\psi)}{\omega_y \, \partial t} = \frac{\partial (-\sin\psi)}{\omega_x \, \partial t} = \frac{\sin\psi \cos\theta \cos\psi}{\sin\theta}. \qquad (2.93)$$

It can be checked that $d\varphi$ and $d\psi$ are also integrable.

**Example 25 ★** *Cardan angles and frequencies.*
The system of Euler angles is singular at $\theta = 0$, and as a consequence, $\varphi$ and $\psi$ become coplanar and indistinguishable. From 12 angle systems of Appendix B, each with certain names, characteristics, advantages, and disadvantages, the rotations about three different axes such as ${}^B A_G = A_{z,\psi} A_{y,\theta} A_{x,\varphi}$ are called Cardan or Bryant angles. The Cardan angle system is not singular at $\theta = 0$, and has some application in mechatronics and attitude analysis of satellites in a central force field.

$$ {}^B A_G = \begin{bmatrix} c\theta c\psi & c\varphi s\psi + s\theta c\psi s\varphi & s\psi s\psi - c\varphi s\theta c\psi \\ -c\theta s\psi & c\varphi c\psi - s\theta s\varphi s\psi & c\psi s\varphi + c\varphi s\theta s\psi \\ s\theta & -c\theta s\varphi & c\theta c\varphi \end{bmatrix} $$

The angular velocity $\omega$ of a rigid body can either be expressed in terms of the components along the axes of $B(Oxyz)$, or in terms of the Cardan frequencies along the axes of the nonorthogonal Cardan frame. The angular velocity in terms of Cardan frequencies is

$$ {}_G\omega_B = \dot{\varphi} A_{z,\psi} A_{y,\theta} \begin{bmatrix} 1 \\ 0 \\ 0 \end{bmatrix} + \dot{\theta} A_{z,\psi} \begin{bmatrix} 0 \\ 1 \\ 0 \end{bmatrix} + \dot{\psi} \begin{bmatrix} 0 \\ 0 \\ 1 \end{bmatrix} \qquad (2.94) $$

therefore,

$$ \begin{bmatrix} \omega_x \\ \omega_y \\ \omega_z \end{bmatrix} = \begin{bmatrix} \cos\theta \cos\psi & \sin\psi & 0 \\ -\cos\theta \sin\psi & \cos\psi & 0 \\ \sin\theta & 0 & 1 \end{bmatrix} \begin{bmatrix} \dot{\varphi} \\ \dot{\theta} \\ \dot{\psi} \end{bmatrix} $$

$$ \begin{bmatrix} \dot{\varphi} \\ \dot{\theta} \\ \dot{\psi} \end{bmatrix} = \begin{bmatrix} \frac{\cos\psi}{\cos\theta} & -\frac{\sin\psi}{\cos\theta} & 0 \\ \sin\psi & \cos\psi & 0 \\ -\tan\theta \cos\psi & \tan\theta \sin\psi & 1 \end{bmatrix} \begin{bmatrix} \omega_x \\ \omega_y \\ \omega_z \end{bmatrix}. $$

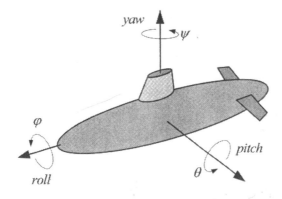

FIGURE 2.18. Local roll-pitch-yaw angles.

*In case of small Cardan angles, we have*

$$^BA_G = \begin{bmatrix} 1 & \psi & -\theta \\ -\psi & 1 & \varphi \\ \theta & -\varphi & 1 \end{bmatrix}$$

*and*

$$\begin{bmatrix} \omega_x \\ \omega_y \\ \omega_z \end{bmatrix} = \begin{bmatrix} 1 & \psi & 0 \\ -\psi & 1 & 0 \\ \theta & 0 & 1 \end{bmatrix} \begin{bmatrix} \dot\varphi \\ \dot\theta \\ \dot\psi \end{bmatrix}.$$

## 2.7   Local Roll-Pitch-Yaw Angles

Rotation about the $x$-axis of the local frame is called *roll* or *bank*, rotation about $y$-axis of the local frame is called *pitch* or *attitude*, and rotation about the $z$-axis of the local frame is called *yaw, spin,* or *heading*. The local roll-pitch-yaw angles are shown in Figure 2.18.

The local *roll-pitch-yaw rotation matrix* is

$$\begin{aligned} ^BA_G &= A_{z,\psi}A_{y,\theta}A_{x,\varphi} \\ &= \begin{bmatrix} c\theta c\psi & c\varphi s\psi + s\theta c\psi s\varphi & s\varphi s\psi - c\varphi s\theta c\psi \\ -c\theta s\psi & c\varphi c\psi - s\theta s\varphi s\psi & c\psi s\varphi + c\varphi s\theta s\psi \\ s\theta & -c\theta s\varphi & c\theta c\varphi \end{bmatrix}. \end{aligned} \quad (2.95)$$

Note the difference between roll-pitch-yaw and Euler angles, although we show both utilizing $\varphi$, $\theta$, and $\psi$.

**Example 26** ★ *Angular velocity and local roll-pitch-yaw rate.*

*Using the roll-pitch-yaw frequencies, the angular velocity of a body $B$ with respect to the global reference frame is*

$$
\begin{aligned}
{}_{G}\boldsymbol{\omega}_B &= \omega_x \hat{\imath} + \omega_y \hat{\jmath} + \omega_z \hat{k} \\
&= \dot{\varphi}\hat{e}_\varphi + \dot{\theta}\hat{e}_\theta + \dot{\psi}\hat{e}_\psi.
\end{aligned}
\tag{2.96}
$$

*Relationships between the components of ${}_{G}\boldsymbol{\omega}_B$ in body frame and roll-pitch-yaw components are found when the local roll unit vector $\hat{e}_\varphi$ and pitch unit vector $\hat{e}_\theta$ are transformed to the body frame. The roll unit vector $\hat{e}_\varphi = \begin{bmatrix} 1 & 0 & 0 \end{bmatrix}^T$ transforms to the body frame after rotation $\theta$ and then rotation $\psi$*

$$
{}^{B}\hat{e}_\varphi = A_{z,\psi} A_{y,\theta} \begin{bmatrix} 1 \\ 0 \\ 0 \end{bmatrix} = \begin{bmatrix} \cos\theta\cos\psi \\ -\cos\theta\sin\psi \\ \sin\theta \end{bmatrix}.
\tag{2.97}
$$

*The pitch unit vector $\hat{e}_\theta = \begin{bmatrix} 0 & 1 & 0 \end{bmatrix}^T$ transforms to the body frame after rotation $\psi$*

$$
{}^{B}\hat{e}_\theta = A_{z,\psi} \begin{bmatrix} 0 \\ 1 \\ 0 \end{bmatrix} = \begin{bmatrix} \sin\psi \\ \cos\psi \\ 0 \end{bmatrix}.
\tag{2.98}
$$

*The yaw unit vector $\hat{e}_\psi = \begin{bmatrix} 0 & 0 & 1 \end{bmatrix}^T$ is already along the local $z$-axis. Hence, ${}_{G}\boldsymbol{\omega}_B$ can be expressed in body frame $Oxyz$ as*

$$
\begin{aligned}
{}^{B}_{G}\boldsymbol{\omega}_B &= \begin{bmatrix} \omega_x \\ \omega_y \\ \omega_z \end{bmatrix} \\[2mm]
&= \dot{\varphi}\begin{bmatrix} \cos\theta\cos\psi \\ -\cos\theta\sin\psi \\ \sin\theta \end{bmatrix} + \dot{\theta}\begin{bmatrix} \sin\psi \\ \cos\psi \\ 0 \end{bmatrix} + \dot{\psi}\begin{bmatrix} 0 \\ 0 \\ 1 \end{bmatrix} \\[2mm]
&= \begin{bmatrix} \cos\theta\cos\psi & \sin\psi & 0 \\ -\cos\theta\sin\psi & \cos\psi & 0 \\ \sin\theta & 0 & 1 \end{bmatrix} \begin{bmatrix} \dot{\varphi} \\ \dot{\theta} \\ \dot{\psi} \end{bmatrix}
\end{aligned}
\tag{2.99}
$$

*and therefore, ${}_{G}\boldsymbol{\omega}_B$ in global frame $OXYZ$ in terms of local roll-pitch-yaw* *frequencies is*

$$
\begin{aligned}
{}^G_G\omega_B &= \begin{bmatrix} \omega_X \\ \omega_Y \\ \omega_Z \end{bmatrix} = {}^B A_G^{-1} \begin{bmatrix} \omega_x \\ \omega_y \\ \omega_z \end{bmatrix} \\[2mm]
&= {}^B A_G^{-1} \begin{bmatrix} \dot\theta\sin\psi + \dot\varphi\cos\theta\cos\psi \\ \dot\theta\cos\psi - \dot\varphi\cos\theta\sin\psi \\ \dot\psi + \dot\varphi\sin\theta \end{bmatrix} \\[2mm]
&= \begin{bmatrix} \dot\varphi + \dot\psi\sin\theta \\ \dot\theta\cos\varphi - \dot\psi\cos\theta\sin\varphi \\ \dot\theta\sin\varphi + \dot\psi\cos\theta\cos\varphi \end{bmatrix} \\[2mm]
&= \begin{bmatrix} 1 & 0 & \sin\theta \\ 0 & \cos\varphi & -\cos\theta\sin\varphi \\ 0 & \sin\varphi & \cos\theta\cos\varphi \end{bmatrix} \begin{bmatrix} \dot\varphi \\ \dot\theta \\ \dot\psi \end{bmatrix} .
\end{aligned} \tag{2.100}
$$

# 2.8   Local Axes Rotation Versus Global Axes Rotation

The global rotation matrix ${}^G Q_B$ is equal to the inverse of the local rotation matrix ${}^B A_G$ and vice versa,

$$
{}^G Q_B = {}^B A_G^{-1} \quad , \quad {}^B A_G = {}^G Q_B^{-1} \tag{2.101}
$$

where

$$
\begin{aligned}
{}^G Q_B &= A_1^{-1} A_2^{-1} A_3^{-1} \cdots A_n^{-1} \tag{2.102} \\
{}^B A_G &= Q_1^{-1} Q_2^{-1} Q_3^{-1} \cdots Q_n^{-1}. \tag{2.103}
\end{aligned}
$$

Also, premultiplication of the global rotation matrix is equal to postmultiplication of the local rotation matrix.

**Proof.** Consider a sequence of global rotations and their resultant global rotation matrix ${}^G Q_B$ to transform a position vector ${}^B\mathbf{r}$ to ${}^G\mathbf{r}$.

$$
{}^G\mathbf{r} = {}^G Q_B \, {}^B\mathbf{r} \tag{2.104}
$$

The global position vector ${}^G\mathbf{r}$ can also be transformed to ${}^B\mathbf{r}$ using a local rotation matrix ${}^B A_G$.

$$
{}^B\mathbf{r} = {}^B A_G \, {}^G\mathbf{r} \tag{2.105}
$$

Combining Equations (2.104) and (2.105) leads to

$$
\begin{aligned}
{}^G\mathbf{r} &= {}^G Q_B \, {}^B A_G \, {}^G\mathbf{r} \tag{2.106} \\
{}^B\mathbf{r} &= {}^B A_G \, {}^G Q_B \, {}^B\mathbf{r} \tag{2.107}
\end{aligned}
$$

and hence,

$$^GQ_B\,^BA_G = {}^BA_G\,^GQ_B = \mathbf{I}. \tag{2.108}$$

Therefore, the global and local rotation matrices are the inverse of each other

$$
\begin{aligned}
^GQ_B &= {}^BA_G^{-1} \\
^GQ_B^{-1} &= {}^BA_G.
\end{aligned}
\tag{2.109}
$$

Assume that $^GQ_B = Q_n \cdots Q_3 Q_2 Q_1$ and $^BA_G = A_n \cdots A_3 A_2 A_1$ then,

$$
\begin{aligned}
^GQ_B &= {}^BA_G^{-1} = A_1^{-1}A_2^{-1}A_3^{-1}\cdots A_n^{-1} & (2.110) \\
^BA_G &= {}^GQ_B^{-1} = Q_1^{-1}Q_2^{-1}Q_3^{-1}\cdots Q_n^{-1} & (2.111)
\end{aligned}
$$

and Equation (2.108) becomes

$$Q_n \cdots Q_2 Q_1 A_n \cdots A_2 A_1 = A_n \cdots A_2 A_1 Q_n \cdots Q_2 Q_1 = \mathbf{I} \tag{2.112}$$

and therefore,

$$
\begin{aligned}
Q_n \cdots Q_3 Q_2 Q_1 &= A_1^{-1}A_2^{-1}A_3^{-1}\cdots A_n^{-1} \\
A_n \cdots A_3 A_2 A_1 &= Q_1^{-1}Q_2^{-1}Q_3^{-1}\cdots Q_n^{-1}
\end{aligned}
\tag{2.113}
$$

or

$$
\begin{aligned}
Q_1^{-1}Q_2^{-1}Q_3^{-1}\cdots Q_n^{-1}Q_n \cdots Q_3 Q_2 Q_1 &= \mathbf{I} & (2.114)\\
A_1^{-1}A_2^{-1}A_3^{-1}\cdots A_n^{-1}A_n \cdots A_3 A_2 A_1 &= \mathbf{I}. & (2.115)
\end{aligned}
$$

Hence, the effect of in order rotations about the global coordinate axes is equivalent to the effect of the same rotations about the local coordinate axes performed in the reverse order. ■

**Example 27** *Global position and postmultiplication of rotation matrix.*

*The local position of a point $P$ after rotation is at $^B\mathbf{r} = \begin{bmatrix} 1 & 2 & 3 \end{bmatrix}^T$. If the local rotation matrix to transform $^G\mathbf{r}$ to $^B\mathbf{r}$ is given as*

$$
^BA_{z,\varphi} =
\begin{bmatrix}
\cos\varphi & \sin\varphi & 0 \\
-\sin\varphi & \cos\varphi & 0 \\
0 & 0 & 1
\end{bmatrix}
=
\begin{bmatrix}
\cos 30 & \sin 30 & 0 \\
-\sin 30 & \cos 30 & 0 \\
0 & 0 & 1
\end{bmatrix},
$$

*then we may find the global position vector $^G\mathbf{r}$ by postmultiplication $^BA_{z,\varphi}$ by the local position vector $^B\mathbf{r}^T$,*

$$
\begin{aligned}
^G\mathbf{r}^T &= {}^B\mathbf{r}^T\,{}^BA_{z,\varphi} \\
&= \begin{bmatrix} 1 & 2 & 3 \end{bmatrix}
\begin{bmatrix}
\cos 30 & \sin 30 & 0 \\
-\sin 30 & \cos 30 & 0 \\
0 & 0 & 1
\end{bmatrix} \\
&= \begin{bmatrix} -0.13 & 2.23 & 3.0 \end{bmatrix}
\end{aligned}
$$

*instead of premultiplication of* $^B A_{z,\varphi}^{-1}$ *by* $^B \mathbf{r}$.

$$
\begin{aligned}
^G\mathbf{r} &= {}^B A_{z,\varphi}^{-1} \, {}^B\mathbf{r} \\
&= \begin{bmatrix} \cos 30 & -\sin 30 & 0 \\ \sin 30 & \cos 30 & 0 \\ 0 & 0 & 1 \end{bmatrix} \begin{bmatrix} 1 \\ 2 \\ 3 \end{bmatrix} = \begin{bmatrix} -0.13 \\ 2.23 \\ 3 \end{bmatrix}
\end{aligned}
$$

## 2.9  General Transformation

Consider a general situation in which two coordinate frames, $G(OXYZ)$ and $B(Oxyz)$ with a common origin $O$, are employed to express the components of a vector $\mathbf{r}$. There is always a *transformation matrix* $^G R_B$ to map the components of $\mathbf{r}$ from the reference frame $B(Oxyz)$ to the other reference frame $G(OXYZ)$.

$$
^G\mathbf{r} = {}^G R_B \, {}^B\mathbf{r} \tag{2.116}
$$

In addition, the inverse map, $^B\mathbf{r} = {}^G R_B^{-1} \, {}^G\mathbf{r}$, can be done by $^B R_G$

$$
^B\mathbf{r} = {}^B R_G \, {}^G\mathbf{r} \tag{2.117}
$$

where,

$$
\left| {}^G R_B \right| = \left| {}^B R_G \right| = 1 \tag{2.118}
$$

and

$$
^B R_G = {}^G R_B^{-1} = {}^G R_B^T. \tag{2.119}
$$

**Proof.** Decomposition of the unit vectors of $G(OXYZ)$ along the axes of $B(Oxyz)$

$$
\begin{aligned}
\hat{I} &= (\hat{I}\cdot\hat{\imath})\hat{\imath} + (\hat{I}\cdot\hat{\jmath})\hat{\jmath} + (\hat{I}\cdot\hat{k})\hat{k} \tag{2.120} \\
\hat{J} &= (\hat{J}\cdot\hat{\imath})\hat{\imath} + (\hat{J}\cdot\hat{\jmath})\hat{\jmath} + (\hat{J}\cdot\hat{k})\hat{k} \tag{2.121} \\
\hat{K} &= (\hat{K}\cdot\hat{\imath})\hat{\imath} + (\hat{K}\cdot\hat{\jmath})\hat{\jmath} + (\hat{K}\cdot\hat{k})\hat{k} \tag{2.122}
\end{aligned}
$$

introduces the transformation matrix $^G R_B$ to map the local frame to the global frame

$$
\begin{bmatrix} \hat{I} \\ \hat{J} \\ \hat{K} \end{bmatrix} = \begin{bmatrix} \hat{I}\cdot\hat{\imath} & \hat{I}\cdot\hat{\jmath} & \hat{I}\cdot\hat{k} \\ \hat{J}\cdot\hat{\imath} & \hat{J}\cdot\hat{\jmath} & \hat{J}\cdot\hat{k} \\ \hat{K}\cdot\hat{\imath} & \hat{K}\cdot\hat{\jmath} & \hat{K}\cdot\hat{k} \end{bmatrix} \begin{bmatrix} \hat{\imath} \\ \hat{\jmath} \\ \hat{k} \end{bmatrix} = {}^G R_B \begin{bmatrix} \hat{\imath} \\ \hat{\jmath} \\ \hat{k} \end{bmatrix} \tag{2.123}
$$

where

$$
{}^{G}R_{B} = \begin{bmatrix} \hat{I}\cdot\hat{\imath} & \hat{I}\cdot\hat{\jmath} & \hat{I}\cdot\hat{k} \\ \hat{J}\cdot\hat{\imath} & \hat{J}\cdot\hat{\jmath} & \hat{J}\cdot\hat{k} \\ \hat{K}\cdot\hat{\imath} & \hat{K}\cdot\hat{\jmath} & \hat{K}\cdot\hat{k} \end{bmatrix}
$$

$$
= \begin{bmatrix} \cos(\hat{I},\hat{\imath}) & \cos(\hat{I},\hat{\jmath}) & \cos(\hat{I},\hat{k}) \\ \cos(\hat{J},\hat{\imath}) & \cos(\hat{J},\hat{\jmath}) & \cos(\hat{J},\hat{k}) \\ \cos(\hat{K},\hat{\imath}) & \cos(\hat{K},\hat{\jmath}) & \cos(\hat{K},\hat{k}) \end{bmatrix}. \qquad (2.124)
$$

Each column of ${}^{G}R_{B}$ is decomposition of a unit vector of the local frame $B(Oxyz)$ in the global frame $G(OXYZ)$

$$
{}^{G}R_{B} = \begin{bmatrix} | & | & | \\ {}^{G}\hat{\imath} & {}^{G}\hat{\jmath} & {}^{G}\hat{k} \\ | & | & | \end{bmatrix} = \begin{bmatrix} | & | & | \\ \hat{\mathbf{r}}_{V_1} & \hat{\mathbf{r}}_{V_2} & \hat{\mathbf{r}}_{V_3} \\ | & | & | \end{bmatrix}. \qquad (2.125)
$$

Similarly, each row of ${}^{G}R_{B}$ is decomposition of a unit vector of the global frame $G(OXYZ)$ in the local frame $B(Oxyz)$

$$
{}^{G}R_{B} = \begin{bmatrix} - & {}^{B}\hat{I} & - \\ - & {}^{B}\hat{J} & - \\ - & {}^{B}\hat{K} & - \end{bmatrix} = \begin{bmatrix} - & \hat{\mathbf{r}}_{H_1} & - \\ - & \hat{\mathbf{r}}_{H_2} & - \\ - & \hat{\mathbf{r}}_{H_3} & - \end{bmatrix}. \qquad (2.126)
$$

The elements of ${}^{G}R_{B}$ are direction cosines of the axes of $G(OXYZ)$ in frame $B(Oxyz)$. This set of nine direction cosines then completely specifies the orientation of the frame $B(Oxyz)$ in the frame $G(OXYZ)$, and can be used to map the coordinates of any point $(x, y, z)$ to its corresponding coordinates $(X, Y, Z)$.

Alternatively, using the method of unit vector decomposition to develop the matrix ${}^{B}R_{G}$ leads to

$$
{}^{B}\mathbf{r} = {}^{B}R_{G}\,{}^{G}\mathbf{r} = {}^{G}R_{B}^{-1}\,{}^{G}\mathbf{r}
$$

$$
{}^{B}R_{G} = \begin{bmatrix} \hat{\imath}\cdot\hat{I} & \hat{\imath}\cdot\hat{J} & \hat{\imath}\cdot\hat{K} \\ \hat{\jmath}\cdot\hat{I} & \hat{\jmath}\cdot\hat{J} & \hat{\jmath}\cdot\hat{K} \\ \hat{k}\cdot\hat{I} & \hat{k}\cdot\hat{J} & \hat{k}\cdot\hat{K} \end{bmatrix}
$$

$$
= \begin{bmatrix} \cos(\hat{\imath},\hat{I}) & \cos(\hat{\imath},\hat{J}) & \cos(\hat{\imath},\hat{K}) \\ \cos(\hat{\jmath},\hat{I}) & \cos(\hat{\jmath},\hat{J}) & \cos(\hat{\jmath},\hat{K}) \\ \cos(\hat{k},\hat{I}) & \cos(\hat{k},\hat{J}) & \cos(\hat{k},\hat{K}) \end{bmatrix} \qquad (2.127)
$$

and shows that the inverse of a transformation matrix is equal to the transpose of the transformation matrix,

$$
{}^{G}R_{B}^{-1} = {}^{G}R_{B}^{T}. \qquad (2.128)
$$

A matrix with condition (2.128) is called *orthogonal*. Orthogonality of $R$ comes from this fact that it maps an orthogonal coordinate frame to another orthogonal coordinate frame.

The transformation matrix $R$ has only three *independent* elements. The constraint equations among the elements of $R$ will be found by applying the orthogonality condition (2.128).

$$^GR_B \cdot {}^GR_B^T = I \qquad (2.129)$$

$$\begin{bmatrix} r_{11} & r_{12} & r_{13} \\ r_{21} & r_{22} & r_{23} \\ r_{31} & r_{32} & r_{33} \end{bmatrix} \begin{bmatrix} r_{11} & r_{21} & r_{31} \\ r_{12} & r_{22} & r_{32} \\ r_{13} & r_{23} & r_{33} \end{bmatrix} = \begin{bmatrix} 1 & 0 & 0 \\ 0 & 1 & 0 \\ 0 & 0 & 1 \end{bmatrix}. \qquad (2.130)$$

Therefore, the dot product of any two different rows of $^GR_B$ is zero, and the dot product of any row of $^GR_B$ with the same row is one.

$$\begin{aligned} r_{11}^2 + r_{12}^2 + r_{13}^2 &= 1 \\ r_{21}^2 + r_{22}^2 + r_{23}^2 &= 1 \\ r_{31}^2 + r_{32}^2 + r_{33}^2 &= 1 \\ r_{11}r_{21} + r_{12}r_{22} + r_{13}r_{23} &= 0 \\ r_{11}r_{31} + r_{12}r_{32} + r_{13}r_{33} &= 0 \\ r_{21}r_{31} + r_{22}r_{32} + r_{23}r_{33} &= 0 \end{aligned} \qquad (2.131)$$

These relations are also true for columns of $^GR_B$, and evidently for rows and columns of $^BR_G$. The orthogonality condition can be summarized in the following equation:

$$\hat{\mathbf{r}}_{H_i} \cdot \hat{\mathbf{r}}_{H_j} = \hat{\mathbf{r}}_{H_i}^T \hat{\mathbf{r}}_{H_j} = \sum_{i=1}^{3} r_{ij}r_{ik} = \delta_{jk} \qquad (j,k=1,2,3) \qquad (2.132)$$

where $r_{ij}$ is the element of row $i$ and column $j$ of the transformation matrix $R$, and $\delta_{jk}$ is the *Kronecker's delta*

$$\delta_{jk} = 1 \text{ if } j = k, \text{ and } \delta_{jk} = 0 \text{ if } j \neq k. \qquad (2.133)$$

Equation (2.132) gives six independent relations satisfied by nine direction cosines. It follows that there are only three independent direction cosines. The independent elements of the matrix $R$ cannot obviously be in the same row or column, or any diagonal.

The determinant of a transformation matrix is equal to one,

$$|^GR_B| = 1 \qquad (2.134)$$

because of Equation (2.129), and noting that

$$\begin{aligned} |^GR_B \cdot {}^GR_B^T| &= |^GR_B| \cdot |^GR_B^T| \\ &= |^GR_B| \cdot |^GR_B| \\ &= |^GR_B|^2 = 1. \end{aligned} \qquad (2.135)$$

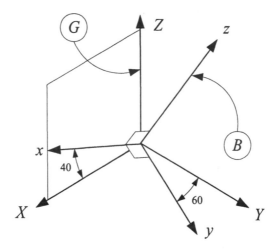

FIGURE 2.19. Body and global coordinate frames of Example 28.

Using linear algebra and row vectors $\hat{\mathbf{r}}_{H_1}, \hat{\mathbf{r}}_{H_2}$, and $\hat{\mathbf{r}}_{H_3}$ of $^{G}R_B$, we know that

$$\left| ^{G}R_B \right| = \hat{\mathbf{r}}_{H_1}^{T} \cdot (\hat{\mathbf{r}}_{H_2} \times \hat{\mathbf{r}}_{H_3}) \tag{2.136}$$

and because the coordinate system is right handed, we have $\hat{\mathbf{r}}_{H_2} \times \hat{\mathbf{r}}_{H_3} = \hat{\mathbf{r}}_{H_1}$ so $\left| ^{G}R_B \right| = \hat{\mathbf{r}}_{H_1}^{T} \cdot \hat{\mathbf{r}}_{H_1} = 1.$ ∎

**Example 28** *Elements of transformation matrix.*

*The position vector* **r** *of a point* $P$ *may be expressed in terms of its components with respect to either* $G\,(OXYZ)$ *or* $B\,(Oxyz)$ *frames. Body and a global coordinate frames are shown in Figure 2.19. If* $^{G}\mathbf{r} = 100\hat{I} - 50\hat{J} + 150\hat{K}$, *and we are looking for components of* **r** *in the* $Oxyz$ *frame, then we have to find the proper rotation matrix* $^{B}R_G$ *first.*

*The row elements of* $^{B}R_G$ *are the direction cosines of the* $Oxyz$ *axes in the* $OXYZ$ *coordinate frame. The* $x$-*axis lies in the* $XZ$ *plane at* 40 deg *from the* $X$-*axis, and the angle between* $y$ *and* $Y$ *is* 60 deg. *Therefore,*

$$
^{B}R_G = \begin{bmatrix} \hat{\imath} \cdot \hat{I} & \hat{\imath} \cdot \hat{J} & \hat{\imath} \cdot \hat{K} \\ \hat{\jmath} \cdot \hat{I} & \hat{\jmath} \cdot \hat{J} & \hat{\jmath} \cdot \hat{K} \\ \hat{k} \cdot \hat{I} & \hat{k} \cdot \hat{J} & \hat{k} \cdot \hat{K} \end{bmatrix}
$$

$$
= \begin{bmatrix} \cos 40 & 0 & \sin 40 \\ \hat{\jmath} \cdot \hat{I} & \cos 60 & \hat{\jmath} \cdot \hat{K} \\ \hat{k} \cdot \hat{I} & \hat{k} \cdot \hat{J} & \hat{k} \cdot \hat{K} \end{bmatrix}
$$

$$
= \begin{bmatrix} 0.766 & 0 & 0.643 \\ \hat{\jmath} \cdot \hat{I} & 0.5 & \hat{\jmath} \cdot \hat{K} \\ \hat{k} \cdot \hat{I} & \hat{k} \cdot \hat{J} & \hat{k} \cdot \hat{K} \end{bmatrix} \tag{2.137}
$$

*and by using* ${}^{B}R_{G}\,{}^{G}R_{B} = {}^{B}R_{G}\,{}^{B}R_{G}^{T} = I$

$$\begin{bmatrix} 0.766 & 0 & 0.643 \\ r_{21} & 0.5 & r_{23} \\ r_{31} & r_{32} & r_{33} \end{bmatrix} \begin{bmatrix} 0.766 & r_{21} & r_{31} \\ 0 & 0.5 & r_{32} \\ 0.643 & r_{23} & r_{33} \end{bmatrix} = \begin{bmatrix} 1 & 0 & 0 \\ 0 & 1 & 0 \\ 0 & 0 & 1 \end{bmatrix} \quad (2.138)$$

*we obtain a set of equations to find the missing elements.*

$$\begin{aligned} 0.766\,r_{21} + 0.643\,r_{23} &= 0 \\ 0.766\,r_{31} + 0.643\,r_{33} &= 0 \\ r_{21}^{2} + r_{23}^{2} + 0.25 &= 1 \\ r_{21}r_{31} + 0.5r_{32} + r_{23}r_{33} &= 0 \\ r_{31}^{2} + r_{32}^{2} + r_{33}^{2} &= 1 \end{aligned} \quad (2.139)$$

*Solving these equations provides the following transformation matrix:*

$$^{B}R_{G} = \begin{bmatrix} 0.766 & 0 & 0.643 \\ 0.557 & 0.5 & -0.663 \\ -0.322 & 0.866 & 0.383 \end{bmatrix} \quad (2.140)$$

*and then we can find the components of* ${}^{B}\mathbf{r}$.

$$\begin{aligned} {}^{B}\mathbf{r} &= {}^{B}R_{G}\,{}^{G}\mathbf{r} \\ &= \begin{bmatrix} 0.766 & 0 & 0.643 \\ 0.557 & 0.5 & -0.663 \\ -0.322 & 0.866 & 0.383 \end{bmatrix} \begin{bmatrix} 100 \\ -50 \\ 150 \end{bmatrix} \\ &= \begin{bmatrix} 173.05 \\ -68.75 \\ -18.05 \end{bmatrix} \end{aligned} \quad (2.141)$$

**Example 29** *Global position, using* ${}^{B}\mathbf{r}$ *and* ${}^{B}R_{G}$.

The position vector $\mathbf{r}$ of a point $P$ may be described in either $G\,(OXYZ)$ or $B\,(Oxyz)$ frames. If ${}^{B}\mathbf{r} = 100\hat{\imath} - 50\hat{\jmath} + 150\hat{k}$, and the following ${}^{B}R_{G}$ is the transformation matrix to map ${}^{G}\mathbf{r}$ to ${}^{B}\mathbf{r}$

$$\begin{aligned} {}^{B}\mathbf{r} &= {}^{B}R_{G}\,{}^{G}\mathbf{r} \\ &= \begin{bmatrix} 0.766 & 0 & 0.643 \\ 0.557 & 0.5 & -0.663 \\ -0.322 & 0.866 & 0.383 \end{bmatrix} {}^{G}\mathbf{r} \end{aligned} \quad (2.142)$$

*then the components of $^G\mathbf{r}$ in $G\,(OXYZ)$ would be*

$$
\begin{aligned}
^G\mathbf{r} &= {}^G R_B {}^B\mathbf{r} \\
&= {}^B R_G^T {}^B\mathbf{r} \\
&= \begin{bmatrix} 0.766 & 0.557 & -0.322 \\ 0 & 0.5 & 0.866 \\ 0.643 & -0.663 & 0.383 \end{bmatrix} \begin{bmatrix} 100 \\ -50 \\ 150 \end{bmatrix} \\
&= \begin{bmatrix} 0.45 \\ 104.9 \\ 154.9 \end{bmatrix}.
\end{aligned}
\tag{2.143}
$$

**Example 30** *Two points transformation matrix.*
*The global position vector of two points, $P_1$ and $P_2$, of a rigid body $B$ are*

$$
^G\mathbf{r}_{P_1} = \begin{bmatrix} 1.077 \\ 1.365 \\ 2.666 \end{bmatrix}
\tag{2.144}
$$

$$
^G\mathbf{r}_{P_2} = \begin{bmatrix} -0.473 \\ 2.239 \\ -0.959 \end{bmatrix}.
\tag{2.145}
$$

*The origin of the body $B\,(Oxyz)$ is fixed on the origin of $G\,(OXYZ)$, and the points $P_1$ and $P_2$ are lying on the local $x$-axis and $y$-axis respectively. To find $^G R_B$, we use the local unit vectors $^G\hat{\imath}$ and $^G\hat{\jmath}$*

$$
^G\hat{\imath} = \frac{^G\mathbf{r}_{P_1}}{|^G\mathbf{r}_{P_1}|} = \begin{bmatrix} 0.338 \\ 0.429 \\ 0.838 \end{bmatrix}
\tag{2.146}
$$

$$
^G\hat{\jmath} = \frac{^G\mathbf{r}_{P_2}}{|^G\mathbf{r}_{P_2}|} = \begin{bmatrix} -0.191 \\ 0.902 \\ -0.387 \end{bmatrix}
\tag{2.147}
$$

*to obtain $^G\hat{k}$*

$$
\begin{aligned}
^G\hat{k} &= \hat{\imath} \times \hat{\jmath} = \tilde{\imath}\,\hat{\jmath} \\
&= \begin{bmatrix} 0 & -0.838 & 0.429 \\ 0.838 & 0 & -0.338 \\ -0.429 & 0.338 & 0 \end{bmatrix} \begin{bmatrix} -0.191 \\ 0.902 \\ -0.387 \end{bmatrix} \\
&= \begin{bmatrix} -0.922 \\ -0.029 \\ 0.387 \end{bmatrix}
\end{aligned}
\tag{2.148}
$$

*where $\tilde{\imath}$ is the skew-symmetric matrix corresponding to $\hat{\imath}$, and $\tilde{\imath}\,\hat{\jmath}$ is an alternative for $\hat{\imath} \times \hat{\jmath}$.*

*Hence, the transformation matrix using the coordinates of two points* $^G\mathbf{r}_{P_1}$ *and* $^G\mathbf{r}_{P_2}$ *would be*

$$^GR_B = \begin{bmatrix} ^G\hat{\imath} & ^G\hat{\jmath} & ^G\hat{k} \end{bmatrix}$$

$$= \begin{bmatrix} 0.338 & -0.191 & -0.922 \\ 0.429 & 0.902 & -0.029 \\ 0.838 & -0.387 & 0.387 \end{bmatrix}. \qquad (2.149)$$

**Example 31** *Length invariant of a position vector.*

*Describing a vector in different frames utilizing rotation matrices does not affect the length and direction properties of the vector. Therefore, length of a vector is an invariant*

$$|\mathbf{r}| = |^G\mathbf{r}| = |^B\mathbf{r}|. \qquad (2.150)$$

*The length invariant property can be shown by*

$$\begin{aligned} |\mathbf{r}|^2 &= {}^G\mathbf{r}^T \, {}^G\mathbf{r} \\ &= \begin{bmatrix} ^GR_B \, {}^B\mathbf{r} \end{bmatrix}^T \, {}^GR_B \, {}^B\mathbf{r} \\ &= {}^B\mathbf{r}^T \, {}^GR_B^T \, {}^GR_B \, {}^B\mathbf{r} \\ &= {}^B\mathbf{r}^T \, {}^B\mathbf{r}. \end{aligned} \qquad (2.151)$$

**Example 32** *Skew symmetric matrices for* $\hat{\imath}$, $\hat{\jmath}$, *and* $\hat{k}$.

*The definition of skew symmetric matrix* $\tilde{a}$ *corresponding to a vector* $\mathbf{a}$ *is defined by*

$$\tilde{a} = \begin{bmatrix} 0 & -a_3 & a_2 \\ a_3 & 0 & -a_1 \\ -a_2 & a_1 & 0 \end{bmatrix}. \qquad (2.152)$$

*Hence,*

$$\tilde{\imath} = \begin{bmatrix} 0 & 0 & 0 \\ 0 & 0 & -1 \\ 0 & 1 & 0 \end{bmatrix} \qquad (2.153)$$

$$\tilde{\jmath} = \begin{bmatrix} 0 & 0 & 1 \\ 0 & 0 & 0 \\ -1 & 0 & 0 \end{bmatrix} \qquad (2.154)$$

$$\tilde{k} = \begin{bmatrix} 0 & -1 & 0 \\ 1 & 0 & 0 \\ 0 & 0 & 0 \end{bmatrix}. \qquad (2.155)$$

**Example 33** *Inverse of Euler angles rotation matrix.*

*Precession-nutation-spin or Euler angle rotation matrix (2.63)*

$$
\begin{aligned}
{}^{B}R_G &= A_{z,\psi}A_{x,\theta}A_{z,\varphi} \\
&= \begin{bmatrix} c\varphi c\psi - c\theta s\varphi s\psi & c\psi s\varphi + c\theta c\varphi s\psi & s\theta s\psi \\ -c\varphi s\psi - c\theta c\psi s\varphi & -s\varphi s\psi + c\theta c\varphi c\psi & s\theta c\psi \\ s\theta s\varphi & -c\varphi s\theta & c\theta \end{bmatrix} \quad (2.156)
\end{aligned}
$$

*must be inverted to be a transformation matrix to map body coordinates to global coordinates.*

$$
\begin{aligned}
{}^{G}R_B &= {}^{B}R_G^{-1} = A_{z,\varphi}^{T}A_{x,\theta}^{T}A_{z,\psi}^{T} \\
&= \begin{bmatrix} c\varphi c\psi - c\theta s\varphi s\psi & -c\varphi s\psi - c\theta c\psi s\varphi & s\theta s\varphi \\ c\psi s\varphi + c\theta c\varphi s\psi & -s\varphi s\psi + c\theta c\varphi c\psi & -c\varphi s\theta \\ s\theta s\psi & s\theta c\psi & c\theta \end{bmatrix} \quad (2.157)
\end{aligned}
$$

The transformation matrix (2.156) is called a *local Euler rotation matrix*, and (2.157) is called a *global Euler rotation matrix*.

**Example 34 ★** *Group property of transformations.*
A set $S$ together with a binary operation $\otimes$ defined on elements of $S$ is called a group $(S, \otimes)$ if it satisfies the following four axioms.

1. **Closure:** If $s_1, s_2 \in S$, then $s_1 \otimes s_2 \in S$.

2. **Identity:** There exists an identity element $s_0$ such that $s_0 \otimes s = s \otimes s_0 = s$ for $\forall s \in S$.

3. **Inverse:** For each $s \in S$, there exists a unique inverse $s^{-1} \in S$ such that $s^{-1} \otimes s = s \otimes s^{-1} = s_0$.

4. **Associativity:** If $s_1, s_2, s_3 \in S$, then $(s_1 \otimes s_2) \otimes s_3 = s_1 \otimes (s_2 \otimes s_3)$.

   Three dimensional coordinate transformations make a group if we define the set of rotation matrices by

$$
S = \left\{ R \in \mathbb{R}^{3\times3} : RR^T = R^T R = \mathbf{I}, |R| = 1 \right\}. \quad (2.158)
$$

Therefore, the elements of the set $S$ are transformation matrices $R_i$, the binary operator $\otimes$ is matrix multiplication, the identity matrix is $\mathbf{I}$, and the inverse of element $R$ is $R^{-1} = R^T$.

$S$ is also a continuous group because

5. The binary matrix multiplication is a continuous operation, and

6. The inverse of any element in $S$ is a continuous function of that element.

*Therefore, S is a **differentiable manifold**. A group that is a differentiable manifold is called a **Lie group**.*

**Example 35** ★ *Transformation with determinant −1.*

*An orthogonal matrix with determinant +1 corresponds to a rotation as described in Equation (2.134). In contrast, an orthogonal matrix with determinant −1 describes a reflection. Moreover it transforms a right-handed coordinate system into a left-handed, and vice versa. This transformation does not correspond to any possible physical action on rigid bodies.*

**Example 36** *Alternative proof for transformation matrix.*

*Starting with an identity*

$$\begin{bmatrix} \hat{\imath} & \hat{\jmath} & \hat{k} \end{bmatrix} \begin{bmatrix} \hat{\imath} \\ \hat{\jmath} \\ \hat{k} \end{bmatrix} = 1 \qquad (2.159)$$

*we may write*

$$\begin{bmatrix} \hat{I} \\ \hat{J} \\ \hat{K} \end{bmatrix} = \begin{bmatrix} \hat{I} \\ \hat{J} \\ \hat{K} \end{bmatrix} \begin{bmatrix} \hat{\imath} & \hat{\jmath} & \hat{k} \end{bmatrix} \begin{bmatrix} \hat{\imath} \\ \hat{\jmath} \\ \hat{k} \end{bmatrix}. \qquad (2.160)$$

*Since matrix multiplication can be performed in any order we find*

$$\begin{bmatrix} \hat{I} \\ \hat{J} \\ \hat{K} \end{bmatrix} = \begin{bmatrix} \hat{I}\cdot\hat{\imath} & \hat{I}\cdot\hat{\jmath} & \hat{I}\cdot\hat{k} \\ \hat{J}\cdot\hat{\imath} & \hat{J}\cdot\hat{\jmath} & \hat{J}\cdot\hat{k} \\ \hat{K}\cdot\hat{\imath} & \hat{K}\cdot\hat{\jmath} & \hat{K}\cdot\hat{k} \end{bmatrix} \begin{bmatrix} \hat{\imath} \\ \hat{\jmath} \\ \hat{k} \end{bmatrix}$$

$$= {}^{G}R_B \begin{bmatrix} \hat{\imath} \\ \hat{\jmath} \\ \hat{k} \end{bmatrix} \qquad (2.161)$$

*where,*

$$ {}^{G}R_B = \begin{bmatrix} \hat{I} \\ \hat{J} \\ \hat{K} \end{bmatrix} \begin{bmatrix} \hat{\imath} & \hat{\jmath} & \hat{k} \end{bmatrix}. \qquad (2.162)$$

*Following the same method we can show that*

$$ {}^{B}R_G = \begin{bmatrix} \hat{\imath} \\ \hat{\jmath} \\ \hat{k} \end{bmatrix} \begin{bmatrix} \hat{I} & \hat{J} & \hat{K} \end{bmatrix}. \qquad (2.163)$$

## 2.10  Active and Passive Transformation

Rotation of a local frame when the position vector ${}^{G}\mathbf{r}$ of a point $P$ is fixed in global frame and does not rotate with the local frame, is called *passive*

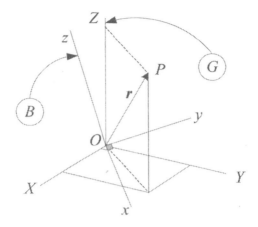

FIGURE 2.20. A position vector **r**, in a local and a global frame.

*transformation.* Alternatively, rotation of a local frame when the position vector $^B\mathbf{r}$ of a point $P$ is fixed in the local frame and rotates with the local frame, is called *active transformation*. Surprisingly, the passive and active transformations are mathematically equivalent. In other words, the rotation matrix for a rotated frame and rotated vector (active transformation) is the same as the rotation matrix for a rotated frame and fixed vector (passive transformation).

**Proof.** Consider a rotated local frame $B(Oxyz)$ with respect to a fixed global frame $G(OXYZ)$, as shown in Figure 2.20. $P$ is a fixed point in the global frame, and so is its global position vector $^G\mathbf{r}$. Position vector of $P$ can be decomposed in either a local or global coordinate frame, denoted by $^B\mathbf{r}$ and $^G\mathbf{r}$ respectively.

The transformation from $^G\mathbf{r}$ to $^B\mathbf{r}$ is equivalent to the required rotation of the body frame $B(Oxyz)$ to be coincided with the global frame $G(OXYZ)$. This is a passive transformation because the local frame cannot move the vector $^G\mathbf{r}$. In a passive transformation, we usually have the coordinates of $P$ in a global frame and we need its coordinates in a local frame; hence, we use the following equation:

$$^B\mathbf{r} = \,^B R_G \,^G\mathbf{r}. \tag{2.164}$$

We may alternatively assume that $B(Oxyz)$ was coincident with $G(OXYZ)$ and the vector $\mathbf{r} = \,^B\mathbf{r}$ was fixed in $B(Oxyz)$, before $B(Oxyz)$ and $^B\mathbf{r}$ move to the new position in $G(OXYZ)$. This is an active transformation and there is a rotation matrix to map the coordinates of $^B\mathbf{r}$ in the local frame to the coordinates of $^G\mathbf{r}$ in global frame. In an active transformation, we usually have the coordinates of $P$ in the local frame and we need its coordinates in the global frame; hence, we use the following equation:

$$^G\mathbf{r} = {}^G R_B\, {}^B\mathbf{r}. \tag{2.165}$$

■

## 2.11  Summary

Two Cartesian coordinate frames with a common origin are convertible based on nine directional cosines of a frame in the other. The conversion of coordinates in the two frames can be cast in matrix transformation

$$^G\mathbf{r} = {}^G R_B\, {}^B\mathbf{r} \tag{2.166}$$

$$\begin{bmatrix} X_2 \\ Y_2 \\ Z_2 \end{bmatrix} = \begin{bmatrix} \hat{I}\cdot\hat{\imath} & \hat{I}\cdot\hat{\jmath} & \hat{I}\cdot\hat{k} \\ \hat{J}\cdot\hat{\imath} & \hat{J}\cdot\hat{\jmath} & \hat{J}\cdot\hat{k} \\ \hat{K}\cdot\hat{\imath} & \hat{K}\cdot\hat{\jmath} & \hat{K}\cdot\hat{k} \end{bmatrix} \begin{bmatrix} x_2 \\ y_2 \\ z_2 \end{bmatrix} \tag{2.167}$$

where,

$$^G R_B = \begin{bmatrix} \cos(\hat{I},\hat{\imath}) & \cos(\hat{I},\hat{\jmath}) & \cos(\hat{I},\hat{k}) \\ \cos(\hat{J},\hat{\imath}) & \cos(\hat{J},\hat{\jmath}) & \cos(\hat{J},\hat{k}) \\ \cos(\hat{K},\hat{\imath}) & \cos(\hat{K},\hat{\jmath}) & \cos(\hat{K},\hat{k}) \end{bmatrix}. \tag{2.168}$$

The transformation matrix $^G R_B$ is orthogonal and therefore its inverse is equal to its transpose

$$^G R_B^{-1} = {}^G R_B^T. \tag{2.169}$$

The orthogonality condition generates six equations between the elements of $^G R_B$ that shows only three elements of $^G R_B$ are independent.

Any rotation can be decomposed into three simpler rotations about the principal coordinate axes in either the $B$ or $G$ frame.

# Exercises

1. Notation and symbols.

   Describe the meaning of

   a- $^G\mathbf{r}$     b- $^G\mathbf{r}_P$     c- $^B\mathbf{r}_P$     d- $^GR_B$     e- $^GR_B^T$     f- $^BR_G$

   g- $^BR_G^{-1}$     h- $^Gd_B$     i- $^2d_1$     j- $Q_X$     k- $Q_{Y,\beta}$     l- $Q_{Y,45}^{-1}$

   m- $\hat{k}$     n- $\hat{J}$     o- $A_{z,\varphi}^T$     p- $\hat{e}_\psi$     q- $\tilde{\imath}$     r- $\mathbf{I}$.

2. Body point and global rotations.

   The point $P$ is at $\mathbf{r}_P = (1,2,1)$ in a body coordinate $B(Oxyz)$. Find the final global position of $P$ after a rotation of 30 deg about the $X$-axis, followed by a 45 deg rotation about the $Z$-axis.

3. Body point after global rotation.

   Find the position of a point $P$ in the local coordinate, if it is moved to $^G\mathbf{r}_P = [1,3,2]^T$ after a 60 deg rotation about the $Z$-axis.

4. Invariant of a vector.

   A point was at $^B\mathbf{r}_P = [1,2,z]^T$. After a rotation of 60 deg about the $X$-axis, followed by a 30 deg rotation about the $Z$-axis, it is at

   $$^G\mathbf{r}_P = \begin{bmatrix} X \\ Y \\ 2.933 \end{bmatrix}.$$

   Find $z$, $X$, and $Y$.

5. Constant length vector.

   Show that the length of a vector will not change by rotation.

   $$\left| {}^G\mathbf{r} \right| = \left| {}^GR_B \, {}^B\mathbf{r} \right|$$

   Show that the distance between two body points will not change by rotation.

   $$\left| {}^B\mathbf{p}_1 - {}^B\mathbf{p}_2 \right| = \left| {}^GR_B \, {}^B\mathbf{p}_1 - {}^GR_B \, {}^B\mathbf{p}_2 \right|$$

6. Repeated global rotations.

   Rotate $^B\mathbf{r}_P = [2,2,3]^T$, 60 deg about the $X$-axis, followed by 30 deg about the $Z$-axis. Then, repeat the sequence of rotations for 60 deg about the $X$-axis, followed by 30 deg about the $Z$-axis. After how many rotations will point $P$ be back to its initial global position?

7. ★ Repeated global rotations.

How many rotations of $\alpha = \pi/m \deg$ about the $X$-axis, followed by $\beta = \pi/n \deg$ about the $Z$-axis are needed to bring a body point to its initial global position, if $m, n \in \mathbb{N}$?

8. Triple global rotations.

Verify the equations in Appendix A.

9. ★ Special triple rotation.

Assume that the first triple rotation in Appendix A brings a body point back to its initial global position. What are the angles $\alpha \neq 0$, $\beta \neq 0$, and $\gamma \neq 0$?

10. ★ Combination of triple rotations.

Any triple rotation in Appendix A can move a body point to its new global position. Assume $\alpha_1, \beta_1$, and $\gamma_1$ for the case $1 - Q_{X,\gamma_1} Q_{Y,\beta_1} Q_{Z,\alpha_1}$ are given. What can $\alpha_2$, $\beta_2$, and $\gamma_2$ be (in terms of $\alpha_1$, $\beta_1$, and $\gamma_1$) to get the same global position if use the case $2 - Q_{Y,\gamma_2} Q_{Z,\beta_2} Q_{X,\alpha_2}$?

11. Global roll-pitch-yaw rotation matrix.

Calculate the global roll-pitch-yaw rotation matrix for $\alpha = 30$, $\beta = 30$, and $\gamma = 30$.

12. Global roll-pitch-yaw rotation angles.

Calculate the role, pitch, and yaw angles for the following rotation matrix:

$$
{}^B R_G = \begin{bmatrix} 0.53 & -0.84 & 0.13 \\ 0.0 & 0.15 & 0.99 \\ -0.85 & -0.52 & 0.081 \end{bmatrix}
$$

13. Body point, local rotation.

What is the global coordinates of a body point at ${}^B r_P = [2, 2, 3]^T$, after a rotation of 60 deg about the $x$-axis?

14. Two local rotations.

Find the global coordinates of a body point at ${}^B r_P = [2, 2, 3]^T$ after a rotation of 60 deg about the $x$-axis followed by 60 deg about the $z$-axis.

15. Triple local rotations.

Verify the equations in Appendix B.

16. Combination of local and global rotations.

Find the final global position of a body point at ${}^B r_P = [10, 10, -10]^T$ after a rotation of 45 deg about the $x$-axis followed by 60 deg about the $Z$-axis.

17. Combination of global and local rotations.

    Find the final global position of a body point at $^B\mathbf{r}_P = [10, 10, -10]^T$ after a rotation of 45 deg about the $X$-axis followed by 60 deg about the $z$-axis.

18. Repeated local rotations.

    Rotate $^B\mathbf{r}_P = [2, 2, 3]^T$, 60 deg about the $x$-axis, followed by 30 deg about the $z$-axis. Then repeat the sequence of rotations for 60 deg about the $x$-axis, followed by 30 deg about the $z$-axis. After how many rotations will point $P$ be back to its initial global position?

19. ★ Repeated local rotations.

    How many rotations of $\alpha = \pi/m$ deg about the $x$-axis, followed by $\beta = \pi/n$ deg about the $z$-axis are needed to bring a body point to its initial global position if $m, n \in \mathbb{N}$?

20. ★ Remaining rotation.

    Find the result of the following sequence of rotations:

    $$^G R_B = A_{z,\varphi}^T A_{y,\theta}^T A_{z,\psi}^T A_{y,-\theta}^T A_{z,-\varphi}^T$$

21. Angles from rotation matrix.

    Find the angles $\varphi$, $\theta$, and $\psi$ if the rotation of the case $1 - A_{x,\psi} A_{y,\theta} A_{z,\varphi}$ in Appendix B is given.

22. Euler angles from rotation matrix.

    Find the Euler angles for the following rotation matrix:

    $$^B R_G = \begin{bmatrix} 0.53 & -0.84 & 0.13 \\ 0.0 & 0.15 & 0.99 \\ -0.85 & -0.52 & 0.081 \end{bmatrix}$$

23. Equivalent Euler angles to two rotations.

    Find the Euler angles corresponding to the rotation matrix $^B R_G = A_{y,45} A_{x,30}$.

24. Equivalent Euler angles to three rotations.

    Find the Euler angles corresponding to the rotation matrix $^B R_G = A_{z,60} A_{y,45} A_{x,30}$.

25. ★ Local and global positions, Euler angles.

    Find the conditions between the Euler angles to transform $^G\mathbf{r}_P = [1, 1, 0]^T$ to $^B\mathbf{r}_P = [0, 1, 1]^T$.

26. ★ Equivalent Euler angles to a triple rotations.

    Find the Euler angles for the rotation matrix of the case

    $$4 - A_{z,\psi'} A_{y,\theta'} A_{x,\varphi'}$$

    in Appendix B.

27. ★ Integrability of Euler frequencies.

    Show that $d\varphi$ and $d\psi$ are integrable, if $\varphi$ and $\psi$ are first and third Euler angles.

28. ★ Cardan angles for Euler angles.

    Find the Cardan angles for a given set of Euler angles.

29. ★ Cardan frequencies for Euler frequencies.

    Find the Euler frequencies in terms of Cardan frequencies.

30. Elements of rotation matrix.

    The elements of rotation matrix $^{G}R_B$ are

    $$^{G}R_B = \begin{bmatrix} \cos(\hat{I},\hat{\imath}) & \cos(\hat{I},\hat{\jmath}) & \cos(\hat{I},\hat{k}) \\ \cos(\hat{J},\hat{\imath}) & \cos(\hat{J},\hat{\jmath}) & \cos(\hat{J},\hat{k}) \\ \cos(\hat{K},\hat{\imath}) & \cos(\hat{K},\hat{\jmath}) & \cos(\hat{K},\hat{k}) \end{bmatrix}.$$

    Find $^{G}R_B$ if $^{G}\mathbf{r}_{P_1} = (0.7071, -1.2247, 1.4142)$ is a point on the $x$-axis, and $^{G}\mathbf{r}_{P_2} = (2.7803, 0.38049, -1.0607)$ is a point on the $y$-axis.

31. Linearly independent vectors.

    A set of vectors $\mathbf{a}_1, \mathbf{a}_2, \cdots, \mathbf{a}_n$ are considered linearly independent if the equation

    $$k_1\mathbf{a}_1 + k_2\mathbf{a}_2 + \cdots + k_n\mathbf{a}_n = 0$$

    in which $k_1, k_2, \cdots, k_n$ are unknown coefficients, has only one solution

    $$k_1 = k_2 = \cdots = k_n = 0.$$

    Verify that the unit vectors of a body frame $B(Oxyz)$, expressed in the global frame $G(OXYZ)$, are linearly independent.

32. Product of orthogonal matrices.

    A matrix $R$ is called orthogonal if $R^{-1} = R^T$ where $\left(R^T\right)_{ij} = R_{ji}$. Prove that the product of two orthogonal matrices is also orthogonal.

33. Vector identity.

    The formula $(a+b)^2 = a^2 + b^2 + 2ab$ for scalars, is equivalent to

    $$(\mathbf{a}+\mathbf{b})^2 = \mathbf{a}\cdot\mathbf{a} + \mathbf{b}\cdot\mathbf{b} + 2\mathbf{a}\cdot\mathbf{b}$$

for vectors. Show that this formula is equal to

$$(\mathbf{a} + \mathbf{b})^2 = \mathbf{a} \cdot \mathbf{a} + \mathbf{b} \cdot \mathbf{b} + 2\,{}^{G}R_B\,\mathbf{a} \cdot \mathbf{b}$$

if $\mathbf{a}$ is a vector in local frame and $\mathbf{b}$ is a vector in global frame.

34. **Rotation as a linear operation.**

    Show that
    $$R\,(\mathbf{a} \times \mathbf{b}) = R\mathbf{a} \times R\mathbf{b}$$

    where $R$ is a rotation matrix and $\mathbf{a}$ and $\mathbf{b}$ are two vectors defined in a coordinate frame.

35. **Scalar triple product.**

    Show that for three arbitrary vectors $\mathbf{a}$, $\mathbf{b}$, and $\mathbf{c}$ we have

    $$\mathbf{a} \cdot (\mathbf{b} \times \mathbf{c}) = (\mathbf{a} \times \mathbf{b}) \cdot \mathbf{c}.$$

# 3

# Orientation Kinematics

## 3.1   Axis-angle Rotation

Two parameters are necessary to define the direction of a line through $O$ and one is necessary to define the amount of rotation of a rigid body about this line. Let the body frame $B(Oxyz)$ rotate $\phi$ about a line indicated by a unit vector $\hat{u}$ with direction cosines $u_1, u_2, u_3$,

$$\hat{u} = u_1\hat{I} + u_2\hat{J} + u_3\hat{K} \tag{3.1}$$

$$\sqrt{u_1^2 + u_2^2 + u_3^2} = 1. \tag{3.2}$$

This is called *axis-angle* representation of a rotation.

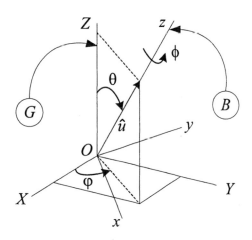

FIGURE 3.1. Axis of rotation $\hat{u}$ when it is coincident with the local $z$-axis.

A transformation matrix $^G R_B$ that maps the coordinates in the local frame $B(Oxyz)$ to the corresponding coordinates in the global frame $G(OXYZ)$,

$$^G\mathbf{r} = {}^G R_B {}^B\mathbf{r} \tag{3.3}$$

is

$$^G R_B = R_{\hat{u},\phi} = \mathbf{I}\cos\phi + \hat{u}\hat{u}^T \operatorname{vers}\phi + \tilde{u}\sin\phi \tag{3.4}$$

$$^G R_B = \begin{bmatrix} u_1^2 \operatorname{vers}\phi + c\phi & u_1 u_2 \operatorname{vers}\phi - u_3 s\phi & u_1 u_3 \operatorname{vers}\phi + u_2 s\phi \\ u_1 u_2 \operatorname{vers}\phi + u_3 s\phi & u_2^2 \operatorname{vers}\phi + c\phi & u_2 u_3 \operatorname{vers}\phi - u_1 s\phi \\ u_1 u_3 \operatorname{vers}\phi - u_2 s\phi & u_2 u_3 \operatorname{vers}\phi + u_1 s\phi & u_3^2 \operatorname{vers}\phi + c\phi \end{bmatrix}$$

$$\tag{3.5}$$

where

$$
\begin{aligned}
\text{vers}\,\phi &= \text{versine}\,\phi \\
&= 1 - \cos\phi \\
&= 2\sin^2\frac{\phi}{2}
\end{aligned}
\tag{3.6}
$$

and $\tilde{u}$ is the skew-symmetric matrix corresponding to the vector $\hat{u}$

$$
\tilde{u} = \begin{bmatrix} 0 & -u_3 & u_2 \\ u_3 & 0 & -u_1 \\ -u_2 & u_1 & 0 \end{bmatrix}.
\tag{3.7}
$$

A matrix $\tilde{u}$ is *skew-symmetric* if

$$
\tilde{u}^T = -\tilde{u}.
\tag{3.8}
$$

The transformation matrix (3.5) is the most general rotation matrix for a local frame rotating with respect to a global frame.

If the axis of rotation (3.1) coincides with a global coordinate axis, then the Equations (2.21), (2.22), or (2.23) will be reproduced.

**Proof.** Interestingly, the effect of rotation $\phi$ about an axis $\hat{u}$ is equivalent to a sequence of rotations about the axes of a local frame in which the local frame is first rotated to bring one of its axes, say the $z$-axis, into coincidence with the rotation axis $\hat{u}$, followed by a rotation $\phi$ about that local axis, then the reverse of the first sequence of rotations.

Figure 3.1 illustrates an axis of rotation $\hat{u} = u_1\hat{I} + u_2\hat{J} + u_3\hat{K}$ , the global frame $G\,(OXYZ)$, and the rotated local frame $B\,(Oxyz)$ when the local $z$-axis is coincident with $\hat{u}$. Based on Figure 3.1, the local frame $B\,(Oxyz)$ undergoes a sequence of rotations $\varphi$ about the $z$-axis and $\theta$ about the $y$-axis to bring the local $z$-axis into coincidence with the rotation axis $\hat{u}$, followed by rotation $\phi$ about $\hat{u}$, and then perform the sequence backward. Therefore, using (2.110), the rotation matrix $^{G}R_B$ to map coordinates in local frame to their coordinates in global frame after rotation $\phi$ about $\hat{u}$ is

$$
\begin{aligned}
^{G}R_B &= {}^{B}R_G^{-1} = {}^{B}R_G^{T} = R_{\hat{u},\phi} \\
&= [A_{z,-\varphi}A_{y,-\theta}A_{z,\phi}A_{y,\theta}A_{z,\varphi}]^{T} \\
&= A_{z,\varphi}^{T}A_{y,\theta}^{T}A_{z,\phi}^{T}A_{y,-\theta}^{T}A_{z,-\varphi}^{T}
\end{aligned}
\tag{3.9}
$$

but

$$
\sin\varphi = \frac{u_2}{\sqrt{u_1^2 + u_2^2}} \qquad \cos\varphi = \frac{u_1}{\sqrt{u_1^2 + u_2^2}}
\tag{3.10}
$$

$$
\sin\theta = \sqrt{u_1^2 + u_2^2} \qquad \cos\theta = u_3
\tag{3.11}
$$

$$
\sin\theta\sin\varphi = u_2 \qquad \sin\theta\cos\varphi = u_1
\tag{3.12}
$$

hence,

$$^{G}R_B = R_{\hat{u},\phi} \tag{3.13}$$

$$= \begin{bmatrix} u_1^2 \text{ vers } \phi + c\phi & u_1 u_2 \text{ vers } \phi - u_3 s\phi & u_1 u_3 \text{ vers } \phi + u_2 s\phi \\ u_1 u_2 \text{ vers } \phi + u_3 s\phi & u_2^2 \text{ vers } \phi + c\phi & u_2 u_3 \text{ vers } \phi - u_1 s\phi \\ u_1 u_3 \text{ vers } \phi - u_2 s\phi & u_2 u_3 \text{ vers } \phi + u_1 s\phi & u_3^2 \text{ vers } \phi + c\phi \end{bmatrix}.$$

The matrix (3.13) can be decomposed to

$$R_{\hat{u},\phi} = \cos\phi \begin{bmatrix} 1 & 0 & 0 \\ 0 & 1 & 0 \\ 0 & 0 & 1 \end{bmatrix}$$

$$+ (1 - \cos\phi) \begin{bmatrix} u_1 \\ u_2 \\ u_3 \end{bmatrix} \begin{bmatrix} u_1 & u_2 & u_3 \end{bmatrix}$$

$$+ \sin\phi \begin{bmatrix} 0 & -u_3 & u_2 \\ u_3 & 0 & -u_1 \\ -u_2 & u_1 & 0 \end{bmatrix} \tag{3.14}$$

to be equal to

$$^{G}R_B = R_{\hat{u},\phi} = \mathbf{I}\cos\phi + \hat{u}\hat{u}^T \text{ vers } \phi + \tilde{u}\sin\phi. \tag{3.15}$$

Equation (3.4) is called the *Rodriguez rotation formula* (or the *Euler-Lexell-Rodriguez formula*). It is sometimes reported in literature as the following equivalent forms:

$$R_{\hat{u},\phi} = \mathbf{I} + (\sin\phi)\,\tilde{u} + (\text{vers }\phi)\,\tilde{u}^2 \tag{3.16}$$

$$R_{\hat{u},\phi} = \left[\mathbf{I} - \hat{u}\hat{u}^T\right]\cos\phi + \tilde{u}\sin\phi + \hat{u}\hat{u}^T \tag{3.17}$$

$$R_{\hat{u},\phi} = -\tilde{u}^2\cos\phi + \tilde{u}\sin\phi + \tilde{u}^2 + \mathbf{I} \tag{3.18}$$

The *inverse of an angle-axis rotation* is

$$^{B}R_G = {}^{G}R_B^T = R_{\hat{u},-\phi}$$

$$= \mathbf{I}\cos\phi + \hat{u}\hat{u}^T \text{ vers }\phi - \tilde{u}\sin\phi. \tag{3.19}$$

It means orientation of $B$ in $G$, when $B$ is rotated $\phi$ about $\hat{u}$, is the same as the orientation of $G$ in $B$, when $B$ is rotated $-\phi$ about $\hat{u}$.

The $3 \times 3$ real orthogonal transformation matrix $R$ is also called a *rotator* and the skew symmetric matrix $\tilde{u}$ is called a *spinor*. We can verify that

$$\tilde{u}\hat{u} = 0 \tag{3.20}$$

$$\mathbf{I} - \hat{u}\hat{u}^T = \tilde{u}^2 \tag{3.21}$$

$$\mathbf{r}^T\tilde{u}\mathbf{r} = 0 \tag{3.22}$$

$$\hat{u} \times \mathbf{r} = \tilde{u}\mathbf{r} = -\tilde{r}\hat{u} = -\mathbf{r} \times \hat{u}. \tag{3.23}$$

**Example 37** *Axis-angle rotation when $\hat{u} = \hat{K}$.*
*If the local frame $B\,(Oxyz)$ rotates about the $Z$-axis, then*

$$\hat{u} = \hat{K} \tag{3.24}$$

*and the transformation matrix (3.5) reduces to*

$$
\begin{aligned}
{}^{G}R_B &= 
\begin{bmatrix}
0\,\mathrm{vers}\,\phi + \cos\phi & 0\,\mathrm{vers}\,\phi - 1\sin\phi & 0\,\mathrm{vers}\,\phi + 0\sin\phi \\
0\,\mathrm{vers}\,\phi + 1\sin\phi & 0\,\mathrm{vers}\,\phi + \cos\phi & 0\,\mathrm{vers}\,\phi - 0\sin\phi \\
0\,\mathrm{vers}\,\phi - 0\sin\phi & 0\,\mathrm{vers}\,\phi + 0\sin\phi & 1\,\mathrm{vers}\,\phi + \cos\phi
\end{bmatrix} \\
&= 
\begin{bmatrix}
\cos\phi & -\sin\phi & 0 \\
\sin\phi & \cos\phi & 0 \\
0 & 0 & 1
\end{bmatrix} \tag{3.25}
\end{aligned}
$$

*which is equivalent to the rotation matrix about the $Z$-axis of global frame in (2.21).*

**Example 38** *Rotation about a rotated local axis.*
*If the body coordinate frame $Oxyz$ rotates $\varphi$ deg about the global $Z$-axis, then the x-axis would be along*

$$
\begin{aligned}
\hat{u}_x &= {}^{G}R_{Z,\varphi}\,\hat{\imath} = 
\begin{bmatrix}
\cos\varphi & -\sin\varphi & 0 \\
\sin\varphi & \cos\varphi & 0 \\
0 & 0 & 1
\end{bmatrix}
\begin{bmatrix}
1 \\ 0 \\ 0
\end{bmatrix} \\
&= 
\begin{bmatrix}
\cos\varphi \\ \sin\varphi \\ 0
\end{bmatrix}. \tag{3.26}
\end{aligned}
$$

*Rotation $\theta$ about $\hat{u}_x = (\cos\varphi)\,\hat{I} + (\sin\varphi)\,\hat{J}$ is defined by Rodriguez's formula (3.5)*

$$
{}^{G}R_{\hat{u}_x,\theta} = 
\begin{bmatrix}
\cos^2\varphi\,\mathrm{vers}\,\theta + \cos\theta & \cos\varphi\sin\varphi\,\mathrm{vers}\,\theta & \sin\varphi\sin\theta \\
\cos\varphi\sin\varphi\,\mathrm{vers}\,\theta & \sin^2\varphi\,\mathrm{vers}\,\theta + \cos\theta & -\cos\varphi\sin\theta \\
-\sin\varphi\sin\theta & \cos\varphi\sin\theta & \cos\theta
\end{bmatrix}.
$$

*Now, rotation $\varphi$ about the global $Z$-axis followed by rotation $\theta$ about the local $x$-axis is transformed by*

$$
\begin{aligned}
{}^{G}R_B &= {}^{G}R_{\hat{u}_x,\theta}\,{}^{G}R_{Z,\varphi} \\
&= 
\begin{bmatrix}
\cos\varphi & -\cos\theta\sin\varphi & \sin\theta\sin\varphi \\
\sin\varphi & \cos\theta\cos\varphi & -\cos\varphi\sin\theta \\
0 & \sin\theta & \cos\theta
\end{bmatrix} \tag{3.27}
\end{aligned}
$$

*that must be equal to $[A_{x,\theta}A_{z,\varphi}]^{-1} = A_{z,\varphi}^{T}A_{x,\theta}^{T}$.*

**Example 39** *Axis and angle of rotation.*

*Given a transformation matrix $^GR_B$ we may obtain the axis $\hat{u}$ and angle $\phi$ of the rotation by considering that*

$$\tilde{u} = \frac{1}{2\sin\phi}\left(^GR_B - {^GR_B^T}\right) \qquad (3.28)$$

$$\cos\phi = \frac{1}{2}\left(\text{tr}\left(^GR_B\right) - 1\right) \qquad (3.29)$$

*add ~~with~~ diagonal of matrix Main*

*because*

$$^GR_B - {^GR_B^T} = \begin{bmatrix} 0 & -2\left(\sin\phi\right)u_3 & 2\left(\sin\phi\right)u_2 \\ 2\left(\sin\phi\right)u_3 & 0 & -2\left(\sin\phi\right)u_1 \\ -2\left(\sin\phi\right)u_2 & 2\left(\sin\phi\right)u_1 & 0 \end{bmatrix}$$

$$= 2\sin\phi \begin{bmatrix} 0 & -u_3 & u_2 \\ u_3 & 0 & -u_1 \\ -u_2 & u_1 & 0 \end{bmatrix}$$

$$= 2\tilde{u}\sin\phi \qquad (3.30)$$

*and*

$$\begin{aligned} \text{tr}\left(^GR_B\right) &= r_{11} + r_{22} + r_{33} \\ &= 3\cos\phi + u_1^2\left(1 - \cos\phi\right) + u_2^2\left(1 - \cos\phi\right) + u_3^2\left(1 - \cos\phi\right) \\ &= 3\cos\phi + u_1^2 + u_2^2 + u_3^2 - \left(u_1^2 + u_2^2 + u_3^2\right)\cos\phi \\ &= 2\cos\phi + 1. \end{aligned} \qquad (3.31)$$

**Example 40** *Axis and angel of a rotation matrix.*

*A body coordinate frame, B, undergoes three Euler rotations $(\varphi, \theta, \psi) = (30, 45, 60)$ deg with respect to a global frame G. The rotation matrix to transform coordinates of B to G is*

$$\begin{aligned} ^GR_B &= {^BR_G^T} = [R_{z,\psi}R_{x,\theta}R_{z,\varphi}]^T \\ &= R_{z,\varphi}^T R_{x,\theta}^T R_{z,\psi}^T \\ &= \begin{bmatrix} 0.126\,83 & -0.926\,78 & 0.353\,55 \\ 0.780\,33 & -0.126\,83 & -0.612\,37 \\ 0.612\,37 & 0.353\,55 & 0.707\,11 \end{bmatrix}. \end{aligned} \qquad (3.32)$$

*The unique angle-axis of rotation for this rotation matrix can then be found by Equations (3.28) and (3.29).*

$$\begin{aligned} \phi &= \cos^{-1}\left(\frac{1}{2}\left(\text{tr}\left(^GR_B\right) - 1\right)\right) \\ &= \cos^{-1}\left(-0.146\,45\right) = 98\,\text{deg} \end{aligned} \qquad (3.33)$$

$$\tilde{u} = \frac{1}{2\sin\phi}\left(^G R_B - {}^G R_B^T\right)$$

$$= \begin{bmatrix} 0.0 & -0.862\,85 & -0.130\,82 \\ 0.862\,85 & 0.0 & -0.488\,22 \\ 0.130\,82 & 0.488\,22 & 0.0 \end{bmatrix} \tag{3.34}$$

$$\hat{u} = \begin{bmatrix} 0.488\,22 \\ -0.130\,82 \\ 0.862\,85 \end{bmatrix} \tag{3.35}$$

As a double check, we may verify the angle-axis rotation formula and derive the same rotation matrix.

$$^G R_B = R_{\hat{u},\phi} = \mathbf{I}\cos\phi + \hat{u}\hat{u}^T \operatorname{vers}\phi + \tilde{u}\sin\phi$$

$$= \begin{bmatrix} 0.126\,82 & -0.926\,77 & 0.353\,54 \\ 0.780\,32 & -0.126\,83 & -0.612\,37 \\ 0.612\,36 & 0.353\,55 & 0.707\,09 \end{bmatrix} \tag{3.36}$$

**Example 41** *Non-uniqueness of angle-axis of rotation.*
The angle-axis representation of rotation is not unique. Rotation $(\theta, \hat{u})$ is equal to rotation $(-\theta, -\hat{u})$, and $(\theta + 2\pi, \hat{u})$.

**Example 42** ★ *Skew-symmetric characteristic of rotation matrix.*
Time derivative of the orthogonality condition of rotation matrix (2.129)

$$\frac{d}{dt}\left(^G R_B^T\, {}^G R_B\right) = {}^G \dot{R}_B^T\, {}^G R_B + {}^G R_B^T\, {}^G \dot{R}_B = 0 \tag{3.37}$$

leads to

$$\left[^G R_B^T\, {}^G \dot{R}_B\right]^T = -{}^G R_B^T\, {}^G \dot{R}_B \tag{3.38}$$

showing that $\left[^G R_B^T\, {}^G \dot{R}_B\right]$ is a skew-symmetric matrix.

If we show the rotation matrix by its elements, $^G R_B = [r_{ij}]$ , then $^G \dot{R}_B = [\dot{r}_{ij}]$.

**Example 43** ★ *Angular velocity vector of a rigid body*
We show the skew-symmetric matrix $\left[^G R_B^T\, {}^G \dot{R}_B\right]$ by

$$\tilde{\omega} = {}^G R_B^T\, {}^G \dot{R}_B \tag{3.39}$$

then, we find the following equation for the time derivative of rotation matrix

$$^G \dot{R}_B = {}^G R_B\, \tilde{\omega} \tag{3.40}$$

where $\omega$ is the vector of angular velocity of the frame $B(Oxyz)$ with respect to frame $G(OXYZ)$, and $\tilde{\omega}$ is its skew-symmetric matrix.

**Example 44** ★ *Differentiating a rotation matrix with respect to a parameter.*

*Suppose that a rotation matrix $R$ is a function of a variable $\tau$; hence, $R = R(\tau)$. To find the differential of $R$ with respect to $\tau$, we use the orthogonality characteristic*

$$RR^T = I \tag{3.41}$$

*and take derivative of both sides*

$$\frac{dR}{d\tau}R^T + R\frac{dR^T}{d\tau} = 0 \tag{3.42}$$

*which can be rewritten in the following form,*

$$\frac{dR}{d\tau}R^T + \left[\frac{dR}{d\tau}R^T\right]^T = 0 \tag{3.43}$$

*showing that $[\frac{dR}{d\tau}R^T]$ is a skew symmetric matrix.*

**Example 45** ★ *Eigenvalues and eigenvectors of $^GR_B$.*

*Consider a rotation matrix $^GR_B$. Applying the rotation on the axis of rotation $\hat{u}$ cannot change its direction*

$$^GR_B\,\hat{u} = \lambda\hat{u} \tag{3.44}$$

*so, the transformation equation implies that*

$$\left|^GR_B - \lambda\mathbf{I}\right| = 0. \tag{3.45}$$

*The characteristic equation of this determinant is*

$$-\lambda^3 + \mathrm{tr}(^GR_B)\lambda^2 - \mathrm{tr}(^GR_B)\lambda + 1 = 0. \tag{3.46}$$

*Factoring the left-hand side, gives*

$$(\lambda - 1)\left[\lambda^2 - \lambda\left(\mathrm{tr}(^GR_B) - 1\right) + 1\right] = 0 \tag{3.47}$$

*and shows that $\lambda_1 = 1$ is always an eigenvalue of $^GR_B$. Hence, there exist a real vector $\hat{u}$, such that every point on the line indicated by vector $\mathbf{n}_1 = \hat{u}$ remains fixed and invariant under transformation $^GR_B$.*

*The rotation angle $\phi$ is defined by*

$$\cos\phi = \frac{1}{2}\left(\mathrm{tr}(^GR_B) - 1\right) = \frac{1}{2}\left(r_{11} + r_{22} + r_{33} - 1\right) \tag{3.48}$$

*then the remaining eigenvalues are*

$$\lambda_2 = e^{i\phi} = \cos\phi + i\sin\phi \tag{3.49}$$

$$\lambda_3 = e^{-i\phi} = \cos\phi - i\sin\phi \tag{3.50}$$

*and their associated eigenvectors are* $\mathbf{v}$ *and* $\bar{\mathbf{v}}$, *where* $\bar{\mathbf{v}}$ *is the complex conjugate of* $\mathbf{v}$. *Since* $^G R_B$ *is orthogonal,* $\mathbf{n}_1, \mathbf{v},$ *and* $\bar{\mathbf{v}}$ *are also orthogonal. The eigenvectors* $\mathbf{v}$ *and* $\bar{\mathbf{v}}$ *span a plane perpendicular to the axis of rotation* $\mathbf{n}_1$. *A real basis for this plane can be found by using the following vectors:*

$$\mathbf{n}_2 = \frac{1}{2}|\mathbf{v} + \bar{\mathbf{v}}| \tag{3.51}$$

$$\mathbf{n}_3 = \frac{i}{2}|\mathbf{v} - \bar{\mathbf{v}}| \tag{3.52}$$

*The basis* $\mathbf{n}_2$ *and* $\mathbf{n}_3$ *transform to*

$$\begin{aligned} ^G R_B\, \mathbf{n}_2 &= \frac{1}{2}|\lambda_2 \mathbf{v} + \lambda_3 \bar{\mathbf{v}}| = \frac{1}{2}\left|e^{i\phi}\mathbf{v} + \overline{e^{i\phi}\mathbf{v}}\right| \\ &= \mathbf{v}\cos\phi + \bar{\mathbf{v}}\sin\phi \end{aligned} \tag{3.53}$$

$$\begin{aligned} ^G R_B\, \mathbf{n}_3 &= \frac{i}{2}|\lambda_2 \mathbf{v} - \lambda_3 \bar{\mathbf{v}}| = \frac{1}{2}\left|e^{i\phi}\mathbf{v} - \overline{e^{i\phi}\mathbf{v}}\right| \\ &= -\mathbf{v}\cos\phi + \bar{\mathbf{v}}\sin\phi. \end{aligned} \tag{3.54}$$

*Therefore, the effect of the transformation* $^G R_B$ *is to rotate vectors in the plane spanned by* $\mathbf{n}_2$ *and* $\mathbf{n}_3$ *through angle* $\phi$ *about* $\mathbf{n}_1$, *while vectors along* $\mathbf{n}_1$ *are invariant.*

## 3.2  ★ Euler Parameters

Assume that $\phi$ is the angle of rotation of a local frame $B(Oxyz)$ about $\hat{u} = u_1\hat{I} + u_2\hat{J} + u_3\hat{K}$ relative to a global frame $G(OXYZ)$. The existence of such an axis of rotation is the analytical representation of the Euler's theorem about rigid body rotation: *the most general displacement of a rigid body with one point fixed is a rotation about some axis.*

To find the axis and angle of rotation we introduce the *Euler parameters* $e_0, e_1, e_2, e_3$ such that $e_0$ is a scalar and $e_1, e_2, e_3$ are components of a vector $\mathbf{e}$,

$$e_0 = \cos\frac{\phi}{2} \tag{3.55}$$

$$\mathbf{e} = e_1\hat{I} + e_2\hat{J} + e_3\hat{K} = \hat{u}\sin\frac{\phi}{2} \tag{3.56}$$

and,

$$e_1^2 + e_2^2 + e_3^2 + e_0^2 = e_0^2 + \mathbf{e}^T\mathbf{e} = 1. \tag{3.57}$$

Then, the transformation matrix $^G R_B$ to satisfy the equation $^G \mathbf{r} = {^G R_B}\, {^B \mathbf{r}}$, can be derived utilizing the Euler parameters

$$^G R_B = R_{\hat{u},\phi} = \left(e_0^2 - \mathbf{e}^2\right) \mathbf{I} + 2\mathbf{e}\,\mathbf{e}^T + 2e_0\tilde{e} \qquad (3.58)$$

$$= \begin{bmatrix} e_0^2 + e_1^2 - e_2^2 - e_3^2 & 2\left(e_1 e_2 - e_0 e_3\right) & 2\left(e_0 e_2 + e_1 e_3\right) \\ 2\left(e_0 e_3 + e_1 e_2\right) & e_0^2 - e_1^2 + e_2^2 - e_3^2 & 2\left(e_2 e_3 - e_0 e_1\right) \\ 2\left(e_1 e_3 - e_0 e_2\right) & 2\left(e_0 e_1 + e_2 e_3\right) & e_0^2 - e_1^2 - e_2^2 + e_3^2 \end{bmatrix}$$

*Rod.*

*matrix*

*w/ Euler*

where $\tilde{e}$ is the *skew-symmetric* matrix corresponding to $\mathbf{e}$ defined below.

$$\tilde{e} = \begin{bmatrix} 0 & -e_3 & e_2 \\ e_3 & 0 & -e_1 \\ -e_2 & e_1 & 0 \end{bmatrix} \qquad (3.59)$$

Euler parameters provide a well-suited, redundant, and nonsingular rotation description for arbitrary and large rotations.

**Proof.** Figure 3.2 depicts a point $P$ of a rigid body with position vector $\mathbf{r}$, and the unit vector $\hat{u}$ along the axis of rotation $ON$ fixed in the global frame. The point moves to $P'$ with position vector $\mathbf{r}'$ after an active rotation $\phi$ about $\hat{u}$. To obtain the relationship between $\mathbf{r}$ and $\mathbf{r}'$, we express $\mathbf{r}'$ by the following vector equation:

$$\mathbf{r}' = \overrightarrow{ON} + \overrightarrow{NQ} + \overrightarrow{QP'}. \qquad (3.60)$$

By investigating Figure 3.2 we may describe Equation (3.60) utilizing $\mathbf{r}$, $\mathbf{r}'$, $\hat{u}$, and $\phi$.

$$\begin{aligned} \mathbf{r}' &= (\mathbf{r} \cdot \hat{u})\,\hat{u} + \hat{u} \times (\mathbf{r} \times \hat{u}) \cos\phi - (\mathbf{r} \times \hat{u}) \sin\phi \\ &= (\mathbf{r} \cdot \hat{u})\,\hat{u} + [\mathbf{r} - (\mathbf{r} \cdot \hat{u})\,\hat{u}]\cos\phi + (\hat{u} \times \mathbf{r})\sin\phi \end{aligned} \qquad (3.61)$$

Rearranging Equation (3.61) leads to a new form of the Rodriguez rotation formula

$$\mathbf{r}' = \mathbf{r}\cos\phi + (1 - \cos\phi)\,(\hat{u} \cdot \mathbf{r})\,\hat{u} + (\hat{u} \times \mathbf{r})\sin\phi. \qquad (3.62)$$

Using the Euler parameters in (3.55) and (3.56), along with the following trigonometric relations

$$\cos\phi = 2\cos^2\frac{\phi}{2} - 1 \qquad (3.63)$$

$$\sin\phi = 2\sin\frac{\phi}{2}\cos\frac{\phi}{2} \qquad (3.64)$$

$$1 - \cos\phi = 2\sin^2\frac{\phi}{2} \qquad (3.65)$$

converts the Rodriguez formula (3.62) to a more useful form,

$$\mathbf{r}' = \mathbf{r}\left(2e_0^2 - 1\right) + 2\mathbf{e}\left(\mathbf{e} \cdot \mathbf{r}\right) + 2e_0\left(\mathbf{e} \times \mathbf{r}\right). \qquad (3.66)$$

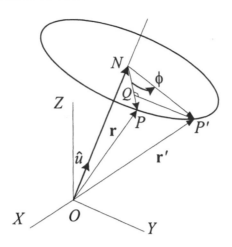

FIGURE 3.2. Axis and angle of rotation.

Using a more compact form and defining new notations for $\mathbf{r}' = {}^G\mathbf{r}$ and $\mathbf{r} = {}^B\mathbf{r}$

$$^G\mathbf{r} = \left(e_0^2 - \mathbf{e}^2\right) {}^B\mathbf{r} + 2\mathbf{e}\left(\mathbf{e}^T\, {}^B\mathbf{r}\right) + 2e_0\left(\tilde{\mathbf{e}}\, {}^B\mathbf{r}\right) \tag{3.67}$$

allows us to factor out the position vector $^B\mathbf{r}$ and extract the Euler parameter transformation matrix $^G R_B$

$$^G R_B = \left(e_0^2 - \mathbf{e}^2\right)\mathbf{I} + 2\mathbf{e}\,\mathbf{e}^T + 2e_0\tilde{\mathbf{e}} \tag{3.68}$$

where

$$^G\mathbf{r} = {}^G R_B\,{}^B\mathbf{r} = R_{\hat{u},\phi}\,{}^B\mathbf{r}. \tag{3.69}$$

∎

**Example 46 ★** *Axis-angle rotation of $^G R_B$.*

*Euler parameters for rotation $\phi = 30\deg$ about $\hat{u} = \left(\hat{I} + \hat{J} + \hat{K}\right)/\sqrt{3}$ are*

$$e_0 = \cos\frac{30}{2} = 0.966 \tag{3.70}$$

$$\mathbf{e} = \hat{u}\sin\frac{\phi}{2} = e_1\hat{I} + e_2\hat{J} + e_3\hat{K}$$

$$= 0.259\left(\hat{I} + \hat{J} + \hat{K}\right) \tag{3.71}$$

*therefore, the corresponding transformation matrix $^G R_B$ is*

$$^G R_B = \begin{bmatrix} 0.866 & -0.366 & 0.635 \\ 0.635 & 0.866 & -0.366 \\ -0.366 & 0.635 & 0.866 \end{bmatrix}. \tag{3.72}$$

*Hence, the global position of a point at $^B r = \begin{bmatrix} 1 & 0 & 0 \end{bmatrix}^T$ is*

$$^G r = {^G R_B}\, {^B r} = \begin{bmatrix} 0.866 \\ 0.635 \\ -0.366 \end{bmatrix}. \tag{3.73}$$

**Example 47** *Axis-angle rotation from $^G R_B$.*
   *Given a transformation matrix $^G R_B$ we may obtain the angle of rotation $\phi$ and the axis of rotation $\hat{u}$ by considering that*

$$\begin{aligned} \cos\phi &= \frac{1}{2}\left(\mathrm{tr}\left(^G R_B\right) - 1\right) \\ &= \frac{1}{2}\left(r_{11} + r_{22} + r_{33} - 1\right) \end{aligned} \tag{3.74}$$

$$\tilde{u} = \frac{1}{2\sin\phi}\left(^G R_B - {^G R_B^T}\right), \tag{3.75}$$

*because*

$$\mathrm{tr}\left(^G R_B\right) = 2\cos\phi + 1 \tag{3.76}$$

*and*

$$^G R_B - {^G R_B^T} = \begin{bmatrix} 0 & -2\left(\sin\phi\right) u_3 & 2\left(\sin\phi\right) u_2 \\ 2\left(\sin\phi\right) u_3 & 0 & -2\left(\sin\phi\right) u_1 \\ -2\left(\sin\phi\right) u_2 & 2\left(\sin\phi\right) u_1 & 0 \end{bmatrix} \tag{3.77}$$

*and therefore,*

$$\hat{u} = \frac{1}{2\sin\phi}\begin{bmatrix} r_{32} - r_{23} \\ r_{13} - r_{31} \\ r_{21} - r_{12} \end{bmatrix}. \tag{3.78}$$

**Example 48** ★ *Euler parameters from $^G R_B$.*
   *Employing a transformation matrix $^G R_B$ we can find the Euler parameters by*

$$e_0^2 = \frac{1}{4}\left(\mathrm{tr}\left(^G R_B\right) + 1\right) = \frac{1}{4}\left(r_{11} + r_{22} + r_{33} + 1\right) \tag{3.79}$$

$$\tilde{e} = \frac{1}{4e_0}\left(^G R_B - {^G R_B^T}\right) \tag{3.80}$$

*because*

$$\mathrm{tr}\left(^G R_B\right) = 3e_0^2 - e_1^2 - e_2^2 - e_3^2 = 4e_0^2 - 1 \tag{3.81}$$

*and*

$$^G R_B - {^G R_B^T} = \begin{bmatrix} 0 & -4e_0 e_3 & 4e_0 e_2 \\ 4e_0 e_3 & 0 & -4e_0 e_1 \\ -4e_0 e_2 & 4e_0 e_1 & 0 \end{bmatrix}. \tag{3.82}$$

**Example 49** ★ *Euler parameters and Euler angles relationship.*

*Comparing the Euler angles rotation matrix (2.63) and the Euler parameter transformation matrix (3.58) we can determine the following relationships between Euler angles and Euler parameters*

$$e_0 = \cos\frac{\theta}{2}\cos\frac{\psi+\varphi}{2} \tag{3.83}$$

$$e_1 = \sin\frac{\theta}{2}\cos\frac{\psi-\varphi}{2} \tag{3.84}$$

$$e_2 = \sin\frac{\theta}{2}\sin\frac{\psi-\varphi}{2} \tag{3.85}$$

$$e_3 = \cos\frac{\theta}{2}\sin\frac{\psi+\varphi}{2} \tag{3.86}$$

*or*

$$\varphi = \cos^{-1}\frac{2\left(e_2e_3 + e_0e_1\right)}{\sin\theta} \tag{3.87}$$

$$\theta = \cos^{-1}\left[2\left(e_0^2 + e_3^2\right) - 1\right] \tag{3.88}$$

$$\psi = \cos^{-1}\frac{-2\left(e_2e_3 - e_0e_1\right)}{\sin\theta}. \tag{3.89}$$

**Example 50** ★ *Rotation matrix for angel of rotation $\phi = k\pi$.*

*When the angle of rotation is $\phi = k\pi$, $k = \pm 1, \pm 3, ...$, then $e_0 = 0$. Therefore, the Euler parameter transformation matrix (3.58) becomes*

$$^G R_B = 2\begin{bmatrix} e_1^2 - \frac{1}{2} & e_1e_2 & e_1e_3 \\ e_1e_2 & e_2^2 - \frac{1}{2} & e_2e_3 \\ e_1e_3 & e_2e_3 & e_3^2 - \frac{1}{2} \end{bmatrix} \tag{3.90}$$

*which is a symmetric matrix and indicates that rotation $\phi = k\pi$, and $\phi = -k\pi$ are equivalent.*

**Example 51** ★ *Vector of infinitesimal rotation.*

*Consider the Rodriguez rotation formula (3.62) for a differential rotation $d\phi$*

$$\mathbf{r}' = \mathbf{r} + (\hat{u} \times \mathbf{r})\,d\phi. \tag{3.91}$$

*In this case the difference between $\mathbf{r}'$ and $\mathbf{r}$ is also very small,*

$$d\mathbf{r} = \mathbf{r}' - \mathbf{r} = d\phi\,\hat{u} \times \mathbf{r} \tag{3.92}$$

*and hence, a differential rotation $d\phi$ about an axis indicated by the unit vector $\hat{u}$ is a vector along $\hat{u}$ with magnitude $d\phi$. Dividing both sides by dt leads to*

$$\dot{\mathbf{r}} = \boldsymbol{\omega} \times \mathbf{r} \tag{3.93}$$

*which represents the global velocity vector of any point in a rigid body rotating about $\hat{u}$.*

**Example 52 ★** *Exponential form of rotation.*

*Consider a point $P$ in the body frame $B$ with a position vector $\mathbf{r}$. If the rigid body has an angular velocity $\boldsymbol{\omega}$, then the velocity of $P$ in the global coordinate frame is*

$$\dot{\mathbf{r}} = \boldsymbol{\omega} \times \mathbf{r} = \tilde{\omega}\mathbf{r}. \tag{3.94}$$

*This is a first-order linear differential equation that may be integrated to give*

$$\mathbf{r}(t) = e^{\tilde{\omega}t}\mathbf{r}(0) \tag{3.95}$$

*where $\mathbf{r}(0)$ is the initial position vector of $P$, and $e^{\tilde{\omega}t}$ is a matrix exponential*

$$e^{\tilde{\omega}t} = \mathbf{I} + \tilde{\omega}t + \frac{(\tilde{\omega}t)^2}{2!} + \frac{(\tilde{\omega}t)^3}{3!} + \cdots \tag{3.96}$$

*The angular velocity $\boldsymbol{\omega}$, has a magnitude $\omega$ and direction indicated by a unit vector $\hat{u}$. Therefore,*

$$\boldsymbol{\omega} = \omega\hat{u} \tag{3.97}$$

$$\tilde{\omega} = \omega\tilde{u} \tag{3.98}$$

$$\tilde{\omega}t = \omega t\tilde{u} = \phi\tilde{u} \tag{3.99}$$

*and hence*

$$e^{\tilde{\omega}t} = e^{\phi\tilde{u}} \tag{3.100}$$

$$= \mathbf{I} + \left(\phi - \frac{\phi^3}{3!} + \frac{\phi^5}{5!} - \cdots\right)\tilde{u} + \left(\frac{\phi^2}{2!} - \frac{\phi^4}{4!} + \frac{\phi^6}{6!} \cdots\right)\tilde{u}^2$$

*or equivalently*

$$e^{\phi\tilde{u}} = \mathbf{I} + \tilde{u}\sin\phi + \tilde{u}^2\left(1 - \cos\phi\right). \tag{3.101}$$

*It is an alternative form of the Rodriguez formula showing that $e^{\phi\tilde{u}}$ is the rotation transformation to map $^B\mathbf{r} = \mathbf{r}(0)$ to $^G\mathbf{r} = \mathbf{r}(t)$.*

**Example 53 ★** *Rotational characteristic of $e^{\phi\tilde{u}}$.*

*To show that $e^{\phi\tilde{u}} \in S$, where $S$ is the set of rotation matrices*

$$S = \left\{R \in \mathbb{R}^{3\times3} : RR^T = \mathbf{I}, |R| = 1\right\} \tag{3.102}$$

*we have to show that $R = e^{\phi\tilde{u}}$ has the orthogonality property $R^T R = I$ and its determinant is $|R| = 1$. The orthogonality can be verified by considering*

$$\left[e^{\phi\tilde{u}}\right]^{-1} = e^{-\phi\tilde{u}} = e^{\phi\tilde{u}^T} = \left[e^{\phi\tilde{u}}\right]^T. \tag{3.103}$$

*Thus $R^{-1} = R^T$ and consequently $RR^T = \mathbf{I}$. From orthogonality, it follows that $|R| = \pm1$, and from continuity of exponential function, it follows that $|e^0| = 1$. Therefore, $|R| = 1$.*

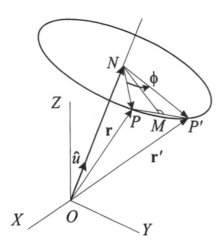

FIGURE 3.3. Illustration of a rotation of a rigid body to derive a new form of the *Rodriguez rotation formula* in Example 55.

**Example 54 ★** $e^{\phi\tilde{u}}$ *is equivalent to the rotation matrix* $^G R_B$.
   *Expanding*

$$e^{\phi\tilde{u}} = \mathbf{I} + \tilde{u}\sin\phi + \tilde{u}^2\left(1 - \cos\phi\right) \tag{3.104}$$

*gives*

$$e^{\phi\tilde{u}} = \begin{bmatrix} u_1^2\,\text{vers}\,\phi + c\phi & u_1u_2\,\text{vers}\,\phi - u_3s\phi & u_1u_3\,\text{vers}\,\phi + u_2s\phi \\ u_1u_2\,\text{vers}\,\phi + u_3s\phi & u_2^2\,\text{vers}\,\phi + c\phi & u_2u_3\,\text{vers}\,\phi - u_1s\phi \\ u_1u_3\,\text{vers}\,\phi - u_2s\phi & u_2u_3\,\text{vers}\,\phi + u_1s\phi & u_3^2\,\text{vers}\,\phi + c\phi \end{bmatrix} \tag{3.105}$$

*which is equal to the axis-angle Equation (3.5), and therefore,*

$$\begin{aligned} e^{\phi\tilde{u}} &= R_{\hat{u},\phi} \\ &= {}^G R_B \\ &= \mathbf{I}\cos\phi + \hat{u}\hat{u}^T\,\text{vers}\,\phi + \tilde{u}\sin\phi. \end{aligned} \tag{3.106}$$

**Example 55 ★** *New form of the Rodriguez rotation formula.*
   *Considering Figure 3.3 we may write*

$$\cos\frac{\phi}{2}\left|\overrightarrow{MP'}\right| = \sin\frac{\phi}{2}\left|\overrightarrow{NM}\right| \tag{3.107}$$

$$\left|\overrightarrow{NM}\right|\overrightarrow{MP'} = \left|\overrightarrow{MP'}\right|\hat{u}\times\overrightarrow{NM} \tag{3.108}$$

*to find*

$$\left(\cos\frac{\phi}{2}\right)\overrightarrow{MP'} = \left(\sin\frac{\phi}{2}\right)\hat{u}\times\overrightarrow{NM}. \tag{3.109}$$

*Now using the following equalities*

$$2\overrightarrow{MP'} = \overrightarrow{NP'} - \overrightarrow{NP} \tag{3.110}$$

$$2\overrightarrow{NM} = \overrightarrow{NP'} + \overrightarrow{NP} \tag{3.111}$$

$$\overrightarrow{NP'} - \overrightarrow{NP} = \mathbf{r'} - \mathbf{r} \tag{3.112}$$

$$\hat{u} \times \left(\overrightarrow{NP'} + \overrightarrow{NP}\right) = \hat{u} \times (\mathbf{r'} + \mathbf{r}) \tag{3.113}$$

*we can write an alternative form of the Rodriguez rotation formula*

$$\cos\frac{\phi}{2}(\mathbf{r'} - \mathbf{r}) = \sin\frac{\phi}{2}\hat{u} \times (\mathbf{r'} + \mathbf{r}) \tag{3.114}$$

*or*

$$(\mathbf{r'} - \mathbf{r}) = \tan\frac{\phi}{2}\hat{u} \times (\mathbf{r'} + \mathbf{r}). \tag{3.115}$$

**Example 56 ★** *Rodriguez vector.*
  *The vector*

$$\mathbf{q} = \tan\frac{\phi}{2}\hat{u} \tag{3.116}$$

*is called the Rodriguez vector. It can be seen that the Euler parameters are related to this vector according to*

$$\mathbf{q} = \frac{\mathbf{e}}{e_0} \tag{3.117}$$

*and*

$$e_0 = \frac{1}{\sqrt{1 + \mathbf{q}^T\mathbf{q}}} = \frac{1}{\sqrt{1 + q^2}} \tag{3.118}$$

$$e_i = \frac{q_i}{\sqrt{1 + \mathbf{q}^T\mathbf{q}}} = \frac{q_i}{\sqrt{1 + q^2}} \qquad i = 1, 2, 3. \tag{3.119}$$

*The Rodriguez formula (3.58) can be converted to a new form based on Rodriguez vector*

$$
\begin{aligned}
{}^{G}R_B &= R_{\hat{u},\phi} = \left(e_0^2 - e^2\right)\mathbf{I} + 2\mathbf{e}\,\mathbf{e}^T + 2e_0\tilde{e} \\
&= \frac{1}{1 + \mathbf{q}^T\mathbf{q}}\left((1 - \mathbf{q}^T\mathbf{q})\mathbf{I} + 2\mathbf{q}\mathbf{q}^T + 2\tilde{q}\right). \tag{3.120}
\end{aligned}
$$

*The combination of two rotations, $\mathbf{q'}$ and $\mathbf{q''}$, is equivalent to a single rotation $\mathbf{q}$, where*

$$\mathbf{q} = \frac{\mathbf{q''} + \mathbf{q'} - \mathbf{q''} \times \mathbf{q'}}{1 - \mathbf{q''} \cdot \mathbf{q'}}. \tag{3.121}$$

**Example 57 ★** *Elements of $^G R_B$.*
  *Introducing* **Levi-Civita density** *or* **permutation symbol**

$$\epsilon_{ijk} = \frac{1}{2}(i-j)(j-k)(k-i) \quad , \quad i,j,k = 1,2,3 \tag{3.122}$$

*and recalling Kronecker delta (2.133)*

$$\delta_{ij} = 1 \ \text{if } j = k, \text{ and } \delta_{jk} = 0 \ \text{if } j \neq k \tag{3.123}$$

*we can redefine the elements of the Rodriguez rotation matrix by*

$$r_{ij} = \delta_{ij}\cos\phi + (1-\cos\phi)\,u_i u_j + \epsilon_{ijk}u_k \sin\phi. \tag{3.124}$$

## 3.3 ★ Determination of Euler Parameters

Assume a transformation matrix is given

$$^G R_B = \begin{bmatrix} r_{11} & r_{12} & r_{13} \\ r_{21} & r_{22} & r_{23} \\ r_{31} & r_{32} & r_{33} \end{bmatrix}. \tag{3.125}$$

It is then possible to find the Euler parameters $e_0, e_1, e_2, e_3$ and indicate the axis and angle of rotation by utilizing one of the following four sets of equations:

$$
\begin{aligned}
e_0 &= \pm\frac{1}{2}\sqrt{1 + r_{11} + r_{22} + r_{33}} \\
e_1 &= \frac{1}{4}\frac{r_{32} - r_{23}}{e_0} \\
e_2 &= \frac{1}{4}\frac{r_{13} - r_{31}}{e_0} \\
e_3 &= \frac{1}{4}\frac{r_{21} - r_{12}}{e_0}
\end{aligned}
\tag{3.126}
$$

$$
\begin{aligned}
e_1 &= \pm\frac{1}{2}\sqrt{1 + r_{11} - r_{22} - r_{33}} \\
e_2 &= \frac{1}{4}\frac{r_{21} + r_{12}}{e_1} \\
e_3 &= \frac{1}{4}\frac{r_{31} + r_{13}}{e_1} \\
e_0 &= \frac{1}{4}\frac{r_{32} + r_{23}}{e_1}
\end{aligned}
\tag{3.127}
$$

$$e_2 = \pm\frac{1}{2}\sqrt{1 - r_{11} + r_{22} - r_{33}}$$

use set

w/ largest

divisor

$$e_3 = \frac{1}{4}\frac{r_{32} + r_{23}}{e_2}$$

$$e_0 = \frac{1}{4}\frac{r_{13} - r_{31}}{e_2}$$

$$e_1 = \frac{1}{4}\frac{r_{21} + r_{12}}{e_2} \tag{3.128}$$

$$e_3 = \pm\frac{1}{2}\sqrt{1 - r_{11} - r_{22} + r_{33}}$$

$$e_0 = \frac{1}{4}\frac{r_{21} - r_{12}}{e_3}$$

$$e_1 = \frac{1}{4}\frac{r_{31} + r_{13}}{e_3}$$

$$e_2 = \frac{1}{4}\frac{r_{32} + r_{23}}{e_3} \tag{3.129}$$

Although Equations (3.126)-(3.129) present four different sets of solutions, their resulting Euler parameters $e_0, e_1, e_2, e_3$ are identical. Therefore, numerical inaccuracies can be minimized by using the set with maximum divisor.

The plus and minus sign indicates that rotation $\phi$ about $\hat{u}$ is equivalent to rotation $-\phi$ about $-\hat{u}$.

**Proof.** Comparing (3.58) and (3.125) shows that for the first set of Equations (3.126), $e_0$ can be found by summing the diagonal elements $r_{11}, r_{22}$, and $r_{33}$ to get tr $\left(^{G}R_B\right) = 4e_0^2 - 1$. To find $e_1$, $e_2$, and $e_3$ we need to simplify $r_{32} - r_{23}$, $r_{13} - r_{31}$, and $r_{21} - r_{12}$, respectively.

The other sets of solutions (3.127)-(3.129) can also be found by comparison. ∎

**Example 58** ★ *Euler parameters from $^{G}R_B$.*

*A transformation matrix is given.*

$$^{G}R_B = \begin{bmatrix} 0.5449 & -0.5549 & 0.6285 \\ 0.3111 & 0.8299 & 0.4629 \\ -0.7785 & -0.0567 & 0.6249 \end{bmatrix} \tag{3.130}$$

*To calculate the corresponding Euler parameters, we use Equation (3.126) and find*

$$\text{tr}\left(^{G}R_B\right) = r_{11} + r_{22} + r_{33}$$
$$= 0.5449 + 0.8299 + 0.6249 = 1.9997 \tag{3.131}$$

*therefore,*

$$e_0 = \sqrt{\left(\text{tr}\left(^G R_B\right) + 1\right)/4} = 0.866 \qquad (3.132)$$

*and*

$$
\begin{aligned}
e_1 &= -0.15 \\
e_2 &= 0.406 \\
e_3 &= 0.25.
\end{aligned} \qquad (3.133)
$$

**Example 59** ★ *Euler parameters when we have one of them.*

*Consider the Euler parameter rotation matrix (3.58) corresponding to rotation $\phi$ about an axis indicated by a unit vector $\hat{u}$. The off diagonal elements of $^G R_B$*

$$
\begin{aligned}
e_0 e_1 &= \frac{1}{4}\left(r_{32} - r_{23}\right) \\
e_0 e_2 &= \frac{1}{4}\left(r_{13} - r_{31}\right) \\
e_0 e_3 &= \frac{1}{4}\left(r_{21} - r_{12}\right) \\
e_1 e_2 &= \frac{1}{4}\left(r_{12} + r_{21}\right) \\
e_1 e_3 &= \frac{1}{4}\left(r_{13} + r_{31}\right) \\
e_2 e_3 &= \frac{1}{4}\left(r_{23} + r_{32}\right) \qquad (3.134)
\end{aligned}
$$

*can be utilized to find $e_i$, $i = 0, 1, 2, 3$ if we have one of them.*

**Example 60** ★ *Euler parameters by the Stanley method.*

*Following an effective method developed by Stanley, we may first find the four $e_i^2$*

work
same
as best
set from
before.

$$
\begin{aligned}
e_0^2 &= \frac{1}{2}\left(1 + \text{tr}\left(^G R_B\right)\right) \\
e_1^2 &= \frac{1}{4}\left(1 + 2r_{11} - \text{tr}\left(^G R_B\right)\right) \\
e_2^2 &= \frac{1}{4}\left(1 + 2r_{22} - \text{tr}\left(^G R_B\right)\right) \\
e_3^2 &= \frac{1}{4}\left(1 + 2r_{33} - \text{tr}\left(^G R_B\right)\right) \qquad (3.135)
\end{aligned}
$$

*and take the positive square root of the largest $e_i^2$. Then the other $e_i$ are found by dividing the appropriate three of the six equations (3.134) by the largest $e_i$.*

# 3.4  ★ Quaternions

A *quaternion q* is defined as a quantity

$$
\begin{aligned}
q &= q_0 + \mathbf{q} \\
&= q_0 + q_1\hat{\imath} + q_2\hat{\jmath} + q_3\hat{k}
\end{aligned}
\tag{3.136}
$$

where $q_0$ is a scalar and $\mathbf{q}$ is a vector. A quaternion can also be shown in a *flag* form

$$
q = q_0 + q_1 i + q_2 j + q_3 k
\tag{3.137}
$$

where, $i, j, k$ are *flags* and defined as follows

$$
\begin{aligned}
i^2 &= j^2 = k^2 = ijk = -1 \tag{3.138} \\
ij &= -ji = k \tag{3.139} \\
jk &= -kj = i \tag{3.140} \\
ki &= -ik = j. \tag{3.141}
\end{aligned}
$$

Addition of two quaternions is a quaternion

$$
\begin{aligned}
q + p &= (q_0 + \mathbf{q}) + (p_0 + \mathbf{p}) \\
&= q_0 + q_1 i + q_2 j + q_3 k + p_0 + p_1 i + p_2 j + p_3 k \\
&= (q_0 + p_0) + (q_1 + p_1) i + (q_2 + p_2) j + (q_3 + p_3) k \tag{3.142}
\end{aligned}
$$

and multiplication of two quaternions is a quaternion defined by

$$
\begin{aligned}
qp &= (q_0 + \mathbf{q}) (p_0 + \mathbf{p}) \\
&= q_0 p_0 + q_0 \mathbf{p} + p_0 \mathbf{q} + \mathbf{q}\mathbf{p} \\
&= q_0 p_0 - \mathbf{q} \cdot \mathbf{p} + q_0 \mathbf{p} + p_0 \mathbf{q} + \mathbf{q} \times \mathbf{p} \\
&= (q_0 p_0 - q_1 p_1 - q_2 p_2 - q_3 p_3) \\
&\quad + (p_0 q_1 + p_1 q_0 - p_2 q_3 + p_3 q_2) i \\
&\quad + (p_0 q_2 + q_0 p_2 + p_1 q_3 - q_1 p_3) j \\
&\quad + (p_0 q_3 - p_1 q_2 + q_0 p_3 + p_2 q_1) k \tag{3.143}
\end{aligned}
$$

where $\mathbf{q}\mathbf{p}$ is cross product minus dot product of $\mathbf{q}$ and $\mathbf{p}$

$$
\mathbf{q}\mathbf{p} = \mathbf{q} \times \mathbf{p} - \mathbf{q} \cdot \mathbf{p}.
\tag{3.144}
$$

Quaternion addition is associative and commutative. Quaternion multiplication is not commutative, however it is associative, and distributes over addition,

$$
\begin{aligned}
(pq) r &= p(qr) \tag{3.145} \\
(p + q) r &= pr + qr. \tag{3.146}
\end{aligned}
$$

A quaternion $q$ has a conjugate $q^*$ where

$$q^* = q_0 - \mathbf{q}$$
$$= q_0 - q_1\hat{\imath} - q_2\hat{\jmath} - q_3\hat{k}. \tag{3.147}$$

Therefore,

$$qq^* = (q_0 + \mathbf{q})(q_0 - \mathbf{q})$$
$$= q_0q_0 + q_0\mathbf{q} - p_0\mathbf{q} - \mathbf{qq}$$
$$= q_0q_0 + \mathbf{q} \cdot \mathbf{q} - \mathbf{q} \times \mathbf{q}$$
$$= q_0^2 + q_1^2 + q_2^2 + q_3^2 \tag{3.148}$$

and then,

$$|q| = \sqrt{qq^*} = \sqrt{q_0^2 + q_1^2 + q_2^2 + q_3^2} \tag{3.149}$$

$$q^{-1} = \frac{1}{q} = \frac{q^*}{|q|^2}. \tag{3.150}$$

If $q$ is a unit quaternion, $|q| = 1$, then $q^{-1} = q^*$.

Let $e(\phi, \hat{u})$ be a unit quaternion, $|e(\phi, \hat{u})| = 1$,

$$e(\phi, \hat{u}) = e_0 + \mathbf{e} \tag{3.151}$$
$$= e_0 + e_1\hat{\imath} + e_2\hat{\jmath} + e_3\hat{k}$$
$$= \cos\frac{\phi}{2} + \sin\frac{\phi}{2}\hat{u}$$

and $\mathbf{r} = 0 + \mathbf{r}$ be a quaternion corresponding to a pure vector $\mathbf{r}$. The vector $\mathbf{r}$ after a rotation $\phi$ about $\hat{u}$ would be

$$\mathbf{r}' = e(\phi, \hat{u})\, \mathbf{r}\, e^*(\phi, \hat{u}) \tag{3.152}$$

equivalent to

$$^G\mathbf{r} = e(\phi, \hat{u})\ ^B\mathbf{r}\ e^*(\phi, \hat{u}). \tag{3.153}$$

Therefore, a rotation $R_{\hat{u}, \phi}$ can be defined by a corresponding quaternion $e(\phi, \hat{u}) = \cos\frac{\phi}{2} + \sin\frac{\phi}{2}\hat{u}$, and consequently, two consecutive rotations $R = R_2 R_1$ are defined by $e(\phi, \hat{u}) = e_2(\phi_2, \hat{u}_2)\, e_1(\phi_1, \hat{u}_1)$.

Note that $e_1(\phi_1, \hat{u}_1)$, $e_2(\phi_2, \hat{u}_2)$, $\cdots$ are quaternion while $e_0$, $e_1$, $e_2$, $e_3$ are Euler parameters.

**Proof.** Employing the quaternion multiplication (3.143) we can write

$$\mathbf{r}e^* = e_0\mathbf{r} + \mathbf{r} \times \mathbf{e}^* - \mathbf{r} \cdot \mathbf{e}^*$$

$$= e_0 \begin{bmatrix} r_1 \\ r_2 \\ r_3 \end{bmatrix} + \begin{bmatrix} r_1 \\ r_2 \\ r_3 \end{bmatrix} \times \begin{bmatrix} -e_1 \\ -e_2 \\ -e_3 \end{bmatrix} + (e_1r_1 + e_2r_2 + e_3r_3)$$

$$= (e_1r_1 + e_2r_2 + e_3r_3) + \begin{bmatrix} e_0r_1 + e_2r_3 - e_3r_2 \\ e_0r_2 - e_1r_3 + e_3r_1 \\ e_0r_3 + e_1r_2 - e_2r_1 \end{bmatrix} \tag{3.154}$$

and therefore,

$$ere^* = e_0 \left(e_1 r_1 + e_2 r_2 + e_3 r_3\right) - \begin{bmatrix} e_1 \\ e_2 \\ e_3 \end{bmatrix} \cdot \begin{bmatrix} e_0 r_1 + e_2 r_3 - e_3 r_2 \\ e_0 r_2 - e_1 r_3 + e_3 r_1 \\ e_0 r_3 + e_1 r_2 - e_2 r_1 \end{bmatrix}$$

$$+ e_0 \begin{bmatrix} e_0 r_1 + e_2 r_3 - e_3 r_2 \\ e_0 r_2 - e_1 r_3 + e_3 r_1 \\ e_0 r_3 + e_1 r_2 - e_2 r_1 \end{bmatrix} + \left(e_1 r_1 + e_2 r_2 + e_3 r_3\right) \begin{bmatrix} e_1 \\ e_2 \\ e_3 \end{bmatrix}$$

$$+ \begin{bmatrix} e_1 \\ e_2 \\ e_3 \end{bmatrix} \times \begin{bmatrix} e_0 r_1 + e_2 r_3 - e_3 r_2 \\ e_0 r_2 - e_1 r_3 + e_3 r_1 \\ e_0 r_3 + e_1 r_2 - e_2 r_1 \end{bmatrix}$$

$$= {}^G R_B \begin{bmatrix} r_1 \\ r_2 \\ r_3 \end{bmatrix} \tag{3.155}$$

which ${}^G R_B$ is equivalent to the Euler parameter transformation matrix (3.58).

$$ {}^G R_B = \begin{bmatrix} e_0^2 + e_1^2 - e_2^2 - e_3^2 & 2\left(e_1 e_2 - e_0 e_3\right) & 2\left(e_0 e_2 + e_1 e_3\right) \\ 2\left(e_0 e_3 + e_1 e_2\right) & e_0^2 - e_1^2 + e_2^2 - e_3^2 & 2\left(c_2 c_3 - e_0 e_1\right) \\ 2\left(e_1 e_3 - e_0 c_2\right) & 2\left(e_0 e_1 + e_2 e_3\right) & e_0^2 - e_1^2 - e_2^2 + e_3^2 \end{bmatrix} \tag{3.156}$$

Using a similar method we can also show that $\mathbf{r} = e^*\left(\phi, \hat{u}\right) \mathbf{r}' \, e\left(\phi, \hat{u}\right)$ is the inverse transformation of $\mathbf{r}' = e\left(\phi, \hat{u}\right) \mathbf{r} \, e^*\left(\phi, \hat{u}\right)$, which is equivalent to ${}^B \mathbf{r} = e^*\left(\phi, \hat{u}\right) {}^G \mathbf{r} \, e\left(\phi, \hat{u}\right)$.

Now assume $e_1\left(\phi_1, \hat{u}_1\right)$ and $e_2\left(\phi_2, \hat{u}_2\right)$ are the quaternions corresponding to the rotation matrix $R_{\hat{u}_1, \phi_1}$ and $R_{\hat{u}_2, \phi_2}$ respectively. The first rotation maps ${}^{B_1} \mathbf{r}$ to ${}^{B_2} \mathbf{r}$, and the second rotation maps ${}^{B_2} \mathbf{r}$ to ${}^{B_3} \mathbf{r}$. Therefore,

$$ {}^{B_2} \mathbf{r} = e_1\left(\phi_1, \hat{u}_1\right) {}^{B_1} \mathbf{r} \, e_1^*\left(\phi_1, \hat{u}_1\right) \tag{3.157}$$

$$ {}^{B_3} \mathbf{r} = e_2\left(\phi_2, \hat{u}_2\right) {}^{B_2} \mathbf{r} \, e_2^*\left(\phi_2, \hat{u}_2\right) \tag{3.158}$$

which implies

$$ {}^{B_3} \mathbf{r} = e_2\left(\phi_2, \hat{u}_2\right) e_1\left(\phi_1, \hat{u}_1\right) {}^{B_1} \mathbf{r} \, e_1^*\left(\phi_1, \hat{u}_1\right) e_2^*\left(\phi_2, \hat{u}_2\right) \tag{3.159}$$

showing that

$$ e\left(\phi, \hat{u}\right) = e_2\left(\phi_2, \hat{u}_2\right) e_1\left(\phi_1, \hat{u}_1\right) \tag{3.160}$$

is the quaternion corresponding to $R = R_2 R_1$. ∎

**Example 61** ★ *Rodriguez rotation formula, using quaternion.*

*We may simplify the Equation (3.155) to have a vectorial form similar to the Rodriguez formula.*

$$\begin{aligned} \mathbf{r}' &= e\left(\phi, \hat{u}\right) \mathbf{r} \, e^*\left(\phi, \hat{u}\right) \\ &= \left(e_0^2 - \mathbf{e} \cdot \mathbf{e}\right) \mathbf{r} + 2 e_0 \left(\mathbf{e} \times \mathbf{r}\right) + 2\mathbf{e} \left(\mathbf{e} \cdot \mathbf{r}\right) \end{aligned} \tag{3.161}$$

**Example 62** ★ *Composition of rotations using quaternions.*

*Using quaternions to represent rotations makes it easy to calculate the composition of rotations. If the quaternion $e_1(\phi_1, \hat{u}_1)$ represents the rotation $R_{\hat{u}_1,\phi_1}$ and $e_2(\phi_2, \hat{u}_2)$ represents $R_{\hat{u}_2,\phi_2}$, then the product*

$$e_2(\phi_2, \hat{u}_2)\, e_1(\phi_1, \hat{u}_1)$$

*represents $R_{\hat{u}_2,\phi_2} R_{\hat{u}_1,\phi_1}$ because*

$$
\begin{aligned}
R_{\hat{u}_2,\phi_2} R_{\hat{u}_1,\phi_1}\mathbf{r} &= e_2(\phi_2, \hat{u}_2)\,(e_1(\phi_1, \hat{u}_1)\,\mathbf{r}\,e_1^*(\phi_1, \hat{u}_1))\,e_2^*(\phi_2, \hat{u}_2) \\
&= (e_2(\phi_2, \hat{u}_2)\,e_1(\phi_1, \hat{u}_1))\,\mathbf{r}\,(e_1^*(\phi_1, \hat{u}_1)\,e_2^*(\phi_2, \hat{u}_2)) \\
&= (e_2(\phi_2, \hat{u}_2)\,e_1(\phi_1, \hat{u}_1))\,\mathbf{r}\,(e_1(\phi_1, \hat{u}_1)\,e_2(\phi_2, \hat{u}_2))^*.
\end{aligned}
$$

**Example 63** ★ *Inner automorphism property of $e(\phi, \hat{u})$.*

*Since $e(\phi, \hat{u})$ is a unit quaternion,*

$$e^*(\phi, \hat{u}) = e^{-1}(\phi, \hat{u}) \tag{3.162}$$

*we may write*

$$^G\mathbf{r} = e(\phi, \hat{u})\,{}^B\mathbf{r}\,e^{-1}(\phi, \hat{u}). \tag{3.163}$$

*In abstract algebra, a mapping of the form $\mathbf{r} = q\,\mathbf{r}\,q^{-1}$, computed by multiplying on the left by an element and on the right by its inverse, is called an* **inner automorphism***. Thus, $^G\mathbf{r}$ is the inner automorphism of $^B\mathbf{r}$ based on the rotation quaternion $e(\phi, \hat{u})$.*

## 3.5  ★ Spinors and Rotators

Finite rotations can be expressed in two general ways: using $3 \times 3$ real orthogonal matrices $R$ called *rotator*, which is an abbreviation for *rotation tensor*; and using $3 \times 3$ real skew-symmetric matrices $\tilde{u}$ called *spinor*, which is an abbreviation for *spin tensor*.

A rotator is a linear operator that maps $^B\mathbf{r}$ to $^G\mathbf{r}$ when a rotation axis $\hat{u}$ and a rotation angle $\phi$, are given.

$$^G\mathbf{r} = {}^G R_B\,{}^B\mathbf{r} \tag{3.164}$$

The spinor $\tilde{u}$ is corresponding to the vector $\hat{u}$, which, along with $\phi$, can be utilized to describe a rotation.

$$^G R_B = \left(\mathbf{I}\cos\phi + \hat{u}\hat{u}^T \operatorname{vers}\phi + \tilde{u}\sin\phi\right) \tag{3.165}$$

For the moment let's forget $|\hat{u}| = \sqrt{u_1^2 + u_2^2 + u_3^2} = 1$ and develop the theory for non-unit vectors indicting the rotation axis. The square of $\tilde{u}$, computed through direct multiplication, is

$$\tilde{u}^2 = \begin{bmatrix} -u_2^2 - u_3^2 & u_1 u_2 & u_1 u_3 \\ u_1 u_2 & -u_1^2 - u_3^2 & u_2 u_3 \\ u_1 u_3 & u_2 u_3 & -u_1^2 - u_2^2 \end{bmatrix}$$

$$= -\tilde{u}\,\tilde{u}^T = -\tilde{u}^T\tilde{u}$$

$$= \hat{u}\hat{u}^T - u^2\mathbf{I}. \tag{3.166}$$

This is a symmetric matrix with $\mathrm{tr}[\tilde{u}^2] = -2|\hat{u}|^2 = -2u^2 = -2(u_1^2 + u_2^2 + u_3^2)$ whose eigenvalues are $0$, $-u^2$, $-u^2$. In other words, $\tilde{u}$ satisfies its own characteristic equation

$$\tilde{u}^2 = -u^2\mathbf{I} \ , \ \tilde{u}^3 = -u^2\tilde{u} \ , \ \cdots \ , \ \tilde{u}^n = -u^2\tilde{u}^{n-2} \ , \ n \geq 3. \tag{3.167}$$

The odd powers of $\tilde{u}$ are skew symmetric with distinct purely imaginary eigenvalues, while even powers of $\tilde{u}$ are symmetric with repeated real eigenvalues.

Spinors and rotators are functions of each other so $R$ must be expandable in a Taylor series of $\tilde{u}$.

$$R = \mathbf{I} + c_1\tilde{u} + c_2\tilde{u}^2 + c_3\tilde{u}^3 + \cdots \tag{3.168}$$

However, because of (3.167), all powers of order three or higher may be eliminated. Therefore, $R$ must be a linear function of $\mathbf{I}$, $\tilde{u}$, and $\tilde{u}^2$.

$$R = \mathbf{I} + a(\lambda\tilde{u}) + b(\lambda\tilde{u})^2 \tag{3.169}$$

$$= \begin{bmatrix} -b\lambda^2(u_2^2 + u_3^2) + 1 & -a\lambda u_3 + b\lambda^2 u_1 u_2 & a\lambda u_2 + b\lambda^2 u_1 u_3 \\ a\lambda u_3 + b\lambda^2 u_1 u_2 & -b\lambda^2(u_1^2 + u_3^2) + 1 & -a\lambda u_1 + b\lambda^2 u_2 u_3 \\ -a\lambda u_2 + b\lambda^2 u_1 u_3 & a\lambda u_1 + b\lambda^2 u_2 u_3 & -b\lambda^2(u_1^2 + u_2^2) + 1 \end{bmatrix}$$

where, $\lambda$ is the spinor normalization factor, while $a = a(\phi, u)$ and $b = b(\phi, u)$ are scalar functions of a rotation angle and an invariant of $\tilde{u}$.

Recalling $u = \sqrt{u_1^2 + u_2^2 + u_3^2} = 1$, Table 3.1 collects some representations of rotator $R$ as a function of the coefficients $a$, $b$, and the spinor $\lambda\tilde{u}$.

Table 3.1 - Rotator $R$ as a function of spinors

| $a$ | $b$ | $\lambda$ | $R$ |
|---|---|---|---|
| $\sin\phi$ | $\sin^2\frac{\phi}{2}$ | $1$ | $\mathbf{I} + \sin\phi\,\tilde{u} + 2\sin^2\frac{\phi}{2}\tilde{u}^2$ |
| $2\cos^2\frac{\phi}{2}$ | $2\cos^2\frac{\phi}{2}$ | $\tan\frac{\phi}{2}$ | $\mathbf{I} + 2\cos^2\frac{\phi}{2}[\tan\frac{\phi}{2}\tilde{u} + \tan^2\frac{\phi}{2}\tilde{u}^2]$ $= [\mathbf{I} + \tan\frac{\phi}{2}\tilde{u}][\mathbf{I} - \tan\frac{\phi}{2}\tilde{u}]^{-1}$ |
| $2\cos\frac{\phi}{2}$ | $2$ | $\sin\frac{\phi}{2}$ | $\mathbf{I} + 2\cos\frac{\phi}{2}\sin\frac{\phi}{2}\tilde{u} + 2\sin^2\frac{\phi}{2}\tilde{u}^2$ |
| $\frac{1}{\phi}\sin\phi$ | $\frac{2}{\phi^2}\sin^2\frac{\phi}{2}$ | $\phi$ | $\mathbf{I} + \sin\phi\,\tilde{u} + 2\sin^2\frac{\phi}{2}\tilde{u}^2$ |

**Proof.** Assuming $\lambda = 1$, we may find $\mathrm{tr}\,R = 1 + 2\cos\phi$, which, because of (3.169), is equal to,

$$\mathrm{tr}\,R = 1 + 2\cos\phi = 3 - 2bu^2 \tag{3.170}$$

and therefore,

$$b = \frac{1 - \cos\phi}{u^2} = \frac{2}{u^2}\sin^2\frac{\phi}{2}. \tag{3.171}$$

Now the orthogonality condition

$$
\begin{aligned}
\mathbf{I} &= R^T R = \left(\mathbf{I} - a\tilde{u} + b\tilde{u}^2\right)\left(\mathbf{I} + a\tilde{u} + b\tilde{u}^2\right) \\
&= \mathbf{I} + (2b - a^2)\tilde{u}^2 + b^2\tilde{u}^4 \\
&= \mathbf{I} + (2b - a^2 - b^2 u^2)\tilde{u}^2
\end{aligned} \tag{3.172}
$$

leads to

$$a = \sqrt{2b - b^2 u^2} = \frac{1}{u}\sin\phi \tag{3.173}$$

and therefore,

$$
\begin{aligned}
R &= \mathbf{I} + \frac{1}{u}\sin\phi\,\tilde{u} + \frac{2}{u^2}\sin^2\frac{\phi}{2}\tilde{u}^2 \\
&= \mathbf{I} + \sin\phi\,\tilde{u} + \operatorname{ver}\phi\,\tilde{u}^2.
\end{aligned} \tag{3.174}
$$

From a numerical viewpoint, the sine-squared form is preferred to avoid the cancellation in computing $1 - \cos\phi$ for small $\phi$. Replacing $a$ and $b$ in (3.169) provides the explicit rotator in terms of $\tilde{u}$ and $\phi$

$$
R = \begin{bmatrix}
u_1^2 + (u_2^2 + u_3^2)c\phi & 2u_1 u_2 s^2\frac{\phi}{2} - u_3 s\phi & 2u_1 u_3 s^2\frac{\phi}{2} + u_2 s\phi \\
2u_1 u_2 s^2\frac{\phi}{2} + u_3 s\phi & u_2^2 + (u_3^2 + u_1^2)c\phi & 2u_2 u_3 s^2\frac{\phi}{2} - u_1 s\phi \\
2u_1 u_3 s^2\frac{\phi}{2} - u_2 s\phi & 2u_2 u_3 s^2\frac{\phi}{2} + u_1 s\phi & u_3^2 + (u_1^2 + u_2^2)c\phi
\end{bmatrix} \tag{3.175}
$$

which is equivalent to Equation (3.13).

If $\lambda \neq 1$ but nonzero, the answers are $a = \frac{1}{\lambda u}\sin\phi$ and $b = \frac{2}{(\lambda u)^2}\sin^2\frac{\phi}{2}$ which do not affect $R$. ■

**Example 64 ★** *Eigenvalues of a spinor.*
*Consider the axis of rotation indicated by*

$$\mathbf{u} = \begin{bmatrix} 6 \\ 2 \\ 3 \end{bmatrix}, \qquad u = 7.$$

*The associated spin matrix and its square are*

$$\tilde{u} = \begin{bmatrix} 0 & -3 & 2 \\ 3 & 0 & -6 \\ -2 & 6 & 0 \end{bmatrix}$$

$$\tilde{u}^2 = \begin{bmatrix} -13 & 12 & 18 \\ 12 & -45 & 6 \\ 18 & 6 & -40 \end{bmatrix}$$

*where the eigenvalues of $\tilde{u}$ are $(0, 7i, -7i)$ while those of of $\tilde{u}^2$ are $(0, -49, -49)$.*

# 3.6 ★ Problems in Representing Rotations

As is evident in this Chapter, there are a number of different methods for representing rotations, however only a few of them are fundamentally distinct. The parameters or coordinates required to completely describe the orientation of a rigid body relative to some reference frames are sometimes called *attitude coordinates*. There are two inherent problems in representing rotations, both related to incontrovertible properties of rotations.

1. Rotations do not commute.

2. Spatial rotations do not topologically allow a smooth mapping in three dimensional Euclidean space.

The non-commutativity of rotations has been reviewed and showed in previous sections. Nonetheless, it is important to obey the order of rotations, although sometimes it seems that we may change the order of rotations and obtain the same result.

The lack of a smooth mapping in three dimensional Euclidean space means we cannot smoothly represent every kind of rotation using one set of three numbers. Any set of three rotational coordinates contains at least one geometrical orientation where the coordinates are singular, and it means at least two coordinates are undefined or not unique. The problem is similar to defining a coordinate system to locate a point on the Earth's surface. Using *longitude* and *latitude* becomes problematic at the north and south poles, where a small displacement can produce a radical change in longitude. We cannot find a superior system because it is not possible to smoothly wrap a sphere with a plane. Similarly, it is not possible to smoothly wrap the space of spatial rotations with three dimensional Euclidean space.

This is the reason why we sometimes describe rotations by using four numbers. We may use only three-number systems and expect to see the resulting singularities, or use four numbers, and cope with the redundancy. The choice depends on the application and method of calculation. For computer applications, the redundancy is not a problem, so most algorithms use representations with extra numbers. However, engineers prefer to work with the minimum set of numbers. Therefore, there is no unique and superior method for representing rotations.

## 3.6.1 Rotation matrix

Rotation matrix representation, derived by determination of directional cosines, is (for many purposes) the most useful representation method of spatial rotations. The two reference frames $G$ and $B$, having a common origin, are defined through orthogonal right-handed sets of unit vectors $\{G\} = \{\hat{I}, \hat{J}, \hat{K}\}$ and $\{B\} = \{\hat{i}, \hat{j}, \hat{k}\}$. The rotation or transformation matrix

between the two frames can simply be found by describing the unit vectors
of one of them in the other.

$$
\begin{aligned}
\hat{I} &= (\hat{I} \cdot \hat{\imath})\hat{\imath} + (\hat{I} \cdot \hat{\jmath})\hat{\jmath} + (\hat{I} \cdot \hat{k})\hat{k} \\
&= \cos(\hat{I}, \hat{\imath})\hat{\imath} + \cos(\hat{I}, \hat{\jmath})\hat{\jmath} + \cos(\hat{I}, \hat{k})\hat{k} \qquad (3.176) \\
\hat{J} &= (\hat{J} \cdot \hat{\imath})\hat{\imath} + (\hat{J} \cdot \hat{\jmath})\hat{\jmath} + (\hat{J} \cdot \hat{k})\hat{k} \\
&= \cos(\hat{J}, \hat{\imath})\hat{\imath} + \cos(\hat{J}, \hat{\jmath})\hat{\jmath} + \cos(\hat{J}, \hat{k})\hat{k} \qquad (3.177) \\
\hat{K} &= (\hat{K} \cdot \hat{\imath})\hat{\imath} + (\hat{K} \cdot \hat{\jmath})\hat{\jmath} + (\hat{K} \cdot \hat{k})\hat{k} \\
&= \cos(\hat{K}, \hat{\imath})\hat{\imath} + \cos(\hat{K}, \hat{\jmath})\hat{\jmath} + \cos(\hat{K}, \hat{k})\hat{k} \qquad (3.178)
\end{aligned}
$$

Therefore, having the rotation matrix $^G R_B$

$$
\begin{aligned}
^G R_B &= \begin{bmatrix} \hat{I} \cdot \hat{\imath} & \hat{I} \cdot \hat{\jmath} & \hat{I} \cdot \hat{k} \\ \hat{J} \cdot \hat{\imath} & \hat{J} \cdot \hat{\jmath} & \hat{J} \cdot \hat{k} \\ \hat{K} \cdot \hat{\imath} & \hat{K} \cdot \hat{\jmath} & \hat{K} \cdot \hat{k} \end{bmatrix} \\
&= \begin{bmatrix} \cos(\hat{I}, \hat{\imath}) & \cos(\hat{I}, \hat{\jmath}) & \cos(\hat{I}, \hat{k}) \\ \cos(\hat{J}, \hat{\imath}) & \cos(\hat{J}, \hat{\jmath}) & \cos(\hat{J}, \hat{k}) \\ \cos(\hat{K}, \hat{\imath}) & \cos(\hat{K}, \hat{\jmath}) & \cos(\hat{K}, \hat{k}) \end{bmatrix} \qquad (3.179)
\end{aligned}
$$

would be enough to find the coordinates of a point in the reference frame
$G$, when its coordinates are given in the reference frame $B$.

$$
^G \mathbf{r} = {}^G R_B \, {}^B \mathbf{r} \qquad (3.180)
$$

The rotation matrices convert the composition of rotations to matrix multiplication. It is simple and convenient especially when rotations are about
the global principal axes, or about the local principal axes.

Orthogonality is the most important and useful property of rotation matrices, which shows that the inverse of a rotation matrix is equivalent to its
transpose, $^G R_B^{-1} = {}^G R_B^T$. The null rotation is represented by the identity
matrix, $\mathbf{I}$.

The primary disadvantage of rotation matrices is that there are so many
numbers, which often make rotation matrices hard to interpret. Numerical
errors may build up until a normalization is necessary.

## 3.6.2  Angle-axis

Angle-axis representation, described by the Rodriguez formula, is a direct
result of the Euler rigid body rotation theorem. In this method, a rotation
is described by the magnitude of rotation, $\phi$, with the positive right-hand
direction about the axis directed by the unit vector, $\hat{u}$.

$$
^G R_B = R_{\hat{u}, \phi} = \mathbf{I} \cos \phi + \hat{u}\hat{u}^T \operatorname{vers} \phi + \tilde{u} \sin \phi \qquad (3.181)
$$

Converting the angle-axis representation to matrix form is simply done by expanding.

$$
{}^G R_B = \begin{bmatrix} u_1^2 \operatorname{vers}\phi + c\phi & u_1 u_2 \operatorname{vers}\phi - u_3 s\phi & u_1 u_3 \operatorname{vers}\phi + u_2 s\phi \\ u_1 u_2 \operatorname{vers}\phi + u_3 s\phi & u_2^2 \operatorname{vers}\phi + c\phi & u_2 u_3 \operatorname{vers}\phi - u_1 s\phi \\ u_1 u_3 \operatorname{vers}\phi - u_2 s\phi & u_2 u_3 \operatorname{vers}\phi + u_1 s\phi & u_3^2 \operatorname{vers}\phi + c\phi \end{bmatrix}
$$

Converting the matrix representation to angle-axis form is shown in Example 39 using matrix manipulation. However, it is sometimes easier if we convert the matrix to a quaternion and then convert the quaternion to angle-axis form.

Angle-axis representation has some problems. First, the rotation axis is indeterminate when $\phi = 0$. Second, the angle-axis representation is a two-to-one mapping system because

$$
R_{-\hat{u},-\phi} = R_{\hat{u},\phi} \tag{3.182}
$$

and it is redundant because for any integer $k$,

$$
R_{\hat{u},\phi+2k\pi} = R_{\hat{u},\phi}. \tag{3.183}
$$

However, both of these problems can be improved to some extent by restricting $\phi$ to some suitable range such as $[0, \pi]$ or $[-\frac{\pi}{2}, \frac{\pi}{2}]$. Finally, angle-axis representation is not efficient to find the composition of two rotations and determine the equivalent angle-axis of rotations.

### 3.6.3  Euler angles

Euler angles are also employed to describe the rotation matrix of rigid bodies utilizing only three numbers.

$$
\begin{aligned}
{}^G R_B &= [R_{z,\psi} R_{x,\theta} R_{z,\varphi}]^T \\
&= \begin{bmatrix} c\varphi c\psi - c\theta s\varphi s\psi & -c\varphi s\psi - c\theta c\psi s\varphi & s\theta s\varphi \\ c\psi s\varphi + c\theta c\varphi s\psi & -s\varphi s\psi + c\theta c\varphi c\psi & -c\varphi s\theta \\ s\theta s\psi & s\theta c\psi & c\theta \end{bmatrix}
\end{aligned} \tag{3.184}
$$

Euler angles and rotation matrices are not generally one-to-one, and also, they are not convenient representations of rotations or of constructing composite rotations. The angles $\varphi$ and $\psi$ are not distinguishable when $\theta \to 0$.

The equivalent rotation matrix is directly obtained by matrix multiplication, while the inverse conversion, from rotation matrix to a set of Euler angles, is not straightforward. It is also not applicable when $\sin\theta = 0$.

$$\theta = \cos^{-1}(r_{33}) \tag{3.185}$$

$$\varphi = -\tan^{-1}\left(\frac{r_{31}}{r_{32}}\right) \tag{3.186}$$

$$\psi = \tan^{-1}\left(\frac{r_{13}}{r_{23}}\right) \tag{3.187}$$

It is possible to use a more efficient method that handles all cases uniformly. The main idea is to work with the sum and difference of $\varphi$ and $\psi$

$$\sigma = \varphi + \psi \tag{3.188}$$

$$\upsilon = \varphi - \psi \tag{3.189}$$

and then,

$$\varphi = \frac{\sigma - \upsilon}{2} \tag{3.190}$$

$$\psi = \frac{\sigma + \upsilon}{2}. \tag{3.191}$$

Therefore,

$$r_{11} + r_{22} = \cos\sigma(1 + \cos\theta) \tag{3.192}$$

$$r_{11} - r_{22} = \cos\upsilon(1 - \cos\theta) \tag{3.193}$$

$$r_{21} - r_{12} = \sin\sigma(1 + \cos\theta) \tag{3.194}$$

$$r_{21} + r_{12} = \sin\upsilon(1 - \cos\theta), \tag{3.195}$$

which leads to

$$\sigma = \tan^{-1}\frac{r_{21} - r_{12}}{r_{11} + r_{22}} \tag{3.196}$$

$$\upsilon = \tan^{-1}\frac{r_{21} + r_{12}}{r_{11} - r_{22}}. \tag{3.197}$$

This approach resolves the problem at $\sin\theta = 0$. At $\theta = 0$, we can find $\sigma$, but $\upsilon$ is undetermined, and at $\theta = \pi$, we can find $\upsilon$, but $\sigma$ is undetermined. The undetermined values are consequence of $\tan^{-1}\frac{0}{0}$. Besides these singularities, both $\sigma$ and $\upsilon$ are uniquely determined. The middle rotation angle, $\theta$, can also be found using $\tan^{-1}$ operator

$$\theta = \tan^{-1}\left(\frac{r_{13}\sin\varphi - r_{23}\cos\varphi}{r_{33}}\right). \tag{3.198}$$

The main advantage of Euler angles is that they use only three numbers. They are integrable, and they provide a good visualization of spatial rotation with no redundancy. Euler angles are used in dynamic analysis of spinning bodies.

The other combinations of Euler angles as well as roll-pitch-yaw angles have the same kind of problems, and similar advantages.

## 3.6.4   Quaternion

Quaternion uses four numbers to represent a rotation according to a special rule for addition and multiplication. Rotation quaternion is a unit quaternion that may be described by Euler parameters, or the axis and angle of rotation.

$$
\begin{aligned}
e\left(\phi, \hat{u}\right) &= e_0 + \mathbf{e} \\
&= e_0 + e_1\hat{\imath} + e_2\hat{\jmath} + e_3\hat{k} \\
&= \cos\frac{\phi}{2} + \sin\frac{\phi}{2}\hat{u}
\end{aligned}
\tag{3.199}
$$

We can also define a $4 \times 4$ matrix to represent a quaternion

$$
\overleftrightarrow{q} =
\begin{bmatrix}
q_0 & -q_1 & -q_2 & -q_3 \\
q_1 & q_0 & -q_3 & q_2 \\
q_2 & q_3 & q_0 & -q_1 \\
q_3 & -q_2 & q_1 & q_0
\end{bmatrix}
\tag{3.200}
$$

which provides the important orthogonality property

$$
\overleftrightarrow{q}^{-1} = \overleftrightarrow{q}^{T}.
\tag{3.201}
$$

The matrix quaternion (3.200) can also be represented by

$$
\overleftrightarrow{q} =
\begin{bmatrix}
q_0 & -\mathbf{q}^T \\
\mathbf{q} & q_0\mathbf{I}_3 - \tilde{q}
\end{bmatrix}
\tag{3.202}
$$

where

$$
\tilde{q} =
\begin{bmatrix}
0 & -q_3 & q_2 \\
q_3 & 0 & -q_1 \\
-q_2 & q_1 & 0
\end{bmatrix}.
\tag{3.203}
$$

Employing the *matrix quaternion,* $\overleftrightarrow{q}$, we can describe the quaternion multiplication by matrix multiplication.

$$
qp = \overleftrightarrow{q}\,p =
\begin{bmatrix}
q_0 & -q_1 & -q_2 & -q_3 \\
q_1 & q_0 & -q_3 & q_2 \\
q_2 & q_3 & q_0 & -q_1 \\
q_3 & -q_2 & q_1 & q_0
\end{bmatrix}
\begin{bmatrix}
p_0 \\
p_1 \\
p_2 \\
p_3
\end{bmatrix}
\tag{3.204}
$$

The matrix description of quaternions ties the quaternion manipulations and matrix manipulations, because if $p$, $q$, and $v$ are three quaternions and

$$
qp = v
\tag{3.205}
$$

then

$$
\overleftrightarrow{q}\,\overleftrightarrow{p} = \overleftrightarrow{v}.
\tag{3.206}
$$

Hence, quaternion representation of transformation between coordinate frames

$$^G\mathbf{r} = e\left(\phi, \hat{u}\right)\ ^B\mathbf{r}\ e^*\left(\phi, \hat{u}\right) \tag{3.207}$$

can also be defined by matrix multiplication

$$\begin{aligned}
\overleftrightarrow{^G\mathbf{r}} &= \overleftrightarrow{e\left(\phi, \hat{u}\right)}\ \overleftrightarrow{^B\mathbf{r}}\ \overleftrightarrow{e^*\left(\phi, \hat{u}\right)} \\
&= \overleftrightarrow{e\left(\phi, \hat{u}\right)}\ \overleftrightarrow{^B\mathbf{r}}\ \overleftrightarrow{e\left(\phi, \hat{u}\right)}^T .
\end{aligned} \tag{3.208}$$

**Proof.** We can use the matrix definition of quaternions and see that

$$\overleftrightarrow{e\left(\phi, \hat{u}\right)} = \begin{bmatrix} e_0 & -e_1 & -e_2 & -e_3 \\ e_1 & e_0 & -e_3 & e_2 \\ e_2 & e_3 & e_0 & -e_1 \\ e_3 & -e_2 & e_1 & e_0 \end{bmatrix} \tag{3.209}$$

$$\overleftrightarrow{^B\mathbf{r}} = \begin{bmatrix} 0 & -\,^Br_1 & -\,^Br_2 & -\,^Br_3 \\ ^Br_1 & 0 & -\,^Br_3 & ^Br_2 \\ ^Br_2 & ^Br_3 & 0 & -\,^Br_1 \\ ^Br_3 & -\,^Br_2 & ^Br_1 & 0 \end{bmatrix} \tag{3.210}$$

$$\overleftrightarrow{e\left(\phi, \hat{u}\right)}^T = \begin{bmatrix} e_0 & e_1 & e_2 & e_3 \\ -e_1 & e_0 & e_3 & -e_2 \\ -e_2 & -e_3 & e_0 & e_1 \\ -e_3 & e_2 & -e_1 & e_0 \end{bmatrix} . \tag{3.211}$$

Therefore,

$$\overleftrightarrow{e\left(\phi, \hat{u}\right)}\ \overleftrightarrow{^B\mathbf{r}}\ \overleftrightarrow{e\left(\phi, \hat{u}\right)}^T = \begin{bmatrix} 0 & -\,^Gr_1 & -\,^Gr_2 & -\,^Gr_3 \\ ^Gr_1 & 0 & -\,^Gr_3 & ^Gr_2 \\ ^Gr_2 & ^Gr_3 & 0 & -\,^Gr_1 \\ ^Gr_3 & -\,^Gr_2 & ^Gr_1 & 0 \end{bmatrix} \tag{3.212}$$

where,

$$\begin{aligned}
^Gr_1 = {}&^Br_1 \left(e_0^2 + e_1^2 - e_2^2 - e_3^2\right) \\
&+ {}^Br_2 \left(2e_1e_2 - 2e_0e_3\right) \\
&+ {}^Br_3 \left(2e_0e_2 + 2e_1e_3\right)
\end{aligned} \tag{3.213}$$

$$\begin{aligned}
^Gr_2 = {}&^Br_1 \left(2e_0e_3 + 2e_1e_2\right) \\
&+ {}^Br_2 \left(e_0^2 - e_1^2 + e_2^2 - e_3^2\right) \\
&+ {}^Br_3 \left(2e_2e_3 - 2e_0e_1\right)
\end{aligned} \tag{3.214}$$

$$\begin{aligned}
^Gr_3 = {}&^Br_1 \left(2e_1e_3 - 2e_0e_2\right) \\
&+ {}^Br_2 \left(2e_0e_1 + 2e_2e_3\right) \\
&+ {}^Br_3 \left(e_0^2 - e_1^2 - e_2^2 + e_3^2\right) .
\end{aligned} \tag{3.215}$$

which are compatible with Equation (3.155). ∎

## 3.6.5 Euler parameters

Euler parameters are the elements of rotation quaternions. Therefore, there is a direct conversion between rotation quaternion and Euler parameters, which in turn are related to angle-axis parameters. We can obtain the axis and angle of rotation $(\phi, \hat{u})$, from Euler parameters or rotation quaternion $e(\phi, \hat{u})$, by

$$\phi = 2\tan^{-1}\frac{|\mathbf{e}|}{e_0} \tag{3.216}$$

$$\hat{u} = \frac{\mathbf{e}}{|\mathbf{e}|}. \tag{3.217}$$

Unit quaternion provides a suitable base for describing spatial rotations, although it needs normalization due to the error pile-up problem. In general, in some applications quaternions offer superior computational efficiency.

It is interesting to know that Euler was the first to derive Rodriguez's formula, while Rodriguez was the first to invent Euler parameters. In addition, Hamilton introduced quaternions, however, Gauss invented them but never published.

**Example 65** ★ *Taylor expansion of rotation matrix.*

*Assume the rotation matrix $R = R(t)$ is a time-dependent transformation between coordinate frames $B$ and $G$. The body frame $B$ is coincident with $G$ at $t = 0$. Therefore, $R(0) = \mathbf{I}$, and we may expand the elements of $R$ in a Taylor series expansion*

$$R(t) = \mathbf{I} + R_1 t + \frac{1}{2!}R_2 t^2 + \frac{1}{3!}R_3 t^3 + \cdots \tag{3.218}$$

*in which $R_i, (i = 1, 2, 3, \cdots)$ is a constant matrix. The rotation matrix $R(t)$ is orthogonal for all $t$, hence,*

$$RR^T = \mathbf{I} \tag{3.219}$$

$$\left(\mathbf{I} + R_1 t + \frac{1}{2!}R_2 t^2 + \cdots\right)\left(\mathbf{I} + R_1^T t + \frac{1}{2!}R_2^T t^2 + \cdots\right) = \mathbf{I}. \tag{3.220}$$

*The coefficient of $t^i, (i = 1, 2, 3, \cdots)$ must vanish on the left-hand side. This gives us*

$$R_1 + R_1^T = 0 \tag{3.221}$$

$$R_2 + 2R_1 R_1^T + R_2^T = 0 \tag{3.222}$$

$$R_3 + 3R_2 R_1^T + 3R_1 R_2^T + R_3^T = 0 \tag{3.223}$$

*or in general*

$$\sum_{i=0}^{n}\binom{n}{i}R_{n-i}R_i^T = 0 \tag{3.224}$$

*where*

$$R_0 = R_0^T = \mathbf{I}. \tag{3.225}$$

*Equation (3.221) shows that $R_1$ is a skew symmetric matrix, and therefore, $R_1 R_1^T = -R_1^2 = C_1$ is symmetric. Now the Equation (3.222)*

$$
\begin{aligned}
R_2 + R_2^T &= -2R_1 R_1^T \\
&= -[R_1 R_1^T + [R_1 R_1^T]^T] \\
&= 2C_1 \tag{3.226}
\end{aligned}
$$

*leads to*

$$
\begin{aligned}
R_2 &= C_1 + [C_1 - R_2^T] \tag{3.227} \\
R_2^T &= C_1 + [C_1 - R_2] \\
&= C_1 + [C_1 - R_2^T]^T \tag{3.228}
\end{aligned}
$$

*that shows $[C_1 - R_2^T]$ is skew symmetric because we must have*

$$[C_1 - R_2^T] + [C_1 - R_2^T]^T = 0. \tag{3.229}$$

*Therefore, the matrix product*

$$[C_1 - R_2^T][C_1 - R_2^T]^T = -[C_1 - R_2^T]^2 \tag{3.230}$$

*is symmetric. The next step shows that*

$$
\begin{aligned}
R_3 + R_3^T &= -3[R_1 R_2^T + R_2 R_1^T] \\
&= -3[R_1 R_2^T + [R_1 R_2^T]^T] \\
&= 3[R_1[R_1^2 - R_2^T] + [R_1^2 - R_2^T]R_1] \\
&= 2C_2 \tag{3.231}
\end{aligned}
$$

*leads to*

$$
\begin{aligned}
R_3 &= C_2 + [C_2 - R_3^T] \tag{3.232} \\
R_3^T &= C_2 + [C_2 - R_3] \\
&= C_2 + [C_2 - R_3^T]^T \tag{3.233}
\end{aligned}
$$

*that shows $[C_2 - R_3^T]$ is skew symmetric because we must have*

$$[C_2 - R_3^T] + [C_2 - R_3^T]^T = 0. \tag{3.234}$$

*Therefore, the matrix product*

$$[C_2 - R_3^T][C_2 - R_3^T]^T = -[C_2 - R_3^T]^2 \tag{3.235}$$

*is also symmetric.*

*Continuing this procedure shows that the expansion of a rotation matrix $R(t)$ around the unit matrix can be written in the form of*

$$
\begin{aligned}
R(t) \;=\; & \mathbf{I} + C_1 t \\
& + \frac{1}{2!} \left[ C_1 + [C_1 - R_2^T] \right] t^2 \\
& + \frac{1}{3!} \left[ C_2 + [C_2 - R_3^T] \right] t^3 + \cdots
\end{aligned}
\tag{3.236}
$$

*where $C_i$ are symmetric and $[C_i - R_{i+1}^T]$ are skew symmetric matrices and*

$$
C_i = \frac{1}{2} \left[ R_{i-1} + R_{1-1}^T \right]. \tag{3.237}
$$

*Therefore, the expansion of an inverse rotation matrix can be written as*

$$
\begin{aligned}
R^T(t) \;=\; & \mathbf{I} + C_1 t \\
& + \frac{1}{2!} \left[ C_1 + [C_1 - R_2^T] \right] t^2 \\
& + \frac{1}{3!} \left[ C_2 + [C_2 - R_3^T] \right] t^3 + \cdots
\end{aligned}
\tag{3.238}
$$

## 3.7 ★ Composition and Decomposition of Rotations

Rotation $\phi_1$ about $\hat{u}_1$ of a rigid body with a fixed point, followed by a rotation $\phi_2$ about $\hat{u}_2$ can be composed to a unique rotation $\phi_3$ about $\hat{u}_3$. In other words, when a rigid body rotates from an initial position to a middle position $^{B_2}\mathbf{r} = {}^{B_2}R_{B_1}\,{}^{B_1}\mathbf{r}$, and then rotates to a final position, $^{B_3}\mathbf{r} = {}^{B_3}R_{B_2}\,{}^{B_2}\mathbf{r}$, the middle position can be skipped to rotate directly to the final position $^{B_3}\mathbf{r} = {}^{B_3}R_{B_1}\,{}^{B_1}\mathbf{r}$.

**Proof.** To show that two successive rotations of a rigid body with a fixed point is equivalent to a single rotation, we start with the Rodriguez rotation formula (3.115) and rewrite it as

$$
(\mathbf{r}' - \mathbf{r}) = \mathbf{w} \times (\mathbf{r}' + \mathbf{r}) \tag{3.239}
$$

where

$$
\mathbf{w} = \tan \frac{\phi}{2} \hat{u} \tag{3.240}
$$

is the Rodriguez vector. Rotation $\mathbf{w}_1$ followed by rotation $\mathbf{w}_2$ are

$$
\begin{aligned}
(\mathbf{r}_2 - \mathbf{r}_1) &= \mathbf{w}_1 \times (\mathbf{r}_2 + \mathbf{r}_1) & (3.241) \\
(\mathbf{r}_3 - \mathbf{r}_2) &= \mathbf{w}_2 \times (\mathbf{r}_3 + \mathbf{r}_2) & (3.242)
\end{aligned}
$$

respectively. The right hand side of the first one is perpendicular to $\mathbf{w}_1$ and the second one is perpendicular to $\mathbf{w}_2$. Hence, dot product of the first one with $\mathbf{w}_1$ and the second one with $\mathbf{w}_2$ show that

$$\mathbf{w}_1 \cdot \mathbf{r}_2 = \mathbf{w}_1 \cdot \mathbf{r}_1 \qquad (3.243)$$

$$\mathbf{w}_2 \cdot \mathbf{r}_3 = \mathbf{w}_2 \cdot \mathbf{r}_2 \qquad (3.244)$$

and cross product of the first one with $\mathbf{w}_2$ and the second one with $\mathbf{w}_1$ show that

$$\begin{aligned}
\mathbf{w}_2 \times (\mathbf{r}_2 - \mathbf{r}_1) - \mathbf{w}_1 \times (\mathbf{r}_3 - \mathbf{r}_2) = {} & \mathbf{w}_1 \left[ \mathbf{w}_2 \cdot (\mathbf{r}_2 + \mathbf{r}_1) \right] \\
& - (\mathbf{w}_1 \cdot \mathbf{w}_2)(\mathbf{r}_2 + \mathbf{r}_1) \\
& - \mathbf{w}_2 \left[ \mathbf{w}_1 \cdot (\mathbf{r}_3 + \mathbf{r}_2) \right] \\
& + (\mathbf{w}_1 \cdot \mathbf{w}_2)(\mathbf{r}_3 + \mathbf{r}_2) . \quad (3.245)
\end{aligned}$$

Rearranging while using Equations (3.243) and (3.244) gives us

$$\begin{aligned}
\mathbf{w}_2 \times (\mathbf{r}_2 - \mathbf{r}_1) - \mathbf{w}_1 \times (\mathbf{r}_3 - \mathbf{r}_2) = {} & (\mathbf{w}_2 \times \mathbf{w}_1) \times (\mathbf{r}_1 + \mathbf{r}_3) \\
& + (\mathbf{w}_1 \cdot \mathbf{w}_2)(\mathbf{r}_3 - \mathbf{r}_1) \quad (3.246)
\end{aligned}$$

which can be written as

$$\begin{aligned}
(\mathbf{w}_1 + \mathbf{w}_2) \times \mathbf{r}_2 = {} & \mathbf{w}_2 \times \mathbf{r}_1 + \mathbf{w}_1 \times \mathbf{r}_3 \\
& + (\mathbf{w}_2 \times \mathbf{w}_1) \times (\mathbf{r}_1 + \mathbf{r}_3) \\
& + (\mathbf{w}_1 \cdot \mathbf{w}_2)(\mathbf{r}_3 - \mathbf{r}_1) . \quad (3.247)
\end{aligned}$$

Adding Equations (3.241) and (3.242) to obtain $(\mathbf{w}_1 + \mathbf{w}_2) \times \mathbf{r}_2$ leads to

$$(\mathbf{w}_1 + \mathbf{w}_2) \times \mathbf{r}_2 = \mathbf{r}_3 - \mathbf{r}_1 - \mathbf{w}_1 \times \mathbf{r}_1 - \mathbf{w}_2 \times \mathbf{r}_3 . \qquad (3.248)$$

Therefore, we obtain the required Rodriguez rotation formula to rotate $\mathbf{r}_1$ to $\mathbf{r}_3$

$$\mathbf{r}_3 - \mathbf{r}_1 = \mathbf{w}_3 \times (\mathbf{r}_3 + \mathbf{r}_1) \qquad (3.249)$$

where

$$\mathbf{w}_3 = \frac{\mathbf{w}_1 + \mathbf{w}_2 + \mathbf{w}_2 \times \mathbf{w}_1}{1 - \mathbf{w}_1 \cdot \mathbf{w}_2} . \qquad (3.250)$$

■

Any rotation $\phi_1$ of a rigid body with a fixed point about $\hat{u}_1$ can be decomposed into three successive rotations about three arbitrary axes $\hat{a}$, $\hat{b}$, and $\hat{c}$ through unique angles $\alpha$, $\beta$, and $\gamma$.

Let $^G R_{\hat{a},\alpha}$, $^G R_{\hat{b},\beta}$, and $^G R_{\hat{c},\gamma}$ be any three in order rotation matrices about non-coaxis non-coplanar unit vectors $\hat{a}$, $\hat{b}$, and $\hat{c}$, through non-vanishing values $\alpha$, $\beta$, and $\gamma$. Then, any other rotation $^G R_{\hat{u},\phi}$ can be expressed in terms of $^G R_{\hat{a},\alpha}$, $^G R_{\hat{b},\beta}$, and $^G R_{\hat{c},\gamma}$

$$^G R_{\hat{u},\phi} = {}^G R_{\hat{c},\gamma} \, {}^G R_{\hat{b},\beta} \, {}^G R_{\hat{a},\alpha} \qquad (3.251)$$

if $\alpha, \beta$, and $\gamma$ are properly chosen numbers.

**Proof.** Using the definition of rotation based on quaternion, we may write

$$^{G}\mathbf{r} = {}^{G}R_{\hat{u},\phi}\,{}^{B}\mathbf{r} = e\,(\phi, \hat{u})\,{}^{B}\mathbf{r}\,e^{*}\,(\phi, \hat{u})\,. \tag{3.252}$$

Let us assume that $\mathbf{r}_1$ indicates the position vector $\mathbf{r}$ before rotation, and $\mathbf{r}_2$, $\mathbf{r}_3$, and $\mathbf{r}_4$ indicate the position vector $\mathbf{r}$ after rotation $R_{\hat{a},\alpha}$, $R_{\hat{b},\beta}$, and $R_{\hat{c},\gamma}$ respectively. Hence,

$$\mathbf{r}_2 = a\mathbf{r}_1 a^{*} \tag{3.253}$$
$$\mathbf{r}_3 = b\mathbf{r}_2 b^{*} \tag{3.254}$$
$$\mathbf{r}_4 = c\mathbf{r}_3 c^{*} \tag{3.255}$$
$$\mathbf{r}_4 = e\mathbf{r}_1 e^{*} \tag{3.256}$$

where

$$a\,(\alpha, \hat{a}) = a_0 + \mathbf{a} = \cos\frac{\alpha}{2} + \sin\frac{\alpha}{2}\hat{a} \tag{3.257}$$

$$b(\beta, \hat{b}) = b_0 + \mathbf{b} = \cos\frac{\beta}{2} + \sin\frac{\beta}{2}\hat{b} \tag{3.258}$$

$$c\,(\gamma, \hat{c}) = c_0 + \mathbf{c} = \cos\frac{\gamma}{2} + \sin\frac{\gamma}{2}\hat{c} \tag{3.259}$$

$$e\,(\phi, \hat{u}) = e_0 + \mathbf{e} = \cos\frac{\phi}{2} + \sin\frac{\phi}{2}\hat{u} \tag{3.260}$$

are quaternions corresponding to $(\alpha, \hat{a})$, $(\beta, \hat{b})$, $(\gamma, \hat{c})$, and $(\phi, \hat{u})$ respectively. We define the following scalars to simplify the equations.

$$\cos\frac{\alpha}{2} = C_1 \quad \cos\frac{\beta}{2} = C_2 \quad \cos\frac{\gamma}{2} = C_3 \quad \cos\frac{\phi}{2} = C \tag{3.261}$$

$$\sin\frac{\alpha}{2} = S_1 \quad \sin\frac{\beta}{2} = S_2 \quad \sin\frac{\gamma}{2} = S_3 \quad \sin\frac{\phi}{2} = S \tag{3.262}$$

$$\frac{b_2 c_3 - b_3 c_2}{S_2 S_3} = f_1 \quad \frac{b_3 c_1 - b_1 c_3}{S_2 S_3} = f_2 \quad \frac{b_1 c_2 - b_2 c_1}{S_2 S_3} = f_3 \tag{3.263}$$

$$\frac{c_2 a_3 - c_3 a_2}{S_3 S_1} = g_1 \quad \frac{a_3 c_1 - a_1 c_3}{S_3 S_1} = g_2 \quad \frac{a_1 c_2 - a_2 c_1}{S_3 S_1} = g_3 \tag{3.264}$$

$$\frac{a_2 b_3 - a_3 b_2}{S_1 S_2} = h_1 \quad \frac{a_3 b_1 - a_1 b_3}{S_1 S_2} = h_2 \quad \frac{a_1 b_2 - a_2 b_1}{S_1 S_2} = h_3 \tag{3.265}$$

$$\mathbf{b} \cdot \mathbf{c} = n_1 S_2 S_3 \tag{3.266}$$
$$\mathbf{c} \cdot \mathbf{a} = n_2 S_3 S_1 \tag{3.267}$$
$$\mathbf{a} \cdot \mathbf{b} = n_3 S_1 S_2 \tag{3.268}$$
$$(\mathbf{a} \times \mathbf{b}) \cdot \mathbf{c} = n_4 S_1 S_2 S_3 \tag{3.269}$$

Direct substitution shows that

$$\mathbf{r}_4 = e\mathbf{r}_1 e^* = c bar_1 a^* b^* c^* \tag{3.270}$$

and therefore,

$$
\begin{aligned}
e &= cba = c\,(b_0 a_0 - \mathbf{b}\cdot\mathbf{a} + b_0\mathbf{a} + a_0\mathbf{b} + \mathbf{b}\times\mathbf{a}) \\
&= c_0 b_0 a_0 - a_0\mathbf{b}\cdot\mathbf{c} - b_0\mathbf{c}\cdot\mathbf{a} - c_0\mathbf{a}\cdot\mathbf{b} + (\mathbf{a}\times\mathbf{b})\cdot\mathbf{c} \\
&\quad + a_0 b_0\mathbf{c} + b_0 c_0\mathbf{a} + c_0 a_0\mathbf{b} \\
&\quad + a_0\,(\mathbf{b}\times\mathbf{c}) + b_0\,(\mathbf{c}\times\mathbf{a}) + c_0\,(\mathbf{b}\times\mathbf{a}) \\
&\quad - (\mathbf{a}\cdot\mathbf{b})\,\mathbf{c} - (\mathbf{b}\cdot\mathbf{c})\,\mathbf{a} + (\mathbf{c}\cdot\mathbf{a})\,\mathbf{b}.
\end{aligned}
\tag{3.271}
$$

Hence,

$$e_0 = c_0 b_0 a_0 - a_0 n_1 S_2 S_3 - b_0 n_2 S_3 S_1 - c_0 n_3 S_1 S_2 + n_4 S_1 S_2 S_3 \tag{3.272}$$

and

$$
\begin{aligned}
\mathbf{e} &= a_0 b_0\mathbf{c} + b_0 c_0\mathbf{a} + c_0 a_0\mathbf{b} \\
&\quad + a_0\,(\mathbf{b}\times\mathbf{c}) + b_0\,(\mathbf{c}\times\mathbf{a}) + c_0\,(\mathbf{b}\times\mathbf{a}) \\
&\quad - n_1 S_2 S_3\mathbf{a} + n_2 S_3 S_1\mathbf{b} - n_3 S_1 S_2\mathbf{c}
\end{aligned}
\tag{3.273}
$$

which generate four equations

$$
\begin{aligned}
C_1 C_2 C_3 - n_1 C_1 S_2 S_3 - n_2 S_1 C_2 S_3 \\
+ n_3 S_1 S_2 C_3 - n_4 S_1 S_2 S_3 = C
\end{aligned}
\tag{3.274}
$$

$$
\begin{aligned}
a_1 S_1 C_2 C_3 + b_1 C_1 S_2 C_3 + c_1 C_1 C_2 S_3 \\
+ f_1 C_1 S_2 S_3 + g_1 S_1 C_2 S_3 + h_1 S_1 S_2 C_3 \\
- n_1 a_1 S_1 S_2 S_3 + n_2 b_1 S_1 S_2 S_3 - n_3 c_1 S_1 S_2 S_3 = u_1 S
\end{aligned}
\tag{3.275}
$$

$$
\begin{aligned}
a_2 S_1 C_2 C_3 + b_2 C_1 S_2 C_3 + c_2 C_1 C_2 S_3 \\
+ f_2 C_1 S_2 S_3 + g_2 S_1 C_2 S_3 + h_2 S_1 S_2 C_3 \\
- n_1 a_2 S_1 S_2 S_3 + n_2 b_2 S_1 S_2 S_3 - n_3 c_2 S_1 S_2 S_3 = u_2 S
\end{aligned}
\tag{3.276}
$$

$$
\begin{aligned}
a_3 S_1 C_2 C_3 + b_{13} C_1 S_2 C_3 + c_3 C_1 C_2 S_3 \\
+ f_3 C_1 S_2 S_3 + g_3 S_1 C_2 S_3 + h_3 S_1 S_2 C_3 \\
- n_1 a_3 S_1 S_2 S_3 + n_2 b_3 S_1 S_2 S_3 - n_3 c_3 S_1 S_2 S_3 = u_3 S
\end{aligned}
\tag{3.277}
$$

Since $e_0^2 + e_1^2 + e_2^2 + e_3^2 = 1$, only the first equation and two out of the others along with

$$C_1 = \sqrt{1 - S_1^2}\ ,\quad C_2 = \sqrt{1 - S_2^2}\ ,\quad C_3 = \sqrt{1 - S_3^2} \tag{3.278}$$

must be utilized to determine $C_1, C_2, C_3, S_1, S_2$, and $S_3$. ∎

**Example 66** ★ *Decomposition of a vector in a non-orthogonal coordinate frame.*

Let $\mathbf{a}, \mathbf{b},$ and $\mathbf{c}$ be any three non-coplanar, non-vanishing vectors; then any other vector $\mathbf{r}$ can be expressed in terms of $\mathbf{a}, \mathbf{b},$ and $\mathbf{c}$

$$\mathbf{r} = u\mathbf{a} + v\mathbf{b} + w\mathbf{c} \tag{3.279}$$

provided $u, v,$ and $w$ are properly chosen numbers. If $(\mathbf{a}, \mathbf{b}, \mathbf{c})$ coordinate system is a Cartesian coordinate system $(\hat{I}, \hat{J}, \hat{K})$, then

$$\mathbf{r} = (\mathbf{r} \cdot \hat{I})\hat{I} + (\mathbf{r} \cdot \hat{J})\hat{J} + (\mathbf{r} \cdot \hat{K})\hat{K}. \tag{3.280}$$

To show this, we start with finding the dot product of Equation (3.279) by $(\mathbf{b} \times \mathbf{c})$

$$\mathbf{r} \cdot (\mathbf{b} \times \mathbf{c}) = u\mathbf{a} \cdot (\mathbf{b} \times \mathbf{c}) + v\mathbf{b} \cdot (\mathbf{b} \times \mathbf{c}) + w\mathbf{c} \cdot (\mathbf{b} \times \mathbf{c}) \tag{3.281}$$

and noting that $(\mathbf{b} \times \mathbf{c})$ is perpendicular to both $\mathbf{b}$ and $\mathbf{c}$, consequently

$$\mathbf{r} \cdot (\mathbf{b} \times \mathbf{c}) = u\mathbf{a} \cdot (\mathbf{b} \times \mathbf{c}). \tag{3.282}$$

Therefore,

$$u = \frac{[\mathbf{rbc}]}{[\mathbf{abc}]} \tag{3.283}$$

where

$$[\mathbf{abc}] = \mathbf{a} \cdot \mathbf{b} \times \mathbf{c} = \mathbf{a} \cdot (\mathbf{b} \times \mathbf{c}) = \begin{vmatrix} a_1 & b_1 & c_1 \\ a_2 & b_2 & c_2 \\ a_3 & b_3 & c_3 \end{vmatrix}. \tag{3.284}$$

Similarly $v$ and $w$ would be

$$v = \frac{[\mathbf{rca}]}{[\mathbf{abc}]} \tag{3.285}$$

$$w = \frac{[\mathbf{rab}]}{[\mathbf{abc}]}. \tag{3.286}$$

Hence,

$$\mathbf{r} = \frac{[\mathbf{rbc}]}{[\mathbf{abc}]}\mathbf{a} + \frac{[\mathbf{rca}]}{[\mathbf{abc}]}\mathbf{b} + \frac{[\mathbf{rab}]}{[\mathbf{abc}]}\mathbf{c} \tag{3.287}$$

that can also be written as

$$\mathbf{r} = \left(\mathbf{r} \cdot \frac{\mathbf{b} \times \mathbf{c}}{[\mathbf{abc}]}\right)\mathbf{a} + \left(\mathbf{r} \cdot \frac{\mathbf{c} \times \mathbf{a}}{[\mathbf{abc}]}\right)\mathbf{b} + \left(\mathbf{r} \cdot \frac{\mathbf{a} \times \mathbf{b}}{[\mathbf{abc}]}\right)\mathbf{c}. \tag{3.288}$$

Multiplying (3.288) by $[\mathbf{abc}]$ gives a symmetric equation

$$[\mathbf{abc}]\,\mathbf{r} - [\mathbf{bcr}]\,\mathbf{a} + [\mathbf{cra}]\,\mathbf{b} - [\mathbf{rab}]\,\mathbf{c} = 0 \tag{3.289}$$

*If the $(\mathbf{a}, \mathbf{b}, \mathbf{c})$ coordinate system is a Cartesian system $(\hat{I}, \hat{J}, \hat{K})$, that is a mutually orthogonal system of unit vectors, then*

$$\left[\hat{I}\hat{J}\hat{K}\right] = 1 \tag{3.290}$$

$$\hat{I} \times \hat{J} = \hat{K} \tag{3.291}$$

$$\hat{J} \times \hat{I} = \hat{K} \tag{3.292}$$

$$\hat{K} \times \hat{I} = \hat{J} \tag{3.293}$$

*and Equation (3.288) becomes*

$$\mathbf{r} = \left(\mathbf{r} \cdot \hat{I}\right) \hat{I} + \left(\mathbf{r} \cdot \hat{J}\right) \hat{J} + \left(\mathbf{r} \cdot \hat{K}\right) \hat{K}. \tag{3.294}$$

## 3.8  Summary

Two Cartesian coordinate frames $B$ and $G$ have a common origin. We may rotate $B$ for $\phi$ rad about a specific axis $\hat{u}$ to transform $B$ to $G$. Having the angle and axis of rotation, the transformation matrix can be calculated by the Rodriguez rotation formula.

$$^{G}R_{B} = R_{\hat{u},\phi} = \mathbf{I}\cos\phi + \hat{u}\hat{u}^{T} \text{ vers } \phi + \tilde{u}\sin\phi \tag{3.295}$$

On the other hand, we can find the angle and axis of rotation by having the transformation matrix $^{G}R_{B}$,

$$\cos\phi = \frac{1}{2}\left(\text{tr}\left(^{G}R_{B}\right) - 1\right) \tag{3.296}$$

$$\tilde{u} = \frac{1}{2\sin\phi}\left(^{G}R_{B} - {^{G}R_{B}^{T}}\right). \tag{3.297}$$

The angle and axis of rotation can also be defined by Euler parameters $e_0, e_1, e_2, e_3$

$$e_0 = \cos\frac{\phi}{2} \tag{3.298}$$

$$\mathbf{e} = e_1\hat{I} + e_2\hat{J} + e_3\hat{K} = \hat{u}\sin\frac{\phi}{2} \tag{3.299}$$

$$^{G}R_{B} = R_{\hat{u},\phi} = \left(e_0^2 - \mathbf{e}^2\right)\mathbf{I} + 2\mathbf{e}\,\mathbf{e}^{T} + 2e_0\tilde{e} \tag{3.300}$$

or unit quaternions

$$\begin{aligned} e\left(\phi, \hat{u}\right) &= e_0 + \mathbf{e} \\ &= e_0 + e_1\hat{i} + e_2\hat{j} + e_3\hat{k} \\ &= \cos\frac{\phi}{2} + \sin\frac{\phi}{2}\hat{u} \end{aligned} \tag{3.301}$$

that provide some advantages for orientation analysis.

# Exercises

1. Notation and symbols.

   Describe the meaning of

   a- $\hat{u}$      b- $\text{vers}\,\phi$   c- $\tilde{u}$      d- $R_{\hat{u},\phi}$   e- $e_0$   f- $\mathbf{e}$

   g- $^{B}R_{G}^{-1}$   h- $d\phi$      i- $q_0 + \mathbf{q}$   j- $ij$      k- $q^{*}$   l- $e\left(\phi, \hat{u}\right)$

   m- $d\phi$      n- $\hat{J}$      o- $\overleftrightarrow{q}$      p- $\tilde{q}$      q- $|q|$   r- $\mathbb{R}^{3 \times 3}$.

2. Invariant axis of rotation.

   Find out if the axis of rotation $\hat{u}$ is fixed in $B(Oxyz)$ or $G(OXYZ)$.

3. $z$-axis-angle rotation matrix.

   Expand

   $$^{G}R_{B} = {}^{B}R_{G}^{-1} = {}^{B}R_{G}^{T} = R_{\hat{u},\phi}$$
   $$= \left[A_{z,-\varphi}\,A_{y,-\theta}\,A_{z,\phi}\,A_{y,\theta}\,A_{z,\varphi}\right]^{T}$$
   $$= A_{z,\varphi}^{T}\,A_{y,\theta}^{T}\,A_{z,\phi}^{T}\,A_{y,-\theta}^{T}\,A_{z,-\varphi}^{T}$$

   and verify the axis-angle rotation matrix

   $$^{G}R_{B} = R_{\hat{u},\phi}$$
   $$= \begin{bmatrix} u_1^2 \text{ vers }\phi + c\phi & u_1 u_2 \text{ vers }\phi - u_3 s\phi & u_1 u_3 \text{ vers }\phi + u_2 s\phi \\ u_1 u_2 \text{ vers }\phi + u_3 s\phi & u_2^2 \text{ vers }\phi + c\phi & u_2 u_3 \text{ vers }\phi - u_1 s\phi \\ u_1 u_3 \text{ vers }\phi - u_2 s\phi & u_2 u_3 \text{ vers }\phi + u_1 s\phi & u_3^2 \text{ vers }\phi + c\phi \end{bmatrix}.$$

4. $x$-axis-angle rotation matrix.

   Find the axis-angle rotation matrix by transforming the $x$-axis on the axis of rotation $\hat{u}$.

5. $y$-axis-angle rotation matrix.

   Find the axis-angle rotation matrix by transforming the $y$-axis on the axis of rotation $\hat{u}$.

6. Axis-angle rotation and Euler angles.

   Find the Euler angles corresponding to the rotation $45\,\text{deg}$ about $\mathbf{u} = \begin{bmatrix} 1 & 1 & 1 \end{bmatrix}^{T}$.

7. Euler angles between two local frames.

   The Euler angles between the coordinate frame $B_1$ and $G$ are $20\,\text{deg}$, $35\,\text{deg}$, and $-40\,\text{deg}$. The Euler angles between the coordinate frame $B_2$ and $G$ are $60\,\text{deg}$, $-30\,\text{deg}$, and $-10\,\text{deg}$. Find the angle and axis of rotation that transforms $B_2$ to $B_1$.

8. ★ Angle and axis of rotation based on Euler angles.

   Compare the Euler angles rotation matrix with the angle-axis rotation matrix and find the angle and axis of rotation based on Euler angles.

9. ★ Euler angles based on angle and axis of rotation.

   Compare the Euler angles rotation matrix with the angle-axis rotation matrix and find the Euler angles based on the angle and axis of rotation.

10. ★ Repeating global-local rotations.

    Rotate ${}^B\mathbf{r}_P = [6, 2, -3]^T$, 60 deg about the $Z$-axis, followed by 30 deg about the $x$-axis. Then repeat the sequence of rotations for 60 deg about the $Z$-axis, followed by 30 deg about the $x$-axis. After how many rotations will point $P$ be back to its initial global position?

11. ★ Repeating global-local rotations.

    How many rotations of $\alpha = \pi/m$ deg about the $X$-axis, followed by $\beta = \pi/k$ deg about the $z$-axis are needed to bring a body point to its initial global position, if $m, k \in \mathbb{N}$?

12. ★ Small rotation angles.

    Show that for very small angles of rotation $\varphi$, $\theta$, and $\psi$ about the axes of the local coordinate frame, the first and third rotations are indistinguishable when they are about the same axis.

13. ★ Inner automorphism properly of $\tilde{\mathbf{a}}$.

    If $R$ is a rotation matrix and $\mathbf{a}$ is a vector, show that

    $$R\tilde{\mathbf{a}}R^T = \widetilde{R\mathbf{a}}.$$

14. ★ Angle-derivative of principal rotation matrices.

    Show that

    $$\frac{dR_{Z,\alpha}}{d\alpha} = \tilde{K}R_{Z,\alpha}$$
    $$\frac{dR_{Y,\beta}}{d\beta} = \tilde{J}R_{Y,\beta}$$
    $$\frac{dR_{X,\gamma}}{d\gamma} = \tilde{I}R_{X,\gamma}.$$

15. ★ Euler angles, Euler parameters.

Compare the Euler angles rotation matrix and Euler parameter transformation matrix and verify the following relationships between Euler angles and Euler parameters

$$e_0 = \cos\frac{\theta}{2}\cos\frac{\psi+\varphi}{2}$$

$$e_1 = \sin\frac{\theta}{2}\cos\frac{\psi-\varphi}{2}$$

$$e_2 = \sin\frac{\theta}{2}\sin\frac{\psi-\varphi}{2}$$

$$e_3 = \cos\frac{\theta}{2}\sin\frac{\psi+\varphi}{2}$$

and

$$\varphi = \cos^{-1}\frac{2\left(e_2e_3 + e_0e_1\right)}{\sin\theta}$$

$$\theta = \cos^{-1}\left[2\left(e_0^2 + e_3^2\right) - 1\right]$$

$$\psi = \cos^{-1}\frac{-2\left(e_2e_3 - e_0e_1\right)}{\sin\theta}.$$

16. ★ Quaternion definition.

Find the unit quaternion $e\left(\phi, \hat{u}\right)$ associated to

$$\hat{u} = \begin{bmatrix} 1/\sqrt{3} \\ 1/\sqrt{3} \\ 1/\sqrt{3} \end{bmatrix}$$

$$\phi = \frac{\pi}{3}$$

and find the result of $e\left(\phi, \hat{u}\right)\hat{\imath}e^*\left(\phi, \hat{u}\right)$.

17. ★ Quaternion product.

Find $pq$, $qp$, $p^*q$, $qp^*$, $p^*p$, $qq^*$, $p^*q^*$, and $p^*rq^*$ if

$$p = 3 + i - 2j + 2k$$
$$q = 2 - i + 2j + 4k$$
$$r = -1 + i + j - 3k.$$

18. ★ Quaternion inverse.

Find $q^{-1}$, $p^{-1}$, $p^{-1}q^{-1}$, $q^{-1}p^*$, $p^*p^{-1}$, $q^{-1}q^*$, and $p^{*-1}q^{*-1}$ if

$$p = 3 + i - 2j + 2k$$
$$q = 2 - i + 2j + 4k.$$

19. ★ Quaternion and angle-axis rotation.

    Find the unit quaternion associated to

    $$
    \begin{aligned}
    p &= 3 + i - 2j + 2k \\
    q &= 2 - i + 2j + 4k
    \end{aligned}
    $$

    and find the angle and axis of rotation for each of the unit quaternion.

20. ★ Unit quaternion and rotation.

    Use the unit quaternion $p$

    $$
    p = \frac{1 + i - j + k}{2}
    $$

    and find the global position of

    $$
    {}^{B}\mathbf{r} = \begin{bmatrix} 2 \\ -2 \\ 6 \end{bmatrix}.
    $$

21. ★ Quaternion matrix.

    Use the unit quaternion matrices associated to

    $$
    \begin{aligned}
    p &= 3 + i - 2j + 2k \\
    q &= 2 - i + 2j + 4k \\
    r &= -1 + i + j - 3k.
    \end{aligned}
    $$

    and find $\overleftrightarrow{p}\,\overleftrightarrow{r}\,\overleftrightarrow{q}$, $\overleftrightarrow{q}\,\overleftrightarrow{p}$, $\overleftrightarrow{p^*}\,\overleftrightarrow{q}$, $\overleftrightarrow{q}\,\overleftrightarrow{p^*}$, $\overleftrightarrow{p^*}\,\overleftrightarrow{p}$, $\overleftrightarrow{q}\,\overleftrightarrow{q^*}$, $\overleftrightarrow{p^*}\,\overleftrightarrow{q^*}$, and $\overleftrightarrow{p^*}\,\overleftrightarrow{r}\,\overleftrightarrow{q^*}$.

22. ★ Euler angles and quaternion.

    Find quaternion components in terms of Euler angles, and Euler angles in terms of quaternion components.

23. Angular velocity vector.

    Use the definition ${}^{G}R_B = [r_{ij}]$ and ${}^{G}\dot{R}_B = [\dot{r}_{ij}]$, and find the angular velocity vector $\boldsymbol{\omega}$, where $\tilde{\omega} = {}^{G}\dot{R}_B\,{}^{G}R_B^T$.

24. ★ *bac-cab* rule.

    Use the Levi-Civita density $\epsilon_{ijk}$ to prove the *bac-cab* rule

    $$
    \mathbf{a} \times (\mathbf{b} \times \mathbf{c}) = \mathbf{b}\,(\mathbf{a} \cdot \mathbf{c}) - \mathbf{c}\,(\mathbf{a} \cdot \mathbf{b}).
    $$

25. ★ *bac-cab* rule application.

    Use the bac-cab rule to show that

    $$\mathbf{a} = \hat{n}\,(\mathbf{a} \cdot \hat{\mathbf{n}}) + \hat{n} \times (\mathbf{a} \times \hat{n})$$

    where $\hat{n}$ is any unit vector. What is the geometric significance of this equation?

26. ★ Two rotations are not enough.

    Show that, in general, it is impossible to move a point $P(X, Y, Z)$ from the initial position $P(X_i, Y_i, Z_i)$ to the final position $P(X_f, Y_f, Z_f)$ only by *two* rotations about the global axes.

27. ★ Three rotations are enough.

    Show that, in general, it is possible to move a point $P(X, Y, Z)$ from the initial position $P(X_i, Y_i, Z_i)$ to the final position $P(X_f, Y_f, Z_f)$ by *three* rotations about different global axes.

28. ★ Closure property.

    Show the closure property of transformation matrices.

29. Sum of two orthogonal matrices.

    Show that the sum of two orthogonal matrices is not, in general, an orthogonal matrix, but their product is.

30. Equivalent cross product.

    Show that if $\mathbf{a} = \begin{bmatrix} a_1 & a_2 & a_3 \end{bmatrix}^T$ and $\mathbf{b} = \begin{bmatrix} b_1 & b_2 & b_3 \end{bmatrix}^T$ are two arbitrary vectors, and

    $$\tilde{\mathbf{a}} = \begin{bmatrix} 0 & -a_3 & a_2 \\ a_3 & 0 & -a_1 \\ -a_2 & a_1 & 0 \end{bmatrix}$$

    is the skew symmetric matrix corresponding to $\mathbf{a}$, then

    $$\tilde{\mathbf{a}}\mathbf{b} = \mathbf{a} \times \mathbf{b}.$$

31. ★ Skew symmetric matrices.

    Use $\mathbf{a} = \begin{bmatrix} a_1 & a_2 & a_3 \end{bmatrix}^T$ and $\mathbf{b} = \begin{bmatrix} b_1 & b_2 & b_3 \end{bmatrix}^T$ to show that:

    a-

    $$\tilde{\mathbf{a}}\mathbf{b} = -\tilde{\mathbf{b}}\mathbf{a}$$

    b-

    $$\widetilde{(\mathbf{a} + \mathbf{b})} = \tilde{\mathbf{a}} + \tilde{\mathbf{b}}$$

    c-

    $$\widetilde{(\tilde{\mathbf{a}}\mathbf{b})} = \mathbf{b}\mathbf{a}^T - \mathbf{a}\mathbf{b}^T$$

32. ★ Rotation matrices identity.

    Show that if $A$, $B$, and $C$ are three rotation matrices then,

    a-

    $$(AB)\,C = A\,(BC) = ABC$$

    b-

    $$(A + B)^T = A^T + B^T$$

    c-

    $$(AB)^T = B^T A^T$$

    d-

    $$\left(A^{-1}\right)^T = \left(A^T\right)^{-1}$$

33. ★ Skew symmetric matrix multiplication.
    Verify that

    a-

    $$\mathbf{a}^T \tilde{\mathbf{a}}^T = -\mathbf{a}^T \tilde{\mathbf{a}} = 0$$

    b-

    $$\tilde{\mathbf{a}}\tilde{\mathbf{b}} = \mathbf{b}\mathbf{a}^T - \mathbf{a}^T\mathbf{b}\,\mathbf{I}$$

34. ★ Skew symmetric matrix derivative.
    Show that

    $$\dot{\tilde{\mathbf{a}}} = \tilde{\dot{\mathbf{a}}}.$$

35. ★ Time derivative of $A = [\mathbf{a}, \tilde{\mathbf{a}}]$.

    Assume that $\mathbf{a}$ is a time dependent vector and $A = [\mathbf{a}, \tilde{\mathbf{a}}]$ is a $3 \times 4$ matrix. What is the time derivative of $C = AA^T$ ?

36. ★ Combined angle-axis rotations.

    The rotation $\phi_1$ about $\hat{u}_1$ followed by rotation $\phi_2$ about $\hat{u}_2$ is equivalent to a rotation $\phi$ about $\hat{u}$. Find the angle $\phi$ and axis $\hat{u}$ in terms of $\phi_1$, $\hat{u}_1$, $\phi_2$, and $\hat{u}_2$.

37. ★ Rodriguez vector.

    Using the Rodriguez rotation formula show that

    $$\mathbf{r}' - \mathbf{r} = \tan\frac{\phi}{2}\hat{u} \times (\mathbf{r}' + \mathbf{r}).$$

38. ★ Equivalent Rodriguez rotation matrices.

    Show that the Rodriguez rotation matrix

    $$^{G}R_B = \mathbf{I}\cos\phi + \hat{u}\hat{u}^{T}\text{ vers }\phi + \tilde{u}\sin\phi$$

    can also be written as

    $$^{G}R_B = \mathbf{I} + (\sin\phi)\,\tilde{u} + (\text{vers }\phi)\,\tilde{u}^{2}.$$

39. ★ Rotation matrix and Rodriguez formula.

    Knowing the alternative definition of the Rodriguez formula

    $$^{G}R_B = \mathbf{I} + (\sin\phi)\,\tilde{u} + (\text{vers }\phi)\,\tilde{u}^{2}$$

    and

    $$\tilde{u}^{2n-1} = (-1)^{n-1}\,\tilde{u}$$
    $$\tilde{u}^{2n} = (-1)^{n-1}\,\tilde{u}^{2}$$

    examine the following equation:

    $$^{G}R_B^{T}\,^{G}R_B = \,^{G}R_B\,^{G}R_B^{T}$$

40. ★ Rodriguez formula application.

    Use the alternative definition of the Rodriguez formula

    $$^{G}R_B = \mathbf{I} + (\sin\phi)\,\tilde{u} + (\text{vers }\phi)\,\tilde{u}^{2}$$

    and find the global position of a body point at

    $$^{B}\mathbf{r} = \begin{bmatrix} 1 & 3 & 4 \end{bmatrix}^{T}$$

    after a rotation of 45 deg about the axis indicated by

    $$\hat{u} = \begin{bmatrix} \frac{1}{\sqrt{3}} & \frac{1}{\sqrt{3}} & \frac{1}{\sqrt{3}} \end{bmatrix}^{T}.$$

41. Axis and angle of rotation.

    Find the axis and angle of rotation for the following transformation matrix:

    $$R = \begin{bmatrix} \frac{3}{4} & \frac{\sqrt{6}}{4} & \frac{1}{4} \\ \frac{-\sqrt{6}}{4} & \frac{2}{4} & \frac{\sqrt{6}}{4} \\ \frac{1}{4} & \frac{-\sqrt{6}}{4} & \frac{3}{4} \end{bmatrix}$$

42. ★ Axis of rotation multiplication.

Show that

$$\tilde{u}^{2k+1} = (-1)^k \, \tilde{u}$$

and

$$\tilde{u}^{2k} = (-1)^k \left( \mathbf{I} - \hat{u}\hat{u}^T \right).$$

43. ★ Stanley method.

Find the Euler parameters of the following rotation matrix based on the Stanley method.

$$
{}^G R_B = \begin{bmatrix}
0.5449 & -0.5549 & 0.6285 \\
0.3111 & 0.8299 & 0.4629 \\
-0.7785 & -0.0567 & 0.6249
\end{bmatrix}
$$

# 4

# Motion Kinematics

## 4.1 Rigid Body Motion

Consider a rigid body with an attached local coordinate frame $B(oxyz)$ moving freely in a fixed global coordinate frame $G(OXYZ)$. The rigid body can rotate in the global frame, while point $o$ of the body frame $B$ can translate relative to the origin $O$ of $G$ as shown in Figure 4.1.

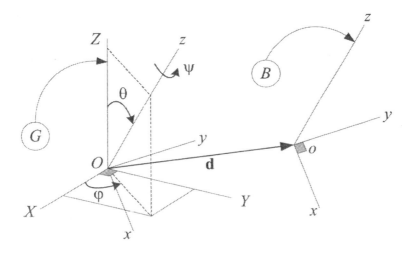

FIGURE 4.1. Rotation and translation of a local frame with respect to a global frame.

If the vector $^G\mathbf{d}$ indicates the position of the moving origin $o$, relative to the fixed origin $O$, then the coordinates of a body point $P$ in local and global frames are related by the following equation:

$$^G\mathbf{r}_P = {}^GR_B\,{}^B\mathbf{r}_P + {}^G\mathbf{d} \tag{4.1}$$

where

$$^G\mathbf{r}_P = \begin{bmatrix} X_P & Y_P & Z_P \end{bmatrix}^T \tag{4.2}$$

$$^B\mathbf{r}_P = \begin{bmatrix} x_P & y_P & z_P \end{bmatrix}^T \tag{4.3}$$

$$^G\mathbf{d} = \begin{bmatrix} X_o & Y_o & Z_o \end{bmatrix}^T. \tag{4.4}$$

The vector $^G\mathbf{d}$ is called the *displacement* or *translation* of $B$ with respect to $G$, and $^GR_B$ is the *rotation matrix* to map $^B\mathbf{r}$ to $^G\mathbf{r}$ when $^G\mathbf{d} = 0$. Such a combination of a rotation and a translation in Equation (4.1) is called *rigid motion*. In other words, the location of a rigid body can be described by the position of the origin $o$ and the orientation of the body frame, with respect to the global frame. Decomposition of a rigid motion into a rotation and a translation is the simplest method for representing spatial displacement. We show the translation by a vector, and the rotation by any of the methods described in the previous Chapter.

**Proof.** Figure 4.1 illustrates a translated and rotated body frame in the global frame. The most general rotation is represented by the *Rodriguez rotation formula* (3.62), which depends on $^B\mathbf{r}_P$, the position vector of a point $P$ measured in the body coordinate frame. In the translation $^G\mathbf{d}$, all points of the body have the same displacement, and therefore, translation of a rigid body is independent of the local position vector $^B\mathbf{r}$. Hence, the most general displacement of a rigid body is represented by the following equation, and has two independent parts: a rotation, and a translation.

$$
\begin{aligned}
^G\mathbf{r} &= \; ^B\mathbf{r}\cos\phi + (1 - \cos\phi)\left(\hat{u} \cdot {}^B\mathbf{r}\right)\hat{u} + \left(\hat{u} \times {}^B\mathbf{r}\right)\sin\phi + {}^G\mathbf{d} \\
&= \; ^GR_B\,{}^B\mathbf{r} + {}^G\mathbf{d}
\end{aligned}
\tag{4.5}
$$

Equation (4.5) shows that the most general displacement of a rigid body is a rotation about an axis and a translation along an axis. The choice of the point of reference $o$ is entirely arbitrary, but when this point is chosen and the body coordinate frame is set up, the rotation and translation are uniquely determined.

Based on translation and rotation, the position of a body can be uniquely determined by six independent parameters: three translation components $X_o$, $Y_o$, $Z_o$; and three rotational components. If a body moves in such a way that its rotational components remain constant, the motion is a *pure translation*; and if it moves in such a way that $X_o$, $Y_o$, and $Z_o$ remain constant, the motion is a *pure rotation*. Therefore, a rigid body has three translational and three rotational degrees of freedom. ∎

**Example 67** *Translation and rotation of a body coordinate frame.*

*A body coordinate frame $B(oxyz)$, that is originally coincident with global coordinate frame $G(OXYZ)$, rotates $45\deg$ about the $X$-axis and translates to $\begin{bmatrix} 3 & 5 & 7 \end{bmatrix}^T$. Then, the global position of a point at $^B\mathbf{r} = \begin{bmatrix} x & y & z \end{bmatrix}^T$ is*

$$
\begin{aligned}
^G\mathbf{r} &= \; ^GR_B\,{}^B\mathbf{r} + {}^G\mathbf{d} \\
&= \begin{bmatrix} 1 & 0 & 0 \\ 0 & \cos 45 & -\sin 45 \\ 0 & \sin 45 & \cos 45 \end{bmatrix} \begin{bmatrix} x \\ y \\ z \end{bmatrix} + \begin{bmatrix} 3 \\ 5 \\ 7 \end{bmatrix} \\
&= (x + 3)\,\hat{I} + (0.707y - 0.707z + 5)\,\hat{J} + (0.707y + 0.707z + 7)\,\hat{K}.
\end{aligned}
$$

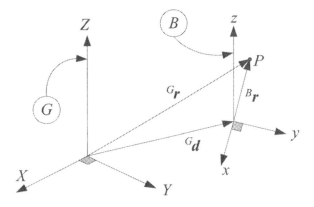

FIGURE 4.2. A translating and rotating body in a global coordinate frame.

**Example 68** *Moving body coordinate frame.*

*Figure 4.2 shows a point $P$ at $^B\mathbf{r}_P = 0.1\hat{\imath} + 0.3\hat{\jmath} + 0.3\hat{k}$ in a body frame $B$, which is rotated 50 deg about the $Z$-axis, and translated $-1$ along $X$, $0.5$ along $Y$, and $0.2$ along the $Z$ axes.*

*The position of $P$ in global coordinate frame is*

$$
\begin{aligned}
^G\mathbf{r} &= \ ^GR_B \ ^B\mathbf{r}_P + \ ^G\mathbf{d} \\
&= \begin{bmatrix} \cos 50 & -\sin 50 & 0 \\ \sin 50 & \cos 50 & 0 \\ 0 & 0 & 1 \end{bmatrix} \begin{bmatrix} 0.1 \\ 0.3 \\ 0.3 \end{bmatrix} + \begin{bmatrix} -1 \\ 0.5 \\ 0.2 \end{bmatrix} \\
&= \begin{bmatrix} -1.166 \\ 0.769 \\ 0.5 \end{bmatrix}.
\end{aligned}
$$

**Example 69** *Rotation of a translated rigid body.*

*Point $P$ of a rigid body $B$ has an initial position vector $^B\mathbf{r}_P$.*

$$
^B\mathbf{r}_P = \begin{bmatrix} 1 & 2 & 3 \end{bmatrix}^T
$$

*If the body rotates 45 deg about the $x$-axis, and then translates to $^G\mathbf{d} = \begin{bmatrix} 4 & 5 & 6 \end{bmatrix}^T$, the final position of $P$ would be*

$$
\begin{aligned}
^G\mathbf{r} &= \ ^BR_{x,45}^T \ ^B\mathbf{r}_P + \ ^G\mathbf{d} \\
&= \begin{bmatrix} 1 & 0 & 0 \\ 0 & \cos 45 & -\sin 45 \\ 0 & \sin 45 & \cos 45 \end{bmatrix} \begin{bmatrix} 1 \\ 2 \\ 3 \end{bmatrix} + \begin{bmatrix} 4 \\ 5 \\ 6 \end{bmatrix} \\
&= \begin{bmatrix} 5.0 \\ 4.23 \\ 9.53 \end{bmatrix}.
\end{aligned}
$$

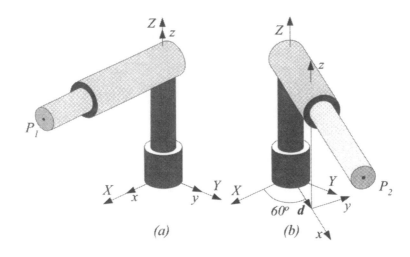

FIGURE 4.3. A polar RP arm.

Note that rotation occurs with the assumption that $^G\mathbf{d} = \mathbf{0}$.

**Example 70** *Arm rotation plus elongation.*

Position vector of point $P_1$ at the tip of an arm shown in Figure 4.3(a) is at $^G\mathbf{r}_{P_1} = {}^B\mathbf{r}_{P_1} = \begin{bmatrix} 1350 & 0 & 900 \end{bmatrix}^T$ mm. The arm rotates 60 deg about the global $Z$-axis, and elongates by $\mathbf{d} = 720.2\,\hat{\imath}$ mm. The final configuration of the arm is shown in Figure 4.3(b).

The new position vector of P is

$$^G\mathbf{r}_{P_2} = {}^GR_B\,{}^B\mathbf{r}_{P_1} + {}^G\mathbf{d}$$

where $^GR_B = R_{Z,60}$ is the rotation matrix to transform $\mathbf{r}_{P_2}$ to $\mathbf{r}_{P_1}$ when $^G\mathbf{d} = 0$,

$$^GR_B = \begin{bmatrix} \cos 60 & -\sin 60 & 0 \\ \sin 60 & \cos 60 & 0 \\ 0 & 0 & 1 \end{bmatrix}$$

and $^G\mathbf{d}$ is the translation vector of o with respect to O in the global frame. The translation vector in the body coordinate frame is $^B\mathbf{d} = \begin{bmatrix} 720.2 & 0 & 0 \end{bmatrix}^T$ so $^G\mathbf{d}$ would be found by a transformation

$$
\begin{aligned}
^G\mathbf{d} &= {}^GR_B\,{}^B\mathbf{d} \\
&= \begin{bmatrix} \cos 60 & -\sin 60 & 0 \\ \sin 60 & \cos 60 & 0 \\ 0 & 0 & 1 \end{bmatrix} \begin{bmatrix} 720.2 \\ 0.0 \\ 0.0 \end{bmatrix} \\
&= \begin{bmatrix} 360.10 \\ 623.71 \\ 0.0 \end{bmatrix}.
\end{aligned}
$$

*Therefore, the final global position of the tip of the arm is at*

$$
\begin{aligned}
{}^{G}\mathbf{r}_{P_2} &= {}^{G}R_B\,{}^{B}\mathbf{r}_{P_1} + {}^{G}\mathbf{d} \\
&= \begin{bmatrix} c60 & -s60 & 0 \\ s60 & c60 & 0 \\ 0 & 0 & 1 \end{bmatrix} \begin{bmatrix} 1350 \\ 0 \\ 900 \end{bmatrix} + \begin{bmatrix} 360.1 \\ 623.7 \\ 0.0 \end{bmatrix} \\
&= \begin{bmatrix} 1035.1 \\ 1792.8 \\ 900.0 \end{bmatrix}.
\end{aligned}
$$

**Example 71** *Composition of transformations.*
   *Assume*

$$
{}^{2}\mathbf{r} = {}^{2}R_1\,{}^{1}\mathbf{r} + {}^{2}\mathbf{d}_1 \tag{4.6}
$$

*indicates the rigid motion of body* $B_1$ *with respect to body* $B_2$, *and*

$$
{}^{G}\mathbf{r} = {}^{G}R_2\,{}^{2}\mathbf{r} + {}^{G}\mathbf{d}_2 \tag{4.7}
$$

*indicates the rigid motion of body* $B_2$ *with respect to frame* $G$. *The composition defines a third rigid motion, which can be described by substituting
the expression for* ${}^{2}\mathbf{r}$ *into the equation for* ${}^{G}\mathbf{r}$.

$$
\begin{aligned}
{}^{G}\mathbf{r} &= {}^{G}R_2\left({}^{2}R_1\,{}^{1}\mathbf{r} + {}^{2}\mathbf{d}_1\right) + {}^{G}\mathbf{d}_2 \\
&= {}^{G}R_2\,{}^{2}R_1\,{}^{1}\mathbf{r} + {}^{G}R_2\,{}^{2}\mathbf{d}_1 + {}^{G}\mathbf{d}_2 \\
&= {}^{G}R_1\,{}^{1}\mathbf{r} + {}^{G}\mathbf{d}_1 \tag{4.8}
\end{aligned}
$$

*Therefore,*

$$
\begin{aligned}
{}^{G}R_1 &= {}^{G}R_2\,{}^{2}R_1 \tag{4.9} \\
{}^{G}\mathbf{d}_1 &= {}^{G}R_2\,{}^{2}\mathbf{d}_1 + {}^{G}\mathbf{d}_2 \tag{4.10}
\end{aligned}
$$

*which shows that the transformation from frame* $B_1$ *to frame* $G$ *can be done
by rotation* ${}^{G}R_1$ *and translation* ${}^{G}\mathbf{d}_1$.

## 4.2   Homogeneous Transformation

As shown in Figure 4.4, an arbitrary point $P$ of a rigid body attached to the
local frame $B$ is denoted by ${}^{B}\mathbf{r}_P$ and ${}^{G}\mathbf{r}_P$ in different frames. The vector
${}^{G}\mathbf{d}$ indicates the position of origin $o$ of the body frame in the global frame.
Therefore, a general motion of a rigid body $B\,(oxyz)$ in the global frame
$G\,(OXYZ)$ is a combination of rotation ${}^{G}R_B$ and translation ${}^{G}\mathbf{d}$.

$$
{}^{G}\mathbf{r} = {}^{G}R_B\,{}^{B}\mathbf{r} + {}^{G}\mathbf{d} \tag{4.11}
$$

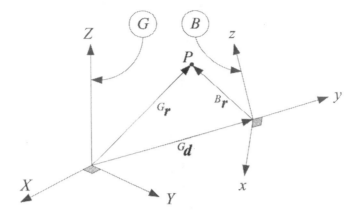

FIGURE 4.4. Representation of a point $P$ in coordinate frames $B$ and $G$.

Using a rotation matrix plus a vector leads us to the use of homogeneous coordinates. Introducing a new $4 \times 4$ *homogeneous transformation matrix* $^{G}T_{B}$, helps us show a rigid motion by a single matrix transformation

$$^{G}\mathbf{r} = {}^{G}T_{B}\,{}^{B}\mathbf{r} \qquad (4.12)$$

where

$$
{}^{G}T_{B} = \begin{bmatrix} r_{11} & r_{12} & r_{13} & X_{o} \\ r_{21} & r_{22} & r_{23} & Y_{o} \\ r_{31} & r_{32} & r_{33} & Z_{o} \\ 0 & 0 & 0 & 1 \end{bmatrix}
$$

$$
\equiv \left[ \begin{array}{ccc|c} & {}^{G}R_{B} & & {}^{G}\mathbf{d} \\ \hline 0 & 0 & 0 & 1 \end{array} \right] \equiv \begin{bmatrix} {}^{G}R_{B} & {}^{G}\mathbf{d} \\ 0 & 1 \end{bmatrix} \qquad (4.13)
$$

and

$$
{}^{G}\mathbf{r} = \begin{bmatrix} X_{P} \\ Y_{P} \\ Z_{P} \\ 1 \end{bmatrix} \qquad (4.14)
$$

$$
{}^{B}\mathbf{r} = \begin{bmatrix} x_{P} \\ y_{P} \\ z_{P} \\ 1 \end{bmatrix} \qquad (4.15)
$$

$$
{}^{G}\mathbf{d} = \begin{bmatrix} X_{o} \\ Y_{o} \\ Z_{o} \\ 1 \end{bmatrix}. \qquad (4.16)
$$

The homogeneous transformation matrix $^GT_B$ is a $4 \times 4$ matrix that maps a homogeneous position vector from one frame to another. This extension of matrix representation of rigid motions is just for simplifying numerical calculations.

Representation of an $n$-component position vector by an $(n+1)$-component vector is called *homogeneous coordinate representation*. The appended element is a *scale factor*, $w$; hence, in general, homogeneous representation of a vector $\mathbf{r} = \begin{bmatrix} x & y & z \end{bmatrix}^T$ is

$$\mathbf{r} = \begin{bmatrix} wx \\ wy \\ wz \\ w \end{bmatrix}. \qquad (4.17)$$

Using homogeneous coordinates shows that the absolute values of the four coordinates are not important. Instead, it is the three ratios, $x/w$, $y/w$, and $z/w$ that are important because,

$$\begin{bmatrix} x \\ y \\ z \\ w \end{bmatrix} = \begin{bmatrix} x/w \\ y/w \\ z/w \\ 1 \end{bmatrix} \qquad (4.18)$$

provided $w \neq 0$, and $w \neq \infty$. Therefore, the homogeneous vector $w\mathbf{r}$ refers to the same point as $\mathbf{r}$ does.

If $w = 1$, then the transformed homogeneous coordinates of a position vector are the same as physical coordinates of the vector and the space is the standard Euclidean space.

Hereafter, if no confusion exists and $w = 1$, we will use the regular vectors, and their homogeneous representation equivalently.

**Proof.** We append a 1 to the coordinates of a point and define *homogeneous position vectors* as follows:

$$^G\mathbf{r}_P = \begin{bmatrix} X_P \\ Y_P \\ Z_P \\ 1 \end{bmatrix} \quad , \quad ^B\mathbf{r}_P = \begin{bmatrix} x_P \\ y_P \\ z_P \\ 1 \end{bmatrix} \quad , \quad ^G\mathbf{d} = \begin{bmatrix} X_o \\ Y_o \\ Z_o \\ 1 \end{bmatrix}.$$

Using the definitions of homogeneous transformation we will find

$$^{G}\mathbf{r}_P = {}^{G}T_B \, {}^{B}\mathbf{r}_P \qquad (4.19)$$

$$\begin{bmatrix} X_P \\ Y_P \\ Z_P \\ 1 \end{bmatrix} = \begin{bmatrix} r_{11} & r_{12} & r_{13} & X_o \\ r_{21} & r_{22} & r_{23} & Y_o \\ r_{31} & r_{32} & r_{33} & Z_o \\ 0 & 0 & 0 & 1 \end{bmatrix} \begin{bmatrix} x_P \\ y_P \\ z_P \\ 1 \end{bmatrix}$$

$$= \begin{bmatrix} X_o + r_{11}x_P + r_{12}y_P + r_{13}z_P \\ Y_o + r_{21}x_P + r_{22}y_P + r_{23}z_P \\ Z_o + r_{31}x_P + r_{32}y_P + r_{33}z_P \\ 1 \end{bmatrix}. \qquad (4.20)$$

However, the standard method reduces to

$$^{G}\mathbf{r}_P = {}^{G}R_B \, {}^{B}\mathbf{r}_p + {}^{G}\mathbf{d} \qquad (4.21)$$

$$\begin{bmatrix} X_P \\ Y_P \\ Z_P \end{bmatrix} = \begin{bmatrix} r_{11} & r_{12} & r_{13} \\ r_{21} & r_{22} & r_{23} \\ r_{31} & r_{32} & r_{33} \end{bmatrix} \begin{bmatrix} x_P \\ y_P \\ z_P \end{bmatrix} + \begin{bmatrix} X_o \\ Y_o \\ Z_o \end{bmatrix}$$

$$= \begin{bmatrix} X_o + r_{11}x_P + r_{12}y_P + r_{13}z_P \\ Y_o + r_{21}x_P + r_{22}y_P + r_{23}z_P \\ Z_o + r_{31}x_P + r_{32}y_P + r_{33}z_P \end{bmatrix}, \qquad (4.22)$$

which is compatible with the definition of homogeneous vector and homogeneous transformation. ■

**Example 72** *Rotation and translation of a body coordinate frame.*
   *A body coordinate frame $B(oxyz)$, that is originally coincident with global coordinate frame $G(OXYZ)$, rotates 45 deg about the $X$-axis and translates to $\begin{bmatrix} 3 & 5 & 7 & 1 \end{bmatrix}^T$. Then, the matrix representation of the global position of a point at $^{B}\mathbf{r} = \begin{bmatrix} x & y & z & 1 \end{bmatrix}^T$ is*

$$^{G}\mathbf{r} = {}^{G}T_B \, {}^{B}\mathbf{r}$$

$$= \begin{bmatrix} X \\ Y \\ Z \\ 1 \end{bmatrix} = \begin{bmatrix} 1 & 0 & 0 & 3 \\ 0 & \cos 45 & -\sin 45 & 5 \\ 0 & \sin 45 & \cos 45 & 7 \\ 0 & 0 & 0 & 1 \end{bmatrix} \begin{bmatrix} x \\ y \\ z \\ 1 \end{bmatrix}.$$

**Example 73** *Decomposition of $^{G}T_B$ into translation and rotation.*
   *Homogeneous transformation matrix $^{G}T_B$ can be decomposed to a matrix multiplication of a pure rotation matrix $^{G}R_B$, and a pure translation matrix*

$^{G}D_B,$

$$^{G}T_B = {}^{G}D_B{}^{G}R_B \tag{4.23}$$

$$= \begin{bmatrix} 1 & 0 & 0 & X_o \\ 0 & 1 & 0 & Y_o \\ 0 & 0 & 1 & Z_o \\ 0 & 0 & 0 & 1 \end{bmatrix} \begin{bmatrix} r_{11} & r_{12} & r_{13} & 0 \\ r_{21} & r_{22} & r_{23} & 0 \\ r_{31} & r_{32} & r_{33} & 0 \\ 0 & 0 & 0 & 1 \end{bmatrix}$$

$$= \begin{bmatrix} r_{11} & r_{12} & r_{13} & X_o \\ r_{21} & r_{22} & r_{23} & Y_o \\ r_{31} & r_{32} & r_{33} & Z_o \\ 0 & 0 & 0 & 1 \end{bmatrix}. \tag{4.24}$$

In other words, a transformation can be achieved by a pure rotation first, followed by a pure translation.

Note that decomposition of a transformation to translation and rotation is not interchangeable

$$^{G}T_B = {}^{G}D_B{}^{G}R_B$$
$$\neq {}^{G}R_B{}^{G}D_B. \tag{4.25}$$

However, according to the definition of $^{G}R_B$ and $^{G}D_B$,

$$^{G}T_B = {}^{G}D_B{}^{G}R_B$$
$$= {}^{G}D_B + {}^{G}R_B - \mathbf{I}$$
$$= {}^{G}R_B + {}^{G}D_B - \mathbf{I}. \tag{4.26}$$

**Example 74** *Pure translation.*

*A rigid body with its coordinate frame $B(oxyz)$, that is originally coincident with global coordinate frame $G(OXYZ)$, translates to*

$$^{G}\mathbf{d} = \begin{bmatrix} d_x \\ d_y \\ d_z \end{bmatrix}.$$

*Then, the motion of the rigid body is a pure translation and the transformation matrix for a point of a rigid body in the global frame is*

$$^{G}\mathbf{r} = {}^{G}T_B{}^{B}\mathbf{r} = {}^{G}D_B{}^{B}\mathbf{r}$$

$$= \begin{bmatrix} X \\ Y \\ Z \\ 1 \end{bmatrix} = \begin{bmatrix} 1 & 0 & 0 & d_x \\ 0 & 1 & 0 & d_y \\ 0 & 0 & 1 & d_z \\ 0 & 0 & 0 & 1 \end{bmatrix} \begin{bmatrix} x \\ y \\ z \\ 1 \end{bmatrix}.$$

**Example 75** *Homogeneous transformation for rotation about global axes.*
    *Using homogeneous representation of rotations about the axes of global coordinate frame, the rotation $\alpha$ about the $Z$-axis is*

$$
{}^{G}T_{B} = R_{Z,\alpha} = \begin{bmatrix} \cos\alpha & -\sin\alpha & 0 & 0 \\ \sin\alpha & \cos\alpha & 0 & 0 \\ 0 & 0 & 1 & 0 \\ 0 & 0 & 0 & 1 \end{bmatrix}. \tag{4.27}
$$

**Example 76** *Rotation about and translation along a global axis.*
    *A point $P$ is located at $(0, 0, 200)$ in a body coordinate frame. If the rigid body rotates 30 deg about the global $X$-axis and the origin of the body frame translates to $(X, Y, Z) = (500, 0, 600)$, then the coordinates of the point in the global frame are*

$$
\begin{bmatrix} 1 & 0 & 0 & 500 \\ 0 & 1 & 0 & 0 \\ 0 & 0 & 1 & 600 \\ 0 & 0 & 0 & 1 \end{bmatrix} \begin{bmatrix} 1 & 0 & 0 & 0 \\ 0 & \cos 30 & -\sin 30 & 0 \\ 0 & \sin 30 & \cos 30 & 0 \\ 0 & 0 & 0 & 1 \end{bmatrix} \begin{bmatrix} 0 \\ 0 \\ 200 \\ 1 \end{bmatrix} = \begin{bmatrix} 500 \\ -100 \\ 773.2 \\ 1 \end{bmatrix}.
$$

**Example 77** *Rotation about and translation along a local axis.*
    *A point $P$ is located at $(0, 0, 200)$ in a body coordinate frame. If the origin of the body frame translates to $(X, Y, Z) = (500, 0, 600)$ and rotates 30 deg about the local $x$-axis, then the coordinates of the point in global frame are*

$$
\begin{bmatrix} 1 & 0 & 0 & 500 \\ 0 & 1 & 0 & 0 \\ 0 & 0 & 1 & 600 \\ 0 & 0 & 0 & 1 \end{bmatrix} \begin{bmatrix} 1 & 0 & 0 & 0 \\ 0 & \cos 30 & \sin 30 & 0 \\ 0 & -\sin 30 & \cos 30 & 0 \\ 0 & 0 & 0 & 1 \end{bmatrix}^{T} \begin{bmatrix} 0 \\ 0 \\ 200 \\ 1 \end{bmatrix} = \begin{bmatrix} 500 \\ -100 \\ 773.2 \\ 1 \end{bmatrix}.
$$

**Example 78** *Translation.*
    *If a body point at*

$$
{}^{B}\mathbf{r} = \begin{bmatrix} -1 & 0 & 2 & 1 \end{bmatrix}^{T}
$$

*is translated to*

$$
{}^{G}\mathbf{r} = \begin{bmatrix} 0 & 10 & -5 & 1 \end{bmatrix}^{T}
$$

*then the corresponding transformation can be found by*

$$
\begin{bmatrix} 2 \\ 10 \\ -5 \\ 1 \end{bmatrix} = \begin{bmatrix} 1 & 0 & 0 & d_X \\ 0 & 1 & 0 & d_Y \\ 0 & 0 & 1 & d_Z \\ 0 & 0 & 0 & 1 \end{bmatrix} \begin{bmatrix} 1 \\ 4 \\ 2 \\ 1 \end{bmatrix}.
$$

*Therefore,*

$$
1 + d_X = 2 \quad , \quad 4 + d_Y = 10 \quad , \quad 2 + d_Z = -5
$$

*and*

$$
d_X = 1 \quad , \quad d_Y = 6 \quad , \quad d_Z = -7.
$$

**Example 79** *Pure rotation and pure translation.*

A set of basic homogeneous transformations for translation along and rotation about $X$, $Y$, and $Z$ axes are given below.

$$
{}^{G}T_B \;=\; D_{X,a} \;=\; \begin{bmatrix} 1 & 0 & 0 & a \\ 0 & 1 & 0 & 0 \\ 0 & 0 & 1 & 0 \\ 0 & 0 & 0 & 1 \end{bmatrix} \tag{4.28}
$$

$$
{}^{G}T_B \;=\; R_{X,\gamma} \;=\; \begin{bmatrix} 1 & 0 & 0 & 0 \\ 0 & \cos\gamma & -\sin\gamma & 0 \\ 0 & \sin\gamma & \cos\gamma & 0 \\ 0 & 0 & 0 & 1 \end{bmatrix} \tag{4.29}
$$

$$
{}^{G}T_B \;=\; D_{Y,b} \;=\; \begin{bmatrix} 1 & 0 & 0 & 0 \\ 0 & 1 & 0 & b \\ 0 & 0 & 1 & 0 \\ 0 & 0 & 0 & 1 \end{bmatrix} \tag{4.30}
$$

$$
{}^{G}T_B \;=\; R_{Y,\beta} \;=\; \begin{bmatrix} \cos\beta & 0 & \sin\beta & 0 \\ 0 & 1 & 0 & 0 \\ -\sin\beta & 0 & \cos\beta & 0 \\ 0 & 0 & 0 & 1 \end{bmatrix} \tag{4.31}
$$

$$
{}^{G}T_B \;=\; D_{Z,c} \;=\; \begin{bmatrix} 1 & 0 & 0 & 0 \\ 0 & 1 & 0 & 0 \\ 0 & 0 & 1 & c \\ 0 & 0 & 0 & 1 \end{bmatrix} \tag{4.32}
$$

$$
{}^{G}T_B \;=\; R_{Z,\alpha} \;=\; \begin{bmatrix} \cos\alpha & -\sin\alpha & 0 & 0 \\ \sin\alpha & \cos\alpha & 0 & 0 \\ 0 & 0 & 1 & 0 \\ 0 & 0 & 0 & 1 \end{bmatrix} \tag{4.33}
$$

**Example 80** *Homogeneous transformation as a consequence of vector addition.*

It is seen in Figure 4.4 that the position of point $P$ can be described by a vector addition

$$
{}^{G}\mathbf{r}_P = {}^{G}\mathbf{d} + {}^{B}\mathbf{r}_P. \tag{4.34}
$$

Since a vector equation is meaningful when all the vectors are described in the same coordinate frame, we need to transform either ${}^{B}\mathbf{r}_P$ to $G$ and ${}^{G}\mathbf{d}$ to $B$. Therefore, the correct vector equation is

$$
{}^{G}\mathbf{r}_P = {}^{G}R_B\,{}^{B}\mathbf{r}_P + {}^{G}\mathbf{d} \tag{4.35}
$$

or

$$
{}^{B}R_G\,{}^{G}\mathbf{r}_P = {}^{B}R_G\,{}^{G}\mathbf{d} + {}^{B}\mathbf{r}_P. \tag{4.36}
$$

*The first one defines a homogenous transformation from B to G,*

$$^G\mathbf{r}_P = {}^G T_B \, {}^B\mathbf{r}_P \tag{4.37}$$

$$^G T_B = \begin{bmatrix} {}^G R_B & {}^G\mathbf{d} \\ 0 & 1 \end{bmatrix} \tag{4.38}$$

*and the second one defines a transformation from G to B.*

$$^B\mathbf{r}_P = {}^B T_G \, {}^G\mathbf{r}_P \tag{4.39}$$

$$^B T_G = \begin{bmatrix} {}^B R_G & -{}^B R_G \, {}^G\mathbf{d} \\ 0 & 1 \end{bmatrix}$$

$$= \begin{bmatrix} {}^G R_B^T & -{}^G R_B^T \, {}^G\mathbf{d} \\ 0 & 1 \end{bmatrix} \tag{4.40}$$

**Example 81** ★ *Point at infinity.*

*Points at infinity have a convenient representation with homogeneous coordinates. Consider the scale factor $w$ as the fourth coordinate of a point, hence the homogeneous representation of the point is given by*

$$\begin{bmatrix} x \\ y \\ z \\ w \end{bmatrix} = \begin{bmatrix} x/w \\ y/w \\ z/w \\ 1 \end{bmatrix}. \tag{4.41}$$

*As $w$ tends to zero, the point tends toward infinity, and we may adapt the convention that the homogeneous coordinate*

$$\begin{bmatrix} x \\ y \\ z \\ 0 \end{bmatrix} \tag{4.42}$$

*represents a point at infinity. More importantly, a point at infinity indicates a direction. In this case, it indicates all lines parallel to the vector $\mathbf{r} = \begin{bmatrix} x & y & z \end{bmatrix}^T$, which intersect at a point at infinity.*

*The homogeneous coordinate transformation of points at infinity introduces a neat decomposition of the homogeneous transformation matrices.*

$$^G T_B = \begin{bmatrix} r_{11} & r_{12} & r_{13} & X_o \\ r_{21} & r_{22} & r_{23} & Y_o \\ r_{31} & r_{32} & r_{33} & Z_o \\ 0 & 0 & 0 & 1 \end{bmatrix} \tag{4.43}$$

*The first three columns have zero as the fourth coordinate. Therefore, they represent points at infinity, which are the directions corresponding to the three coordinate axes. The fourth column has one as the fourth coordinate, and represents the location of the coordinate frame origin.*

**Example 82** ★ *The most general homogeneous transformation.*

*The homogeneous transformation (4.13) is a special case of the general homogeneous transformation. The most general homogeneous transformation, which has been extensively used in the field of computer graphics, takes the form*

$$
{}^AT_B = \left[ \begin{array}{c|c} {}^AR_B \, (3 \times 3) & {}^A\mathbf{d} \, (3 \times 1) \\ \hline p \, (1 \times 3) & w \, (1 \times 1) \end{array} \right] \tag{4.44}
$$

$$
= \left[ \begin{array}{c|c} rotation & translation \\ \hline perspective & scale\ factor \end{array} \right].
$$

*For the purpose of robotics, we always take the last row vector of* $[T]$ *to be* $(0,0,0,1)$. *However, the more general form of (4.44) could be useful, for example, when a graphical simulator or a vision system is added to the overall robotic system.*

*The upper left* $3 \times 3$ *submatrix* ${}^AR_B$ *denotes the orientation of a moving frame* $B$ *with respect to a reference frame* $A$. *The upper right* $3 \times 1$ *submatrix* ${}^A\mathbf{d}$ *denotes the position of the origin of the moving frame* $B$ *relative to the reference frame* $A$. *The lower left* $1 \times 3$ *submatrix* $p$ *denotes a perspective transformation, and the lower right element* $w$ *is a scaling factor.*

## 4.3  Inverse Homogeneous Transformation

The advantage of simplicity to work with homogeneous transformation matrices come with the penalty of losing the orthogonality property. If we show ${}^GT_B$ by

$$
{}^GT_B = \left[ \begin{array}{cc} \mathbf{I} & {}^G\mathbf{d} \\ 0 & 1 \end{array} \right] \left[ \begin{array}{cc} {}^GR_B & 0 \\ 0 & 1 \end{array} \right]
$$

$$
= \left[ \begin{array}{cc} {}^GR_B & {}^G\mathbf{d} \\ 0 & 1 \end{array} \right] \tag{4.45}
$$

then,

$$
{}^BT_G = {}^GT_B^{-1}
$$

$$
= \left[ \begin{array}{cc} {}^GR_B & {}^G\mathbf{d} \\ 0 & 1 \end{array} \right]^{-1}
$$

$$
= \left[ \begin{array}{cc} {}^GR_B^T & -{}^GR_B^T \, {}^G\mathbf{d} \\ 0 & 1 \end{array} \right] \tag{4.46}
$$

showing that,

$$
{}^GT_B^{-1} \, {}^GT_B = \mathbf{I}_4. \tag{4.47}
$$

However, a transformation matrix is not orthogonal and its inverse is not equal to its transpose

$$
{}^GT_B^{-1} \neq {}^GT_B^T. \tag{4.48}
$$

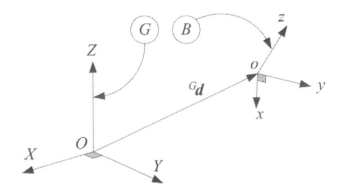

FIGURE 4.5. Illustration of a rotated and translated body frame $B(oxyz)$ with respect to the global frame $G(OXYZ)$.

**Proof. 1.** Figure 4.5 depicts a rotated and translated body frame $B(oxyz)$ with respect to the global frame $G(OXYZ)$. Transformation of the coordinates of a point $P$ from the global frame to the body frame is $^BT_G$, that is the inverse of the transformation $^GT_B$.

Let's start with the expression of $^Gr$ and the definition of $^GT_B$ for mapping $^Br$ to $^Gr$

$$^Gr = {}^GR_B {}^Br + {}^Gd \tag{4.49}$$

$$\begin{bmatrix} {}^Gr \\ 1 \end{bmatrix} = \begin{bmatrix} {}^GR_B & {}^Gd \\ 0 & 1 \end{bmatrix} \begin{bmatrix} {}^Br \\ 1 \end{bmatrix} \tag{4.50}$$

and find

$$\begin{aligned} {}^Br &= {}^GR_B^{-1}({}^Gr - {}^Gd) \\ &= {}^GR_B^T {}^Gr - {}^GR_B^T {}^Gd \end{aligned} \tag{4.51}$$

to express the transformation matrix $^BT_G$ for mapping $^Gr$ to $^Br$

$$^BT_G = \begin{bmatrix} {}^GR_B^T & -{}^GR_B^T {}^Gd \\ 0 & 1 \end{bmatrix}. \tag{4.52}$$

∎

**Proof. 2.** $^BT_G$ can also be found by a geometric expression. Using the inverse of the rotation matrix $^GR_B$

$$^GR_B^{-1} = {}^GR_B^T = {}^BR_G \tag{4.53}$$

and describing the reverse of $^Gd$ in the body coordinate frame to indicate the origin of the global frame with respect to the origin of the body frame

$$^Bd = {}^BR_G {}^Gd = {}^GR_B^T {}^Gd \tag{4.54}$$

allows us to define the homogeneous transformation $^BT_G$

$$^BT_G = \begin{bmatrix} ^BR_G & -^B\mathbf{d} \\ 0 & 1 \end{bmatrix}$$

$$= \begin{bmatrix} ^GR_B^T & -^GR_B^T \, ^G\mathbf{d} \\ 0 & 1 \end{bmatrix}. \tag{4.55}$$

We use the notation

$$^BT_G = {}^GT_B^{-1} \tag{4.56}$$

and remember that the inverse of a homogeneous transformation matrix must be calculated according to Equation (4.46), and not by regular inversion of a $4 \times 4$ matrix. This notation is consistent with the multiplication of a $T$ matrix by its inverse $T^{-1}$, because

$$^GT_B^{-1} \, {}^GT_B = \mathbf{I}_4. \tag{4.57}$$

■

**Example 83** *Inverse of a homogeneous transformation matrix.*
   *Assume that*

$$^GT_B = \begin{bmatrix} 0.643 & -0.766 & 0 & -1 \\ 0.766 & 0.643 & 0 & 0.5 \\ 0 & 0 & 1 & 0.2 \\ 0 & 0 & 0 & 1 \end{bmatrix} = \begin{bmatrix} ^GR_B & ^G\mathbf{d} \\ 0 & 1 \end{bmatrix}$$

*then*

$$^GR_B = \begin{bmatrix} 0.643 & -0.766 & 0 \\ 0.766 & 0.643 & 0 \\ 0 & 0 & 1 \end{bmatrix}$$

$$^G\mathbf{d} = \begin{bmatrix} -1 \\ 0.5 \\ 0.2 \end{bmatrix}$$

*and therefore,*

$$^BT_G = {}^GT_B^{-1} = \begin{bmatrix} ^GR_B^T & -^GR_B^T \, ^G\mathbf{d} \\ 0 & 1 \end{bmatrix}$$

$$= \begin{bmatrix} 0.643 & 0.766 & 0 & 0.26 \\ -0.766 & 0.643 & 0 & -1.087 \\ 0 & 0 & 1 & -0.2 \\ 0 & 0 & 0 & 1 \end{bmatrix}.$$

**Example 84** *Transformation matrix and coordinate of points.*
   *It is possible and sometimes convenient to describe a rigid body motion in terms of known displacement of specified points fixed in the body.*

*Assume A, B, C, and D are four points with the following coordinates at two different positions:*

$$A_1(2,4,1) \quad , \quad B_1(2,6,1) \quad , \quad C_1(1,5,1) \quad , \quad D_1(3,5,2)$$

$$A_2(5,1,1) \quad , \quad B_2(7,1,1) \quad , \quad C_2(6,2,1) \quad , \quad D_2(6,2,3)$$

*There must be a transformation matrix T to map the initial positions to the final,*

$$[T] \begin{bmatrix} 2 & 2 & 1 & 3 \\ 4 & 6 & 5 & 5 \\ 1 & 1 & 2 & 2 \\ 1 & 1 & 1 & 1 \end{bmatrix} = \begin{bmatrix} 5 & 7 & 6 & 6 \\ 1 & 1 & 2 & 2 \\ 1 & 1 & 1 & 3 \\ 1 & 1 & 1 & 1 \end{bmatrix}$$

*and hence*

$$[T] = \begin{bmatrix} 5 & 7 & 6 & 6 \\ 1 & 1 & 2 & 2 \\ 1 & 1 & 1 & 3 \\ 1 & 1 & 1 & 1 \end{bmatrix} \begin{bmatrix} 2 & 2 & 1 & 3 \\ 4 & 6 & 5 & 5 \\ 1 & 1 & 2 & 2 \\ 1 & 1 & 1 & 1 \end{bmatrix}^{-1}$$

$$= \begin{bmatrix} 5 & 7 & 6 & 6 \\ 1 & 1 & 2 & 2 \\ 1 & 1 & 1 & 3 \\ 1 & 1 & 1 & 1 \end{bmatrix} \begin{bmatrix} 0 & -1/2 & -1/2 & 7/2 \\ 0 & 1/2 & -1/2 & -3/2 \\ -1/2 & 0 & 1/2 & 1/2 \\ 1/2 & 0 & 1/2 & -3/2 \end{bmatrix}$$

$$= \begin{bmatrix} 0 & 1 & 0 & 1 \\ 0 & 0 & 1 & 0 \\ 1 & 0 & 1 & -2 \\ 0 & 0 & 0 & 1 \end{bmatrix}.$$

**Example 85** *Quick inverse transformation.*

*For numerical calculation, it is more practical to decompose a transformation matrix into translation times rotation, and take advantage of the inverse of matrix multiplication.*

*Consider a transformation matrix*

$$[T] = \begin{bmatrix} r_{11} & r_{12} & r_{13} & r_{14} \\ r_{21} & r_{22} & r_{23} & r_{24} \\ r_{31} & r_{32} & r_{33} & r_{34} \\ 0 & 0 & 0 & 1 \end{bmatrix} = [D][R] \qquad (4.58)$$

$$= \begin{bmatrix} 1 & 0 & 0 & r_{14} \\ 0 & 1 & 0 & r_{24} \\ 0 & 0 & 1 & r_{34} \\ 0 & 0 & 0 & 1 \end{bmatrix} \begin{bmatrix} r_{11} & r_{12} & r_{13} & 0 \\ r_{21} & r_{22} & r_{23} & 0 \\ r_{31} & r_{32} & r_{33} & 0 \\ 0 & 0 & 0 & 1 \end{bmatrix}$$

*therefore,*

$$T^{-1} = [DR]^{-1} = R^{-1}D^{-1} = R^T D^{-1} \tag{4.59}$$

$$= \begin{bmatrix} r_{11} & r_{21} & r_{31} & 0 \\ r_{12} & r_{22} & r_{32} & 0 \\ r_{13} & r_{23} & r_{33} & 0 \\ 0 & 0 & 0 & 1 \end{bmatrix} \begin{bmatrix} 1 & 0 & 0 & -r_{14} \\ 0 & 1 & 0 & -r_{24} \\ 0 & 0 & 1 & -r_{34} \\ 0 & 0 & 0 & 1 \end{bmatrix}$$

$$= \begin{bmatrix} r_{11} & r_{21} & r_{31} & -r_{11}r_{14} - r_{21}r_{24} - r_{31}r_{34} \\ r_{12} & r_{22} & r_{32} & -r_{12}r_{14} - r_{22}r_{24} - r_{32}r_{34} \\ r_{13} & r_{23} & r_{33} & -r_{13}r_{14} - r_{23}r_{24} - r_{33}r_{34} \\ 0 & 0 & 0 & 1 \end{bmatrix}.$$

**Example 86 ★** *Inverse of the general homogeneous transformation.*
Let $[T]$ *be defined as a general matrix combined by four submatrices* $[A]$, $[B]$, $[C]$, *and* $[D]$.

$$[T] = \begin{bmatrix} A & B \\ C & D \end{bmatrix} \tag{4.60}$$

*Then, the inverse is given by*

$$T^{-1} = \begin{bmatrix} A^{-1} + A^{-1}BE^{-1}CA^{-1} & -A^{-1}BE^{-1} \\ -E^{-1}CA^{-1} & E^{-1} \end{bmatrix} \tag{4.61}$$

*where*

$$[E] = D - CA^{-1}B. \tag{4.62}$$

*In the case of the most general homogeneous transformation,*

$$^AT_B = \left[ \begin{array}{c|c} ^AR_B\,(3 \times 3) & ^A\mathbf{d}\,(3 \times 1) \\ \hline p\,(1 \times 3) & w\,(1 \times 1) \end{array} \right] \tag{4.63}$$

$$= \left[ \begin{array}{c|c} \text{rotation} & \text{translation} \\ \hline \text{perspective} & \text{scale factor} \end{array} \right]$$

*we have*

$$[T] = {}^AT_B \tag{4.64}$$
$$[A] = {}^AR_B \tag{4.65}$$
$$[B] = {}^A\mathbf{d} \tag{4.66}$$
$$[C] = \begin{bmatrix} p_1 & p_2 & p_3 \end{bmatrix} \tag{4.67}$$
$$[D] = [w_{1\times1}] = w \tag{4.68}$$

*and therefore,*

$$
\begin{aligned}
[E] &= [E_{1\times 1}] = E \\
&= w - \begin{bmatrix} p_1 & p_2 & p_3 \end{bmatrix}{}^A R_B^{-1}\,{}^A\mathbf{d} \\
&= w - \begin{bmatrix} p_1 & p_2 & p_3 \end{bmatrix}{}^A R_B^T\,{}^A\mathbf{d} \\
&= 1 - p_1 (d_1 r_{11} + d_2 r_{21} + d_3 r_{31}) \\
&\quad + p_2 (d_1 r_{12} + d_2 r_{22} + d_3 r_{32}) \\
&\quad + p_3 (d_1 r_{13} + d_2 r_{23} + d_3 r_{33})
\end{aligned}
\tag{4.69}
$$

$$
-E^{-1}CA^{-1} = \frac{1}{E}\begin{bmatrix} g_1 & g_2 & g_3 \end{bmatrix}
\tag{4.70}
$$

$$
\begin{aligned}
g_1 &= p_1 r_{11} + p_2 r_{12} + p_3 r_{13} \\
g_2 &= p_1 r_{21} + p_2 r_{22} + p_3 r_{23} \\
g_3 &= p_1 r_{31} + p_2 r_{32} + p_3 r_{33}
\end{aligned}
$$

$$
-A^{-1}BE^{-1} = -\frac{1}{E}\begin{bmatrix} d_1 r_{11} + d_2 r_{21} + d_3 r_{31} \\ d_1 r_{12} + d_2 r_{22} + d_3 r_{32} \\ d_1 r_{13} + d_2 r_{23} + d_3 r_{33} \end{bmatrix}
\tag{4.71}
$$

$$
\begin{aligned}
A^{-1} + A^{-1}BE^{-1}CA^{-1} &= [F] \\
&= \begin{bmatrix} f_{11} & f_{12} & f_{13} \\ f_{21} & f_{22} & f_{23} \\ f_{31} & f_{32} & r_{33} \end{bmatrix}
\end{aligned}
\tag{4.72}
$$

*where*

$$
f_{11} = r_{11} + \frac{1}{E}(p_1 r_{11} + p_2 r_{12} + p_3 r_{13})(d_1 r_{11} + d_2 r_{21} + d_3 r_{31})
$$

$$
f_{12} = r_{21} + \frac{1}{E}(p_1 r_{21} + p_2 r_{22} + p_3 r_{23})(d_1 r_{11} + d_1 r_{21} + d_1 r_{31})
$$

$$
f_{13} = r_{31} + \frac{1}{E}(p_1 r_{31} + p_2 r_{32} + p_3 r_{33})(d_1 r_{11} + d_2 r_{21} + d_3 r_{31})
$$

$$
f_{21} = r_{12} + \frac{1}{E}(p_1 r_{11} + p_2 r_{12} + p_3 r_{13})(d_1 r_{12} + d_2 r_{22} + d_3 r_{32})
$$

$$
f_{22} = r_{22} + \frac{1}{E}(p_1 r_{21} + p_2 r_{22} + p_3 r_{23})(d_1 r_{12} + d_2 r_{22} + d_3 r_{32})
$$

$$
f_{23} = r_{32} + \frac{1}{E}(p_1 r_{31} + p_2 r_{32} + p_3 r_{33})(d_1 r_{12} + d_2 r_{22} + d_3 r_{32})
$$

$$
f_{31} = r_{13} + \frac{1}{E}(p_1 r_{11} + p_2 r_{12} + p_3 r_{13})(d_1 r_{13} + d_2 r_{23} + d_3 r_{33})
$$

$$
f_{32} = r_{23} + \frac{1}{E}(p_1 r_{21} + p_2 r_{22} + p_3 r_{23})(d_1 r_{13} + d_2 r_{23} + d_3 r_{33})
$$

$$
f_{33} = r_{33} + \frac{1}{E}(p_1 r_{31} + p_2 r_{32} + p_3 r_{33})(d_1 r_{13} + d_2 r_{23} + d_3 r_{33}).
$$

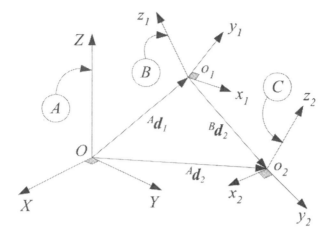

FIGURE 4.6. Three coordinate frames to analyze compound transformations.

*In the case of a coordinate homogeneous transformation*

$$[T] \quad = \quad {}^{A}T_{B} \tag{4.73}$$
$$[A] \quad = \quad {}^{A}R_{B} \tag{4.74}$$
$$[B] \quad = \quad {}^{A}\mathbf{d} \tag{4.75}$$
$$[C] \quad = \quad [\ 0 \quad 0 \quad 0\ ] \tag{4.76}$$
$$[D] \quad = \quad [1] \tag{4.77}$$

*we have*

$$[E] = [1] \tag{4.78}$$

*and therefore,*

$${}^{A}T_{B}^{-1} = \begin{bmatrix} {}^{A}R_{B}^{T} & -{}^{A}R_{B}^{T}\,{}^{A}\mathbf{d} \\ 0 & 1 \end{bmatrix}. \tag{4.79}$$

## 4.4 Compound Homogeneous Transformation

Figure 4.6 shows three reference frames: $A, B$, and $C$.

The transformation matrices to transform coordinates from frame $B$ to $A$, and from frame $C$ to $B$ are

$${}^{A}T_{B} \quad = \quad \begin{bmatrix} {}^{A}R_{B} & {}^{A}\mathbf{d}_{1} \\ 0 & 1 \end{bmatrix} \tag{4.80}$$

$${}^{B}T_{C} \quad = \quad \begin{bmatrix} {}^{B}R_{C} & {}^{B}\mathbf{d}_{2} \\ 0 & 1 \end{bmatrix}. \tag{4.81}$$

Hence, the transformation matrix from $C$ to $A$ is

$$
\begin{aligned}
^A T_C &= \ ^A T_B \, ^B T_C \\
&= \begin{bmatrix} ^A R_B & ^A \mathbf{d}_1 \\ 0 & 1 \end{bmatrix} \begin{bmatrix} ^B R_C & ^B \mathbf{d}_2 \\ 0 & 1 \end{bmatrix} \\
&= \begin{bmatrix} ^A R_B \, ^B R_C & ^A R_B \, ^B \mathbf{d}_2 + ^A \mathbf{d}_1 \\ 0 & 1 \end{bmatrix} \\
&= \begin{bmatrix} ^A R_C & ^A \mathbf{d}_2 \\ 0 & 1 \end{bmatrix}
\end{aligned}
\tag{4.82}
$$

and therefore, the inverse transformation is

$$
\begin{aligned}
^C T_A &= \begin{bmatrix} ^B R_C^T \, ^A R_B^T & - ^B R_C^T \, ^A R_B^T \left[ ^A R_B \, ^B \mathbf{d}_2 + ^A \mathbf{d}_1 \right] \\ 0 & 1 \end{bmatrix} \\
&= \begin{bmatrix} ^B R_C^T \, ^A R_B^T & - ^B R_C^T \, ^B \mathbf{d}_2 - ^A R_C^T \, ^A \mathbf{d}_1 \\ 0 & 1 \end{bmatrix} \\
&= \begin{bmatrix} ^A R_C^T & - ^A R_C^T \, ^A \mathbf{d}_2 \\ 0 & 1 \end{bmatrix} .
\end{aligned}
\tag{4.83}
$$

The value of homogeneous coordinates are better appreciated when several displacements occur in succession which, for instance, can be written as

$$
^G T_4 = \ ^G T_1 \, ^1 T_2 \, ^2 T_3 \, ^3 T_4
\tag{4.84}
$$

rather than

$$
\begin{aligned}
^G R_4 \, ^4 \mathbf{r}_P \ &+ ^G \mathbf{d}_4 \\
&= \ ^G R_1 \left( ^1 R_2 \left( ^2 R_3 \left( ^3 R_4 \, ^4 \mathbf{r}_P + ^3 \mathbf{d}_4 \right) + ^2 \mathbf{d}_3 \right) + ^1 \mathbf{d}_2 \right) + ^G \mathbf{d}_1 .
\end{aligned}
\tag{4.85}
$$

**Example 87** *Homogeneous transformation for multiple frames.*
*Figure 4.7 depicts a point $P$ in a local frame $B_2 \, (x_2 y_2 z_2)$. The coordinates of $P$ in the global frame $G(OXYZ)$ can be found by using the homogeneous transformation matrices.*
*The position of $P$ in frame $B_2 \, (x_2 y_2 z_2)$ is indicated by $^2 \mathbf{r}_P$. Therefore, its position in frame $B_1 \, (x_1 y_1 z_1)$ is*

$$
\begin{bmatrix} x_1 \\ y_1 \\ z_1 \\ 1 \end{bmatrix} = \begin{bmatrix} ^1 R_2 & ^1 \mathbf{d}_2 \\ 0 & 1 \end{bmatrix} \begin{bmatrix} x_2 \\ y_2 \\ z_2 \\ 1 \end{bmatrix}
\tag{4.86}
$$

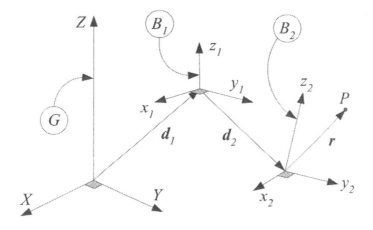

FIGURE 4.7. Point $P$ in a local frame $B_2$ $(x_2y_2z_2)$.

*and therefore, its position in the global frame $G(OXYZ)$ would be*

$$
\begin{bmatrix} X \\ Y \\ Z \\ 1 \end{bmatrix} = \begin{bmatrix} {}^GR_1 & {}^G\mathbf{d}_1 \\ 0 & 1 \end{bmatrix} \begin{bmatrix} x_1 \\ y_1 \\ z_1 \\ 1 \end{bmatrix}
$$

$$
= \begin{bmatrix} {}^GR_2 & {}^G\mathbf{d}_1 \\ 0 & 1 \end{bmatrix} \begin{bmatrix} {}^1R_2 & {}^1\mathbf{d}_2 \\ 0 & 1 \end{bmatrix} \begin{bmatrix} x_2 \\ y_2 \\ z_2 \\ 1 \end{bmatrix}
$$

$$
= \begin{bmatrix} {}^GR_2\,{}^GR_2 & {}^GR_2\,{}^1\mathbf{d}_2 + {}^G\mathbf{d}_1 \\ 0 & 1 \end{bmatrix} \begin{bmatrix} x_2 \\ y_2 \\ z_2 \\ 1 \end{bmatrix}. \qquad (4.87)
$$

**Example 88** *Rotation about an axis not going through origin.*

*The homogeneous transformation matrix can represent rotations about an axis going through a point different from the origin. Figure 4.8 indicates an angle of rotation, $\phi$, around the axis $\hat{u}$, passing through a point $P$ apart from the origin.*

*We set a local frame $B$ at point $P$ parallel to the global frame $G$. Then, a rotation around $\hat{u}$ can be expressed as a translation along $-\mathbf{d}$, to bring the body fame $B$ to the global frame $G$, followed by a rotation about $\hat{u}$ and*

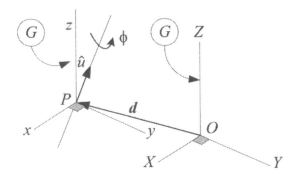

FIGURE 4.8. Rotation about an axis not going through origin.

*a reverse translation along* **d**

$$
{}^G T_B = D_{\hat{d},d}\, R_{\hat{u},\phi}\, D_{\hat{d},-d}
$$
$$
= \begin{bmatrix} \mathbf{I} & \mathbf{d} \\ 0 & 1 \end{bmatrix} \begin{bmatrix} R_{\hat{u},\phi} & 0 \\ 0 & 1 \end{bmatrix} \begin{bmatrix} \mathbf{I} & -\mathbf{d} \\ 0 & 1 \end{bmatrix}
$$
$$
= \begin{bmatrix} R_{\hat{u},\phi} & \mathbf{d} - R_{\hat{u},\phi}\mathbf{d} \\ 0 & 1 \end{bmatrix} \tag{4.88}
$$

*where,*

$$
R_{\hat{u},\phi} = \begin{bmatrix} u_1^2 \operatorname{vers} \phi + c\phi & u_1 u_2 \operatorname{vers} \phi - u_3 s\phi & u_1 u_3 \operatorname{vers} \phi + u_2 s\phi \\ u_1 u_2 \operatorname{vers} \phi + u_3 s\phi & u_2^2 \operatorname{vers} \phi + c\phi & u_2 u_3 \operatorname{vers} \phi - u_1 s\phi \\ u_1 u_3 \operatorname{vers} \phi - u_2 s\phi & u_2 u_3 \operatorname{vers} \phi + u_1 s\phi & u_3^2 \operatorname{vers} \phi + c\phi \end{bmatrix} \tag{4.89}
$$

*and*

$$
\mathbf{d} - R_{\hat{u},\phi}\mathbf{d} = \tag{4.90}
$$
$$
\begin{bmatrix} d_1(1-u_1^2) \operatorname{vers} \phi - u_1 \operatorname{vers} \phi \, (d_2 u_2 + d_3 u_3) + s\phi \, (d_2 u_3 - d_3 u_2) \\ d_2(1-u_2^2) \operatorname{vers} \phi - u_2 \operatorname{vers} \phi \, (d_3 u_3 + d_1 u_1) + s\phi \, (d_3 u_1 - d_1 u_3) \\ d_3(1-u_3^2) \operatorname{vers} \phi - u_3 \operatorname{vers} \phi \, (d_1 u_1 + d_2 u_2) + s\phi \, (d_1 u_2 - d_2 u_1) \end{bmatrix}.
$$

**Example 89** *End effector of an RPR robot in a global frame.*

Point $P$ indicates the tip point of the last arm of the robot shown in Figure 4.9. Position vector of $P$ in frame $B_2\,(x_2 y_2 z_2)$ is ${}^2\mathbf{r}_P$. Frame $B_2\,(x_2 y_2 z_2)$ can rotate about $z_2$ and slide along $y_1$. Frame $B_1\,(x_1 y_1 z_1)$ can rotate about the $Z$-axis of the global frame $G(OXYZ)$ while its origin is at ${}^G\mathbf{d}_1$. The position of $P$ in $G(OXYZ)$ would then be at

$$
{}^G\mathbf{r} = {}^G R_1\,{}^1 R_2\,{}^2\mathbf{r}_P + {}^G R_1\,{}^1\mathbf{d}_2 + {}^G\mathbf{d}_1
$$
$$
= {}^G T_1\,{}^1 T_2\,{}^2\mathbf{r}_P
$$
$$
= {}^G T_2\,{}^2\mathbf{r}_P \tag{4.91}
$$

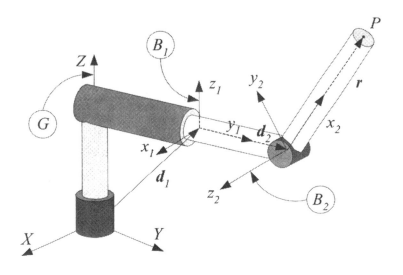

FIGURE 4.9. An RPR manipulator robot.

*where*

$$
{}^1T_2 = \begin{bmatrix} {}^1R_2 & {}^1\mathbf{d}_2 \\ 0 & 1 \end{bmatrix} \tag{4.92}
$$

$$
{}^GT_1 = \begin{bmatrix} {}^GR_1 & {}^G\mathbf{d}_1 \\ 0 & 1 \end{bmatrix} \tag{4.93}
$$

*and*

$$
{}^GT_2 = \begin{bmatrix} {}^GR_1{}^1R_2 & {}^GR_1{}^1\mathbf{d}_2 + {}^G\mathbf{d}_1 \\ 0 & 1 \end{bmatrix}. \tag{4.94}
$$

**Example 90** *End-effector of a SCARA robot in a global frame.*
 *Figure 4.10 depicts an* $R\|R\|P$ *(SCARA) robot with a global coordinate frame* $G(OXYZ)$ *attached to the base link along with the coordinate frames* $B_1(o_1x_1y_1z_1)$ *and* $B_2(o_2x_2y_2z_2)$ *attached to link (1) and the tip of link (3).*

*The transformation matrix, which is utilized to map points in frame* $B_1$ *to the base frame* $G$ *is*

$$
{}^GT_{B_1} = \begin{bmatrix} \cos\theta_1 & -\sin\theta_1 & 0 & l_1\cos\theta_1 \\ \sin\theta_1 & \cos\theta_1 & 0 & l_1\sin\theta_1 \\ 0 & 0 & 1 & 0 \\ 0 & 0 & 0 & 1 \end{bmatrix} \tag{4.95}
$$

*and the transformation matrix that is utilized to map points in frame* $B_2$

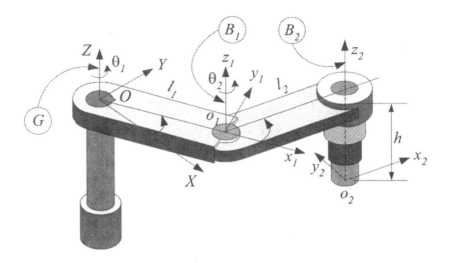

FIGURE 4.10. The SCARA robot of Example 90.

*to the frame $B_1$ is*

$$^{B_1}T_{B_2} = \begin{bmatrix} \cos\theta_2 & -\sin\theta_2 & 0 & l_2\cos\theta_2 \\ \sin\theta_2 & \cos\theta_2 & 0 & l_2\sin\theta_2 \\ 0 & 0 & 1 & -h \\ 0 & 0 & 0 & 1 \end{bmatrix}. \qquad (4.96)$$

*Therefore, the transformation matrix to map points in the end-effector frame $B_2$ to the base frame $G$ is*

$$^{G}T_{B_2} = {}^{G}T_{B_1}\,{}^{B_1}T_{B_2} \qquad (4.97)$$

$$= \begin{bmatrix} c(\theta_1+\theta_2) & -s(\theta_1+\theta_2) & 0 & l_1 c\theta_1 + l_2 c(\theta_1+\theta_2) \\ s(\theta_1+\theta_2) & c(\theta_1+\theta_2) & 0 & l_1 s\theta_1 + l_2 s(\theta_1+\theta_2) \\ 0 & 0 & 1 & -h \\ 0 & 0 & 0 & 1 \end{bmatrix}.$$

*The origin of the last frame is at $^{B_2}\mathbf{r}_{o_2} = \begin{bmatrix} 0 & 0 & 0 & 1 \end{bmatrix}^T$, in its local frame. Hence, the position of $o_2$ in the base coordinate frame is at*

$$^{G}\mathbf{r}_2 = {}^{G}T_{B_2}\,{}^{B_2}\mathbf{r}_{o_2}$$

$$= \begin{bmatrix} l_1\cos\theta_1 + l_2\cos(\theta_1+\theta_2) \\ l_1\sin\theta_1 + l_2\sin(\theta_1+\theta_2) \\ -h \\ 1 \end{bmatrix}. \qquad (4.98)$$

**Example 91** *Object manipulation.*

  *The geometry of a rigid body may be represented by an array containing the homogeneous coordinates of some specific points of the body described*

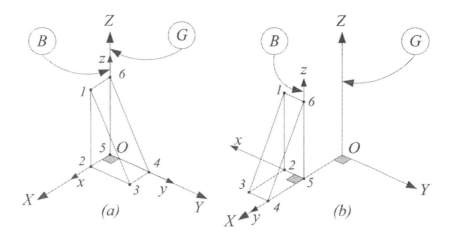

FIGURE 4.11. Describing the motion of a rigid body in terms of some body points.

in a local coordinate frame. The specific points are usually the corners, if there are any.

Figure 4.11(a) illustrates the configuration of a wedge. The coordinates of the corners of the wedge in its body frame $B$ are collected in the matrix $P$.

$$
{}^{B}P = \begin{bmatrix} 1 & 1 & 1 & 0 & 0 & 0 \\ 0 & 0 & 3 & 3 & 0 & 0 \\ 4 & 0 & 0 & 0 & 0 & 4 \\ 1 & 1 & 1 & 1 & 1 & 1 \end{bmatrix} \tag{4.99}
$$

The configuration of the wedge after a rotation of $-90\,\deg$ about the $Z$-axis and a translation of three units along the $X$-axis is shown in Figure 4.11(b). The new coordinates of its corners in the global frame $G$ are found by multiplying the corresponding transformation matrix by the $P$ matrix.

$$
\begin{aligned}
{}^{G}T_{B} &= D_{X,3}R_{Z,90} \\
&= \begin{bmatrix} 1 & 0 & 0 & 3 \\ 0 & 1 & 0 & 0 \\ 0 & 0 & 1 & 0 \\ 0 & 0 & 0 & 1 \end{bmatrix} \begin{bmatrix} \cos-90 & -\sin-90 & 0 & 0 \\ \sin-90 & \cos-90 & 0 & 0 \\ 0 & 0 & 1 & 0 \\ 0 & 0 & 0 & 1 \end{bmatrix} \\
&= \begin{bmatrix} 0 & 1 & 0 & 3 \\ -1 & 0 & 0 & 0 \\ 0 & 0 & 1 & 0 \\ 0 & 0 & 0 & 1 \end{bmatrix}. \tag{4.100}
\end{aligned}
$$

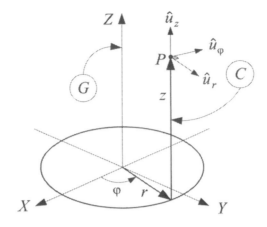

FIGURE 4.12. Cylindical coordinates of a point $P$.

*Therefore, the global coordinates of corners 1 to 6 after motion are*

$$
\begin{aligned}
{}^{G}P &= {}^{G}T_{B}\,{}^{B}P \\[4pt]
&= \begin{bmatrix} 0 & 1 & 0 & 3 \\ -1 & 0 & 0 & 0 \\ 0 & 0 & 1 & 0 \\ 0 & 0 & 0 & 1 \end{bmatrix}\begin{bmatrix} 1 & 1 & 1 & 0 & 0 & 0 \\ 0 & 0 & 3 & 3 & 0 & 0 \\ 4 & 0 & 0 & 0 & 0 & 4 \\ 1 & 1 & 1 & 1 & 1 & 1 \end{bmatrix} \\[4pt]
&= \begin{bmatrix} 3 & 3 & 6 & 6 & 3 & 3 \\ -1 & -1 & -1 & 0 & 0 & 0 \\ 4 & 0 & 0 & 0 & 0 & 4 \\ 1 & 1 & 1 & 1 & 1 & 1 \end{bmatrix}.
\end{aligned}
\tag{4.101}
$$

**Example 92** *Cylindrical coordinates.*

*There are situation in which we wish to specify the position of a robot end-effector in cylindrical coordinates. A set of cylindrical coordinates, as shown in Figure 4.12, can be achieved by a translation $r$ along the $X$-axis, followed by a rotation $\varphi$ about the $Z$-axis, and finally a translation $z$ along the $Z$-axis.*

*Therefore, the homogeneous transformation matrix for going from cylin-*

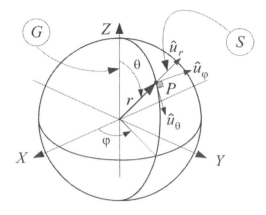

FIGURE 4.13. Spherical coordinates of a point $P$.

*drical coordinates $C(Or\varphi z)$ to Cartesian coordinates $G(OXYZ)$ is*

$$
\begin{aligned}
{}^{G}T_{C} &= D_{Z,z}\, R_{Z,\varphi}\, D_{X,r} \\
&= \begin{bmatrix} 1 & 0 & 0 & 0 \\ 0 & 1 & 0 & 0 \\ 0 & 0 & 1 & z \\ 0 & 0 & 0 & 1 \end{bmatrix} \begin{bmatrix} \cos\varphi & -\sin\varphi & 0 & 0 \\ \sin\varphi & \cos\varphi & 0 & 0 \\ 0 & 0 & 1 & 0 \\ 0 & 0 & 0 & 1 \end{bmatrix} \begin{bmatrix} 1 & 0 & 0 & r \\ 0 & 1 & 0 & 0 \\ 0 & 0 & 1 & 0 \\ 0 & 0 & 0 & 1 \end{bmatrix} \\
&= \begin{bmatrix} \cos\varphi & -\sin\varphi & 0 & r\cos\varphi \\ \sin\varphi & \cos\varphi & 0 & r\sin\varphi \\ 0 & 0 & 1 & z \\ 0 & 0 & 0 & 1 \end{bmatrix}.
\end{aligned}
\tag{4.102}
$$

*As an example, consider a point $P$ at $(1, \frac{\pi}{3}, 2)$ in a cylindrical coordinate frame. Then, the Cartesian coordinates of $P$ would be*

$$
\begin{bmatrix} \cos\frac{\pi}{3} & -\sin\frac{\pi}{3} & 0 & \cos\frac{\pi}{3} \\ \sin\frac{\pi}{3} & \cos\frac{\pi}{3} & 0 & \sin\frac{\pi}{3} \\ 0 & 0 & 1 & 2 \\ 0 & 0 & 0 & 1 \end{bmatrix} \begin{bmatrix} 0 \\ 0 \\ 0 \\ 1 \end{bmatrix} = \begin{bmatrix} 0.5 \\ 0.866 \\ 2.0 \\ 1.0 \end{bmatrix}.
$$

**Example 93** *Spherical coordinates.*

*A set of spherical coordinates, as shown in Figure 4.13, can be achieved by a translation $r$ along the $Z$-axis, followed by a rotation $\theta$ about the $Y$-axis, and finally a rotation $\varphi$ about the $Z$-axis.*

*Therefore, the homogeneous transformation matrix for going from spher-*

*ical coordinates $S(Or\theta\varphi)$ to Cartesian coordinates $G(OXYZ)$ is*

$$
^{G}T_{S} = R_{Z,\varphi}\, R_{Y,\theta}\, D_{Z,r}
$$

$$
= \begin{bmatrix}
\cos\theta\cos\varphi & -\sin\varphi & \cos\varphi\sin\theta & r\cos\varphi\sin\theta \\
\cos\theta\sin\varphi & \cos\varphi & \sin\theta\sin\varphi & r\sin\theta\sin\varphi \\
-\sin\theta & 0 & \cos\theta & r\cos\theta \\
0 & 0 & 0 & 1
\end{bmatrix} \quad (4.103)
$$

*where*

$$
R_{Z,\varphi} = \begin{bmatrix}
\cos\varphi & -\sin\varphi & 0 & 0 \\
\sin\varphi & \cos\varphi & 0 & 0 \\
0 & 0 & 1 & 0 \\
0 & 0 & 0 & 1
\end{bmatrix} \quad (4.104)
$$

$$
R_{Y,\theta} = \begin{bmatrix}
\cos\theta & 0 & \sin\theta & 0 \\
0 & 1 & 0 & 0 \\
-\sin\theta & 0 & \cos\theta & 0 \\
0 & 0 & 0 & 1
\end{bmatrix} \quad (4.105)
$$

$$
D_{Z,r} = \begin{bmatrix}
1 & 0 & 0 & 0 \\
0 & 1 & 0 & 0 \\
0 & 0 & 1 & r \\
0 & 0 & 0 & 1
\end{bmatrix}. \quad (4.106)
$$

*As an example, consider a point $P$ at $(2, \frac{\pi}{3}, \frac{\pi}{3})$ in a spherical coordinate frame. Then, the Cartesian coordinates of $P$ would be*

$$
\begin{bmatrix}
\cos\frac{\pi}{3}\cos\frac{\pi}{3} & -\sin\frac{\pi}{3} & \cos\frac{\pi}{3}\sin\frac{\pi}{3} & 2\cos\frac{\pi}{3}\sin\frac{\pi}{3} \\
\cos\frac{\pi}{3}\sin\frac{\pi}{3} & \cos\frac{\pi}{3} & \sin\frac{\pi}{3}\sin\frac{\pi}{3} & 2\sin\frac{\pi}{3}\sin\frac{\pi}{3} \\
-\sin\frac{\pi}{3} & 0 & \cos\frac{\pi}{3} & 2\cos\frac{\pi}{3} \\
0 & 0 & 0 & 1
\end{bmatrix}
\begin{bmatrix} 0 \\ 0 \\ 0 \\ 1 \end{bmatrix}
=
\begin{bmatrix} 0.866 \\ 1.5 \\ 1.0 \\ 1.0 \end{bmatrix}.
$$

## 4.5  ★ Screw Coordinates

Any rigid body motion can be produced by a single translation along an axis combined with a unique rotation about that axis. This is called *Chasles theorem*. Such a motion is called *screw*. Consider the *screw* motion illustrated in Figure 4.14. Point $P$ rotates about the screw axis indicated by $\hat{u}$ and simultaneously translates along the same axis. Hence, any point on the *screw axis* moves along the axis, while any point off the axis moves along a *helix*.

The angular rotation of the rigid body about the screw is called *twist*. *Pitch* of a screw, $p$, is the ratio of *translation*, $h$, to *rotation*, $\phi$.

$$
p = \frac{h}{\phi} \quad (4.107)
$$

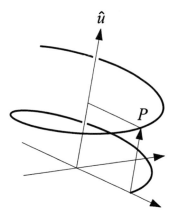

FIGURE 4.14. A screw motion is translation along a line combined with a rotation about the line.

In other words, the rectilinear distance through which the rigid body translates parallel to the axis of screw for a unit rotation is called pitch. Note that if $p > 0$, then the screw is *right-handed*, whereas if $p < 0$, it is *left-handed*.

A screw is shown by $\check{s}(h, \phi, \hat{u}, s)$ and is indicated by a unit vector $\hat{u}$, a *location* vector $s$, a twist angle $\phi$, and a translation $h$ (or pitch $p$). The location vector $s$ indicates the global position of a point on the screw axis. The twist angle $\phi$, the twist axis $\hat{u}$, and the pitch $p$ (or translation $h$) are called *screw parameters*.

Homogeneous matrix representation of a screw transformation is a combination of rotation and translation about the screw axis. If $\hat{u}$ passes through the origin of the coordinate frame, then $s = 0$ and the screw motion is called *central screw* $\check{s}(h, \phi, \hat{u})$.

The screw is basically another transformation method to describe the motion of a rigid body. A linear displacement along an axis combined with an angular displacement about the same axis arises repeatedly in robotic application.

For a central screw we have

$$^{G}\check{s}_{B}(h, \phi, \hat{u}) = D_{\hat{u},h}\, R_{\hat{u},\phi} \tag{4.108}$$

where,

$$D_{\hat{u},h} = \begin{bmatrix} 1 & 0 & 0 & hu_1 \\ 0 & 1 & 0 & hu_2 \\ 0 & 0 & 1 & hu_3 \\ 0 & 0 & 0 & 1 \end{bmatrix} \tag{4.109}$$

$$R_{\hat{u},\phi} =$$
$$\begin{bmatrix} u_1^2 \operatorname{vers}\phi + c\phi & u_1u_2 \operatorname{vers}\phi - u_3s\phi & u_1u_3 \operatorname{vers}\phi + u_2s\phi & 0 \\ u_1u_2 \operatorname{vers}\phi + u_3s\phi & u_2^2 \operatorname{vers}\phi + c\phi & u_2u_3 \operatorname{vers}\phi - u_1s\phi & 0 \\ u_1u_3 \operatorname{vers}\phi - u_2s\phi & u_2u_3 \operatorname{vers}\phi + u_1s\phi & u_3^2 \operatorname{vers}\phi + c\phi & 0 \\ 0 & 0 & 0 & 1 \end{bmatrix}$$
$$(4.110)$$

and hence,

$$^G\check{s}_B(h,\phi,\hat{u}) =$$
$$\begin{bmatrix} u_1^2 \operatorname{vers}\phi + c\phi & u_1u_2 \operatorname{vers}\phi - u_3s\phi & u_1u_3 \operatorname{vers}\phi + u_2s\phi & hu_1 \\ u_1u_2 \operatorname{vers}\phi + u_3s\phi & u_2^2 \operatorname{vers}\phi + c\phi & u_2u_3 \operatorname{vers}\phi - u_1s\phi & hu_2 \\ u_1u_3 \operatorname{vers}\phi - u_2s\phi & u_2u_3 \operatorname{vers}\phi + u_1s\phi & u_3^2 \operatorname{vers}\phi + c\phi & hu_3 \\ 0 & 0 & 0 & 1 \end{bmatrix}.$$
$$(4.111)$$

As a result, a central screw transformation matrix includes the pure or fundamental translations and rotations as special cases because a pure translation corresponds to $\phi = 0$, and a pure rotation corresponds to $h = 0$ (or $p = \infty$). A *reverse central screw* is defined as $\check{s}(-h, -\phi, \hat{u})$.

When the screw is not central and $\hat{u}$ is not passing through the origin, a screw motion to move $\mathbf{p}$ to $\mathbf{p}''$ is denoted by

$$\begin{aligned} \mathbf{p}'' &= (\mathbf{p} - \mathbf{s})\cos\phi + (1 - \cos\phi)(\hat{u} \cdot (\mathbf{p} - \mathbf{s}))\hat{u} \\ &\quad + (\hat{u} \times (\mathbf{p} - \mathbf{s}))\sin\phi + \mathbf{s} + h\hat{u} \end{aligned} \tag{4.112}$$

or

$$\begin{aligned} \mathbf{p}'' &= {}^GR_B(\mathbf{p} - \mathbf{s}) + \mathbf{s} + h\hat{u} \\ &= {}^GR_B\mathbf{p} + \mathbf{s} - {}^GR_B\mathbf{s} + h\hat{u} \end{aligned} \tag{4.113}$$

and therefore,

$$\mathbf{p}'' = \check{s}(h,\phi,\hat{u},\mathbf{s})\mathbf{p} = [T]\mathbf{p} \tag{4.114}$$

where

$$\begin{aligned} [T] &= \begin{bmatrix} {}^GR_B & {}^G\mathbf{s} - {}^GR_B{}^G\mathbf{s} + h\hat{u} \\ 0 & 1 \end{bmatrix} \\ &= \begin{bmatrix} {}^GR_B & {}^G\mathbf{d} \\ 0 & 1 \end{bmatrix}. \end{aligned} \tag{4.115}$$

The vector $^G\mathbf{s}$, called *location vector,* is the global position of the body frame before screw motion. The vectors $\mathbf{p}''$ and $\mathbf{p}$ are global positions of a point $P$ after and before screw, as shown in Figure 4.15.

The screw axis is indicated by the unit vector $\hat{u}$. Now a body point $P$ moves from its first position to its second position $P'$ by a rotation about $\hat{u}$. Then it moves to $P''$ by a translation $h$ parallel to $\hat{u}$. The initial position of $P$ is pointed by $\mathbf{p}$ and its final position is pointed by $\mathbf{p}''$.

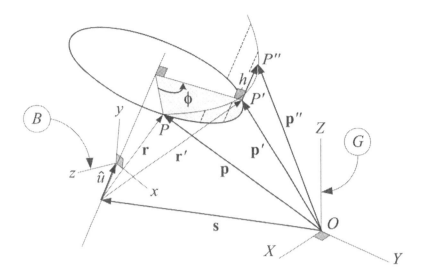

FIGURE 4.15. Screw motion of a rigid body.

In some references, screw is defined as a line with a pitch. To indicate a motion, they need to add the angle of twist. So, screw is the helical path of motion, and twist is the actual motion. However, in this text we define a screw as a four variable function $\check{s}(h, \phi, \hat{u}, \mathbf{s})$ to indicate the motion. A screw has a line of action $\hat{u}$ at $^G\mathbf{s}$, a twist $\phi$, and a translation $h$.

The instantaneous screw axis was first used by Mozzi (1763) although Chasles (1830) is credited with this discovery.

**Proof.** The angle-axis rotation formula (3.4) relates $\mathbf{r}'$ and $\mathbf{r}$, which are position vectors of $P$ after and before rotation $\phi$ about $\hat{u}$ when $\mathbf{s} = 0$, $h = 0$.

$$\mathbf{r}' = \mathbf{r} \cos \phi + (1 - \cos \phi)(\hat{u} \cdot \mathbf{r}) \hat{u} + (\hat{u} \times \mathbf{r}) \sin \phi \qquad (4.116)$$

However, when the screw axis does not pass through the origin of $G(OXYZ)$, then $\mathbf{r}'$ and $\mathbf{r}$ must accordingly be substituted with the following equations:

$$\mathbf{r} = \mathbf{p} - \mathbf{s} \qquad (4.117)$$
$$\mathbf{r}' = \mathbf{p}'' - \mathbf{s} - h\hat{u} \qquad (4.118)$$

where $\mathbf{r}'$ is a vector after rotation and hence in $G$ coordinate frame, and $\mathbf{r}$ is a vector before rotation and hence in $B$ coordinate frame.

Therefore, the relationship between the new and old positions of the body point $P$ after a screw motion is

$$\begin{aligned} \mathbf{p}'' = {} & (\mathbf{p} - \mathbf{s}) \cos \phi + (1 - \cos \phi)(\hat{u} \cdot (\mathbf{p} - \mathbf{s})) \hat{u} \\ & + (\hat{u} \times (\mathbf{p} - \mathbf{s})) \sin \phi + (\mathbf{s} + h\hat{u}). \end{aligned} \qquad (4.119)$$

Equation (4.119) is the *Rodriguez formula* for the most general rigid body motion. Defining new notations $^G\mathbf{p} = \mathbf{p}''$ and $^B\mathbf{p} = \mathbf{p}$ and also noting that $\mathbf{s}$ indicates a point on the rotation axis and therefore rotation does not affect $\mathbf{s}$, we may factor out $^B\mathbf{p}$ and write the Rodriguez formula in the following form

$$
^G\mathbf{p} = \left(\mathbf{I}\cos\phi + \hat{u}\hat{u}^T(1-\cos\phi) + \tilde{u}\sin\phi\right)\,^B\mathbf{p}
$$
$$
- \left(\mathbf{I}\cos\phi + \hat{u}\hat{u}^T(1-\cos\phi) + \tilde{u}\sin\phi\right)\,^G\mathbf{s} + \,^G\mathbf{s} + h\hat{u} \quad (4.120)
$$

which can be rearranged to show that a screw can be represented by a homogeneous transformation

$$
^G\mathbf{p} = \,^GR_B\,^B\mathbf{p} + \,^G\mathbf{s} - \,^GR_B\,^G\mathbf{s} + h\hat{u} \quad (4.121)
$$
$$
= \,^GR_B\,^B\mathbf{p} + \,^G\mathbf{d}
$$
$$
= \,^GT_B\,^B\mathbf{p}
$$

$$
^GT_B = \,^G\check{s}_B(h,\phi,\hat{u},\mathbf{s}) \quad (4.122)
$$
$$
= \begin{bmatrix} ^GR_B & \,^G\mathbf{s} - \,^GR_B\,^G\mathbf{s} + h\hat{u} \\ 0 & 1 \end{bmatrix} = \begin{bmatrix} ^GR_B & \,^G\mathbf{d} \\ 0 & 1 \end{bmatrix}
$$

where,

$$
^GR_B = \mathbf{I}\cos\phi + \hat{u}\hat{u}^T(1-\cos\phi) + \tilde{u}\sin\phi \quad (4.123)
$$
$$
^G\mathbf{d} = \left((\mathbf{I}-\hat{u}\hat{u}^T)(1-\cos\phi) - \tilde{u}\sin\phi\right)\,^G\mathbf{s} + h\hat{u}. \quad (4.124)
$$

Direct substitution shows that

$$
^GR_B = \begin{bmatrix} u_1^2\,\mathrm{vers}\,\phi + c\phi & u_1u_2\,\mathrm{vers}\,\phi - u_3s\phi & u_1u_3\,\mathrm{vers}\,\phi + u_2s\phi \\ u_1u_2\,\mathrm{vers}\,\phi + u_3s\phi & u_2^2\,\mathrm{vers}\,\phi + c\phi & u_2u_3\,\mathrm{vers}\,\phi - u_1s\phi \\ u_1u_3\,\mathrm{vers}\,\phi - u_2s\phi & u_2u_3\,\mathrm{vers}\,\phi + u_1s\phi & u_3^2\,\mathrm{vers}\,\phi + c\phi \end{bmatrix}
$$
$$
(4.125)
$$
$$
^G\mathbf{d} = \begin{bmatrix} hu_1 - u_1(s_3u_3 + s_2u_2 + s_1u_1)\,\mathrm{vers}\,\phi + (s_2u_3 - s_3u_2)s\phi + s_1 \\ hu_2 - u_2(s_3u_3 + s_2u_2 + s_1u_1)\,\mathrm{vers}\,\phi + (s_3u_1 - s_1u_3)s\phi + s_2 \\ hu_3 - u_3(s_3u_3 + s_2u_2 + s_1u_1)\,\mathrm{vers}\,\phi + (s_1u_2 - s_2u_1)s\phi + s_3 \end{bmatrix}.
$$
$$
(4.126)
$$

This representation of a rigid motion requires six independent parameters, namely one for rotation angle $\phi$, one for translation $h$, two for screw axis $\hat{u}$, and two for location vector $^G\mathbf{s}$. It is because three components of $\hat{u}$ are related to each other according to

$$
\hat{u}^T\hat{u} = 1 \quad (4.127)
$$

and the location vector $^G\mathbf{s}$ can locate any arbitrary point on the screw axis. It is convenient to choose the point where it has the minimum distance from $O$ to make $^G\mathbf{s}$ perpendicular to $\hat{u}$. Let's indicate the *shortest location vector* by $^G\mathbf{s}_0$, then there is a constraint among the components of the location vector

$$^G\mathbf{s}_0^T \, \hat{u} = 0. \tag{4.128}$$

If $\mathbf{s} = 0$ then the screw axis passes through the origin of $G$ and (4.122) reduces to (4.111).

The screw parameters $\phi$ and $h$, together with the screw axis $\hat{u}$ and location vector $^G\mathbf{s}$, completely define a rigid motion of $B(oxyz)$ in $G(OXYZ)$. It means, given the screw parameters and screw axis, we can find the elements of the transformation matrix by Equations (4.125) and (4.126). On the other hand, given the transformation matrix $^GT_B$, we can find the screw angle and axis by

$$\begin{aligned}
\cos\phi &= \frac{1}{2}\left(\mathrm{tr}\left(^GR_B\right) - 1\right) \\
&= \frac{1}{2}\left(\mathrm{tr}\left(^GT_B\right) - 2\right) \\
&= \frac{1}{2}\left(r_{11} + r_{22} + r_{33} - 1\right)
\end{aligned} \tag{4.129}$$

$$\tilde{u} = \frac{1}{2\sin\phi}\left(^GR_B - {}^GR_B^T\right) \tag{4.130}$$

hence,

$$\hat{u} = \frac{1}{2\sin\phi}\begin{bmatrix} r_{32} - r_{23} \\ r_{13} - r_{31} \\ r_{21} - r_{12} \end{bmatrix}. \tag{4.131}$$

To find all the required screw parameters, we must find $h$ and coordinates of one point on the screw axis. Since the points on the screw axis are invariant under the rotation, we must have

$$\begin{bmatrix} r_{11} & r_{12} & r_{13} & r_{14} \\ r_{21} & r_{22} & r_{23} & r_{24} \\ r_{31} & r_{32} & r_{33} & r_{34} \\ 0 & 0 & 0 & 1 \end{bmatrix} \begin{bmatrix} X \\ Y \\ Z \\ 1 \end{bmatrix} = \begin{bmatrix} 1 & 0 & 0 & hu_1 \\ 0 & 1 & 0 & hu_2 \\ 0 & 0 & 1 & hu_3 \\ 0 & 0 & 0 & 1 \end{bmatrix} \begin{bmatrix} X \\ Y \\ Z \\ 1 \end{bmatrix} \tag{4.132}$$

where $(X, Y, Z)$ are coordinates of points on the screw axis.

As a sample point, we may find the intersection point of the screw line with $YZ$-plane, by setting $X_s = 0$ and searching for $\mathbf{s} = \begin{bmatrix} 0 & Y_s & Z_s \end{bmatrix}^T$. Therefore,

$$\begin{bmatrix} r_{11} - 1 & r_{12} & r_{13} & r_{14} - hu_1 \\ r_{21} & r_{22} - 1 & r_{23} & r_{24} - hu_2 \\ r_{31} & r_{32} & r_{33} - 1 & r_{34} - hu_3 \\ 0 & 0 & 0 & 0 \end{bmatrix} \begin{bmatrix} 0 \\ Y_s \\ Z_s \\ 1 \end{bmatrix} = \begin{bmatrix} 0 \\ 0 \\ 0 \\ 0 \end{bmatrix} \tag{4.133}$$

which generates three equations to be solved for $Y_s$, $Z_s$, and $h$.

$$\begin{bmatrix} h \\ Y_s \\ Z_s \end{bmatrix} = \begin{bmatrix} u_1 & -r_{12} & -r_{13} \\ u_2 & 1-r_{22} & -r_{23} \\ u_3 & -r_{32} & 1-r_{33} \end{bmatrix}^{-1} \begin{bmatrix} r_{14} \\ r_{24} \\ r_{34} \end{bmatrix} \tag{4.134}$$

Now we can find the shortest location vector $^G s_0$ by

$$^G s_0 = s - (s \cdot \hat{u})\hat{u}. \tag{4.135}$$

■

**Example 94 ★** *Central screw transformation of a base unit vector.*

*Consider two initially coincident frames $G(OXYZ)$ and $B(oxyz)$. The body performs a screw motion along the Y-axis for $h = 2$ and $\phi = 90 \deg$. The position of a body point at $\begin{bmatrix} 1 & 0 & 0 & 1 \end{bmatrix}^T$ can be found by applying the central screw transformation.*

$$\check{s}(h, \phi, \hat{u}) = \check{s}(2, \frac{\pi}{2}, \hat{J}) \tag{4.136}$$

$$= D(2\hat{J}) R(\hat{J}, \frac{\pi}{2})$$

$$= \begin{bmatrix} 1 & 0 & 0 & 0 \\ 0 & 1 & 0 & 2 \\ 0 & 0 & 1 & 0 \\ 0 & 0 & 0 & 1 \end{bmatrix} \begin{bmatrix} 0 & 0 & 1 & 0 \\ 0 & 1 & 0 & 0 \\ -1 & 0 & 0 & 0 \\ 0 & 0 & 0 & 1 \end{bmatrix}$$

$$= \begin{bmatrix} 0 & 0 & 1 & 0 \\ 0 & 1 & 0 & 2 \\ -1 & 0 & 0 & 0 \\ 0 & 0 & 0 & 1 \end{bmatrix}$$

*Therefore,*

$$^G \hat{i} = \check{s}(2, \frac{\pi}{2}, \hat{J}) \, ^B \hat{i} \tag{4.137}$$

$$= \begin{bmatrix} 0 & 0 & 1 & 0 \\ 0 & 1 & 0 & 2 \\ -1 & 0 & 0 & 0 \\ 0 & 0 & 0 & 1 \end{bmatrix} \begin{bmatrix} 1 \\ 0 \\ 0 \\ 1 \end{bmatrix}$$

$$= \begin{bmatrix} 0 & 2 & -1 & 1 \end{bmatrix}^T.$$

*The pitch of this screw is*

$$p = \frac{h}{\phi} = \frac{2}{\frac{\pi}{2}} = \frac{4}{\pi} = 1.2732 \ \ unit/rad. \tag{4.138}$$

**Example 95 ★** *Screw transformation of a point.*

*Consider two initially parallel frames $G(OXYZ)$ and $B(oxyz)$. The body performs a screw motion along $X = 2$ and parallel to the $Y$-axis for $h = 2$ and $\phi = 90\,\text{deg}$. Therefore, the body coordinate frame is at location $\mathbf{s} = \begin{bmatrix} 2 & 0 & 0 \end{bmatrix}^T$. The position of a body point at $^B\mathbf{r} = \begin{bmatrix} 3 & 0 & 0 & 1 \end{bmatrix}^T$ can be found by applying the screw transformation, which is*

$$^GT_B = \begin{bmatrix} ^GR_B & \mathbf{s} - {}^GR_B\,\mathbf{s} + h\hat{u} \\ 0 & 1 \end{bmatrix} \tag{4.139}$$

$$= \begin{bmatrix} 0 & 0 & 1 & 2 \\ 0 & 1 & 0 & 2 \\ -1 & 0 & 0 & 2 \\ 0 & 0 & 0 & 1 \end{bmatrix}$$

*because,*

$$^GR_B = \begin{bmatrix} 0 & 0 & 1 \\ 0 & 1 & 0 \\ -1 & 0 & 0 \end{bmatrix}$$

$$\mathbf{s} = \begin{bmatrix} 2 \\ 0 \\ 0 \end{bmatrix}$$

$$\hat{u} = \begin{bmatrix} 0 \\ 1 \\ 0 \end{bmatrix}.$$

*Therefore, the position vector of $^G\mathbf{r}$ would then be*

$$^G\mathbf{r} = {}^GT_B\,{}^B\mathbf{r} \tag{4.140}$$

$$= \begin{bmatrix} 0 & 0 & 1 & 2 \\ 0 & 1 & 0 & 2 \\ -1 & 0 & 0 & 2 \\ 0 & 0 & 0 & 1 \end{bmatrix} \begin{bmatrix} 3 \\ 0 \\ 0 \\ 1 \end{bmatrix} = \begin{bmatrix} 2 \\ 2 \\ -1 \\ 1 \end{bmatrix}.$$

**Example 96 ★** *Rotation of a vector.*

*Transformation equation $^G\mathbf{r} = {}^GR_B\,{}^B\mathbf{r}$ and Rodriguez rotation formula (3.4) describe the rotation of any vector fixed in a rigid body. However, the vector can conveniently be described in terms of two points fixed in the body to derive the screw equation.*

*A reference point $P_1$ with position vector $\mathbf{r}_1$ at the tail, and a point $P_2$ with position vector $\mathbf{r}_2$ at the head, define a vector in the rigid body. Then the transformation equation between body and global frames can be written as*

$$^G(\mathbf{r}_2 - \mathbf{r}_1) = {}^GR_B\,{}^B(\mathbf{r}_2 - \mathbf{r}_1). \tag{4.141}$$

*Assume the original and final positions of the reference point $P_1$ are along the rotation axis. Equation (4.141) can then be rearranged in a form suitable* *for calculating coordinates of the new position of point $P_2$ in a transformation matrix form*

$$
\begin{aligned}
{}^G\mathbf{r}_2 &= {}^GR_B\,{}^B(\mathbf{r}_2 - \mathbf{r}_1) + {}^G\mathbf{r}_1 \\
&= {}^GR_B\,{}^B\mathbf{r}_2 + {}^G\mathbf{r}_1 - {}^GR_B\,{}^B\mathbf{r}_1 \\
&= {}^GT_B\,{}^B\mathbf{r}_2
\end{aligned}
\tag{4.142}
$$

*where*

$$
{}^GT_B = \begin{bmatrix} {}^GR_B & {}^G\mathbf{r}_1 - {}^GR_B\,{}^B\mathbf{r}_1 \\ 0 & 1 \end{bmatrix}.
\tag{4.143}
$$

*It is compatible with screw motion (4.122) for $h = 0$.*

**Example 97** ★ *Special cases for screw determination.*
    *There are two special cases for screws. The first one occurs when $r_{11} = r_{22} = r_{33} = 1$, then, $\phi = 0$ and the motion is a pure translation $h$ parallel to $\hat{u}$, where,*

$$
\hat{u} = \frac{r_{14} - s_1}{h}\hat{I} + \frac{r_{24} - s_2}{h}\hat{J} + \frac{r_{34} - s_3}{h}\hat{K}.
\tag{4.144}
$$

*Since there is no unique screw axis in this case, we cannot locate any specific point on the screw axis.*
    *The second special case occurs when $\phi = 180\,\mathrm{deg}$. In this case*

$$
\hat{u} = \begin{bmatrix} \sqrt{\frac{1}{2}(r_{11}+1)} \\ \sqrt{\frac{1}{2}(r_{22}+1)} \\ \sqrt{\frac{1}{2}(r_{33}+1)} \end{bmatrix}
\tag{4.145}
$$

*however, $h$ and $(X, Y, Z)$ can again be calculated from (4.134).*

**Example 98** ★ *Rotation and translation in a plane.*
    *Assume a plane is displaced from position 1 to position 2 according to Figure 4.16.*
    *New coordinates of $Q_2$ are*

$$
\begin{aligned}
\mathbf{r}_{Q_2} &= {}^2R_1\,(\mathbf{r}_{Q_1} - \mathbf{r}_{P_1}) + \mathbf{r}_{P_2} \\
&= \begin{bmatrix} \cos 58 & -\sin 58 & 0 \\ \sin 58 & \cos 58 & 0 \\ 0 & 0 & 1 \end{bmatrix}\left(\begin{bmatrix} 3 \\ 1 \\ 0 \end{bmatrix} - \begin{bmatrix} 1 \\ 1 \\ 0 \end{bmatrix}\right) + \begin{bmatrix} 4 \\ 1.5 \\ 0 \end{bmatrix} \\
&= \begin{bmatrix} 1.06 \\ 1.696 \\ 0 \end{bmatrix} + \begin{bmatrix} 4 \\ 1.5 \\ 0 \end{bmatrix} = \begin{bmatrix} 5.06 \\ 3.196 \\ 0.0 \end{bmatrix}
\end{aligned}
\tag{4.146}
$$

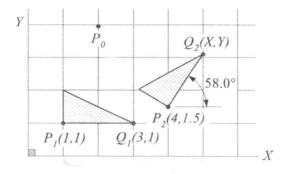

FIGURE 4.16. Motion in a plane.

*or equivalently*

$$\mathbf{r}_{Q_2} = {}^2T_1 \, \mathbf{r}_{Q_1} \qquad (4.147)$$

$$= \begin{bmatrix} {}^2R_1 & \mathbf{r}_{P_2} - {}^2R_1 \, \mathbf{r}_{P_1} \\ 0 & 1 \end{bmatrix} \mathbf{r}_{Q_1}$$

$$= \begin{bmatrix} \cos 58 & -\sin 58 & 0 & 4.318 \\ \sin 58 & \cos 58 & 0 & 0.122 \\ 0 & 0 & 1 & 0 \\ 0 & 0 & 0 & 1 \end{bmatrix} \begin{bmatrix} 3 \\ 1 \\ 0 \\ 1 \end{bmatrix}$$

$$= \begin{bmatrix} 5.06 \\ 3.196 \\ 0 \\ 1 \end{bmatrix}.$$

**Example 99** ★ *Pole of planar motion.*

*In the planar motion of a rigid body, going from position 1 to position 2, there is always one point in the plane of motion that does not change its position. Hence, the body can be considered as having rotated about this point, which is known as the finite rotation pole. The transformation matrix can be used to locate the pole. Figure 4.16 depicts a planar motion of a triangle. To locate the pole of motion $P_0(X_0, Y_0)$ we need the transformation of the motion. Using the data given in Figure 4.16 we have*

$$
{}^2T_1 = \begin{bmatrix} {}^2R_1 & \mathbf{r}_{P_2} - {}^2R_1 \, \mathbf{r}_{P_1} \\ 0 & 1 \end{bmatrix} \qquad (4.148)
$$

$$
= \begin{bmatrix} c\alpha & -s\alpha & 0 & -c\alpha + s\alpha + 4 \\ s\alpha & c\alpha & 0 & -c\alpha - s\alpha + 3.5 \\ 0 & 0 & 1 & 0 \\ 0 & 0 & 0 & 1 \end{bmatrix}.
$$

*The pole would be conserved under the transformation. Therefore,*

$$\mathbf{r}_{P_0} = {}^2T_1\, \mathbf{r}_{P_0} \tag{4.149}$$

$$\begin{bmatrix} X_0 \\ Y_0 \\ 0 \\ 1 \end{bmatrix} = \begin{bmatrix} \cos\alpha & -\sin\alpha & 0 & -\cos\alpha + \sin\alpha + 4 \\ \sin\alpha & \cos\alpha & 0 & -\cos\alpha - \sin\alpha + 1.5 \\ 0 & 0 & 1 & 0 \\ 0 & 0 & 0 & 1 \end{bmatrix} \begin{bmatrix} X_0 \\ Y_0 \\ 0 \\ 1 \end{bmatrix}$$

*which for $\alpha = 58\deg$ provides*

$$X_0 = -1.5\sin\alpha + 1 - 4\cos\alpha = 2.049$$
$$Y_0 = 4\sin\alpha + 1 - 1.5\cos\alpha = 3.956.$$

**Example 100 ★** *Determination of screw parameters.*

*We are able to determine screw parameters when we have the original and final position of three non-colinear points of a rigid body. Assume $\mathbf{p}_0$, $\mathbf{q}_0$, and $\mathbf{r}_0$ denote the position of points $P$, $Q$, and $R$ before the screw motion, and $\mathbf{p}_1$, $\mathbf{q}_1$, and $\mathbf{r}_1$ denote their positions after the screw motion.*

*To determine screw parameters, $\phi$, $\hat{u}$, $h$, and $\mathbf{s}$, we should solve the following three simultaneous Rodriguez equations:*

$$\mathbf{p}_1 - \mathbf{p}_0 = \tan\frac{\phi}{2}\hat{u} \times (\mathbf{p}_1 + \mathbf{p}_0 - 2\mathbf{s}) + h\hat{u} \tag{4.150}$$

$$\mathbf{q}_1 - \mathbf{q}_0 = \tan\frac{\phi}{2}\hat{u} \times (\mathbf{q}_1 + \mathbf{q}_0 - 2\mathbf{s}) + h\hat{u} \tag{4.151}$$

$$\mathbf{r}_1 - \mathbf{r}_0 = \tan\frac{\phi}{2}\hat{u} \times (\mathbf{r}_1 + \mathbf{r}_0 - 2\mathbf{s}) + h\hat{u} \tag{4.152}$$

*We start with subtracting Equation (4.152) from (4.150) and (4.151).*

$$(\mathbf{p}_1 - \mathbf{p}_0) - (\mathbf{r}_1 - \mathbf{r}_0) = \tan\frac{\phi}{2}\hat{u} \times [(\mathbf{p}_1 + \mathbf{p}_0) - (\mathbf{r}_1 - \mathbf{r}_0)] \tag{4.153}$$

$$(\mathbf{q}_1 - \mathbf{q}_0) - (\mathbf{r}_1 - \mathbf{r}_0) = \tan\frac{\phi}{2}\hat{u} \times [(\mathbf{q}_1 + \mathbf{q}_0) - (\mathbf{r}_1 - \mathbf{r}_0)] \tag{4.154}$$

*Now multiplying both sides of (4.153) by $[(\mathbf{q}_1 - \mathbf{q}_0) - (\mathbf{r}_1 - \mathbf{r}_0)]$ which is perpendicular to $\hat{u}$*

$$[(\mathbf{q}_1 - \mathbf{q}_0) - (\mathbf{r}_1 - \mathbf{r}_0)] \times [(\mathbf{p}_1 - \mathbf{p}_0) - (\mathbf{r}_1 - \mathbf{r}_0)]$$
$$= \tan\tfrac{\phi}{2}[(\mathbf{q}_1 - \mathbf{q}_0) - (\mathbf{r}_1 - \mathbf{r}_0)] \times \{\hat{u} \times [(\mathbf{p}_1 + \mathbf{p}_0) - (\mathbf{r}_1 - \mathbf{r}_0)]\} \tag{4.155}$$

*gives us*

$$[(\mathbf{q}_1 - \mathbf{q}_0) - (\mathbf{r}_1 - \mathbf{r}_0)] \times (\mathbf{p}_1 + \mathbf{p}_0) - (\mathbf{r}_1 - \mathbf{r}_0)$$
$$= \tan\tfrac{\phi}{2}[(\mathbf{q}_1 - \mathbf{q}_0) - (\mathbf{r}_1 - \mathbf{r}_0)] \cdot (\mathbf{p}_1 + \mathbf{p}_0) - (\mathbf{r}_1 - \mathbf{r}_0)\,\hat{u} \tag{4.156}$$

*and therefore, the rotation angle can be found by equating $\tan\frac{\phi}{2}$ and the norm of the right-hand side of the following equation:*

$$\tan\frac{\phi}{2}\hat{u} = \frac{[(\mathbf{q}_1 - \mathbf{q}_0) - (\mathbf{r}_1 - \mathbf{r}_0)] \times (\mathbf{p}_1 + \mathbf{p}_0) - (\mathbf{r}_1 - \mathbf{r}_0)}{[(\mathbf{q}_1 - \mathbf{q}_0) - (\mathbf{r}_1 - \mathbf{r}_0)] \cdot (\mathbf{p}_1 + \mathbf{p}_0) - (\mathbf{r}_1 - \mathbf{r}_0)} \tag{4.157}$$

*To find* $\mathbf{s}$, *we may start with the cross product of* $\hat{u}$ *with Equation (4.150).*

$$\hat{u} \times \mathbf{p}_1 - \mathbf{p}_0 = \hat{u} \times \left[ \tan \frac{\phi}{2} \hat{u} \times (\mathbf{p}_1 + \mathbf{p}_0 - 2\mathbf{s}) + h\hat{u} \right] \qquad (4.158)$$

$$= \tan \frac{\phi}{2} \{ [\hat{u} \cdot (\mathbf{p}_1 + \mathbf{p}_0)] \hat{u} - (\mathbf{p}_1 + \mathbf{p}_0) + 2 [\mathbf{s} - (\hat{u} \cdot \mathbf{s}) \hat{u}] \}$$

*Note that* $\mathbf{s} - (\hat{u} \cdot \mathbf{s}) \hat{u}$ *is the component of* $\mathbf{s}$ *perpendicular to* $\hat{u}$, *where* $\mathbf{s}$ *is a vector from the origin of the global frame* $G(OXYZ)$ *to an arbitrary point on the screw axis. This perpendicular component indicates a vector with the shortest distance between* $O$ *and* $\hat{u}$. *Let's assume* $\mathbf{s}_0$ *is the name of the shortest* $\mathbf{s}$. *Therefore,*

$$\mathbf{s}_0 = \mathbf{s} - (\hat{u} \cdot \mathbf{s}) \hat{u}$$

$$= \frac{1}{2} \left[ \frac{\hat{u} \times \mathbf{p}_1 - \mathbf{p}_0}{\tan \frac{\phi}{2}} - [\hat{u} \cdot (\mathbf{p}_1 + \mathbf{p}_0)] \hat{u} + \mathbf{p}_1 + \mathbf{p}_0 \right]. \qquad (4.159)$$

*The last parameter of the screw is the pitch* $h$, *which can be found from any one of the Equations (4.150), (4.151), or (4.152).*

$$h = \hat{u} \cdot (\mathbf{p}_1 - \mathbf{p}_0)$$
$$= \hat{u} \cdot (\mathbf{q}_1 - \mathbf{q}_0)$$
$$= \hat{u} \cdot (\mathbf{r}_1 - \mathbf{r}_0) \qquad (4.160)$$

**Example 101** ★ *Alternative derivation of screw transformation.*

*Assume the screw axis does not pass through the origin of* $G$. *If* $^G\mathbf{s}$ *is the position vector of some point on the axis* $\hat{u}$, *then we can derive the matrix representation of screw* $\check{s}(h, \phi, \hat{u}, \mathbf{s})$ *by translating the screw axis back to the origin, performing the central screw motion, and translating the line back to its original position.*

$$\check{s}(h, \phi, \hat{u}, \mathbf{s}) = D(^G\mathbf{s}) \, \check{s}(h, \phi, \hat{u}) \, D(-^G\mathbf{s})$$

$$= D(^G\mathbf{s}) \, D(h\hat{u}) \, R(\hat{u}, \phi) \, D(-^G\mathbf{s})$$

$$= \begin{bmatrix} \mathbf{I} & ^G\mathbf{s} \\ 0 & 1 \end{bmatrix} \begin{bmatrix} ^GR_B & h\hat{u} \\ 0 & 1 \end{bmatrix} \begin{bmatrix} \mathbf{I} & -^G\mathbf{s} \\ 0 & 1 \end{bmatrix}$$

$$= \begin{bmatrix} ^GR_B & ^G\mathbf{s} - {^GR_B} \, {^G\mathbf{s}} + h\hat{u} \\ 0 & 1 \end{bmatrix} \qquad (4.161)$$

**Example 102** ★ *Rotation about an off-center axis.*

*Rotation of a rigid body about an axis indicated by* $\hat{u}$ *and passing through a point at* $^G\mathbf{s}$, *where* $^G\mathbf{s} \times \hat{u} \neq 0$ *is a rotation about an off-center axis. The transformation matrix associated with an off-center rotation can be obtained from the screw transformation by setting* $h = 0$. *Therefore, an off-center rotation transformation is*

$$^GT_B = \begin{bmatrix} ^GR_B & ^G\mathbf{s} - {^GR_B} \, {^G\mathbf{s}} \\ 0 & 1 \end{bmatrix}. \qquad (4.162)$$

**Example 103 ★** *Principal central screw.*

There are three principal central screws, namely the x-screw, y-screw, and z-screw, which are

$$\check{s}(h_Z, \alpha, \hat{K}) = \begin{bmatrix} \cos\alpha & -\sin\alpha & 0 & 0 \\ \sin\alpha & \cos\alpha & 0 & 0 \\ 0 & 0 & 1 & p_Z\,\alpha \\ 0 & 0 & 0 & 1 \end{bmatrix} \qquad (4.163)$$

$$\check{s}(h_Y, \beta, \hat{J}) = \begin{bmatrix} \cos\beta & 0 & \sin\beta & 0 \\ 0 & 1 & 0 & p_Y\,\beta \\ -\sin\beta & 0 & \cos\beta & 0 \\ 0 & 0 & 0 & 1 \end{bmatrix} \qquad (4.164)$$

$$\check{s}(h_X, \gamma, \hat{I}) = \begin{bmatrix} 1 & 0 & 0 & p_X\,\gamma \\ 0 & \cos\gamma & -\sin\gamma & 0 \\ 0 & \sin\gamma & \cos\gamma & 0 \\ 0 & 0 & 0 & 1 \end{bmatrix}. \qquad (4.165)$$

**Example 104 ★** *Proof of Chasles theorem.*

Let $[T]$ be an arbitrary spatial displacement, and decompose it into a rotation $R$ about $\hat{u}$ and a translation $D$.

$$[T] = [D][R] \qquad (4.166)$$

We may also decompose the translation $[D]$ into two components $[D_\parallel]$ and $[D_\perp]$, parallel and perpendicular to $\hat{u}$, respectively.

$$[T] = [D_\parallel][D_\perp][R] \qquad (4.167)$$

Now $[D_\perp][R]$ is a planar motion, and is therefore equivalent to some rotation $[R'] = [D_\perp][R]$ about an axis parallel to the rotation axis $\hat{u}$. This yields the decomposition $[T] = [D_\parallel][R']$. This decomposition completes the proof, since the axis of $[D_\parallel]$ can be taken equal to $\hat{u}$.

**Example 105 ★** *Every rigid motion is a screw.*

To show that any proper rigid motion can be considered as a screw motion, we must show that a homogeneous transformation matrix

$$^G T_B = \begin{bmatrix} ^G R_B & ^G\mathbf{d} \\ 0 & 1 \end{bmatrix} \qquad (4.168)$$

can be written in the form

$$^G T_B = \begin{bmatrix} ^G R_B & (\mathbf{I} - {}^G R_B)\mathbf{s} + h\hat{u} \\ 0 & 1 \end{bmatrix}. \qquad (4.169)$$

This problem is then equivalent to the following equation to find $h$ and $\hat{u}$.

$$^G\mathbf{d} = (\mathbf{I} - {}^G R_B)\mathbf{s} + h\hat{u} \qquad (4.170)$$

*The matrix* $[\mathbf{I} - {}^G R_B]$ *is singular because* ${}^G R_B$ *always has 1 as an eigenvalue. This eigenvalue corresponds to* $\hat{u}$ *as eigenvector. Therefore,*

$$[\mathbf{I} - {}^G R_B]\hat{u} = [\mathbf{I} - {}^G R_B^T]\hat{u} = 0 \tag{4.171}$$

*and an inner product shows that*

$$\begin{aligned} \hat{u} \cdot {}^G\mathbf{d} &= \hat{u} \cdot [\mathbf{I} - {}^G R_B]\,\mathbf{s} + \hat{u} \cdot h\hat{u} \\ &= [\mathbf{I} - {}^G R_B]\,\hat{u} \cdot \mathbf{s} + \hat{u} \cdot h\hat{u} \end{aligned}$$

*which leads to*

$$h = \hat{u} \cdot {}^G\mathbf{d}. \tag{4.172}$$

*Now we may use* $h$ *to find* $\mathbf{s}$

$$\mathbf{s} = [\mathbf{I} - {}^G R_B]^{-1} ({}^G\mathbf{d} - h\hat{u}). \tag{4.173}$$

**Example 106** ★ *Classification of motions of a rigid body.*

*Imagine a body coordinate frame* $B(oxyz)$ *is moving with respect to a global frame* $G(OXYZ)$. *Point* $P$ *with position vector* $\mathbf{p}_B = [\begin{array}{ccc} p_1 & p_2 & p_3 \end{array}]^T$ *is an arbitrary point in* $B(oxyz)$. *The possible motions of* $B(oxyz)$ *and the required transformation between frames can be classified as:*

1. *Rotation* $\phi$ *about an axis passing through the origin and indicating by the unit vector* $\hat{u} = [\begin{array}{ccc} u_1 & u_2 & u_3 \end{array}]^T$

$$ {}^G\mathbf{p} = {}^G R_B \, {}^B\mathbf{p} \tag{4.174}$$

   *where* ${}^G R_B$ *comes from the Rodriguez's rotation formula (3.4).*

2. *Translation by* ${}^G\mathbf{d} = [\begin{array}{ccc} d_1 & d_2 & d_3 \end{array}]^T$ *plus a rotation* ${}^G R_B$.

$$ {}^G\mathbf{p} = {}^G R_B \, {}^B\mathbf{p} + {}^G\mathbf{d} \tag{4.175}$$

3. *Rotation* $\phi$ *about an axis on the unit vector* $\hat{u} = [\begin{array}{ccc} u_1 & u_2 & u_3 \end{array}]^T$ *passing through an arbitrary point indicated by* ${}^G\mathbf{s} = [\begin{array}{ccc} s_1 & s_2 & s_3 \end{array}]^T$.

$$ {}^G\mathbf{p} = {}^G R_B \, {}^B\mathbf{p} + {}^G\mathbf{s} - {}^G R_B \, {}^G\mathbf{s} \tag{4.176}$$

4. *Screw motion with angle* $\phi$ *and displacement* $h$, *about and along an axis directed by* $\hat{u} = [\begin{array}{ccc} u_1 & u_2 & u_3 \end{array}]^T$ *passing through an arbitrary point indicated by* ${}^G\mathbf{s} = [\begin{array}{ccc} s_1 & s_2 & s_3 \end{array}]^T$.

$$ {}^G\mathbf{p} = {}^G R_B \, {}^B\mathbf{p} + {}^G\mathbf{s} - {}^G R_B \, {}^G\mathbf{s} + h\hat{u} \tag{4.177}$$

5. *Reflection*

(a) *in the xy-plane;*

$$^G\mathbf{p} = {}^GR_B\,{}^B\mathbf{p}_{(-z)} \tag{4.178}$$

*where*

$$\mathbf{p}_{(-z)} = \begin{bmatrix} p_1 \\ p_2 \\ -p_3 \end{bmatrix}. \tag{4.179}$$

(b) *in the yz-plane;*

$$^G\mathbf{p} = {}^GR_B\,{}^B\mathbf{p}_{(-x)} \tag{4.180}$$

*where*

$$\mathbf{p}_{(-z)} = \begin{bmatrix} -p_1 \\ p_2 \\ p_3 \end{bmatrix}. \tag{4.181}$$

(c) *in the xz-plane;*

$$^G\mathbf{p} = {}^GR_B\,{}^B\mathbf{p}_{(-y)} \tag{4.182}$$

*where*

$$\mathbf{p}_{(-z)} = \begin{bmatrix} p_1 \\ -p_2 \\ p_3 \end{bmatrix}. \tag{4.183}$$

(d) *in a plane with equation $u_1 x + u_2 y + u_3 z + h = 0$;*

$$
\begin{aligned}
^G\mathbf{p} &= \frac{1}{u_1^2 + u_2^2 + u_3^2}\left({}^GR_B\,{}^B\mathbf{p} - 2h\hat{u}\right) \\
&= {}^GR_B\,{}^B\mathbf{p} - 2h\hat{u}
\end{aligned} \tag{4.184}
$$

*where*

$$^GR_B = \begin{bmatrix} -u_1^2 + u_2^2 + u_3^2 & -2u_1 u_2 & -2u_3 u_1 \\ -2u_2 u_1 & u_1^2 - u_2^2 + u_3^2 & -2u_2 u_3 \\ -2u_1 u_3 & -2u_3 u_2 & u_1^2 + u_2^2 - u_3^2 \end{bmatrix}. \tag{4.185}$$

(e) *in a plane going through point $\begin{bmatrix} s_1 & s_2 & s_3 \end{bmatrix}^T$ and normal to $\hat{u} = \begin{bmatrix} u_1 & u_2 & u_3 \end{bmatrix}^T$*

$$^G\mathbf{p} = {}^GR_B\,{}^B\mathbf{p} + {}^G\mathbf{s} - {}^GR_B\,{}^G\mathbf{s} \tag{4.186}$$

*where $^GR_B$ is as in (4.185).*

## 4.6 ★ Inverse Screw

Inverse of the screw $\check{s}(h, \phi, \hat{u}, \mathbf{s})$ is defined by

$$
\begin{aligned}
{}^{G}\check{s}_{B}^{-1}(h, \phi, \hat{u}, \mathbf{s}) &= {}^{B}\check{s}_{G}(h, \phi, \hat{u}, \mathbf{s}) \\
&= \begin{bmatrix} {}^{G}R_{B}^{T} & {}^{G}\mathbf{s} - {}^{G}R_{B}^{T}\,{}^{G}\mathbf{s} - h\hat{u} \\ 0 & 1 \end{bmatrix}
\end{aligned}
\tag{4.187}
$$

where $\hat{u}$ is a unit vector indicating the axis of screw, $\mathbf{s}$ is the location vector of a point on the axis of screw, $\phi$ is the screw angle, and $h$ is the screw translation. If the screw is central, the axis of screw passes through the origin and $\mathbf{s} = 0$. Therefore, the inverse of a central screw is

$$
{}^{G}\check{s}_{B}^{-1}(h, \phi, \hat{u}) = \begin{bmatrix} {}^{G}R_{B}^{T} & -h\hat{u} \\ 0 & 1 \end{bmatrix}.
\tag{4.188}
$$

**Proof.** The homogeneous matrix expression of a screw $\check{s}(h, \phi, \hat{u}, \mathbf{s})$ is

$$
\begin{aligned}
{}^{G}T_{B} &= {}^{G}\check{s}_{B}(h, \phi, \hat{u}, \mathbf{s}) \\
&= \begin{bmatrix} {}^{G}R_{B} & {}^{G}\mathbf{s} - {}^{G}R_{B}\,{}^{G}\mathbf{s} + h\hat{u} \\ 0 & 1 \end{bmatrix}.
\end{aligned}
\tag{4.189}
$$

A homogeneous matrix

$$
{}^{G}T_{B} = \begin{bmatrix} {}^{G}R_{B} & {}^{G}\mathbf{d} \\ 0 & 1 \end{bmatrix}
\tag{4.190}
$$

can be inverted according to

$$
\begin{aligned}
{}^{B}T_{G} &= {}^{G}T_{B}^{-1} \\
&= \begin{bmatrix} {}^{G}R_{B}^{T} & -{}^{G}R_{B}^{T}\,{}^{G}\mathbf{d} \\ 0 & 1 \end{bmatrix}.
\end{aligned}
\tag{4.191}
$$

To show the correctness of Equation (4.187), we need to calculate $-{}^{G}R_{B}^{T}\,{}^{G}\mathbf{d}$:

$$
\begin{aligned}
-{}^{G}R_{B}^{T}\,{}^{G}\mathbf{d} &= -{}^{G}R_{B}^{T}\left({}^{G}\mathbf{s} - {}^{G}R_{B}\,{}^{G}\mathbf{s} + h\hat{u}\right) \\
&= -{}^{G}R_{B}^{T}\,{}^{G}\mathbf{s} + {}^{G}R_{B}^{T}\,{}^{G}R_{B}\,{}^{G}\mathbf{s} - {}^{G}R_{B}^{T}\, h\hat{u} \\
&= -{}^{G}R_{B}^{T}\,{}^{G}\mathbf{s} + {}^{G}\mathbf{s} - {}^{G}R_{B}^{T}\, h\hat{u}.
\end{aligned}
\tag{4.192}
$$

Since $\hat{u}$ is an invariant vector in both coordinate frames $B$ and $G$, we have

$$
\hat{u} = {}^{G}R_{B}\,\hat{u} = {}^{G}R_{B}^{T}\,\hat{u}
\tag{4.193}
$$

and therefore,

$$
-{}^{G}R_{B}^{T}\,{}^{G}\mathbf{d} = {}^{G}\mathbf{s} - {}^{G}R_{B}^{T}\,{}^{G}\mathbf{s} - h\hat{u}.
\tag{4.194}
$$

This completes the inversion of a general screw:

$$^G\breve{s}_B^{-1}(h, \phi, \hat{u}, \mathbf{s}) = \begin{bmatrix} ^G R_B^T & ^G\mathbf{s} - {}^G R_B^T\, {}^G\mathbf{s} - h\hat{u} \\ 0 & 1 \end{bmatrix}. \qquad (4.195)$$

If the screw is central, the location vector is zero and the inverse of the screw simplifies to

$$^G\breve{s}_B^{-1}(h, \phi, \hat{u}) = \begin{bmatrix} ^G R_B^T & -h\hat{u} \\ 0 & 1 \end{bmatrix}. \qquad (4.196)$$

Since the inversion of a rotation matrix $^G R_B = R_{\hat{u},\phi}$ can be found by a rotation $-\phi$ about $\hat{u}$

$$^G R_B^{-1} = {}^G R_B^T = {}^B R_G = R_{\hat{u},-\phi} \qquad (4.197)$$

the inversion of a screw can also be interpreted as a rotation $-\phi$ about $\hat{u}$, plus a translation $-h$ along $\hat{u}$.

$$^G\breve{s}_B^{-1}(h, \phi, \hat{u}, \mathbf{s}) = \breve{s}(-h, -\phi, \hat{u}, \mathbf{s}) \qquad (4.198)$$

■

**Example 107** *Checking the inversion formula.*
   *Employing the inversion screw formula, we must have*

$$\breve{s}(h, \phi, \hat{u}, \mathbf{s})\, {}^G\breve{s}_B^{-1}(h, \phi, \hat{u}, \mathbf{s}) = \mathbf{I}_4. \qquad (4.199)$$

*It can be checked by*

$$\begin{bmatrix} ^G R_B & ^G\mathbf{s} - {}^G R_B\, {}^G\mathbf{s} + h\hat{u} \\ 0 & 1 \end{bmatrix} \begin{bmatrix} ^G R_B^T & ^G\mathbf{s} - {}^G R_B^T\, {}^G\mathbf{s} - h\hat{u} \\ 0 & 1 \end{bmatrix}$$

$$= \begin{bmatrix} \mathbf{I}_3 & ^G R_B\left({}^G\mathbf{s} - {}^G R_B^T\, {}^G\mathbf{s} - h\hat{u}\right) + \left({}^G\mathbf{s} - {}^G R_B\, {}^G\mathbf{s} + h\hat{u}\right) \\ 0 & 1 \end{bmatrix}$$

$$= \begin{bmatrix} \mathbf{I}_3 & ^G R_B\, {}^G\mathbf{s} - {}^G\mathbf{s} - h\, {}^G R_B\, \hat{u} + {}^G\mathbf{s} - {}^G R_B\, {}^G\mathbf{s} + h\hat{u} \\ 0 & 1 \end{bmatrix}$$

$$= \begin{bmatrix} \mathbf{I}_3 & 0 \\ 0 & 1 \end{bmatrix} = \mathbf{I}_4. \qquad (4.200)$$

# 4.7  ★ Compound Screw Transformation

Assume $^1\breve{s}_2(h_1, \phi_1, \hat{u}_1, \mathbf{s}_1)$ is a screw motion to move from coordinate frame $B_2$ to $B_1$ and $^G\breve{s}_1(h_0, \phi_0, \hat{u}_0, \mathbf{s}_0)$ is a screw motion to move from coordinate frame $B_1$ to $G$. Then, the screw motion to move from $B_2$ to $G$ is

$$^G\breve{s}_2(h, \phi, \hat{u}, \mathbf{s}) = {}^G\breve{s}_1(h_0, \phi_0, \hat{u}_0, \mathbf{s}_0)\, {}^1\breve{s}_2(h_1, \phi_1, \hat{u}_1, \mathbf{s}_1) \qquad (4.201)$$

$$= \begin{bmatrix} ^G R_2 & ^G R_1(\mathbf{I} - {}^1 R_2)\,\mathbf{s}_1 + (\mathbf{I} - {}^G R_1)\,\mathbf{s}_0 + h_1\, {}^G R_1 \hat{u}_1 + h_0 \hat{u}_0 \\ 0 & 1 \end{bmatrix}.$$

**Proof.** Direct substitution for ${}^1s_2(h_1, \phi_1, \hat{u}_1)$ and ${}^Gs_1(h_0, \phi_0, \hat{u}_0)$

$$
{}^G\check{s}_1(h_0, \phi_0, \hat{u}_0, s_0) = \begin{bmatrix} {}^GR_1 & s_0 - {}^GR_1 s_0 + h_0\hat{u}_0 \\ 0 & 1 \end{bmatrix} \quad (4.202)
$$

$$
{}^1\check{s}_2(h_1, \phi_1, \hat{u}_1, s_1) = \begin{bmatrix} {}^1R_2 & s_1 - {}^1R_2 s_1 + h_1\hat{u}_1 \\ 0 & 1 \end{bmatrix} \quad (4.203)
$$

shows that

$$
{}^G\check{s}_2(h, \phi, \hat{u}, s) \quad (4.204)
$$

$$
= \begin{bmatrix} {}^GR_1 & s_0 - {}^GR_1 s_0 + h_0\hat{u}_0 \\ 0 & 1 \end{bmatrix} \begin{bmatrix} {}^1R_2 & s_1 - {}^1R_2 s_1 + h_1\hat{u}_1 \\ 0 & 1 \end{bmatrix}
$$

$$
= \begin{bmatrix} {}^GR_2 & {}^GR_1\left(s_1 - {}^1R_2 s_1 + h_1\hat{u}_1\right) + s_0 - {}^GR_1 s_0 + h_0\hat{u}_0 \\ 0 & 1 \end{bmatrix}
$$

$$
= \begin{bmatrix} {}^GR_2 & {}^GR_1(\mathbf{I} - {}^1R_2) s_1 + (\mathbf{I} - {}^GR_1) s_0 + h_1\,{}^GR_1\hat{u}_1 + h_0\hat{u}_0 \\ 0 & 1 \end{bmatrix}
$$

where

$$
{}^GR_2 = {}^GR_1\,{}^1R_2. \quad (4.205)
$$

To find the screw parameters of the equivalent screw ${}^G\check{s}_2(h, \phi, \hat{u}, s)$, we start obtaining $\hat{u}$ and $\phi$ from ${}^GR_2$ based on (4.131) and (4.129). Then, utilizing (4.172) and (4.173) we can find $h$ and $s$

$$
h = \hat{u} \cdot {}^G\mathbf{d} \quad (4.206)
$$

$$
s = \left[\mathbf{I} - {}^GR_2\right]^{-1} \left({}^G\mathbf{d} - h\hat{u}\right) \quad (4.207)
$$

where

$$
\begin{aligned}
{}^G\mathbf{d} &= {}^GR_1(\mathbf{I} - {}^1R_2) s_1 + (\mathbf{I} - {}^GR_1) s_0 + h_1\,{}^GR_1\hat{u}_1 + h_0\hat{u}_0 \\
&= \left({}^GR_1 - {}^GR_2\right) s_1 + {}^GR_1\left(h_1\hat{u}_1 - s_0\right) + s_0 + h_0\hat{u}_0. \quad (4.208)
\end{aligned}
$$

∎

**Example 108 ★** *Exponential representation of a screw.*

*To compute a rigid body motion associated with a screw, consider the motion of point P in Figure 4.15. The final position of the point can be given by*

$$
\begin{aligned}
\mathbf{p}'' &= s + e^{\phi\tilde{u}}\mathbf{r} + h\hat{u} \quad (4.209) \\
&= s + e^{\phi\tilde{u}}(\mathbf{p} - s) + h\hat{u} \\
&= [T]\mathbf{p}
\end{aligned}
$$

*where $[T]$ is the exponential representation of screw motion*

$$
[T] = \begin{bmatrix} e^{\phi\tilde{u}} & (\mathbf{I} - e^{\phi\tilde{u}}) s + h\hat{u} \\ 0 & 1 \end{bmatrix}. \quad (4.210)
$$

*The exponential screw transformation matrix (4.210) is based on the exponential form of the Rodriguez formula (3.101)*

$$e^{\phi\tilde{u}} = I + \tilde{u}\sin\phi + \tilde{u}^2(1 - \cos\phi).\tag{4.211}$$

*Therefore, the combination of two screws can also be found by*

$$[T] = T_1 T_2$$

$$= \begin{bmatrix} e^{\phi_1\tilde{u}_1} & (\mathbf{I} - e^{\phi_1\tilde{u}_1})\,\mathbf{s}_1 + h_1\hat{u}_1 \\ 0 & 1 \end{bmatrix} \begin{bmatrix} e^{\phi_2\tilde{u}_2} & (\mathbf{I} - e^{\phi_2\tilde{u}_2})\,\mathbf{s}_2 + h_2\hat{u}_2 \\ 0 & 1 \end{bmatrix}$$

$$= \begin{bmatrix} e^{\phi_1\tilde{u}_1 + \phi_2\tilde{u}_2} & {}^G\mathbf{d} \\ 0 & 1 \end{bmatrix}\tag{4.212}$$

*where*

$${}^G\mathbf{d} = \left(e^{\phi_1\tilde{u}_1} - e^{\phi_1\tilde{u}_1 + \phi_2\tilde{u}_2}\right)\mathbf{s}_2 + e^{\phi_1\tilde{u}_1}\left(h_2\hat{u}_2 - \mathbf{s}_1\right) + \mathbf{s}_1 + h_1\hat{u}_1.\tag{4.213}$$

**Example 109 ★** *Combination of two principal central screws.*

*Combination of every two principal central screws can be found by matrix multiplication. As an example, a screw motion about $Y$ followed by another screw motion about $X$ is*

$$\check{s}(h_X, \gamma, \hat{I})\,\check{s}(h_Y, \beta, \hat{J})\tag{4.214}$$

$$= \begin{bmatrix} 1 & 0 & 0 & \gamma p_X \\ 0 & c\gamma & -s\gamma & 0 \\ 0 & s\gamma & c\gamma & 0 \\ 0 & 0 & 0 & 1 \end{bmatrix} \begin{bmatrix} c\beta & 0 & s\beta & 0 \\ 0 & 1 & 0 & \beta p_Y \\ -s\beta & 0 & c\beta & 0 \\ 0 & 0 & 0 & 1 \end{bmatrix}$$

$$= \begin{bmatrix} \cos\beta & 0 & \sin\beta & \gamma p_X \\ \sin\beta\sin\gamma & \cos\gamma & -\cos\beta\sin\gamma & \beta p_Y\cos\gamma \\ -\cos\gamma\sin\beta & \sin\gamma & \cos\beta\cos\gamma & \beta p_Y\sin\gamma \\ 0 & 0 & 0 & 1 \end{bmatrix}.$$

*Screw combination is not commutative and therefore,*

$$\check{s}(h_X, \gamma, \hat{I})\,\check{s}(h_Y, \beta, \hat{J}) \neq \check{s}(h_Y, \beta, \hat{J})\,\check{s}(h_X, \gamma, \hat{I}).\tag{4.215}$$

**Example 110 ★** *Decomposition of a screw.*

*Every general screw can be decomposed to three principal central screws.*

$$\begin{aligned}{}^G\check{s}_B(h, \phi, \hat{u}, \mathbf{s}) &= \begin{bmatrix} {}^G R_B & \mathbf{s} - {}^G R_B\mathbf{s} + h\hat{u} \\ 0 & 1 \end{bmatrix} \\ &= \check{s}(h_X, \gamma, \hat{I})\,\check{s}(h_Y, \beta, \hat{J})\,\check{s}(h_Z, \alpha, \hat{K})\end{aligned}\tag{4.216}$$

*Therefore,*

$${}^G R_B = \begin{bmatrix} c\alpha c\beta & -c\beta s\alpha & s\beta \\ c\gamma s\alpha + c\alpha s\beta s\gamma & c\alpha c\gamma - s\alpha s\beta s\gamma & -c\beta s\gamma \\ s\alpha s\gamma - c\alpha c\gamma s\beta & c\alpha s\gamma + c\gamma s\alpha s\beta & c\beta c\gamma \end{bmatrix}\tag{4.217}$$

*and*

$$\mathbf{s} - {}^{G}R_B\,\mathbf{s} + h\hat{u} = \begin{bmatrix} \gamma p_X + \alpha p_Z \sin\beta \\ \beta p_Y \cos\gamma - \alpha p_Z \cos\beta \sin\gamma \\ \beta p_Y \sin\gamma + \alpha p_Z \cos\beta \cos\gamma \end{bmatrix} = \begin{bmatrix} d_X \\ d_Y \\ d_Z \end{bmatrix}. \quad (4.218)$$

*The twist angles* $\alpha, \beta, \gamma$ *can be found from* ${}^{G}R_B$, *then, the pitches* $p_X$, $p_Y$, $p_Z$ *can be found as follows*

$$p_Z = \frac{d_Z \cos\gamma - d_Y \sin\gamma}{\alpha \cos\beta} \quad (4.219)$$

$$p_Y = \frac{d_Z \sin\gamma + d_Y \cos\gamma}{\beta} \quad (4.220)$$

$$p_X = \frac{d_X}{\gamma} - \frac{d_Z \cos\gamma - d_Y \sin\gamma}{\gamma \cos\beta}\sin\beta. \quad (4.221)$$

**Example 111 ★** *Decomposition of a screw to principal central screws.*

In general, there are six different independent combinations of triple principal central screws and therefore, there are six different methods to decompose a general screw into a combination of principal central screws. They are:

$$\begin{array}{l} 1 - \check{s}(h_X, \gamma, \hat{I})\,\check{s}(h_Y, \beta, \hat{J})\,\check{s}(h_Z, \alpha, \hat{K}) \\ 2 - \check{s}(h_Y, \beta, \hat{J})\,\check{s}(h_Z, \alpha, \hat{K})\,\check{s}(h_X, \gamma, \hat{I}) \\ 3 - \check{s}(h_Z, \alpha, \hat{K})\,\check{s}(h_X, \gamma, \hat{I})\,\check{s}(h_Y, \beta, \hat{J}) \\ 4 - \check{s}(h_Z, \alpha, \hat{K})\,\check{s}(h_Y, \beta, \hat{J})\,\check{s}(h_X, \gamma, \hat{I}) \\ 5 - \check{s}(h_Y, \beta, \hat{J})\,\check{s}(h_X, \gamma, \hat{I})\,\check{s}(h_Z, \alpha, \hat{K}) \\ 6 - \check{s}(h_X, \gamma, \hat{I})\,\check{s}(h_Z, \alpha, \hat{K})\,\check{s}(h_Y, \beta, \hat{J}) \end{array} \quad (4.222)$$

The expanded form of the six combinations of principal central screws are presented in Appendix C.

## 4.8 ★ The Plücker Line Coordinate

The most common coordinate set for showing a line is the *Plücker coordinates*. Analytical representation of a line in space can be found if we have the position of two different points of that line. Assume $P_1(X_1, Y_1, Z_1)$ and $P_2(X_2, Y_2, Z_2)$ at $\mathbf{r}_1$ and $\mathbf{r}_2$ are two different points on the line $l$ as shown in Figure 4.17.

Using the position vectors $\mathbf{r}_1$ and $\mathbf{r}_2$, the equation of line $l$ can be defined by six elements of two vectors

$$l = \begin{bmatrix} \hat{u} \\ \rho \end{bmatrix} = \begin{bmatrix} L \\ M \\ N \\ P \\ Q \\ R \end{bmatrix} \quad (4.223)$$

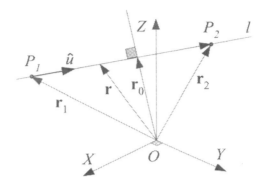

FIGURE 4.17. A line indicated by two points.

referred to as *Plücker coordinates* of the *directed* line, where,

$$\hat{u} = \frac{\mathbf{r}_2 - \mathbf{r}_1}{|\mathbf{r}_2 - \mathbf{r}_1|} = L\hat{I} + M\hat{J} + N\hat{K} \qquad (4.224)$$

is a unit vector along the line referred to as *direction vector*, and

$$\boldsymbol{\rho} = \mathbf{r}_1 \times \hat{u} = P\hat{I} + Q\hat{J} + R\hat{K} \qquad (4.225)$$

is the *moment vector* of $\hat{u}$ about the origin.

The Plücker method is a canonical representation of line definition and therefore is more efficient than the other methods such as parametric form $l(t) = \mathbf{r}_1 + t\hat{u}$, point and direction form $(\mathbf{r}_1, \hat{u})$, or two-point representation $(\mathbf{r}_1, \mathbf{r}_2)$.

**Proof.** The unit vector $\hat{u}$, whose direction is along the line connecting $P_1$ and $P_2$, is

$$
\begin{aligned}
\hat{u} &= \frac{\mathbf{r}_2 - \mathbf{r}_1}{|\mathbf{r}_2 - \mathbf{r}_1|} \\
&= L\hat{I} + M\hat{J} + N\hat{K} \\
&= \frac{X_2 - X_1}{d}\hat{I} + \frac{Y_2 - Y_1}{d}\hat{J} + \frac{Z_2 - Z_1}{d}\hat{K} \qquad (4.226)
\end{aligned}
$$

where,

$$L^2 + M^2 + N^2 = 1 \qquad (4.227)$$

and the distance between $P_1$ and $P_2$ is

$$d = \sqrt{(X_2 - X_1)^2 + (Y_2 - Y_1)^2 + (Z_2 - Z_1)^2}. \qquad (4.228)$$

If $\mathbf{r}$ represents a vector from the origin $O$ to a point on the line $l$, then the vector $\mathbf{r} - \mathbf{r}_1$ is parallel to $\hat{u}$ and therefore, the equation of the line can be written as

$$(\mathbf{r} - \mathbf{r}_1) \times \hat{u} = 0 \qquad (4.229)$$

or equivalently
$$\mathbf{r} \times \hat{u} = \rho \qquad (4.230)$$
where
$$\rho = \mathbf{r_1} \times \hat{u}$$
is the moment of the direction vector $\hat{u}$ about $O$. Furthermore, because vectors $\rho$ and $\hat{u}$ are perpendicular, there is a constraint among their components
$$\hat{u} \cdot \rho = 0. \qquad (4.231)$$
Expanding (4.225) yields
$$\rho = \begin{vmatrix} \hat{I} & \hat{J} & \hat{K} \\ X_1 & Y_1 & Z_1 \\ L & M & N \end{vmatrix} = P\hat{I} + Q\hat{J} + R\hat{K} \qquad (4.232)$$
where
$$\begin{aligned} P &= Y_1 N - Z_1 M \\ Q &= Z_1 L - X_1 N \\ R &= X_1 M - Y_1 L \end{aligned} \qquad (4.233)$$
and therefore the orthogonality condition (4.231) can be expressed as
$$LP + MQ + NR = 0. \qquad (4.234)$$

The Plücker coordinates of the line (4.223) have four independent coordinates because of two constraints: (4.227) and (4.234).

Our arrangement of Plücker coordinates in the form of (4.223) is the *line arrangement* and is called *ray coordinates*, however sometimes the reverse order in *axis arrangement* $l = \begin{bmatrix} \rho & \hat{u} \end{bmatrix}^T$ is also used by some other textbooks. In either case, a vertical line $\begin{bmatrix} \hat{u} & | & \rho \end{bmatrix}^T$ or a semi-colon $\begin{bmatrix} \hat{u} & ; & \rho \end{bmatrix}^T$ may be utilized to separate the first three elements from the last three. Both arrangements can be used in kinematics efficiently.

The Plücker line coordinates $\begin{bmatrix} \hat{u} & \rho \end{bmatrix}^T$ are homogeneous because Equation (4.225) shows that the coordinates $\begin{bmatrix} w\hat{u} & w\rho \end{bmatrix}^T$, where $w \in \mathbb{R}$, determines the same line.

Force - moment, angular velocity - translational velocity, and rigid motion act like a line vector and can be described in Plücker coordinates. ∎

**Example 112 ★** *Plücker coordinates of a line connecting two points.*
*Plücker line coordinates of the line connecting points $P_1(1, 0, 0)$ and $P_2(0, 1, 1)$ are*
$$\begin{aligned} l &= \begin{bmatrix} \hat{u} \\ \rho \end{bmatrix} \\ &= \begin{bmatrix} -1 & 1 & 1 & 0 & -1 & 1 \end{bmatrix}^T \end{aligned}$$

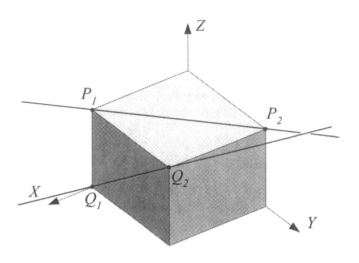

FIGURE 4.18. A unit cube.

*because*

$$\sqrt{3}\hat{u} = \frac{\mathbf{r}_2 - \mathbf{r}_1}{|\mathbf{r}_2 - \mathbf{r}_1|} = -\hat{I} + \hat{J} + \hat{K}$$

*and*

$$\sqrt{3}\rho = \mathbf{r}_1 \times \sqrt{3}\hat{u} = -\hat{J} + \hat{K}.$$

**Example 113 ★** *Plücker coordinates of diagonals of a cube.*

Figure 4.18 depicts a unit cube and two lines on diagonals of two adjacent faces. Line $l_1$ connecting corners $P_1(1,0,1)$ and $P_2(0,1,1)$ is

$$l_1 = \begin{bmatrix} \hat{u}_1 \\ \rho_1 \end{bmatrix}$$

$$= \begin{bmatrix} -\frac{\sqrt{2}}{2} & \frac{\sqrt{2}}{2} & 0 & -\frac{\sqrt{2}}{2} & -\frac{\sqrt{2}}{2} & \frac{\sqrt{2}}{2} \end{bmatrix}^T$$

*because*

$$\hat{u}_1 = \frac{\mathbf{p}_2 - \mathbf{p}_1}{|\mathbf{p}_2 - \mathbf{p}_1|} = \frac{-\hat{I} + \hat{J}}{\sqrt{2}}$$

*and*

$$\rho_1 = \mathbf{p}_1 \times \hat{u}_1 = \frac{-\hat{I} - \hat{J} + \hat{K}}{\sqrt{2}}.$$

Line $l_2$ connecting corners $Q_1(1,0,0)$ and $Q_2(1,1,1)$ is

$$l_2 = \begin{bmatrix} \hat{u}_2 \\ \rho_2 \end{bmatrix}$$

$$= \begin{bmatrix} 0 & \frac{\sqrt{2}}{2} & \frac{\sqrt{2}}{2} & 0 & -\frac{\sqrt{2}}{2} & \frac{\sqrt{2}}{2} \end{bmatrix}^T$$

*because*

$$\hat{u}_2 = \frac{\mathbf{q}_2 - \mathbf{q}_1}{|\mathbf{q}_2 - \mathbf{q}_1|} = \frac{\hat{J} + \hat{K}}{\sqrt{2}}$$

*and*

$$\rho_2 = \mathbf{q}_1 \times \hat{u}_2 = \frac{-\hat{J} + \hat{K}}{\sqrt{2}}.$$

**Example 114** ★ *Grassmanian matrix to show Plücker coordinates.*
   It can be verified that the Grassmanian matrix for coordinates of two points

$$\begin{bmatrix} w_1 & X_1 & Y_1 & Z_1 \\ w_2 & X_2 & Y_2 & Z_2 \end{bmatrix} \tag{4.235}$$

*is a short notation for the Plücker coordinates if we define*

$$L = \begin{vmatrix} w_1 & X_1 \\ w_2 & X_2 \end{vmatrix} \quad , \quad P = \begin{vmatrix} Y_1 & Z_1 \\ Y_2 & Z_2 \end{vmatrix}$$

$$M = \begin{vmatrix} w_1 & Y_1 \\ w_2 & Y_2 \end{vmatrix} \quad , \quad Q = \begin{vmatrix} Z_1 & X_1 \\ Z_2 & X_2 \end{vmatrix} \tag{4.236}$$

$$N = \begin{vmatrix} w_1 & Z_1 \\ w_2 & Z_2 \end{vmatrix} \quad , \quad R = \begin{vmatrix} X_1 & Y_1 \\ X_2 & Y_2 \end{vmatrix}.$$

**Example 115** ★ *Ray-axis arrangement transformation.*
   It can be verified that the ray arrangement of Plücker coordinates,

$$l_{ray} = \begin{bmatrix} \hat{u} \\ \rho \end{bmatrix} \tag{4.237}$$

*can be transformed to the axis arrangement,*

$$l_{axis} = \begin{bmatrix} \rho \\ \hat{u} \end{bmatrix} \tag{4.238}$$

*and vice versa utilizing the following 6 × 6 transformation matrix:*

$$\begin{bmatrix} \rho \\ \hat{u} \end{bmatrix} = \Delta \begin{bmatrix} \hat{u} \\ \rho \end{bmatrix} \tag{4.239}$$

$$\Delta = \begin{bmatrix} 0 & 0 & 0 & 1 & 0 & 0 \\ 0 & 0 & 0 & 0 & 1 & 0 \\ 0 & 0 & 0 & 0 & 0 & 1 \\ 1 & 0 & 0 & 0 & 0 & 0 \\ 0 & 1 & 0 & 0 & 0 & 0 \\ 0 & 0 & 1 & 0 & 0 & 0 \end{bmatrix} = \begin{bmatrix} 0 & I_3 \\ I_3 & 0 \end{bmatrix} \tag{4.240}$$

*The transformation matrix $\Delta$ is symmetric and satisfies the following equations:*

$$\Delta^2 = \Delta\Delta = I \tag{4.241}$$

$$\Delta^T = \Delta \tag{4.242}$$

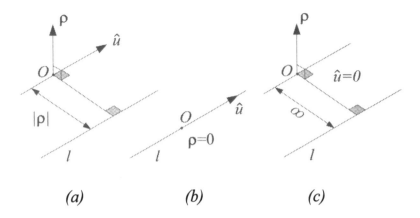

FIGURE 4.19. Three cases of Plücker coordinates; (a) general case, (b) line through origin, (c) line at infinity.

**Example 116** ★ *Classification of Plücker coordinates.*

*There are three cases of Plücker coordinates as shown in Figure 4.19. They are:* **general case, line through origin,** *and* **line at infinity.**

*The general case of* $l = \begin{bmatrix} \hat{u} & \rho \end{bmatrix}^T$, *illustrated in Figure 4.19(a), is when both $\hat{u}$ and $\rho$ are non-zero. The direction vector $\hat{u}$ is parallel to the line, $\rho$ is normal to the plane including the origin and the line, and $|\rho|$ gives the distance from the origin to the line.*

*Line through origin* $l = \begin{bmatrix} \hat{u} & \mathbf{0} \end{bmatrix}^T$, *illustrated in Figure 4.19(b), is when the line passes through the origin and $\rho$ is zero.*

*Line at infinity* $l = \begin{bmatrix} 0 & \rho \end{bmatrix}^T$, *illustrated in Figure 4.19(c), is when the distance of the line from origin tends to infinity. In this case we may assume $\hat{u}$ is zero. When the line is at infinity, it is better to redefine the Plücker coordinates by normalizing the moment vector*

$$l = \begin{bmatrix} \dfrac{\hat{u}}{|\rho|} & \dfrac{\rho}{|\rho|} \end{bmatrix}^T. \tag{4.243}$$

*Therefore, the direction components of the line tends to zero with the distance, while the moment components remain finite.*

*By noting that the line at infinity can be assumed as the intersection of the set of all planes perpendicular to the moment vector, it is evident that the moment vector completely specifies the line.*

*No line is defined by the zero Plücker coordinates* $\begin{bmatrix} 0 & 0 & 0 & 0 & 0 & 0 \end{bmatrix}^T$.

**Example 117** ★ *Transformation of a line vector.*

*Consider the line ${}^B l$ in Figure 4.20 that is defined in a local frame $B(oxyz)$ by*

$$^B l = \begin{bmatrix} {}^B\hat{u} \\ {}^B\rho \end{bmatrix} = \begin{bmatrix} {}^B\hat{u} \\ {}^B\mathbf{r}_P \times {}^B\hat{u} \end{bmatrix} \tag{4.244}$$

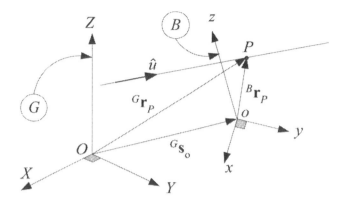

FIGURE 4.20. A line vector in $B$ and $G$ frames.

*where $\hat{u}$ is a unit vector parallel to the line $l$, and $P$ is any point on the line. The Plücker coordinates of the line in the global frame $G(OXYZ)$ is expressed by*

$$^Gl = \begin{bmatrix} {}^G\hat{u} \\ {}^G\rho \end{bmatrix} = \begin{bmatrix} {}^G\hat{u} \\ {}^G\mathbf{r}_P \times {}^G\hat{u} \end{bmatrix} \tag{4.245}$$

*where*

$$^G\hat{u} = {}^GR_B\,{}^B\hat{u} \tag{4.246}$$

*and $^G\rho$ is the moment of $^G\hat{u}$ about $O$*

$$
\begin{aligned}
^G\rho &= {}^G\mathbf{r}_P \times {}^G\hat{u} \\
&= ({}^G\mathbf{s}_o + {}^GR_B\,{}^B\mathbf{r}_P) \times {}^GR_B\,{}^B\hat{u} \\
&= {}^G\mathbf{s}_o \times {}^GR_B\,{}^B\hat{u} + {}^GR_B({}^B\mathbf{r}_P \times {}^B\hat{u}) \\
&= {}^G\mathbf{s}_o \times {}^GR_B\,{}^B\hat{u} + {}^GR_B\,{}^B\rho \\
&= {}^G\tilde{\mathbf{s}}_o\,{}^GR_B\,{}^B\hat{u} + {}^GR_B\,{}^B\rho.
\end{aligned}
\tag{4.247}
$$

*Therefore, the $6 \times 1$ Plücker coordinates $\begin{bmatrix} \hat{u} & \rho \end{bmatrix}^T$ for a line vector can be transformed from a frame $B$ to another frame $G$*

$$^Gl = {}^G\Gamma_B\,{}^Bl \tag{4.248}$$

$$\begin{bmatrix} {}^G\hat{u} \\ {}^G\rho \end{bmatrix} = {}^G\Gamma_B \begin{bmatrix} {}^B\hat{u} \\ {}^B\rho \end{bmatrix} \tag{4.249}$$

*by a $6 \times 6$ transformation matrix defined as*

$$^G\Gamma_B = \begin{bmatrix} {}^GR_B & 0 \\ {}^G\tilde{\mathbf{s}}_o\,{}^GR_B & {}^GR_B \end{bmatrix} \tag{4.250}$$

*where*

$$^GR_B = \begin{bmatrix} r_{11} & r_{12} & r_{13} \\ r_{21} & r_{22} & r_{23} \\ r_{31} & r_{32} & r_{33} \end{bmatrix} \tag{4.251}$$

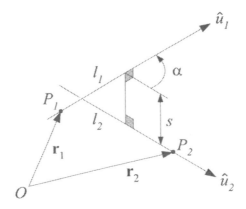

FIGURE 4.21. Illustration of two skew lines.

$$
{}^{G}\tilde{\mathbf{s}}_{o} = \begin{bmatrix} 0 & -s_3 & s_2 \\ s_3 & 0 & -s_1 \\ -s_2 & s_1 & 0 \end{bmatrix} \tag{4.252}
$$

$$
{}^{G}\tilde{\mathbf{s}}_{o}\,{}^{G}R_B = \begin{bmatrix} s_2 r_{31} - s_3 r_{21} & s_2 r_{32} - s_3 r_{22} & s_2 r_{33} - s_3 r_{23} \\ -s_1 r_{31} + s_3 r_{11} & -s_1 r_{32} + s_3 r_{12} & -s_1 r_{33} + s_3 r_{13} \\ s_1 r_{21} - s_2 r_{11} & s_1 r_{22} - s_2 r_{12} & s_1 r_{23} - s_2 r_{13} \end{bmatrix}. \tag{4.253}
$$

# 4.9 ★ The Geometry of Plane and Line

Plücker coordinates introduces a suitable method to define the moment between two lines, shortest distance between two lines, and the angle between two lines.

## 4.9.1 ★ Moment

Consider two arbitrary lines $l_1 = \begin{bmatrix} \hat{u}_1 & \boldsymbol{\rho}_1 \end{bmatrix}^T$ and $l_2 = \begin{bmatrix} \hat{u}_2 & \boldsymbol{\rho}_2 \end{bmatrix}^T$ as shown in Figure 4.21. Points $P_1$ and $P_2$ on $l_1$ and $l_2$ are indicated by vectors $\mathbf{r}_1$ and $\mathbf{r}_2$ respectively. Direction vectors of the lines are $\hat{u}_1$ and $\hat{u}_2$.

The moment of the line $l_2$ about $P_1$ is

$$
(\mathbf{r}_2 - \mathbf{r}_1) \times \hat{u}_2
$$

and we can define the moment of the line $l_2$ about $l_1$ by

$$
l_2 \times l_1 = \hat{u}_1 \cdot (\mathbf{r}_2 - \mathbf{r}_1) \times \hat{u}_2 \tag{4.254}
$$

which, because of $\hat{u}_1 \cdot \mathbf{r}_1 \times \hat{u}_2 = \hat{u}_2 \cdot \hat{u}_1 \times \mathbf{r}_1$, simplifies to

$$
l_2 \times l_1 = \hat{u}_1 \cdot \boldsymbol{\rho}_2 + \hat{u}_2 \cdot \boldsymbol{\rho}_1. \tag{4.255}
$$

The *reciprocal product* or *virtual product* of two lines described by Plücker coordinates is defined as

$$l_2 \times l_1 = \begin{bmatrix} \hat{u}_2 \\ \rho_2 \end{bmatrix} \otimes \begin{bmatrix} \hat{u}_1 \\ \rho_1 \end{bmatrix}$$

$$= \hat{u}_2 \cdot \rho_1 + \hat{u}_1 \cdot \rho_2. \qquad (4.256)$$

The reciprocal product is commutative and gives the moment between two directed lines.

### 4.9.2   ★ Angle and Distance

If $s$ is the shortest distance between two lines $l_1 = \begin{bmatrix} \hat{u}_1 & \rho_1 \end{bmatrix}^T$ and $l_2 = \begin{bmatrix} \hat{u}_2 & \rho_2 \end{bmatrix}^T$, and $\alpha \in [0, \pi]$ is the angle between $l_1$ and $l_2$, then

$$\sin \alpha = |\hat{u}_2 \times \hat{u}_1| \qquad (4.257)$$

and

$$
\begin{aligned}
s &= \frac{1}{\sin \alpha} |\hat{u}_2 \cdot \rho_1 + \hat{u}_1 \cdot \rho_2| \\
&= \frac{1}{\sin \alpha} \left| \begin{bmatrix} \hat{u}_2 \\ \rho_2 \end{bmatrix} \otimes \begin{bmatrix} \hat{u}_1 \\ \rho_1 \end{bmatrix} \right| \\
&= \frac{1}{\sin \alpha} |l_2 \times l_1|. \qquad (4.258)
\end{aligned}
$$

Therefore, two lines $l_1$ and $l_2$ intersect if and only if their reciprocal product is zero. Two parallel lines may be assumed to intersect at infinity. The distance expression does not work for parallel lines.

### 4.9.3   ★ Plane and Line

The equation of a plane $^G\pi$ having a normal unit vector $\hat{n} = n_1 \hat{I} + n_2 \hat{J} + n_3 \hat{K}$ is

$$n_1 X + n_2 Y + n_3 Z = s \qquad (4.259)$$

where $s$ is the minimum distance of the plane to the origin $O$. We may indicate a plane using a homogeneous representation,

$$\pi = \begin{bmatrix} n_1 & n_2 & n_3 & s \end{bmatrix}^T \qquad (4.260)$$

and write the condition $\pi^T \cdot \mathbf{r} = 0$ for a point $\mathbf{r} = \begin{bmatrix} X & Y & Z & w \end{bmatrix}^T$ to be in the plane by

$$\pi^T \cdot \mathbf{r} = \begin{bmatrix} n_1 & n_2 & n_3 & s \end{bmatrix} \begin{bmatrix} X \\ Y \\ Z \\ w \end{bmatrix} = 0. \qquad (4.261)$$

Moreover, the $w = 0$ indicates all points at infinity, and $s = 0$ indicates all planes containing the origin.

The intersection of $\pi$-plane with the $X$-axis, or the $X$-intercept, is $X = -s/n_1$, the $Y$-intercept is $Y = -s/n_2$, and the $Z$-intercept is $Z = -s/n_3$. The plane is perpendicular to $XY$-plane if $n_3 = 0$. It is perpendicular to the $X$-axis if $n_2 = n_3 = 0$. There are similar conditions for the other planes and axes. If $(X_0, Y_0, Z_0)$ is a point in the plane (4.260), then

$$n_1(X - X_0) + n_2(Y - Y_0) + n_3(Z - Z_0) = s. \qquad (4.262)$$

The distance of a point $\begin{bmatrix} X & Y & Z & w \end{bmatrix}^T$ from the origin is

$$d = \sqrt{\frac{X^2 + Y^2 + Z^2}{w^2}} \qquad (4.263)$$

while the distance of a plane $\pi = \begin{bmatrix} m_1 & m_2 & m_3 & s \end{bmatrix}^T$ from the origin is

$$s = \sqrt{\frac{s^2}{m_1^2 + m_2^2 + m_3^2}}. \qquad (4.264)$$

The equation of a line connecting two points $P_1(X_1, Y_1, Z_1)$ and $P_2(X_2, Y_2, Z_2)$ at $\mathbf{r}_1$ and $\mathbf{r}_2$ can also be expressed by

$$l = \mathbf{r}_1 + m(\mathbf{r}_2 - \mathbf{r}_1) \qquad (4.265)$$

and the distance of a point $P(X, Y, Z)$ from any point on $l$ is given by

$$\begin{aligned} s^2 \;=\; & (X_1 + m(X_2 - X_1) - X)^2 \\ & + (Y_1 + m(Y_2 - Y_1) - Y)^2 \\ & + (Z_1 + m(Z_2 - Z_1) - Z)^2 \end{aligned} \qquad (4.266)$$

which is minimum for

$$m = -\frac{(X_2 - X_1)(X_1 - X) + (Y_2 - Y_1)(Y_1 - Y) + (Z_2 - Z_1)(Z_1 - Z)}{(X_2 - X_1)^2 + (Y_2 - Y_1)^2 + (Z_2 - Z_1)^2}. \qquad (4.267)$$

To find the minimum distance of the origin we set $X = Y = Z = 0$.

**Example 118** ★ *Angle and distance between two diagonals of a cube.*

*The Plücker coordinates of the two diagonals of the unit cube shown in Figure 4.18 are*

$$\begin{aligned} l_1 \;&=\; \begin{bmatrix} \hat{u}_1 \\ \rho_1 \end{bmatrix} \\ &=\; \begin{bmatrix} -\frac{\sqrt{2}}{2} & \frac{\sqrt{2}}{2} & 0 & -\frac{\sqrt{2}}{2} & -\frac{\sqrt{2}}{2} & \frac{\sqrt{2}}{2} \end{bmatrix}^T \end{aligned}$$

$$l_2 = \begin{bmatrix} \hat{u}_2 \\ \rho_2 \end{bmatrix}$$

$$= \begin{bmatrix} 0 & \frac{\sqrt{2}}{2} & \frac{\sqrt{2}}{2} & 0 & -\frac{\sqrt{2}}{2} & \frac{\sqrt{2}}{2} \end{bmatrix}^T.$$

*The angle between $l_1$ and $l_2$ is*

$$\alpha = \sin^{-1}|\hat{u}_2 \times \hat{u}_1| = \sin^{-1}\frac{\sqrt{3}}{2} = 60\,\text{deg}$$

*and the distance between them is*

$$s = \frac{1}{\sin\alpha}\left\|\begin{bmatrix} \hat{u}_1 \\ \rho_1 \end{bmatrix} \otimes \begin{bmatrix} \hat{u}_2 \\ \rho_2 \end{bmatrix}\right\|$$

$$= \frac{1}{\sin\alpha}|\hat{u}_1 \cdot \rho_2 + \hat{u}_2 \cdot \rho_1|$$

$$= \frac{2}{\sqrt{3}}|-0.5| = \frac{1}{\sqrt{3}}.$$

**Example 119** *Distance of a point from a line.*

*The equation of the line connecting two points $r_1 = \begin{bmatrix} -1 & 2 & 1 & 1 \end{bmatrix}^T$ and $r_2 = \begin{bmatrix} 1 & -2 & -1 & 1 \end{bmatrix}^T$ is*

$$l = r_1 + m(r_2 - r_1)$$

$$= \begin{bmatrix} -1 + 2m \\ 2 - 4m \\ 1 - 2m \\ 1 \end{bmatrix}.$$

*Now the distance between point $r_3 = \begin{bmatrix} 1 & 1 & 0 & 1 \end{bmatrix}^T$ and $l$ is given by*

$$\begin{aligned} s^2 &= (X_1 + m(X_2 - X_1) - X_3)^2 \\ &\quad + (Y_1 + m(Y_2 - Y_1) - Y_3)^2 \\ &\quad + (Z_1 + m(Z_2 - Z_1) - Z_3)^2 \\ &= 24m^2 - 20m + 6 \end{aligned}$$

*which is minimum for*

$$m = \frac{5}{12}.$$

*Thus, the point on the line at a minimum distance from $r_3$ is*

$$r = \begin{bmatrix} -1/6 \\ 1/3 \\ 1/6 \\ 1 \end{bmatrix}.$$

**Example 120** *Distance between two lines.*

The distance between the line connecting $r_1 = \begin{bmatrix} -1 & 2 & 1 & 1 \end{bmatrix}^T$ and $r_2 = \begin{bmatrix} 1 & -2 & -1 & 1 \end{bmatrix}^T$ is

$$
\begin{aligned}
l &= r_1 + m(r_2 - r_1) \\
&= \begin{bmatrix} -1 + 2m \\ 2 - 4m \\ 1 - 2m \\ 1 \end{bmatrix}
\end{aligned}
$$

and the line connecting $r_3 = \begin{bmatrix} 1 & 1 & 0 & 1 \end{bmatrix}^T$ and $r_4 = \begin{bmatrix} 0 & -1 & 2 & 1 \end{bmatrix}^T$ is

$$
\begin{aligned}
l &= r_1 + m(r_2 - r_1) \\
&= \begin{bmatrix} 1 - n \\ 1 - 2n \\ 2n \\ 1 \end{bmatrix}.
\end{aligned}
$$

The distance between two arbitrary points on the lines is

$$
s^2 = \left(-1 + 2m - 1 + n^2\right) + (2 - 4m - 1 + 2n)^2 + (1 - 2m - 2n)^2.
$$

The minimum distance is found by minimizing $s^2$ with respect to $m$ and $n$.

$$
m = 0.443 \quad , \quad n = 0.321
$$

Thus, the two points on the lines at a minimum distance apart are at

$$
r_m = \begin{bmatrix} -0.114 \\ 0.228 \\ 0.114 \\ 1 \end{bmatrix}
$$

$$
r_n = \begin{bmatrix} 0.679 \\ 0.358 \\ 0.642 \\ 1 \end{bmatrix}.
$$

**Example 121** ★ *Intersection condition for two lines.*

If two lines $l_1 = \begin{bmatrix} \hat{u}_1 & \rho_1 \end{bmatrix}^T$ and $l_2 = \begin{bmatrix} \hat{u}_2 & \rho_2 \end{bmatrix}^T$ intersect, and the position of their common point is at $r$, then

$$
\begin{aligned}
\rho_1 &= r \times \hat{u}_1 \\
\rho_2 &= r \times \hat{u}_2
\end{aligned}
$$

*and therefore*

$$\boldsymbol{\rho}_1 \cdot \hat{u}_2 = (\mathbf{r} \times \hat{u}_1) \cdot \hat{u}_2 = \mathbf{r} \cdot (\hat{u}_1 \times \hat{u}_2)$$
$$\boldsymbol{\rho}_2 \cdot \hat{u}_1 = (\mathbf{r} \times \hat{u}_2) \cdot \hat{u}_1 = \mathbf{r} \cdot (\hat{u}_2 \times \hat{u}_1)$$

*which implies*

$$\hat{u}_1 \cdot \boldsymbol{\rho}_2 + \hat{u}_2 \cdot \boldsymbol{\rho}_1 = 0$$

*or equivalently*

$$\begin{bmatrix} \hat{u}_1 \\ \boldsymbol{\rho}_1 \end{bmatrix} \otimes \begin{bmatrix} \hat{u}_2 \\ \boldsymbol{\rho}_2 \end{bmatrix} = 0.$$

**Example 122 ★** *Plücker coordinates of the axis of rotation.*

*Consider a homogenous transformation matrix corresponding to a rotation $\alpha$ about $Z$, along with a translation in $XY$-plane*

$$^G T_B = \begin{bmatrix} r_{11} & r_{12} & 0 & X_o \\ r_{21} & r_{22} & 0 & Y_0 \\ 0 & 0 & 1 & 0 \\ 0 & 0 & 0 & 1 \end{bmatrix} \tag{4.268}$$

*which must be equal to*

$$^G T_B = \begin{bmatrix} \cos\alpha & -\sin\alpha & 0 & X_o \\ \sin\alpha & \cos\alpha & 0 & Y_o \\ 0 & 0 & 1 & 0 \\ 0 & 0 & 0 & 1 \end{bmatrix}. \tag{4.269}$$

*The angle of rotation can be obtained by comparison*

$$\alpha = \tan^{-1} \frac{r_{21}}{r_{11}}. \tag{4.270}$$

*The pole of rotation can be found by searching for a point that has the same coordinates in both frames*

$$\begin{bmatrix} r_{11} & r_{12} & 0 & X_o \\ r_{21} & r_{22} & 0 & Y_o \\ 0 & 0 & 1 & 0 \\ 0 & 0 & 0 & 1 \end{bmatrix} \begin{bmatrix} X_p \\ Y_p \\ 0 \\ 1 \end{bmatrix} = \begin{bmatrix} X_p \\ Y_p \\ 0 \\ 1 \end{bmatrix} \tag{4.271}$$

*that can be written as*

$$\begin{bmatrix} r_{11} - 1 & r_{12} & 0 & X_o \\ r_{21} & r_{22} - 1 & 0 & Y_o \\ 0 & 0 & 1 & 0 \\ 0 & 0 & 0 & 1 \end{bmatrix} \begin{bmatrix} X_p \\ Y_p \\ 0 \\ 1 \end{bmatrix} = \begin{bmatrix} 0 \\ 0 \\ 0 \\ 1 \end{bmatrix}. \tag{4.272}$$

*The solutions of these equations are*

$$X_p = \frac{1}{2}X_o - \frac{1}{2}\frac{r_{21}}{1 - r_{11}}Y_o$$

$$= \frac{1}{2}X_o - \frac{1}{2}\frac{\sin\alpha}{\text{vers}\,\alpha}Y_o \tag{4.273}$$

$$Y_p = \frac{1}{2}Y_o + \frac{1}{2}\frac{r_{21}}{1 - r_{11}}X_o$$

$$= \frac{1}{2}Y_o + \frac{1}{2}\frac{\sin\alpha}{\text{vers}\,\alpha}X_o. \tag{4.274}$$

*The Plücker line coordinates* $l = \begin{bmatrix} \hat{u} & \rho \end{bmatrix}^T$ *of the pole axis is then equal to*

$$l = \begin{bmatrix} 0 & 0 & 1 & Y_p & -X_p & 0 \end{bmatrix}^T. \tag{4.275}$$

## 4.10   ★ Screw and Plücker Coordinate

Consider a screw $\check{s}(h, \phi, \hat{u}, \mathbf{s})$ whose line of action is given by Plücker coordinates $\begin{bmatrix} \hat{u} & \rho \end{bmatrix}^T$, and whose pitch is $p = \frac{h}{\phi}$. The screw can be defined by a set of Plücker coordinates

$$\check{s}(h, \phi, \hat{u}, \mathbf{s}) = \begin{bmatrix} \hat{u} \\ \boldsymbol{\xi} \end{bmatrix}$$

$$= \begin{bmatrix} \hat{u} \\ \rho + p\hat{u} \end{bmatrix}$$

$$= \begin{bmatrix} \phi\hat{u} \\ \phi\rho + h\hat{u} \end{bmatrix}. \tag{4.276}$$

If the pitch is infinite $p = \infty$, then the screw reduces to a pure translation, or equivalently, a line at infinity

$$\check{s}(h, 0, \hat{u}, \mathbf{r}) = \begin{bmatrix} 0 \\ h\hat{u} \end{bmatrix}. \tag{4.277}$$

A zero pitch screw $p = 0$ corresponds to a pure rotation, then the screw coordinates are identical to the Plücker coordinates of the screw line.

$$\check{s}(0, \phi, \hat{u}, \mathbf{s}) = \begin{bmatrix} \phi\hat{u} \\ \phi\rho \end{bmatrix} = \begin{bmatrix} \hat{u} \\ \rho \end{bmatrix} \tag{4.278}$$

A central screw is defined by a line through origin.

$$
\begin{aligned}
\breve{s}(h, \phi, \hat{u}) &= \breve{s}(h, \phi, \hat{u}, 0) \\
&= \begin{bmatrix} \hat{u} \\ p\hat{u} \end{bmatrix} \\
&= \begin{bmatrix} \phi\hat{u} \\ h\hat{u} \end{bmatrix} \\
&= D(h\hat{u})\, R(\hat{u}, \phi)
\end{aligned} \tag{4.279}
$$

Screw coordinates for differential screw motion is useful in velocity analysis of robots. Consider a screw axis $l$, an angular velocity $\boldsymbol{\omega} = \omega\hat{u} = \dot\phi\hat{u}$ about $l$, and a velocity $\mathbf{v}$ along $l$. If the location vector $\mathbf{s}$ is the position of a point on $l$, then the Plücker coordinates of the line $l$ are

$$
l = \begin{bmatrix} \hat{u} \\ \boldsymbol{\rho} \end{bmatrix} = \begin{bmatrix} \hat{u} \\ \mathbf{s} \times \hat{u} \end{bmatrix}. \tag{4.280}
$$

The pitch of screw is

$$
p = \frac{|\mathbf{v}|}{|\boldsymbol{\omega}|} \tag{4.281}
$$

and the direction of screw is defined by

$$
\hat{u} = \frac{\boldsymbol{\omega}}{|\boldsymbol{\omega}|} \tag{4.282}
$$

so the instantaneous screw coordinates $\breve{v}(p, \omega, \hat{u}, \mathbf{s})$ are

$$
\begin{aligned}
\breve{v}(p, \omega, \hat{u}, \mathbf{r}) &= \begin{bmatrix} \omega\hat{u} & \dfrac{\mathbf{r} \times \boldsymbol{\omega} + |\mathbf{v}|\,\boldsymbol{\omega}}{|\boldsymbol{\omega}|} \end{bmatrix}^T \\
&= \begin{bmatrix} \omega\hat{u} \\ \mathbf{s} \times \hat{u} + \mathbf{v} \end{bmatrix} \\
&= \begin{bmatrix} \boldsymbol{\omega} \\ \mathbf{s} \times \hat{u} + p\boldsymbol{\omega} \end{bmatrix} \\
&= \begin{bmatrix} \boldsymbol{\omega} \\ \boldsymbol{\rho} + p\boldsymbol{\omega} \end{bmatrix}. 
\end{aligned} \tag{4.283}
$$

**Example 123 ★** *Pitch of an instantaneous screw.*
*The pitch of an instantaneous screw, defined by Plücker coordinates, can be found by*

$$
p = \hat{u} \cdot \boldsymbol{\xi} \tag{4.284}
$$

*because two Plücker vectors are orthogonal, $\hat{u} \cdot \boldsymbol{\rho} = 0$, and therefore,*

$$
\begin{aligned}
\phi\hat{u} \cdot \phi\boldsymbol{\xi} &= \phi\hat{u} \cdot (\boldsymbol{\rho} + \phi p\hat{u}) \\
&= (\phi\hat{u} \cdot \boldsymbol{\rho} + \phi^2 p) \\
&= \phi^2 p.
\end{aligned} \tag{4.285}
$$

**Example 124 ★** *Nearest point on a screw axis to the origin.*

*The point on the instantaneous screw axis, nearest to the origin, is indicated by the following position vector:*

$$\mathbf{s}_0 = \phi\hat{u} \times \phi\boldsymbol{\xi} \tag{4.286}$$

## 4.11 Summary

Arbitrary motion of a body $B$ with respect to another body $G$ is called rigid motion and can be expressed by

$$^G\mathbf{r}_P = {}^GR_B\,{}^B\mathbf{r}_P + {}^G\mathbf{d} \tag{4.287}$$

where,

$$^G\mathbf{r}_P = \begin{bmatrix} X_P & Y_P & Z_P \end{bmatrix}^T \tag{4.288}$$

$$^B\mathbf{r}_P = \begin{bmatrix} x_P & y_P & z_P \end{bmatrix}^T \tag{4.289}$$

$$^G\mathbf{d} = \begin{bmatrix} X_o & Y_o & Z_o \end{bmatrix}^T. \tag{4.290}$$

The vector $^G\mathbf{d}$ is translation of $B$ with respect to $G$, and $^GR_B$ is the rotation matrix to map $^B\mathbf{r}$ to $^G\mathbf{r}$ when $^G\mathbf{d} = 0$.

By introducing the homogeneous coordinate representation for a point

$$\mathbf{r} = \begin{bmatrix} wx \\ wy \\ wz \\ w \end{bmatrix} = \begin{bmatrix} x/w \\ y/w \\ z/w \\ 1 \end{bmatrix} \tag{4.291}$$

we may combine the rotation and translation of a rigid motion in a $4 \times 4$ homogeneous transformation matrix and show the coordinate transformation by

$$^G\mathbf{r} = {}^GT_B\,{}^B\mathbf{r}$$

where,

$$^GT_B = \begin{bmatrix} {}^GR_B & {}^G\mathbf{d} \\ 0 & 1 \end{bmatrix}. \tag{4.292}$$

Since the homogeneous transformation matrix $^GT_B$ is not orthogonal, its inverse obeys a specific rule

$$^GT_B^{-1} = {}^BT_G = \begin{bmatrix} {}^GR_B^T & -{}^GR_B^T\,{}^G\mathbf{d} \\ 0 & 1 \end{bmatrix} \tag{4.293}$$

to be consistent with

$$^GT_B^{-1}\,{}^GT_B = I_4. \tag{4.294}$$

The rigid motion can also be expressed with screw coordinate $\check{s}(h, \phi, \hat{u}, \mathbf{s})$ and screw transformation

$$^G\mathbf{r} = {}^G\check{s}_B(h, \phi, \hat{u}, \mathbf{s}) \, {}^B\mathbf{r} \tag{4.295}$$

$$\check{s}(h, \phi, \hat{u}, \mathbf{s}) = \begin{bmatrix} {}^GR_B & {}^G\mathbf{s} - {}^GR_B \, {}^G\mathbf{s} + h\hat{u} \\ 0 & 1 \end{bmatrix}. \tag{4.296}$$

The screw $\check{s}(h, \phi, \hat{u}, \mathbf{s})$ is indicated by screw parameters; a unit vector on the axis of rotation $\hat{u}$, a location vector $\mathbf{s}$, a twist angle $\phi$, and a translation $h$ (or pitch $p$). The location vector $\mathbf{s}$ indicates the global position of a point on the screw axis. When $\mathbf{s} = 0$, then, $\hat{u}$ passes through the origin of the coordinate frame and the screw motion is called central screw $\check{s}(h, \phi, \hat{u})$. Every screw can be decomposed into three principal central screws about the three axes of the coordinate frame $G$.

A rigid motion can be expressed more effectively by screw and Plücker coordinates of directed lines.

$$\check{s}(h, \phi, \hat{u}, \mathbf{s}) = \begin{bmatrix} \hat{u} \\ \xi \end{bmatrix} = \begin{bmatrix} \hat{u} \\ \rho + p\hat{u} \end{bmatrix} = \begin{bmatrix} \phi\hat{u} \\ \phi\rho + h\hat{u} \end{bmatrix} \tag{4.297}$$

$$\rho = \mathbf{r} \times \hat{u} \tag{4.298}$$

# Exercises

1. Notation and symbols.

   Describe the meaning of

   a- $^2\mathbf{d}_1$      b- $^GT_B^{-1}$   c- $^BT_G$   d- $^GD_B$      e- $^G\check{s}_B(h,\phi,\hat{u})$

   f- $D_{\hat{u},h}$      g- $\check{s}^{-1}$      h- $^G\Gamma_B$   i- $\begin{bmatrix} \hat{u} \\ \rho \end{bmatrix}$      j- $^G\check{s}_B(2,\frac{\pi}{3},\hat{K})$

   k- $p=\frac{h}{\phi}$   l- $^G\mathbf{s}_0^T$      m- $\check{s}^{-1}$      n- $\begin{bmatrix} \hat{u} \\ \boldsymbol{\xi} \end{bmatrix}$      o- $\check{s}(h,\phi,\hat{u},\mathbf{s})$.

2. Global position in a rigid body motion.

   We move the body coordinate frame $B$ to

   $$^G\mathbf{d} = \begin{bmatrix} 4 & -3 & 7 \end{bmatrix}^T.$$

   Find $^G\mathbf{r}_P$ if the local position of a point is

   $$^B\mathbf{r}_P = \begin{bmatrix} 7 & 3 & 2 \end{bmatrix}^T$$

   and the orientation of $B$ with respect to the global frame $G$ can be found by a rotation 45 deg about the $X$-axis, and then 45 deg about the $Y$-axis.

3. Rotation matrix compatibility.

   It is not possible to find $^GR_B$ from the equation of rigid body motion

   $$^G\mathbf{r}_P = {}^GR_B\,{}^B\mathbf{r}_P + {}^G\mathbf{d}$$

   if $^G\mathbf{r}_P$, $^B\mathbf{r}_P$, and $^G\mathbf{d}$ are given. Explain why and find the required conditions to be able to find $^GR_B$.

4. Global position with local rotation in a rigid body motion.

   Assume a body coordinate frame $B$ is at

   $$^G\mathbf{d} = \begin{bmatrix} 4 & -3 & 7 \end{bmatrix}^T.$$

   Find $^G\mathbf{r}_P$ if the local position of a point is

   $$^B\mathbf{r}_P = \begin{bmatrix} 7 & 3 & 2 \end{bmatrix}^T$$

   and the orientation of $B$ with respect to the global frame $G$ can be found by a rotation 45 deg about the $x$-axis, and then 45 deg about the $y$-axis.

5. Global and local rotation in a rigid body motion.

    A body coordinate frame $B$ is translated to

    $$^G\mathbf{d} = [\ 4 \ \ -3 \ \ 7\ ]^T.$$

    Find $^G\mathbf{r}_P$ if the local position of a point is

    $$^B\mathbf{r}_P = [\ 7 \ \ 3 \ \ 2\ ]^T$$

    and the orientation of $B$ with respect to the global frame $G$ can be found by a rotation 45 deg about the $X$-axis, then 45 deg about the $y$-axis, and finally 45 deg about the $z$-axis.

6. Combination of rigid motions.

    The frame $B_1$ is rotated 35 deg about the $z_1$-axis and translated to

    $$^{B_2}\mathbf{d} = [\ -40 \ \ 30 \ \ 20\ ]^T$$

    with respect to another frame $B_2$. The orientation of the frame $B_2$ in the global frame $G$ can be found by a rotation of 55 deg about

    $$\mathbf{u} = [\ 2 \ \ -3 \ \ 4\ ].$$

    Calculate $^G\mathbf{d}_1$, and $^GR_1$.

7. Rotation submatrix in a homogeneous transformation matrix.

    Find the missing elements in this homogeneous transformation matrix:

    $$[T] = \begin{bmatrix} ? & 0 & ? & 4 \\ 0.707 & ? & ? & 3 \\ ? & ? & 0 & 1 \\ 0 & 0 & 0 & 1 \end{bmatrix}$$

8. Angle and axis of rotation.

    Find the angle and axis of rotation for $[T]$ and $T^{-1}$.

    $$[T] = \begin{bmatrix} 0.866 & -0.5 & 0 & 4 \\ 0.5 & 0.866 & 0 & 3 \\ 0 & 0 & 1 & 1 \\ 0 & 0 & 0 & 1 \end{bmatrix}$$

9. Combination of homogeneous transformations.

    Assume that the origin of the frames $B_1$, and $B_2$ are at

    $$^2\mathbf{d}_1 = [\ -10 \ \ 20 \ \ -20\ ]^T$$
    $$^G\mathbf{d}_2 = [\ -15 \ \ -20 \ \ 30\ ]^T.$$

The orientation of $B_1$ in $B_2$ can be found by a rotation of $60 \deg$ about

$$^2\mathbf{u} = [\ 1 \quad -2 \quad 4\ ]$$

and the orientation of $B_2$ in the global frame $G$ can be found by a rotation of $30 \deg$ about

$$^G\mathbf{u} = [\ 4 \quad 3 \quad -4\ ].$$

Calculate the transformation matrix $^GT_2$ and $^GT_2^{-1}$

10. Rotation about an axis not going through origin.

    Find the global position of a body point at

    $$^B\mathbf{r}_P = [\ 7 \quad 3 \quad 2\ ]^T$$

    after a rotation of $30 \deg$ about an axis parallel to

    $$^G\mathbf{u} = [\ 4 \quad 3 \quad -4\ ]$$

    and passing through a point at $(3,3,3)$.

11. Inversion of a square matrix.

    Knowing that the inverse of a $2 \times 2$ matrix

    $$[A] = \begin{bmatrix} a & b \\ c & d \end{bmatrix} \tag{4.299}$$

    is

    $$A^{-1} = \begin{bmatrix} -\dfrac{d}{-ad+bc} & \dfrac{b}{-ad+bc} \\ \dfrac{c}{-ad+bc} & -\dfrac{a}{-ad+bc} \end{bmatrix}, \tag{4.300}$$

    use the inverse method of splitting a matrix $[T]$ into

    $$[T] = \begin{bmatrix} A & B \\ C & D \end{bmatrix}$$

    and applying the inverse technique (4.61), verify Equation (4.300), and calculate the inverse of

    $$[A] = \begin{bmatrix} 1 & 2 & 3 & 4 \\ 12 & 13 & 14 & 5 \\ 11 & 16 & 15 & 6 \\ 10 & 9 & 8 & 7 \end{bmatrix}.$$

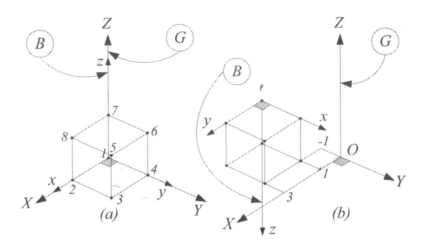

FIGURE 4.22. A cube, before and after a rigid motion.

12. Combination of rotations about non-central axes.

    Consider a rotation 30 deg about an axis at the global point $(3,0,0)$ followed by another rotation 30 deg about an axis at the global point $(0,3,0)$. Find the final global coordinates of

    $$^B\mathbf{r}_P = \begin{bmatrix} 1 & 1 & 0 \end{bmatrix}^T$$

    to

    $$^G\mathbf{r}_P = \begin{bmatrix} \sqrt{2} & 0 & 3 \end{bmatrix}^T.$$

13. Transformation matrix from body points.

    Figure 4.22 (a) shows a cube at initial configurations. Label the corners of the cube, at the final configuration shown in Figure 4.22 (b), and find the associated homogeneous transformation matrix. The length of each side of the cube is 2.

14. ★ Central screw.

    Find the central screw that moves the point

    $$^B\mathbf{r}_P = \begin{bmatrix} 1 & 0 & 0 \end{bmatrix}^T$$

    to

    $$^G\mathbf{r}_P = \begin{bmatrix} 0 & 1 & 4 \end{bmatrix}^T.$$

15. ★ Screw motion.

    Find the global position of

    $$^B\mathbf{r}_P = \begin{bmatrix} 1 & 0 & 0 & 1 \end{bmatrix}^T$$

after a screw motion $^G \check{s}_B(h, \phi, \hat{u}, s) = {}^G \check{s}_B(4, 60 \deg, \hat{u}, s)$ where

$$^G \mathbf{s} = \begin{bmatrix} 3 & 0 & 0 \end{bmatrix}^T$$
$$^G \mathbf{u} = \begin{bmatrix} 1 & 1 & 1 \end{bmatrix}^T.$$

16. ★ Pole of a planar motion.

Find the pole position of a planar motion if we have the coordinates of two body points, before and after the motion, as given below.

$$P_1 (1, 1, 1) \qquad P_2(5, 2, 1)$$
$$Q_1 (4, 1, 1) \qquad Q_2(7, 2 + \sqrt{5}, 1)$$

17. ★ Screw parameters

Find the global coordinates of the body points

$$P_1 (5, 0, 0)$$
$$Q_1 (5, 5, 0)$$
$$R_1 (0, 5, 0)$$

after a rotation of 30 deg about the $x$-axis followed by a rotation of 40 deg about an axis parallel to the vector

$$^G \mathbf{u} = \begin{bmatrix} 4 & 3 & -4 \end{bmatrix}$$

and passing through a global point at $(0, 3, 0)$. Use the coordinates of the points and calculate the screw parameters that are equivalent to the two rotations.

18. ★ Non-central rotation.

Find the global coordinates of a point at

$$^B \mathbf{r}_P = \begin{bmatrix} 10 & 20 & -10 \end{bmatrix}^T$$

when the body frame rotates about

$$^G \mathbf{u} = \begin{bmatrix} 1 & 1 & 1 \end{bmatrix}^T$$

which passes through a point at $(1-, -4, 2)$.

19. ★ Equivalent screw.

Calculate the transformation matrix $^G T_B$ for a rotation of 30 deg about the $x$-axis followed by a translation parallel to

$$^G \mathbf{s} = \begin{bmatrix} 3 & 2 & -1 \end{bmatrix}^T$$

and then a rotation of 40 deg about an axis parallel to the vector

$$^G \mathbf{u} = \begin{bmatrix} 2 & -1 & 1 \end{bmatrix}^T.$$

Find the screw parameters of $^G T_B$.

20. ★ Central screw decomposition.

Find a triple central screws for the case 1 in Appendix C

$$^G\check{s}_B(h, \phi, \hat{u}, \mathbf{s}) = \check{s}(h_X, \gamma, \hat{I})\,\check{s}(h_Y, \beta, \hat{J})\,\check{s}(h_Z, \alpha, \hat{K})$$

to get the same screw as

$$^G\check{s}_B(h, \phi, \hat{u}, \mathbf{s}) = {}^G\check{s}_B(4, 60, \hat{u}, \mathbf{s})$$

where

$$^G\mathbf{s} = \begin{bmatrix} 3 & 0 & 0 \end{bmatrix}^T$$
$$^G\mathbf{u} = \begin{bmatrix} 1 & 1 & 1 \end{bmatrix}^T.$$

21. ★ Central screw composition.

What is the final position of a point at

$$^B\mathbf{r}_P = \begin{bmatrix} 10 & 0 & 0 \end{bmatrix}^T$$

after the central screw $\check{s}(4, 30\,\mathrm{deg}, \hat{J})$ followed by $\check{s}(2, 30\,\mathrm{deg}, \hat{I})$ and $\check{s}(-6, 120\,\mathrm{deg}, \hat{K})$?

22. ★ Screw composition.

Find the final position of a point at

$$^B\mathbf{r}_P = \begin{bmatrix} 10 & 0 & 0 \end{bmatrix}^T$$

after a screw

$$^1\check{s}_2(h_1, \phi_1, \hat{u}_1, \mathbf{s}_1) = {}^1\check{s}_2\left(1, 40\,\mathrm{deg}, \begin{bmatrix} 1/\sqrt{3} \\ -1/\sqrt{3} \\ 1/\sqrt{3} \end{bmatrix}, \begin{bmatrix} 2 \\ 3 \\ -1 \end{bmatrix}\right)$$

followed by

$$^G\check{s}_1(h_0, \phi_0, \hat{u}_0, \mathbf{s}_0) = {}^G\check{s}_1\left(-1, 45\,\mathrm{deg}, \begin{bmatrix} 1/9 \\ 4/9 \\ 4/9 \end{bmatrix}, \begin{bmatrix} -3 \\ 1 \\ 5 \end{bmatrix}\right).$$

23. ★ Plücker line coordinate.

Find the missing numbers

$$l = \begin{bmatrix} 1/3 & 1/5 & ? & ? & 2 & -1 \end{bmatrix}^T.$$

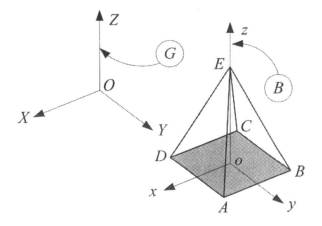

FIGURE 4.23. A pyramid.

24. ★ Plücker lines.

Find the Plücker lines for $AE$, $BE$, $CE$, $DE$ in the local coordinate $B$, and calculate the angle between $AE$, and the $Z$-axis in the pyramid shown in Figure 4.23. The local coordinate of the edges are

$$A(1,1,0)$$
$$B(-1,1,0)$$
$$C(-1,-1,0)$$
$$D(1,-1,0)$$
$$E(0,0,3).$$

Transforms the Plücker lines $AE$, $BE$, $CE$, $DE$ to the global coordinate $G$. The global position of $o$ is at

$$o(1,10,2).$$

25. ★ Angle between two lines.

Find the angle between $OE$, $OD$, of the pyramid shown in Figure 4.23. The coordinates of the points are

$$D(1,-1,0)$$
$$E(0,0,3).$$

26. ★ Distance from the origin.

The equation of a plane is given as

$$4X - 5Y - 12Z - 1 = 0.$$

Determine the perpendicular distance of the plane from the origin.

# 5

# Forward Kinematics

## 5.1 Denavit-Hartenberg Notation

A series robot with $n$ joints will have $n+1$ links. Numbering of links starts from (0) for the immobile grounded *base link* and increases sequentially up to $(n)$ for the *end-effector* link. Numbering of joints starts from 1, for the joint connecting the first movable link to the base link, and increases sequentially up to $n$. Therefore, the link $(i)$ is connected to its *lower link* $(i-1)$ at its *proximal end* by joint $i$ and is connected to its *upper link* $(i+1)$ at its *distal end* by joint $i+1$, as shown in Figure 5.1.

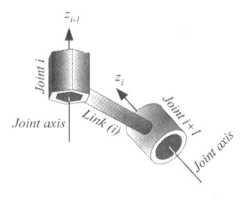

FIGURE 5.1. Link $(i)$ and its beginning joint $i-1$ and its end joint $i$.

Figure 5.2 shows links $(i-1), (i)$, and $(i+1)$ of a serial robot, along with joints $i-1, i$, and $i+1$. Every joint is indicated by its axis, which may be translational or rotational. To relate the kinematic information of robot components, we rigidly attach a local coordinate frame $B_i$ to each link $(i)$ at joint $i+1$ based on the following *standard method*, known as *Denavit-Hartenberg* (DH) method.

1. The $z_i$-axis is aligned with the $i+1$ joint axis.

   All joints, without exception, are represented by a $z$-axis. We always start with identifying the $z_i$-axes. The positive direction of the $z_i$-axis is arbitrary. Identifying the joint axes for revolute joints is obvious, however, a prismatic joint may pick any axis parallel to the direction of translation. By assigning the $z_i$-axes, the pairs of links on either side of each joint, and also two joints on either side of each link are also identified.

2. The $x_i$-axis is defined along the *common normal* between the $z_{i-1}$ and $z_i$ axes, pointing from the $z_{i-1}$ to the $z_i$-axis.

   In general, the $z$-axes may be skew lines, however there is always one line mutually perpendicular to any two skew lines, called the common normal. The common normal has the shortest distance between the two skew lines.

   When the two $z$-axes are parallel, there are an infinite number of common normals. In that case, we pick the common normal that is collinear with the common normal of the previous joints.

   When the two $z$-axes are intersecting, there is no common normal between them. In that case, we assign the $x_i$-axis perpendicular to the plane formed by the two $z$-axes in the direction of $z_{i-1} \times z_i$.

   In case the two $z$-axes are collinear, the only nontrivial arrangement of joints is either P∥R or R∥P. Hence, we assign the $x_i$-axis such that we have the joint variable equal to $\theta_i = 0$ in the rest position of the robot.

3. The $y_i$-axis is determined by the right-hand rule, $y_i = z_i \times x_i$.

   Generally speaking, we assign reference frames to each link so that one of the three coordinate axes $x_i$, $y_i$, or $z_i$ (usually $x_i$) is aligned along the axis of the distal joint.

By applying the DH method, the origin $o_i$ of the frame $B_i(o_i, x_i, y_i, z_i)$, attached to the link $(i)$, is placed at the intersection of the $i + 1$-th joint axis with the common normal between the $z_{i-1}$ and $z_i$ axes.

A DH coordinate frame is identified by four parameters: $a_i$, $\alpha_i$, $\theta_i$, and $d_i$.

1. *Link length* $a_i$ is the distance between $z_{i-1}$ and $z_i$ axes along the $x_i$-axis. $a_i$ is the *kinematic length* of link $(i)$.

2. *Link twist* $\alpha_i$ is the required rotation of the $z_{i-1}$-axis about the $x_i$-axis to become parallel to the $z_i$-axis.

3. *Joint distance* $d_i$ is the distance between $x_{i-1}$ and $x_i$ axes along the $z_{i-1}$-axis. Joint distance is also called *link offset*.

4. *Joint angle* $\theta_i$ is the required rotation of $x_{i-1}$-axis about the $z_{i-1}$-axis to become parallel to the $x_i$-axis.

DH frame parameters of the links in Figure 5.2 are illustrated in Figure 5.3. The parameters $\theta_i$ and $d_i$ are called *joint parameters*, since they define

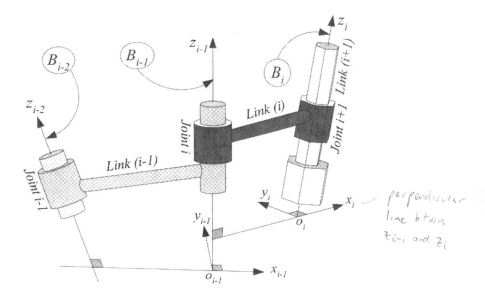

FIGURE 5.2. Links $(i-1)$, $(i)$, and $(i+1)$ along with coordinate frames $B_i$ and $B_{i+1}$.

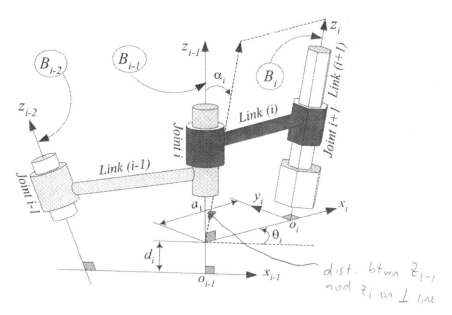

FIGURE 5.3. DH parameters $a_i, \alpha_i, d_i, \theta_i$ defined for joint $i$ and link $(i)$.

the relative position of two adjacent links connected at joint $i$. In a given design for a robot, each joint is revolute or prismatic. Thus, for each joint, it will always be the case that either $\theta_i$ or $d_i$ is fixed and the other is variable. For a revolute joint (R) at joint $i$, the value of $d_i$ is fixed, while $\theta_i$ is the unique joint variable. For a prismatic joint (P), the value of $\theta_i$ is fixed and $d_i$ is the only joint variable. The variable parameter either $\theta_i$ or $d_i$ is called the *joint variable*. The joint parameters $\theta_i$ and $d_i$ define a screw motion because $\theta_i$ is a rotation about the $z_{i-1}$-axis, and $d_i$ is a translation along the $z_{i-1}$-axis.

The parameters $\alpha_i$ and $a_i$ are called *link parameters*, because they define relative positions of joints $i$ and $i+1$ at two ends of link $(i)$. The link twist $\alpha_i$ is the angle of rotation $z_{i-1}$-axis about $x_i$ to become parallel with the $z_i$-axis. The other link parameter, $a_i$, is the translation along the $x_i$-axis to bring the $z_{i-1}$-axis on the $z_i$-axis. The link parameters $\alpha_i$ and $a_i$ define a screw motion because $\alpha_i$ is a rotation about the $x_i$-axis, and $a_i$ is a translation along the $x_i$-axis.

In other words, we can move the $z_{i-1}$-axis to the $z_i$-axis by a central screw $\check{s}(a_i, \alpha_i, \hat{\imath})$, and move the $x_{i-1}$-axis to the $x_i$-axis by a central screw $\check{s}(d_i, \theta_i, \hat{k}_{i-1})$.

**Example 125** *Simplification comments for the DH method.*

*There are some comments to simplify the application of the DH frame method.*

1. *Showing only $z$ and $x$ axes is sufficient to identify a coordinate frame. Drawing is made clearer by not showing $y$ axes.*

2. *Certain parameters of the frames attached to the first and last links need not be defined. If the first and last joints are R, then*

$$a_0 = 0 \quad , \quad a_n = 0 \tag{5.1}$$
$$\alpha_0 = 0 \quad , \quad \alpha_n = 0. \tag{5.2}$$

*In these cases, the zero position for $\theta_1$, and $\theta_n$ can be chosen arbitrarily, and link offsets can be set to zero.*

$$d_1 = 0 \quad , \quad d_n = 0 \tag{5.3}$$

*If the first and last joints are P, then*

$$\theta_1 = 0 \quad , \quad \theta_n = 0 \tag{5.4}$$

*and the zero position for $d_1$, and $d_n$ can be chosen arbitrarily, but generally to make as many parameters as possible to zero.*

*If the final joint $n$ is R, we choose $x_n$ to align with $x_{n-1}$ when $\theta_n = 0$ and the origin of $B_n$ is chosen such that $d_n = 0$. If the final joint $n$ is P, we choose $x_n$ such that $\theta_n = 0$ and the origin of $B_n$ is chosen at the intersection of $x_{n-1}$ and joint axis $n$ that $d_n = 0$.*

3. *Each link, except the base and the last, is a binary link and is connected to two other links.*

4. *The parameters $a_i$ and $\alpha_i$ are determined by the geometric design of the robot and are always constant. The distance $d_i$ is the offset of the frame $B_i$ with respect to $B_{i-1}$ along the $z_{i-1}$-axis. Since $a_i$ is a length, $a_i \geq 0$.*

5. *The angles $\alpha_i$ and $\theta_i$ are directional. Positive direction is determined by the right-hand rule and the direction of $x_i$ and $z_{i-1}$ respectively.*

6. *For industrial robots, the link twist angle, $\alpha_i$, is usually a multiple of $\pi/2$ radians.*

7. *The DH coordinate frames are not unique because the direction of $z_i$-axes are arbitrary.*

8. *The base frame $B_0(x_0, y_0, z_0) = G(X, Y, Z)$ is the global frame for an immobile robot. It is convenient to choose the Z-axis along the axis of joint 1, and set the origin O where the axes of the G frame are colinear or parallel with the axes of the $B_1$ frame at rest position.*

9. *The configuration of a robot at which all the joint variables are zero is called the* **home configuration** *or* **rest position,** *which is the reference for all motions of a robot. The best rest position is where it makes as many axes parallel to each other and coplanar as possible.*

10. *For convenience, we can relax the strict DH definition for the direction of $x_i$, so that it points from $z_i$ to $z_{i-1}$ and still obtains a valid DH parameterization. The direction flip of $x_i$ is to set a more convenient reference frame when most of the joint parameters are zero.*

11. *A DH parameter table helps to establish a systematic link frame. As shown in Table 5.1, a DH table has five columns for frame index and four DH parameters. The rows of the four DH parameters for each frame will be filled by constant parameters and the joint variable. The joint variable can be found by considering what frames and links will move with each varying active joint.*

*Table 5.1 - DH parameter table for establishment of link frames.*

| Frame No. | $a_i$ | $\alpha_i$ | $d_i$ | $\theta_i$ |
|-----------|-------|------------|-------|------------|
| 1 | $a_1$ | $\alpha_1$ | $d_1$ | $\theta_1$ |
| 2 | $a_2$ | $\alpha_2$ | $d_2$ | $\theta_2$ |
| ......... | ........ | ........ | ........ | ........ |
| $j$ | $a_j$ | $\alpha_j$ | $d_j$ | $\theta_j$ |
| ......... | ........ | ........ | ........ | ........ |
| $n$ | $a_n$ | $\alpha_n$ | $d_n$ | $\theta_n$ |

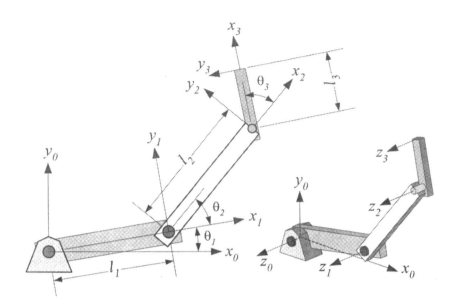

FIGURE 5.4. Illustration of a 3R planar manipulator robot and DH frames of each link.

**Example 126** *DH table and coordinate frames for 3R planar manipulator.*
*R stands for revolute, hence, an $R\|R\|R$ manipulator is a planar robot with three parallel revolute joints. Figure 5.4 illustrates a 3R planar manipulator robot. The DH table can be filled as shown in Table 5.2, and the link coordinate frames can be set up as shown in the Figure.*

*Table 5.2 - DH table for the 3R planar manipulator robot of Figure 5.4.*

| Frame No. | $a_i$ | $\alpha_i$ | $d_i$ | $\theta_i$ |
|-----------|-------|------------|-------|------------|
| 1 | $l_1$ | 0 | 0 | $\theta_1$ |
| 2 | $l_2$ | 0 | 0 | $\theta_2$ |
| 3 | $l_3$ | 0 | 0 | $\theta_3$ |

**Example 127** *Coordinate frames for a 3R PUMA robot.*
*A PUMA manipulator shown in Figure 5.5 has $R\vdash R\|R$ main revolute joints, ignoring the structure of the end-effector of the robot. Coordinate frames attached to the links of the robot are indicated in the Figure and tabulated in Table 5.3.*

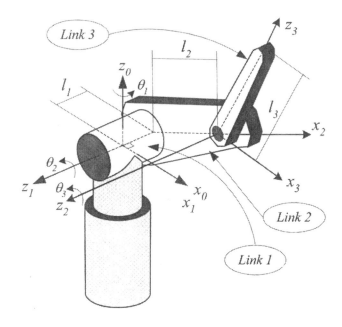

FIGURE 5.5. 3R PUMA manipulator and links coordinate frame.

Table 5.3 - DH table for the 3R PUMA manipulator shown in Figure 5.5.

| Frame No. | $a_i$ | $\alpha_i$ | $d_i$ | $\theta_i$ |
|-----------|-------|------------|-------|------------|
| 1 | 0 | 90 deg | 0 | $\theta_1$ |
| 2 | $l_2$ | 0 | $l_1$ | $\theta_2$ |
| 3 | 0 | 0 | 0 | $\theta_3$ |

The joint axes of an $R \perp R \| R$ manipulator are called waist $z_0$, shoulder $z_1$, and elbow $z_2$. Typically, the joint axes $z_1$ and $z_2$ are parallel, and perpendicular to $z_0$.

**Example 128** *Stanford arm.*

*A schematic illustration of the Stanford arm, which is a spherical robot $R \vdash R \vdash P$ attached to a spherical wrist $R \vdash R \vdash R$, is shown in Figure 5.6 and tabulated in Table 5.3.*

*The DH parameters of the Stanford arm are tabulated in Table 5.4. This robot has 6 DOF: $\theta_1$, $\theta_2$, $d_3$, $\theta_4$, $\theta_5$, $\theta_6$.*

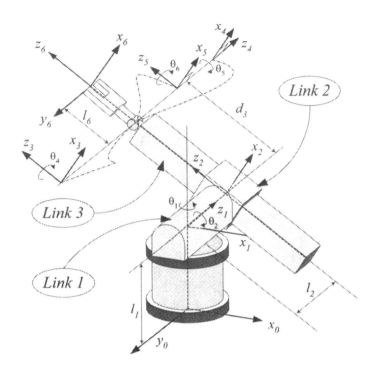

FIGURE 5.6. Stanford arm R⊢R⊢P:R⊢R⊢R.

Table 5.4 - DH table for Stanford arm shown in Figure 5.6.

| Frame No. | $a_i$ | $\alpha_i$ | $d_i$ | $\theta_i$ |
|-----------|-------|------------|-------|------------|
| 1 | 0 | $-90\deg$ | $l_1$ | $\theta_1$ |
| 2 | 0 | $90\deg$ | $l_2$ | $\theta_2$ |
| 3 | 0 | 0 | $d_3$ | 0 |
| 4 | 0 | $-90\deg$ | 0 | $\theta_4$ |
| 5 | 0 | $90\deg$ | 0 | $\theta_5$ |
| 6 | 0 | 0 | $l_6$ | $\theta_6$ |

**Example 129** *Special coordinate frames.*

*In a robotic manipulator, some frames have special names.*

*The* **base frame** $B_0$ *or* $G$ *is the grounded link on which the robot is installed. It is in the base frame that every kinematic information must be calculated because the departure point, path of motion, and the arrival point of the end effector are defined in this frame. The base frame is illustrated in Figure 5.7.*

*The* **station frame** $S$, *also called* **world frame** *or* **universe frame**, *is the frame in which the kinematics of the robot are calculated. Station frame is important when several robots are installed in a workshop, or the robot*

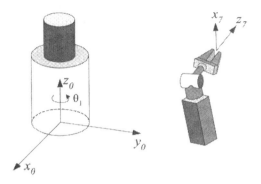

FIGURE 5.7. Base and tool frames.

*is mobile.*

The **wrist frames** $W_i$ *are installed on the wrist point where the hand is attached to the last arm of a robot. The hands of industrial robots are usually attached to a wrist mechanism with three orthogonal revolute axes.*

The **tool frame** $T$ *is attached to the end of any tool the robot is holding. It may also be called the **end-effector frame**, or **final frame**. When the hand is empty, the tool frame is chosen with its origin between the fingertips of the robot. The tool frame is illustrated in Figure 5.7.*

The **goal frame** $F$ *is the location where the robot is to move the tool. The goal frame is specified relative to the station frame.*

**Example 130** *Transforming the $B_{i-1}$ to $B_i$.*

*Two neighbor coordinate frames can be brought into coincidence by several sequences of translations and rotations. The following prescribed set of two rotations and two translations is a straightforward method to move the frame $B_{i-1}$ to coincide with the frame $B_i$.*

1. *Translate frame $B_{i-1}$ along the $z_{i-1}$-axis by distance $d_i$.*

2. *Rotate frame $B_{i-1}$ through $\theta_i$ around the $z_{i-1}$-axis.*

3. *Translate frame $B_{i-1}$ along the $x_i$-axis by distance $a_i$.*

4. *Rotate frame $B_{i-1}$ through $\alpha_i$ about the $x_i$-axis.*

*However, to make a transformation matrix we should start from two coincident coordinate frames $B_i$ and $B_{i-1}$ at the current configuration of $B_{i-1}$, then move $B_i$ to come to its present position. Therefore, we must follow the following sequence of motions:*

1. *Rotate frame $B_i$ through $\alpha_i$ around the $x_{i-1}$-axis.*

2. *Translate frame $B_i$ along the $x_{i-1}$-axis by distance $a_i$.*

3. *Rotate frame $B_i$ through $\theta_i$ about the $z_{i-1}$-axis.*

4. *Translate frame $B_i$ along the $z_{i-1}$-axis by distance $d_i$.*

Note that during these maneuvers, $B_{i-1}$ acts as the global coordinate frame for the local coordinate frame $B_i$, and these motions are about and along the global axes.

**Example 131** *Shortcomings of the Denavit-Hartenberg method.*
The Denavit-Hartenberg method for describing link coordinate frames is neither unique nor the best method. The drawbacks of the DH method are

1. The successive coordinate axes are defined in such a way that the origin $o_i$ and axis $x_i$ are defined on the common perpendicular to adjacent link axes. This may be a difficult task depending on the geometry of the links, and may produce singularity.

2. The DH notation cannot be extended to ternary and compound links.

## 5.2 Transformation Between Two Adjacent Coordinate Frames

The coordinate frame $B_i$ is fixed to the link $(i)$ and the coordinate frame $B_{i-1}$ is fixed to the link $(i-1)$. Based on the Denavit-Hartenberg convention, the transformation matrix $^{i-1}T_i$ to transform coordinate frames $B_i$ to $B_{i-1}$ is represented as a product of four basic transformations using the parameters of link $(i)$ and joint $i$.

$$
^{i-1}T_i \;=\; D_{z_{i-1},d_i}\, R_{z_{i-1},\theta_i}\, D_{x_{i-1},a_i}\, R_{x_{i-1},\alpha_i} \tag{5.5}
$$

$$
= \begin{bmatrix} \cos\theta_i & -\sin\theta_i\,\cos\alpha_i & \sin\theta_i\,\sin\alpha_i & a_i\,\cos\theta_i \\ \sin\theta_i & \cos\theta_i\,\cos\alpha_i & -\cos\theta_i\,\sin\alpha_i & a_i\,\sin\theta_i \\ 0 & \sin\alpha_i & \cos\alpha_i & d_i \\ 0 & 0 & 0 & 1 \end{bmatrix}
$$

where

$$
R_{x_{i-1},\alpha_i} = \begin{bmatrix} 1 & 0 & 0 & 0 \\ 0 & \cos\alpha_i & -\sin\alpha_i & 0 \\ 0 & \sin\alpha_i & \cos\alpha_i & 0 \\ 0 & 0 & 0 & 1 \end{bmatrix} \tag{5.6}
$$

$$
D_{x_{i-1},a_i} = \begin{bmatrix} 1 & 0 & 0 & a_i \\ 0 & 1 & 0 & 0 \\ 0 & 0 & 1 & 0 \\ 0 & 0 & 0 & 1 \end{bmatrix} \tag{5.7}
$$

$$R_{z_{i-1},\theta_i} = \begin{bmatrix} \cos\theta_i & -\sin\theta_i & 0 & 0 \\ \sin\theta_i & \cos\theta_i & 0 & 0 \\ 0 & 0 & 1 & 0 \\ 0 & 0 & 0 & 1 \end{bmatrix}. \tag{5.8}$$

$$D_{z_{i-1},d_i} = \begin{bmatrix} 1 & 0 & 0 & 0 \\ 0 & 1 & 0 & 0 \\ 0 & 0 & 1 & d_i \\ 0 & 0 & 0 & 1 \end{bmatrix}. \tag{5.9}$$

Therefore the transformation equation from coordinate frame $B_i(x_i, y_i, z_i)$, to its previous coordinate frame $B_{i-1}(x_{i-1}, y_{i-1}, z_{i-1})$, is

$$\begin{bmatrix} x_{i-1} \\ y_{i-1} \\ z_{i-1} \\ 1 \end{bmatrix} = {}^{i-1}T_i \begin{bmatrix} x_i \\ y_i \\ z_i \\ 1 \end{bmatrix} \tag{5.10}$$

where,

$$^{i-1}T_i = \begin{bmatrix} \cos\theta_i & -\sin\theta_i\cos\alpha_i & \sin\theta_i\sin\alpha_i & a_i\cos\theta_i \\ \sin\theta_i & \cos\theta_i\cos\alpha_i & -\cos\theta_i\sin\alpha_i & a_i\sin\theta_i \\ 0 & \sin\alpha_i & \cos\alpha_i & d_i \\ 0 & 0 & 0 & 1 \end{bmatrix}. \tag{5.11}$$

This $4 \times 4$ matrix may be partitioned into two submatrices, which represent a unique rotation combined with a unique translation to produce the same rigid motion required to move from $B_i$ to $B_{i-1}$,

$$^{i-1}T_i = \begin{bmatrix} ^{i-1}R_i & ^{i-1}\mathbf{d}_i \\ 0 & 1 \end{bmatrix} \tag{5.12}$$

where

$$^{i-1}R_i = \begin{bmatrix} \cos\theta_i & -\sin\theta_i\cos\alpha_i & \sin\theta_i\sin\alpha_i \\ \sin\theta_i & \cos\theta_i\cos\alpha_i & -\cos\theta_i\sin\alpha_i \\ 0 & \sin\alpha_i & \cos\alpha_i \end{bmatrix} \tag{5.13}$$

and

$$^{i-1}\mathbf{d}_i = \begin{bmatrix} a_i\cos\theta_i \\ a_i\sin\theta_i \\ d_i \end{bmatrix}. \tag{5.14}$$

The inverse of the homogeneous transformation matrix (5.11) is

$$^iT_{i-1} = {}^{i-1}T_i^{-1} \tag{5.15}$$

$$= \begin{bmatrix} \cos\theta_i & \sin\theta_i & 0 & -a_i \\ -\sin\theta_i\cos\alpha_i & \cos\theta_i\cos\alpha_i & \sin\alpha_i & -d_i\sin\alpha_i \\ \sin\theta_i\sin\alpha_i & -\cos\theta_i\sin\alpha_i & \cos\alpha_i & -d_i\cos\alpha_i \\ 0 & 0 & 0 & 1 \end{bmatrix}.$$

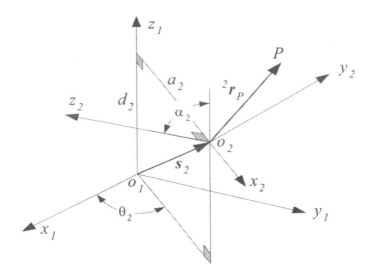

FIGURE 5.8. Two coordinate frames based on Denevit-Hartenberg rules.

**Proof. 1**

Assume that the coordinate frames $B_2(x_2, y_2, z_2)$ and $B_1(x_1, y_1, z_1)$ in Figure 5.8 are set up based on Denavit-Hartenberg rules.

The position vector of point $P$ can be found in frame $B_1(x_1, y_1, z_1)$ using $^2\mathbf{r}_P$ and $^1\mathbf{s}_2$

$$^1\mathbf{r}_P = {}^1R_2 {}^2\mathbf{r}_P + {}^1\mathbf{s}_2 \tag{5.16}$$

which, in homogeneous coordinate representation, is equal to

$$
\begin{bmatrix} x_1 \\ y_1 \\ z_1 \\ 1 \end{bmatrix} =
\begin{bmatrix}
\cos(\hat{\imath}_2, \hat{\imath}_1) & \cos(\hat{\jmath}_2, \hat{\imath}_1) & \cos(\hat{k}_2, \hat{\imath}_1) & s_{2_x} \\
\cos(\hat{\imath}_2, \hat{\jmath}_1) & \cos(\hat{\jmath}_2, \hat{\jmath}_1) & \cos(\hat{k}_2, \hat{\jmath}_1) & s_{2_y} \\
\cos(\hat{\imath}_2, \hat{k}_1) & \cos(\hat{\jmath}_2, \hat{k}_1) & \cos(\hat{k}_2, \hat{k}_1) & s_{2_z} \\
0 & 0 & 0 & 1
\end{bmatrix}
\begin{bmatrix} x_2 \\ y_2 \\ z_2 \\ 1 \end{bmatrix} . \tag{5.17}
$$

Using the parameters introduced in Figure 5.8, Equation (5.17) becomes

$$
\begin{bmatrix} x_1 \\ y_1 \\ z_1 \\ 1 \end{bmatrix} =
\begin{bmatrix}
c\theta_2 & -s\theta_2 c\alpha_2 & s\theta_2 s\alpha_2 & a_2 c\theta_2 \\
s\theta_2 & c\theta_2 c\alpha_2 & -c\theta_2 s\alpha_2 & a_2 s\theta_2 \\
0 & s\alpha_2 & c\alpha_2 & d_2 \\
0 & 0 & 0 & 1
\end{bmatrix}
\begin{bmatrix} x_2 \\ y_2 \\ z_2 \\ 1 \end{bmatrix} . \tag{5.18}
$$

If we substitute the coordinate frame $B_1$ by $B_{i-1}(x_{i-1}, y_{i-1}, z_{i-1})$, and $B_2$ by $B_i(x_i, y_i, z_i)$, then we may rewrite the above equation in the required form

$$
\begin{bmatrix} x_{i-1} \\ y_{i-1} \\ z_{i-1} \\ 1 \end{bmatrix} =
\begin{bmatrix}
c\theta_i & -s\theta_i c\alpha_i & s\theta_i s\alpha_i & a_i c\theta_i \\
s\theta_i & c\theta_i c\alpha_i & -c\theta_i s\alpha_i & a_i s\theta_i \\
0 & s\alpha_i & c\alpha_i & d_i \\
0 & 0 & 0 & 1
\end{bmatrix}
\begin{bmatrix} x_i \\ y_i \\ z_i \\ 1 \end{bmatrix} . \tag{5.19}
$$

Following the inversion rule of homogeneous transformation matrix

$$^{i-1}T_i = \begin{bmatrix} ^GR_B & ^G\mathbf{s} \\ 0 & 1 \end{bmatrix} \qquad (5.20)$$

$$^{i-1}T_i^{-1} = \begin{bmatrix} ^GR_B^T & -^GR_B^T\,^G\mathbf{s} \\ 0 & 1 \end{bmatrix} \qquad (5.21)$$

we also find the required inverse transformation (5.15).

$$\begin{bmatrix} x_i \\ y_i \\ z_i \\ 1 \end{bmatrix} = \begin{bmatrix} c\theta_i & s\theta_i & 0 & -a_i \\ -s\theta_i c\alpha_i & c\theta_i c\alpha_i & s\alpha_i & -d_i s\alpha_i \\ s\theta_i s\alpha_i & -c\theta_i s\alpha_i & c\alpha_i & -d_i c\alpha_i \\ 0 & 0 & 0 & 1 \end{bmatrix} \begin{bmatrix} x_{i-1} \\ y_{i-1} \\ z_{i-1} \\ 1 \end{bmatrix} \qquad (5.22)$$

■

## Proof. 2

An alternative method to find $^iT_{i-1}$ is to follow the sequence of translations and rotations that brings the frame $B_{i-1}$ to the present configuration starting from a coincident position with the frame $B_i$. Note that working with two frames can also be equivalently described by $B \equiv B_i$ and $G \equiv B_{i-1}$, so all the following rotations and translations are about and along the local coordinate frame axes. Inspecting Figure 5.8 shows that:

1. Frame $B_{i-1}$ is translated along the local $z_i$-axis by distance $-d_i$,

2. The displaced frame $B_{i-1}$ is rotated through $-\theta_i$ about the local $z_i$-axis,

3. The displaced frame $B_{i-1}$ is translated along the local $x_i$-axis by distance $-a_i$, and

4. The displaced frame $B_{i-1}$ is rotated through $-\alpha_i$ about the local $x_i$-axis.

Following these displacement sequences, the transformation matrix $^iT_{i-1}$ would be

$$^iT_{i-1} = R_{x_i,-\alpha_i}\, D_{x_i,-a_i}\, R_{z_i,-\theta_i}\, D_{z_i,-d_i} \qquad (5.23)$$

$$= \begin{bmatrix} \cos\theta_i & \sin\theta_i & 0 & -a_i \\ -\sin\theta_i\cos\alpha_i & \cos\theta_i\cos\alpha_i & \sin\alpha_i & -d_i\sin\alpha_i \\ \sin\theta_i\sin\alpha_i & -\cos\theta_i\sin\alpha_i & \cos\alpha_i & -d_i\cos\alpha_i \\ 0 & 0 & 0 & 1 \end{bmatrix}$$

where

$$D_{z_i,-d_i} = \begin{bmatrix} 1 & 0 & 0 & 0 \\ 0 & 1 & 0 & 0 \\ 0 & 0 & 1 & -d_i \\ 0 & 0 & 0 & 1 \end{bmatrix} \qquad (5.24)$$

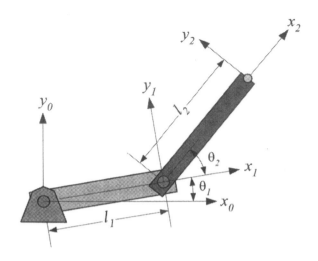

FIGURE 5.9. Illustration of a 2R planar manipulator robot and DH frames of each link.

$$R_{z_i,-\theta_i} = \begin{bmatrix} \cos\theta_i & -\sin\theta_i & 0 & 0 \\ \sin\theta_i & \cos\theta_i & 0 & 0 \\ 0 & 0 & 1 & 0 \\ 0 & 0 & 0 & 1 \end{bmatrix} \tag{5.25}$$

$$D_{x_i,-a_i} = \begin{bmatrix} 1 & 0 & 0 & -a_i \\ 0 & 1 & 0 & 0 \\ 0 & 0 & 1 & 0 \\ 0 & 0 & 0 & 1 \end{bmatrix} \tag{5.26}$$

$$R_{x_i,-\alpha_i} = \begin{bmatrix} 1 & 0 & 0 & 0 \\ 0 & \cos\alpha_i & -\sin\alpha_i & 0 \\ 0 & \sin\alpha_i & \cos\alpha_i & 0 \\ 0 & 0 & 0 & 1 \end{bmatrix}. \tag{5.27}$$

Using $^{i-1}T_i = {}^iT_{i-1}^{-1}$ we find

$$\begin{aligned} {}^{i-1}T_i &= {}^iT_{i-1}^{-1} \tag{5.28} \\ &= \begin{bmatrix} \cos\theta_i & -\sin\theta_i\cos\alpha_i & \sin\theta_i\sin\alpha_i & a_i\cos\theta_i \\ \sin\theta_i & \cos\theta_i\cos\alpha_i & -\cos\theta_i\sin\alpha_i & a_i\sin\theta_i \\ 0 & \sin\alpha_i & \cos\alpha_i & d_i \\ 0 & 0 & 0 & 1 \end{bmatrix}. \end{aligned}$$

∎

**Example 132** *DH transformation matrices for a 2R planar manipulator.*
*Figure 5.9 illustrates an R∥R planar manipulator and its DH link coordinate frames.*

*Table 5.5 - DH table for 2R planar manipulator shown in Figure 5.9.*

| Frame No. | $a_i$ | $\alpha_i$ | $d_i$ | $\theta_i$ |
|-----------|-------|------------|-------|------------|
| 1 | $l_1$ | 0 | 0 | $\theta_1$ |
| 2 | $l_2$ | 0 | 0 | $\theta_2$ |

Based on the DH Table 5.5 we can find the transformation matrices from frame $B_i$ to frame $B_{i-1}$ by direct substitution of DH parameters in Equation 5.11. Therefore,

$$^1T_2 = \begin{bmatrix} \cos\theta_2 & -\sin\theta_2 & 0 & l_2\cos\theta_2 \\ \sin\theta_2 & \cos\theta_2 & 0 & l_2\sin\theta_2 \\ 0 & 0 & 1 & 0 \\ 0 & 0 & 0 & 1 \end{bmatrix} \tag{5.29}$$

$$^0T_1 = \begin{bmatrix} \cos\theta_1 & -\sin\theta_1 & 0 & l_1\cos\theta_1 \\ \sin\theta_1 & \cos\theta_1 & 0 & l_1\sin\theta_1 \\ 0 & 0 & 1 & 0 \\ 0 & 0 & 0 & 1 \end{bmatrix} \tag{5.30}$$

and consequently

$$\begin{aligned} ^0T_2 &= {}^0T_1\,{}^1T_2 \\ &= \begin{bmatrix} c(\theta_1+\theta_2) & -s(\theta_1+\theta_2) & 0 & l_1c\theta_1 + l_2c(\theta_1+\theta_2) \\ s(\theta_1+\theta_2) & c(\theta_1+\theta_2) & 0 & l_1s\theta_1 + l_2s(\theta_1+\theta_2) \\ 0 & 0 & 1 & 0 \\ 0 & 0 & 0 & 1 \end{bmatrix}. \end{aligned} \tag{5.31}$$

**Example 133** *Link with $R\|R$ or $R\|P$ joints.*

When the proximal joint of link $(i)$ is revolute and the distal joint is either revolute or prismatic, and the joint axes at two ends are parallel as shown in Figure 5.10, then $\alpha_i = 0\deg$ (or $\alpha_i = 180\deg$), $a_i$ is the distance between the joint axes, and $\theta_i$ is the only variable parameter. The joint distance $d_i = cte$ is the distance between the origin of $B_i$ and $B_{i-1}$ along $z_i$ however we usually set $x_iy_i$ and $x_{i-1}y_{i-1}$ coplanar to have $d_i = 0$. The $x_i$ and $x_{i-1}$ are parallel for an $R\|R$ link at rest position.

Therefore, the transformation matrix $^{i-1}T_i$ for a link with $\alpha_i = 0$ and $R\|R$ or $R\|P$ joints, known as $R\|R(0)$ or $R\|P(0)$, is

$$^{i-1}T_i = \begin{bmatrix} \cos\theta_i & -\sin\theta_i & 0 & a_i\cos\theta_i \\ \sin\theta_i & \cos\theta_i & 0 & a_i\sin\theta_i \\ 0 & 0 & 1 & d_i \\ 0 & 0 & 0 & 1 \end{bmatrix} \tag{5.32}$$

while for a link with $\alpha_i = 180\deg$ and $R\|R$ or $R\|P$ joints, known as

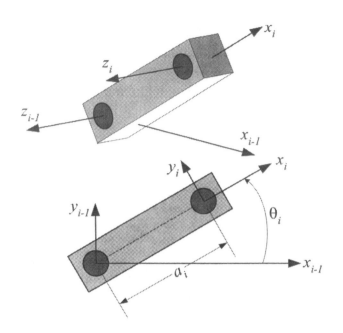

FIGURE 5.10. A R‖R(0) link.

$R\|R(180)$ *or* $R\|P(180)$, *is*

$$
^{i-1}T_i =
\begin{bmatrix}
\cos\theta_i & \sin\theta_i & 0 & a_i\cos\theta_i \\
\sin\theta_i & -\cos\theta_i & 0 & a_i\sin\theta_i \\
0 & 0 & -1 & d_i \\
0 & 0 & 0 & 1
\end{bmatrix}.
\tag{5.33}
$$

**Example 134** *Link with $R\perp R$ or $R\perp P$ joints.*

*When the proximal joint of link $(i)$ is revolute and the distal joint is either revolute or prismatic, and the joint axes at two ends are perpendicular as shown in Figure 5.11, then $\alpha_i = 90\,\text{deg}$ (or $\alpha_i = -90\,\text{deg}$), $a_i$ is the distance between the joint axes on $x_i$, and $\theta_i$ is the only variable parameter. The joint distance $d_i = cte$ is the distance between the origin of $B_i$ and $B_{i-1}$ along $z_i$. However we usually set $x_iy_i$ and $x_{i-1}y_{i-1}$ coplanar to have $d_i = 0$.*

*The $R\perp R$ link is made by twisting the $R\|R$ link about its center line $x_{i-1}$-axis by 90 deg. The $x_i$ and $x_{i-1}$ are parallel for an $R\perp R$ link at rest position.*

*Therefore, the transformation matrix $^{i-1}T_i$ for a link with $\alpha_i = 90\,\text{deg}$ and $R\perp R$ or $R\perp P$ joints, known as $R\perp R(90)$ or $R\perp P(90)$, is*

$$
^{i-1}T_i =
\begin{bmatrix}
\cos\theta_i & 0 & \sin\theta_i & a_i\cos\theta_i \\
\sin\theta_i & 0 & -\cos\theta_i & a_i\sin\theta_i \\
0 & 1 & 0 & d_i \\
0 & 0 & 0 & 1
\end{bmatrix}
\tag{5.34}
$$

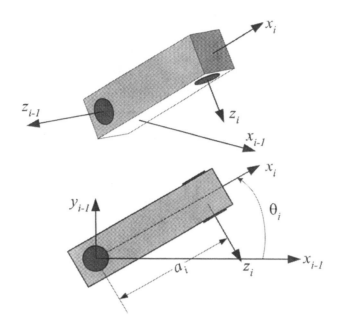

FIGURE 5.11. A $R{\perp}R(90)$ link.

*while for a link with $\alpha_i = -90\,\mathrm{deg}$ and $R{\perp}R$ or $R{\perp}P$ joints, known as $R{\perp}R(-90)$ or $R{\perp}P(-90)$, is*

$$
{}^{i-1}T_i = \begin{bmatrix} \cos\theta_i & 0 & -\sin\theta_i & a_i\cos\theta_i \\ \sin\theta_i & 0 & \cos\theta_i & a_i\sin\theta_i \\ 0 & -1 & 0 & d_i \\ 0 & 0 & 0 & 1 \end{bmatrix}. \tag{5.35}
$$

**Example 135** *Link with $R{\vdash}R$ or $R{\vdash}P$ joints.*

*When the proximal joint of link $(i)$ is revolute and the distal joint is either revolute or prismatic, and the joint axes at two ends are intersecting orthogonal, as shown in Figure 5.12, then $\alpha_i = 90\,\mathrm{deg}$ (or $\alpha_i = -90\,\mathrm{deg}$), $a_i = 0$, $d_i = cte$ is the distance between the coordinates origin on $z_i$, and $\theta_i$ is the only variable parameter. Note that it is possible to have or assume $d_i = 0$. The $x_i$ and $x_{i-1}$ of an $R{\vdash}R$ link at rest position are coincident (when $d_i = 0$) or parallel (when $d_i \neq 0$).*

*Therefore, the transformation matrix ${}^{i-1}T_i$ for a link with $\alpha_i = 90\,\mathrm{deg}$ and $R{\vdash}R$ or $R{\vdash}P$ joints, known as $R{\vdash}R(90)$ or $R{\vdash}P(90)$, is*

$$
{}^{i-1}T_i = \begin{bmatrix} \cos\theta_i & 0 & \sin\theta_i & 0 \\ \sin\theta_i & 0 & -\cos\theta_i & 0 \\ 0 & 1 & 0 & d_i \\ 0 & 0 & 0 & 1 \end{bmatrix} \tag{5.36}
$$

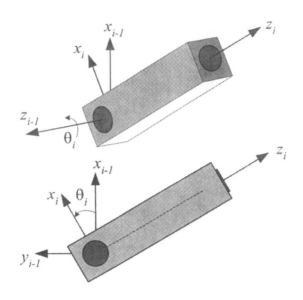

FIGURE 5.12. An R⊢R(90) link.

*while for a link with $\alpha_i = -90\deg$ and R⊢R or R⊢P joints, known as R⊢R(−90) or R⊢P(−90), is*

$$
{}^{i-1}T_i = \begin{bmatrix} \cos\theta_i & 0 & -\sin\theta_i & 0 \\ \sin\theta_i & 0 & \cos\theta_i & 0 \\ 0 & -1 & 0 & d_i \\ 0 & 0 & 0 & 1 \end{bmatrix}.
\tag{5.37}
$$

**Example 136** *Link with $P\|R$ or $P\|P$ joints.*

*When the proximal joint of link $(i)$ is prismatic and its distal joint is either revolute or prismatic, and the joint axes at two ends are parallel as shown in Figure 5.13, then $\alpha_i = 0\deg$ (or $\alpha_i = 180\deg$), $\theta_i = 0$, $a_i$ is the distance between the joint axes on $x_i$, and $d_i$ is the only variable parameter. Note that it is possible to have $a_i = 0$.*

*Therefore, the transformation matrix ${}^{i-1}T_i$ for a link with $\alpha_i = 0$ and $P\|R$ or $P\|P$ joints, known as $P\|R(0)$ or $P\|P(0)$, is*

$$
{}^{i-1}T_i = \begin{bmatrix} 1 & 0 & 0 & a_i \\ 0 & 1 & 0 & 0 \\ 0 & 0 & 1 & d_i \\ 0 & 0 & 0 & 1 \end{bmatrix}
\tag{5.38}
$$

*while for a link with $\alpha_i = 180\deg$ and $P\|R$ or $P\|P$ joints, known as*

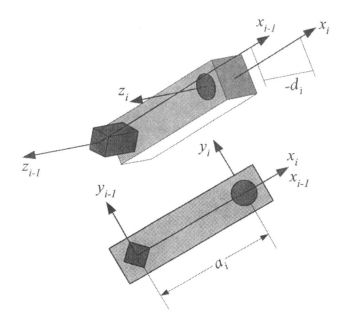

FIGURE 5.13. A $P\|R(0)$ link.

$P\|R(180)$ *or* $P\|P(180)$, *is*

$$^{i-1}T_i = \begin{bmatrix} 1 & 0 & 0 & a_i \\ 0 & -1 & 0 & 0 \\ 0 & 0 & -1 & d_i \\ 0 & 0 & 0 & 1 \end{bmatrix}. \tag{5.39}$$

*The origin of the $B_{i-1}$ frame can be chosen at any point on the $z_{i-1}$-axis or parallel to $z_{i-1}$-axis arbitrarily. One simple setup is to locate the origin $o_i$ of a prismatic joint at the previous origin $o_{i-1}$. This sets $a_i = 0$ and furthermore, sets the initial value of the joint variable $d_i = 0$, which will vary when $o_i$ slides up and down parallel to the $z_{i-1}$-axis.*

**Example 137** *Link with $P\perp R$ or $P\perp P$ joints.*

*When the proximal joint of link $(i)$ is prismatic and its distal joint is either revolute or prismatic, with orthogonal axes as shown in Figure 5.14, then $\alpha_i = 90$ deg (or $\alpha_i = -90$ deg), $\theta_i = 0$, $a_i$ is the distance between the joint axes on $x_i$, and $d_i$ is the only variable parameter.*

*Therefore, the transformation matrix $^{i-1}T_i$ for a link with $\alpha_i = 90$ deg and $P\perp R$ or $P\perp P$ joints, known as $P\perp R(90)$ or $P\perp P(90)$, is*

$$^{i-1}T_i = \begin{bmatrix} 1 & 0 & 0 & a_i \\ 0 & 0 & -1 & 0 \\ 0 & 1 & 0 & d_i \\ 0 & 0 & 0 & 1 \end{bmatrix} \tag{5.40}$$

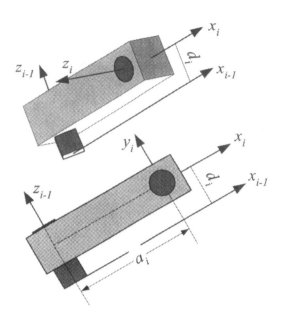

FIGURE 5.14. A $P \perp R(90)$ link.

*while for a link with $\alpha_i = -90$ deg and $P \perp R$ or $P \perp P$ joints, known as $P \perp R(-90)$ or $P \perp P(-90)$, is*

$$
{}^{i-1}T_i = \begin{bmatrix} 1 & 0 & 0 & a_i \\ 0 & 0 & 1 & 0 \\ 0 & -1 & 0 & d_i \\ 0 & 0 & 0 & 1 \end{bmatrix}. \tag{5.41}
$$

**Example 138** *Link with $P\vdash R$ or $P\vdash P$ joints.*

When the proximal joint of link $(i)$ is prismatic and the distal joint is either revolute or prismatic, and the joint axes at two ends are intersecting orthogonal as shown in Figure 5.15, then $\alpha_i = 90$ deg (or $\alpha_i = -90$ deg), $\theta_i = 0$, $a_i = 0$, and $d_i$ is the only variable parameter. Note that $x_i$ must be perpendicular to the plane of $z_{i-1}$ and $z_i$, and it is possible to have $a_i \neq 0$.

Therefore, the transformation matrix ${}^{i-1}T_i$ for a link with $\alpha_i = 90$ deg and $P\vdash R$ or $P\vdash P$ joints, known as $P\vdash R(90)$ or $P\vdash P(90)$, is

$$
{}^{i-1}T_i = \begin{bmatrix} 1 & 0 & 0 & 0 \\ 0 & 0 & -1 & 0 \\ 0 & 1 & 0 & d_i \\ 0 & 0 & 0 & 1 \end{bmatrix} \tag{5.42}
$$

*while for a link with $\alpha_i = -90$ deg and $P\vdash R$ or $P\vdash P$ joints, known as*

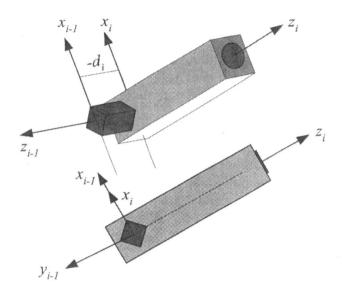

FIGURE 5.15. A P⊢R(90) link.

P⊢R(−90) or P⊢P(−90), is

$$
{}^{i-1}T_i = \begin{bmatrix} 1 & 0 & 0 & 0 \\ 0 & 0 & 1 & 0 \\ 0 & -1 & 0 & d_i \\ 0 & 0 & 0 & 1 \end{bmatrix}. \tag{5.43}
$$

**Example 139** *Classification of industrial robot links.*

*A robot link is identified by its joints at both ends, which determines the transformation matrix to go from the distal joint coordinate frame $B_i$ to the proximal joint coordinate frame $B_{i-1}$. There are 12 types of links to make an industrial robot. The transformation matrix for each type depends solely on the proximal joint, and angle between the z-axes. The 12 types of*

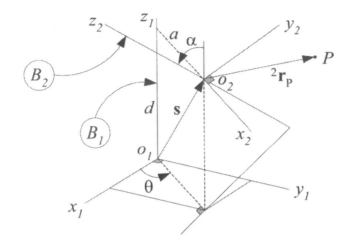

FIGURE 5.16. Alternative method to derive the Denavit-Hartenberg coordinate transformation.

*transformation matrices are:*

| | | | |
|---|---|---|---|
| 1 | $R\|R(0)$ | *or* | $R\|P(0)$ |
| 2 | $R\|R(180)$ | *or* | $R\|P(180)$ |
| 3 | $R\perp R(90)$ | *or* | $R\perp P(90)$ |
| 4 | $R\perp R(-90)$ | *or* | $R\perp P(-90)$ |
| 5 | $R\vdash R(90)$ | *or* | $R\vdash P(90)$ |
| 6 | $R\vdash R(-90)$ | *or* | $R\vdash P(-90)$ |
| 7 | $P\|R(0)$ | *or* | $P\|P(0)$ |
| 8 | $P\|R(180)$ | *or* | $P\|P(180)$ |
| 9 | $P\perp R(90)$ | *or* | $P\perp P(90)$ |
| 10 | $P\perp R(-90)$ | *or* | $P\perp P(-90)$ |
| 11 | $P\vdash R(90)$ | *or* | $P\vdash P(90)$ |
| 12 | $P\vdash R(-90)$ | *or* | $P\vdash P(-90)$ |

**Example 140** *DH coordinate transformation based on vector addition.*

*The DH transformation from a coordinate frame to the other can also be described by a vector addition. The coordinates of a point P in frame $B_1$, as shown in Figure 5.16, are given by the vector equation*

$$\overrightarrow{O_1P} = \overrightarrow{O_2P} + \overrightarrow{O_1O_2} \tag{5.44}$$

*where*

$$^{B_1}\overrightarrow{O_1O_2} = \begin{bmatrix} s_1 & s_2 & s_3 \end{bmatrix}^T \tag{5.45}$$

$$^{B_1}\overrightarrow{O_1P} = \begin{bmatrix} x_1 & y_1 & z_1 \end{bmatrix}^T \tag{5.46}$$

$$^{B_2}\overrightarrow{O_2P} = \begin{bmatrix} x_2 & y_2 & z_2 \end{bmatrix}^T. \tag{5.47}$$

*However, they must be expressed in the same coordinate frame, using cosines of the angles between axes of the two coordinate frames.*

$$
\begin{aligned}
x_1 &= x_2 \cos(x_2, x_1) + y_2 \cos(y_2, x_1) + z_2 \cos(z_2, x_1) + s_1 \\
y_1 &= x_2 \cos(x_2, y_1) + y_2 \cos(y_2, y_1) + z_2 \cos(z_2, y_1) + s_2 \\
z_1 &= x_2 \cos(x_2, z_1) + y_2 \cos(y_2, z_1) + z_2 \cos(z_2, z_1) + s_3 \\
1 &= x_2(0) + y_2(0) + z_2(0) + 1
\end{aligned}
\tag{5.48}
$$

*The transformation (5.48) can be rearranged to be described with the homogeneous matrix transformation*

$$
\begin{bmatrix} x_1 \\ y_1 \\ z_1 \\ 1 \end{bmatrix} =
\begin{bmatrix}
\cos(x_2, x_1) & \cos(y_2, x_1) & \cos(z_2, x_1) & s_1 \\
\cos(x_2, y_1) & \cos(y_2, y_1) & \cos(z_2, y_1) & s_2 \\
\cos(x_2, z_1) & \cos(y_2, z_1) & \cos(z_2, z_1) & s_3 \\
0 & 0 & 0 & 1
\end{bmatrix}
\begin{bmatrix} x_2 \\ y_2 \\ z_2 \\ 1 \end{bmatrix}.
$$

*In Figure 5.16 the axis $x_2$ has been selected such that it lies along the shortest common perpendicular between axes $z_1$ and $z_2$. The axis $y_2$ completes a right-handed set of coordinate axes. Other parameters are defined as follows:*

1. *a is the distance between axes $z_1$ and $z_2$.*

2. *$\alpha$ is the twist angle that screws the $z_1$-axis into the $z_2$-axis along a.*

3. *d is the distance from the $x_1$-axis to the $x_2$-axis.*

4. *$\theta$ is the angle that screws the $x_1$-axis into the $x_2$-axis along d.*

*Using these definitions, the homogeneous transformation matrix becomes,*

$$
\begin{bmatrix} x_1 \\ y_1 \\ z_1 \\ 1 \end{bmatrix} =
\begin{bmatrix}
\cos\theta & -\sin\theta\cos\alpha & -\sin\theta\sin\alpha & a\cos\theta \\
\sin\theta & \cos\theta\cos\alpha & \cos\theta\sin\alpha & a\sin\theta \\
0 & -\sin\alpha & \cos\alpha & d \\
0 & 0 & 0 & 1
\end{bmatrix}
\begin{bmatrix} x_2 \\ y_2 \\ z_2 \\ 1 \end{bmatrix}
$$

*or*

$$
{}^1\mathbf{r}_P = {}^1T_2 \; {}^2\mathbf{r}_P
\tag{5.49}
$$

*where*

$$
{}^1T_2 = (a, \alpha, d, \theta).
\tag{5.50}
$$

*The parameters $a, \alpha, \theta, d$ define the configuration of $B_2$ with respect to $B_1$ and belong to $B_2$. Hence, in general, the parameters $a_i, \alpha_i, \theta_i, d_i$ define the configuration of $B_i$ with respect to $B_{i-1}$ and belong to $B_i$.*

$$
{}^{i-1}T_i = (a_i, \alpha_i, d_i, \theta_i).
\tag{5.51}
$$

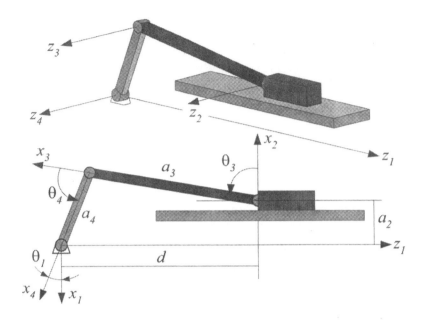

FIGURE 5.17. A planar slider-crank linkage, making a closed loop or parallel mechanism.

**Example 141** *The same DH transformation matrix.*

*Because in DH method of setting coordinate frames, a translation $D$ and a rotation $R$ are about and along one axis, it is immaterial if we apply the translation $D$ first and then the rotation $R$ or vice versa. Therefore, we can change the order of $D$ and $R$ about and along the same axis and obtain the same DH transformation matrix 5.11. Therefore,*

$$
\begin{aligned}
^{i-1}T_i &= D_{z_i,d_i}\, R_{z_i,\theta_i}\, D_{x_i,a_i}\, R_{x_i,\alpha_i} \\
&= R_{z_i,\theta_i}\, D_{z_i,d_i}\, D_{x_i,a_i}\, R_{x_i,\alpha_i} \\
&= D_{z_i,d_i}\, R_{z_i,\theta_i}\, R_{x_i,\alpha_i}\, D_{x_i,a_i} \\
&= R_{z_i,\theta_i}\, D_{z_i,d_i}\, R_{x_i,\alpha_i}\, D_{x_i,a_i}.
\end{aligned}
\tag{5.52}
$$

**Example 142** ★ *DH application for a slider-crank planar linkage.*

*For a closed loop robot or mechanism there would also be a connection between the first and last links. So, the DH convention will not be satisfied by this connection. Figure 5.17 depicts a planar slider-crank linkage $R \perp P \vdash R \| R \| R$ and DH coordinate frames installed on each link.*

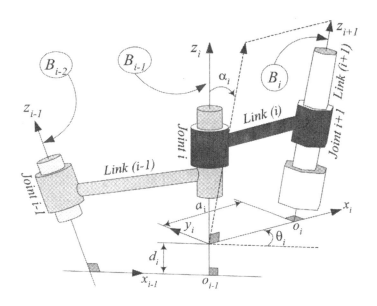

FIGURE 5.18. Non-standard definition of DH parameters $a_i, \alpha_i, d_i, \theta_i$ defined for joint $i$ and link $(i)$.

Table 5.6 - DH table for slider-crank planar linkage shown in Figure 5.17.

| Frame No. | $a_i$ | $\alpha_i$ | $d_i$ | $\theta_i$ |
|-----------|-------|------------|-------|------------|
| 1 | $a_2$ | $-90\deg$ | $d$ | $180\deg$ |
| 2 | $a_3$ | 0 | 0 | $\theta_3$ |
| 3 | $a_4$ | 0 | 0 | $\theta_4$ |
| 4 | 0 | $-90\deg$ | 0 | $\theta_1$ |

Applying a loop transformation leads to

$$[T] = {}^1T_2 \ {}^2T_3 \ {}^3T_4 \ {}^4T_1 = \mathbf{I}_4 \qquad (5.53)$$

where the transformation matrix $[T]$ contains elements that are functions of $a_2, d, a_3, \theta_3, a_4, \theta_4,$ and $\theta_1$. The parameters $a_2, a_3,$ and $a_4$ are constant while $d, \theta_3, \theta_4,$ and $\theta_1$ are variable. Assuming $\theta_1$ is input and specified, we may solve for other unknown variables $\theta_3, \theta_4,$ and $d$ by equating corresponding elements of $[T]$ and $\mathbf{I}$.

**Example 143** ★ *Non-standard Denavit-Hartenberg notation.*
The Denavit-Hartenberg notation presented in this section is the standard DH method. However, we may adopt a different set of DH parameters, simply by setting the link coordinate frame $B_i$ at proximal joint $i$ instead of the distal joint $i + 1$ as shown in Figure 5.18. The $z_i$-axis is along the

axis of joint $i$ and the $x_i$-axis is along the common normal of the $z_i$ and $z_{i+1}$ axes, directed from $z_i$ to $z_{i+1}$ axes. The $y_i$-axis makes the $B_i$ frame a right-handed coordinate frame.

The parameterization of this shift of coordinate frames are:

1. $a_i$ is the distance between the $z_i$ and $z_{i+1}$ axes along the $x_i$-axis.

2. $\alpha_i$ is the angle from $z_i$ to $z_{i+1}$ axes about the $x_i$-axis.

3. $d_i$ is the distance between the $x_{i-1}$ and $x_i$ axes along the $z_i$-axis.

4. $\theta_i$ is the angle from the $x_{i-1}$ and $x_i$ axes about the $z_i$-axis.

The transformation matrix from $B_{i-1}$ to $B_i$ utilizing the non-standard DH method, is made of two rotations and two translations about and along the local coordinate axes of $B_{i-1}$. 1-Rotate $\alpha_{i-1}$ about $x_{i-1}$. 2-Translate $a_{i-1}$ along $x_{i-1}$. 3-Translate $d_i$ along $z_{i-1}$. 4-Rotate $\theta_i$ about $z_{i-1}$.

$$
{}^iT_{i-1} = R_{z_{i-1},\theta_i}\, D_{z_{i-1},-d_i}\, D_{x_{i-1},a_{i-1}}\, R_{x_{i-1},-\alpha_{i-1}} \tag{5.54}
$$

$$
= \begin{bmatrix}
\cos\theta_i & \sin\theta_i\cos\alpha_{i-1} & \sin\theta_i\sin\alpha_{i-1} & -a_{i-1}\cos\theta_i \\
-\sin\theta_i & \cos\theta_i\cos\alpha_{i-1} & \cos\theta_i\sin\alpha_{i-1} & a_{i-1}\sin\theta_i \\
0 & -\sin\alpha_{i-1} & \cos\alpha_{i-1} & -d_i \\
0 & 0 & 0 & 1
\end{bmatrix}
$$

Therefore, the transformation matrix from the $B_i$ to $B_{i-1}$ for the non-standard DH method is

$$
{}^{i-1}T_i = \begin{bmatrix}
\cos\theta_i & -\sin\theta_i & 0 & a_{i-1} \\
\sin\theta_i\cos\alpha_{i-1} & \cos\theta_i\cos\alpha_{i-1} & -\sin\alpha_{i-1} & -d_i\sin\alpha_{i-1} \\
\sin\theta_i\sin\alpha_{i-1} & \cos\theta_i\sin\alpha_{i-1} & \cos\alpha_{i-1} & d_i\cos\alpha_{i-1} \\
0 & 0 & 0 & 1
\end{bmatrix}.
$$
$$\tag{5.55}$$

An advantage of the non-standard DH method is that the rotation $\theta_i$ is around the $z_i$-axis and the joint number is the same as the coordinate number. Actuation force, which is exerted at joint $i$, is also at the same place as the coordinate frame $B_i$. Addressing the link's geometrical characteristics, such as center of gravity, are more natural in this system.

A disadvantage of the non-standard DH method is that the transformation matrix is a mix of $i-1$ and $i$ indices.

Applying the standard or non-standard notation is a personal preference, since both can be applied effectively.

**Example 144** ★ *Hayati-Roberts notation and singularity of DH notation.*
In DH notation, the common normal is not well defined when the two joint axes are parallel. In this condition, the DH notation has a **singularity**, because a small change in the spatial positions of the parallel joint axes can cause a large change in the DH coordinate representation of their relative position.

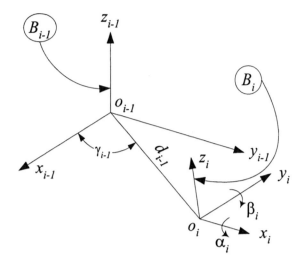

FIGURE 5.19. Hayati-Roberts (HR) notation to avoid the singularity in the DH method.

*The Hayati-Roberts (HR) notation is another convention to represent subsequent links. HR avoids the coordinate singularity in the DH method for the case of parallel lines. In the HR method, the direction of the $z_i$-axis is defined in the $B_{i-1}$ frame using roll and pitch angles $\alpha_i$ and $\beta_i$ as shown in Figure 5.19. The origin of the $B_i$ frame is chosen to lie in the $x_{i-1}y_{i-1}$-plane where the distance $d_i$ is measured between $o_{i-1}$ and $o_i$.*

*Similar to DH convention, there is no unique HR convention concerning the freedom in choosing the angle of rotations. Furthermore, although the HR method can eliminate the parallel joint axes' singularity, it has its own singularities when the $z_i$-axis is parallel to either the $x_{i-1}$ or $y_{i-1}$ axes, or when $z_i$ intersects the origin of the $B_{i-1}$ frame.*

**Example 145 ★** *Parametrically Continuous Convention.*

*There exists another alternative method for coordinate transformation called parametrically continuous convention (PC). The PC method represents the $z_i$-axis by two steps:*

1. *Direction of the $z_i$-axis is defined by two direction cosines, $\alpha_i$ and $\beta_i$, with respect to the axes $x_{i-1}$ and $y_{i-1}$.*

2. *Position of the $z_i$-axis is defined by two distance parameters, $l_i$ and $m_i$ to indicate the $x_{i-1}$ and $y_{i-1}$ coordinates of the intersection of $z_i$ in the $x_{i-1}y_{i-1}$-plane from the origin and perpendicular to the $z_i$-axis.*

*The PC homogeneous matrix to transform $B_{i-1}$ coordinates to $B_i$ is*

$$
{}^{i-1}T_i =
\begin{bmatrix}
1 - \frac{\alpha_i^2}{1+\gamma_i} & -\frac{\alpha_i\beta_i}{1+\gamma_i} & \alpha_i & l_i \\
-\frac{\alpha_i\beta_i}{1+\gamma_i} & 1 - \frac{\beta_i^2}{1+\gamma_i} & \beta_i & m_i \\
-\alpha_i & -\beta_i & \gamma_i & 0 \\
0 & 0 & 0 & 1
\end{bmatrix}
\tag{5.56}
$$

*where*

$$
\gamma_i = \sqrt{1 - \alpha_i^2 - \beta_i^2}. \tag{5.57}
$$

*The PC notation uses four parameters to indicate the position and orientation of the $B_{i-1}$ frame in the $B_i$ frame. Since we need six parameters in general, there must be two conditions:*

1. *The origin of the $B_i$ frame must lie on the common normal and be along the joint axis.*

2. *The $x_i$-axis must lie along the common normal.*

## 5.3    Forward Position Kinematics of Robots

The *forward* or *direct kinematics* is the transformation of kinematic information from the robot joint variable space to the Cartesian coordinate space of the final frame. So, the problem of finding the end-effector position and orientation for a given set of joint variables is the forward kinematics problem. This problem can be solved by determining transformation matrices ${}^0T_i$ to describe the kinematic information of link $(i)$ in the base link coordinate frame. The traditional way of producing forward kinematic equations for robotic manipulators is to proceed link by link using the Denavit-Hartenberg notations and frames. Hence, the forward kinematics is basically transformation matrix manipulation.

For a six DOF robot, six DH transformation matrices, one for each link, are required to transform the final coordinates to the base coordinates. The last frame attached to the final frame is usually set at the center of the gripper as shown in Figure 5.20. For a given set of joint variables, the transformation matrices ${}^{i-1}T_i$ are uniquely determined. Therefore, the position and orientation of the end-effector is also a unique function of the joint variables.

The kinematic information includes: position, velocity, acceleration, and jerk. However, forward kinematics generally refers to the position analysis. So the forward position kinematics is equivalent to a determination of a combined transformation matrix

$$
{}^0T_n = {}^0T_1(q_1)\, {}^1T_2(q_2)\, {}^2T_3(q_3)\, {}^3T_4(q_4) \cdots {}^{n-1}T_n(q_n) \tag{5.58}
$$

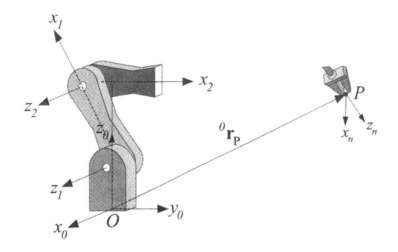

FIGURE 5.20. The position of the final frame in the base frame.

to find the coordinates of a point P in the base coordinate frame, when its coordinates are given in the final frame.

$$^0\mathbf{r}_P = {}^0T_n\, {}^n\mathbf{r}_P \tag{5.59}$$

**Example 146** *3R planar manipulator forward kinematics.*

*Figure 5.21 illustrates an $R\|R\|R$ planar manipulator. Utilizing the DH parameters indicated in Example 126 and applying Equation (5.11), we can find the transformation matrices $^{i-1}T_i$ for $i = 3, 2, 1$. It is also possible to use the transformation matrix (5.32) of Example 133, since links (1) and (2) are both $R\|R(0)$.*

$$^2T_3 = \begin{bmatrix} \cos\theta_3 & -\sin\theta_3 & 0 & l_3\cos\theta_3 \\ \sin\theta_3 & \cos\theta_3 & 0 & l_3\sin\theta_3 \\ 0 & 0 & 1 & 0 \\ 0 & 0 & 0 & 1 \end{bmatrix} \tag{5.60}$$

$$^1T_2 = \begin{bmatrix} \cos\theta_2 & -\sin\theta_2 & 0 & l_2\cos\theta_2 \\ \sin\theta_2 & \cos\theta_2 & 0 & l_2\sin\theta_2 \\ 0 & 0 & 1 & 0 \\ 0 & 0 & 0 & 1 \end{bmatrix} \tag{5.61}$$

$$^0T_1 = \begin{bmatrix} \cos\theta_1 & -\sin\theta_1 & 0 & l_1\cos\theta_1 \\ \sin\theta_1 & \cos\theta_1 & 0 & l_1\sin\theta_1 \\ 0 & 0 & 1 & 0 \\ 0 & 0 & 0 & 1 \end{bmatrix} \tag{5.62}$$

*Therefore, the transformation matrix to relate the end-effector frame to the*

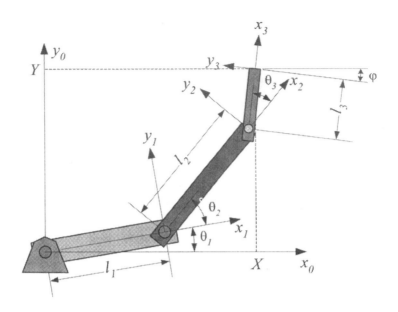

FIGURE 5.21. A R∥R∥R planar manipulator.

*base frame is*

$$
\begin{aligned}
{}^0T_3 \;=\;& {}^0T_1 \, {}^1T_2 \, {}^2T_3 \\
=\;& \begin{bmatrix}
\cos(\theta_1 + \theta_2 + \theta_3) & -\sin(\theta_1 + \theta_2 + \theta_3) & 0 & r_{14} \\
\sin(\theta_1 + \theta_2 + \theta_3) & \cos(\theta_1 + \theta_2 + \theta_3) & 0 & r_{24} \\
0 & 0 & 1 & 0 \\
0 & 0 & 0 & 1
\end{bmatrix} \quad (5.63)
\end{aligned}
$$

$$
\begin{aligned}
r_{14} &= l_1 \cos\theta_1 + l_2 \cos(\theta_1 + \theta_2) + l_3 \cos(\theta_1 + \theta_2 + \theta_3) \\
r_{24} &= l_1 \sin\theta_1 + l_2 \sin(\theta_1 + \theta_2) + l_3 \sin(\theta_1 + \theta_2 + \theta_3).
\end{aligned}
$$

*The position of the origin of the frame $B_3$, which is the tip point of the robot, is at*

$$
{}^0T_3 \begin{bmatrix} 0 \\ 0 \\ 0 \\ 1 \end{bmatrix} =
\begin{bmatrix}
l_1 c\theta_1 + l_2 c(\theta_1 + \theta_2) + l_3 c(\theta_1 + \theta_2 + \theta_3) \\
l_1 s\theta_1 + l_2 s(\theta_1 + \theta_2) + l_3 s(\theta_1 + \theta_2 + \theta_3) \\
0 \\
1
\end{bmatrix}. \quad (5.64)
$$

*It means we can find the coordinate of the tip point in the base Cartesian coordinate frame if we have the geometry of the robot and all joint variables.*

$$
\begin{aligned}
X &= l_1 \cos\theta_1 + l_2 \cos(\theta_1 + \theta_2) + l_3 \cos(\theta_1 + \theta_2 + \theta_3) \quad (5.65) \\
Y &= l_1 \sin\theta_1 + l_2 \sin(\theta_1 + \theta_2) + l_3 \sin(\theta_1 + \theta_2 + \theta_3). \quad (5.66)
\end{aligned}
$$

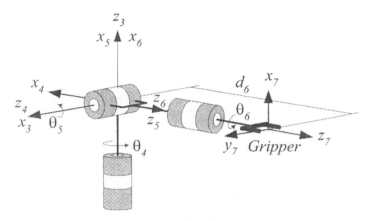

*Attached to forearm*

FIGURE 5.22. Spherical wrist kinematics.

The rest position of the manipulator is lying on the $x_0$-axis where $\theta_1 = 0$, $\theta_2 = 0$, $\theta_3 = 0$ because, $^0T_3$ becomes

$$^0T_3 = \begin{bmatrix} 1 & 0 & 0 & l_1 + l_2 + l_3 \\ 0 & 1 & 0 & 0 \\ 0 & 0 & 1 & 0 \\ 0 & 0 & 0 & 1 \end{bmatrix}. \tag{5.67}$$

**Example 147** *Spherical wrist forward kinematics.*

*A spherical wrist is made by three R⊢R links with zero lengths and zero offset where their joint axes are mutually orthogonal and intersecting at a point called the* **wrist point***. Spherical wrist is a common wrist configuration in robotic applications. Figure 5.22 illustrates a schematic of a spherical wrist configuration. It is made of an R⊢R(−90) link, attached to another R⊢R(90) link, that finally is attached to a spinning gripper R∥R(0). The gripper coordinate frame $B_7$ is always parallel to $B_6$ and is attached at a distance $d_6$ from the wrist point.*

*The wrist will be attached to the final link of a manipulator, which is usually link (3). The coordinate of the final link is usually set up at the wrist point or at the previous joint and acts as the ground link for the wrist mechanism. So, the first frame in a wrist mechanism is labeled 3, because we will be adding the wrist at the end of a manipulator. Then the $z_3$-axis will point along the forearm of the manipulator that is the first wrist joint axis. We chose $x_5 = z_4 \times z_5$ and $\alpha = −90$ deg because when $\theta_5 = 0$ we wish for the hand to point straight up from the forearm.*

*Utilizing the transformation matrix (5.37) for link (4), (5.36) for link (5), $R_{Z,\theta_6}$ for link (6), and a $D_{Z,d_6}$ for frame $B_7$, we will find the following* *individual transformation matrices:*

$$
{}^{3}T_4 = \begin{bmatrix} \cos\theta_4 & 0 & -\sin\theta_4 & 0 \\ \sin\theta_4 & 0 & \cos\theta_4 & 0 \\ 0 & -1 & 0 & 0 \\ 0 & 0 & 0 & 1 \end{bmatrix}
\tag{5.68}
$$

$$
{}^{4}T_5 = \begin{bmatrix} \cos\theta_5 & 0 & \sin\theta_5 & 0 \\ \sin\theta_5 & 0 & -\cos\theta_5 & 0 \\ 0 & 1 & 0 & 0 \\ 0 & 0 & 0 & 1 \end{bmatrix}
\tag{5.69}
$$

$$
{}^{5}T_6 = \begin{bmatrix} \cos\theta_6 & -\sin\theta_6 & 0 & 0 \\ \sin\theta_6 & \cos\theta_6 & 0 & 0 \\ 0 & 0 & 1 & 0 \\ 0 & 0 & 0 & 1 \end{bmatrix}
\tag{5.70}
$$

$$
{}^{6}T_7 = \begin{bmatrix} 1 & 0 & 0 & 0 \\ 0 & 1 & 0 & 0 \\ 0 & 0 & 1 & d_6 \\ 0 & 0 & 0 & 1 \end{bmatrix}.
\tag{5.71}
$$

*The matrix ${}^{3}T_6 = {}^{3}T_4\,{}^{4}T_5\,{}^{5}T_6$ provides the wrist's orientation in the fore-beam coordinate frame $B_3$*

$$
\begin{aligned}
{}^{3}T_6 &= {}^{3}T_4\,{}^{4}T_5\,{}^{5}T_6 \\[4pt]
&= \begin{bmatrix} c\theta_4 c\theta_5 c\theta_6 - s\theta_4 s\theta_6 & -c\theta_6 s\theta_4 - c\theta_4 c\theta_5 s\theta_6 & c\theta_4 s\theta_5 & 0 \\ c\theta_4 s\theta_6 + c\theta_5 c\theta_6 s\theta_4 & c\theta_4 c\theta_6 - c\theta_5 s\theta_4 s\theta_6 & s\theta_4 s\theta_5 & 0 \\ -c\theta_6 s\theta_5 & s\theta_5 s\theta_6 & c\theta_5 & 0 \\ 0 & 0 & 0 & 1 \end{bmatrix}
\end{aligned}
\tag{5.72}
$$

*and the following transformation matrix provides the configuration of the tool frame $B_7$ in the forebeam coordinate frame $B_3$.*

$$
\begin{aligned}
{}^{3}T_7 &= {}^{3}T_4\,{}^{4}T_5\,{}^{5}T_6\,{}^{6}T_7 \\[4pt]
&= \begin{bmatrix} c\theta_4 c\theta_5 c\theta_6 - s\theta_4 s\theta_6 & -c\theta_6 s\theta_4 - c\theta_4 c\theta_5 s\theta_6 & c\theta_4 s\theta_5 & d_6 c\theta_4 s\theta_5 \\ c\theta_4 s\theta_6 + c\theta_5 c\theta_6 s\theta_4 & c\theta_4 c\theta_6 - c\theta_5 s\theta_4 s\theta_6 & s\theta_4 s\theta_5 & d_6 s\theta_4 s\theta_5 \\ -c\theta_6 s\theta_5 & s\theta_5 s\theta_6 & c\theta_5 & d_6 c\theta_5 \\ 0 & 0 & 0 & 1 \end{bmatrix}
\end{aligned}
\tag{5.73}
$$

*Sometimes it is easier to define a compact ${}^{5}T_6$ to include rotation $\theta_6$ and translation $d_6$ by*

$$
{}^{5}T_6 = \begin{bmatrix} \cos\theta_6 & -\sin\theta_6 & 0 & 0 \\ \sin\theta_6 & \cos\theta_6 & 0 & 0 \\ 0 & 0 & 1 & d_6 \\ 0 & 0 & 0 & 1 \end{bmatrix}.
\tag{5.74}
$$

*Employing a compact $^5T_6$ reduces the number of matrices, and therefore the number of numerical calculations. This is the reason we sometimes call $B_6$ the end-effector frame. Such an approach looks simpler, however, note that the position of the wrist point is invariant in a robot structure, but the position of the end-effector at $d_6$ depends on application and the tool being used.*

*The transformation matrix at rest position, where $\theta_4 = 0, \theta_5 = 0, \theta_6 = 0$, is*

$$^3T_7 = \begin{bmatrix} 1 & 0 & 0 & 0 \\ 0 & 1 & 0 & 0 \\ 0 & 0 & 1 & d_6 \\ 0 & 0 & 0 & 1 \end{bmatrix}. \tag{5.75}$$

*Figures 5.23 and 5.24 depict a better illustration of a robot hand's links and joints. The common origin of frames $B_4$, $B_5$, and $B_6$ is at the wrist point. The final frame, which is called the tool or end-effector frame, is denoted by three vectors, $\mathbf{a}$, $\mathbf{s}$, $\mathbf{n}$, and is set at a symmetric point between the fingers of an empty hand or at the tip of the tools hold by the hand. The vector $\mathbf{n}$ is called **tilt** and is the normal vector perpendicular to the fingers or jaws. The vector $\mathbf{s}$ is called **twist** and is the slide vector showing the direction of fingers opening. The vector $\mathbf{a}$ is called **turn** and is the approach vector perpendicular to the palm of the hand.*

*The placement of internal links' coordinate frames is predetermined by the DH method, however, for the end link the placement of the tool's frame $B_n$ is somehow arbitrary and not clear. This arbitrariness may be resolved through simplifying choices or by placement at a distinguished location in the gripper. It is easier to work with the coordinate system $B_n$ if $z_n$ is made coincident with $z_{n-1}$. This choice sets $a_i = 0$ and $\alpha_i = 0$.*

**Example 148** $R\vdash R\|R$ *articulated arm forward kinematics.*

*Consider an $R\vdash R\|R$ arm as shown schematically in Figure 5.25. To develop the forward kinematics of the robot, the DH parameter table of the robot at rest position is set up as indicated in Table 5.7. The rest position of the robot could be set at any configuration where the joint axes $z_1$, $z_2$, $z_3$ are coplanar, however two positions are mostly being used. The first position is where $x_1$ is coplanar with $z_1$, $z_2$, $z_3$, and the second position is where $x_0$ is coplanar with $z_1$, $z_2$, $z_3$.*

*Table 5.7 - DH parameter table for setting up the link frames.*

| Frame No. | $a_i$ | $\alpha_i$ | $d_i$ | $\theta_i$ |
|-----------|-------|------------|-------|------------|
| 1 | 0 | $-90\deg$ | $d_1$ | $\theta_1$ |
| 2 | $l_2$ | 0 | $d_2$ | $\theta_2$ |
| 3 | 0 | $90\deg$ | $l_3$ | $\theta_3$ |

*We recommend applying the link-joints classification in Examples 133 to 138. Therefore, we must be able to determine the type of link-joints*

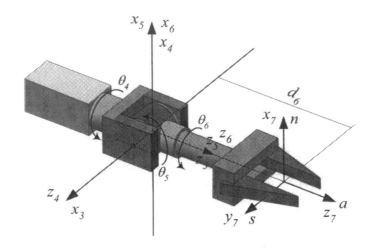

FIGURE 5.23. Hand of a robot at rest position.

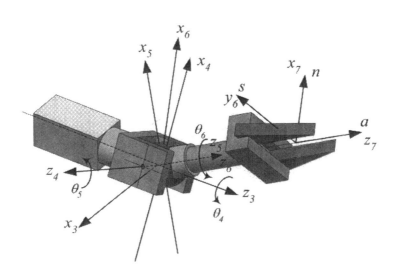

FIGURE 5.24. Hand of a robot in motion.

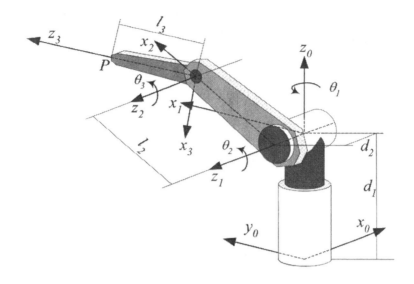

FIGURE 5.25. R⊢R∥R articulated arm.

*combination as shown in Table 5.8.*

*Table 5.8 - Link classification for set-up of the link frames.*

| Link No. | Type |
|----------|------|
| 1 | R⊢R(−90) |
| 2 | R∥R(0) |
| 3 | R⊢R(90) |

*Therefore, the successive transformation matrices have the following expressions:*

$$
{}^{0}T_1 = \begin{bmatrix} \cos\theta_1 & 0 & -\sin\theta_1 & 0 \\ \sin\theta_1 & 0 & \cos\theta_1 & 0 \\ 0 & -1 & 0 & d_1 \\ 0 & 0 & 0 & 1 \end{bmatrix}
\tag{5.76}
$$

$$
{}^{1}T_2 = \begin{bmatrix} \cos\theta_2 & -\sin\theta_2 & 0 & l_2\cos\theta_2 \\ \sin\theta_2 & \cos\theta_2 & 0 & l_2\sin\theta_2 \\ 0 & 0 & 1 & d_2 \\ 0 & 0 & 0 & 1 \end{bmatrix}
\tag{5.77}
$$

$$
{}^{2}T_3 = \begin{bmatrix} \cos\theta_3 & 0 & \sin\theta_3 & 0 \\ \sin\theta_3 & 0 & -\cos\theta_3 & 0 \\ 0 & 1 & 0 & 0 \\ 0 & 0 & 0 & 1 \end{bmatrix}.
\tag{5.78}
$$

*To express the complete transformation*

$$
{}^{0}T_3 = {}^{0}T_1\,{}^{1}T_2\,{}^{2}T_3
\tag{5.79}
$$

*we only need to find the result of a matrix multiplication. Therefore,*

$$^0T_3 = {}^0T_1 \, {}^1T_2 \, {}^2T_3$$

$$= \begin{bmatrix} r_{11} & r_{12} & r_{13} & r_{14} \\ r_{21} & r_{22} & r_{23} & r_{24} \\ r_{31} & r_{32} & r_{33} & r_{34} \\ 0 & 0 & 0 & 1 \end{bmatrix} \tag{5.80}$$

*where*

$$r_{11} = \cos\theta_1 \cos(\theta_2 + \theta_3) \tag{5.81}$$
$$r_{21} = \sin\theta_1 \cos(\theta_2 + \theta_3) \tag{5.82}$$
$$r_{31} = -\sin(\theta_2 + \theta_3) \tag{5.83}$$

$$r_{12} = -\sin\theta_1 \tag{5.84}$$
$$r_{22} = \cos\theta_1 \tag{5.85}$$
$$r_{32} = 0 \tag{5.86}$$

$$r_{13} = \cos\theta_1 \sin(\theta_2 + \theta_3) \tag{5.87}$$
$$r_{23} = \sin\theta_1 \sin(\theta_2 + \theta_3) \tag{5.88}$$
$$r_{33} = \cos(\theta_2 + \theta_3) \tag{5.89}$$

$$r_{14} = l_2 \cos\theta_1 \cos\theta_2 - d_2 \sin\theta_1 \tag{5.90}$$
$$r_{24} = l_2 \cos\theta_2 \sin\theta_1 + d_2 \cos\theta_1 \tag{5.91}$$
$$r_{34} = d_1 - l_2 \sin\theta_2. \tag{5.92}$$

The tip point $P$ of the third arm is at $\begin{bmatrix} 0 & 0 & l_3 \end{bmatrix}^T$ in $B_3$. So, its position in the base frame would be at

$$^0\mathbf{r}_P = {}^0T_3 \, {}^3\mathbf{r}_P = {}^0T_3 \begin{bmatrix} 0 \\ 0 \\ l_3 \\ 1 \end{bmatrix}$$

$$= \begin{bmatrix} -d_2 s\theta_1 + l_2 c\theta_1 c\theta_2 + l_3 c\theta_1 s (\theta_2 + \theta_3) \\ d_2 c\theta_1 + l_2 c\theta_2 s\theta_1 + l_3 s\theta_1 s (\theta_2 + \theta_3) \\ d_1 - l_2 s\theta_2 + l_3 c (\theta_2 + \theta_3) \\ 1 \end{bmatrix}. \tag{5.93}$$

The transformation matrix at rest position, where $\theta_1 = 0$, $\theta_2 = 0$, $\theta_3 = 0$, is

$$^0T_3 = \begin{bmatrix} 1 & 0 & 0 & l_2 \\ 0 & 1 & 0 & d_2 \\ 0 & 0 & 1 & d_1 \\ 0 & 0 & 0 & 1 \end{bmatrix}. \tag{5.94}$$

Therefore, at rest position, $x_0$, $x_1$ are parallel, and $z_1$, $z_2$, $z_3$ are coplanar.

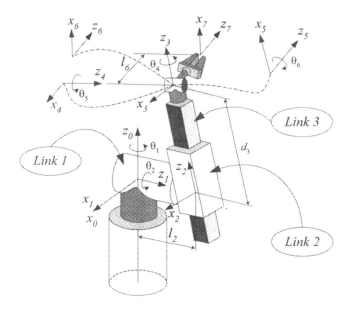

FIGURE 5.26. A spherical robot made by a spherical manipulator equipped with a spherical wrist.

**Example 149** *Spherical robot forward kinematics.*

*Figure 5.26 illustrates a spherical manipulator attached with a spherical wrist to make an R⊢R⊢P robot. The associated DH table is shown in Table 5.9.*

Table 5.9 - DH parameter table for Stanford arm.

| Frame No. | $a_i$ | $\alpha_i$ | $d_i$ | $\theta_i$ |
|-----------|-------|-----------|-------|-----------|
| 1 | 0 | −90 deg | 0 | $\theta_1$ |
| 2 | 0 | 90 deg | $l_2$ | $\theta_2$ |
| 3 | 0 | 0 | $d_3$ | 0 |
| 4 | 0 | −90 deg | 0 | $\theta_4$ |
| 5 | 0 | 90 deg | 0 | $\theta_5$ |
| 6 | 0 | 0 | 0 | $\theta_6$ |

*However, it is recommended to apply the link-joints classification in Examples 133 to 138. Therefore, we must be able to determine the type of link-joint combinations as shown in Table 5.10.*

Table 5.10 - DH parameter table for setting up the link frames.

| Link No. | Type |
|:---:|:---:|
| 1 | R⊢R(−90) |
| 2 | R⊢P(90) |
| 3 | P‖R(0) |
| 4 | R⊢R(−90) |
| 5 | R⊢R(90) |
| 6 | R‖R(0) |

Employing the associated transformation matrices for moving from $B_i$ to $B_{i-1}$ shows that

$$
{}^0T_1 = \begin{bmatrix} \cos\theta_1 & 0 & -\sin\theta_1 & 0 \\ \sin\theta_1 & 0 & \cos\theta_1 & 0 \\ 0 & -1 & 0 & 0 \\ 0 & 0 & 0 & 1 \end{bmatrix} \tag{5.95}
$$

$$
{}^1T_2 = \begin{bmatrix} \cos\theta_2 & 0 & \sin\theta_2 & 0 \\ \sin\theta_2 & 0 & -\cos\theta_2 & 0 \\ 0 & 1 & 0 & l_2 \\ 0 & 0 & 0 & 1 \end{bmatrix} \tag{5.96}
$$

$$
{}^2T_3 = \begin{bmatrix} 1 & 0 & 0 & 0 \\ 0 & 1 & 0 & 0 \\ 0 & 0 & 1 & d_3 \\ 0 & 0 & 0 & 1 \end{bmatrix} \tag{5.97}
$$

$$
{}^3T_4 = \begin{bmatrix} \cos\theta_4 & 0 & -\sin\theta_4 & 0 \\ \sin\theta_4 & 0 & \cos\theta_4 & 0 \\ 0 & -1 & 0 & 0 \\ 0 & 0 & 0 & 1 \end{bmatrix} \tag{5.98}
$$

$$
{}^4T_5 = \begin{bmatrix} \cos\theta_5 & 0 & \sin\theta_5 & 0 \\ \sin\theta_5 & 0 & -\cos\theta_5 & 0 \\ 0 & 1 & 0 & 0 \\ 0 & 0 & 0 & 1 \end{bmatrix} \tag{5.99}
$$

$$
{}^5T_6 = \begin{bmatrix} \cos\theta_6 & -\sin\theta_6 & 0 & 0 \\ \sin\theta_6 & \cos\theta_6 & 0 & 0 \\ 0 & 0 & 1 & 0 \\ 0 & 0 & 0 & 1 \end{bmatrix}. \tag{5.100}
$$

Therefore, the wrist point's position in the global coordinate frame is at

$$
\begin{aligned}
{}^0T_6 &= {}^0T_1\,{}^1T_2\,{}^2T_3\,{}^3T_4\,{}^4T_5\,{}^5T_6 \\
&= \begin{bmatrix} r_{11} & r_{12} & r_{13} & r_{14} \\ r_{21} & r_{22} & r_{23} & r_{24} \\ r_{31} & r_{32} & r_{33} & r_{34} \\ 0 & 0 & 0 & 1 \end{bmatrix}
\end{aligned} \tag{5.101}
$$

*where,*

$$
\begin{aligned}
r_{11} &= s\theta_6 \left(-c\theta_4 s\theta_1 - c\theta_1 c\theta_2 s\theta_4\right) \\
&\quad + c\theta_6 \left(-c\theta_1 s\theta_2 s\theta_5 + c\theta_5 \left(-s\theta_1 s\theta_4 + c\theta_1 c\theta_2 c\theta_4\right)\right) \quad (5.102) \\
r_{21} &= s\theta_6 \left(c\theta_1 c\theta_4 - c\theta_2 s\theta_1 s\theta_4\right) \\
&\quad + c\theta_6 \left(-s\theta_1 s\theta_2 s\theta_5 + c\theta_5 \left(c\theta_1 s\theta_4 + c\theta_2 c\theta_4 s\theta_1\right)\right) \quad (5.103) \\
r_{31} &= s\theta_2 s\theta_4 s\theta_6 + c\theta_6 \left(-c\theta_2 s\theta_5 - c\theta_4 c\theta_5 s\theta_2\right) \quad (5.104)
\end{aligned}
$$

$$
\begin{aligned}
r_{12} &= c\theta_6 \left(-c\theta_4 s\theta_1 - c\theta_1 c\theta_2 s\theta_4\right) \\
&\quad - s\theta_6 \left(-c\theta_1 s\theta_2 s\theta_5 + c\theta_5 \left(-s\theta_1 s\theta_4 + c\theta_1 c\theta_2 c\theta_4\right)\right) \quad (5.105) \\
r_{22} &= c\theta_6 \left(c\theta_1 c\theta_4 - c\theta_2 s\theta_1 s\theta_4\right) \\
&\quad - s\theta_6 \left(-s\theta_1 s\theta_2 s\theta_5 + c\theta_5 \left(c\theta_1 s\theta_4 + c\theta_2 c\theta_4 s\theta_1\right)\right) \quad (5.106) \\
r_{32} &= c\theta_6 s\theta_2 s\theta_4 - s\theta_6 \left(-c\theta_2 s\theta_5 - c\theta_4 c\theta_5 s\theta_2\right) \quad (5.107)
\end{aligned}
$$

$$
\begin{aligned}
r_{13} &= c\theta_1 c\theta_5 s\theta_2 + s\theta_5 \left(-s\theta_1 s\theta_4 + c\theta_1 c\theta_2 c\theta_4\right) \quad (5.108) \\
r_{23} &= c\theta_5 s\theta_1 s\theta_2 + s\theta_5 \left(c\theta_1 s\theta_4 + c\theta_2 c\theta_4 s\theta_1\right) \quad (5.109) \\
r_{33} &= c\theta_2 c\theta_5 - c\theta_4 s\theta_2 s\theta_5 \quad (5.110)
\end{aligned}
$$

$$
\begin{aligned}
r_{14} &= -l_2 s\theta_1 + d_3 c\theta_1 s\theta_2 \quad (5.111) \\
r_{24} &= l_2 c\theta_1 + d_3 s\theta_1 s\theta_2 \quad (5.112) \\
r_{34} &= d_3 c\theta_2. \quad (5.113)
\end{aligned}
$$

*The end-effector kinematics can be solved by multiplying the position of the tool frame $B_7$ with respect to the wrist point, by $^0T_6$*

$$
^0T_7 = {}^0T_6\, {}^6T_7 \quad (5.114)
$$

*where*

$$
^6T_7 = \begin{bmatrix} 1 & 0 & 0 & 0 \\ 0 & 1 & 0 & 0 \\ 0 & 0 & 1 & d_6 \\ 0 & 0 & 0 & 1 \end{bmatrix}. \quad (5.115)
$$

**Example 150** *Checking the robot transformation matrix.*

To check the correctness of the final transformation matrix to map the coordinates in tool frame into the base frame, we may set the joint variables at a specific rest position. Let us substitute the joint rotational angles of the spherical robot analyzed in Example 149 equal to zero.

$$
\theta_1 = 0, \; \theta_2 = 0, \; \theta_3 = 0, \; \theta_4 = 0, \; \theta_5 = 0 \quad (5.116)
$$

*Therefore, the transformation matrix (5.101) would be*

$$
\begin{aligned}
{}^0T_6 &= {}^0T_1\,{}^1T_2\,{}^2T_3\,{}^3T_4\,{}^4T_5\,{}^5T_6 \\
&= \begin{bmatrix} 1 & 0 & 0 & 0 \\ 0 & 1 & 0 & l_2 \\ 0 & 0 & 1 & d_3 \\ 0 & 0 & 0 & 1 \end{bmatrix}
\end{aligned}
\tag{5.117}
$$

*that correctly indicates the origin of the tool frame in robot's stretched-up configuration, at*

$$
{}^G\mathbf{r}_{o_6} = \begin{bmatrix} 0 \\ l_2 \\ d_3 \end{bmatrix}.
\tag{5.118}
$$

**Example 151** *Assembling a wrist mechanism to a manipulator.*

*Consider a robot made by mounting the hand shown in Figure 5.23, to the tip point of the articulated arm shown in Figure 5.25. The resulting robot would have six DOF to reach any point within the working space in a desired orientation. The robot's forward kinematics can be found by combining the wrist transformation matrix (5.73) and the manipulator transformation matrix (5.80).*

$$
\begin{aligned}
{}^0T_7 &= T_{arm}\,T_{wrist}\,{}^6T_7 \tag{5.119} \\
&= {}^0T_3\,{}^3T_6\,{}^6T_7
\end{aligned}
$$

*The wrist transformation matrix ${}^3T_7$ has been analyzed in Example 147, and the arm transformation matrix ${}^0T_3$ has been found in Example 148. However, since we are attaching the wrist at point P of the frame $B_3$, the transformation matrix ${}^3T_4$ in (5.68) must include this joint distance. So, we substitute matrix (5.68) with*

$$
{}^3T_4 = \begin{bmatrix} \cos\theta_4 & 0 & -\sin\theta_4 & 0 \\ \sin\theta_4 & 0 & \cos\theta_4 & 0 \\ 0 & -1 & 0 & l_3 \\ 0 & 0 & 0 & 1 \end{bmatrix}
\tag{5.120}
$$

*to find*

$$
\begin{aligned}
T_{wrist} &= {}^3T_6 \tag{5.121} \\
&= \begin{bmatrix}
-s\theta_4 s\theta_6 + c\theta_4 c\theta_5 c\theta_6 & -c\theta_6 s\theta_4 - c\theta_4 c\theta_5 s\theta_6 & c\theta_4 s\theta_5 & 0 \\
c\theta_4 s\theta_6 + c\theta_5 c\theta_6 s\theta_4 & c\theta_4 c\theta_6 - c\theta_5 s\theta_4 s\theta_6 & s\theta_4 s\theta_5 & 0 \\
-c\theta_6 s\theta_5 & s\theta_5 s\theta_6 & c\theta_5 & l_3 \\
0 & 0 & 0 & 1
\end{bmatrix}.
\end{aligned}
$$

*The rest position of the robot can be checked to be at*

$$
{}^0T_7 = \begin{bmatrix} 1 & 0 & 0 & l_2 \\ 0 & 1 & 0 & d_2 + l_3 \\ 0 & 0 & 1 & d_1 + d_6 \\ 0 & 0 & 0 & 1 \end{bmatrix}
\tag{5.122}
$$

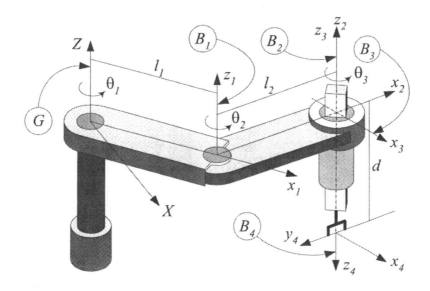

FIGURE 5.27. An R∥R∥R∥P SCARA manipulator robot.

*because*

$$
{}^6T_7 = \begin{bmatrix} 1 & 0 & 0 & 0 \\ 0 & 1 & 0 & 0 \\ 0 & 0 & 1 & d_6 \\ 0 & 0 & 0 & 1 \end{bmatrix}.
$$

(5.123)

**Example 152** *SCARA robot forward kinematics.*

*Consider the R∥R∥R∥P robot shown in Figure 5.27. The forward kinematics of the robot can be solved by obtaining individual transformation matrices ${}^{i-1}T_i$. The first link is an R∥R(0) link, which has the following transformation matrix:*

$$
{}^0T_1 = \begin{bmatrix} \cos\theta_1 & -\sin\theta_1 & 0 & l_1\cos\theta_1 \\ \sin\theta_1 & \cos\theta_1 & 0 & l_1\sin\theta_1 \\ 0 & 0 & 1 & 0 \\ 0 & 0 & 0 & 1 \end{bmatrix}
$$

(5.124)

*The second link is also an R∥R(0) link*

$$
{}^1T_2 = \begin{bmatrix} \cos\theta_2 & -\sin\theta_2 & 0 & l_2\cos\theta_2 \\ \sin\theta_2 & \cos\theta_2 & 0 & l_2\sin\theta_2 \\ 0 & 0 & 1 & 0 \\ 0 & 0 & 0 & 1 \end{bmatrix}.
$$

(5.125)

*The third link is an $R\|R(0)$ with zero length*

$$
{}^2T_3 = \begin{bmatrix} \cos\theta_3 & -\sin\theta_3 & 0 & 0 \\ \sin\theta_3 & \cos\theta_3 & 0 & 0 \\ 0 & 0 & 1 & 0 \\ 0 & 0 & 0 & 1 \end{bmatrix} \tag{5.126}
$$

*and finally the fourth link is an $R\|P(0)$*

$$
{}^3T_4 = \begin{bmatrix} 1 & 0 & 0 & 0 \\ 0 & 1 & 0 & 0 \\ 0 & 0 & 1 & d \\ 0 & 0 & 0 & 1 \end{bmatrix}. \tag{5.127}
$$

*Therefore, the configuration of the end-effector in the base coordinate frame is*

$$
\begin{aligned}
{}^0T_4 &= {}^0T_1\,{}^1T_2\,{}^2T_3\,{}^3T_4 \tag{5.128} \\
&= \begin{bmatrix} c(\theta_1+\theta_2+\theta_3) & -s(\theta_1+\theta_2+\theta_3) & 0 & l_1 c\theta_1 + l_2 c(\theta_1+\theta_2) \\ s(\theta_1+\theta_2+\theta_3) & c(\theta_1+\theta_2+\theta_3) & 0 & l_1 s\theta_1 + l_2 s(\theta_1+\theta_2) \\ 0 & 0 & 1 & d \\ 0 & 0 & 0 & 1 \end{bmatrix}
\end{aligned}
$$

*that shows the rest position of the robot $\theta_1 = \theta_2 = \theta_3 = d = 0$ is at*

$$
{}^0T_4 = \begin{bmatrix} 1 & 0 & 0 & l_1 + l_2 \\ 0 & 1 & 0 & 0 \\ 0 & 0 & 1 & 0 \\ 0 & 0 & 0 & 1 \end{bmatrix}. \tag{5.129}
$$

**Example 153** *Space station remote manipulator system (SSRMS).*

*Shuttle remote manipulator system (SRMS), also known as (SSRMS), is an arm and a hand attached to the Shuttle or space station. It is utilized for several purposes such as: satellite deployment, construction of a space station, transporting a crew member at the end of the arm, surveying and inspecting the outside of the station using a camera.*

*A simplified model of the SRMS, schematically shown in Figure 5.28, has the characteristics indicated in Table 5.11.*

Table 5.11- Space station's robot arm characteristics.

| Length | 14.22m |
|---|---|
| Diameter | 38.1cm |
| Weight | 1336kg |
| Number of joints | Seven |
| Handling capacity | 116000kg (in space) |
| Max velocity of end of arm | Carrying nothing : 37cm/s |
|  | Full capacity : 1.2cm/s |
| Max rotational speed | Approx. 4 deg /s |

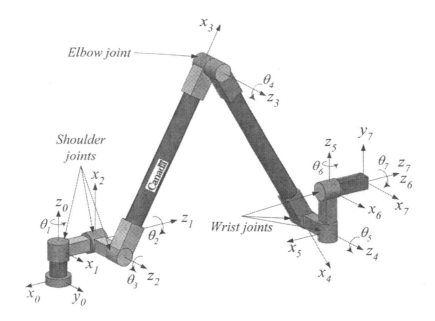

FIGURE 5.28. Illustration of the space station remote manipulator system.

Table 5.12 - DH parameters for SRMS.

| Frame No. | $a_i$ | $\alpha_i$ | $d_i$ | $\theta_i$ |
|-----------|-------|------------|-------|------------|
| 1 | 0 | −90 deg | 380mm | $\theta_1$ |
| 2 | 0 | −90 deg | 1360mm | $\theta_2$ |
| 3 | 7110mm | 0 | 570mm | $\theta_3$ |
| 4 | 7110mm | 0 | 475mm | $\theta_4$ |
| 5 | 0 | 90 deg | 570mm | $\theta_5$ |
| 6 | 0 | −90 deg | 635mm | $\theta_6$ |
| 7 | 0 | 0 | $d_7$ | 0 |

Consider the numerical values of joints offset and links' length as tabulated in Table 5.12. Utilizing these values and indicating the type of each link enables us to determine the required transformation matrices to solve the forward kinematics problem.

Links (1) and (2) are R⊢R(−90), and therefore,

$$
{}^0T_1 = \begin{bmatrix}
\cos\theta_1 & 0 & -\sin\theta_1 & 0 \\
\sin\theta_1 & 0 & \cos\theta_1 & 0 \\
0 & -1 & 0 & d_1 \\
0 & 0 & 0 & 1
\end{bmatrix} \tag{5.130}
$$

$$
{}^{1}T_2 =
\begin{bmatrix}
\cos\theta_2 & 0 & -\sin\theta_2 & 0 \\
\sin\theta_2 & 0 & \cos\theta_2 & 0 \\
0 & -1 & 0 & d_2 \\
0 & 0 & 0 & 1
\end{bmatrix}.
\tag{5.131}
$$

*Links (3) and (4) are $R\|R(0)$, hence*

$$
{}^{2}T_3 =
\begin{bmatrix}
\cos\theta_3 & -\sin\theta_3 & 0 & a_3\cos\theta_3 \\
\sin\theta_3 & \cos\theta_3 & 0 & a_3\sin\theta_3 \\
0 & 0 & 1 & d_3 \\
0 & 0 & 0 & 1
\end{bmatrix}
\tag{5.132}
$$

$$
{}^{3}T_4 =
\begin{bmatrix}
\cos\theta_4 & -\sin\theta_4 & 0 & a_4\cos\theta_4 \\
\sin\theta_4 & \cos\theta_4 & 0 & a_4\sin\theta_4 \\
0 & 0 & 1 & d_4 \\
0 & 0 & 0 & 1
\end{bmatrix}.
\tag{5.133}
$$

*Link (5) is $R \vdash R(90)$, and link (6) is $R \vdash R(-90)$, therefore*

$$
{}^{4}T_5 =
\begin{bmatrix}
\cos\theta_5 & 0 & \sin\theta_5 & 0 \\
\sin\theta_5 & 0 & -\cos\theta_5 & 0 \\
0 & 1 & 0 & d_5 \\
0 & 0 & 0 & 1
\end{bmatrix}
\tag{5.134}
$$

$$
{}^{5}T_6 =
\begin{bmatrix}
\cos\theta_6 & 0 & -\sin\theta_6 & 0 \\
\sin\theta_6 & 0 & \cos\theta_6 & 0 \\
0 & -1 & 0 & d_6 \\
0 & 0 & 0 & 1
\end{bmatrix}.
\tag{5.135}
$$

*Finally link (7) is $R \vdash R(0)$ and the coordinate frame attached to the end-effector has a translation $d_7$ with respect to the coordinate frame $B_6$.*

$$
{}^{6}T_7 =
\begin{bmatrix}
\cos\theta_7 & -\sin\theta_7 & 0 & 0 \\
\sin\theta_7 & \cos\theta_7 & 0 & 0 \\
0 & 0 & 1 & d_7 \\
0 & 0 & 0 & 1
\end{bmatrix}
\tag{5.136}
$$

*The forward kinematics of SSRMS can be found by direct multiplication of $^{i-1}T_i$, $(i = 1, 2, \cdots, 7)$.*

$$
{}^{0}T_7 = {}^{0}T_1\,{}^{1}T_2\,{}^{2}T_3\,{}^{3}T_4\,{}^{4}T_5\,{}^{5}T_6\,{}^{6}T_7
\tag{5.137}
$$

## 5.4    ★ Coordinate Transformation Using Screws

It is possible and easier to use screws to describe a transformation matrix between two adjacent coordinate frames $B_i$ and $B_{i-1}$. We can move $B_{i-1}$

to $B_i$ by a central screw $\check{s}(a_i, \alpha_i, \hat{\imath}_{i-1})$ followed by another central screw $\check{s}(d_i, \theta_i, \hat{k}_{i-1})$.

$$
\begin{aligned}
{}^{i-1}T_i &= \check{s}(d_i, \theta_i, \hat{k}_{i-1})\,\check{s}(a_i, \alpha_i, \hat{\imath}_{i-1}) \qquad\qquad\qquad (5.138) \\[4pt]
&= \begin{bmatrix}
\cos\theta_i & -\sin\theta_i \cos\alpha_i & \sin\theta_i \sin\alpha_i & a_i \cos\theta_i \\
\sin\theta_i & \cos\theta_i \cos\alpha_i & -\cos\theta_i \sin\alpha_i & a_i \sin\theta_i \\
0 & \sin\alpha_i & \cos\alpha_i & d_i \\
0 & 0 & 0 & 1
\end{bmatrix}
\end{aligned}
$$

**Proof.** The central screw $\check{s}(a_i, \alpha_i, \hat{\imath}_{i-1})$ is

$$
\begin{aligned}
\check{s}(a_i, \alpha_i, \hat{\imath}_{i-1}) &= D(a_i, \hat{\imath}_{i-1})R(\hat{\imath}_{i-1}, \alpha_i) \qquad\qquad\qquad (5.139) \\[4pt]
&= D_{x_{i-1}, a_i} \, R_{x_{i-1}, \alpha_i} \\[4pt]
&= \begin{bmatrix}
1 & 0 & 0 & a_i \\
0 & 1 & 0 & 0 \\
0 & 0 & 1 & 0 \\
0 & 0 & 0 & 1
\end{bmatrix}
\begin{bmatrix}
1 & 0 & 0 & 0 \\
0 & \cos\alpha_i & -\sin\alpha_i & 0 \\
0 & \sin\alpha_i & \cos\alpha_i & 0 \\
0 & 0 & 0 & 1
\end{bmatrix} \\[4pt]
&= \begin{bmatrix}
1 & 0 & 0 & a_i \\
0 & \cos\alpha_i & -\sin\alpha_i & 0 \\
0 & \sin\alpha_i & \cos\alpha_i & 0 \\
0 & 0 & 0 & 1
\end{bmatrix}
\end{aligned}
$$

and the central screw $\check{s}(d_i, \theta_i, \hat{k}_{i-1})$ is

$$
\begin{aligned}
\check{s}(d_i, \theta_i, \hat{k}_{i-1}) &= D(d_i, \hat{k}_{i-1})R(\hat{k}_{i-1}, \theta_i) \qquad\qquad\qquad (5.140) \\[4pt]
&= D_{z_{i-1}, d_i} \, R_{z_{i-1}, \theta_i} \\[4pt]
&= \begin{bmatrix}
1 & 0 & 0 & 0 \\
0 & 1 & 0 & 0 \\
0 & 0 & 1 & d_i \\
0 & 0 & 0 & 1
\end{bmatrix}
\begin{bmatrix}
\cos\theta_i & -\sin\theta_i & 0 & 0 \\
\sin\theta_i & \cos\theta_i & 0 & 0 \\
0 & 0 & 1 & 0 \\
0 & 0 & 0 & 1
\end{bmatrix} \\[4pt]
&= \begin{bmatrix}
\cos\theta_i & -\sin\theta_i & 0 & 0 \\
\sin\theta_i & \cos\theta_i & 0 & 0 \\
0 & 0 & 1 & d_i \\
0 & 0 & 0 & 1
\end{bmatrix}.
\end{aligned}
$$

Therefore, the transformation matrix ${}^{i-1}T_i$ made by two screw motions would be

$$^{i-1}T_i \;=\; \check{s}(d_i, \theta_i, \hat{k}_{i-1})\, \check{s}(a_i, \alpha_i, \hat{i}_{i-1}) \tag{5.141}$$

$$= \begin{bmatrix} \cos\theta_i & -\sin\theta_i & 0 & 0 \\ \sin\theta_i & \cos\theta_i & 0 & 0 \\ 0 & 0 & 1 & d_i \\ 0 & 0 & 0 & 1 \end{bmatrix} \begin{bmatrix} 1 & 0 & 0 & a_i \\ 0 & \cos\alpha_i & -\sin\alpha_i & 0 \\ 0 & \sin\alpha_i & \cos\alpha_i & 0 \\ 0 & 0 & 0 & 1 \end{bmatrix}$$

$$= \begin{bmatrix} \cos\theta_i & -\cos\alpha_i \sin\theta_i & \sin\theta_i \sin\alpha_i & a_i \cos\theta_i \\ \sin\theta_i & \cos\theta_i \cos\alpha_i & -\cos\theta_i \sin\alpha_i & a_i \sin\theta_i \\ 0 & \sin\alpha_i & \cos\alpha_i & d_i \\ 0 & 0 & 0 & 1 \end{bmatrix}.$$

The resultant transformation matrix $^{i-1}T_i$ is equivalent to a general screw whose parameters can be found based on Equations (4.125) and (4.126).

The twist of screw, $\phi$, can be computed based on Equation (4.129)

$$\cos\phi \;=\; \frac{1}{2}\left(\mathrm{tr}\left(^G R_B\right) - 1\right)$$

$$= \frac{1}{2}(\cos\theta_i + \cos\theta_i \cos\alpha_i + \cos\alpha_i - 1) \tag{5.142}$$

and the axis of screw, $\hat{u}$, can be found by using Equation (4.131)

$$\tilde{u} \;=\; \frac{1}{2\sin\phi}\left(^G R_B - {}^G R_B^T\right)$$

$$= \frac{1}{2s\phi}\begin{bmatrix} c\theta_i & -c\alpha_i s\theta_i & s\theta_i s\alpha_i \\ s\theta_i & c\theta_i c\alpha_i & -c\theta_i s\alpha_i \\ 0 & s\alpha_i & c\alpha_i \end{bmatrix} - \begin{bmatrix} c\theta_i & s\theta_i & 0 \\ -c\alpha_i s\theta_i & c\theta_i c\alpha_i & s\alpha_i \\ s\theta_i s\alpha_i & -c\theta_i s\alpha_i & c\alpha_i \end{bmatrix}$$

$$= \frac{1}{2s\phi}\begin{bmatrix} 0 & -s\theta_i - c\alpha_i s\theta_i & s\theta_i s\alpha_i \\ s\theta_i + c\alpha_i s\theta_i & 0 & -s\alpha_i - c\theta_i s\alpha_i \\ -s\theta_i s\alpha_i & s\alpha_i + c\theta_i s\alpha_i & 0 \end{bmatrix} \tag{5.143}$$

and therefore,

$$\hat{u} = \frac{1}{2s\phi}\begin{bmatrix} \sin\alpha_i + \cos\theta_i \sin\alpha_i \\ \sin\theta_i \sin\alpha_i \\ \sin\theta_i + \cos\alpha_i \sin\theta_i \end{bmatrix}. \tag{5.144}$$

The translation parameter, $h$, and the position vector of a point on the screw axis, for instance $\begin{bmatrix} 0 & y_{i-1} & z_{i-1} \end{bmatrix}$, can be found based on Equation (4.134).

$$
\begin{bmatrix} h \\ y_{i-1} \\ y_{i-1} \end{bmatrix} = \begin{bmatrix} u_1 & -r_{12} & -r_{13} \\ u_2 & 1-r_{22} & -r_{23} \\ u_3 & -r_{32} & 1-r_{33} \end{bmatrix}^{-1} \begin{bmatrix} r_{14} \\ r_{24} \\ r_{34} \end{bmatrix} \tag{5.145}
$$

$$
= \frac{1}{2s\phi} \begin{bmatrix} s\alpha_i + c\theta_i s\alpha_i & -s\theta_i\, c\alpha_i & s\theta_i\, s\alpha_i \\ s\theta_i s\alpha_i & 1-c\theta_i\, c\alpha_i & -c\theta_i\, s\alpha_i \\ s\theta_i + c\alpha_i s\theta_i & s\alpha_i & c\alpha_i \end{bmatrix}^{-1} \begin{bmatrix} a_i\, c\theta_i \\ a_i\, s\theta_i \\ d_i \end{bmatrix}
$$

■

**Example 154** ★ *Classification of industrial robot link by screws.*

    *There are 12 different configurations that are mostly used for industrial robots. Each type has its own class of geometrical configuration and transformation. Each class is identified by its joints at both ends, and has its own transformation matrix to go from the distal joint coordinate frame $B_i$ to the proximal joint coordinate frame $B_{i-1}$. The transformation matrix of each class depends solely on the proximal joint, and the angle between z-axes. The screw expression for two arbitrary coordinate frames is*

$$
{}^{i-1}T_i = \breve{s}(d_i, \theta_i, \hat{k}_{i-1})\, \breve{s}(a_i, \alpha_i, \hat{\imath}_{i-1}) \tag{5.146}
$$

*where*

$$
\breve{s}(d_i, \theta_i, \hat{k}_{i-1}) = D(d_i, \hat{k}_{i-1}) R(\hat{k}_{i-1}, \theta_i) \tag{5.147}
$$
$$
\breve{s}(a_i, \alpha_i, \hat{\imath}_{i-1}) = D(a_i, \hat{\imath}_{i-1}) R(\hat{\imath}_{i-1}, \alpha_i). \tag{5.148}
$$

*The screw expression of the frame transformation can be simplified for each class according to Table 5.13.*

Table 5.13 - Classification of industrial robot link by screws.

| No. | Type | of | Link | ${}^{i-1}T_i =$ |
|---|---|---|---|---|
| 1 | $R\Vert R(0)$ | or | $R\Vert P(0)$ | $\breve{s}(0,\theta_i,\hat{k}_{i-1})\,\breve{s}(a_i,0,\hat{\imath}_{i-1})$ |
| 2 | $R\Vert R(180)$ | or | $R\Vert P(180)$ | $\breve{s}(0,\theta_i,\hat{k}_{i-1})\,\breve{s}(a_i,2\pi,\hat{\imath}_{i-1})$ |
| 3 | $R\perp R(90)$ | or | $R\perp P(90)$ | $\breve{s}(0,\theta_i,\hat{k}_{i-1})\,\breve{s}(a_i,\pi,\hat{\imath}_{i-1})$ |
| 4 | $R\perp R(-90)$ | or | $R\perp P(-90)$ | $\breve{s}(0,\theta_i,\hat{k}_{i-1})\,\breve{s}(a_i,-\pi,\hat{\imath}_{i-1})$ |
| 5 | $R\vdash R(90)$ | or | $R\vdash P(90)$ | $\breve{s}(0,\theta_i,\hat{k}_{i-1})\,\breve{s}(0,\pi,\hat{\imath}_{i-1})$ |
| 6 | $R\vdash R(-90)$ | or | $R\vdash P(-90)$ | $\breve{s}(0,\theta_i,\hat{k}_{i-1})\,\breve{s}(0,-\pi,\hat{\imath}_{i-1})$ |
| 7 | $P\Vert R(0)$ | or | $P\Vert P(0)$ | $\breve{s}(d_i,0,\hat{k}_{i-1})\,\breve{s}(a_i,0,\hat{\imath}_{i-1})$ |
| 8 | $P\Vert R(180)$ | or | $P\Vert P(180)$ | $\breve{s}(d_i,0,\hat{k}_{i-1})\,\breve{s}(a_i,2\pi,\hat{\imath}_{i-1})$ |
| 9 | $P\perp R(90)$ | or | $P\perp P(90)$ | $\breve{s}(d_i,0,\hat{k}_{i-1})\,\breve{s}(a_i,\pi,\hat{\imath}_{i-1})$ |
| 10 | $P\perp R(-90)$ | or | $P\perp P(-90)$ | $\breve{s}(d_i,0,\hat{k}_{i-1})\,\breve{s}(a_i,-\pi,\hat{\imath}_{i-1})$ |
| 11 | $P\vdash R(90)$ | or | $P\vdash P(90)$ | $\breve{s}(d_i,0,\hat{k}_{i-1})\,\breve{s}(0,\pi,\hat{\imath}_{i-1})$ |
| 12 | $P\vdash R(-90)$ | or | $P\vdash P(-90)$ | $\breve{s}(d_i,0,\hat{k}_{i-1})\,\breve{s}(0,-\pi,\hat{\imath}_{i-1})$ |

*As an example, we may examine the first class*

$$
\begin{aligned}
{}^{i-1}T_i &= \check{s}(0, \theta_i, \hat{k}_{i-1})\, \check{s}(a_i, 0, \hat{\imath}_{i-1}) \\
&= D(0, \hat{k}_{i-1})R(\hat{k}_{i-1}, \theta_i)D(a_i, \hat{\imath}_{i-1})R(\hat{\imath}_{i-1}, 0) \\
&= \begin{bmatrix} \cos\theta_i & -\sin\theta_i & 0 & 0 \\ \sin\theta_i & \cos\theta_i & 0 & 0 \\ 0 & 0 & 1 & 0 \\ 0 & 0 & 0 & 1 \end{bmatrix}
\begin{bmatrix} 1 & 0 & 0 & a_i \\ 0 & 1 & 0 & 0 \\ 0 & 0 & 1 & 0 \\ 0 & 0 & 0 & 1 \end{bmatrix} \\
&= \begin{bmatrix} \cos\theta_i & -\sin\theta_i & 0 & a_i\cos\theta_i \\ \sin\theta_i & \cos\theta_i & 0 & a_i\sin\theta_i \\ 0 & 0 & 1 & 0 \\ 0 & 0 & 0 & 1 \end{bmatrix}
\end{aligned}
$$

*and find the same result as Equation (5.32).*

**Example 155** ★ *Spherical robot forward kinematics based on screws.*
*Application of screws in forward kinematics can be done by determining the class of each link and applying the associated screws. The class of links for the spherical robot shown in Figure 5.26, are indicated in Table 5.14.*

*Table 5.14 - Screw transformation for the spherical robot shown in Figure 5.26.*

| Link No. | Class | Screw transformation |
|----------|-------|----------------------|
| 1 | R⊢R(−90) | ${}^{0}T_1 = \check{s}(0, \theta_i, \hat{k}_{i-1})\, \check{s}(0, -\pi, \hat{\imath}_{i-1})$ |
| 2 | R⊢P(90) | ${}^{1}T_2 = \check{s}(0, \theta_i, \hat{k}_{i-1})\, \check{s}(0, \pi, \hat{\imath}_{i-1})$ |
| 3 | P∥R(0) | ${}^{2}T_3 = \check{s}(d_i, 0, \hat{k}_{i-1})\, \check{s}(a_i, 0, \hat{\imath}_{i-1})$ |
| 4 | R⊢R(−90) | ${}^{3}T_4 = \check{s}(0, \theta_i, \hat{k}_{i-1})\, \check{s}(0, -\pi, \hat{\imath}_{i-1})$ |
| 5 | R⊢R(90) | ${}^{4}T_5 = \check{s}(0, \theta_i, \hat{k}_{i-1})\, \check{s}(0, \pi, \hat{\imath}_{i-1})$ |
| 6 | R∥R(0) | ${}^{5}T_6 = \check{s}(0, \theta_i, \hat{k}_{i-1})\, \check{s}(a_i, 0, \hat{\imath}_{i-1})$ |

*Therefore, the configuration of the end-effector frame of the spherical robot in the base frame is*

$$
\begin{aligned}
{}^{0}T_6 &= {}^{0}T_1\, {}^{1}T_2\, {}^{2}T_3\, {}^{3}T_4\, {}^{4}T_5\, {}^{5}T_6 \\
&= \check{s}(0, \theta_i, \hat{k}_{i-1})\, \check{s}(0, -\pi, \hat{\imath}_{i-1})\, \check{s}(0, \theta_i, \hat{k}_{i-1})\, \check{s}(0, \pi, \hat{\imath}_{i-1}) \\
&\quad \times \check{s}(d_i, 0, \hat{k}_{i-1})\, \check{s}(a_i, 0, \hat{\imath}_{i-1})\, \check{s}(0, \theta_i, \hat{k}_{i-1})\, \check{s}(0, -\pi, \hat{\imath}_{i-1}) \\
&\quad \times \check{s}(0, \theta_i, \hat{k}_{i-1})\, \check{s}(0, \pi, \hat{\imath}_{i-1})\, \check{s}(0, \theta_i, \hat{k}_{i-1})\, \check{s}(a_i, 0, \hat{\imath}_{i-1}). \quad (5.149)
\end{aligned}
$$

**Example 156** ★ *Plücker coordinate of a central screw.*
*Utilizing Plücker coordinates we can define a central screw*

$$
\check{s}(h, \phi, \hat{u}) = \begin{bmatrix} \phi\hat{u} \\ h\hat{u} \end{bmatrix} \quad (5.150)
$$

*which is equal to*

$$
\begin{bmatrix} \phi\hat{u} \\ h\hat{u} \end{bmatrix} = D(h\hat{u})\, R(\hat{u}, \phi). \quad (5.151)
$$

**Example 157** ★ *Plücker coordinate for the central screw* $\check{s}(a_i, \alpha_i, \hat{\imath}_{i-1})$.

The central screw $\check{s}(a_i, \alpha_i, \hat{\imath}_{i-1})$ can also be describe by a proper Plücker coordinate.

$$\check{s}(a_i, \alpha_i, \hat{\imath}_{i-1}) = \begin{bmatrix} \alpha_i\,\hat{\imath}_{i-1} \\ a_i\,\hat{\imath}_{i-1} \end{bmatrix} \tag{5.152}$$

$$= D(a_i, \hat{\imath}_{i-1}) R(\hat{\imath}_{i-1}, \alpha_i)$$

Similarly, the central screw $\check{s}(d_i, \theta_i, \hat{k}_{i-1})$ can also be described by a proper Plücker coordinate.

$$\check{s}(d_i, \theta_i, \hat{k}_{i-1}) = \begin{bmatrix} \theta_i\,\hat{k}_{i-1} \\ d_i\,\hat{k}_{i-1} \end{bmatrix} \tag{5.153}$$

$$= D(d_i, \hat{k}_{i-1}) R(\hat{k}_{i-1}, \theta_i)$$

**Example 158** ★ *Intersecting two central screws.*

Two lines (and therefore two screws) are intersecting if their reciprocal product is zero. We can check that the reciprocal product of the screws $\check{s}(a_i, \alpha_i, \hat{\imath}_{i-1})$ and $\check{s}(d_i, \theta_i, \hat{k}_{i-1})$ is zero.

$$\check{s}(d_i, \theta_i, \hat{k}_{i-1}) \times \check{s}(a_i, \alpha_i, \hat{\imath}_{i-1}) = \begin{bmatrix} \theta_i\,\hat{k}_{i-1} \\ d_i\,\hat{k}_{i-1} \end{bmatrix} \otimes \begin{bmatrix} \alpha_i\,\hat{\imath}_{i-1} \\ a_i\,\hat{\imath}_{i-1} \end{bmatrix}$$

$$= \theta_i\,\hat{k}_{i-1} \cdot a_i\,\hat{\imath}_{i-1} + \alpha_i\,\hat{\imath}_{i-1} \cdot \theta_i\,\hat{k}_{i-1}$$

$$= 0. \tag{5.154}$$

## 5.5 ★ Sheth Method

The Sheth method can overcome the limitation of the DH method for higher order links, by introducing a number of frames equal to the number of joints on the link. It also provides more flexibility to specify the link geometry compared to the DH method.

In Sheth method, we define a coordinate frame at each joint of a link, so an $n$ joint robot would have $2n$ frames. Figure 5.29 shows the case of a binary link $(i)$ where a first frame $(x_i, y_i, z_i)$ is attached at the *origin* of the link and a second frame $(u_i, v_i, w_i)$ to the *end* of the link. The assignment of the origin joint and the end joint are arbitrary, however it is easier if they are in the direction of base-to-tool frames.

To describe the geometry, first we locate the joint axes by $z_i$ and $w_i$, and then determine the common perpendicular to both joint axes $z_i$ and $w_i$. The common normal is indicated by a unit vector $\hat{n}_i$. Specifying the link geometry requires six parameters, and are determined as follows:

1. $a_i$ is the distance from $z_i$ to $w_i$, measured along $\hat{n}_i$. It is the kinematic distance between $z_i$ and $w_i$.

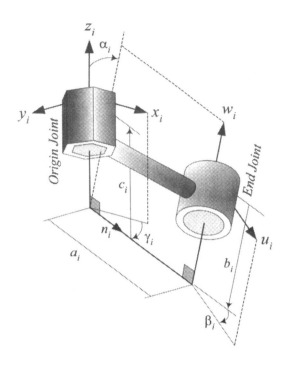

FIGURE 5.29. Sheth method for defining the origin and end coordinate frames on a binary link.

2. $b_i$ is the distance from $\hat{n}_i$ to $u_i$, measured along $w_i$. It is the elevation of the $w_i$-axis.

3. $c_i$ is the distance from $x_i$ to $\hat{n}_i$, measured along $z_i$. It is the elevation of the $z_i$-axis.

4. $\alpha_i$ is the angle made by axes $z_i$ and $w_i$, measured positively from $z_i$ to $w_i$ about $\hat{n}_i$.

5. $\beta_i$ is the angle made by axes $\hat{n}_i$ and $u_i$, measured positively from $\hat{n}_i$ to $u_i$ about $w_i$.

6. $\gamma_i$ is the angle made by axes $x_i$ and $\hat{n}_i$, measured positively from $x_i$ to $\hat{n}_i$ about $z_i$.

The Sheth parameters generate a homogeneous transformation matrix

$$^{o}T_e = {}^{o}T_e\,(a_i, b_i, c_i, \alpha_i, \beta_i, \gamma_i) \tag{5.155}$$

to map the end coordinate frame $B_e$ to the origin coordinate frame $B_o$

$$\begin{bmatrix} x_i \\ y_i \\ z_i \\ 1 \end{bmatrix} = {}^oT_e = \begin{bmatrix} u_i \\ v_i \\ w_i \\ 1 \end{bmatrix} \tag{5.156}$$

where ${}^oT_e$ denotes the Sheth transformation matrix

$$ {}^oT_e = \begin{bmatrix} r_{11} & r_{12} & r_{13} & r_{14} \\ r_{21} & r_{22} & r_{23} & r_{24} \\ r_{31} & r_{32} & r_{33} & r_{34} \\ 0 & 0 & 0 & 1 \end{bmatrix} \tag{5.157}$$

where,

$$ r_{11} = \cos\beta_i \cos\gamma_i - \cos\alpha_i \sin\beta_i \sin\gamma_i \tag{5.158}$$
$$ r_{21} = \cos\beta_i \sin\gamma_i + \cos\alpha_i \cos\gamma_i \sin\beta_i \tag{5.159}$$
$$ r_{31} = \sin\alpha_i \sin\beta_i \tag{5.160}$$

$$ r_{12} = -\cos\gamma_i \sin\beta_i - \cos\alpha_i \cos\beta_i \sin\gamma_i \tag{5.161}$$
$$ r_{22} = -\sin\beta_i \sin\gamma_i + \cos\alpha_i \cos\beta_i \cos\gamma_i \tag{5.162}$$
$$ r_{32} = \cos\beta_i \sin\alpha_i \tag{5.163}$$

$$ r_{13} = \sin\alpha_i \sin\gamma_i \tag{5.164}$$
$$ r_{23} = -\cos\gamma_i \sin\alpha_i \tag{5.165}$$
$$ r_{33} = \cos\alpha_i \tag{5.166}$$

$$ r_{14} = a_i \cos\gamma_i + b_i \sin\alpha_i \sin\gamma_i \tag{5.167}$$
$$ r_{24} = a_i \sin\gamma_i - b_i \cos\gamma_i \sin\alpha_i \tag{5.168}$$
$$ r_{34} = c_i + b_i \cos\alpha_i. \tag{5.169}$$

The Sheth transformation matrix for two coordinate frames at a joint is simplified to a translation for a prismatic joint, and a rotation about the $Z$-axis for a revolute joint.

**Proof.** The homogenous transformation matrix to provide the coordinates in $B_i$, when the coordinates in $B_j$ are given, is

$$ {}^oT_e = D_{z_i,c_i}\, R_{z_i,\gamma_i}\, D_{x_i,a_i}\, R_{x_i,\alpha_i}\, D_{z_i,b_i}\, R_{z_i,\beta_i}. \tag{5.170}$$

Employing the associated transformation matrices

$$ R_{z_i,\beta_i} = \begin{bmatrix} \cos\beta_i & -\sin\beta_i & 0 & 0 \\ \sin\beta_i & \cos\beta_i & 0 & 0 \\ 0 & 0 & 1 & 0 \\ 0 & 0 & 0 & 1 \end{bmatrix} \tag{5.171}$$

$$D_{z_i,b_i} = \begin{bmatrix} 1 & 0 & 0 & 0 \\ 0 & 1 & 0 & 0 \\ 0 & 0 & 1 & b_i \\ 0 & 0 & 0 & 1 \end{bmatrix} \tag{5.172}$$

$$R_{x_i,\alpha_i} = \begin{bmatrix} 1 & 0 & 0 & 0 \\ 0 & \cos\alpha_i & -\sin\alpha_i & 0 \\ 0 & \sin\alpha_i & \cos\alpha_i & 0 \\ 0 & 0 & 0 & 1 \end{bmatrix} \tag{5.173}$$

$$D_{x_i,a_i} = \begin{bmatrix} 1 & 0 & 0 & a_i \\ 0 & 1 & 0 & 0 \\ 0 & 0 & 1 & 0 \\ 0 & 0 & 0 & 1 \end{bmatrix} \tag{5.174}$$

$$R_{z_i,\gamma_i} = \begin{bmatrix} \cos\gamma_i & -\sin\gamma_i & 0 & 0 \\ \sin\gamma_i & \cos\gamma_i & 0 & 0 \\ 0 & 0 & 1 & 0 \\ 0 & 0 & 0 & 1 \end{bmatrix} \tag{5.175}$$

$$D_{z_i,c_i} = \begin{bmatrix} 1 & 0 & 0 & 0 \\ 0 & 1 & 0 & 0 \\ 0 & 0 & 1 & c_i \\ 0 & 0 & 0 & 1 \end{bmatrix} \tag{5.176}$$

we can verify that the Sheth transformation matrix is

$$^{o}T_e = \begin{bmatrix} c\beta_i c\gamma_i & -c\gamma_i s\beta_i & & a_i c\gamma_i \\ -c\alpha_i s\beta_i s\gamma_i & -c\alpha_i c\beta_i s\gamma_i & s\alpha_i s\gamma_i & +b_i s\alpha_i s\gamma_i \\ & & & \\ c\beta_i s\gamma_i & -s\beta_i s\gamma_i & & a_i s\gamma_i \\ +c\alpha_i c\gamma_i s\beta_i & +c\alpha_i c\beta_i c\gamma_i & -c\gamma_i s\alpha_i & -b_i c\gamma_i s\alpha_i \\ & & & \\ s\alpha_i s\beta_i & c\beta_i s\alpha_i & c\alpha_i & c_i + b_i c\alpha_i \\ 0 & 0 & 0 & 1 \end{bmatrix}. \tag{5.177}$$

■

**Example 159 ★** *Sheth transformation matrix at revolute and prismatic joints.*

Two links connected by a revolute joint are shown in Figure 5.30 (a). The coordinate frames of the two links at the common joint are set such that the axes $z_i$ and $w_j$ coincide with the rotation axis, and both frames have the same origin. The Sheth parameters are $a = 0$, $b = 0$, $c = 0$, $\alpha = 0$, $\beta = 0$, $\gamma = \theta$ , and therefore,

$$^{i}T_j = \begin{bmatrix} \cos\theta & -\sin\theta & 0 & 0 \\ \sin\theta & \cos\theta & 0 & 0 \\ 0 & 0 & 1 & 0 \\ 0 & 0 & 0 & 1 \end{bmatrix}. \tag{5.178}$$

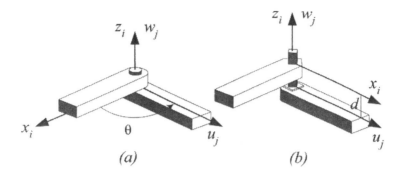

FIGURE 5.30. Illustration of (a) a revolute joint and (b) a prosmatic joint to define Sheth coordinate transformation.

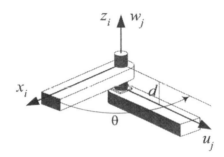

FIGURE 5.31. Illustration of a cylindrical joint to define Sheth coordinate transformation.

*Two links connected by a prismatic joint are illustrated in Figure 5.30 (b). The Sheth variable at this joint is $d$ along the joint axis. The coordinate frames of the two links at the common joint are set such that the axes $z_i$ and $w_j$ coincide with the translational axis, and axes $x_i$ and $u_j$ are chosen parallel in the same direction. The Sheth parameters are $a = 0$, $b = 0$, $c = d$, $\alpha = 0$, $\beta = 0$, $\gamma = 0$, and therefore,*

$$
{}^iT_j = \begin{bmatrix} 1 & 0 & 0 & 0 \\ 0 & 1 & 0 & 0 \\ 0 & 0 & 1 & d \\ 0 & 0 & 0 & 1 \end{bmatrix}. \tag{5.179}
$$

**Example 160** ★ *Sheth transformation matrix at a cylindrical joint.*

*A cylindrical joint provides two DOF, a rotational and a translational about the same axis. Two links connected by a cylindrical joint are shown in Figure 5.31. The transformation matrix for a cylindrical joint can be described by combining a revolute and a prismatic joint. Therefore, the*

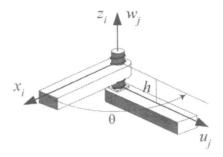

FIGURE 5.32. Illustration of a screw joint to define Sheth coordinate transformation.

*Sheth parameters are* $a = 0$, $b = 0$, $c = d$, $\alpha = 0$, $\beta = 0$, $\gamma = \theta$ , *and*

$$
{}^iT_j = \begin{bmatrix} \cos\theta & -\sin\theta & 0 & 0 \\ \sin\theta & \cos\theta & 0 & 0 \\ 0 & 0 & 1 & d \\ 0 & 0 & 0 & 1 \end{bmatrix}.
\tag{5.180}
$$

**Example 161** ★ *Sheth transformation matrix at a screw joint.*
    *A screw joint, as shown in Figure 5.32, provides a proportional rotation and translation motion, which has one DOF. The relationship between translation h and rotation $\theta$ is called **pitch of screw** and is defined by*

$$
p = \frac{h}{\theta}.
\tag{5.181}
$$

*The transformation for a screw joint may be expressed in terms of the relative rotation $\theta$*

$$
{}^iT_j = \begin{bmatrix} \cos\theta & -\sin\theta & 0 & 0 \\ \sin\theta & \cos\theta & 0 & 0 \\ 0 & 0 & 1 & p\theta \\ 0 & 0 & 0 & 1 \end{bmatrix}
\tag{5.182}
$$

*or displacement h*

$$
{}^iT_j = \begin{bmatrix} \cos\frac{h}{p} & -\sin\frac{h}{p} & 0 & 0 \\ \sin\frac{h}{p} & \cos\frac{h}{p} & 0 & 0 \\ 0 & 0 & 1 & h \\ 0 & 0 & 0 & 1 \end{bmatrix}.
\tag{5.183}
$$

    *The coordinate frames are installed on the two connected links at the screw joint such that the axes $w_j$ and $z_i$ are aligned along the screw axis, and the axes $u_j$ and $x_i$ coincide at rest position.*

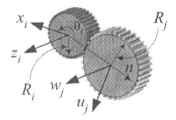

FIGURE 5.33. Illustration of a gear joint to define Sheth coordinate transformation.

**Example 162** ★ *Sheth transformation matrix at a gear joint.*

 *The Sheth method can also be utilized to describe the relative motion of two links connected by a gear joint. A gear joint, as shown in Figure 5.33, provides a proportional rotation, which has 1 DOF. The axes of rotations indicate the axes $w_j$ and $z_i$, and their common perpendicular shows the $\hat{n}$ vector. Then, the Sheth parameters are defined as $a = R_i + R_j$, $b = 0$, $c = 0$, $\alpha = 0$, $\beta = \theta_j$, $\gamma = \theta_i$ , to have*

$$
{}^iT_j =
\begin{bmatrix}
\cos(1+\varepsilon)\,\theta_i & -\sin(1+\varepsilon)\,\theta_i & 0 & R_j\,(1+\varepsilon)\cos\theta_i \\
\sin(1+\varepsilon)\,\theta_i & \cos(1+\varepsilon)\,\theta_i & 0 & R_j\,(1+\varepsilon)\sin\theta_i \\
0 & 0 & 1 & 0 \\
0 & 0 & 0 & 1
\end{bmatrix}
\tag{5.184}
$$

*where*

$$
\varepsilon = \frac{R_i}{R_j}.
\tag{5.185}
$$

## 5.6  Summary

Forward kinematics describes the end-effector coordinate frame in the base coordinate frame of a robot when the joint variables are given.

$$
{}^0\mathbf{r}_P = {}^0T_n\,{}^n\mathbf{r}_P
\tag{5.186}
$$

For an $n$ link serial robot, it is equivalent to finding the transformation matrix ${}^0T_n$ as a function of joint variables

$$
{}^0T_n = {}^0T_1(q_1)\,{}^1T_2(q_2)\,{}^2T_3(q_3)\,{}^3T_4(q_4) \cdots {}^{n-1}T_n(q_n).
\tag{5.187}
$$

There is a special rule for installing the coordinate frames attached to each robot's link called the standard Denavit-Hartenberg convention. Based on the DH rule, each transformation matrix ${}^{i-1}T_i$ from the coordinate frame

$B_i$ to $B_{i-1}$ can be expressed by four parameters; link length $a_i$, link twist $\alpha_i$, joint distance $d_i$, and joint angle $\theta_i$.

$$
{}^{i-1}T_i =
\begin{bmatrix}
\cos\theta_i & -\sin\theta_i \cos\alpha_i & \sin\theta_i \sin\alpha_i & a_i \cos\theta_i \\
\sin\theta_i & \cos\theta_i \cos\alpha_i & -\cos\theta_i \sin\alpha_i & a_i \sin\theta_i \\
0 & \sin\alpha_i & \cos\alpha_i & d_i \\
0 & 0 & 0 & 1
\end{bmatrix}
\tag{5.188}
$$

However, for most industrial robots, the link transformation matrix ${}^{i-1}T_i$ can be classified into 12 simple types.

| 1 | R‖R(0) | or | R‖P(0) |
|---|---|---|---|
| 2 | R‖R(180) | or | R‖P(180) |
| 3 | R⊥R(90) | or | R⊥P(90) |
| 4 | R⊥R(−90) | or | R⊥P(−90) |
| 5 | R⊢R(90) | or | R⊢P(90) |
| 6 | R⊢R(−90) | or | R⊢P(−90) |
| 7 | P‖R(0) | or | P‖P(0) |
| 8 | P‖R(180) | or | P‖P(180) |
| 9 | P⊥R(90) | or | P⊥P(90) |
| 10 | P⊥R(−90) | or | P⊥P(−90) |
| 11 | P⊢R(90) | or | P⊢P(90) |
| 12 | P⊢R(−90) | or | P⊢P(−90) |

Most industrial robots are made of a 3 DOF manipulator equipped with a 3 DOF spherical wrist. The transformation matrix ${}^{0}T_7$ can be decomposed into three submatrices ${}^{0}T_3$, ${}^{3}T_6$ and ${}^{6}T_7$.

$$
{}^{0}T_6 = {}^{0}T_3 \, {}^{3}T_6 \, {}^{6}T_7
\tag{5.189}
$$

The matrix ${}^{0}T_3$ positions the wrist point and depends on the manipulator links and joints. The matrix ${}^{3}T_6$ is the wrist transformation matrix and ${}^{6}T_7$ is the tools transformation matrix. Decomposing ${}^{0}T_7$ into submatrices enables us to make the forward kinematics modular.

# Exercises

1. Notation and symbols.

   Describe the meaning of

   a- $\theta_i$    b- $\alpha_i$    c- $\begin{bmatrix} \phi\hat{u} \\ h\hat{u} \end{bmatrix}$    d- $\check{s}(d_i, \theta_i, \hat{k}_{i-1})$    e- $R\|R(180)$

   f- $^{i-1}T_i$    g- $\otimes$    h- $\begin{bmatrix} \theta_i\,\hat{k}_{i-1} \\ d_i\,\hat{k}_{i-1} \end{bmatrix}$    i- $\check{s}(h, \phi, \hat{u})$    j- $P\bot R(-90)$

   k- $d_i$    l- $a_i$    m- $\begin{bmatrix} \alpha_i\,\hat{i}_{i-1} \\ a_i\,\hat{i}_{i-1} \end{bmatrix}$    n- $\check{s}(a_i, \alpha_i, \hat{i}_{i-1})$    o- $R\vdash R(-90)$.

2. A 4R planar manipulator.

   For the 4R planar manipulator, shown in Figure 5.34, find the

   (a) DH table

   (b) link-type table

   (c) individual frame transformation matrices $^{i-1}T_i$, $i = 1, 2, 3, 4$

   (d) global coordinates of the end-effector

   (e) orientation of the end-effector $\varphi$.

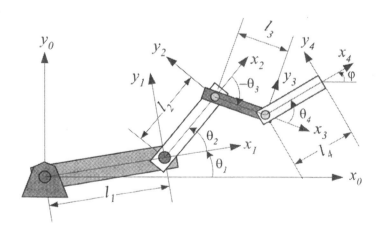

FIGURE 5.34. A 4R planar manipulator.

3. A one-link $R\vdash R(-90)$ arm.

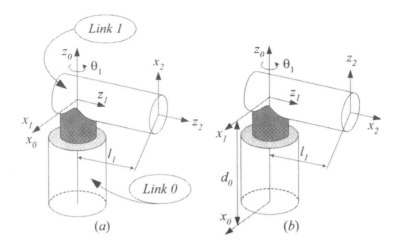

FIGURE 5.35. A one-link R⊢R($-90$) manipulator.

For the one-link R⊢R($-90$) manipulator shown in Figure 5.35 (a) and
(b), find the transformation matrices $^0T_1$, $^1T_2$, and $^0T_2$. Compare the
transformation matrix $^1T_2$ for both frame installations.

4. A 2R planar manipulator.

   Determine the link's transformation matrices $^0T_1$, $^1T_2$, and $^0T_2$ for
   the 2R planar manipulator shown in Figure 5.36.

5. A polar manipulator.

   Determine the link's transformation matrices $^1T_2$, $^2T_3$, and $^1T_3$ for
   the polar manipulator shown in Figure 5.37.

6. A planar Cartesian manipulator.

   Determine the link's transformation matrices $^1T_2$, $^2T_3$, and $^1T_3$ for
   the planar Cartesian manipulator shown in Figure 5.38.

7. Modular articulated manipulators.

   Most of the industrial robots are modular. Some of them are manu-
   factured by attaching a 2 DOF manipulator to a one-link R⊢R($-90$)
   arm. Articulated manipulators are made by attaching a 2R planar
   manipulator, such as the one shown in Figure 5.36, to a one-link
   R⊢R($-90$) manipulator shown in Figure 5.35 (a). Attach the 2R manipulator to the one-link R⊢R($-90$) arm and make an articulated
   manipulator. Make the required changes into the coordinate frames
   of Exercises 3 and 4 to find the link's transformation matrices of the
   articulated manipulator. Examine the rest position of the manipulator.

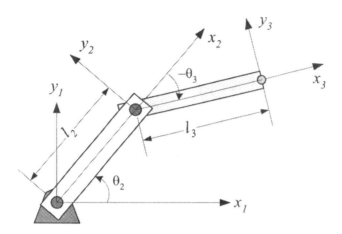

FIGURE 5.36. A 2R planar manipulator.

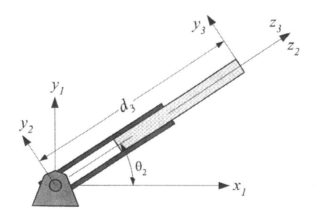

FIGURE 5.37. A 2 DOF polar manipulator.

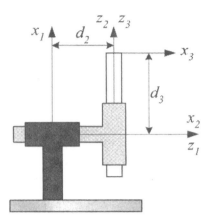

FIGURE 5.38. A 2 DOF Cartesian manipulator.

8. Modular spherical manipulators.

   Spherical manipulators are made by attaching a polar manipulator shown in Figure 5.37, to a one-link R⊢R(−90) manipulator shown in Figure 5.35 (b). Attach the polar manipulator to the one-link R⊢R(−90) arm and make a spherical manipulator. Make the required changes to the coordinate frames Exercises 3 and 5 to find the link's transformation matrices of the spherical manipulator. Examine the rest position of the manipulator.

9. Modular cylindrical manipulators.

   Cylindrical manipulators are made by attaching a 2 DOF Cartesian manipulator shown in Figure 5.38, to a one-link R⊢R(−90) manipulator shown in Figure 5.35 (a). Attach the 2 DOF Cartesian manipulator to the one-link R⊢R(−90) arm and make a cylindrical manipulator. Make the required changes into the coordinate frames of Exercises 3 and 6 to find the link's transformation matrices of the cylindrical manipulator. Examine the rest position of the manipulator.

10. Disassembled spherical wrist.

    A spherical wrist has three revolute joints in such a way that their joint axes intersect at a common point, called the wrist point. Each revolute joint of the wrist attaches two links. Disassembled links of a spherical wrist are shown in Figure 5.39. Define the required DH coordinate frames to the links in (a), (b), and (c) consistently. Find the transformation matrices $^3T_4$ for (a), $^4T_5$ for (b), and $^5T_6$ for (c).

11. Assembled spherical wrist.

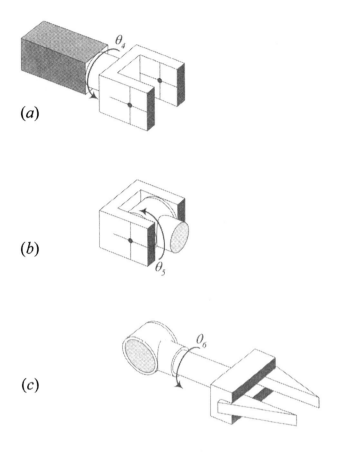

FIGURE 5.39. Disassembled links of a spherical wrist.

Label the coordinate frames attached to the spherical wrist in Figure 5.40 according to the frames that you installed in Exercise 10. Determine the transformation matrices $^3T_6$ and $^3T_7$ for the wrist.

12. Articulated robots.

Attach the spherical wrist of Exercise 11 to the articulated manipulator of Exercise 7 and make a 6 DOF articulated robot. Change your DH coordinate frames in the exercises accordingly and solve the forward kinematics problem of the robot.

13. Spherical robots.

Attach the spherical wrist of Exercise 11 to the spherical manipulator of Exercise 8 and make a 6 DOF spherical robot. Change your DH coordinate frames in the exercises accordingly and solve the forward kinematics problem of the robot.

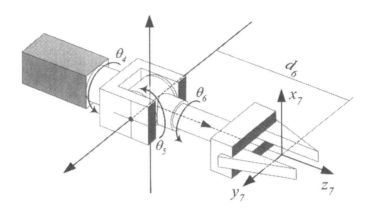

FIGURE 5.40. Assembled spherical wrist.

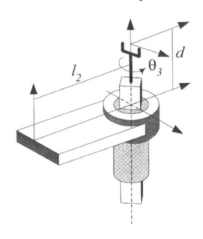

FIGURE 5.41. A 2 DOF R∥P manipulator

14. Cylindrical robots.

    Attach the spherical wrist of Exercise 11 to the cylindrical manip-
    ulator of Exercise 9 and make a 6 DOF cylindrical robot. Change
    your DH coordinate frames in the exercises accordingly and solve the
    forward kinematics problem of the robot.

15. An R∥P manipulator.

    Figure 5.41 shows a 2 DOF R∥P manipulator. The end-effector of the
    manipulator can slide on a line and rotate about the same line. Label
    the coordinate frames installed on the links of the manipulator and
    determine the transformation matrix of the end-effector to the base
    link.

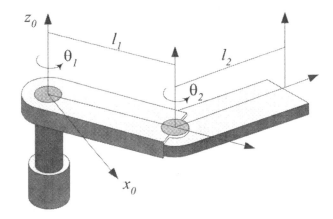

FIGURE 5.42. A 2R manipulator, acting in a horizontal plane.

16. Horizontal 2R manipulator

    Figure 5.41 illustrates a 2R planar manipulator that acts in a horizontal plane. Label the coordinate frames and determine the transformation matrix of the end-effector in the base frame.

17. SCARA manipulator.

    A SCARA robot can be made by attaching a 2 DOF R∥P manipulator to a 2R planar manipulator. Attach the 2 DOF R∥P manipulator of Exercise 15 to the 2R horizontal manipulator of Exercise 16 and make a SCARA manipulator. Solve the forward kinematics problem for the manipulator.

18. ★ SCARA robot with a spherical wrist.

    Attach the spherical wrist of Exercise 11 to the SCARA manipulator of Exercise 17 and make a 7 DOF robot. Change your DH coordinate frames in the exercises accordingly and solve the forward kinematics problem of the robot.

19. ★ Modular articulated manipulators by screws.

    Solve Exercise 7 by screws.

20. ★ Modular spherical manipulators by screws.

    Solve Exercise 8 by screws.

21. ★ Modular cylindrical manipulators by screws.

    Solve Exercise 9 by screws.

22. ★ Spherical wrist kinematics by screws.

    Solve Exercise 11 by screws.

23. ★ Modular SCARA manipulator by screws.

    Solve Exercise 15, 16, and 17 by screws.

24. ★ Space station remote manipulator system.

    Attach a spherical wrist to the SSRMS and make a 10 DOF robot. Solve the forward kinematics of the robot by matrix and screw methods.

# 6

# Inverse Kinematics

## 6.1 Decoupling Technique

Determination of joint variables in terms of the end-effector position and orientation is called *inverse kinematics*. Mathematically, inverse kinematics is searching for the elements of vector $\mathbf{q}$ when a transformation is given as a function of the joint variables $q_1, q_2, q_3, \cdots$.

$$^0T_n = {}^0T_1(q_1)\,{}^1T_2(q_2)\,{}^2T_3(q_3)\,{}^3T_4(q_4) \cdots {}^{n-1}T_n(q_n) \qquad (6.1)$$

Computer controlled robots are usually actuated in the joint variable space, however objects to be manipulated are usually expressed in the global coordinate frame. Therefore, carrying kinematic information, back and forth, between joint space and Cartesian space, is a need in robot applications. To control the configuration of the end-effector to reach an object, the inverse kinematics solution must be solved. Hence, we need to know what the required values of joint variables are, to reach a desired point in a desired orientation.

The result of forward kinematics of a 6 DOF robot is a $4 \times 4$ transformation matrix

$$
\begin{aligned}
{}^0T_6 &= {}^0T_1\,{}^1T_2\,{}^2T_3\,{}^3T_4\,{}^4T_5\,{}^5T_6 \\
&= \begin{bmatrix} r_{11} & r_{12} & r_{13} & r_{14} \\ r_{21} & r_{22} & r_{23} & r_{24} \\ r_{31} & r_{32} & r_{33} & r_{34} \\ 0 & 0 & 0 & 1 \end{bmatrix}
\end{aligned} \qquad (6.2)
$$

where 12 elements are trigonometric functions of six unknown joint variables. However, since the upper left $3 \times 3$ submatrix of (6.2) is a rotation matrix, only three elements of them are independent. This is because of the orthogonality condition (2.129). Hence, only six equations out of the 12 equations of (6.2) are independent.

Trigonometric functions inherently provide multiple solutions. Therefore, multiple configurations of the robot are expected when the six equations are solved for the unknown joint variables.

It is possible to decouple the inverse kinematics problem into two subproblems, known as *inverse position* and *inverse orientation* kinematics. The practical consequence of such a decoupling is the allowance to break the problem into two independent problems, each with only three unknown parameters. Following the decoupling principle, the overall transformation matrix of a robot can be decomposed to a translation and a rotation

$$
\begin{aligned}
{}^0T_6 \;&=\; {}^0D_6\,{}^0R_6 \\
&=\; \begin{bmatrix} \mathbf{I} & {}^0\mathbf{d}_6 \\ 0 & 1 \end{bmatrix} \begin{bmatrix} {}^0R_6 & \mathbf{0} \\ 0 & 1 \end{bmatrix} \\
&=\; \begin{bmatrix} {}^0R_6 & {}^0\mathbf{d}_6 \\ 0 & 1 \end{bmatrix}.
\end{aligned}
\tag{6.3}
$$

The translation matrix ${}^0D_6$ can be solved for wrist position variables, and the rotation matrix ${}^0R_6$ can be solved for wrist orientation variables.

**Proof.** Most robots have a wrist made of three revolute joints with intersecting and orthogonal axes at the wrist point. Taking advantage of having a spherical wrist, we can decouple the kinematics of the wrist and manipulator by decomposing the overall forward kinematics transformation matrix ${}^0T_6$ into the wrist orientation and wrist position

$$
\begin{aligned}
{}^0T_6 \;&=\; {}^0T_3\,{}^3T_6 \\
&=\; \begin{bmatrix} {}^0R_3 & {}^0\mathbf{d}_6 \\ 0 & 1 \end{bmatrix} \begin{bmatrix} {}^3R_6 & \mathbf{0} \\ 0 & 1 \end{bmatrix}
\end{aligned}
\tag{6.4}
$$

where the wrist orientation matrix is

$$
{}^3R_6 = {}^0R_3^T\,{}^0R_6 = {}^0R_3^T \begin{bmatrix} r_{11} & r_{12} & r_{13} \\ r_{21} & r_{22} & r_{23} \\ r_{31} & r_{32} & r_{33} \end{bmatrix}
\tag{6.5}
$$

and the wrist position vector is

$$
{}^0\mathbf{d}_6 = \begin{bmatrix} r_{14} \\ r_{24} \\ r_{34} \end{bmatrix}.
\tag{6.6}
$$

The wrist position vector includes the manipulator joint variables only. Hence, to solve the inverse kinematics of such a robot, we must solve ${}^0T_3$ for position of the wrist point, and then solve ${}^3T_6$ for orientation of the wrist.

The components of the wrist position vector ${}^0\mathbf{d}_6 = \mathbf{d}_w$ provides three equations for the three unknown manipulator joint variables. Solving $\mathbf{d}_w$, for manipulator joint variables, leads to calculating ${}^3R_6$ from (6.5). Then, the wrist orientation matrix ${}^3R_6$ can be solved for wrist joint variables.

In case we include the tool coordinate frame in forward kinematics, the decomposition must be done according to the following equation to exclude the effect of tool distance $d_6$ from the robot's kinematics.

$$
\begin{aligned}
{}^{0}T_7 &= {}^{0}T_3\,{}^{3}T_7 \\
&= {}^{0}T_3\,{}^{3}T_6\,{}^{6}T_7 \\
&= \begin{bmatrix} {}^{0}R_3 & \mathbf{d}_w \\ 0 & 1 \end{bmatrix} \begin{bmatrix} {}^{3}R_6 & 0 \\ 0 & 1 \end{bmatrix} \begin{bmatrix} \mathbf{I} & \begin{matrix} 0 \\ 0 \\ d_6 \end{matrix} \\ 0 & 1 \end{bmatrix}
\end{aligned}
\tag{6.7}
$$

In this case, inverse kinematics starts from determination of ${}^{0}T_6$, which can be found by

$$
\begin{aligned}
{}^{0}T_6 &= {}^{0}T_7\,{}^{6}T_7^{-1} \tag{6.8} \\
&= {}^{0}T_7 \begin{bmatrix} 1 & 0 & 0 & 0 \\ 0 & 1 & 0 & 0 \\ 0 & 0 & 1 & d_6 \\ 0 & 0 & 0 & 1 \end{bmatrix}^{-1} \\
&= {}^{0}T_7 \begin{bmatrix} 1 & 0 & 0 & 0 \\ 0 & 1 & 0 & 0 \\ 0 & 0 & 1 & -d_6 \\ 0 & 0 & 0 & 1 \end{bmatrix}.
\end{aligned}
$$

∎

**Example 163** *Inverse kinematics of an articulated robot.*

*The forward kinematics of the articulated robot, illustrated in Figure 6.1, was found in Example 151, where the overall transformation matrix of the end-effector was found, based on the wrist and arm transformation matrices.*

$$
\begin{aligned}
{}^{0}T_7 &= T_{arm}T_{wrist} \tag{6.9} \\
&= {}^{0}T_3\,{}^{3}T_7
\end{aligned}
$$

*The wrist transformation matrix $T_{wrist}$ is described in (5.73) and the manipulator transformation matrix, $T_{arm}$ is found in (5.80). However, according to a new setup coordinate frame, as shown in Figure 6.1, we have a 6R robot with a six links configuration*

| 1 | $R \vdash R(90)$ |
|---|---|
| 2 | $R \| R(0)$ |
| 3 | $R \vdash R(90)$ |
| 4 | $R \vdash R(-90)$ |
| 5 | $R \vdash R(90)$ |
| 6 | $R \| R(0)$ |

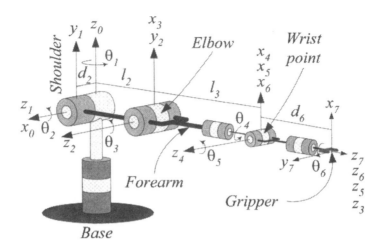

FIGURE 6.1. A 6 DOF articulated manipulator.

and a displacement $T_{Z,d_6}$. Therefore, the individual links' transformation matrices are

$$
^0T_1 = \begin{bmatrix} \cos\theta_1 & 0 & \sin\theta_1 & 0 \\ \sin\theta_1 & 0 & -\cos\theta_1 & 0 \\ 0 & 1 & 0 & 0 \\ 0 & 0 & 0 & 1 \end{bmatrix} \tag{6.10}
$$

$$
^1T_2 = \begin{bmatrix} \cos\theta_2 & -\sin\theta_2 & 0 & l_2\cos\theta_2 \\ \sin\theta_2 & \cos\theta_2 & 0 & l_2\sin\theta_2 \\ 0 & 0 & 1 & d_2 \\ 0 & 0 & 0 & 1 \end{bmatrix} \tag{6.11}
$$

$$
^2T_3 = \begin{bmatrix} \cos\theta_3 & 0 & \sin\theta_3 & 0 \\ \sin\theta_3 & 0 & -\cos\theta_3 & 0 \\ 0 & 1 & 0 & 0 \\ 0 & 0 & 0 & 1 \end{bmatrix} \tag{6.12}
$$

$$
^3T_4 = \begin{bmatrix} \cos\theta_4 & 0 & -\sin\theta_4 & 0 \\ \sin\theta_4 & 0 & \cos\theta_4 & 0 \\ 0 & -1 & 0 & l_3 \\ 0 & 0 & 0 & 1 \end{bmatrix} \tag{6.13}
$$

$$
^4T_5 = \begin{bmatrix} \cos\theta_5 & 0 & \sin\theta_5 & 0 \\ \sin\theta_5 & 0 & -\cos\theta_5 & 0 \\ 0 & 1 & 0 & 0 \\ 0 & 0 & 0 & 1 \end{bmatrix} \tag{6.14}
$$

$$^5T_6 = \begin{bmatrix} \cos\theta_6 & -\sin\theta_6 & 0 & 0 \\ \sin\theta_6 & \cos\theta_6 & 0 & 0 \\ 0 & 0 & 1 & 0 \\ 0 & 0 & 0 & 1 \end{bmatrix} \tag{6.15}$$

$$^6T_7 = \begin{bmatrix} 1 & 0 & 0 & 0 \\ 0 & 1 & 0 & 0 \\ 0 & 0 & 1 & d_6 \\ 0 & 0 & 0 & 1 \end{bmatrix} \tag{6.16}$$

and the tool transformation matrix in the base coordinate frame is

$$
\begin{aligned}
^0T_7 &= \ ^0T_1\,^1T_2\,^2T_3\,^3T_4\,^4T_5\,^5T_6\,^6T_7 \tag{6.17} \\
&= \ ^0T_3\,^3T_6\,^6T_7 \\
&= \begin{bmatrix} t_{11} & t_{12} & t_{13} & t_{14} \\ t_{21} & t_{22} & t_{23} & t_{24} \\ t_{31} & t_{32} & t_{33} & t_{34} \\ 0 & 0 & 0 & 1 \end{bmatrix}
\end{aligned}
$$

where

$$^0T_3 = \begin{bmatrix} c\theta_1 c(\theta_2+\theta_3) & s\theta_1 & c\theta_1 s(\theta_2+\theta_3) & l_2 c\theta_1 c\theta_2 + d_2 s\theta_1 \\ s\theta_1 c(\theta_2+\theta_3) & -c\theta_1 & s\theta_1 s(\theta_2+\theta_3) & l_2 c\theta_2 s\theta_1 - d_2 c\theta_1 \\ s(\theta_2+\theta_3) & 0 & -c(\theta_2+\theta_3) & l_2 s\theta_2 \\ 0 & 0 & 0 & 1 \end{bmatrix} \tag{6.18}$$

$$^3T_6 = \begin{bmatrix} c\theta_4 c\theta_5 c\theta_6 - s\theta_4 s\theta_6 & -c\theta_6 s\theta_4 - c\theta_4 c\theta_5 s\theta_6 & c\theta_4 s\theta_5 & 0 \\ c\theta_5 c\theta_6 s\theta_4 + c\theta_4 s\theta_6 & c\theta_4 c c\theta_6 - c\theta_5 s\theta_4 s\theta_6 & s\theta_4 s\theta_5 & 0 \\ -c\theta_6 s\theta_5 & s\theta_5 s\theta_6 & c\theta_5 & l_3 \\ 0 & 0 & 0 & 1 \end{bmatrix} \tag{6.19}$$

and

$$
\begin{aligned}
t_{11} &= c\theta_1 \left( c\,(\theta_2+\theta_3)\,(c\theta_4 c\theta_5 c\theta_6 - s\theta_4 s\theta_6) - c\theta_6 s\theta_5 s\,(\theta_2+\theta_3) \right) \\
&\quad + s\theta_1 \left( c\theta_4 s\theta_6 + c\theta_5 c\theta_6 s\theta_4 \right) \tag{6.20} \\
t_{21} &= s\theta_1 \left( c\,(\theta_2+\theta_3)\,(-s\theta_4 s\theta_6 + c\theta_4 c\theta_5 c\theta_6) - c\theta_6 s\theta_5 s\,(\theta_2+\theta_3) \right) \\
&\quad - c\theta_1 \left( c\theta_4 s\theta_6 + c\theta_5 c\theta_6 s\theta_4 \right) \tag{6.21} \\
t_{31} &= s\,(\theta_2+\theta_3)\,(c\theta_4 c\theta_5 c\theta_6 - s\theta_4 s\theta_6) + c\theta_6 s\theta_5 c\,(\theta_2+\theta_3) \tag{6.22}
\end{aligned}
$$

$$
\begin{aligned}
t_{12} &= c\theta_1 \left( s\theta_5 s\theta_6 s\,(\theta_2+\theta_3) - c\,(\theta_2+\theta_3)\,(c\theta_6 s\theta_4 + c\theta_4 c\theta_5 s\theta_6) \right) \\
&\quad + s\theta_1 \left( c\theta_4 c\theta_6 - c\theta_5 s\theta_4 s\theta_6 \right) \tag{6.23} \\
t_{22} &= s\theta_1 \left( s\theta_5 s\theta_6 s\,(\theta_2+\theta_3) - c\,(\theta_2+\theta_3)\,(c\theta_6 s\theta_4 + c\theta_4 c\theta_5 s\theta_6) \right) \\
&\quad + c\theta_1 \left( -c\theta_4 c\theta_6 + c\theta_5 s\theta_4 s\theta_6 \right) \tag{6.24} \\
t_{32} &= -s\theta_5 s\theta_6 c\,(\theta_2+\theta_3) - s\,(\theta_2+\theta_3)\,(c\theta_6 s\theta_4 + c\theta_4 c\theta_5 s\theta_6) \tag{6.25}
\end{aligned}
$$

$$t_{13} = s\theta_1 s\theta_4 s\theta_5 + c\theta_1 \left( c\theta_5 s \left( \theta_2 + \theta_3 \right) + c\theta_4 s\theta_5 c \left( \theta_2 + \theta_3 \right) \right) \quad (6.26)$$

$$t_{23} = -c\theta_1 s\theta_4 s\theta_5 + s\theta_1 \left( c\theta_5 s \left( \theta_2 + \theta_3 \right) + c\theta_4 s\theta_5 c \left( \theta_2 + \theta_3 \right) \right) \quad (6.27)$$

$$t_{33} = c\theta_4 s\theta_5 s \left( \theta_2 + \theta_3 \right) - c\theta_5 c \left( \theta_2 + \theta_3 \right) \quad (6.28)$$

$$t_{14} = d_6 \left( s\theta_1 s\theta_4 s\theta_5 + c\theta_1 \left( c\theta_4 s\theta_5 c \left( \theta_2 + \theta_3 \right) + c\theta_5 s \left( \theta_2 + \theta_3 \right) \right) \right)$$
$$+ l_3 c\theta_1 s \left( \theta_2 + \theta_3 \right) + d_2 s\theta_1 + l_2 c\theta_1 c\theta_2 \quad (6.29)$$

$$t_{24} = d_6 \left( -c\theta_1 s\theta_4 s\theta_5 + s\theta_1 \left( c\theta_4 s\theta_5 c \left( \theta_2 + \theta_3 \right) + c\theta_5 s \left( \theta_2 + \theta_3 \right) \right) \right)$$
$$+ s\theta_1 s \left( \theta_2 + \theta_3 \right) l_3 - d_2 c\theta_1 + l_2 c\theta_2 s\theta_1 \quad (6.30)$$

$$t_{34} = d_6 \left( c\theta_4 s\theta_5 s \left( \theta_2 + \theta_3 \right) - c\theta_5 c \left( \theta_2 + \theta_3 \right) \right)$$
$$+ l_2 s\theta_2 + l_3 c \left( \theta_2 + \theta_3 \right). \quad (6.31)$$

*Solution of the inverse kinematics problem starts with the wrist position vector* $\mathbf{d}$, *which is* $\begin{bmatrix} t_{14} & t_{24} & t_{34} \end{bmatrix}^T$ *of* $^0T_7$ *for* $d_6 = 0$

$$\mathbf{d} = \begin{bmatrix} c\theta_1 \left( l_3 s \left( \theta_2 + \theta_3 \right) + l_2 c\theta_2 \right) + d_2 s\theta_1 \\ s\theta_1 \left( l_3 s \left( \theta_2 + \theta_3 \right) + l_2 c\theta_2 \right) - d_2 c\theta_1 \\ l_3 c \left( \theta_2 + \theta_3 \right) + l_2 s\theta_2 \end{bmatrix} = \begin{bmatrix} d_x \\ d_y \\ d_z \end{bmatrix}. \quad (6.32)$$

*Theoretically, we must be able to solve Equation (6.32) for the three joint variables* $\theta_1$, $\theta_2$, *and* $\theta_3$. *It can be seen that*

$$d_x \sin \varphi_1 - d_y \cos \theta_1 = d_2 \quad (6.33)$$

*which provides*

$$\theta_1 = 2 \operatorname{atan2}(d_x \pm \sqrt{d_x^2 + d_y^2 - d_2^2}, d_2 - d_y). \quad (6.34)$$

*Equation (6.34) has two solutions for* $d_x^2 + d_y^2 > d_2^2$, *one solution for* $d_x^2 + d_y^2 = d_2^2$, *and no real solution for* $d_x^2 + d_y^2 < d_2^2$.
*Combining the first two elements of* $\mathbf{d}$ *gives*

$$l_3 \sin \left( \theta_2 + \theta_3 \right) = \pm \sqrt{d_x^2 + d_y^2 - d_2^2} - l_2 \cos \theta_2 \quad (6.35)$$

*then, the third element of* $\mathbf{d}$ *may be utilized to find*

$$l_3^2 = \left( \pm \sqrt{d_x^2 + d_y^2 - d_2^2} - l_2 \cos \theta_2 \right)^2 + (d_z - l_2 \sin \theta_2)^2 \quad (6.36)$$

*which can be rearranged to the following form*

$$a \cos \theta_2 + b \sin \theta_2 = c \quad (6.37)$$

$$a = 2 l_2 \sqrt{d_x^2 + d_y^2 - d_2^2} \quad (6.38)$$

$$b = 2 l_2 d_z \quad (6.39)$$

$$c = d_x^2 + d_y^2 + d_z^2 - d_2^2 + l_2^2 - l_3^2. \quad (6.40)$$

*with two solutions*

$$\theta_2 = \text{atan2}(\frac{c}{r}, \pm\sqrt{1 - \frac{c^2}{r^2}}) - \text{atan2}(a, b) \tag{6.41}$$

$$\theta_2 = \text{atan2}(\frac{c}{r}, \pm\sqrt{r^2 - c^2}) - \text{atan2}(a, b) \tag{6.42}$$

$$r^2 = a^2 + b^2. \tag{6.43}$$

*Summing the squares of the elements of* **d** *gives*

$$d_x^2 + d_y^2 + d_z^2 = d_2^2 + l_2^2 + l_3^2 + 2l_2l_3 \sin(2\theta_2 + \theta_3) \tag{6.44}$$

*that provides*

$$\theta_3 = \arcsin\left(\frac{d_x^2 + d_y^2 + d_z^2 - d_2^2 - l_2^2 - l_3^2}{2l_2l_3}\right) - 2\theta_2. \tag{6.45}$$

*Having $\theta_1$, $\theta_2$, and $\theta_3$ means we can find the wrist point in space. However, because the joint variables in $^0T_3$ and in $^3T_6$ are independent, we should find the orientation of the end-effector by solving $^3T_6$ or $^3R_6$ for $\theta_4$, $\theta_5$, and $\theta_6$.*

$$^3R_6 = \begin{bmatrix} c\theta_4 c\theta_5 c\theta_6 - s\theta_4 s\theta_6 & -c\theta_6 s\theta_4 - c\theta_4 c\theta_5 s\theta_6 & c\theta_4 s\theta_5 \\ c\theta_5 c\theta_6 s\theta_4 + c\theta_4 s\theta_6 & c\theta_4 c c\theta_6 - c\theta_5 s\theta_4 s\theta_6 & s\theta_4 s\theta_5 \\ -c\theta_6 s\theta_5 & s\theta_5 s\theta_6 & c\theta_5 \end{bmatrix}$$

$$= \begin{bmatrix} s_{11} & s_{12} & s_{13} \\ s_{21} & s_{22} & s_{23} \\ s_{31} & s_{32} & s_{33} \end{bmatrix} \tag{6.46}$$

*The angles $\theta_4$, $\theta_5$, and $\theta_6$ can be found by examining elements of $^3R_6$*

$$\theta_4 = \text{atan2}(s_{23}, s_{13}) \tag{6.47}$$

$$\theta_5 = \text{atan2}\left(\sqrt{s_{13}^2 + s_{23}^2}, s_{33}\right) \tag{6.48}$$

$$\theta_6 = \text{atan2}(s_{32}, -s_{31}). \tag{6.49}$$

**Example 164** *Solution of equation $a\cos\theta + b\sin\theta = c$.*
*The first type of trigonometric equation*

$$a\cos\theta + b\sin\theta = c \tag{6.50}$$

*can be solved by introducing two new variables $r$ and $\phi$ such that*

$$a = r\sin\phi \tag{6.51}$$

$$b = r\cos\phi \tag{6.52}$$

*and*

$$r = \sqrt{a^2 + b^2} \tag{6.53}$$

$$\phi = \text{atan2}(a, b). \tag{6.54}$$

*Substituting the new variables show that*

$$\sin(\phi + \theta) = \frac{c}{r} \tag{6.55}$$

$$\cos(\phi + \theta) = \pm\sqrt{1 - \frac{c^2}{r^2}}. \tag{6.56}$$

*Hence, the solutions of the problem are*

$$\theta = \text{atan2}(\frac{c}{r}, \pm\sqrt{1 - \frac{c^2}{r^2}}) - \text{atan2}(a, b) \tag{6.57}$$

*and*

$$\theta = \text{atan2}(\frac{c}{r}, \pm\sqrt{r^2 - c^2}) - \text{atan2}(a, b). \tag{6.58}$$

*Therefore, the equation $a\cos\theta + b\sin\theta = c$ has two solutions if $r^2 = a^2 + b^2 > c^2$, one solution if $r^2 = c^2$, and no solution if $r^2 < c^2$.*

**Example 165** *Meaning of the function $\tan_2^{-1}\frac{y}{x} = \text{atan2}(y, x)$.*

*In robotic calculation, specially in solving inverse kinematics problems, we need to find an angle based on the $\sin$ and $\cos$ functions of the angle. However, $\tan^{-1}$ cannot show the effect of the individual sign for the numerator and denominator. It always represents an angle in the first or fourth quadrant. To overcome this problem and determine the joint angles in the correct quadrant, the atan2 function is introduced.*

$$
\begin{aligned}
\text{atan2}(y, x) &= \tan^{-1}\frac{y}{x} & \text{if} & \quad y > 0 \\
&= \tan^{-1}\frac{y}{x} + \pi\,\text{sign}\,y & \text{if} & \quad y < 0 \quad (6.59) \\
&= \frac{\pi}{2}\,\text{sign}\,x & \text{if} & \quad y = 0
\end{aligned}
$$

(handwritten: ✗ wrong)

*In this text, whether it has been mentioned or not, wherever $\tan^{-1}\frac{y}{x}$ is used, it must be calculated based on atan2$(y, x)$.*

## 6.2   Inverse Transformation Technique

Assume we have the transformation matrix $^0T_6$ indicating the global position and the orientation of the end-effector of a 6 DOF robot in the base frame $B_0$. Furthermore, assume the geometry and individual transformation matrices $^0T_1(q_1)$, $^1T_2(q_2)$, $^2T_3(q_3)$, $^3T_4(q_4)$, $^4T_5(q_5)$, and $^5T_6(q_6)$ are given as functions of joint variables.

According to forward kinematics,

$$^0T_6 = {}^0T_1\,{}^1T_2\,{}^2T_3\,{}^3T_4\,{}^4T_5\,{}^5T_6 \tag{6.60}$$

$$= \begin{bmatrix} r_{11} & r_{12} & r_{13} & r_{14} \\ r_{21} & r_{22} & r_{23} & r_{24} \\ r_{31} & r_{32} & r_{33} & r_{34} \\ 0 & 0 & 0 & 1 \end{bmatrix}.$$

We can solve the inverse kinematics problem by solving the following equations for the unknown joint variables:

$$^1T_6 = {}^0T_1^{-1}\,{}^0T_6 \tag{6.61}$$

$$^2T_6 = {}^1T_2^{-1}\,{}^0T_1^{-1}\,{}^0T_6 \tag{6.62}$$

$$^3T_6 = {}^2T_3^{-1}\,{}^1T_2^{-1}\,{}^0T_1^{-1}\,{}^0T_6 \tag{6.63}$$

$$^4T_6 = {}^3T_4^{-1}\,{}^2T_3^{-1}\,{}^1T_2^{-1}\,{}^0T_1^{-1}\,{}^0T_6 \tag{6.64}$$

$$^5T_6 = {}^4T_5^{-1}\,{}^3T_4^{-1}\,{}^2T_3^{-1}\,{}^1T_2^{-1}\,{}^0T_1^{-1}\,{}^0T_6 \tag{6.65}$$

$$I = {}^5T_6^{-1}\,{}^4T_5^{-1}\,{}^3T_4^{-1}\,{}^2T_3^{-1}\,{}^1T_2^{-1}\,{}^0T_1^{-1}\,{}^0T_6 \tag{6.66}$$

**Proof.** We multiply both sides of the transformation matrix $^0T_6$ by $^0T_1^{-1}$ to obtain

$$^0T_1^{-1}\,{}^0T_6 = {}^0T_1^{-1}\left({}^0T_1\,{}^1T_2\,{}^2T_3\,{}^3T_4\,{}^4T_5\,{}^5T_6\right)$$

$$= {}^1T_6. \tag{6.67}$$

Note that $^0T_1^{-1}$ is the mathematical inverse of the $4 \times 4$ matrix $^0T_1$, and not an inverse transformation. So, $^0T_1^{-1}$ must be calculated by a mathematical matrix inversion.

The left-hand side of the Equation (6.67) is a function of $q_1$. However, the elements of the matrix $^1T_6$ on the right-hand side are either zero, constant, or functions of $q_2$, $q_3$, $q_4$, $q_5$, and $q_6$. The zero or constant elements of the right-hand side provides the required algebraic equation to be solved for $q_1$.

Then, we multiply both sides of (6.67) by $^1T_2^{-1}$ to obtain

$$^1T_2^{-1}\,{}^0T_1^{-1}\,{}^0T_6 = {}^1T_2^{-1}\,{}^0T_1^{-1}\left({}^0T_1\,{}^1T_2\,{}^2T_3\,{}^3T_4\,{}^4T_5\,{}^5T_6\right)$$

$$= {}^2T_6. \tag{6.68}$$

The left-hand side of the this equation is a function of $q_2$, while the elements of the matrix $^2T_6$, on the right hand side, are either zero, constant, or functions of $q_3$, $q_4$, $q_5$, and $q_6$. Equating the associated element, with constant or zero elements on the right-hand side, provides the required algebraic equation to be solved for $q_2$.

Following this procedure, we can find the joint variables $q_3$, $q_4$, $q_5$, and $q_6$ by using the following equalities respectively.

$$^2T_3^{-1}\,{}^1T_2^{-1}\,{}^0T_1^{-1}\,{}^0T_6$$
$$= {}^2T_3^{-1}\,{}^1T_2^{-1}\,{}^0T_1^{-1}\left({}^0T_1\,{}^1T_2\,{}^2T_3\,{}^3T_4\,{}^4T_5\,{}^5T_6\right) \tag{6.69}$$
$$= {}^3T_6.$$

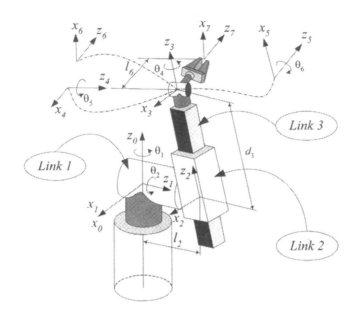

FIGURE 6.2. A spherical robot, made of a spherical manipulator attached to a spherical wrist.

$$
\begin{aligned}
&^3T_4^{-1}\,{}^2T_3^{-1}\,{}^1T_2^{-1}\,{}^0T_1^{-1}\,{}^0T_6 \\
&= {}^3T_4^{-1}\,{}^2T_3^{-1}\,{}^1T_2^{-1}\,{}^0T_1^{-1}\left({}^0T_1\,{}^1T_2\,{}^2T_3\,{}^3T_4\,{}^4T_5\,{}^5T_6\right) \\
&= {}^4T_6.
\end{aligned}
\qquad (6.70)
$$

$$
\begin{aligned}
&^4T_5^{-1}\,{}^3T_4^{-1}\,{}^2T_3^{-1}\,{}^1T_2^{-1}\,{}^0T_1^{-1}\,{}^0T_6 \\
&= {}^4T_5^{-1}\,{}^3T_4^{-1}\,{}^2T_3^{-1}\,{}^1T_2^{-1}\,{}^0T_1^{-1}\left({}^0T_1\,{}^1T_2\,{}^2T_3\,{}^3T_4\,{}^4T_5\,{}^5T_6\right) \\
&= {}^5T_6.
\end{aligned}
\qquad (6.71)
$$

$$
\begin{aligned}
&^5T_6^{-1}\,{}^4T_5^{-1}\,{}^3T_4^{-1}\,{}^2T_3^{-1}\,{}^1T_2^{-1}\,{}^0T_1^{-1}\,{}^0T_6 \\
&= {}^5T_6^{-1}\,{}^4T_5^{-1}\,{}^3T_4^{-1}\,{}^2T_3^{-1}\,{}^1T_2^{-1}\,{}^0T_1^{-1}\left({}^0T_1\,{}^1T_2\,{}^2T_3\,{}^3T_4\,{}^4T_5\,{}^5T_6\right) \\
&= \mathbf{I}.
\end{aligned}
\qquad (6.72)
$$

The *inverse transformation technique* may sometimes be called *Pieper technique*. ∎

**Example 166** *Inverse kinematics for a spherical robot.*
*Transformation matrices of the spherical robot shown in Figure 6.2 are*

$$
^0T_1 = \begin{bmatrix} c\theta_1 & 0 & -s\theta_1 & 0 \\ s\theta_1 & 0 & c\theta_1 & 0 \\ 0 & -1 & 0 & 0 \\ 0 & 0 & 0 & 1 \end{bmatrix}
\qquad
^1T_2 = \begin{bmatrix} c\theta_2 & 0 & s\theta_2 & 0 \\ s\theta_2 & 0 & -c\theta_2 & 0 \\ 0 & 1 & 0 & l_2 \\ 0 & 0 & 0 & 1 \end{bmatrix}
$$

$$^2T_3 = \begin{bmatrix} 1 & 0 & 0 & 0 \\ 0 & 1 & 0 & 0 \\ 0 & 0 & 1 & d_3 \\ 0 & 0 & 0 & 1 \end{bmatrix} \qquad ^3T_4 = \begin{bmatrix} c\theta_4 & 0 & -s\theta_4 & 0 \\ s\theta_4 & 0 & c\theta_4 & 0 \\ 0 & -1 & 0 & 0 \\ 0 & 0 & 0 & 1 \end{bmatrix}$$

$$^4T_5 = \begin{bmatrix} c\theta_5 & 0 & s\theta_5 & 0 \\ s\theta_5 & 0 & -c\theta_5 & 0 \\ 0 & 1 & 0 & 0 \\ 0 & 0 & 0 & 1 \end{bmatrix} \qquad ^5T_6 = \begin{bmatrix} c\theta_6 & -s\theta_6 & 0 & 0 \\ s\theta_6 & c\theta_6 & 0 & 0 \\ 0 & 0 & 1 & 0 \\ 0 & 0 & 0 & 1 \end{bmatrix} \quad (6.73)$$

*Therefore, the position and orientation of the end-effector for a set of joint variables, which solves the forward kinematics problem, can be found by matrix multiplication*

$$
\begin{aligned}
^0T_6 &= {}^0T_1\,{}^1T_2\,{}^2T_3\,{}^3T_4\,{}^4T_5\,{}^5T_6 \\
&= \begin{bmatrix} r_{11} & r_{12} & r_{13} & r_{14} \\ r_{21} & r_{22} & r_{23} & r_{24} \\ r_{31} & r_{32} & r_{33} & r_{34} \\ 0 & 0 & 0 & 1 \end{bmatrix}
\end{aligned}
\qquad (6.74)
$$

*where the elements of $^0T_6$ are the same as the elements of the matrix in Equation (5.101).*

*Multiplying both sides of the (6.74) by $^0T_1^{-1}$ provides*

$$
\begin{aligned}
^0T_1^{-1}\,{}^0T_6 &= \begin{bmatrix} \cos\theta_1 & \sin\theta_1 & 0 & 0 \\ 0 & 0 & -1 & 0 \\ -\sin\theta_1 & \cos\theta_1 & 0 & 0 \\ 0 & 0 & 0 & 1 \end{bmatrix} \begin{bmatrix} r_{11} & r_{12} & r_{13} & r_{14} \\ r_{21} & r_{22} & r_{23} & r_{24} \\ r_{31} & r_{32} & r_{33} & r_{34} \\ 0 & 0 & 0 & 1 \end{bmatrix} \\
&= \begin{bmatrix} f_{11} & f_{12} & f_{13} & f_{14} \\ f_{21} & f_{22} & f_{23} & f_{24} \\ f_{31} & f_{32} & f_{33} & f_{34} \\ 0 & 0 & 0 & 1 \end{bmatrix}
\end{aligned}
\qquad (6.75)
$$

*where*

$$f_{1i} = r_{1i}\cos\theta_1 + r_{2i}\sin\theta_1 \qquad (6.76)$$
$$f_{2i} = -r_{3i} \qquad (6.77)$$
$$f_{3i} = r_{2i}\cos\theta_1 - r_{1i}\sin\theta_1 \qquad (6.78)$$
$$i = 1, 2, 3, 4.$$

*Based on the given transformation matrices, we find that*

$$
\begin{aligned}
^1T_6 &= {}^1T_2\,{}^2T_3\,{}^3T_4\,{}^4T_5\,{}^5T_6 \\
&= \begin{bmatrix} f_{11} & f_{12} & f_{13} & f_{14} \\ f_{21} & f_{22} & f_{23} & f_{24} \\ f_{31} & f_{32} & f_{33} & f_{34} \\ 0 & 0 & 0 & 1 \end{bmatrix}
\end{aligned}
\qquad (6.79)
$$

$$f_{11} = -c\theta_2 s\theta_4 s\theta_6 + c\theta_6 \left(-s\theta_2 s\theta_5 + c\theta_2 c\theta_4 c\theta_5\right) \tag{6.80}$$
$$f_{21} = -s\theta_2 s\theta_4 s\theta_6 + c\theta_6 \left(c\theta_2 s\theta_5 + c\theta_4 c\theta_5 s\theta_2\right) \tag{6.81}$$
$$f_{31} = c\theta_4 s\theta_6 + c\theta_5 c\theta_6 s\theta_4 \tag{6.82}$$

$$f_{12} = -c\theta_2 c\theta_6 s\theta_4 - s\theta_6 \left(-s\theta_2 s\theta_5 + c\theta_2 c\theta_4 c\theta_5\right) \tag{6.83}$$
$$f_{22} = -c\theta_6 s\theta_2 s\theta_4 - s\theta_6 \left(c\theta_2 s\theta_5 + c\theta_4 c\theta_5 s\theta_2\right) \tag{6.84}$$
$$f_{32} = c\theta_4 c\theta_6 - c\theta_5 s\theta_4 s\theta_6 \tag{6.85}$$

$$f_{13} = c\theta_5 s\theta_2 + c\theta_2 c\theta_4 s\theta_5 \tag{6.86}$$
$$f_{23} = -c\theta_2 c\theta_5 + c\theta_4 s\theta_2 s\theta_5 \tag{6.87}$$
$$f_{33} = s\theta_4 s\theta_5 \tag{6.88}$$

$$f_{14} = d_3 s\theta_2 \tag{6.89}$$
$$f_{24} = -d_3 c\theta_2 \tag{6.90}$$
$$f_{34} = l_2. \tag{6.91}$$

*The only constant element of the matrix (6.79) is $f_{34} = l_2$, therefore,*

$$r_{24} \cos\theta_1 - r_{14} \sin\theta_1 = l_2. \tag{6.92}$$

*This kind of trigonometric equation frequently appears in robotic inverse kinematics, which has a systematic method of solution. We assume*

$$r_{14} = r\cos\phi \tag{6.93}$$
$$r_{24} = r\sin\phi \tag{6.94}$$

*where*

$$r = \sqrt{r_{14}^2 + r_{24}^2} \tag{6.95}$$

$$\phi = \tan^{-1}\frac{r_{24}}{r_{14}} \tag{6.96}$$

*and therefore, Equation (6.92) becomes*

$$\frac{l_2}{r} = \sin\phi\cos\theta_1 - \cos\phi\sin\theta_1 \tag{6.97}$$
$$= \sin(\phi - \theta_1) \tag{6.98}$$

*showing that*

$$\pm\sqrt{1 - (l_2/r)^2} = \cos(\phi - \theta_1). \tag{6.99}$$

*Hence, the solution of Equation (6.92) for $\theta_1$ is*

$$\theta_1 = \tan^{-1}\frac{r_{24}}{r_{14}} - \tan^{-1}\frac{l_2}{\pm\sqrt{r^2 - l_2^2}}. \tag{6.100}$$

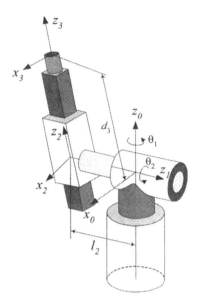

FIGURE 6.3. Left shoulder configuration of a spherical robot.

*The $(-)$ sign corresponds to a left shoulder configuration of the robot, as shown in Figure 6.3, and the $(+)$ sign corresponds to the right shoulder configuration.*

*The elements $f_{14}$ and $f_{24}$ of matrix (6.79) are functions of $\theta_2$ and $\theta_2$ only.*

$$f_{14} = d_3 \sin\theta_2 = r_{14}\cos\theta_1 + r_{24}\sin\theta_1 \qquad (6.101)$$
$$f_{24} = -d_3 \cos\theta_2 = -r_{34} \qquad (6.102)$$

*Hence, it is possible to use them and find $\theta_2$*

$$\theta_2 = \tan^{-1}\frac{r_{14}\cos\theta_1 + r_{24}\sin\theta_1}{r_{34}} \qquad (6.103)$$

*where $\theta_1$ must be substituted from (6.100).*
*In the next step, we find the third joint variable $d_3$ from*

$$^1T_2^{-1}\,{}^0T_1^{-1}\,{}^0T_6 = {}^2T_6 \qquad (6.104)$$

*where*

$$^1T_2^{-1} = \begin{bmatrix} \cos\theta_2 & \sin\theta_2 & 0 & 0 \\ 0 & 0 & 1 & -l_2 \\ \sin\theta_2 & -\cos\theta_2 & 0 & 0 \\ 0 & 0 & 0 & 1 \end{bmatrix} \qquad (6.105)$$

*and*

$$^2T_6 = \begin{bmatrix} -s\theta_4 s\theta_6 + c\theta_4 c\theta_5 c\theta_6 & -c\theta_6 s\theta_4 - c\theta_4 c\theta_5 s\theta_6 & c\theta_4 s\theta_5 & 0 \\ c\theta_4 s\theta_6 + c\theta_5 c\theta_6 s\theta_4 & c\theta_4 c\theta_6 - c\theta_5 s\theta_4 s\theta_6 & s\theta_4 s\theta_5 & 0 \\ -c\theta_6 s\theta_5 & s\theta_5 s\theta_6 & c\theta_5 & d_3 \\ 0 & 0 & 0 & 1 \end{bmatrix}.$$

$$(6.106)$$

*Employing the elements of the matrices on both sides of Equation (6.104) shows that the element (3, 4) can be utilized to find $d_3$.*

$$d_3 = r_{34} \cos\theta_2 + r_{14} \cos\theta_1 \sin\theta_2 + r_{24} \sin\theta_1 \sin\theta_2 \qquad (6.107)$$

*Since there is no other element in Equation (6.104) to be a function of another single variable, we move to the next step and evaluate $\theta_4$ from*

$$^3T_4^{-1}\, ^2T_3^{-1}\, ^1T_2^{-1}\, ^0T_1^{-1}\, ^0T_6 = \,^4T_6 \qquad (6.108)$$

*because $^2T_3^{-1}\, ^1T_2^{-1}\, ^0T_1^{-1}\, ^0T_6 = \,^3T_6$ provides no new equation. Evaluating $^4T_6$*

$$^4T_6 = \begin{bmatrix} \cos\theta_5 \cos\theta_6 & -\cos\theta_5 \sin\theta_6 & \sin\theta_5 & 0 \\ \cos\theta_6 \sin\theta_5 & -\sin\theta_5 \sin\theta_6 & -\cos\theta_5 & 0 \\ \sin\theta_6 & \cos\theta_6 & 0 & 0 \\ 0 & 0 & 0 & 1 \end{bmatrix} \qquad (6.109)$$

*and the left-hand side of (6.108) utilizing*

$$^2T_3^{-1} = \begin{bmatrix} 1 & 0 & 0 & 0 \\ 0 & 1 & 0 & 0 \\ 0 & 0 & 1 & -d_3 \\ 0 & 0 & 0 & 1 \end{bmatrix} \qquad (6.110)$$

*and*

$$^3T_4^{-1} = \begin{bmatrix} \cos\theta_4 & \sin\theta_4 & 0 & 0 \\ 0 & 0 & -1 & 0 \\ -\sin\theta_4 & \cos\theta_4 & 0 & 0 \\ 0 & 0 & 0 & 1 \end{bmatrix} \qquad (6.111)$$

*shows that*

$$^3T_4^{-1}\, ^2T_3^{-1}\, ^1T_2^{-1}\, ^0T_1^{-1}\, ^0T_6 = \begin{bmatrix} g_{11} & g_{12} & g_{13} & g_{14} \\ g_{21} & g_{22} & g_{23} & g_{24} \\ g_{31} & g_{32} & g_{33} & g_{34} \\ 0 & 0 & 0 & 1 \end{bmatrix} \qquad (6.112)$$

*where*

$$g_{1i} = -r_{3i}c\theta_4 s\theta_2 + r_{2i}(c\theta_1 s\theta_4 + c\theta_2 c\theta_4 s\theta_1)$$
$$+r_{1i}(-s\theta_1 s\theta_4 + c\theta_1 c\theta_2 c\theta_4) \tag{6.113}$$

$$g_{2i} = d_3\delta_{4i} - r_{31}c\theta_2 - r_{11}c\theta_1 s\theta_2 - r_{21}s\theta_1 s\theta_2 \tag{6.114}$$

$$g_{3i} = r_{31}s\theta_2 s\theta_4 + r_{21}(c\theta_1 c\theta_4 - c\theta_2 s\theta_1 s\theta_4)$$
$$+r_{11}(-c\theta_4 s\theta_1 - c\theta_1 c\theta_2 s\theta_4) \tag{6.115}$$

$$i = 1, 2, 3, 4.$$

*The symbol $\delta_{4i}$ indicates the Kronecker delta and is equal to*

$$\delta_{4i} \begin{aligned} &= 1 \quad if \quad i = 4 \\ &= 0 \quad if \quad i \neq 4. \end{aligned} \tag{6.116}$$

*Therefore, we can find $\theta_4$ by equating the element $(3,3)$, $\theta_5$ by equating the elements $(1,3)$ or $(2,3)$, and $\theta_6$ by equating the elements $(3,1)$ or $(3,2)$. Starting from element $(3,3)$*

$$r_{13}(-c\theta_4 s\theta_1 - c\theta_1 c\theta_2 s\theta_4) + r_{23}(c\theta_1 c\theta_4 - c\theta_2 s\theta_1 s\theta_4) + r_{33}s\theta_2 s\theta_4 = 0 \tag{6.117}$$

*we find $\theta_4$*

$$\theta_4 = \tan^{-1} \frac{-r_{13}s\theta_1 + r_{23}c\theta_1}{c\theta_2(r_{13}c\theta_1 + r_{23}s\theta_1) - r_{33}s\theta_2} \tag{6.118}$$

*which, based on the second value of $\theta_1$, can also be equal to*

$$\theta_4 = \frac{\pi}{2} + \tan^{-1} \frac{-r_{13}s\theta_1 + r_{23}c\theta_1}{c\theta_2(r_{13}c\theta_1 + r_{23}s\theta_1) - r_{33}s\theta_2}. \tag{6.119}$$

*Now we use elements $(1,3)$ and $(2,3)$,*

$$\sin\theta_5 = r_{23}(\cos\theta_1 \sin\theta_4 + \cos\theta_2 \cos\theta_4 \sin\theta_1) - r_{33}\cos\theta_4 \sin\theta_2$$
$$+r_{13}(\cos\theta_1 \cos\theta_2 \cos\theta_4 - \sin\theta_1 \sin\theta_4) \tag{6.120}$$

$$-\cos\theta_5 = -r_{33}\cos\theta_2 - r_{13}\cos\theta_1 \sin\theta_2 - r_{23}\sin\theta_1 \sin\theta_2 \tag{6.121}$$

*to find $\theta_5$*

$$\theta_5 = \tan^{-1} \frac{\sin\theta_5}{\cos\theta_5}. \tag{6.122}$$

*Finally, $\theta_6$ can be found from the elements $(3,1)$ and $(3,2)$*

$$\sin\theta_6 = r_{31}\sin\theta_2 \sin\theta_4 + r_{21}(\cos\theta_1 \cos\theta_4 - \cos\theta_2 \sin\theta_1 \sin\theta_4)$$
$$+r_{11}(-\cos\theta_4 \sin\theta_1 - \cos\theta_1 \cos\theta_2 \sin\theta_4) \tag{6.123}$$

$$\cos\theta_6 = r_{32}\sin\theta_2 \sin\theta_4 + r_{22}(\cos\theta_1 \cos\theta_4 - \cos\theta_2 \sin\theta_1 \sin\theta_4)$$
$$+r_{12}(-\cos\theta_4 \sin\theta_1 - \cos\theta_1 \cos\theta_2 \sin\theta_4) \tag{6.124}$$

$$\theta_6 = \tan^{-1} \frac{\sin\theta_6}{\cos\theta_6}. \tag{6.125}$$

**Example 167** *Inverse of Euler angles transformation matrix.*
*The global rotation matrix based on Euler angles has been found in Equation (2.64).*

$$
\begin{aligned}
{}^{G}R_B &= [A_{z,\psi}\,A_{x,\theta}\,A_{z,\varphi}]^T \\
&= R_{Z,\varphi}\,R_{X,\theta}\,R_{Z,\psi} \\
&= \begin{bmatrix}
c\varphi c\psi - c\theta s\varphi s\psi & -c\varphi s\psi - c\theta c\psi s\varphi & s\theta s\varphi \\
c\psi s\varphi + c\theta c\varphi s\psi & -s\varphi s\psi + c\theta c\varphi c\psi & -c\varphi s\theta \\
s\theta s\psi & s\theta c\psi & c\theta
\end{bmatrix} \\
&= \begin{bmatrix}
r_{11} & r_{12} & r_{13} \\
r_{21} & r_{22} & r_{23} \\
r_{31} & r_{32} & r_{33}
\end{bmatrix}
\end{aligned} \tag{6.126}
$$

*Premultiplying* ${}^{G}R_B$ *by* $R_{Z,\varphi}^{-1}$, *gives*

$$
\begin{aligned}
&\begin{bmatrix}
\cos\varphi & \sin\varphi & 0 \\
-\sin\varphi & \cos\varphi & 0 \\
0 & 0 & 1
\end{bmatrix} {}^{G}R_B \\
&= \begin{bmatrix}
r_{11}c\varphi + r_{21}s\varphi & r_{12}c\varphi + r_{22}s\varphi & r_{13}c\varphi + r_{23}s\varphi \\
r_{21}c\varphi - r_{11}s\varphi & r_{22}c\varphi - r_{12}s\varphi & r_{23}c\varphi - r_{13}s\varphi \\
r_{31} & r_{32} & r_{33}
\end{bmatrix} \\
&= \begin{bmatrix}
\cos\psi & -\sin\psi & 0 \\
\cos\theta\sin\psi & \cos\theta\cos\psi & -\sin\theta \\
\sin\theta\sin\psi & \sin\theta\cos\psi & \cos\theta
\end{bmatrix}.
\end{aligned} \tag{6.127}
$$

*Equating the elements* $(1,3)$ *of both sides*

$$
r_{13}\cos\varphi + r_{23}\sin\varphi = 0 \tag{6.128}
$$

*gives*

$$
\varphi = \operatorname{atan2}\left(r_{13}, -r_{23}\right). \tag{6.129}
$$

*Having* $\varphi$ *helps us to find* $\psi$ *by using elements* $(1,1)$ *and* $(1,2)$

$$
\begin{aligned}
\cos\psi &= r_{11}\cos\varphi + r_{21}\sin\varphi \\
-\sin\psi &= r_{12}\cos\varphi + r_{22}\sin\varphi
\end{aligned}
\tag{6.130}
\tag{6.131}
$$

*therefore,*

$$
\psi = \operatorname{atan2}\frac{-r_{12}\cos\varphi - r_{22}\sin\varphi}{r_{11}\cos\varphi + r_{21}\sin\varphi}. \tag{6.132}
$$

*In the next step, we may postmultiply $^G R_B$ by $R_{Z,\psi}^{-1}$, to provide*

$$
^G R_B \begin{bmatrix} \cos\psi & \sin\psi & 0 \\ -\sin\psi & \cos\psi & 0 \\ 0 & 0 & 1 \end{bmatrix}
$$

$$
= \begin{bmatrix} r_{11}c\psi - r_{12}s\psi & r_{12}c\psi + r_{11}s\psi & r_{13} \\ r_{21}c\psi - r_{22}s\psi & r_{22}c\psi + r_{21}s\psi & r_{23} \\ r_{31}c\psi - r_{32}s\psi & r_{32}c\psi + r_{31}s\psi & r_{33} \end{bmatrix}
$$

$$
= \begin{bmatrix} \cos\varphi & -\cos\theta\sin\varphi & \sin\theta\sin\varphi \\ \sin\varphi & \cos\theta\cos\varphi & -\cos\varphi\sin\theta \\ 0 & \sin\theta & \cos\theta \end{bmatrix}. \tag{6.133}
$$

*The elements $(1,1)$ on both sides make an equation to find $\psi$.*

$$
r_{31}\cos\psi - r_{31}\sin\psi = 0 \tag{6.134}
$$

*Therefore, it is possible to find $\psi$ from the following equation:*

$$
\psi = \mathrm{atan2}\,(r_{31}, r_{31})\,. \tag{6.135}
$$

*Finally, $\theta$ can be found using elements $(3,2)$ and $(3,3)$*

$$
r_{32}c\psi + r_{31}s\psi = \sin\theta \tag{6.136}
$$
$$
r_{33} = \cos\theta \tag{6.137}
$$

*which give*

$$
\theta = \mathrm{atan2}\,\frac{r_{32}\cos\psi + r_{31}\sin\psi}{r_{33}}. \tag{6.138}
$$

**Example 168** *Inverse kinematics for a 2R planar manipulator.*

*Figure 5.9 illustrates a 2R planar manipulator with two $R\|R$ links according to the coordinate frames setup shown in the figure.*

*The forward kinematics of the manipulator was found to be*

$$
^0 T_2 = {}^0 T_1\, {}^1 T_2 \tag{6.139}
$$

$$
= \begin{bmatrix} c(\theta_1 + \theta_2) & -s(\theta_1 + \theta_2) & 0 & l_1 c\theta_1 + l_2 c(\theta_1 + \theta_2) \\ s(\theta_1 + \theta_2) & c(\theta_1 + \theta_2) & 0 & l_1 s\theta_1 + l_2 s(\theta_1 + \theta_2) \\ 0 & 0 & 1 & 0 \\ 0 & 0 & 0 & 1 \end{bmatrix}.
$$

*The inverse kinematics of planar robots are generally easier to find analytically. In this case, we can see that the position of the tip point of the manipulator is at*

$$
\begin{bmatrix} X \\ Y \end{bmatrix} = \begin{bmatrix} l_1 \cos\theta_1 + l_2 \cos(\theta_1 + \theta_2) \\ l_1 \sin\theta_1 + l_2 \sin(\theta_1 + \theta_2) \end{bmatrix} \tag{6.140}
$$

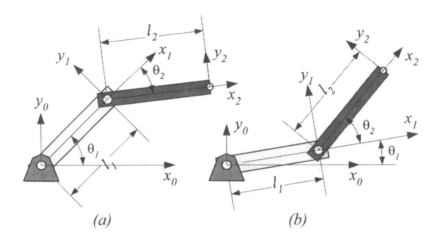

FIGURE 6.4. Illustration of a 2R planar manipulator in two possible configurations: (a) elbow up and (b) elbow down.

*therefore*

$$X^2 + Y^2 = l_1^2 + l_2^2 - 2l_1l_2 \cos\theta_2 \tag{6.141}$$

*and*

$$\cos\theta_2 = \frac{X^2 + Y^2 - l_1^2 - l_2^2}{-2l_1l_2}. \tag{6.142}$$

*However, we should avoid using arcsin and arccos because of the inaccuracy. So, we employ the half angle formula*

$$\tan^2\frac{\theta}{2} = \frac{1 - \cos\theta}{1 + \cos\theta} \tag{6.143}$$

*to find $\theta_2$ using an atan2 function*

$$\theta_2 = \pm 2\,\text{atan2}\sqrt{\frac{(l_1 + l_2)^2 - (X + Y)^2}{(X^2 + Y^2) - (l_1^2 + l_2^2)}} \tag{6.144}$$

*The $\pm$ is because of the square root, which generates two solutions. These two solutions are called elbow up and elbow down, as shown in Figure 6.4. The first joint variable $\theta_1$ can be found from*

$$\theta_1 = \text{atan2}\frac{X(l_1 + l_2\cos\theta_2) + Yl_2\sin\theta_2}{Y(l_1 + l_2\cos\theta_2) - Xl_2\sin\theta_2}. \tag{6.145}$$

*The two different solutions for $\theta_1$ correspond to the elbow up and elbow down configurations.*

**Example 169** ★ *Inverse kinematics and nonstandard DH frames.*
*Consider a 3 DOF planar manipulator shown in Figure 5.4. The non-standard DH transformation matrices of the manipulator are*

$$
{}^0T_1 = \begin{bmatrix} \cos\theta_1 & -\sin\theta_1 & 0 & 0 \\ \sin\theta_1 & \cos\theta_1 & 0 & 0 \\ 0 & 0 & 1 & 0 \\ 0 & 0 & 0 & 1 \end{bmatrix} \tag{6.146}
$$

$$
{}^1T_2 = \begin{bmatrix} \cos\theta_2 & -\sin\theta_2 & 0 & l_1 \\ \sin\theta_2 & \cos\theta_2 & 0 & 0 \\ 0 & 0 & 1 & 0 \\ 0 & 0 & 0 & 1 \end{bmatrix} \tag{6.147}
$$

$$
{}^2T_3 = \begin{bmatrix} \cos\theta_3 & -\sin\theta_3 & 0 & l_2 \\ \sin\theta_3 & \cos\theta_3 & 0 & 0 \\ 0 & 0 & 1 & 0 \\ 0 & 0 & 0 & 1 \end{bmatrix} \tag{6.148}
$$

$$
{}^3T_4 = \begin{bmatrix} 1 & 0 & 0 & l_3 \\ 0 & 1 & 0 & 0 \\ 0 & 0 & 1 & 0 \\ 0 & 0 & 0 & 1 \end{bmatrix}. \tag{6.149}
$$

*Solution of the inverse kinematics problem is a mathematical problem and none of the standard or nonstandard DH methods for defining link frames provide any simplicity.*

*To calculate the inverse kinematics, we start with calculating the forward kinematics transformation matrix* ${}^0T_4$

$$
\begin{aligned}
{}^0T_4 &= {}^0T_1 \, {}^1T_2 \, {}^2T_3 \, {}^3T_4 \tag{6.150} \\
&= \begin{bmatrix} \cos\theta_{123} & -\sin\theta_{123} & 0 & l_1\cos\theta_1 + l_2\cos\theta_{12} + l_3\cos\theta_{123} \\ \sin\theta_{123} & \cos\theta_{123} & 0 & l_1\sin\theta_1 + l_2\sin\theta_{12} + l_3\sin\theta_{123} \\ 0 & 0 & 1 & 0 \\ 0 & 0 & 0 & 1 \end{bmatrix} \\
&= \begin{bmatrix} r_{11} & r_{12} & r_{13} & r_{14} \\ r_{21} & r_{22} & r_{23} & r_{24} \\ r_{31} & r_{32} & r_{33} & r_{34} \\ 0 & 0 & 0 & 1 \end{bmatrix}
\end{aligned}
$$

*where we used the following short notation to simplify the equation.*

$$
\theta_{ijk} = \theta_i + \theta_j + \theta_k \tag{6.151}
$$

*Examining the matrix* ${}^0T_4$ *indicates that*

$$
\theta_{123} = \text{atan2}\,(r_{21}, r_{11}). \tag{6.152}
$$

*The next equation*

$$^0T_4\,^3T_4^{-1} = {}^0T_1\,^1T_2\,^2T_3 \tag{6.153}$$

$$\begin{bmatrix} r_{11} & r_{12} & 0 & r_{14} - l_3 r_{11} \\ r_{21} & r_{22} & 0 & r_{24} - l_3 r_{21} \\ 0 & 0 & 1 & 0 \\ 0 & 0 & 0 & 1 \end{bmatrix} = \begin{bmatrix} c\theta_{123} & -s\theta_{123} & 0 & l_1 c\theta_1 + l_2 c\theta_{12} \\ s\theta_{123} & c\theta_{123} & 0 & l_1 s\theta_1 + l_2 s\theta_{12} \\ 0 & 0 & 1 & 0 \\ 0 & 0 & 0 & 1 \end{bmatrix}$$

*shows that*

$$\theta_2 = \arccos \frac{f_1^2 + f_2^2 - l_1^2 - l_2^2}{2 l_1 l_2} \tag{6.154}$$

$$\theta_1 = \text{atan2}\,(f_2 f_3 - f_1 f_4\,,\ f_1 f_3 + f_2 f_4)$$

*where*

$$\begin{aligned} f_1 &= r_{14} - l_3 r_{11} \\ &= c\theta_1\,(l_2 c\theta_2 + l_1) - s\theta_1\,(l_2 s\theta_2) \\ &= c\theta_1 f_3 - s\theta_1 f_4 \end{aligned} \tag{6.155}$$

$$\begin{aligned} f_2 &= r_{24} - l_3 r_{21} \\ &= s\theta_1\,(l_2 c\theta_2 + l_1) + c\theta_1\,(l_2 s\theta_2) \\ &= s\theta_1 f_3 + c\theta_1 f_4. \end{aligned} \tag{6.156}$$

*Finally, the angle $\theta_3$ is*

$$\theta_3 = \theta_{123} - \theta_1 - \theta_2. \tag{6.157}$$

## 6.3   Iterative Technique

The inverse kinematics problem can be interpreted as searching for the solution $q_k$ of a set of nonlinear algebraic equations

$$\begin{aligned} ^0T_n &= \mathbf{T}(\mathbf{q}) \tag{6.158} \\ &= {}^0T_1(q_1)\,^1T_2(q_2)\,^2T_3(q_3)\,^3T_4(q_4) \cdots {}^{n-1}T_n(q_n) \\ &= \begin{bmatrix} r_{11} & r_{12} & r_{13} & r_{14} \\ r_{21} & r_{22} & r_{23} & r_{24} \\ r_{31} & r_{32} & r_{33} & r_{34} \\ 0 & 0 & 0 & 1 \end{bmatrix} \end{aligned}$$

or

$$r_{ij} = r_{ij}(q_k) \qquad k = 1, 2, \cdots n. \tag{6.159}$$

where $n$ is the number of DOF. However, maximum $m = 6$ out of 12 equations of (6.158) are independent and can be utilized to solve for joint variables $q_k$. The functions $\mathbf{T}(\mathbf{q})$ are transcendental, which are given explicitly based on forward kinematic analysis.

Numerous methods are available to find the zeros of Equation (6.158). However, the methods are, in general, *iterative*. The most common method is known as the *Newton-Raphson method*.

In the *iterative technique*, to solve the kinematic equations

$$\mathbf{T}(\mathbf{q}) = 0 \qquad (6.160)$$

for variables $\mathbf{q}$, we start with an initial guess

$$\mathbf{q}^{\star} = \mathbf{q} + \delta\mathbf{q} \qquad (6.161)$$

for the joint variables. Using the forward kinematics, we can determine the configuration of the end-effector frame for the guessed joint variables.

$$\mathbf{T}^{\star} = \mathbf{T}(\mathbf{q}^{\star}) \qquad (6.162)$$

The difference between the configuration calculated with the forward kinematics and the desired configuration represents an *error*, called *residue*, which must be minimized.

$$\delta\mathbf{T} = \mathbf{T} - \mathbf{T}^{\star} \qquad (6.163)$$

A first order Taylor expansion of the set of equations is

$$\mathbf{T} = \mathbf{T}(\mathbf{q}^{\star} + \delta\mathbf{q}) \qquad (6.164)$$
$$= \mathbf{T}(\mathbf{q}^{\star}) + \frac{\partial\mathbf{T}}{\partial\mathbf{q}}\delta\mathbf{q} + O(\delta\mathbf{q}^2). \qquad (6.165)$$

Assuming $\delta\mathbf{q} \ll \mathbf{I}$ allows us to work with a set of linear equations

$$\delta\mathbf{T} = \mathbf{J}\,\delta\mathbf{q} \qquad (6.166)$$

where $\mathbf{J}$ is the Jacobian matrix of the set of equations

$$\mathbf{J}(\mathbf{q}) = \left[\frac{\partial T_i}{\partial q_j}\right] \qquad (6.167)$$

that implies

$$\delta\mathbf{q} = \mathbf{J}^{-1}\,\delta\mathbf{T}. \qquad (6.168)$$

Therefore, the unknown variables $\mathbf{q}$ are

$$\mathbf{q} = \mathbf{q}^{\star} + \mathbf{J}^{-1}\,\delta\mathbf{T}. \qquad (6.169)$$

We may use the values obtained by (6.169) as a new approximation to repeat the calculations and find newer values. Repeating the methods can be summarized in the following iterative equation to converge to the exact value of the variables.

$$q^{(i+1)} = q^{(i)} + J^{-1}(q^{(i)}) \, \delta T(q^{(i)}) \tag{6.170}$$

This iteration technique can be set in an algorithm for easier numerical calculations.

**Algorithm 6.1.** Inverse kinematics iteration technique.

1. *Set the initial counter $i = 0$.*

2. *Find or guess an initial estimate $q^{(0)}$.*

3. *Calculate the residue $\delta T(q^{(i)}) = J(q^{(i)}) \, \delta q^{(i)}$.*

   *If every element of $T(q^{(i)})$ or its norm $\|T(q^{(i)})\|$ is less than a tolerance, $\|T(q^{(i)})\| < \epsilon$ then terminate the iteration. The $q^{(i)}$ is the desired solution.*

4. *Calculate $q^{(i+1)} = q^{(i)} + J^{-1}(q^{(i)}) \, \delta T(q^{(i)})$.*

5. *Set $i = i + 1$ and return to step 3.*

The tolerance $\epsilon$ can equivalently be set up on variables

$$q^{(i+1)} - q^{(i)} < \epsilon \tag{6.171}$$

or on Jacobian

$$J - I < \epsilon. \tag{6.172}$$

**Example 170** *Inverse kinematics for a 2R planar manipulator.*
   *In Example 168 we have seen that the tip point of a 2R planar manipulator can be described by*

$$\begin{bmatrix} X \\ Y \end{bmatrix} = \begin{bmatrix} l_1 c\theta_1 + l_2 c\,(\theta_1 + \theta_2) \\ l_1 s\theta_1 + l_2 s\,(\theta_1 + \theta_2) \end{bmatrix}. \tag{6.173}$$

*To solve the inverse kinematics of the manipulator and find the joint coordinates for a known position of the tip point, we define*

$$q = \begin{bmatrix} \theta_1 \\ \theta_2 \end{bmatrix} \tag{6.174}$$

$$T = \begin{bmatrix} X \\ Y \end{bmatrix} \tag{6.175}$$

*therefore, the Jacobian of the equations is*

$$\mathbf{J}(\mathbf{q}) = \left[ \frac{\partial T_i}{\partial q_j} \right] = \left[ \begin{array}{cc} \frac{\partial X}{\partial \theta_1} & \frac{\partial X}{\partial \theta_2} \\ \frac{\partial Y}{\partial \theta_1} & \frac{\partial Y}{\partial \theta_2} \end{array} \right]$$

$$= \left[ \begin{array}{cc} -l_1 \sin\theta_1 - l_2 \sin(\theta_1 + \theta_2) & -l_2 \sin(\theta_1 + \theta_2) \\ l_1 \cos\theta_1 + l_2 \cos(\theta_1 + \theta_2) & l_2 \cos(\theta_1 + \theta_2) \end{array} \right]. \quad (6.176)$$

*The inverse of the Jacobian is*

$$\mathbf{J}^{-1} = \frac{-1}{l_1 l_2 s\theta_2} \left[ \begin{array}{cc} -l_2 c(\theta_1 + \theta_2) & -l_2 s(\theta_1 + \theta_2) \\ l_1 c\theta_1 + l_2 c(\theta_1 + \theta_2) & l_1 s\theta_1 + l_2 s(\theta_1 + \theta_2) \end{array} \right] \quad (6.177)$$

*and therefore, the iterative formula (6.170) is set up as*

$$\left[ \begin{array}{c} \theta_1 \\ \theta_2 \end{array} \right]^{(i+1)} = \left[ \begin{array}{c} \theta_1 \\ \theta_2 \end{array} \right]^{(i)} + \mathbf{J}^{-1} \left( \left[ \begin{array}{c} X \\ Y \end{array} \right] - \left[ \begin{array}{c} X \\ Y \end{array} \right]^{(i)} \right). \quad (6.178)$$

*Let's assume*

$$l_1 = l_2 = 1$$

$$\mathbf{T} = \left[ \begin{array}{c} X \\ Y \end{array} \right] = \left[ \begin{array}{c} 1 \\ 1 \end{array} \right]$$

*and start from a guess value*

$$\mathbf{q}^{(0)} = \left[ \begin{array}{c} \theta_1 \\ \theta_2 \end{array} \right]^{(0)} = \left[ \begin{array}{c} \pi/3 \\ -\pi/3 \end{array} \right]$$

*for which*

$$\delta \mathbf{T} = \left[ \begin{array}{c} 1 \\ 1 \end{array} \right] - \left[ \begin{array}{c} \cos\pi/3 + \cos(\pi/3 + -\pi/3) \\ \sin\pi/3 + \sin(\pi/3 + -\pi/3) \end{array} \right]$$

$$= \left[ \begin{array}{c} 1 \\ 1 \end{array} \right] - \left[ \begin{array}{c} \frac{3}{2} \\ \frac{1}{2}\sqrt{3} \end{array} \right]$$

$$= \left[ \begin{array}{c} -\frac{1}{2} \\ -\frac{1}{2}\sqrt{3} + 1 \end{array} \right].$$

*The Jacobian and its inverse for these values are*

$$\mathbf{J} = \left[ \begin{array}{cc} -\frac{1}{2}\sqrt{3} & 0 \\ \frac{3}{2} & 1 \end{array} \right]$$

$$\mathbf{J}^{-1} = \left[ \begin{array}{cc} -\frac{2}{3}\sqrt{3} & 0 \\ \sqrt{3} & 1 \end{array} \right]$$

*and therefore,*

$$
\begin{bmatrix} \theta_1 \\ \theta_2 \end{bmatrix}^{(1)} = \begin{bmatrix} \theta_1 \\ \theta_2 \end{bmatrix}^{(0)} + \mathbf{J}^{-1}\, \delta \mathbf{T}
$$

$$
= \begin{bmatrix} \pi/3 \\ -\pi/3 \end{bmatrix} + \begin{bmatrix} -\frac{2}{3}\sqrt{3} & 0 \\ \sqrt{3} & 1 \end{bmatrix}\begin{bmatrix} -\frac{1}{2} \\ -\frac{1}{2}\sqrt{3} + 1 \end{bmatrix}
$$

$$
= \begin{bmatrix} 1.624\,5 \\ -1.779\,2 \end{bmatrix}.
$$

Based on the iterative technique, we can find the following values and find the solution in a few iterations.

Iteration 1.

$$
\mathbf{J} = \begin{bmatrix} -\frac{1}{2}\sqrt{3} & 0 \\ \frac{3}{2} & 1 \end{bmatrix}
$$

$$
\delta \mathbf{T} = \begin{bmatrix} -\frac{1}{2} \\ -\frac{1}{2}\sqrt{3} + 1 \end{bmatrix}
$$

$$
\mathbf{q}^{(1)} = \begin{bmatrix} 1.624\,5 \\ -1.779\,2 \end{bmatrix}
$$

Iteration 2.

$$
\mathbf{J} = \begin{bmatrix} -0.844 & 0.154 \\ 0.934 & 0.988 \end{bmatrix}
$$

$$
\delta \mathbf{T} = \begin{bmatrix} 6.516 \times 10^{-2} \\ 0.155\,53 \end{bmatrix}
$$

$$
\mathbf{q}^{(2)} = \begin{bmatrix} 1.583 \\ -1.582 \end{bmatrix}
$$

Iteration 3.

$$
\mathbf{J} = \begin{bmatrix} -1.00 & -.433 \times 10^{-3} \\ .988 & .999 \end{bmatrix}
$$

$$
\delta \mathbf{T} = \begin{bmatrix} .119 \times 10^{-1} \\ -.362 \times 10^{-3} \end{bmatrix}
$$

$$
\mathbf{q}^{(3)} = \begin{bmatrix} 1.570795886 \\ -1.570867014 \end{bmatrix}
$$

Iteration 4.

$$
\mathbf{J} = \begin{bmatrix} -1.000 & 0.0 \\ 0.998\,50 & 1.0 \end{bmatrix}
$$

$$
\delta \mathbf{T} = \begin{bmatrix} -.438 \times 10^{-6} \\ .711 \times 10^{-4} \end{bmatrix}
$$

$$
\mathbf{q}^{(4)} = \begin{bmatrix} 1.570796329 \\ -1.570796329 \end{bmatrix}
$$

The result of the fourth iteration $\mathbf{q}^{(4)}$ is close enough to the exact value $\mathbf{q} = \begin{bmatrix} \pi/2 & -\pi/2 \end{bmatrix}^T$.

# 6.4 ★ Comparison of the Inverse Kinematics Techniques

## 6.4.1 ★ Existence and Uniqueness of Solution

It is clear that when the desired tool frame position $^0\mathbf{d}_7$ is outside the working space of the robot, there can not be any real solution for the joint variables of the robot. In this condition, the overall resultant of the terms under square root signs would be negative. Furthermore, even when the tool frame position $^0\mathbf{d}_7$ is within the working space, there may be some tool orientations $^0\mathbf{R}_7$ that are not achievable without breaking joint constraints and violating one or more joint variable limits. Therefore, existing solutions for inverse kinematics problem generally depends on the geometric configuration of the robot.

The normal case is when the number of joints is six. Then, provided that no DOF is redundant and the configuration assigned to the end-effectors of the robot lies within the workspace, the inverse kinematics solution exists in finite numbers. The different solutions correspond to possible configurations to reach the same end-effector configuration.

Generally speaking, when the solution of the inverse kinematics of a robot exists, they are not unique. Multiple solutions appear because a robot can reach to a point within the working space in different configurations. Every set of solutions is associated to a particular configuration. The elbow-up and elbow-down configuration of the 2R manipulator in Example 168 is a simple example.

The multiplicity of the solution depends on the number of joints of the manipulator and their type. The fact that a manipulator has multiple solutions may cause problems since the system has to be able to select one of them. The criteria on which to base a decision may vary, but a very reasonable choice consists of choosing the closest solution to the current configuration.

When the number of joints is less than six, no solution exists unless freedom is reduced in the same time in the task space, for example, by constraining the tool orientation to certain directions.

When the number of joints exceeds six, the structure becomes redundant and an infinite number of solutions exists to reach the same end-effector configuration within the robot workspace. Redundancy of the robot architecture is an interesting feature for systems installed in a highly constrained environment. From the kinematic point of view, the difficulty lies in formulating the environment constraints in mathematical form, to ensure the uniqueness of the solution to the inverse kinematic problem.

## 6.4.2  ★ Inverse Kinematics Techniques

The inverse kinematics problem of robots can be solved by several methods, such as *decoupling, inverse transformation, iterative, screw algebra, dual matrices, dual quaternions,* and *geometric techniques.* The decoupling and inverse transform technique using $4 \times 4$ homogeneous transformation matrices suffers from the fact that the solution does not clearly indicate how to select the appropriate solution from multiple possible solutions for a particular configuration. Thus, these techniques rely on the skills and intuition of the engineer. The iterative solution method often requires a vast amount of computation and moreover, it does not guarantee convergence to the correct solution. It is especially weak when the robot is close to the singular and degenerate configurations. The iterative solution method also lacks a method for selecting the appropriate solution from multiple possible solutions.

Although the set of nonlinear trigonometric equations is typically not possible to be solved analytically, there are some robot structures that are *solvable* analytically. The sufficient condition of solvability is when the 6 DOF robot has three consecutive revolute joints with axes intersecting in one point. The other property of inverse kinematics is ambiguity of a solution in singular points. However, when closed-form solutions to the arm equation can be found, they are seldom unique.

**Example 171** ★ *Iteration technique and n-m relationship.*
*1- Iteration method when $n = m$.*

*When the number of joint variables $n$ is equal to the number of independent equations generated in forward kinematics $m$, then provided that the Jacobian matrix remains non singular, the linearized equation*

$$\delta \mathbf{T} = \mathbf{J} \, \delta \mathbf{q} \tag{6.179}$$

*has a unique set of solutions and therefore, the Newton-Raphson technique may be utilized to solve the inverse kinematics problem.*

*The cost of the procedure depends on the number of iterations to be performed, which depends upon different parameters such as the distance between the estimated and effective solutions, and the condition number of the Jacobian matrix at the solution. Since the solution to the inverse kinematics problem is not unique, it may generate different configurations according to the choice of the estimated solution. No convergence may be observed if the initial estimate of the solution falls outside the convergence domain of the algorithm.*

*2- Iteration method when $n > m$.*

*When the number of joint variables $n$ is more than the number of independent equations $m$, then the problem is an overdetermined case for which no solution exists in general because the number of joints is not enough to generate an arbitrary configuration for the end-effector. A solution can be generated, which minimizes the position error.*

*3- Iteration method when n < m.*

*When the number of joint variables n is less than the number of independent equations m, then the problem is a redundant case for which an infinite number of solutions are generally available.*

# 6.5  ★ Singular Configuration

*if $\theta_2 = 0$ or $180$ or $360$ (not allowed to go to this config)*

Generally speaking, for any robot, redundant or not, it is possible to discover some configurations, called *singular configurations*, in which the number of DOF of the end-effector is inferior to the dimension in which it generally operates. Singular configurations happen when:

1. Two axes of prismatic joints become parallel

2. Two axes of revolute joints become identical.

At singular positions, the end-effector loses one or more degrees of freedom, since the kinematic equations become linearly dependent or certain solutions become undefined. Singular positions must be avoided as the velocities required to move the end-effector become theoretically infinite.

The singular configurations can be determined from the Jacobian matrix. The Jacobian matrix $\mathbf{J}$ relates the infinitesimal displacements of the end-effector

$$\delta \mathbf{X} = [\delta X_1, \cdots \delta X_m] \tag{6.180}$$

to the infinitesimal joint variables

$$\delta \mathbf{q} = [\delta q_1, \cdots \delta q_n] \tag{6.181}$$

and has thus dimension $m \times n$, where $n$ is the number of joints, and $m$ is the number of end-effector DOF.

When $n$ is larger than $m$ and $\mathbf{J}$ has full rank, then there are $m - n$ redundancies in the system to which $m - n$ arbitrary variables correspond.

The Jacobian matrix $\mathbf{J}$ also determines the relationship between end-effector velocities $\dot{\mathbf{X}}$ and joint velocities $\dot{\mathbf{q}}$

$$\dot{\mathbf{X}} = \mathbf{J} \dot{\mathbf{q}}. \tag{6.182}$$

This equation can be interpreted as a linear mapping from an $m$-dimensional vector space $\mathbf{X}$ to an $n$-dimensional vector space $\mathbf{q}$. The subspace $\mathbb{R}(\mathbf{J})$ is the *range space* of the linear mapping, and represents all the possible end-effector velocities that can be generated by the $n$ joints in the current configuration. $\mathbf{J}$ has full row-rank, which means that the system does not present any singularity in that configuration, then the range space $\mathbb{R}(\mathbf{J})$ covers the entire vector space $\mathbf{X}$. Otherwise, there exists at least one direction in which the end-effector cannot be moved.

The null space $\mathbb{N}(\mathbf{J})$ represents the solutions of $\mathbf{J}\dot{\mathbf{q}} = 0$. Therefore, any vector $\dot{\mathbf{q}} \in \mathbb{N}(\mathbf{J})$ does not generate any motion for the end-effector.

If the manipulator has full rank, the dimension of the null space is then equal to the number $m - n$ of redundant DOF. When $\mathbf{J}$ is degenerate, the dimension of $\mathbb{R}(\mathbf{J})$ decreases and the dimension of the null space increases by the same amount. Therefore,

$$\dim \mathbb{R}(\mathbf{J}) + \dim \mathbb{N}(\mathbf{J}) = n. \tag{6.183}$$

Configurations in which the Jacobian no longer has full rank, corresponds to singularities of the robot, which are generally of two types:

1. *Workspace boundary singularities* are those occurring when the manipulator is fully stretched out or folded back on itself. In this case, the end effector is near or at the workspace boundary.

2. *Workspace interior singularities* are those occurring away from the boundary. In this case, generally two or more axes line up.

Mathematically, singularity configurations can be found by calculating the conditions that make

$$|\mathbf{J}| = 0 \tag{6.184}$$

or

$$\left|\mathbf{J}\mathbf{J}^T\right| = 0. \tag{6.185}$$

Identification and avoidance of singularity configurations are very important in robotics. Some of the main reasons are:

1. Certain directions of motion may be unattainable.

2. Some of the joint velocities are infinite.

3. Some of the joint torques are infinite.

4. There will not exist a unique solution to the inverse kinematics problem.

Detecting the singular configurations using the Jacobian determinant may be a tedious task for complex robots. However, for robots having a spherical wrist, it is possible to split the singularity detection problem into two separate problems:

1. Arm singularities resulting from the motion of the manipulator arms.

2. Wrist singularities resulting from the motion of the wrist.

## 6.6   Summary

Inverse kinematics refers to determining the joint variables of a robot for a given position and orientation of the end-effector frame. The forward kinematics of a 6 DOF robot generates a $4 \times 4$ transformation matrix

$$
\begin{aligned}
{}^0T_6 &= {}^0T_1\,{}^1T_2\,{}^2T_3\,{}^3T_4\,{}^4T_5\,{}^5T_6 \\[2mm]
&= \begin{bmatrix} {}^0R_6 & {}^0\mathbf{d}_6 \\ 0 & 1 \end{bmatrix} = \begin{bmatrix} r_{11} & r_{12} & r_{13} & r_{14} \\ r_{21} & r_{22} & r_{23} & r_{24} \\ r_{31} & r_{32} & r_{33} & r_{34} \\ 0 & 0 & 0 & 1 \end{bmatrix}
\end{aligned} \tag{6.186}
$$

where only six elements out of the 12 elements of ${}^0T_6$ are independent. Therefore, the inverse kinematics reduces to finding the six independent elements for a given ${}^0T_6$ matrix.

Decoupling, inverse transformation, and iterative techniques are three applied methods for solving the inverse kinematics problem. In decoupling technique, the inverse kinematics of a robot with a spherical wrist can be decoupled into two subproblems: inverse position and inverse orientation kinematics. Practically, the tools transformation matrix ${}^0T_7$ is decomposed into three submatrices ${}^0T_3$, ${}^3T_6$, and ${}^6T_7$.

$$
{}^0T_6 = {}^0T_3\,{}^3T_6\,{}^6T_7 \tag{6.187}
$$

The matrix ${}^0T_3$ positions the wrist point and depends on the three manipulator joints' variables. The matrix ${}^3T_6$ is the wrist transformation matrix and the ${}^6T_7$ is the tools transformation matrix.

In inverse transformation technique, we extract equations with only one unknown from the following matrix equations, step by step.

$$
\begin{aligned}
{}^1T_6 &= {}^0T_1^{-1}\,{}^0T_6 & (6.188) \\
{}^2T_6 &= {}^1T_2^{-1}\,{}^0T_1^{-1}\,{}^0T_6 & (6.189) \\
{}^3T_6 &= {}^2T_3^{-1}\,{}^1T_2^{-1}\,{}^0T_1^{-1}\,{}^0T_6 & (6.190) \\
{}^4T_6 &= {}^3T_4^{-1}\,{}^2T_3^{-1}\,{}^1T_2^{-1}\,{}^0T_1^{-1}\,{}^0T_6 & (6.191) \\
{}^5T_6 &= {}^4T_5^{-1}\,{}^3T_4^{-1}\,{}^2T_3^{-1}\,{}^1T_2^{-1}\,{}^0T_1^{-1}\,{}^0T_6 & (6.192) \\
\mathbf{I} &= {}^5T_6^{-1}\,{}^4T_5^{-1}\,{}^3T_4^{-1}\,{}^2T_3^{-1}\,{}^1T_2^{-1}\,{}^0T_1^{-1}\,{}^0T_6 & (6.193)
\end{aligned}
$$

The iterative technique is a numerical method seeking to find the joint variable vector $\mathbf{q}$ for a set of equations $\mathbf{T}(\mathbf{q}) = 0$.

# Exercises

1. Notation and symbols.

   Describe the meaning of

   a- atan2 $(a, b)$    b- $^0T_n$    c- $\mathbf{T(q)}$    d- $w$    e- $\mathbf{q}$    f- $\mathbf{J}$.

2. 3R planar manipulator inverse kinematics.

   Figure 5.21 illustrates an R‖R‖R planar manipulator. The forward kinematics of the manipulator generates the following matrices. Solve the inverse kinematics and find $\theta_1, \theta_2, \theta_3$.

$$^2T_3 = \begin{bmatrix} \cos\theta_3 & -\sin\theta_3 & 0 & l_3\cos\theta_3 \\ \sin\theta_3 & \cos\theta_3 & 0 & l_3\sin\theta_3 \\ 0 & 0 & 1 & 0 \\ 0 & 0 & 0 & 1 \end{bmatrix}$$

$$^1T_2 = \begin{bmatrix} \cos\theta_2 & -\sin\theta_2 & 0 & l_2\cos\theta_2 \\ \sin\theta_2 & \cos\theta_2 & 0 & l_2\sin\theta_2 \\ 0 & 0 & 1 & 0 \\ 0 & 0 & 0 & 1 \end{bmatrix}$$

$$^0T_1 = \begin{bmatrix} \cos\theta_1 & -\sin\theta_1 & 0 & l_1\cos\theta_1 \\ \sin\theta_1 & \cos\theta_1 & 0 & l_1\sin\theta_1 \\ 0 & 0 & 1 & 0 \\ 0 & 0 & 0 & 1 \end{bmatrix}$$

3. Spherical wrist inverse kinematics.

   Figure 5.22 illustrates a schematic of a spherical wrist with following transformation matrices. Assume that the frame $B_3$ is the base frame. Solve the inverse kinematics and find $\theta_4, \theta_5, \theta_6$.

$$^3T_4 = \begin{bmatrix} \cos\theta_4 & 0 & -\sin\theta_4 & 0 \\ \sin\theta_4 & 0 & \cos\theta_4 & 0 \\ 0 & -1 & 0 & 0 \\ 0 & 0 & 0 & 1 \end{bmatrix}$$

$$^4T_5 = \begin{bmatrix} \cos\theta_5 & 0 & \sin\theta_5 & 0 \\ \sin\theta_5 & 0 & -\cos\theta_5 & 0 \\ 0 & 1 & 0 & 0 \\ 0 & 0 & 0 & 1 \end{bmatrix}$$

$$^5T_6 = \begin{bmatrix} \cos\theta_6 & -\sin\theta_6 & 0 & 0 \\ \sin\theta_6 & \cos\theta_6 & 0 & 0 \\ 0 & 0 & 1 & 0 \\ 0 & 0 & 0 & 1 \end{bmatrix}$$

4. **SCARA robot inverse kinematics.**

Consider the R‖R‖R‖P robot shown in Figure 5.27 with the following forward kinematics solution. Solve the inverse kinematics and find $\theta_1$, $\theta_2$, $\theta_3$, $d$.

$$^0T_1 = \begin{bmatrix} \cos\theta_1 & -\sin\theta_1 & 0 & l_1\cos\theta_1 \\ \sin\theta_1 & \cos\theta_1 & 0 & l_1\sin\theta_1 \\ 0 & 0 & 1 & 0 \\ 0 & 0 & 0 & 1 \end{bmatrix}$$

$$^1T_2 = \begin{bmatrix} \cos\theta_2 & -\sin\theta_2 & 0 & l_2\cos\theta_2 \\ \sin\theta_2 & \cos\theta_2 & 0 & l_2\sin\theta_2 \\ 0 & 0 & 1 & 0 \\ 0 & 0 & 0 & 1 \end{bmatrix}$$

$$^2T_3 = \begin{bmatrix} \cos\theta_3 & -\sin\theta_3 & 0 & 0 \\ \sin\theta_3 & \cos\theta_3 & 0 & 0 \\ 0 & 0 & 1 & 0 \\ 0 & 0 & 0 & 1 \end{bmatrix}$$

$$^3T_4 = \begin{bmatrix} 1 & 0 & 0 & 0 \\ 0 & 1 & 0 & 0 \\ 0 & 0 & 1 & d \\ 0 & 0 & 0 & 1 \end{bmatrix}$$

$$\begin{aligned} ^0T_4 &= {}^0T_1\,{}^1T_2\,{}^2T_3\,{}^3T_4 \\ &= \begin{bmatrix} c\theta_{123} & -s\theta_{123} & 0 & l_1 c\theta_1 + l_2 c\theta_{12} \\ s\theta_{123} & c\theta_{123} & 0 & l_1 s\theta_1 + l_2 s\theta_{12} \\ 0 & 0 & 1 & d \\ 0 & 0 & 0 & 1 \end{bmatrix} \end{aligned}$$

$$\theta_{123} = \theta_1 + \theta_2 + \theta_3$$
$$\theta_{12} = \theta_1 + \theta_2$$

5. **R⊢R‖R articulated arm inverse kinematics.**

Figure 5.25 illustrates 3 DOF R⊢R‖R manipulator. Use the following transformation matrices and solve the inverse kinematics for $\theta_1$, $\theta_2$, $\theta_3$.

$$^0T_1 = \begin{bmatrix} \cos\theta_1 & 0 & -\sin\theta_1 & 0 \\ \sin\theta_1 & 0 & \cos\theta_1 & 0 \\ 0 & -1 & 0 & d_1 \\ 0 & 0 & 0 & 1 \end{bmatrix}$$

$$^1T_2 = \begin{bmatrix} \cos\theta_2 & -\sin\theta_2 & 0 & l_2\cos\theta_2 \\ \sin\theta_2 & \cos\theta_2 & 0 & l_2\sin\theta_2 \\ 0 & 0 & 1 & d_2 \\ 0 & 0 & 0 & 1 \end{bmatrix}$$

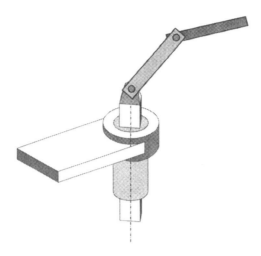

FIGURE 6.5. A PRRR manipulator.

$$
^{2}T_{3} = \begin{bmatrix} \cos\theta_3 & 0 & \sin\theta_3 & 0 \\ \sin\theta_3 & 0 & -\cos\theta_3 & 0 \\ 0 & 1 & 0 & 0 \\ 0 & 0 & 0 & 1 \end{bmatrix}
$$

6. Kinematics of a PRRR manipulator.

A PRRR manipulator is shown in Figure 6.5. Set up the links's co-ordinate frame according to standard DH rules. Determine the class of each link. Find the links' transformation matrices. Calculate the forward kinematics of the manipulator. Solve the inverse kinematics problem for the manipulator.

7. ★ Space station remote manipulator system inverse kinematics.

Shuttle remote manipulator system (SRMS) is shown in Figure 5.28 schematically. The forward kinematics of the robot provides the following transformation matrices. Solve the inverse kinematics for the SRMS.

$$
^{0}T_{1} = \begin{bmatrix} \cos\theta_1 & 0 & -\sin\theta_1 & 0 \\ \sin\theta_1 & 0 & \cos\theta_1 & 0 \\ 0 & -1 & 0 & d_1 \\ 0 & 0 & 0 & 1 \end{bmatrix}
$$

$$
^{1}T_{2} = \begin{bmatrix} \cos\theta_2 & 0 & -\sin\theta_2 & 0 \\ \sin\theta_2 & 0 & \cos\theta_2 & 0 \\ 0 & -1 & 0 & d_2 \\ 0 & 0 & 0 & 1 \end{bmatrix}
$$

$$
^2T_3 = \begin{bmatrix} \cos\theta_3 & -\sin\theta_3 & 0 & a_3\cos\theta_3 \\ \sin\theta_3 & \cos\theta_3 & 0 & a_3\sin\theta_3 \\ 0 & 0 & 1 & d_3 \\ 0 & 0 & 0 & 1 \end{bmatrix}
$$

$$
^3T_4 = \begin{bmatrix} \cos\theta_4 & -\sin\theta_4 & 0 & a_4\cos\theta_4 \\ \sin\theta_4 & \cos\theta_4 & 0 & a_4\sin\theta_4 \\ 0 & 0 & 1 & d_4 \\ 0 & 0 & 0 & 1 \end{bmatrix}
$$

$$
^4T_5 = \begin{bmatrix} \cos\theta_5 & 0 & \sin\theta_5 & 0 \\ \sin\theta_5 & 0 & -\cos\theta_5 & 0 \\ 0 & 1 & 0 & d_5 \\ 0 & 0 & 0 & 1 \end{bmatrix}
$$

$$
^5T_6 = \begin{bmatrix} \cos\theta_6 & 0 & -\sin\theta_6 & 0 \\ \sin\theta_6 & 0 & \cos\theta_6 & 0 \\ 0 & -1 & 0 & d_6 \\ 0 & 0 & 0 & 1 \end{bmatrix}
$$

$$
^6T_7 = \begin{bmatrix} \cos\theta_7 & -\sin\theta_7 & 0 & 0 \\ \sin\theta_7 & \cos\theta_7 & 0 & 0 \\ 0 & 0 & 1 & d_7 \\ 0 & 0 & 0 & 1 \end{bmatrix}
$$

# 7

# Angular Velocity

## 7.1 Angular Velocity Vector and Matrix

Consider a rotating rigid body $B(Oxyz)$ with a fixed point $O$ in a reference frame $G(OXYZ)$ as shown in Figure 7.1. The motion of the body can be described by a time varying rotation transformation matrix between the global and body frames to map the instantaneous coordinates of any fixed point in body frame $B$ into their coordinates in the global frame $G$

$$^G\mathbf{r}(t) = {^GR_B(t)} \, {^B\mathbf{r}}. \tag{7.1}$$

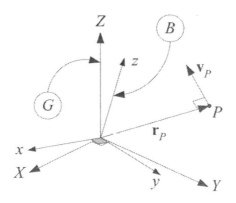

FIGURE 7.1. A rotating rigid body $B(Oxyz)$ with a fixed point $O$ in a global frame $G(OXYZ)$.

The velocity of a body point in the global frame is

$$
\begin{aligned}
^G\dot{\mathbf{r}}(t) &= {^G\mathbf{v}(t)} & (7.2)\\
&= {^G\dot{R}_B(t)} \, {^B\mathbf{r}} & (7.3)\\
&= {_G\tilde{\omega}_B} \, {^G\mathbf{r}(t)} & (7.4)\\
&= {_G\boldsymbol{\omega}_B} \times {^G\mathbf{r}(t)} & (7.5)
\end{aligned}
$$

where $_G\boldsymbol{\omega}_B$ is the *angular velocity vector* of $B$ with respect to $G$. It is equal to a rotation with *angular rate* $\dot{\phi}$ about an *instantaneous axis of rotation* $\hat{u}$.

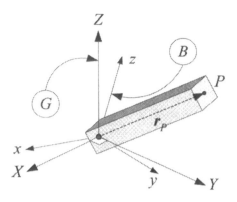

FIGURE 7.2. A body fixed point $P$ at $^B\mathbf{r}$ in the rotating body frame $B$.

$$\boldsymbol{\omega} = \begin{bmatrix} \omega_1 \\ \omega_2 \\ \omega_3 \end{bmatrix} = \dot{\phi}\,\hat{u} \qquad (7.6)$$

The angular velocity vector is associated with a skew symmetric matrix $_G\tilde{\omega}_B$ called the *angular velocity matrix*

$$\tilde{\omega} = \begin{bmatrix} 0 & -\omega_3 & \omega_2 \\ \omega_3 & 0 & -\omega_1 \\ -\omega_2 & \omega_1 & 0 \end{bmatrix} \qquad (7.7)$$

where

$$
\begin{aligned}
_G\tilde{\omega}_B &= {}^G\dot{R}_B\,{}^GR_B^T & (7.8) \\
&= \dot{\phi}\,\tilde{u}. & (7.9)
\end{aligned}
$$

**Proof.** Consider a rigid body with a fixed point $O$ and an attached frame $B(Oxyz)$ as shown in Figure 7.2. The body frame $B$ is initially coincident with the global frame $G$. Therefore, the position vector of a body point $P$ is

$$^G\mathbf{r}(t_0) = {}^B\mathbf{r}. \qquad (7.10)$$

The global time derivative of $^G\mathbf{r}$ is

$$
\begin{aligned}
^G\mathbf{v} &= {}^G\dot{\mathbf{r}} \\
&= \frac{{}^Gd}{dt}\,{}^G\mathbf{r}(t) \\
&= \frac{{}^Gd}{dt}\left[{}^GR_B(t)\,{}^B\mathbf{r}\right] \\
&= \frac{{}^Gd}{dt}\left[{}^GR_B(t)\,{}^G\mathbf{r}(t_0)\right] \\
&= {}^G\dot{R}_B(t)\,{}^B\mathbf{r}. & (7.11)
\end{aligned}
$$

Eliminating $^B\mathbf{r}$ between (7.1) and (7.11) determines the velocity of the point in global frame

$$^G\mathbf{v} = {^G\dot{R}_B(t)} \, {^GR_B^T(t)} \, {^G\mathbf{r}(t)}. \tag{7.12}$$

We denote the coefficient of $^G\mathbf{r}(t)$ by $\tilde{\omega}$

$$_G\tilde{\omega}_B = {^G\dot{R}_B} \, {^GR_B^T} \tag{7.13}$$

and write the Equation (7.12) as

$$^G\mathbf{v} = {_G\tilde{\omega}_B} \, {^G\mathbf{r}(t)} \tag{7.14}$$

or as

$$^G\mathbf{v} = {_G\omega_B} \times {^G\mathbf{r}(t)}. \tag{7.15}$$

The time derivative of the orthogonality condition, $^GR_B \, {^GR_B^T} = \mathbf{I}$, introduces an important identity

$$^G\dot{R}_B \, {^GR_B^T} + {^GR_B} \, {^G\dot{R}_B^T} = 0 \tag{7.16}$$

which can be utilized to show that $_G\tilde{\omega}_B = [^G\dot{R}_B \, {^GR_B^T}]$ is a skew symmetric matrix, because

$$^GR_B \, {^G\dot{R}_B^T} = \left[^G\dot{R}_B \, {^GR_B^T}\right]^T. \tag{7.17}$$

The vector $_G^G\omega_B$ is called the *instantaneous angular velocity* of the body $B$ relative to the global frame $G$ as seen from the $G$ frame.

Since a vectorial equation can be expressed in any coordinate frame, we may use any of the following expressions for the velocity of a body point in body or global frames

$$\begin{aligned} _G^G\mathbf{v}_P &= {_G^G\omega_B} \times {^G\mathbf{r}_P} & (7.18) \\ _G^B\mathbf{v}_P &= {_G^B\omega_B} \times {^B\mathbf{r}_P} & (7.19) \end{aligned}$$

where $_G^G\mathbf{v}_P$ is the global velocity of point $P$ expressed in global frame and $_G^B\mathbf{v}_P$ is the global velocity of point $P$ expressed in body frame.

$$\begin{aligned} _G^G\mathbf{v}_P &= {^GR_B} \, {_G^B\mathbf{v}_P} \\ &= {^GR_B} \left({_G^B\omega_B} \times {^B\mathbf{r}_P}\right) \end{aligned} \tag{7.20}$$

$_G^G\mathbf{v}_P$ and $_G^B\mathbf{v}_P$ can be converted to each other using a rotation matrix

$$\begin{aligned} _G^B\mathbf{v}_P &= {^GR_B^T} \, {_G^G\mathbf{v}_P} & (7.21) \\ &= {^GR_B^T} \, {_G\tilde{\omega}_B} \, {_G^G\mathbf{r}_P} \\ &= {^GR_B^T} \, {^G\dot{R}_B} \, {^GR_B^T} \, {_G^G\mathbf{r}_P} \\ &= {^GR_B^T} \, {^G\dot{R}_B} \, {_G^B\mathbf{r}_P}. & (7.22) \end{aligned}$$

showing that

$$\tensor[_G^B]{\tilde{\omega}}{_B} = {}^G R_B^T \, {}^G \dot{R}_B \tag{7.23}$$

which is called the *instantaneous angular velocity* of $B$ relative to the global frame $G$ as seen from the $B$ frame. From the definitions of $\tensor[_G]{\tilde{\omega}}{_B}$ and $\tensor[_G^B]{\tilde{\omega}}{_B}$ we are able to transform the two angular velocity matrices by

$$\tensor[_G]{\tilde{\omega}}{_B} = {}^G R_B \, \tensor[_G^B]{\tilde{\omega}}{_B} \, {}^G R_B^T \tag{7.24}$$

$$\tensor[_G^B]{\tilde{\omega}}{_B} = {}^G R_B^T \, \tensor[_G]{\tilde{\omega}}{_B} \, {}^G R_B \tag{7.25}$$

or equivalently

$$^G \dot{R}_B = \tensor[_G]{\tilde{\omega}}{_B} \, {}^G R_B \tag{7.26}$$

$$^G \dot{R}_B = {}^G R_B \, \tensor[_G^B]{\tilde{\omega}}{_B} \tag{7.27}$$

$$\tensor[_G]{\tilde{\omega}}{_B} \, {}^G R_B = {}^G R_B \, \tensor[_G^B]{\tilde{\omega}}{_B}. \tag{7.28}$$

The angular velocity of $B$ in $G$ is negative of the angular velocity of $G$ in $B$ if both are expressed in the same coordinate frame.

$$\tensor[_G^G]{\tilde{\omega}}{_B} = -\tensor[_B^G]{\tilde{\omega}}{_G} \tag{7.29}$$

$$\tensor[_G^B]{\tilde{\omega}}{_B} = -\tensor[_B^B]{\tilde{\omega}}{_G}. \tag{7.30}$$

$\tensor[_G]{\omega}{_B}$ and can always be expressed in the form

$$\tensor[_G]{\omega}{_B} = \omega \hat{u} \tag{7.31}$$

where $\hat{u}$ is a unit vector parallel to $\tensor[_G]{\omega}{_B}$ and indicates the *instantaneous axis of rotation*.

Using the Rodriguez rotation formula (3.5) we can show that

$$\dot{R}_{\hat{u},\phi} = \dot{\phi} \, \tilde{u} \, R_{\hat{u},\phi} \tag{7.32}$$

and therefore

$$\tilde{\omega} = \dot{\phi} \, \tilde{u} \tag{7.33}$$

or equivalently

$$\tensor[_G]{\tilde{\omega}}{_B} = \lim_{\phi \to 0} \frac{{}^G d}{dt} R_{\hat{u},\phi} \tag{7.34}$$

$$= \lim_{\phi \to 0} \frac{{}^G d}{dt} \left( -\tilde{u}^2 \cos\phi + \tilde{u} \sin\phi + \tilde{u}^2 + \mathbf{I} \right)$$

$$= \dot{\phi} \, \tilde{u}$$

and therefore

$$\omega = \dot{\phi} \, \hat{u}. \tag{7.35}$$

■

**Example 172** *Rotation of a body point about a global axis.*

The slab shown in Figure 2.4 is turning about the $Z$-axis with $\dot\alpha = 10\,\deg/s$. The global velocity of the corner point $P(5, 30, 10)$, when the slab is at $\alpha = 30\,\deg$, is

$$
{}^G\mathbf{v}_P = {}^G\dot{R}_B(t)\,{}^B\mathbf{r}_P \tag{7.36}
$$

$$
= \frac{{}^Gd}{dt}\left(\begin{bmatrix} \cos\alpha & -\sin\alpha & 0 \\ \sin\alpha & \cos\alpha & 0 \\ 0 & 0 & 1 \end{bmatrix}\right)\begin{bmatrix} 5 \\ 30 \\ 10 \end{bmatrix}
$$

$$
= \dot\alpha \begin{bmatrix} -\sin\alpha & -\cos\alpha & 0 \\ \cos\alpha & -\sin\alpha & 0 \\ 0 & 0 & 0 \end{bmatrix}\begin{bmatrix} 5 \\ 30 \\ 10 \end{bmatrix}
$$

$$
= \frac{10\pi}{180}\begin{bmatrix} -\sin\frac{\pi}{6} & -\cos\frac{\pi}{6} & 0 \\ \cos\frac{\pi}{6} & -\sin\frac{\pi}{6} & 0 \\ 0 & 0 & 0 \end{bmatrix}\begin{bmatrix} 5 \\ 30 \\ 10 \end{bmatrix} = \begin{bmatrix} -4.97 \\ -1.86 \\ 0 \end{bmatrix}
$$

at this moment, the point $P$ is at

$$
{}^G\mathbf{r}_P = {}^GR_B\,{}^B\mathbf{r}_P \tag{7.37}
$$

$$
= \begin{bmatrix} \cos\frac{\pi}{6} & -\sin\frac{\pi}{6} & 0 \\ \sin\frac{\pi}{6} & \cos\frac{\pi}{6} & 0 \\ 0 & 0 & 1 \end{bmatrix}\begin{bmatrix} 5 \\ 30 \\ 10 \end{bmatrix} = \begin{bmatrix} -10.67 \\ 28.48 \\ 10 \end{bmatrix}.
$$

**Example 173** *Rotation of a global point about a global axis.*

The corner $P$ of the slab shown in Figure 2.4, is at ${}^B\mathbf{r}_P = \begin{bmatrix} 5 & 30 & 10 \end{bmatrix}^T$. When it is turned $\alpha = 30\,\deg$ about the $Z$-axis, the global position of $P$ is

$$
{}^G\mathbf{r}_P = {}^GR_B\,{}^B\mathbf{r}_P \tag{7.38}
$$

$$
= \begin{bmatrix} \cos\frac{\pi}{6} & -\sin\frac{\pi}{6} & 0 \\ \sin\frac{\pi}{6} & \cos\frac{\pi}{6} & 0 \\ 0 & 0 & 1 \end{bmatrix}\begin{bmatrix} 5 \\ 30 \\ 10 \end{bmatrix} = \begin{bmatrix} -10.67 \\ 28.48 \\ 10 \end{bmatrix}.
$$

If the slab is turning with $\dot\alpha = 10\,\deg/s$, the global velocity of the point $P$ would be

$$
{}^G\mathbf{v}_P = {}^G\dot{R}_B\,{}^GR_B^T\,{}^G\mathbf{r}_P \tag{7.39}
$$

$$
= \frac{10\pi}{180}\begin{bmatrix} -s\frac{\pi}{6} & -c\frac{\pi}{6} & 0 \\ c\frac{\pi}{6} & -s\frac{\pi}{6} & 0 \\ 0 & 0 & 0 \end{bmatrix}\begin{bmatrix} c\frac{\pi}{6} & -s\frac{\pi}{6} & 0 \\ s\frac{\pi}{6} & c\frac{\pi}{6} & 0 \\ 0 & 0 & 1 \end{bmatrix}^T\begin{bmatrix} -10.67 \\ 28.48 \\ 10 \end{bmatrix}
$$

$$
= \begin{bmatrix} -4.97 \\ -1.86 \\ 0 \end{bmatrix}.
$$

**Example 174** *Principal angular velocities.*
  *The principal rotational matrices about the axes $X$, $Y$, and $Z$ are*

$$R_{X,\gamma} = \begin{bmatrix} 1 & 0 & 0 \\ 0 & \cos\gamma & -\sin\gamma \\ 0 & \sin\gamma & \cos\gamma \end{bmatrix} \tag{7.40}$$

$$R_{Y,\beta} = \begin{bmatrix} \cos\beta & 0 & \sin\beta \\ 0 & 1 & 0 \\ -\sin\beta & 0 & \cos\beta \end{bmatrix} \tag{7.41}$$

$$R_{Z,\alpha} = \begin{bmatrix} \cos\alpha & -\sin\alpha & 0 \\ \sin\alpha & \cos\alpha & 0 \\ 0 & 0 & 1 \end{bmatrix}. \tag{7.42}$$

*and hence, their time derivatives are*

$$\dot{R}_{X,\gamma} = \dot{\gamma} \begin{bmatrix} 0 & 0 & 0 \\ 0 & -\sin\gamma & -\cos\gamma \\ 0 & \cos\gamma & -\sin\gamma \end{bmatrix} \tag{7.43}$$

$$\dot{R}_{Y,\beta} = \dot{\beta} \begin{bmatrix} -\sin\beta & 0 & \cos\beta \\ 0 & 0 & 0 \\ -\cos\beta & 0 & -\sin\beta \end{bmatrix} \tag{7.44}$$

$$\dot{R}_{Z,\alpha} = \dot{\alpha} \begin{bmatrix} -\sin\alpha & -\cos\alpha & 0 \\ \cos\alpha & -\sin\alpha & 0 \\ 0 & 0 & 0 \end{bmatrix}. \tag{7.45}$$

*Therefore, the principal angular velocity matrices about axes $X$, $Y$, and $Z$ are*

$$_G\tilde{\omega}_X = \dot{R}_{X,\gamma} R_{X,\gamma}^T = \dot{\gamma} \begin{bmatrix} 0 & 0 & 0 \\ 0 & 0 & -1 \\ 0 & 1 & 0 \end{bmatrix} \tag{7.46}$$

$$_G\tilde{\omega}_Y = \dot{R}_{Y,\beta} R_{Y,\beta}^T = \dot{\beta} \begin{bmatrix} 0 & 0 & 1 \\ 0 & 0 & 0 \\ -1 & 0 & 0 \end{bmatrix} \tag{7.47}$$

$$_G\tilde{\omega}_Z = \dot{R}_{Z,\alpha} R_{Z,\alpha}^T = \dot{\alpha} \begin{bmatrix} 0 & -1 & 0 \\ 1 & 0 & 0 \\ 0 & 0 & 0 \end{bmatrix} \tag{7.48}$$

*which are equivalent to*

$$_G\tilde{\omega}_X = \dot{\gamma}\tilde{I} \tag{7.49}$$
$$_G\tilde{\omega}_Y = \dot{\beta}\tilde{J} \tag{7.50}$$
$$_G\tilde{\omega}_Z = \dot{\alpha}\tilde{K} \tag{7.51}$$

*and therefore, the principal angular velocity vectors are*

$$_G\boldsymbol{\omega}_X \;=\; \omega_X\,\hat{I} = \dot{\gamma}\hat{I} \tag{7.52}$$

$$_G\boldsymbol{\omega}_Y \;=\; \omega_Y\,\hat{J} = \dot{\beta}\hat{J} \tag{7.53}$$

$$_G\boldsymbol{\omega}_Z \;=\; \omega_Z\,\hat{K} = \dot{\alpha}\hat{K}. \tag{7.54}$$

*Utilizing the same technique, we can find the following principal angular velocity matrices about the local axes.*

$$_G^B\tilde{\omega}_x = R_{x,\psi}^T\,\dot{R}_{x,\psi} = \dot{\psi}\begin{bmatrix} 0 & 0 & 0 \\ 0 & 0 & -1 \\ 0 & 1 & 0 \end{bmatrix} = \dot{\psi}\,\tilde{\imath} \tag{7.55}$$

$$_G^B\tilde{\omega}_y = R_{y,\theta}^T\dot{R}_{y,\theta} = \dot{\theta}\begin{bmatrix} 0 & 0 & 1 \\ 0 & 0 & 0 \\ -1 & 0 & 0 \end{bmatrix} = \dot{\theta}\,\tilde{\jmath} \tag{7.56}$$

$$_G^B\tilde{\omega}_z = R_{z,\varphi}^T\dot{R}_{z,\varphi} = \dot{\varphi}\begin{bmatrix} 0 & -1 & 0 \\ 1 & 0 & 0 \\ 0 & 0 & 0 \end{bmatrix} = \dot{\varphi}\,\tilde{k} \tag{7.57}$$

**Example 175** *Decomposition of an angular velocity vector.*
   *Every angular velocity vector can be decomposed to three principal angular velocity vectors.*

$$\begin{aligned}
_G\boldsymbol{\omega}_B \;&=\; \left(_G\boldsymbol{\omega}_B \cdot \hat{I}\right)\hat{I} + \left(_G\boldsymbol{\omega}_B \cdot \hat{J}\right)\hat{J} + \left(_G\boldsymbol{\omega}_B \cdot \hat{K}\right)\hat{K} \tag{7.58}\\
&=\; \omega_X\,\hat{I} + \omega_Y\,\hat{J} + \omega_Z\,\hat{K} \\
&=\; \dot{\gamma}\hat{I} + \dot{\beta}\hat{J} + \dot{\alpha}\hat{K} \\
&=\; \boldsymbol{\omega}_X + \boldsymbol{\omega}_Y + \boldsymbol{\omega}_Z
\end{aligned}$$

**Example 176** *Combination of angular velocities.*
   *Starting from a combination of rotations*

$$^0R_2 = {}^0R_1\,{}^1R_2 \tag{7.59}$$

*and taking derivative, we find*

$$^0\dot{R}_2 = {}^0\dot{R}_1\,{}^1R_2 + {}^0R_1\,{}^1\dot{R}_2. \tag{7.60}$$

*Now, substituting the derivative of rotation matrices with*

$$^0\dot{R}_2 \;=\; {}_0\tilde{\omega}_2\,{}^0R_2 \tag{7.61}$$

$$^0\dot{R}_1 \;=\; {}_0\tilde{\omega}_1\,{}^0R_1 \tag{7.62}$$

$$^1\dot{R}_2 \;=\; {}_1\tilde{\omega}_2\,{}^1R_2 \tag{7.63}$$

*results in*

$$
\begin{aligned}
_0\tilde{\omega}_2 \, {}^0R_2 &= {}_0\tilde{\omega}_1 \, {}^0R_1 \, {}^1R_2 + {}^0R_1 \, {}_1\tilde{\omega}_2 \, {}^1R_2 \\
&= {}_0\tilde{\omega}_1 \, {}^0R_2 + {}^0R_1 \, {}_1\tilde{\omega}_2 \, {}^0R_1^T \, {}^0R_1 \, {}^1R_2 \\
&= {}_0\tilde{\omega}_1 \, {}^0R_2 + {}_1^0\tilde{\omega}_2 \, {}^0R_2
\end{aligned} \tag{7.64}
$$

*where*

$$
{}^0R_1 \, {}_1\tilde{\omega}_2 \, {}^0R_1^T = {}_1^0\tilde{\omega}_2. \tag{7.65}
$$

*Therefore, we find*

$$
_0\tilde{\omega}_2 = {}_0\tilde{\omega}_1 + {}_1^0\tilde{\omega}_2 \tag{7.66}
$$

*which indicates that the angular velocities may be added relatively*

$$
_0\boldsymbol{\omega}_2 = {}_0\boldsymbol{\omega}_1 + {}_1^0\boldsymbol{\omega}_2. \tag{7.67}
$$

*This result also holds for any number of angular velocities*

$$
\begin{aligned}
_0\boldsymbol{\omega}_n &= {}_0\boldsymbol{\omega}_1 + {}_1^0\boldsymbol{\omega}_2 + {}_2^0\boldsymbol{\omega}_3 + \cdots + {}_{n-1}^{\;\;0}\boldsymbol{\omega}_n \tag{7.68} \\
&= \sum_{i=1}^{n} {}_{i-1}^{\;\;0}\boldsymbol{\omega}_i.
\end{aligned}
$$

**Example 177** *Angular velocity in terms of Euler frequencies.*

The angular velocity vector can be expressed by Euler frequencies as described in Chapter 2. Therefore,

$$
\begin{aligned}
{}_G^B\boldsymbol{\omega}_B &= \omega_x \hat{\imath} + \omega_y \hat{\jmath} + \omega_z \hat{k} \\
&= \dot{\varphi}\hat{e}_\varphi + \dot{\theta}\hat{e}_\theta + \dot{\psi}\hat{e}_\psi \\
&= \dot{\varphi}\begin{bmatrix} \sin\theta\sin\psi \\ \sin\theta\cos\psi \\ \cos\theta \end{bmatrix} + \dot{\theta}\begin{bmatrix} \cos\psi \\ -\sin\psi \\ 0 \end{bmatrix} + \dot{\psi}\begin{bmatrix} 0 \\ 0 \\ 1 \end{bmatrix} \\
&= \begin{bmatrix} \sin\theta\sin\psi & \cos\psi & 0 \\ \sin\theta\cos\psi & -\sin\psi & 0 \\ \cos\theta & 0 & 1 \end{bmatrix}\begin{bmatrix} \dot{\varphi} \\ \dot{\theta} \\ \dot{\psi} \end{bmatrix}
\end{aligned} \tag{7.69}
$$

*and*

$$
\begin{aligned}
{}_G^G\boldsymbol{\omega}_B &= {}^BR_G^{-1} \, {}_G^B\boldsymbol{\omega}_B \\
&= {}^BR_G^{-1}\begin{bmatrix} \dot{\varphi}\sin\theta\sin\psi + \dot{\theta}\cos\psi \\ \dot{\varphi}\sin\theta\cos\psi - \dot{\theta}\sin\psi \\ \dot{\varphi}\cos\theta + \dot{\psi} \end{bmatrix} \\
&= \begin{bmatrix} 0 & \cos\varphi & \sin\theta\sin\varphi \\ 0 & \sin\varphi & -\cos\varphi\sin\theta \\ 1 & 0 & \cos\theta \end{bmatrix}\begin{bmatrix} \dot{\varphi} \\ \dot{\theta} \\ \dot{\psi} \end{bmatrix}
\end{aligned} \tag{7.70}
$$

*where the inverse of the Euler transformation matrix is*

$$^{B}R_{G}^{-1} = \begin{bmatrix} c\varphi c\psi - c\theta s\varphi s\psi & -c\varphi s\psi - c\theta c\psi s\varphi & s\theta s\varphi \\ c\psi s\varphi + c\theta c\varphi s\psi & -s\varphi s\psi + c\theta c\varphi c\psi & -c\varphi s\theta \\ s\theta s\psi & s\theta c\psi & c\theta \end{bmatrix}. \qquad (7.71)$$

**Example 178** *Angular velocity in terms of rotation frequencies.*

Appendices A and B show the 12 global and 12 local axes' triple rotations. Utilizing those equations, we are able to find the angular velocity matrix and vector of a rigid body in terms of rotation frequencies. As an example, consider the Euler angles transformation matrix in case 9, Appendix B.

$$^{B}R_{G} = R_{z,\psi}R_{x,\theta}R_{z,\varphi} \qquad (7.72)$$

*The angular velocity matrix is then equal to*

$$\begin{aligned}
_{B}\tilde{\omega}_{G} &= {}^{B}\dot{R}_{G}{}^{B}R_{G}^{T} \\
&= \left( \dot{\varphi}\, R_{z,\psi}R_{x,\theta}\frac{dR_{z,\varphi}}{dt} + \dot{\theta}\, R_{z,\psi}\frac{dR_{x,\theta}}{dt}R_{z,\varphi} + \dot{\psi}\,\frac{dR_{z,\psi}}{dt}R_{x,\theta}R_{z,\varphi} \right) \\
&\quad \times (R_{z,\psi}R_{x,\theta}R_{z,\varphi})^{T} \\
&= \dot{\varphi}\, R_{z,\psi}R_{x,\theta}\frac{dR_{z,\varphi}}{dt}R_{z,\varphi}^{T}R_{x,\theta}^{T}R_{z,\psi}^{T} \\
&\quad + \dot{\theta}\, R_{z,\psi}\frac{dR_{x,\theta}}{dt}R_{x,\theta}^{T}R_{z,\psi}^{T} \\
&\quad + \dot{\psi}\,\frac{dR_{z,\psi}}{dt}R_{z,\psi}^{T} \qquad (7.73)
\end{aligned}$$

*which, in matrix form, is*

$$\begin{aligned}
_{B}\tilde{\omega}_{G} &= \dot{\varphi}\begin{bmatrix} 0 & \cos\theta & -\sin\theta\cos\psi \\ -\cos\theta & 0 & \sin\theta\sin\psi \\ \sin\theta\cos\psi & -\sin\theta\sin\psi & 0 \end{bmatrix} \\
&\quad + \dot{\theta}\begin{bmatrix} 0 & 0 & \sin\psi \\ 0 & 0 & \cos\psi \\ -\sin\psi & -\cos\psi & 0 \end{bmatrix} + \dot{\psi}\begin{bmatrix} 0 & 1 & 0 \\ -1 & 0 & 0 \\ 0 & 0 & 0 \end{bmatrix} \qquad (7.74)
\end{aligned}$$

*or*

$$_{B}\tilde{\omega}_{G} = \begin{bmatrix} 0 & \dot{\psi} + \dot{\varphi}c\theta & \dot{\theta}s\psi - \dot{\varphi}s\theta c\psi \\ -\dot{\psi} - \dot{\varphi}c\theta & 0 & \dot{\theta}c\psi + \dot{\varphi}s\theta s\psi \\ -\dot{\theta}s\psi + \dot{\varphi}s\theta c\psi & -\dot{\theta}c\psi - \dot{\varphi}s\theta s\psi & 0 \end{bmatrix}. \qquad (7.75)$$

*The corresponding angular velocity vector is*

$$\begin{aligned}
_{B}\omega_{G} &= -\begin{bmatrix} \dot{\theta}c\psi + \dot{\varphi}s\theta s\psi \\ -\dot{\theta}s\psi + \dot{\varphi}s\theta c\psi \\ \dot{\psi} + \dot{\varphi}c\theta \end{bmatrix} \\
&= -\begin{bmatrix} \sin\theta\sin\psi & \cos\psi & 0 \\ \sin\theta\cos\psi & -\sin\psi & 0 \\ \cos\theta & 0 & 1 \end{bmatrix}\begin{bmatrix} \dot{\varphi} \\ \dot{\theta} \\ \dot{\psi} \end{bmatrix}. \qquad (7.76)
\end{aligned}$$

*However,*

$$
\begin{aligned}
{}_{B}^{B}\tilde{\omega}_G &= -{}_{G}^{B}\tilde{\omega}_B & (7.77) \\
{}_{B}^{B}\omega_G &= -{}_{G}^{B}\omega_B & (7.78)
\end{aligned}
$$

*and therefore,*

$$
{}_{G}^{B}\omega_B = \begin{bmatrix} \sin\theta\sin\psi & \cos\psi & 0 \\ \sin\theta\cos\psi & -\sin\psi & 0 \\ \cos\theta & 0 & 1 \end{bmatrix} \begin{bmatrix} \dot{\varphi} \\ \dot{\theta} \\ \dot{\psi} \end{bmatrix}. \tag{7.79}
$$

**Example 179** ★ *Coordinate transformation of angular velocity.*
   *Angular velocity ${}_{1}\omega_2$ of coordinate frame $B_2$ with respect to $B_1$ and ex-pressed in $B_1$ can be expressed in base coordinate frame $B_0$ according to*

$$
{}^{0}R_1\,{}_{1}\tilde{\omega}_2\,{}^{0}R_1^{T} = {}_{1}^{0}\tilde{\omega}_2. \tag{7.80}
$$

*To show this equation, it is enough to apply both sides on an arbitrary vector ${}^{0}\mathbf{r}$. Therefore, the left-hand side would be*

$$
\begin{aligned}
{}^{0}R_1\,{}_{1}\tilde{\omega}_2\,{}^{0}R_1^{T}\,{}^{0}\mathbf{r} &= {}^{0}R_1\,{}_{1}\tilde{\omega}_2\,{}^{1}R_0\,{}^{0}\mathbf{r} \\
&= {}^{0}R_1\,{}_{1}\tilde{\omega}_2\,{}^{1}\mathbf{r} \\
&= {}^{0}R_1\left({}_{1}\omega_2 \times {}^{1}\mathbf{r}\right) \\
&= {}^{0}R_1\,{}_{1}\omega_2 \times {}^{0}R_1\,{}^{1}\mathbf{r} \\
&= {}_{1}^{0}\omega_2 \times {}^{0}\mathbf{r} \tag{7.81}
\end{aligned}
$$

*which is equal to the right-hand side after applying on the vector ${}^{0}\mathbf{r}$*

$$
{}_{1}^{0}\tilde{\omega}_2\,{}^{0}\mathbf{r} = {}_{1}^{0}\omega_2 \times {}^{0}\mathbf{r}. \tag{7.82}
$$

**Example 180** ★ *Time derivative of unit vectors.*
   *Using Equation (7.19) we can define the time derivative of unit vectors of a body coordinate frame $B(\hat{\imath}, \hat{\jmath}, \hat{k})$, rotating in the global coordinate frame $G(\hat{I}, \hat{J}, \hat{K})$*

$$
\frac{{}^{G}d\hat{\imath}}{dt} = {}_{G}^{B}\omega_B \times \hat{\imath} \tag{7.83}
$$

$$
\frac{{}^{G}d\hat{\jmath}}{dt} = {}_{G}^{B}\omega_B \times \hat{\jmath} \tag{7.84}
$$

$$
\frac{{}^{G}d\hat{k}}{dt} = {}_{G}^{B}\omega_B \times \hat{k}. \tag{7.85}
$$

**Example 181 ★** *Angular velocity in terms of quaternion and Euler parameters.*

*The angular velocity vector can also be expressed by Euler parameters. Starting from the unit quaternion representation of a finite rotation*

$$
\begin{aligned}
{}^{G}\mathbf{r} &= e(t)\,{}^{B}\mathbf{r}\,e^{*}(t) \\
&= e(t)\,{}^{B}\mathbf{r}\,e^{-1}(t)
\end{aligned}
\tag{7.86}
$$

*where*

$$
\begin{aligned}
e &= e_0 + \mathbf{e} \tag{7.87} \\
e^{*} &= e^{-1} = e_0 - \mathbf{e} \tag{7.88}
\end{aligned}
$$

*we can find*

$$
\begin{aligned}
{}^{G}\dot{\mathbf{r}} &= \dot{e}\,{}^{B}\mathbf{r}\,e^{*} + e\,{}^{B}\mathbf{r}\,\dot{e}^{*} \\
&= \dot{e}\,e^{*}\,{}^{G}\mathbf{r} + {}^{G}\mathbf{r}\,e\,\dot{e}^{*} \\
&= 2\dot{e}\,e^{*}\,{}^{G}\mathbf{r}
\end{aligned}
\tag{7.89}
$$

*and therefore, the angular velocity quaternion is*

$$
{}_{G}\boldsymbol{\omega}_{B} = 2\dot{e}\,e^{*}.
\tag{7.90}
$$

*We have used the orthogonality property of unit quaternion*

$$
e\,e^{-1} = e\,e^{*} = 1
\tag{7.91}
$$

*which provides*

$$
\dot{e}\,e^{*} + e\,\dot{e}^{*} = 0.
\tag{7.92}
$$

*The angular velocity quaternion can be expanded using quaternion products to find the angular velocity components based on Euler parameters.*

$$
\begin{aligned}
{}_{G}\boldsymbol{\omega}_{B} &= 2\dot{e}\,e^{*} = 2\,(\dot{e}_0 + \dot{\mathbf{e}})\,(e_0 - \mathbf{e}) \\
&= 2\,(\dot{e}_0 e_0 + e_0\dot{\mathbf{e}} - \dot{e}_0\mathbf{e} + \dot{\mathbf{e}}\cdot\mathbf{e} - \dot{\mathbf{e}}\times\mathbf{e}) \\
&= 2\begin{bmatrix} 0 \\ e_0\dot{e}_1 - e_1\dot{e}_0 + e_2\dot{e}_3 - e_3\dot{e}_2 \\ e_0\dot{e}_2 - e_2\dot{e}_0 - e_1\dot{e}_3 + e_3\dot{e}_1 \\ e_0\dot{e}_3 + e_1\dot{e}_2 - e_2\dot{e}_1 - e_3\dot{e}_0 \end{bmatrix} \\
&= 2\begin{bmatrix} \dot{e}_0 & -\dot{e}_1 & -\dot{e}_2 & -\dot{e}_3 \\ \dot{e}_1 & \dot{e}_0 & -\dot{e}_3 & \dot{e}_2 \\ \dot{e}_2 & \dot{e}_3 & \dot{e}_0 & -\dot{e}_1 \\ \dot{e}_3 & -\dot{e}_2 & \dot{e}_1 & \dot{e}_0 \end{bmatrix}\begin{bmatrix} e_0 \\ -e_1 \\ -e_2 \\ -e_3 \end{bmatrix}
\end{aligned}
\tag{7.93}
$$

*The scalar component of the angular velocity quaternion is zero because*

$$
\dot{e}_0 e_0 + \dot{\mathbf{e}}\cdot\mathbf{e} = \dot{e}_0 e_0 + e_1\dot{e}_1 + e_2\dot{e}_2 + e_3\dot{e}_3 = 0.
\tag{7.94}
$$

*The angular velocity vector can also be defined as a quaternion*

$$\overleftrightarrow{_G\omega_B} = 2\overleftrightarrow{\dot{e}}\ \overleftrightarrow{e^*} \tag{7.95}$$

*to be utilized for definition of the derivative of a rotation quaternion*

$$\overleftrightarrow{\dot{e}} = \frac{1}{2}\overleftrightarrow{_G\omega_B}\ \overleftrightarrow{e}. \tag{7.96}$$

*A coordinate transformation can transform the angular velocity into a body coordinate frame*

$$\begin{aligned}
{}^B_G\omega_B &= e^*\ {}^G_G\omega_B\ e \tag{7.97}\\
&= 2e^*\ \dot{e} \tag{7.98}\\
&= 2\begin{bmatrix} e_0 & e_1 & e_2 & e_3 \\ -e_1 & e_0 & e_3 & -e_2 \\ -e_2 & -e_3 & e_0 & e_1 \\ -e_3 & -e_2 & e_1 & e_0 \end{bmatrix}\begin{bmatrix} \dot{e}_0 \\ \dot{e}_1 \\ \dot{e}_2 \\ \dot{e}_3 \end{bmatrix} \tag{7.99}
\end{aligned}$$

*and therefore,*

$$\overleftrightarrow{_G^B\omega_B} = 2\overleftrightarrow{e^*}\ \overleftrightarrow{\dot{e}} \tag{7.100}$$

$$\overleftrightarrow{\dot{e}} = \frac{1}{2}\overleftrightarrow{e}\ \overleftrightarrow{_G^B\omega_B}. \tag{7.101}$$

**Example 182 ★** *Differential of Euler parameters.*
   The rotation matrix $^GR_B$ based on Euler parameters is given in Equation (3.58)

$$\begin{aligned}
^GR_B &= \begin{bmatrix} e_0^2 + e_1^2 - e_2^2 - e_3^2 & 2\left(e_1e_2 - e_0e_3\right) & 2\left(e_0e_2 + e_1e_3\right) \\ 2\left(e_0e_3 + e_1e_2\right) & e_0^2 - e_1^2 + e_2^2 - e_3^2 & 2\left(e_2e_3 - e_0e_1\right) \\ 2\left(e_1e_3 - e_0e_2\right) & 2\left(e_0e_1 + e_2e_3\right) & e_0^2 - e_1^2 - e_2^2 + e_3^2 \end{bmatrix} \\[2mm]
&= \begin{bmatrix} r_{11} & r_{12} & r_{13} \\ r_{21} & r_{22} & r_{23} \\ r_{31} & r_{32} & r_{33} \end{bmatrix} \tag{7.102}
\end{aligned}$$

*and the individual parameters can be found from any set of Equations (3.126) to (3.129). The first set indicates that*

$$e_0 = \pm\frac{1}{2}\sqrt{1 + r_{11} + r_{22} + r_{33}} \tag{7.103}$$

$$e_1 = \frac{1}{4}\frac{r_{32} - r_{23}}{e_0} \tag{7.104}$$

$$e_2 = \frac{1}{4}\frac{r_{13} - r_{31}}{e_0} \tag{7.105}$$

$$e_3 = \frac{1}{4}\frac{r_{21} - r_{12}}{e_0} \tag{7.106}$$

*and therefore,*

$$\dot{e}_0 = \frac{\dot{r}_{11} + \dot{r}_{22} + \dot{r}_{33}}{8e_0} \tag{7.107}$$

$$\dot{e}_1 = \frac{1}{4e_0^2}\left((\dot{r}_{32} - \dot{r}_{23})\,e_0 - (r_{32} - r_{23})\,\dot{e}_0\right) \tag{7.108}$$

$$\dot{e}_2 = \frac{1}{4e_0^2}\left((\dot{r}_{13} - \dot{r}_{31})\,e_0 - (r_{13} - r_{31})\,\dot{e}_0\right) \tag{7.109}$$

$$\dot{e}_3 = \frac{1}{4e_0^2}\left((\dot{r}_{21} - \dot{r}_{12})\,e_0 - (r_{21} - r_{12})\,\dot{e}_0\right). \tag{7.110}$$

*We may use the differential of the transformation matrix*

$$^G\dot{R}_B = {}_G\tilde{\omega}_B\,{}^G R_B$$

*to show that*

$$\dot{e}_0 = \frac{1}{2}\left(-e_1\omega_1 - e_2\omega_2 - e_3\omega_3\right) \tag{7.111}$$

$$\dot{e}_1 = \frac{1}{2}\left(e_0\omega_1 + e_2\omega_3 - e_3\omega_2\right) \tag{7.112}$$

$$\dot{e}_2 = \frac{1}{2}\left(e_0\omega_2 - e_1\omega_3 - e_3\omega_1\right) \tag{7.113}$$

$$\dot{e}_3 = \frac{1}{2}\left(e_0\omega_3 + e_1\omega_2 - e_2\omega_1\right). \tag{7.114}$$

*Similarly we can find $\dot{e}_1$, $\dot{e}_2$, and $\dot{e}_3$, however the final result can be set in a matrix form*

$$\begin{bmatrix} \dot{e}_0 \\ \dot{e}_1 \\ \dot{e}_2 \\ \dot{e}_3 \end{bmatrix} = \frac{1}{2} \begin{bmatrix} 0 & -\omega_1 & -\omega_2 & -\omega_3 \\ \omega_1 & 0 & \omega_3 & -\omega_2 \\ \omega_2 & -\omega_3 & 0 & \omega_1 \\ \omega_3 & \omega_2 & -\omega_1 & 0 \end{bmatrix} \begin{bmatrix} e_0 \\ e_1 \\ e_2 \\ e_3 \end{bmatrix} \tag{7.115}$$

*or*

$$\begin{bmatrix} \dot{e}_0 \\ \dot{e}_1 \\ \dot{e}_2 \\ \dot{e}_3 \end{bmatrix} = \frac{1}{2} \begin{bmatrix} e_0 & -e_3 & -e_2 & -e_1 \\ e_1 & e_0 & -e_3 & e_2 \\ e_2 & e_1 & e_0 & -e_3 \\ e_3 & -e_2 & e_1 & e_0 \end{bmatrix} \begin{bmatrix} 0 \\ \omega_1 \\ \omega_2 \\ \omega_3 \end{bmatrix}. \tag{7.116}$$

**Example 183 ★** *Elements of the angular velocity matrix.*
   *Utilizing the permutation symbol introduced in (3.122)*

$$\epsilon_{ijk} = \frac{1}{2}(i - j)(j - k)(k - i) \quad , \quad i, j, k = 1, 2, 3 \tag{7.117}$$

*allows us to find the elements of the angular velocity matrix, $\tilde{\omega}$, when the angular velocity vector, $\boldsymbol{\omega} = \begin{bmatrix} \omega_1 & \omega_2 & \omega_3 \end{bmatrix}^T$, is given.*

$$\tilde{\omega}_{ij} = \epsilon_{ijk}\,\omega_k \tag{7.118}$$

## 7.2  Time Derivative and Coordinate Frames

The time derivative of a vector depends on the coordinate frame in which we are taking the derivative. The time derivative of a vector $\mathbf{r}$ in the global frame is called *G-derivative* and is denoted by

$$\frac{{}^G d}{dt}\mathbf{r}$$

while the time derivative of the vector in the body frame is called the *B-derivative* and is denoted by

$$\frac{{}^B d}{dt}\mathbf{r}.$$

The left superscript on the derivative symbol indicates the frame in which the derivative is taken, and hence, its unit vectors are considered constant. Therefore, the derivative of ${}^B\mathbf{r}_P$ in $B$ and the derivative of ${}^G\mathbf{r}_P$ in $G$ are

$$\frac{{}^B d}{dt}{}^B\mathbf{r}_P \;=\; {}^B\dot{\mathbf{r}}_P = {}^B\mathbf{v}_P = \dot{x}\,\hat{\imath} + \dot{y}\,\hat{\jmath} + \dot{z}\,\hat{k} \tag{7.119}$$

$$\frac{{}^G d}{dt}{}^G\mathbf{r}_P \;=\; {}^G\dot{\mathbf{r}}_P = {}^G\mathbf{v}_P = \dot{X}\,\hat{I} + \dot{Y}\,\hat{J} + \dot{Z}\,\hat{K}. \tag{7.120}$$

It is also possible to find the $G$-derivative of ${}^B\mathbf{r}_P$ and the $B$-derivative of ${}^G\mathbf{r}_P$. We use Equation (7.19) for the global velocity of a body fixed point $P$, expressed in body frame

$$\genfrac{}{}{0pt}{}{B}{G}\mathbf{v}_P = \genfrac{}{}{0pt}{}{B}{G}\boldsymbol{\omega}_B \times {}^B\mathbf{r}_P \tag{7.121}$$

to define the $G$-derivative of a body vector ${}^B\mathbf{r}_P$ by

$$\genfrac{}{}{0pt}{}{B}{G}\mathbf{v}_P = \frac{{}^G d}{dt}{}^B\mathbf{r}_P \tag{7.122}$$

and similarly, a $B$-derivative of a global vector ${}^G\mathbf{r}_P$ by

$$\genfrac{}{}{0pt}{}{G}{B}\mathbf{v}_P = \frac{{}^B d}{dt}{}^G\mathbf{r}_P. \tag{7.123}$$

When point $P$ is moving in frame $B$ while $B$ is rotating in $G$, the $G$-derivative of ${}^B\mathbf{r}_P(t)$ is defined by

$$\frac{{}^G d}{dt}{}^B\mathbf{r}_P(t) \;=\; {}^B\dot{\mathbf{r}}_P + \genfrac{}{}{0pt}{}{B}{G}\boldsymbol{\omega}_B \times {}^B\mathbf{r}_P$$

$$=\; \genfrac{}{}{0pt}{}{B}{G}\dot{\mathbf{r}}_P \tag{7.124}$$

and the $B$-derivative of ${}^G\mathbf{r}_P$ is defined by

$$\frac{{}^B d}{dt}{}^G\mathbf{r}_P(t) \;=\; {}^G\dot{\mathbf{r}}_P - {}_G\boldsymbol{\omega}_B \times {}^G\mathbf{r}_P$$

$$=\; \genfrac{}{}{0pt}{}{G}{B}\dot{\mathbf{r}}_P. \tag{7.125}$$

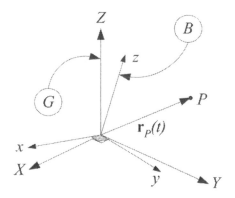

FIGURE 7.3. A moving body point $P$ at $^B\mathbf{r}(t)$ in the rotating body frame $B$.

**Proof.** Let $G(OXYZ)$ with unit vectors $\hat{I}$, $\hat{J}$, and $\hat{K}$ be the global coordinate frame, and let $B(Oxyz)$ with unit vectors $\hat{i}$, $\hat{j}$, and $\hat{k}$ be a body coordinate frame. The position vector of a moving point $P$, as shown in Figure 7.3, can be expressed in the body and global frames

$$\begin{align}
^B\mathbf{r}_P(t) &= x(t)\,\hat{i} + y(t)\,\hat{j} + z(t)\,\hat{k} \tag{7.126}\\
^G\mathbf{r}_P(t) &= X(t)\,\hat{I} + Y(t)\,\hat{J} + Z(t)\,\hat{K}. \tag{7.127}
\end{align}$$

The time derivative of $^B\mathbf{r}_P$ in $B$ and $^G\mathbf{r}_P$ in $G$ are

$$\frac{^Bd}{dt}\,^B\mathbf{r}_P = {}^B\dot{\mathbf{r}}_P = {}^B\mathbf{v}_P = \dot{x}\,\hat{i} + \dot{y}\,\hat{j} + \dot{z}\,\hat{k} \tag{7.128}$$

$$\frac{^Gd}{dt}\,^G\mathbf{r}_P = {}^G\dot{\mathbf{r}}_P = {}^G\mathbf{v}_P = \dot{X}\,\hat{I} + \dot{Y}\,\hat{J} + \dot{Z}\,\hat{K} \tag{7.129}$$

because the unit vectors of $B$ in Equation (7.126) and the unit vectors of $G$ in Equation (7.127) are considered constant.

Using the definition (7.122), we can find the $G$-derivative of the position vector $^B\mathbf{r}_P$ as

$$\begin{align}
\frac{^Gd}{dt}\,^B\mathbf{r}_P &= \frac{^Gd}{dt}\left(x\hat{i} + y\hat{j} + z\hat{k}\right)\\
&= \dot{x}\,\hat{i} + \dot{y}\,\hat{j} + \dot{z}\,\hat{k} + x\frac{^Gd\hat{i}}{dt} + y\frac{^Gd\hat{j}}{dt} + z\frac{^Gd\hat{k}}{dt}\\
&= {}^B\dot{\mathbf{r}}_P + {}^B_G\omega_B \times \left(x\hat{i} + y\hat{j} + z\hat{k}\right)\\
&= {}^B\dot{\mathbf{r}}_P + {}^B_G\omega_B \times {}^B\mathbf{r}_P\\
&= \frac{^Bd}{dt}\,^B\mathbf{r}_P + {}^B_G\omega_B \times {}^B\mathbf{r}_P. \tag{7.130}
\end{align}$$

We achieved this result because the $x$, $y$, and $z$ components of $^B\mathbf{r}_P$ are scalar. Scalars are invariant with respect to frame transformations. Therefore, if $x$ is a scalar then,

$$\frac{^Gd}{dt}x = \frac{^Bd}{dt}x = \dot{x}. \tag{7.131}$$

The $B$-derivative of $^G\mathbf{r}_P$ is

$$
\begin{aligned}
\frac{^Bd}{dt}{}^G\mathbf{r}_P &= \frac{^Bd}{dt}\left(X\hat{I} + Y\hat{J} + Z\hat{K}\right) \\
&= \dot{X}\hat{I} + \dot{Y}\hat{J} + \dot{Z}\hat{K} + X\frac{^Bd\hat{I}}{dt} + Y\frac{^Bd\hat{J}}{dt} + Z\frac{^Bd\hat{K}}{dt} \\
&= {}^G\dot{\mathbf{r}}_P + {}^G_B\boldsymbol{\omega}_G \times {}^G\mathbf{r}_P
\end{aligned}
\tag{7.132}
$$

and therefore,

$$\frac{^Bd}{dt}{}^G\mathbf{r}_P = {}^G\dot{\mathbf{r}}_P - {}_G\boldsymbol{\omega}_B \times {}^G\mathbf{r}_P. \tag{7.133}$$

The angular velocity of $B$ relative to $G$ is a vector quantity and can be expressed in either frames.

$$
\begin{aligned}
{}^G_G\boldsymbol{\omega}_B &= \omega_X\hat{I} + \omega_Y\hat{J} + \omega_Z\hat{K} \tag{7.134} \\
{}^B_G\boldsymbol{\omega}_B &= \omega_x\hat{i} + \omega_y\hat{j} + \omega_z\hat{k}. \tag{7.135}
\end{aligned}
$$

∎

**Example 184** *Time derivative of a moving point in B.*

*Consider a local frame $B$, rotating in $G$ by $\dot{\alpha}$ about the $Z$-axis, and a moving point at $^B\mathbf{r}_P(t) = t\hat{i}$ . Therefore,*

$$
\begin{aligned}
{}^G\mathbf{r}_P &= {}^GR_B\,{}^B\mathbf{r}_P = R_{Z,\alpha}(t)\,{}^B\mathbf{r}_P \\
&= \begin{bmatrix} \cos\alpha & -\sin\alpha & 0 \\ \sin\alpha & \cos\alpha & 0 \\ 0 & 0 & 1 \end{bmatrix}\begin{bmatrix} t \\ 0 \\ 0 \end{bmatrix} \\
&= t\cos\alpha\hat{I} + t\sin\alpha\hat{J}.
\end{aligned}
\tag{7.136}
$$

*The angular velocity matrix is*

$$_G\tilde{\omega}_B = {}^G\dot{R}_B\,{}^GR_B^T = \dot{\alpha}\tilde{K} \tag{7.137}$$

*that gives*

$$_G\boldsymbol{\omega}_B = \dot{\alpha}\hat{K}. \tag{7.138}$$

*It can also be verified that*

$$_G^B\tilde{\omega}_B = {}^GR_B^T\,{}_G\tilde{\omega}_B\,{}^GR_B = \dot{\alpha}\tilde{k} \tag{7.139}$$

*and therefore,*

$$_G^B\boldsymbol{\omega}_B = \dot{\alpha}\hat{k}. \tag{7.140}$$

*Now we can find the following derivatives*

$$\frac{^Bd}{dt} \, ^B\mathbf{r}_P \;=\; ^B\dot{\mathbf{r}}_P$$

$$= \; \hat{\imath} \qquad\qquad (7.141)$$

$$\frac{^Gd}{dt} \, ^G\mathbf{r}_P \;=\; ^G\dot{\mathbf{r}}_P$$

$$= \; (\cos\alpha - t\dot\alpha\sin\alpha)\,\hat{I} + (\sin\alpha + t\dot\alpha\cos\alpha)\,\hat{J}. \quad (7.142)$$

*For the mixed derivatives we start with*

$$\frac{^Gd}{dt} \, ^B\mathbf{r}_P \;=\; \frac{^Bd}{dt} \, ^B\mathbf{r}_P + \, ^B_G\boldsymbol\omega_B \times \, ^B\mathbf{r}_P$$

$$= \begin{bmatrix} 1 \\ 0 \\ 0 \end{bmatrix} + \dot\alpha \begin{bmatrix} 0 \\ 0 \\ 1 \end{bmatrix} \times \begin{bmatrix} t \\ 0 \\ 0 \end{bmatrix}$$

$$= \begin{bmatrix} 1 \\ t\dot\alpha \\ 0 \end{bmatrix} = \hat{\imath} + t\dot\alpha\hat{\jmath} = \, ^B_G\dot{\mathbf{r}}_P \qquad (7.143)$$

*which is the global velocity of P expressed in B. We may, however, transform $^B_G\dot{\mathbf{r}}_P$ to the global frame and find the global velocity expressed in G.*

$$^G\dot{\mathbf{r}}_P \;=\; ^GR_B \, ^B_G\dot{\mathbf{r}}_P$$

$$= \begin{bmatrix} \cos\alpha & -\sin\alpha & 0 \\ \sin\alpha & \cos\alpha & 0 \\ 0 & 0 & 1 \end{bmatrix} \begin{bmatrix} 1 \\ t\dot\alpha \\ 0 \end{bmatrix}$$

$$= \begin{bmatrix} \cos\alpha - t\dot\alpha\sin\alpha \\ \sin\alpha + t\dot\alpha\cos\alpha \\ 0 \end{bmatrix}$$

$$= \; (\cos\alpha - t\dot\alpha\sin\alpha)\,\hat{I} + (\sin\alpha + t\dot\alpha\cos\alpha)\,\hat{J} \qquad (7.144)$$

*The next derivative is*

$$\frac{^Bd}{dt} \, ^G\mathbf{r}_P \;=\; ^G\dot{\mathbf{r}}_P - \, _G\boldsymbol\omega_B \times \, ^G\mathbf{r}_P$$

$$= \begin{bmatrix} \cos\alpha - t\dot\alpha\sin\alpha \\ \sin\alpha + t\dot\alpha\cos\alpha \\ 0 \end{bmatrix} - \dot\alpha \begin{bmatrix} 0 \\ 0 \\ 1 \end{bmatrix} \times \begin{bmatrix} t\cos\alpha \\ t\sin\alpha \\ 0 \end{bmatrix}$$

$$= \begin{bmatrix} \cos\alpha \\ \sin\alpha \\ 0 \end{bmatrix} = (\cos\alpha)\,\hat{I} + (\sin\alpha)\,\hat{J}$$

$$= \; ^G_B\dot{\mathbf{r}}_P \qquad\qquad (7.145)$$

*which is the velocity of P relative to B and expressed in G. To express this velocity in B we apply a frame transformation*

$$
\begin{aligned}
{}^B\dot{\mathbf{r}}_P &= {}^G R_B^T \, {}^G\dot{\mathbf{r}}_P \\[2mm]
&= \begin{bmatrix} \cos\alpha & -\sin\alpha & 0 \\ \sin\alpha & \cos\alpha & 0 \\ 0 & 0 & 1 \end{bmatrix}^T \begin{bmatrix} \cos\alpha \\ \sin\alpha \\ 0 \end{bmatrix} \\[2mm]
&= \begin{bmatrix} 1 \\ 0 \\ 0 \end{bmatrix} = \hat{\imath}.
\end{aligned}
\tag{7.146}
$$

*Sometimes it is more applied if we transform the vector to the same frame in which we are taking the derivative and then apply the differential operator. Therefore,*

$$
\begin{aligned}
\frac{{}^G d}{dt} {}^B\mathbf{r}_P &= \frac{{}^G d}{dt} \left( {}^G R_B \, {}^B\mathbf{r}_P \right) = \frac{{}^G d}{dt} \begin{bmatrix} t\cos\alpha \\ t\sin\alpha \\ 0 \end{bmatrix} \\[2mm]
&= \begin{bmatrix} \cos\alpha - t\dot{\alpha}\sin\alpha \\ \sin\alpha + t\dot{\alpha}\cos\alpha \\ 0 \end{bmatrix}
\end{aligned}
\tag{7.147}
$$

*and*

$$
\begin{aligned}
\frac{{}^B d}{dt} {}^G\mathbf{r}_P &= \frac{{}^B d}{dt} \left( {}^G R_B^T \, {}^G\mathbf{r}_P \right) \\[2mm]
&= \frac{{}^B d}{dt} \begin{bmatrix} t \\ 0 \\ 0 \end{bmatrix} = \begin{bmatrix} 1 \\ 0 \\ 0 \end{bmatrix}.
\end{aligned}
\tag{7.148}
$$

**Example 185** *Orthogonality of position and velocity vectors.*
*If the position vector of a body point in global frame is denoted by $\mathbf{r}$ then*

$$
\frac{d\mathbf{r}}{dt} \cdot \mathbf{r} = 0.
\tag{7.149}
$$

*To show this property we may take a derivative from*

$$
\mathbf{r} \cdot \mathbf{r} = r^2
\tag{7.150}
$$

*and find*

$$
\begin{aligned}
\frac{d}{dt}(\mathbf{r}\cdot\mathbf{r}) &= \frac{d\mathbf{r}}{dt}\cdot\mathbf{r} + \mathbf{r}\cdot\frac{d\mathbf{r}}{dt} \\[2mm]
&= 2\frac{d\mathbf{r}}{dt}\cdot\mathbf{r} \\[2mm]
&= 0.
\end{aligned}
\tag{7.151}
$$

*The Equation (7.149) is correct in every coordinate frame and for every constant length vector, as long as the vector and the derivative are expressed in the same coordinate frame.*

**Example 186** ★ *Derivative transformation formula.*
*The global velocity of a fixed point in the body coordinate frame $B\,(Oxyz)$ can be found by Equation (7.5). Now consider a point $P$ that can move in $B\,(Oxyz)$. In this case, the body position vector $^B\mathbf{r}_P$ is not constant, and therefore, the global velocity of such a point expressed in $B$ is*

$$\frac{^Gd}{dt}\,^B\mathbf{r}_P \;=\; \frac{^Bd}{dt}\,^B\mathbf{r}_P + {}^B_G\boldsymbol{\omega}_B \times {}^B\mathbf{r}_P \tag{7.152}$$

$$\;=\; {}^B_G\dot{\mathbf{r}}_P. \tag{7.153}$$

*Sometimes the result of Equation (7.152) is utilized to define transformation of the differential operator from a body to a global coordinate frame*

$$\frac{^Gd}{dt}\square \;=\; \frac{^Bd}{dt}\square + {}^B_G\boldsymbol{\omega}_B \times {}^B_G\square \tag{7.154}$$

$$\;=\; {}^B_G\dot{\square} \tag{7.155}$$

*however, special attention must be paid to the coordinate frame in which the vector $\square$ and the final result are expressed. The final result is ${}^B_G\dot{\square}$ showing the global $(G)$ time derivative expressed in body frame $(B)$. The vector $\square$ might be any vector such as position, velocity, angular velocity, momentum, angular velocity, or even a time-varying force vector.*
*The Equation (7.154) is called the* **derivative transformation formula** *and relates the time derivative of a vector as it would be seen from frame $G$ to its derivative as seen in frame $B$. The derivative transformation formula (7.154) is more general and can be applied to every vector for derivative transformation between every two relatively moving coordinate frames.*

**Example 187** ★ *Differential equation for rotation matrix.*
*Equation (7.8) for defining the angular velocity matrix may be written as a first-order differential equation*

$$\frac{d}{dt}\,{}^GR_B - {}^GR_B\,{}_G\tilde{\omega}_B = 0. \tag{7.156}$$

*The solution of the equation confirms the exponential definition of the rotation matrix as*

$$^GR_B = e^{\tilde{\omega}t} \tag{7.157}$$

*or*

$$\tilde{\omega}t = \dot{\phi}\tilde{u} = \ln\left(^GR_B\right). \tag{7.158}$$

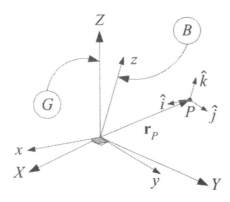

FIGURE 7.4. A body coordinate frame moving with a fixed point in the global coordinate frame.

**Example 188 ★** *Acceleration of a body point in the global frame.*
    *The angular acceleration vector of a rigid body $B(Oxyz)$ in the global frame $G(OXYZ)$ is denoted by $_G\alpha_B$ and is defined as the global time derivative of $_G\omega_B$.*

$$_G\alpha_B = \frac{^Gd}{dt}\,_G\omega_B \tag{7.159}$$

*Using this definition, the acceleration of a fixed body point in the global frame is*

$$
\begin{aligned}
^G\mathbf{a}_P &= \frac{^Gd}{dt}\left(_G\omega_B \times {}^G\mathbf{r}_P\right) \\
&= {}_G\alpha_B \times {}^G\mathbf{r}_P + {}_G\omega_B \times \left({}_G\omega_B \times {}^G\mathbf{r}_P\right). \tag{7.160}
\end{aligned}
$$

**Example 189 ★** *Alternative definition of angular velocity vector.*
    *The angular velocity vector of a rigid body $B(\hat{\imath},\hat{\jmath},\hat{k})$ in global frame $G(\hat{I},\hat{J},\hat{K})$ can also be defined by*

$$_G^B\omega_B = \hat{\imath}(\frac{^Gd\hat{\jmath}}{dt}\cdot\hat{k}) + \hat{\jmath}(\frac{^Gd\hat{k}}{dt}\cdot\hat{\imath}) + \hat{k}(\frac{^Gd\hat{\imath}}{dt}\cdot\hat{\jmath}). \tag{7.161}$$

**Proof.** *Consider a body coordinate frame $B$ moving with a fixed point in the global coordinate frame $G$. The fixed point of the body is taken as the origin of both coordinate frames, as shown in Figure 7.4. In order to describe the motion of the body, it is sufficient to describe the motion of the local unit vectors $\hat{\imath}$, $\hat{\jmath}$, $\hat{k}$ . Let $\mathbf{r}_P$ be the position vector of a body point $P$. Then, $^B\mathbf{r}_P$ is a vector with constant components.*

$$^B\mathbf{r}_P = x\hat{\imath} + y\hat{\jmath} + z\hat{k} \tag{7.162}$$

*When the body moves, it is only the unit vectors $\hat{\imath}$, $\hat{\jmath}$, and $\hat{k}$ that vary relative to the global coordinate frame. Therefore, the vector of differential displacement is*

$$d\mathbf{r}_P = x\,d\hat{\imath} + y\,d\hat{\jmath} + z\,d\hat{k} \tag{7.163}$$

*which can also be expressed by*

$$d\mathbf{r}_P = (d\mathbf{r}_P \cdot \hat{\imath})\,\hat{\imath} + (d\mathbf{r}_P \cdot \hat{\jmath})\,\hat{\jmath} + \left(d\mathbf{r}_P \cdot \hat{k}\right)\hat{k}. \tag{7.164}$$

*Substituting (7.163) in the right-hand side of (7.164) results in*

$$
\begin{aligned}
d\mathbf{r}_P \;=\;& \left(x\hat{\imath} \cdot d\hat{\imath} + y\hat{\imath} \cdot d\hat{\jmath} + z\hat{\imath} \cdot d\hat{k}\right)\hat{\imath} \\
&+ \left(x\hat{\jmath} \cdot d\hat{\imath} + y\hat{\jmath} \cdot d\hat{\jmath} + z\hat{\jmath} \cdot d\hat{k}\right)\hat{\jmath} \\
&+ \left(x\hat{k} \cdot d\hat{\imath} + y\hat{k} \cdot d\hat{\jmath} + z\hat{k} \cdot d\hat{k}\right)\hat{k}. 
\end{aligned} \tag{7.165}
$$

*Utilizing the unit vectors' relationships*

$$
\begin{aligned}
\hat{\jmath} \cdot d\hat{\imath} &= -\hat{\imath} \cdot d\hat{\jmath} & (7.166) \\
\hat{k} \cdot d\hat{\jmath} &= -\hat{\jmath} \cdot d\hat{k} & (7.167) \\
\hat{\imath} \cdot d\hat{k} &= -\hat{k} \cdot d\hat{\imath} & (7.168) \\
\hat{\imath} \cdot d\hat{\imath} &= \hat{\jmath} \cdot d\hat{\jmath} = \hat{k} \cdot d\hat{k} = 0 & (7.169) \\
\hat{\imath} \cdot \hat{\jmath} &= \hat{\jmath} \cdot \hat{k} = \hat{k} \cdot \hat{\imath} = 0 & (7.170) \\
\hat{\imath} \cdot \hat{\imath} &= \hat{\jmath} \cdot \hat{\jmath} = \hat{k} \cdot \hat{k} = 1 & (7.171)
\end{aligned}
$$

*the $d\mathbf{r}_P$ reduces to*

$$
\begin{aligned}
d\mathbf{r}_P \;=\;& \left(z\hat{\imath} \cdot d\hat{k} - y\hat{\jmath} \cdot d\hat{\imath}\right)\hat{\imath} \\
&+ \left(x\hat{\jmath} \cdot d\hat{\imath} - z\hat{k} \cdot d\hat{\jmath}\right)\hat{\jmath} \\
&+ \left(y\hat{k} \cdot d\hat{\jmath} - x\hat{\imath} \cdot d\hat{k}\right)\hat{k}. 
\end{aligned} \tag{7.172}
$$

*This equation can be rearranged to be expressed as a vector product*

$$d\mathbf{r}_P = \left((\hat{k} \cdot d\hat{\jmath})\hat{\imath} + (\hat{\imath} \cdot d\hat{k})\hat{\jmath} + (\hat{\jmath} \cdot d\hat{\imath})\hat{k}\right) \times \left(x\hat{\imath} + y\hat{\jmath} + z\hat{k}\right) \tag{7.173}$$

*or*

$$^{B}_{G}\dot{\mathbf{r}}_P = \left((\hat{k} \cdot \frac{^{G}d\hat{\jmath}}{dt})\hat{\imath} + (\hat{\imath} \cdot \frac{^{G}d\hat{k}}{dt})\hat{\jmath} + (\hat{\jmath} \cdot \frac{^{G}d\hat{\imath}}{dt})\hat{k}\right) \times \left(x\hat{\imath} + y\hat{\jmath} + z\hat{k}\right).$$

*Comparing this result with*

$$\dot{\mathbf{r}}_P = {}_{G}\boldsymbol{\omega}_B \times \mathbf{r}_P$$

*shows that*

$$_G^B\omega_B = \hat{\imath}\left(\frac{^Gd\hat{\jmath}}{dt}\cdot\hat{k}\right) + \hat{\jmath}\left(\frac{^Gd\hat{k}}{dt}\cdot\hat{\imath}\right) + \hat{k}\left(\frac{^Gd\hat{\imath}}{dt}\cdot\hat{\jmath}\right). \tag{7.174}$$

∎

**Example 190** ★ *Alternative proof for angular velocity definition (7.161).*
*The angular velocity definition presented in Equation (7.161) can also be shown by direct substitution for $^GR_B$ in the angular velocity matrix $_G^B\tilde{\omega}_B$*

$$_G^B\tilde{\omega}_B = {}^GR_B^T\,{}^G\dot{R}_B. \tag{7.175}$$

*Therefore,*

$$
_G^B\tilde{\omega}_B =
\begin{bmatrix}
\hat{\imath}\cdot\hat{I} & \hat{\imath}\cdot\hat{J} & \hat{\imath}\cdot\hat{K} \\
\hat{\jmath}\cdot\hat{I} & \hat{\jmath}\cdot\hat{J} & \hat{\jmath}\cdot\hat{K} \\
\hat{k}\cdot\hat{I} & \hat{k}\cdot\hat{J} & \hat{k}\cdot\hat{K}
\end{bmatrix}
\frac{^Gd}{dt}
\begin{bmatrix}
\hat{I}\cdot\hat{\imath} & \hat{I}\cdot\hat{\jmath} & \hat{I}\cdot\hat{k} \\
\hat{J}\cdot\hat{\imath} & \hat{J}\cdot\hat{\jmath} & \hat{J}\cdot\hat{k} \\
\hat{K}\cdot\hat{\imath} & \hat{K}\cdot\hat{\jmath} & \hat{K}\cdot\hat{k}
\end{bmatrix}
$$
$$
=
\begin{bmatrix}
\left(\hat{\imath}\cdot\frac{^Gd\hat{\imath}}{dt}\right) & \left(\hat{\imath}\cdot\frac{^Gd\hat{\jmath}}{dt}\right) & \left(\hat{\imath}\cdot\frac{^Gd\hat{k}}{dt}\right) \\
\left(\hat{\jmath}\cdot\frac{^Gd\hat{\imath}}{dt}\right) & \left(\hat{\jmath}\cdot\frac{^Gd\hat{\jmath}}{dt}\right) & \left(\hat{\jmath}\cdot\frac{^Gd\hat{k}}{dt}\right) \\
\left(\hat{k}\cdot\frac{^Gd\hat{\imath}}{dt}\right) & \left(\hat{k}\cdot\frac{^Gd\hat{\jmath}}{dt}\right) & \left(\hat{k}\cdot\frac{^Gd\hat{k}}{dt}\right)
\end{bmatrix}
\tag{7.176}
$$

*which shows that*

$$
_G^B\omega_B =
\begin{bmatrix}
\left(\frac{^Gd\hat{\jmath}}{dt}\cdot\hat{k}\right) \\
\left(\frac{^Gd\hat{k}}{dt}\cdot\hat{\imath}\right) \\
\left(\frac{^Gd\hat{\imath}}{dt}\cdot\hat{\jmath}\right)
\end{bmatrix}.
\tag{7.177}
$$

**Example 191** ★ *Second derivative.*
*In general, $^Gd\mathbf{r}/dt$ is a variable vector in $G(OXYZ)$ and in any other coordinate frame such as $B(oxyz)$. Therefore, it can be differentiated in either coordinate frames $G$ or $B$. However, the order of differentiating is important. In general,*

$$\frac{^Bd}{dt}\frac{^Gd\mathbf{r}}{dt} \neq \frac{^Gd}{dt}\frac{^Bd\mathbf{r}}{dt}. \tag{7.178}$$

*As an example, consider a rotating body coordinate frame about the $Z$-axis, and a variable vector as*

$$^G\mathbf{r} = t\hat{I}. \tag{7.179}$$

*Therefore,*

$$\frac{^G d\mathbf{r}}{dt} = {}^G\dot{\mathbf{r}} = \hat{I} \tag{7.180}$$

*and hence,*

$$
\begin{aligned}
{}^B\left(\frac{^G d\mathbf{r}}{dt}\right) &= {}^B_G\dot{\mathbf{r}} = R^T_{Z,\varphi}\left[\hat{I}\right]\\
&= \begin{bmatrix} \cos\varphi & \sin\varphi & 0\\ -\sin\varphi & \cos\varphi & 0\\ 0 & 0 & 1 \end{bmatrix}\begin{bmatrix} 1\\0\\0 \end{bmatrix}\\
&= \cos\varphi\hat{\imath} - \sin\varphi\hat{\jmath}
\end{aligned} \tag{7.181}
$$

*which provides*

$$\frac{^B d}{dt}\frac{^G d\mathbf{r}}{dt} = -\dot{\varphi}\sin\varphi\hat{\imath} - \dot{\varphi}\cos\varphi\hat{\jmath} \tag{7.182}$$

*and*

$$^G\left(\frac{^B d}{dt}\frac{^G d\mathbf{r}}{dt}\right) = -\dot{\varphi}\hat{J}. \tag{7.183}$$

*Now*

$$^B\mathbf{r} = R^T_{Z,\varphi}\left[t\hat{I}\right] = t\cos\varphi\hat{\imath} - t\sin\varphi\hat{\jmath} \tag{7.184}$$

*that provides*

$$\frac{^B d\mathbf{r}}{dt} = (-t\dot{\varphi}\sin\varphi + \cos\varphi)\hat{\imath} - (\sin\varphi + t\dot{\varphi}\cos\varphi)\hat{\jmath} \tag{7.185}$$

*and*

$$
\begin{aligned}
{}^G\left(\frac{^B d\mathbf{r}}{dt}\right) &= {}^G_B\dot{\mathbf{r}}\\
&= R_{Z,\varphi}\left((-t\dot{\varphi}\sin\varphi + \cos\varphi)\hat{\imath} - (\sin\varphi + t\dot{\varphi}\cos\varphi)\hat{\jmath}\right)\\
&= \begin{bmatrix} \cos\varphi & -\sin\varphi & 0\\ \sin\varphi & \cos\varphi & 0\\ 0 & 0 & 1 \end{bmatrix}\begin{bmatrix} -t\dot{\varphi}\sin\varphi + \cos\varphi\\ -\sin\varphi - t\dot{\varphi}\cos\varphi\\ 0 \end{bmatrix}\\
&= \hat{I} - t\dot{\varphi}\hat{J}
\end{aligned} \tag{7.186}
$$

*which shows*

$$
\frac{^G d}{dt}\frac{^B d\mathbf{r}}{dt} = -(\dot{\varphi} + t\ddot{\varphi})\hat{J} \tag{7.187}
$$

$$\neq \frac{^B d}{dt}\frac{^G d\mathbf{r}}{dt}.$$

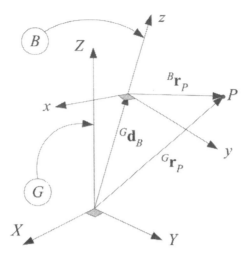

FIGURE 7.5. A rigid body with an attached coordinate frame $B\,(oxyz)$ moving freely in a global coordinate frame $G(OXYZ)$.

## 7.3  Rigid Body Velocity

Consider a rigid body with an attached local coordinate frame $B\,(oxyz)$ moving freely in a fixed global coordinate frame $G(OXYZ)$, as shown in Figure 7.5. The rigid body can rotate in the global frame, while the origin of the body frame $B$ can translate relative to the origin of $G$. The coordinates of a body point $P$ in local and global frames are related by the following equation:

$$^G\mathbf{r}_P = {}^G R_B\, {}^B\mathbf{r}_P + {}^G\mathbf{d}_B \qquad (7.188)$$

where $^G\mathbf{d}_B$ indicates the position of the moving origin $o$ relative to the fixed origin $O$.

The velocity of the point $P$ in $G$ is

$$
\begin{aligned}
^G\mathbf{v}_P &= {}^G\dot{\mathbf{r}}_P \\
&= {}^G\dot{R}_B\, {}^B\mathbf{r}_P + {}^G\dot{\mathbf{d}}_B \\
&= {}_G\tilde{\omega}_B\, {}^G_B\mathbf{r}_P + {}^G\dot{\mathbf{d}}_B \\
&= {}_G\tilde{\omega}_B\left({}^G\mathbf{r}_P - {}^G\mathbf{d}_B\right) + {}^G\dot{\mathbf{d}}_B \\
&= {}_G\omega_B \times \left({}^G\mathbf{r}_P - {}^G\mathbf{d}_B\right) + {}^G\dot{\mathbf{d}}_B.
\end{aligned}
\qquad (7.189)
$$

**Proof.** Direct differentiating shows

$$
\begin{aligned}
{}^G\mathbf{v}_P &= \frac{{}^Gd}{dt}\, {}^G\mathbf{r}_P = {}^G\dot{\mathbf{r}}_P \\
&= \frac{{}^Gd}{dt}\left({}^GR_B\,{}^B\mathbf{r}_P + {}^G\mathbf{d}_B\right) \\
&= {}^G\dot{R}_B\,{}^B\mathbf{r}_P + {}^G\dot{\mathbf{d}}_B.
\end{aligned}
\tag{7.190}
$$

The local position vector ${}^B\mathbf{r}_P$ can be substituted from (7.188) to obtain

$$
\begin{aligned}
{}^G\mathbf{v}_P &= {}^G\dot{R}_B\,{}^GR_B^T\left({}^G\mathbf{r}_P - {}^G\mathbf{d}_B\right) + {}^G\dot{\mathbf{d}}_B \\
&= {}_G\tilde{\omega}_B\left({}^G\mathbf{r}_P - {}^G\mathbf{d}_B\right) + {}^G\dot{\mathbf{d}}_B \\
&= {}_G\omega_B \times \left({}^G\mathbf{r}_P - {}^G\mathbf{d}_B\right) + {}^G\dot{\mathbf{d}}_B.
\end{aligned}
\tag{7.191}
$$

It may also be written using relative position vector

$$
{}^G\mathbf{v}_P = {}_G\omega_B \times {}^G_B\mathbf{r}_P + {}^G\dot{\mathbf{d}}_B.
\tag{7.192}
$$

■

**Example 192** *Geometric interpretation of rigid body velocity.*
*Figure 7.6 illustrates a body point P of a moving rigid body. The global velocity of the point P*

$$
{}^G\mathbf{v}_P = {}_G\omega_B \times {}^G_B\mathbf{r}_P + {}^G\dot{\mathbf{d}}_B
$$

*is a vector addition of rotational and translational velocities, both expressed in the global frame. At the moment, the body frame is assumed to be coincident with the global frame, and the body frame has a velocity ${}^G\dot{\mathbf{d}}_B$ with respect to the global frame. The translational velocity ${}^G\dot{\mathbf{d}}_B$ is a common property for every point of the body, but the rotational velocity ${}_G\omega_B \times {}^G_B\mathbf{r}_P$ differs for different points of the body.*

**Example 193** *Velocity of a moving point in a moving body frame.*
*Assume that point P in Figure 7.5 is moving in the frame B, indicating by a time varying position vector ${}^B\mathbf{r}_P(t)$. The global velocity of P is a composition of the velocity of P in B, rotation of B relative to G, and velocity of B relative to G.*

$$
\begin{aligned}
\frac{{}^Gd}{dt}\,{}^G\mathbf{r}_P &= \frac{{}^Gd}{dt}\left({}^G\mathbf{d}_B + {}^GR_B\,{}^B\mathbf{r}_P\right) \\
&= \frac{{}^Gd}{dt}\,{}^G\mathbf{d}_B + \frac{{}^Gd}{dt}\left({}^GR_B\,{}^B\mathbf{r}_P\right) \\
&= {}^G\dot{\mathbf{d}}_B + {}^G_B\dot{\mathbf{r}}_P + {}_G\omega_B \times {}^G_B\mathbf{r}_P
\end{aligned}
\tag{7.193}
$$

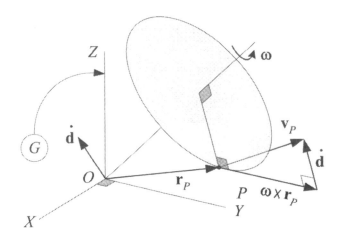

FIGURE 7.6. Geometric interpretation of rigid body velocity.

**Example 194** *Velocity of a body point in multiple coordinate frames.*

Consider three frames, $B_0$, $B_1$ and $B_2$, as shown in Figure 7.7. The velocity of point $P$ should be measured and expressed in a coordinate frame. If the point is stationary in a frame, say $B_2$, then the time derivative of $^2\mathbf{r}_P$ in $B_2$ is zero. If the frame $B_2$ is moving relative to the frame $B_1$, then, the time derivative of $^1\mathbf{r}_P$ is a combination of the rotational component due to rotation of $B_2$ relative to $B_1$ and the velocity of $B_2$ relative to $B_1$. In forward velocity kinematics of robots, the velocities must be measured in the base frame $B_0$. Therefore, the velocity of point $P$ in the base frame is a combination of the velocity of $B_2$ relative to $B_1$ and the velocity of $B_1$ relative to $B_0$.

The global coordinate of the body point $P$ is

$$^0\mathbf{r}_P = {}^0\mathbf{d}_1 + {}^0_1\mathbf{d}_2 + {}^0_2\mathbf{r}_P \qquad (7.194)$$
$$= {}^0\mathbf{d}_1 + {}^0R_1\,{}^1\mathbf{d}_2 + {}^0R_2\,{}^2\mathbf{r}_P. \qquad (7.195)$$

*Therefore, the velocity of point $P$ can be found by combining the relative velocities*

$$^0\dot{\mathbf{r}}_P = {}^0\dot{\mathbf{d}}_1 + ({}^0\dot{R}_1\,{}^1\mathbf{d}_2 + {}^0R_1\,{}^1\dot{\mathbf{d}}_2) + {}^0\dot{R}_2\,{}^2\mathbf{r}_P$$
$$= {}^0\dot{\mathbf{d}}_1 + {}^0_0\boldsymbol{\omega}_1 \times {}^0_1\mathbf{d}_2 + {}^0R_1\,{}^1\dot{\mathbf{d}}_2 + {}^0_0\boldsymbol{\omega}_2 \times {}^0_2\mathbf{r}_P \qquad (7.196)$$

*Most of the time, it is better to use a relative velocity method and write*

$$^0_0\mathbf{v}_P = {}^0_0\mathbf{v}_1 + {}^0_1\mathbf{v}_2 + {}^0_2\mathbf{v}_P \qquad (7.197)$$

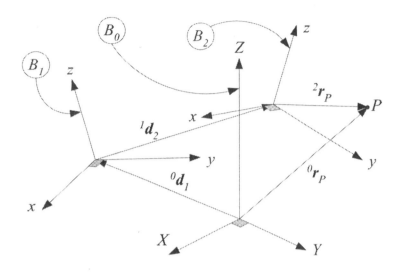

FIGURE 7.7. A rigid body coordinate frame $B_2$ is moving in a frame $B_1$ that is moving in the base coordinate frame $B_0$.

*because*

$$
{}^0_0\mathbf{v}_1 = {}^0_0\dot{\mathbf{d}}_1 \tag{7.198}
$$

$$
{}^0_1\mathbf{v}_2 = {}^0_0\omega_1 \times {}^0_1\mathbf{d}_2 + {}^0R_1\,{}^1\dot{\mathbf{d}}_2 \tag{7.199}
$$

$$
{}^0_2\mathbf{v}_P = {}^0_0\omega_2 \times {}^0_2\mathbf{r}_P \tag{7.200}
$$

*and therefore,*

$$
{}^0\mathbf{v}_P = {}^0\dot{\mathbf{d}}_1 + {}^0_0\omega_1 \times {}^0_1\mathbf{d}_2 + {}^0R_1\,{}^1\dot{\mathbf{d}}_2 + {}^0_0\omega_2 \times {}^0_2\mathbf{r}_P. \tag{7.201}
$$

**Example 195**  *Velocity vectors are free vectors.*

*Velocity vectors are free, so to express them in different coordinate frames we need only to premultiply them by a rotation matrix. Hence, considering ${}^k_j\mathbf{v}_i$ as the velocity of the origin of the $B_i$ coordinate frame with respect to the origin of the frame $B_j$ expressed in frame $B_k$, we can write*

$$
{}^k_j\mathbf{v}_i = -{}^k_i\mathbf{v}_j \tag{7.202}
$$

*and*

$$
{}^k_j\mathbf{v}_i = {}^kR_m\,{}^m_j\mathbf{v}_i \tag{7.203}
$$

*and therefore,*

$$
\frac{{}^id}{dt}\,{}^i_i\mathbf{r}_P = {}^i\mathbf{v}_P = {}^i_j\mathbf{v}_P + {}^i_i\omega_j \times {}^i_j\mathbf{r}_P. \tag{7.204}
$$

**Example 196 ★** *Zero velocity points.*

   To answer whether there is a point with zero velocity at each time, we may utilize Equation (7.189) and write

$$G\tilde{\omega}_B \left( {}^G\mathbf{r}_0 - {}^G\mathbf{d}_B \right) + {}^G\dot{\mathbf{d}}_B = 0 \tag{7.205}$$

to search for ${}^G\mathbf{r}_0$ which refers to a point with zero velocity

$$ {}^G\mathbf{r}_0 = {}^G\mathbf{d}_B - G\tilde{\omega}_B^{-1} \, {}^G\dot{\mathbf{d}}_B \tag{7.206}$$

however, the skew symmetric matrix $G\tilde{\omega}_B$ is singular and has no inverse. In other words, there is no general solution for Equation (7.205).

   If we restrict ourselves to planar motions, say $XY$-plane, then $G\omega_B = \omega\hat{K}$ and $G\tilde{\omega}_B^{-1} = 1/\omega$. Hence, in 2D space there is a point at any time with zero velocity at position ${}^G\mathbf{r}_0$ given by

$$ {}^G\mathbf{r}_0(t) = {}^G\mathbf{d}_B(t) - \frac{1}{\omega} \, {}^G\dot{\mathbf{d}}_B(t). \tag{7.207}$$

The zero velocity point is called the **pole** or **instantaneous center of rotation**. The position of the pole is generally a function of time and the path of its motion is called a **centroid**.

**Example 197 ★** *Eulerian and Lagrangian view points.*

   When a variable quantity is measured within the stationary global coordinate frame, it is called absolute or the **Lagrangian** viewpoint. On the other hand, when the variable is measured within a moving body coordinate frame, it is called relative or the **Eulerian** viewpoint.

   In 2D planar motion of a rigid body, there is always a pole of zero velocity at

$$ {}^G\mathbf{r}_0 = {}^G\mathbf{d}_B - \frac{1}{\omega} \, {}^G\dot{\mathbf{d}}_B. \tag{7.208}$$

The position of the pole in the body coordinate frame can be found by substituting for ${}^G\mathbf{r}$ from (7.188)

$$ {}^G R_B \, {}^B\mathbf{r}_0 + {}^G\mathbf{d}_B = {}^G\mathbf{d}_B - G\tilde{\omega}_B^{-1} \, {}^G\dot{\mathbf{d}}_B \tag{7.209}$$

and solving for the position of the zero velocity point in the body coordinate frame ${}^B\mathbf{r}_0$.

$$
\begin{aligned}
{}^B\mathbf{r}_0 &= -{}^G R_B^T \, G\tilde{\omega}_B^{-1} \, {}^G\dot{\mathbf{d}}_B \\
&= -{}^G R_B^T \left[ {}^G\dot{R}_B \, {}^G R_B^T \right]^{-1} {}^G\dot{\mathbf{d}}_B \\
&= -{}^G R_B^T \left[ {}^G R_B \, {}^G\dot{R}_B^{-1} \right] {}^G\dot{\mathbf{d}}_B \\
&= -{}^G\dot{R}_B^{-1} \, {}^G\dot{\mathbf{d}}_B \tag{7.210}
\end{aligned}
$$

   Therefore, ${}^G\mathbf{r}_0$ indicates the path of motion of the pole in the global frame, while ${}^B\mathbf{r}_0$ indicates the same path in the body frame. The ${}^G\mathbf{r}_0$ refers to Lagrangian centroid and ${}^B\mathbf{r}_0$ refers to Eulerian centroid.

**Example 198 ★** *Screw axis and screw motion.*

*The screw axis may be defined as a line for a moving rigid body $B$ whose points $P$ have velocity parallel to the angular velocity vector ${}_G\omega_B = \omega\hat{u}$. Such points satisfy*

$$
\begin{aligned}
{}^G\mathbf{v}_P &= {}_G\omega_B \times \left({}^G\mathbf{r}_P - {}^G\mathbf{d}_B\right) + {}^G\dot{\mathbf{d}}_B \\
&= p\,{}_G\omega_B.
\end{aligned}
\tag{7.211}
$$

*where, $p$ is a scalar. Since ${}_G\omega_B$ is perpendicular to ${}_G\omega_B \times \left({}^G\mathbf{r} - {}^G\mathbf{d}\right)$, a dot product of Equation (7.211) by ${}_G\omega_B$ yields*

$$
p = \frac{1}{\omega^2}\left({}_G\omega_B \cdot {}^G\dot{\mathbf{d}}_B\right).
\tag{7.212}
$$

*Introducing a parameter $m$ to indicate different points of the line, the equation of the screw axis is defined by*

$$
{}^G\mathbf{r}_P = {}^G\mathbf{d}_B + \frac{1}{\omega^2}\left({}_G\omega_B \times {}^G\dot{\mathbf{d}}_B\right) + m\,{}_G\omega_B
\tag{7.213}
$$

*because if we have $\mathbf{a} \times \mathbf{x} = \mathbf{b}$, and $\mathbf{a} \cdot \mathbf{b} = 0$, then $\mathbf{x} = -\mathbf{a}^{-2}(\mathbf{a} \times \mathbf{b}) + m\mathbf{a}$. In our case, ${}_G\omega_B \times \left({}^G\mathbf{r}_P - {}^G\mathbf{d}_B\right) - p\,{}_G\omega_B - {}^G\dot{\mathbf{d}}_B$, and $\left({}^G\mathbf{r}_P - {}^G\mathbf{d}_B\right)$ is perpendicular to ${}_G\omega_B \times \left({}^G\mathbf{r}_P - {}^G\mathbf{d}_B\right)$, hence to $(p\,{}_G\omega_B - {}^G\dot{\mathbf{d}}_B)$.*

*Therefore, there exists at any time a line $s$ in space, parallel to ${}_G\omega_B$, which is the locus of points whose velocity is parallel to ${}_G\omega_B$.*

*If $\mathbf{s}$ is the position vector of a point on $s$, then*

$$
{}_G\omega_B \times \left({}^G\mathbf{s} - {}^G\mathbf{d}_B\right) = p\,{}_G\omega_B - {}^G\dot{\mathbf{d}}_B
\tag{7.214}
$$

*and the velocity of any point out of $\mathbf{s}$ is*

$$
{}^G\mathbf{v} = {}_G\omega_B \times \left({}^G\mathbf{r} - {}^G\mathbf{s}\right) + p\,{}_G\omega_B
\tag{7.215}
$$

*which expresses that at any time the velocity of a body point can be decomposed into perpendicular and parallel components to the angular velocity vector ${}_G\omega_B$. Therefore, the motion of any point of a rigid body is a screw. The parameter $p$ is the ratio of translation velocity to rotation velocity, and is called **pitch**. In general, $\mathbf{s}$, ${}_G\omega_B$, and $p$ may be functions of time.*

## 7.4  Velocity Transformation Matrix

Consider the motion of a rigid body $B$ in the global coordinate frame $G$, as shown in Figure 7.5. Assume the body frame $B(oxyz)$ is coincident at some initial time $t_0$ with the global frame $G(OXYZ)$. At any time $t \neq t_0$, $B$ is not necessarily coincident with $G$ and therefore, the homogeneous transformation matrix ${}^GT_B(t)$ is time varying.

The global position vector $^G\mathbf{r}_P(t)$ of a point $P$ of the rigid body is a function of time, but its local position vector $^B\mathbf{r}_P$ is a constant, which is equal to $^G\mathbf{r}_P(t_0)$.

$$^B\mathbf{r}_P \equiv {^G}\mathbf{r}_P(t_0) \tag{7.216}$$

The velocity of point $P$ on the rigid body $B$ as seen in the reference frame $G$ is obtained by differentiating the position vector $^G\mathbf{r}(t)$ in the reference frame $G$

$$^G\mathbf{v}_P = \frac{d}{dt}\,{^G}\mathbf{r}_P(t) = {^G}\dot{\mathbf{r}}_P \tag{7.217}$$

where $^G\dot{\mathbf{r}}_P$ denotes the differentiation of the quantity $^G\mathbf{r}_P(t)$ in the reference frame $G$.

The velocity of a body point in global coordinate frame can be found by applying a homogeneous transformation matrix

$$^G\mathbf{v}(t) = {^G}V_B\,{^G}\mathbf{r}(t) \tag{7.218}$$

where $^GV_B$ is the *velocity transformation matrix*

$$
\begin{aligned}
^GV_B &= {^G}\dot{T}_B\,{^G}T_B^{-1} &\tag{7.219}\\[4pt]
&= \begin{bmatrix} {^G}\dot{R}_B\,{^G}R_B^T & {^G}\dot{\mathbf{d}}_B - {^G}\dot{R}_B\,{^G}R_B^T\,{^G}\mathbf{d}_B \\ 0 & 0 \end{bmatrix} &\tag{7.220}\\[4pt]
&= \begin{bmatrix} {_G}\tilde{\omega}_B & {^G}\dot{\mathbf{d}}_B - {_G}\tilde{\omega}_B\,{^G}\mathbf{d}_B \\ 0 & 0 \end{bmatrix}. &\tag{7.221}
\end{aligned}
$$

**Proof.** Based on homogeneous coordinate transformation, we have

$$^G\mathbf{r}_P(t) = {^G}T_B(t)\,{^B}\mathbf{r}_P = {^G}T_B(t)\,{^G}\mathbf{r}_P(t_0) \tag{7.222}$$

and therefore,

$$
\begin{aligned}
^G\mathbf{v}_P &= \frac{^Gd}{dt}\,[{^G}T_B\,{^B}\mathbf{r}_P] = {^G}\dot{T}_B\,{^B}\mathbf{r}_P &\tag{7.223}\\[4pt]
&= \begin{bmatrix} \frac{^Gd}{dt}\,{^G}R_B & \frac{^Gd}{dt}\,{^G}\mathbf{d}_B \\ 0 & 0 \end{bmatrix}{^B}\mathbf{r}_P \\[4pt]
&= \begin{bmatrix} {^G}\dot{R}_B & {^G}\dot{\mathbf{d}}_B \\ 0 & 0 \end{bmatrix}{^B}\mathbf{r}_P
\end{aligned}
$$

Substituting for $^B\mathbf{r}_P$ from Equation (7.222), gives

$$
\begin{aligned}
^G\mathbf{v}_P &= {^G}\dot{T}_B\,{^G}T_B^{-1}\,{^G}\mathbf{r}_P(t) \\[4pt]
&= \begin{bmatrix} {^G}\dot{R}_B & {^G}\dot{\mathbf{d}}_B \\ 0 & 0 \end{bmatrix}\begin{bmatrix} {^G}R_B^T & -{^G}R_B^T\,{^G}\mathbf{d}_B \\ 0 & 1 \end{bmatrix}{^G}\mathbf{r}_P(t) \\[4pt]
&= \begin{bmatrix} {^G}\dot{R}_B\,{^G}R_B^T & {^G}\dot{\mathbf{d}}_B - {^G}\dot{R}_B\,{^G}R_B^T\,{^G}\mathbf{d}_B \\ 0 & 0 \end{bmatrix}{^G}\mathbf{r}_P(t) \\[4pt]
&= \begin{bmatrix} {_G}\tilde{\omega}_B & {^G}\dot{\mathbf{d}}_B - {_G}\tilde{\omega}_B\,{^G}\mathbf{d}_B \\ 0 & 0 \end{bmatrix}{^G}\mathbf{r}_P(t). &\tag{7.224}
\end{aligned}
$$

Thus, the velocity of any point $P$ of the rigid body $B$ in the reference frame $G$ can be obtained by premultiplying the position vector of the point $P$ in $G$ with the *velocity transformation matrix*, $^G V_B$,

$$^G \mathbf{v}_P(t) = {}^G V_B \, {}^G \mathbf{r}_P(t) \tag{7.225}$$

where,

$$
\begin{aligned}
^G V_B &= {}^G \dot{T}_B \, {}^G T_B^{-1} \\
&= \begin{bmatrix} _G \tilde{\omega}_B & {}^G \dot{\mathbf{d}}_B - {}_G \tilde{\omega}_B \, {}^G \mathbf{d}_B \\ 0 & 0 \end{bmatrix} \\
&= \begin{bmatrix} _G \tilde{\omega}_B & {}^G \mathbf{v}_B \\ 0 & 0 \end{bmatrix}
\end{aligned} \tag{7.226}
$$

and

$$_G \tilde{\omega}_B = {}^G \dot{R}_B \, {}^G R_B^T \tag{7.227}$$

$$
\begin{aligned}
^G \mathbf{v}_B &= {}^G \dot{\mathbf{d}}_B - {}^G \dot{R}_B \, {}^G R_B^T \, {}^G \mathbf{d}_B \\
&= {}^G \dot{\mathbf{d}}_B - {}_G \tilde{\omega}_B \, {}^G \mathbf{d}_B \\
&= {}^G \dot{\mathbf{d}}_B - {}_G \boldsymbol{\omega}_B \times {}^G \mathbf{d}_B.
\end{aligned} \tag{7.228}
$$

The *velocity transformation matrix* $^G V_B$ may be assumed as a matrix operator that provides the global velocity of any point attached to $B(oxyz)$. It consists of the angular velocity matrix $_G \tilde{\omega}_B$ and the frame velocity $^G \dot{\mathbf{d}}_B$ both described in the global frame $G(OXYZ)$. The matrix $^G V_B$ depends on six parameters: the three components of the angular velocity vector $_G \boldsymbol{\omega}_B$ and the three components of the frame velocity $^G \dot{\mathbf{d}}_B$. Sometimes it is convenient to introduce a $6 \times 1$ vector called *velocity transformation vector* to simplify numerical calculations.

$$_G \mathbf{t}_B = \begin{bmatrix} ^G \mathbf{v}_B \\ _G \boldsymbol{\omega}_B \end{bmatrix} = \begin{bmatrix} ^G \dot{\mathbf{d}}_B - {}_G \tilde{\omega}_B \, {}^G \mathbf{d}_B \\ _G \boldsymbol{\omega}_B \end{bmatrix} \tag{7.229}$$

In analogy to the two representations of the angular velocity, the velocity of body $B$ in reference frame $G$ can be represented either as the velocity transformation matrix $^G V_B$ in (7.226) or as the velocity transformation vector $_G \mathbf{t}_B$ in (7.229). The velocity transformation vector represents a noncommensurate vector since the dimension of $_G \boldsymbol{\omega}_B$ and $^G \mathbf{v}_B$ differ. ∎

**Example 199** *Velocity transformation matrix based on coordinate transformation matrix.*

*The velocity transformation matrix can be found based on a coordinate transformation matrix. Starting from*

$$^G \mathbf{r}(t) = {}^G T_B \, {}^B \mathbf{r} = \begin{bmatrix} ^G R_B & {}^G \mathbf{d} \\ 0 & 1 \end{bmatrix} {}^B \mathbf{r} \tag{7.230}$$

*and taking the derivative, shows that*

$$^G\mathbf{v} = \frac{^Gd}{dt}\left[^GT_B\,^B\mathbf{r}\right] = {^G\dot{T}_B}\,^B\mathbf{r} \tag{7.231}$$

$$= \begin{bmatrix} ^G\dot{R}_B & ^G\dot{\mathbf{d}} \\ 0 & 0 \end{bmatrix}\,^B\mathbf{r}$$

*however,*

$$^B\mathbf{r} = {^GT_B^{-1}}\,^G\mathbf{r} \tag{7.232}$$

*and therefore,*

$$^G\mathbf{v} = \begin{bmatrix} ^G\dot{R}_B & ^G\dot{\mathbf{d}} \\ 0 & 0 \end{bmatrix}{^GT_B^{-1}}\,^G\mathbf{r} \tag{7.233}$$

$$= \begin{bmatrix} ^G\dot{R}_B & ^G\dot{\mathbf{d}} \\ 0 & 0 \end{bmatrix}\begin{bmatrix} ^GR_B^T & -{^GR_B^T}\,^G\mathbf{d} \\ 0 & 1 \end{bmatrix}\,^G\mathbf{r}$$

$$= \begin{bmatrix} ^G\dot{R}_B\,^GR_B^T & ^G\dot{\mathbf{d}} - {^G\dot{R}_B}\,^GR_B^T\,^G\mathbf{d} \\ 0 & 0 \end{bmatrix}\,^G\mathbf{r}$$

$$= {^GV_B}\,^G\mathbf{r}.$$

**Example 200** *Inverse of a velocity transformation matrix.*

*Transformation from a body frame to the global frame is given by Equation (4.46)*

$$^GT_B^{-1} = \begin{bmatrix} ^GR_B^T & -{^GR_B^T}\,^G\mathbf{d} \\ 0 & 1 \end{bmatrix}. \tag{7.234}$$

*Following the same principle, we may introduce the inverse velocity transformation matrix by*

$$^BV_G = {^GV_B^{-1}}$$

$$= \begin{bmatrix} \left(^G\dot{R}_B\,^GR_B^T\right)^{-1} & -\left(^G\dot{R}_B\,^GR_B^T\right)^{-1}\left(^G\dot{\mathbf{d}} - {^G\dot{R}_B}\,^GR_B^T\,^G\mathbf{d}\right) \\ 0 & 0 \end{bmatrix}$$

$$= \begin{bmatrix} ^GR_B\,^G\dot{R}_B^{-1} & -{^GR_B}\,^G\dot{R}_B^{-1}\left(^G\dot{\mathbf{d}} - {^G\dot{R}_B}\,^GR_B^T\,^G\mathbf{d}\right) \\ 0 & 0 \end{bmatrix}$$

$$= \begin{bmatrix} ^GR_B\,^G\dot{R}_B^{-1} & -{^GR_B}\,^G\dot{R}_B^{-1}\,^G\dot{\mathbf{d}} + {^G\mathbf{d}} \\ 0 & 0 \end{bmatrix} \tag{7.235}$$

*to have*

$$^GV_B\,^GV_B^{-1} = \mathbf{I}. \tag{7.236}$$

*Therefore, having the velocity vector of a body point $^G\mathbf{v}_P$ and the velocity transformation matrix $^GV_B$ we can find the global position of the point by*

$$^G\mathbf{r}_P = {^GV_B^{-1}}\,^G\mathbf{v}_P. \tag{7.237}$$

**Example 201** *Velocity transformation matrix in body frame.*
*The velocity transformation matrix $^G V_B$ defined in the global frame $G$ is described by*

$$^G V_B = \begin{bmatrix} ^G \dot{R}_B \, ^G R_B^T & ^G \dot{\mathbf{d}} - \, ^G \dot{R}_B \, ^G R_B^T \, ^G \mathbf{d} \\ 0 & 0 \end{bmatrix} \tag{7.238}$$

*However, the velocity transformation matrix can be expressed in the body coordinate frame $B$ as well*

$$
\begin{aligned}
{}_G^B V_B &= \, ^G T_B^{-1} \, ^G \dot{T}_B \tag{7.239}\\
&= \begin{bmatrix} ^G R_B^T & -\, ^G R_B^T \, ^G \mathbf{d} \\ 0 & 1 \end{bmatrix} \begin{bmatrix} ^G \dot{R}_B & ^G \dot{\mathbf{d}} \\ 0 & 0 \end{bmatrix} \\
&= \begin{bmatrix} ^G R_B^T \, ^G \dot{R}_B & ^G R_B^T \, ^G \dot{\mathbf{d}} \\ 0 & 0 \end{bmatrix} \\
&= \begin{bmatrix} {}_G^B \omega_B & ^B \dot{\mathbf{d}} \\ 0 & 0 \end{bmatrix}
\end{aligned}
$$

*where ${}_G^B \omega_B$ is the angular velocity vector of $B$ with respect to $G$ expressed in $B$, and $^B \dot{\mathbf{d}}$ is the velocity of the origin of $B$ in $G$ expressed in $B$.*
*It is also possible to use a matrix multiplication to find the velocity transformation matrix in the body coordinate frame.*

$$
\begin{aligned}
{}_G^B \mathbf{v}_P &= \, ^G T_B^{-1} \, ^G \mathbf{v}_P \\
&= \, ^G T_B^{-1} \, ^G \dot{T}_B \, ^B \mathbf{r}_P \\
&= \, {}_G^B V_B \, ^B \mathbf{r}_P \tag{7.240}
\end{aligned}
$$

*Using the definition of (7.219) and (7.239) we are able to transform the velocity transformation matrices between the $B$ and $G$ frames.*

$$^G V_B = \, ^G T_B \, {}_G^B V_B \, ^G T_B^{-1}. \tag{7.241}$$

*It can also be useful if we define the time derivative of the transformation matrix by*

$$^G \dot{T}_B = \, ^G V_B \, ^G T_B \tag{7.242}$$

*or*

$$^G \dot{T}_B = \, ^G T_B \, {}_G^B V_B. \tag{7.243}$$

*Similarly, we may define a velocity transformation matrix from link $(i)$ to $(i-1)$ by*

$$^{i-1} V_i = \begin{bmatrix} ^{i-1} \dot{R}_i \, ^{i-1} R_i^T & ^{i-1} \dot{\mathbf{d}} - \, ^{i-1} \dot{R}_i \, ^{i-1} R_i^T \, ^{i-1} \mathbf{d} \\ 0 & 0 \end{bmatrix} \tag{7.244}$$

*and*

$$^i_{i-1} V_i = \begin{bmatrix} ^{i-1} R_i^T \, ^{i-1} \dot{R}_i & ^{i-1} R_i^T \, ^{i-1} \dot{\mathbf{d}} \\ 0 & 0 \end{bmatrix}. \tag{7.245}$$

**Example 202** *Motion with a point fixed.*

When a point of a rigid body is fixed to the global frame, it is convenient to set the origins of the moving coordinate frame $B(Oxyz)$ and the global coordinate frame $G(OXYZ)$ on the fixed point. Under these conditions,

$$^G\mathbf{d}_B = 0 \quad , \quad ^G\dot{\mathbf{d}}_B = 0$$

and Equation (7.224) reduces to

$$^G\mathbf{v}_P = {_G}\tilde{\omega}_B \, ^G\mathbf{r}_P(t) = {_G}\omega_B \times {^G}\mathbf{r}_P(t). \tag{7.246}$$

**Example 203** *Velocity in spherical coordinates*

The homogeneous transformation matrix from the spherical coordinates $S(Or\theta\varphi)$ to Cartesian coordinates $G(OXYZ)$ is found as

$$^GT_S \;=\; R_{Z,\varphi}\,R_{Y,\theta}\,D_{Z,r} = \begin{bmatrix} ^GR_B & ^G\mathbf{d} \\ 0 & 1 \end{bmatrix}$$

$$= \begin{bmatrix} \cos\theta\cos\varphi & -\sin\varphi & \cos\varphi\sin\theta & r\cos\varphi\sin\theta \\ \cos\theta\sin\varphi & \cos\varphi & \sin\theta\sin\varphi & r\sin\theta\sin\varphi \\ -\sin\theta & 0 & \cos\theta & r\cos\theta \\ 0 & 0 & 0 & 1 \end{bmatrix}. \tag{7.247}$$

Time derivative of $^GT_S$ shows that

$$\begin{aligned}
^G\dot{T}_S &= {^G}V_S \, ^GT_S \\
&= \begin{bmatrix} {_G}\tilde{\omega}_S & ^G\mathbf{v}_S \\ 0 & 0 \end{bmatrix} {^G}T_S \tag{7.248} \\
&= \begin{bmatrix} 0 & -\varphi' & \theta'\cos\varphi & r'\cos\varphi\sin\theta \\ \varphi' & 0 & \theta'\sin\varphi & r'\sin\theta\sin\varphi \\ -\theta'\cos\varphi & -\theta'\sin\varphi & 0 & r'\cos\theta \\ 0 & 0 & 0 & 0 \end{bmatrix} {^G}T_B.
\end{aligned}$$

## 7.5 Derivative of a Homogeneous Transformation Matrix

The velocity transformation matrix can be found directly from the homogeneous link transformation matrix. According to forward kinematics, there is a $4 \times 4$ homogeneous transformation matrix to move between every two coordinate frames.

$$^GT_B = \begin{bmatrix} ^GR_B & ^G\mathbf{d} \\ 0 & 1 \end{bmatrix} = \begin{bmatrix} r_{11} & r_{12} & r_{13} & r_{14} \\ r_{21} & r_{22} & r_{23} & r_{24} \\ r_{31} & r_{32} & r_{33} & r_{34} \\ 0 & 0 & 0 & 1 \end{bmatrix} \tag{7.249}$$

When the elements of the transformation matrix are time varying, its derivative is

$$
\frac{^G dT}{dt} = {}^G \dot{T}_B = 
\begin{bmatrix}
\dfrac{dr_{11}}{dt} & \dfrac{dr_{12}}{dt} & \dfrac{dr_{13}}{dt} & \dfrac{dr_{14}}{dt} \\
\dfrac{dr_{21}}{dt} & \dfrac{dr_{22}}{dt} & \dfrac{dr_{23}}{dt} & \dfrac{dr_{24}}{dt} \\
\dfrac{dr_{31}}{dt} & \dfrac{dr_{32}}{dt} & \dfrac{dr_{33}}{dt} & \dfrac{dr_{34}}{dt} \\
0 & 0 & 0 & 0
\end{bmatrix}.
\tag{7.250}
$$

The time derivative of the transformation matrix can be arranged to be proportional to the transformation matrix

$$
{}^G \dot{T}_B = {}^G V_B \, {}^G T_B
\tag{7.251}
$$

where ${}^G V_B$ is a $4 \times 4$ homogeneous matrix called velocity transformation matrix or *velocity operator matrix* and is equal to

$$
\begin{aligned}
{}^G V_B &= {}^G \dot{T}_B \, {}^G T_B^{-1} \\
&= \begin{bmatrix} {}^G \dot{R}_B \, {}^G R_B^T & {}^G \dot{d} - {}^G \dot{R}_B \, {}^G R_B^T \, {}^G d \\ 0 & 1 \end{bmatrix}.
\end{aligned}
\tag{7.252}
$$

The homogeneous matrix and its derivative based on the velocity transformation matrix are useful in forward velocity kinematics. The ${}^{i-1} \dot{T}_i$ for two links connected by a revolute joint is

$$
{}^{i-1} \dot{T}_i = \dot{\theta}_i 
\begin{bmatrix}
-\sin\theta_i & -\cos\theta_i \cos\alpha_i & \cos\theta_i \sin\alpha_i & -a_i \sin\theta_i \\
\cos\theta_i & -\sin\theta_i \cos\alpha_i & \sin\theta_i \sin\alpha_i & a_i \cos\theta_i \\
0 & 0 & 0 & 0 \\
0 & 0 & 0 & 0
\end{bmatrix}
\tag{7.253}
$$

and for two links connected by a prismatic joint is

$$
{}^{i-1} \dot{T}_i = 
\begin{bmatrix}
0 & 0 & 0 & 0 \\
0 & 0 & 0 & 0 \\
0 & 0 & 0 & \dot{d}_i \\
0 & 0 & 0 & 0
\end{bmatrix}.
\tag{7.254}
$$

The associated velocity transformation matrix for a revolute joint is

$$
{}^{i-1} V_i = \dot{\theta}_i \, \Delta_R = \dot{\theta}_i 
\begin{bmatrix}
0 & -1 & 0 & 0 \\
1 & 0 & 0 & 0 \\
0 & 0 & 0 & 0 \\
0 & 0 & 0 & 0
\end{bmatrix}
\tag{7.255}
$$

and for a prismatic joint is

$$
{}^{i-1} V_i = \dot{d}_i \, \Delta_P = \dot{d}_i 
\begin{bmatrix}
0 & 0 & 0 & 0 \\
0 & 0 & 0 & 0 \\
0 & 0 & 0 & 1 \\
0 & 0 & 0 & 0
\end{bmatrix}.
\tag{7.256}
$$

**Proof.** Since any transformation matrix can be decomposed into a rotation and translation

$$[T] = \begin{bmatrix} R_{\hat{u},\phi} & \mathbf{d} \\ 0 & 1 \end{bmatrix} = \begin{bmatrix} \mathbf{I} & \mathbf{d} \\ 0 & 1 \end{bmatrix} \begin{bmatrix} R_{\hat{u},\phi} & 0 \\ 0 & 1 \end{bmatrix}$$

$$= [D] [R] \tag{7.257}$$

we can find $\dot{T}$ as

$$\dot{T} = \begin{bmatrix} \dot{R}_{\hat{u},\phi} & \dot{\mathbf{d}} \\ 0 & 0 \end{bmatrix} = \begin{bmatrix} \mathbf{I} & \dot{\mathbf{d}} \\ 0 & 1 \end{bmatrix} \begin{bmatrix} \dot{R}_{\hat{u},\phi} & 0 \\ 0 & 1 \end{bmatrix} - \mathbf{I}$$

$$= \left[\mathbf{I} + \dot{D}\right] \left[\mathbf{I} + \dot{R}\right] - \mathbf{I} \tag{7.258}$$

$$= [V] [T]$$

where $[V]$ is the velocity transformation matrix described as

$$[V] = \dot{T}\, T^{-1}$$

$$= \begin{bmatrix} \dot{R}_{\hat{u},\phi} & \dot{\mathbf{d}} \\ 0 & 0 \end{bmatrix} \begin{bmatrix} R_{\hat{u},\phi}^T & -R_{\hat{u},\phi}^T \mathbf{d} \\ 0 & 1 \end{bmatrix}$$

$$= \begin{bmatrix} \dot{R}_{\hat{u},\phi} R_{\hat{u},\phi}^T & \dot{\mathbf{d}} - \dot{R}_{\hat{u},\phi} R_{\hat{u},\phi}^T \mathbf{d} \\ 0 & 1 \end{bmatrix}$$

$$= \begin{bmatrix} \tilde{\omega} & \dot{\mathbf{d}} - \tilde{\omega}\,\mathbf{d} \\ 0 & 1 \end{bmatrix}. \tag{7.259}$$

The transformation matrix between two neighbor coordinate frames of a robot is described in Equation (5.11) based on the DH parameters,

$$^{i-1}T_i = \begin{bmatrix} \cos\theta_i & -\sin\theta_i \cos\alpha_i & \sin\theta_i \sin\alpha_i & a_i \cos\theta_i \\ \sin\theta_i & \cos\theta_i \cos\alpha_i & -\cos\theta_i \sin\alpha_i & a_i \sin\theta_i \\ 0 & \sin\alpha_i & \cos\alpha_i & d_i \\ 0 & 0 & 0 & 1 \end{bmatrix}. \tag{7.260}$$

Direct differentiating shows that in case two links are connected via a revolute joint, $\theta_i$ is the only variable of the DH matrix, and therefore,

$$^{i-1}\dot{T}_i = \dot{\theta}_i \begin{bmatrix} -\sin\theta_i & -\cos\theta_i \cos\alpha_i & \cos\theta_i \sin\alpha_i & -a_i \sin\theta_i \\ \cos\theta_i & -\sin\theta_i \cos\alpha_i & \sin\theta_i \sin\alpha_i & a_i \cos\theta_i \\ 0 & 0 & 0 & 0 \\ 0 & 0 & 0 & 0 \end{bmatrix}$$

$$= \dot{\theta}_i \begin{bmatrix} 0 & -1 & 0 & 0 \\ 1 & 0 & 0 & 0 \\ 0 & 0 & 0 & 0 \\ 0 & 0 & 0 & 0 \end{bmatrix}\; {}^{i-1}T_i = \dot{\theta}_i\, \Delta_R\; {}^{i-1}T_i. \tag{7.261}$$

which shows that the *revolute velocity transformation matrix* is

$$^{i-1}V_i = \dot{\theta}_i \, \Delta_R = \dot{\theta}_i \begin{bmatrix} 0 & -1 & 0 & 0 \\ 1 & 0 & 0 & 0 \\ 0 & 0 & 0 & 0 \\ 0 & 0 & 0 & 0 \end{bmatrix}. \tag{7.262}$$

However, if the two links are connected via a prismatic joint, $d_i$ is the only variable of the DH matrix, and therefore,

$$^{i-1}\dot{T}_i = \dot{d}_i \begin{bmatrix} 0 & 0 & 0 & 0 \\ 0 & 0 & 0 & 0 \\ 0 & 0 & 0 & 1 \\ 0 & 0 & 0 & 0 \end{bmatrix} = \dot{d}_i \, \Delta_P \, ^{i-1}T_i \tag{7.263}$$

which shows that the *prismatic velocity transformation matrix* is

$$^{i-1}V_i = \dot{d}_i \, \Delta_P = \dot{d}_i \begin{bmatrix} 0 & 0 & 0 & 0 \\ 0 & 0 & 0 & 0 \\ 0 & 0 & 0 & 1 \\ 0 & 0 & 0 & 0 \end{bmatrix}. \tag{7.264}$$

The $\Delta_R$ and $\Delta_P$ are revolute and prismatic *velocity coefficient matrices* with some application in velocity analysis of robots. ∎

**Example 204** *Velocity of frame $B_i$ in $B_0$.*

The velocity of the frame $B_i$ attached to the link $(i)$ with respect to the base coordinate frame $B_0$ can be found by differentiating $^0\mathbf{d}_i$ in the base frame.

$$\begin{aligned}
^0\mathbf{v}_i &= \frac{^0d}{dt} \, ^0\mathbf{d}_i = \frac{^0d}{dt} \left( ^0T_i \, ^i\mathbf{d}_i \right) \\
&= ^0\dot{T}_1 \, ^1T_2 \cdots \, ^{i-1}T_i \, ^i\mathbf{d}_i + ^0T_1 \, ^1\dot{T}_2 \, ^2T_3 \cdots \, ^{i-1}T_i \, ^i\mathbf{d}_i \\
&\quad + ^0T_1 \cdots \, ^{i-1}\dot{T}_i \, ^i\mathbf{d}_i \\
&= \left[ \sum_{j=1}^{i} \frac{\partial\, ^0T_i}{\partial q_j} \dot{q}_j \right] \, ^i\mathbf{d}_i \tag{7.265}
\end{aligned}$$

However, the partial derivatives $\partial\, ^{i-1}T_i / \partial q_i$ can be found by utilizing the velocity coefficient matrices $\Delta_i$, which is either $\Delta_R$ or $\Delta_P$.

$$\frac{\partial\, ^{i-1}T_i}{\partial q_i} = \Delta_i \, ^{i-1}T_i. \tag{7.266}$$

Hence,

$$\frac{\partial\, ^0T_i}{\partial q_j} = \begin{cases} ^0T_1 \, ^1T_2 \cdots \, ^{j-2}T_{j-1} \, \Delta_j \, ^{j-1}T_j \cdots \, ^{i-1}T_i & \text{for } j \le i \\ 0 & \text{for } j > i. \end{cases} \tag{7.267}$$

**Example 205** *Differential rotation and translation.*

*Assume the angle of rotation about the axis $\hat{u}$ is too small and indicated by $d\phi$, then the differential rotation matrix is*

$$\mathbf{I} + dR_{\hat{u},\phi} = \mathbf{I} + R_{\hat{u},d\phi} = \begin{bmatrix} 1 & -u_3 d\phi & u_2 d\phi & 0 \\ u_3 d\phi & 1 & -u_1 d\phi & 0 \\ -u_2 d\phi & +u_1 d\phi & 1 & 0 \\ 0 & 0 & 0 & 1 \end{bmatrix} \quad (7.268)$$

*because when $\phi \ll 1$, then,*

$$\begin{aligned} \sin \phi &\simeq d\phi \\ \cos \phi &\simeq 1 \\ \text{vers } \phi &\simeq 0. \end{aligned}$$

*Differential translation* $d\mathbf{d} = d(d_x\hat{I} + d_y\hat{J} + d_z\hat{K})$ *is shown by a differential translation matrix*

$$\mathbf{I} + dD = \begin{bmatrix} 1 & 0 & 0 & dd_x \\ 0 & 1 & 0 & dd_y \\ 0 & 0 & 1 & dd_z \\ 0 & 0 & 0 & 0 \end{bmatrix} \quad (7.269)$$

*and therefore,*

$$\begin{aligned} dT &= [\mathbf{I} + dD][\mathbf{I} + dR] - \mathbf{I} \\ &= \begin{bmatrix} 0 & -d\phi u_3 & d\phi u_2 & dd_x \\ d\phi u_3 & 0 & -d\phi u_1 & dd_y \\ -d\phi u_2 & d\phi u_1 & 0 & dd_z \\ 0 & 0 & 0 & 0 \end{bmatrix}. \end{aligned} \quad (7.270)$$

**Example 206** *Combination of principal differential rotations.*

*The differential rotation about $X, Y, Z$ are*

$$R_{X,d\gamma} = \begin{bmatrix} 1 & 0 & 0 & 0 \\ 0 & 1 & -d\gamma & 0 \\ 0 & d\gamma & 1 & 0 \\ 0 & 0 & 0 & 1 \end{bmatrix} \quad (7.271)$$

$$R_{Y,d\beta} = \begin{bmatrix} 1 & 0 & d\beta & 0 \\ 0 & 1 & 0 & 0 \\ -d\beta & 0 & 1 & 0 \\ 0 & 0 & 0 & 1 \end{bmatrix} \quad (7.272)$$

$$R_{Z,d\alpha} = \begin{bmatrix} 1 & -d\alpha & 0 & 0 \\ d\alpha & 1 & 0 & 0 \\ 0 & 0 & 1 & 0 \\ 0 & 0 & 0 & 1 \end{bmatrix} \quad (7.273)$$

*therefore, the combination of the principal differential rotation matrices about axes $X$, $Y$, and $Z$ is*

$$[\mathbf{I} + R_{X,d\gamma}]\,[\mathbf{I} + R_{Y,d\beta}]\,[\mathbf{I} + R_{Z,d\alpha}]$$

$$= \begin{bmatrix} 1 & 0 & 0 & 0 \\ 0 & 1 & -d\gamma & 0 \\ 0 & d\gamma & 1 & 0 \\ 0 & 0 & 0 & 1 \end{bmatrix} \begin{bmatrix} 1 & 0 & d\beta & 0 \\ 0 & 1 & 0 & 0 \\ -d\beta & 0 & 1 & 0 \\ 0 & 0 & 0 & 1 \end{bmatrix} \begin{bmatrix} 1 & -d\alpha & 0 & 0 \\ d\alpha & 1 & 0 & 0 \\ 0 & 0 & 1 & 0 \\ 0 & 0 & 0 & 1 \end{bmatrix}$$

$$= \begin{bmatrix} 1 & -d\alpha & d\beta & 0 \\ d\alpha & 1 & -d\gamma & 0 \\ -d\beta & d\gamma & 1 & 0 \\ 0 & 0 & 0 & 1 \end{bmatrix}$$

$$= [\mathbf{I} + R_{Z,d\alpha}]\,[\mathbf{I} + R_{Y,d\beta}]\,[\mathbf{I} + R_{X,d\gamma}]. \tag{7.274}$$

*The combination of differential rotations is commutative.*

**Example 207** *Differential of a transformation matrix.*
   *Assume a transformation matrix is given as*

$$T = \begin{bmatrix} 0 & 0 & 1 & 4 \\ 1 & 0 & 0 & 4 \\ 0 & 1 & 0 & 4 \\ 0 & 0 & 0 & 1 \end{bmatrix} \tag{7.275}$$

*subject to a differential rotation and differential translation given by*

$$d\phi\hat{u} = \begin{bmatrix} 0.1 & 0.2 & 0.3 \end{bmatrix} \tag{7.276}$$
$$d\mathbf{d} = \begin{bmatrix} 0.6 & 0.4 & 0.2 \end{bmatrix}. \tag{7.277}$$

*Then, the differential transformation matrix $dT$ is*

$$\begin{aligned} dT &= [\mathbf{I} + dD]\,[\mathbf{I} + dR] - \mathbf{I} \\ &= \begin{bmatrix} 0 & -0.3 & 0.2 & 0.6 \\ 0.3 & 0 & -0.1 & 0.4 \\ -0.2 & 0.1 & 0 & 0.2 \\ 0 & 0 & 0 & 0 \end{bmatrix}. \end{aligned} \tag{7.278}$$

**Example 208** *Derivative of rotation matrix.*
   *Based on the Rodriguez formula, the angle-axis rotation matrix is*

$$R_{\hat{u},\phi} = \mathbf{I}\cos\phi + \hat{u}\hat{u}^T\,\text{vers}\,\phi + \tilde{u}\sin\phi \tag{7.279}$$

*therefore, the time rate of the Rodriguez formula is*

$$\begin{aligned} \dot{R}_{\hat{u},\phi} &= -\dot{\phi}\sin\phi\,\mathbf{I} + \hat{u}\hat{u}^T\,\dot{\phi}\sin\phi + \tilde{u}\dot{\phi}\cos\phi \\ &= \dot{\phi}\tilde{u}R_{\hat{u},\phi}. \end{aligned} \tag{7.280}$$

## 7.6  Summary

When a body coordinate frame $B$ and a global frame $G$ have a common origin and frame $B$ rotates continuously with respect to frame $G$, the rotation matrix $^{G}R_{B}$ is time dependent.

$$^{G}\mathbf{r}(t) = {}^{G}R_{B}(t)\,{}^{B}\mathbf{r} \tag{7.281}$$

Then, the global velocity of a point in $B$ is

$$^{G}\dot{\mathbf{r}}(t) = {}^{G}\mathbf{v}(t) = {}^{G}\dot{R}_{B}(t)\,{}^{B}\mathbf{r} = {}_{G}\tilde{\omega}_{B}\,{}^{G}\mathbf{r}(t) \tag{7.282}$$

where $_{G}\tilde{\omega}_{B}$ is the skew symmetric angular velocity matrix

$$_{G}\tilde{\omega}_{B} = {}^{G}\dot{R}_{B}\,{}^{G}R_{B}^{T} = \begin{bmatrix} 0 & -\omega_{3} & \omega_{2} \\ \omega_{3} & 0 & -\omega_{1} \\ -\omega_{2} & \omega_{1} & 0 \end{bmatrix}. \tag{7.283}$$

The matrix $_{G}\tilde{\omega}_{B}$ is associated with the angular velocity vector $_{G}\boldsymbol{\omega}_{B} = \dot{\phi}\hat{u}$, which is equal to an angular rate $\dot{\phi}$ about the instantaneous axis of rotation $\hat{u}$. Angular velocities of connected links of a robot may be added relatively to find the angular velocity of the end link in the base frame

$$_{0}\boldsymbol{\omega}_{n} = {}_{0}\boldsymbol{\omega}_{1} + {}_{1}^{0}\boldsymbol{\omega}_{2} + {}_{2}^{0}\boldsymbol{\omega}_{3} + \cdots + {}_{n-1}^{0}\boldsymbol{\omega}_{n} = \sum_{i=1}^{n} {}_{i-1}^{0}\boldsymbol{\omega}_{i}. \tag{7.284}$$

To work with angular velocities of relatively moving links, we need to follow the rules of relative derivatives in body and global coordinate frames.

$$\frac{^{B}d}{dt}\,{}^{B}\mathbf{r}_{P} = {}^{B}\dot{\mathbf{r}}_{P} = {}^{B}\mathbf{v}_{P} = \dot{x}\,\hat{\imath} + \dot{y}\,\hat{\jmath} + \dot{z}\,\hat{k} \tag{7.285}$$

$$\frac{^{G}d}{dt}\,{}^{G}\mathbf{r}_{P} = {}^{G}\dot{\mathbf{r}}_{P} = {}^{G}\mathbf{v}_{P} = \dot{X}\,\hat{I} + \dot{Y}\,\hat{J} + \dot{Z}\,\hat{K} \tag{7.286}$$

$$\frac{^{G}d}{dt}\,{}^{B}\mathbf{r}_{P}(t) = {}^{B}\dot{\mathbf{r}}_{P} + {}_{G}^{B}\boldsymbol{\omega}_{B} \times {}^{B}\mathbf{r}_{P} = {}_{G}^{B}\dot{\mathbf{r}}_{P} \tag{7.287}$$

$$\frac{^{B}d}{dt}\,{}^{G}\mathbf{r}_{P}(t) = {}^{G}\dot{\mathbf{r}}_{P} - {}_{G}\boldsymbol{\omega}_{B} \times {}^{G}\mathbf{r}_{P} = {}_{B}^{G}\dot{\mathbf{r}}_{P}. \tag{7.288}$$

The global velocity of a point $P$ in a moving frame $B$ at

$$^{G}\mathbf{r}_{P} = {}^{G}R_{B}\,{}^{B}\mathbf{r}_{P} + {}^{G}\mathbf{d}_{B}$$

is

$$\begin{aligned} ^{G}\mathbf{v}_{P} &= {}^{G}\dot{\mathbf{r}}_{P} \\ &= {}_{G}\tilde{\omega}_{B}\left({}^{G}\mathbf{r}_{P} - {}^{G}\mathbf{d}_{B}\right) + {}^{G}\dot{\mathbf{d}}_{B} \\ &= {}_{G}\boldsymbol{\omega}_{B} \times \left({}^{G}\mathbf{r}_{P} - {}^{G}\mathbf{d}_{B}\right) + {}^{G}\dot{\mathbf{d}}_{B}. \end{aligned} \tag{7.289}$$

The velocity relationship for a body $B$ having a continues rigid motion in $G$ may also be expressed by a homogeneous velocity transformation matrix $^G V_B$

$$^G \mathbf{v}(t) = {}^G V_B \, {}^G \mathbf{r}(t) \tag{7.290}$$

where,

$$^G V_B = {}^G \dot{T}_B \, {}^G T_B^{-1} = \begin{bmatrix} {}_G \tilde{\omega}_B & {}^G \dot{\mathbf{d}}_B - {}_G \tilde{\omega}_B \, {}^G \mathbf{d}_B \\ 0 & 0 \end{bmatrix}. \tag{7.291}$$

# Exercises

1. Notation and symbols.

   Describe the meaning of

   a- $_G\omega_B$     b- $_B\omega_G$     c- $_G^G\omega_B$     d- $_G^B\omega_B$     e- $_B^B\omega_G$     f- $_B^G\omega_G$

   g- $_2^0\omega_1$     h- $_2^2\omega_1$     i- $_2^3\omega_1$     j- $^G\dot{R}_B$     k- $_2^0\tilde{\omega}_1$     l- $_j^k\omega_i$.

   m- $^G\mathbf{r}_P(t)$     n- $_2^G\mathbf{v}_P$     o- $\Delta_R$     p- $\Delta_P$     q- $\dfrac{^Gd}{dt}$     r- $\dfrac{^Bd}{dt}$

   s- $\dfrac{^Gd}{dt}{}^G\mathbf{r}_P$     t- $\dfrac{^Gd}{dt}{}^B\mathbf{r}_P$     u- $\dfrac{^Bd}{dt}{}^B\mathbf{r}_P$     v- $^G\dot{\mathbf{r}}_P$     w- $^G\dot{\mathbf{d}}_P$     x- $^G\mathbf{V}_B$

2. Local position, global velocity.

   A body is turning about the $Z$-axis at a constant angular rate $\dot{\alpha} = 2\,\text{rad/sec}$. Find the global velocity of a point at

   $$^B\mathbf{r} = \begin{bmatrix} 5 \\ 30 \\ 10 \end{bmatrix}.$$

3. Global position, constant angular velocity.

   A body is turning about the $Z$-axis at a constant angular rate $\dot{\alpha} = 2\,\text{rad/s}$. Find the global position of a point at

   $$^B\mathbf{r} = \begin{bmatrix} 5 \\ 30 \\ 10 \end{bmatrix}$$

   after $t = 3\,\text{sec}$ if the body and global coordinate frames were coincident at $t = 0\,\text{sec}$.

4. Turning about $x$-axis.

   Find the angular velocity matrix when the body coordinate frame is turning $35\,\text{deg/sec}$ at $45\,\text{deg}$ about the $x$-axis.

5. Combined rotation and angular velocity.

   Find the rotation matrix for a body frame after $30\,\text{deg}$ rotation about the $Z$-axis, followed by $30\,\text{deg}$ about the $X$-axis, and then $90\,\text{deg}$ about the $Y$-axis. Then calculate the angular velocity of the body if it is turning with $\dot{\alpha} = 20\,\text{deg/sec}$, $\dot{\beta} = -40\,\text{deg/sec}$, and $\dot{\gamma} = 55\,\text{deg/sec}$ about the $Z$, $Y$, and $X$ axes respectively.

6. Angular velocity, expressed in body frame.

   The point $P$ is at $\mathbf{r}_P = (1, 2, 1)$ in a body coordinate $B(Oxyz)$. Find $_G^B\tilde{\omega}_B$ when the body frame is turned 30 deg about the $X$-axis at a rate $\dot{\gamma} = 75\,\mathrm{deg}/\mathrm{sec}$, followed by 45 deg about the $Z$-axis at a rate $\dot{\alpha} = 25\,\mathrm{deg}/\mathrm{sec}$.

7. Global roll-pitch-yaw angular velocity.

   Calculate the angular velocity for a global roll-pitch-yaw rotation of $\alpha = 30\,\mathrm{deg}$, $\beta = 30\,\mathrm{deg}$, and $\gamma = 30\,\mathrm{deg}$ with $\dot{\alpha} = 20\,\mathrm{deg}/\mathrm{sec}$, $\dot{\beta} = -20\,\mathrm{deg}/\mathrm{sec}$, and $\dot{\gamma} = 20\,\mathrm{deg}/\mathrm{sec}$.

8. Roll-pitch-yaw angular velocity.

   Find $_G^B\tilde{\omega}_B$ and $_G\tilde{\omega}_B$ for the role, pitch, and yaw rates equal to $\dot{\alpha} = 20\,\mathrm{deg}/\mathrm{sec}$, $\dot{\beta} = -20\,\mathrm{deg}/\mathrm{sec}$, and $\dot{\gamma} = 20\,\mathrm{deg}/\mathrm{sec}$ respectively, and having the following rotation matrix:

$$^B R_G = \begin{bmatrix} 0.53 & -0.84 & 0.13 \\ 0.0 & 0.15 & 0.99 \\ -0.85 & -0.52 & 0.081 \end{bmatrix}$$

9. Angular velocity from the Rodriguez formula.

   We may find the time derivative of $^G R_B = R_{\hat{u}, \phi}$ by

$$^G \dot{R}_B = \frac{d}{dt}{}^G R_B = \dot{\phi}\frac{d}{d\phi}{}^G R_B.$$

   Use the Rodriguez rotation formula and find $_G\tilde{\omega}_B$ and $_G^B\tilde{\omega}_B$.

10. Skew symmetric matrix

    Show that any square matrix can be expressed as the sum of a symmetric and skew symmetric matrix.

$$\begin{aligned} A &= B + C \\ B &= \frac{1}{2}(A + A) \\ C &= \frac{1}{2}(A - A) \end{aligned}$$

11. A rotating slider.

    Figure 7.8 illustrates a slider link on a rotating arm. Calculate

$$\frac{^G d\hat{\imath}}{dt}, \quad \frac{^G d\hat{\jmath}}{dt}, \quad \frac{^G d\hat{k}}{dt}$$
$$\frac{^G d^2\hat{\imath}}{dt^2}, \quad \frac{^G d^2\hat{\jmath}}{dt^2}, \quad \frac{^G d^2\hat{k}}{dt^2}$$

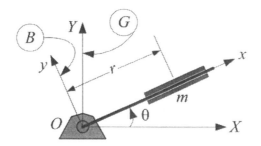

FIGURE 7.8. A slider on a rotating bar.

and find $^{B}\mathbf{v}$ and $^{B}\mathbf{a}$ of $m$ at $CM$ of the slider by using the rule of mixed derivative

$$\frac{^{G}d}{dt}\left(\frac{^{B}d}{dt}\mathbf{r}\right) = \frac{^{B}d}{dt}\left(\frac{^{B}d}{dt}\mathbf{r}\right) + {}_{G}\boldsymbol{\omega}_{B} \times \left(\frac{^{B}d}{dt}\mathbf{r}\right).$$

12. ★ Differentiating in local and global frames.

Consider a local point at $^{B}\mathbf{r}_{P} = t\hat{\imath} + \hat{\jmath}$. The local frame $B$ is rotating in $G$ by $\dot{\alpha}$ about the $Z$-axis. Calculate $\frac{^{B}d}{dt}\,^{B}\mathbf{r}_{P}$, $\frac{^{G}d}{dt}\,^{G}\mathbf{r}_{P}$, $\frac{^{B}d}{dt}\,^{G}\mathbf{r}_{P}$, and $\frac{^{G}d}{dt}\,^{B}\mathbf{r}_{P}$.

13. ★ Skew symmetric identity for angular velocity.

Show that

$$R\tilde{\omega}R^{T} = \widetilde{R\omega}.$$

14. ★ Transformation of angular velocity exponents.

Show that

$${}_{G}^{B}\tilde{\omega}_{B}^{n} = {}^{G}R_{B}^{T}\,{}_{G}^{G}\tilde{\omega}_{B}^{n}\,{}^{G}R_{B}.$$

15. ★ An angular velocity matrix identity.

Show that

$$\tilde{\omega}^{2k+1} = (-1)^{k}\,\omega^{2k}\,\tilde{\omega}$$

and

$$\tilde{\omega}^{2k} = (-1)^{k}\,\omega^{2(k-1)}\left(\omega^{2}\,\mathbf{I} - \boldsymbol{\omega}\boldsymbol{\omega}^{T}\right).$$

16. ★ Angular velocity and global triple rotation frequencies.

Find the angular velocity in terms of the 12 cases of triple global rotations in Appendix A.

17. ★ Angular velocity and local triple rotation frequencies.

Find the angular velocity in terms of the 12 cases of triple local rotations in Appendix B.

# 8

# Velocity Kinematics

## 8.1  Rigid Link Velocity

The angular velocity of link $(i)$ in the global coordinate frame $B_0$ is a summation of global angular velocities of the links $(j)$, for $j \leq i$

$$_0^0\omega_i = \sum_{j=1}^{i} {}_{j-1}^0\omega_j \qquad (8.1)$$

where

$$_{j-1}^0\omega_j = \begin{cases} \dot{\theta}_j \, {}^0\hat{k}_{j-1} & \text{if joint } i \text{ is R} \\ 0 & \text{if joint } i \text{ is P}. \end{cases} \qquad (8.2)$$

The velocity of the origin of $B_i$ attached to link $(i)$ in the base coordinate frame is

$$_{i-1}^0\dot{\mathbf{d}}_i = \begin{cases} {}_0^0\omega_i \times {}_{i-1}^0\mathbf{d}_i & \text{if joint } i \text{ is R} \\ \dot{d}_i \, {}^0\hat{k}_{i-1} + {}_0^0\omega_i \times {}_{i-1}^0\mathbf{d}_i & \text{if joint } i \text{ is P} \end{cases} \qquad (8.3)$$

where $\theta$ and $d$ are DH parameters, and $\mathbf{d}$ is a frame's origin position vector.

Therefore, if a robot has $n$ links, the global angular velocity of the final coordinate frame is

$$_0^0\omega_n = \sum_{i=1}^{n} {}_{i-1}^0\omega_i \qquad (8.4)$$

and the global velocity vector of the last link's coordinate frame is

$$_0^0\dot{\mathbf{d}}_n = \sum_{i=1}^{n} {}_{i-1}^0\dot{\mathbf{d}}_i. \qquad (8.5)$$

**Proof.** According to the DH definition, the position vector of the coordinate frame $B_i$ with respect to $B_{i-1}$ is

$$_{i-1}^0\mathbf{d}_i = d_i \, {}^0\hat{k}_{i-1} + a_i \, {}^0\hat{\imath}_i \qquad (8.6)$$

which depends on joint variables $q_j$, for $j \leq i$, and therefore, ${}_{i-1}^0\dot{\mathbf{d}}_i$ is a function of joint velocities $\dot{q}_j$, for $j \leq i$ .

Assume that every joint of a robot except joint $i$ is locked. Then, the angular velocity of link $(i)$ connecting via a revolute joint to link $(i-1)$ is

$$_{i-1}^0\omega_i = \dot{\theta}_i \, {}^0\hat{k}_{i-1} \qquad \text{if the only movable joint } i \text{ is R.} \qquad (8.7)$$

However, if the link $(i)$ and $(i-1)$ are connecting via a prismatic joint then,

$$_{i-1}^{0}\boldsymbol{\omega}_i = 0 \qquad \text{if the only movable joint } i \text{ is P.} \qquad (8.8)$$

The relative position vector (8.6) shows that the velocity of link $(i)$ connecting via a revolute joint to the link $(i-1)$ is

$$
\begin{aligned}
_{i-1}^{0}\dot{\mathbf{d}}_i &= \dot{\theta}_i \; {}^{0}\hat{k}_{i-1} \times a_i \; {}^{0}\hat{\imath}_i \\
&= {}_{i-1}^{0}\boldsymbol{\omega}_i \times {}_{i-1}^{0}\mathbf{d}_i \qquad \text{if the only movable joint } i \text{ is R.} \quad (8.9)
\end{aligned}
$$

We could substitute $a_i \, {}^{0}\hat{\imath}_i$ with ${}_{i-1}^{0}\mathbf{d}_i$ because the $x_i$-axis is turning about the $z_{i-1}$-axis with angular velocity $\dot{\theta}_i$ and therefore, $\dot{\theta}_i \, {}^{0}\hat{k}_{i-1} \times d_i \, {}^{0}\hat{k}_{i-1} = 0$. However, if the link $(i)$ and $(i-1)$ are connecting via a prismatic joint then,

$$_{i-1}^{0}\dot{\mathbf{d}}_i = \dot{d}_i \; {}^{0}\hat{k}_{i-1} \qquad \text{if the only movable joint } i \text{ is P.} \qquad (8.10)$$

Now assume that all lower joints $j \le i$ are moving. Then, the angular velocity of link $(i)$ in the base coordinate frame is

$$
{}_{0}^{0}\boldsymbol{\omega}_i = \sum_{j=1}^{i} {}_{j-1}^{0}\boldsymbol{\omega}_j
$$

or

$$
{}_{0}^{0}\boldsymbol{\omega}_i =
\begin{cases}
\displaystyle\sum_{j=1}^{i} \dot{\theta}_i \; {}^{0}\hat{k}_{i-1} & \text{if joint } j \text{ is R} \\[2mm]
0 & \text{if joint } j \text{ is P}
\end{cases}
\qquad (8.11)
$$

which can be written in a recursive form

$$
{}_{0}^{0}\boldsymbol{\omega}_i = {}_{0}^{0}\boldsymbol{\omega}_{i-1} + {}_{i-1}^{0}\boldsymbol{\omega}_i
\qquad (8.12)
$$

or

$$
{}_{0}^{0}\boldsymbol{\omega}_i =
\begin{cases}
{}_{0}^{0}\boldsymbol{\omega}_{i-1} + \dot{\theta}_i \; {}^{0}\hat{k}_{i-1} & \text{if joint } i \text{ is R} \\
{}_{0}^{0}\boldsymbol{\omega}_{i-1} & \text{if joint } i \text{ is P.}
\end{cases}
\qquad (8.13)
$$

The velocity of the origin of the link $(i)$ in the base coordinate frame is

$$
{}_{i-1}^{0}\dot{\mathbf{d}}_i =
\begin{cases}
{}_{0}^{0}\boldsymbol{\omega}_i \times {}_{i-1}^{0}\mathbf{d}_i & \text{if joint } i \text{ is R} \\
\dot{d}_i \, {}^{0}\hat{k}_{i-1} + {}_{0}^{0}\boldsymbol{\omega}_i \times {}_{i-1}^{0}\mathbf{d}_i & \text{if joint } i \text{ is P.}
\end{cases}
\qquad (8.14)
$$

The translation and angular velocities of the last link of a $n$ link robot is then a direct application of these results. ∎

**Example 209** *Serial rigid links angular velocity.*
    *Consider a serial manipulator with $n$ links and $n$ revolute joints. The global angular velocity of link $(i)$ in terms of the angular velocity of its previous links is*

$$
{}^{0}\boldsymbol{\omega}_i = {}^{0}\boldsymbol{\omega}_{i-1} + \dot{\theta}_i \, {}^{0}\hat{k}_{i-1}
\qquad (8.15)
$$

*or in general*

$$^0\omega_i = \sum_{j=1}^{i} \dot{\theta}_j \, ^0\hat{k}_{j-1} \tag{8.16}$$

*because*

$$^0_{i-1}\omega_i = \dot{\theta}_i \, ^0\hat{k}_{i-1} \tag{8.17}$$

$$^0_{i-2}\omega_{i-1} = \dot{\theta}_{i-1} \, ^0\hat{k}_{i-2} \tag{8.18}$$

$$
\begin{aligned}
^0_{i-2}\omega_i &= \, ^0_{i-2}\omega_{i-1} + \, ^0_{i-1}\omega_i \\
&= \, ^0_{i-2}\omega_{i-1} + \dot{\theta}_i \, ^0\hat{k}_{i-1} \\
&= \dot{\theta}_{i-1} \, ^0\hat{k}_{i-2} + \dot{\theta}_i \, ^0\hat{k}_{i-1}
\end{aligned} \tag{8.19}
$$

*and therefore,*

$$
\begin{aligned}
^0\omega_i &= \sum_{j=1}^{i-1} \, ^0_{j-1}\omega_j + \dot{\theta}_i \, ^0\hat{k}_{i-1} \\
&= \sum_{j=1}^{i-1} \dot{\theta}_j \, ^0\hat{k}_{j-1} + \dot{\theta}_i \, ^0\hat{k}_{i-1} \\
&= \sum_{j=1}^{i} \dot{\theta}_j \, ^0\hat{k}_{j-1}.
\end{aligned} \tag{8.20}
$$

**Example 210** *Serial rigid links translational velocity.*
   *The global angular velocity of link (i) in a serial manipulator in terms of the angular velocity of its previous links is*

$$^0v_i = \, ^0v_{i-1} + \, ^0_{i-1}\omega_i \times \, ^0_{i-1}d_i \tag{8.21}$$

*where*

$$^0v_i = \, ^0\dot{d}_i \tag{8.22}$$

*or in general*

$$^0v_i = \sum_{j=1}^{i} \left( ^0\hat{k}_{j-1} \times \, ^0_{i-1}d_i \right) \dot{\theta}_j \tag{8.23}$$

*because*

$$^0_{i-1}v_i = \, _0\omega_i \times \, ^0_{i-1}d_i \tag{8.24}$$

$$^0_{i-2}v_{i-1} = \, _0\omega_{i-1} \times \, ^0_{i-2}d_{i-1} \tag{8.25}$$

$$
\begin{aligned}
^0_{i-2}v_i &= \, ^0_{i-2}v_{i-1} + \, ^0_{i-1}v_i \\
&= \, ^0_{i-2}v_{i-1} + \, _0\omega_i \times \, ^0_{i-1}d_i \\
&= \, _0\omega_{i-1} \times \, ^0_{i-2}d_{i-1} + \, _0\omega_i \times \, ^0_{i-1}d_i \\
&= \dot{\theta}_{i-1} \, ^0\hat{k}_{i-2} \times \, ^0_{i-2}d_{i-1} + \dot{\theta}_i \, ^0\hat{k}_{i-1} \times \, ^0_{i-1}d_i
\end{aligned} \tag{8.26}
$$

*and therefore,*

$$^0\mathbf{v}_i = \sum_{j=1}^{i-1} {}_{j-1}^{0}\mathbf{v}_j + {}_{i-1}^{0}\mathbf{v}_i$$

$$= \sum_{j=1}^{i} \left( {}^0\hat{k}_{j-1} \times {}_{i-1}^{0}\mathbf{d}_i \right) \dot{\theta}_j . \tag{8.27}$$

## 8.2 Forward Velocity Kinematics and the Jacobian Matrix

The *forward velocity kinematics* solves the problem of relating joint velocities, $\dot{\mathbf{q}}$, to the end-effector velocities $\dot{\mathbf{X}}$

$$\dot{\mathbf{X}} = \mathbf{J}\,\dot{\mathbf{q}} \tag{8.28}$$

where $\mathbf{q}$ is the *joint variable vector*, and $\dot{\mathbf{q}}$ is the *joint velocity vector*

$$\mathbf{q} = \begin{bmatrix} q_n & q_n & q_n & \cdots & q_n \end{bmatrix}^T \tag{8.29}$$

$$\dot{\mathbf{q}} = \begin{bmatrix} \dot{q}_n & \dot{q}_n & \dot{q}_n & \cdots & \dot{q}_n \end{bmatrix}^T . \tag{8.30}$$

In addition, $\mathbf{X}$ is the *end-effector configuration vector* and $\dot{\mathbf{X}}$ is the *end-effector configuration velocity vector.*

$$\begin{aligned} \mathbf{X} &= \begin{bmatrix} X_n & Y_n & Z_n & \varphi_n & \theta_n & \psi_n \end{bmatrix}^T \\ &= \begin{bmatrix} {}^0\mathbf{d}_n \\ {}_0\phi_n \end{bmatrix} . \end{aligned} \tag{8.31}$$

$$\begin{aligned} \dot{\mathbf{X}} &= \begin{bmatrix} \dot{X}_n & \dot{Y}_n & \dot{Z}_n & \omega_{Xn} & \omega_{Yn} & \omega_{Zn} \end{bmatrix}^T \\ &= \begin{bmatrix} {}^0\dot{\mathbf{d}}_n \\ {}^0\boldsymbol{\omega}_n \end{bmatrix} = \begin{bmatrix} {}^0\mathbf{v}_n \\ {}^0\boldsymbol{\omega}_n \end{bmatrix} \end{aligned} \tag{8.32}$$

This problem is solved by finding the $6 \times n$ transformation matrix $\mathbf{J}(\mathbf{q})$ where $n$ is the number of joint variables. The matrix $\mathbf{J}(\mathbf{q})$ is called *Jacobian*. It is possible to break the forward velocity problem into two sub problems to find the translation and angular velocities of the end frame independently.

$$^0\mathbf{v}_n = \mathbf{J}_D\,\dot{\mathbf{q}} \tag{8.33}$$

$$^0\boldsymbol{\omega}_n = \mathbf{J}_R\,\dot{\mathbf{q}} \tag{8.34}$$

$$\dot{\mathbf{X}} = \begin{bmatrix} {}^0\mathbf{v}_n \\ {}^0\boldsymbol{\omega}_n \end{bmatrix} = \begin{bmatrix} \mathbf{J}_D \\ \mathbf{J}_R \end{bmatrix} \dot{\mathbf{q}} = \mathbf{J}\,\dot{\mathbf{q}} \tag{8.35}$$

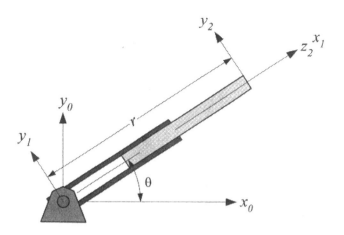

FIGURE 8.1. A planar polar manipulator.

From the forward kinematics we know that

$$^0T_n(\mathbf{q}) = \begin{bmatrix} ^0R_n & ^0\mathbf{d}_n \\ 0 & 1 \end{bmatrix} \tag{8.36}$$

and therefore,

$$\mathbf{J}(\mathbf{q}) = \frac{\partial T(\mathbf{q})}{\partial \mathbf{q}} \tag{8.37}$$

**Example 211** *Jacobian matrix for a planar polar manipulator.*
*Figure 8.1 illustrates a planar polar manipulator with the following forward kinematics.*

$$
\begin{aligned}
^0T_2 &= {}^0T_1\,{}^1T_2 \\
&= \begin{bmatrix} \cos\theta & -\sin\theta & 0 & 0 \\ \sin\theta & \cos\theta & 0 & 0 \\ 0 & 0 & 1 & 0 \\ 0 & 0 & 0 & 1 \end{bmatrix} \begin{bmatrix} 1 & 0 & 0 & r \\ 0 & 1 & 0 & 0 \\ 0 & 0 & 1 & 0 \\ 0 & 0 & 0 & 1 \end{bmatrix} \\
&= \begin{bmatrix} \cos\theta & -\sin\theta & 0 & r\cos\theta \\ \sin\theta & \cos\theta & 0 & r\sin\theta \\ 0 & 0 & 1 & 0 \\ 0 & 0 & 0 & 1 \end{bmatrix}
\end{aligned} \tag{8.38}
$$

*The tip point of the manipulator is at*

$$\begin{bmatrix} X \\ Y \end{bmatrix} = \begin{bmatrix} r\cos\theta \\ r\sin\theta \end{bmatrix} \tag{8.39}$$

*and therefore, its velocity is*

$$\begin{bmatrix} \dot{X} \\ \dot{Y} \end{bmatrix} = \begin{bmatrix} \cos\theta & -r\sin\theta \\ \sin\theta & r\cos\theta \end{bmatrix} \begin{bmatrix} \dot{r} \\ \dot{\theta} \end{bmatrix} \tag{8.40}$$

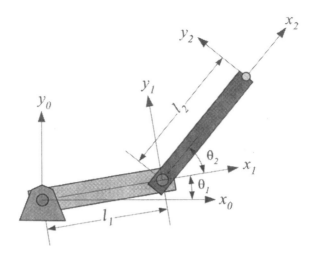

FIGURE 8.2. A 2R planar manipulator.

*which shows that*

$$J_D = \begin{bmatrix} \cos\theta & -r\sin\theta \\ \sin\theta & r\cos\theta \end{bmatrix}.$$  (8.41)

**Example 212** *Jacobian matrix for the 2R planar manipulator.*

*A 2R planar manipulator with two R‖R links was illustrated in Figure 5.9 and is shown in Figure 8.2 again. The manipulator has been analyzed in Example 132 for forward kinematics, and in Example 168 for inverse kinematics.*

*The angular velocity of links (1) and (2) are*

$$\begin{align}
{}_0\boldsymbol{\omega}_1 &= \dot{\theta}_1\,{}^0\hat{k}_0 \tag{8.42}\\
{}_1^0\boldsymbol{\omega}_2 &= \dot{\theta}_2\,{}^0\hat{k}_1 \tag{8.43}\\
{}_0\boldsymbol{\omega}_2 &= {}_0\boldsymbol{\omega}_1 + {}_1^0\boldsymbol{\omega}_2\\
&= \left(\dot{\theta}_1 + \dot{\theta}_2\right){}^0\hat{k}_0 \tag{8.44}
\end{align}$$

*and*

$$\begin{align}
{}_0\boldsymbol{\omega}_2 &= {}_0\boldsymbol{\omega}_1 + {}_1^0\boldsymbol{\omega}_2\\
&= \left(\dot{\theta}_1 + \dot{\theta}_2\right){}^0\hat{k}_0 \tag{8.45}
\end{align}$$

*and the global velocity of the tip position of the manipulator is*

$$
\begin{aligned}
{}^0\dot{\mathbf{d}}_2 &= {}^0\dot{\mathbf{d}}_1 + {}^0_1\dot{\mathbf{d}}_2 \\
&= {}_0\boldsymbol{\omega}_1 \times {}^0\mathbf{d}_1 + {}_0\boldsymbol{\omega}_2 \times {}^0_1\mathbf{d}_2 \\
&= \dot{\theta}_1\,{}^0\hat{k}_0 \times l_1\,{}^0\hat{\imath}_1 + \left(\dot{\theta}_1 + \dot{\theta}_2\right)\,{}^0\hat{k}_0 \times l_2\,{}^0\hat{\imath}_2 \\
&= l_1\dot{\theta}_1\,{}^0\hat{\jmath}_1 \times l_2\left(\dot{\theta}_1 + \dot{\theta}_2\right)\,{}^0\hat{\jmath}_2.
\end{aligned}
\tag{8.46}
$$

*The unit vectors ${}^0\hat{\jmath}_1$ and ${}^0\hat{\jmath}_2$ can be found by using the coordinate transformation method,*

$$
\begin{aligned}
{}^0\hat{\jmath}_1 &= R_{z,\theta_1}\,{}^1\hat{\jmath}_1 \\
&= \begin{bmatrix} \cos\theta_1 & -\sin\theta_1 & 0 \\ \sin\theta_1 & \cos\theta_1 & 0 \\ 0 & 0 & 1 \end{bmatrix} \begin{bmatrix} 0 \\ 1 \\ 0 \end{bmatrix} \\
&= \begin{bmatrix} -\sin\theta_1 \\ \cos\theta_1 \\ 0 \end{bmatrix}
\end{aligned}
\tag{8.47}
$$

$$
\begin{aligned}
{}^0\hat{\jmath}_2 &= R_{z,\theta_1+\theta_2}\,{}^2\hat{\jmath}_2 \\
&= \begin{bmatrix} \cos(\theta_1+\theta_2) & -\sin(\theta_1+\theta_2) & 0 \\ \sin(\theta_1+\theta_2) & \cos(\theta_1+\theta_2) & 0 \\ 0 & 0 & 1 \end{bmatrix} \begin{bmatrix} 0 \\ 1 \\ 0 \end{bmatrix} \\
&= \begin{bmatrix} -\sin(\theta_1+\theta_2) \\ \cos(\theta_1+\theta_2) \\ 0 \end{bmatrix}.
\end{aligned}
\tag{8.48}
$$

*Substituting back shows that*

$$
{}^0\dot{\mathbf{d}}_2 = l_1\dot{\theta}_1 \begin{bmatrix} -\sin\theta_1 \\ \cos\theta_1 \\ 0 \end{bmatrix} \times l_2\left(\dot{\theta}_1 + \dot{\theta}_2\right) \begin{bmatrix} -\sin(\theta_1+\theta_2) \\ \cos(\theta_1+\theta_2) \\ 0 \end{bmatrix}
\tag{8.49}
$$

*which can be rearranged to have*

$$
\begin{aligned}
\begin{bmatrix} \dot{X} \\ \dot{Y} \end{bmatrix} &= \begin{bmatrix} -l_1 s\theta_1 - l_2 s\left(\theta_1 + \theta_2\right) & -l_2 s\left(\theta_1 + \theta_2\right) \\ l_1 c\theta_1 + l_2 c\left(\theta_1 + \theta_2\right) & l_2 c\left(\theta_1 + \theta_2\right) \end{bmatrix} \begin{bmatrix} \dot{\theta}_1 \\ \dot{\theta}_2 \end{bmatrix} \\
&= \mathbf{J}_D \begin{bmatrix} \dot{\theta}_1 \\ \dot{\theta}_2 \end{bmatrix}.
\end{aligned}
\tag{8.50}
$$

*Taking advantage of the structural simplicity of the 2R manipulator, we may find its Jacobian simpler. The forward kinematics of the manipulator was found as*

$$^0T_2 = {^0T_1}\,{^1T_2} \tag{8.51}$$

$$= \begin{bmatrix} c\,(\theta_1 + \theta_2) & -s\,(\theta_1 + \theta_2) & 0 & l_1 c\theta_1 + l_2 c\,(\theta_1 + \theta_2) \\ s\,(\theta_1 + \theta_2) & c\,(\theta_1 + \theta_2) & 0 & l_1 s\theta_1 + l_2 s\,(\theta_1 + \theta_2) \\ 0 & 0 & 1 & 0 \\ 0 & 0 & 0 & 1 \end{bmatrix}$$

*which shows the tip position $^0_0\mathbf{d}_2$ of the manipulator is at*

$$\begin{bmatrix} X \\ Y \end{bmatrix} = \begin{bmatrix} l_1 \cos\theta_1 + l_2 \cos(\theta_1 + \theta_2) \\ l_1 \sin\theta_1 + l_2 \sin(\theta_1 + \theta_2) \end{bmatrix}. \tag{8.52}$$

*Direct differentiating gives*

$$\begin{bmatrix} \dot{X} \\ \dot{Y} \end{bmatrix} = \begin{bmatrix} -l_1 \dot{\theta}_1 \sin\theta_1 - l_2 \left(\dot{\theta}_1 + \dot{\theta}_2\right) \sin(\theta_1 + \theta_2) \\ l_1 \dot{\theta}_1 \cos\theta_1 + l_2 \left(\dot{\theta}_1 + \dot{\theta}_2\right) \cos(\theta_1 + \theta_2) \end{bmatrix} \tag{8.53}$$

*which can be rearranged in a matrix form*

$$\begin{bmatrix} \dot{X} \\ \dot{Y} \end{bmatrix} = \begin{bmatrix} -l_1 s\theta_1 - l_2 s\,(\theta_1 + \theta_2) & -l_2 s\,(\theta_1 + \theta_2) \\ l_1 c\theta_1 + l_2 c\,(\theta_1 + \theta_2) & l_2 c\,(\theta_1 + \theta_2) \end{bmatrix} \begin{bmatrix} \dot{\theta}_1 \\ \dot{\theta}_2 \end{bmatrix} \tag{8.54}$$

*or*

$$\dot{\mathbf{X}} = \mathbf{J}_D\,\dot{\boldsymbol{\theta}}. \tag{8.55}$$

**Example 213** *Columns of the Jacobian for the 2R manipulator.*

*The Jacobian of the 2R planar manipulator can be found systematically using the column-by-column method. The global position vector of the coordinate frames are*

$$^1_1\mathbf{d}_2 = l_2\,{^0\hat{\imath}_2} \tag{8.56}$$

$$^0\mathbf{d}_2 = l_1\,{^0\hat{\imath}_1} + l_2\,{^0\hat{\imath}_2} \tag{8.57}$$

*and therefore,*

$$\begin{aligned} ^0\dot{\mathbf{d}}_2 &= {_0\boldsymbol{\omega}_1} \times {^0\mathbf{d}_2} + {_1^0\boldsymbol{\omega}_2} \times {_1^0\mathbf{d}_2} \\ &= \dot{\theta}_1\,{^0\hat{k}_0} \times \left(l_1\,{^0\hat{\imath}_1} + l_2\,{^0\hat{\imath}_2}\right) + \dot{\theta}_2\,{^0\hat{k}_1} \times l_2\,{^0\hat{\imath}_2} \\ &= \begin{bmatrix} {^0\hat{k}_0} \times \left(l_1\,{^0\hat{\imath}_1} + l_2\,{^0\hat{\imath}_2}\right) & {^0\hat{k}_1} \times l_2\,{^0\hat{\imath}_2} \end{bmatrix} \begin{bmatrix} \dot{\theta}_1 \\ \dot{\theta}_2 \end{bmatrix} \end{aligned} \tag{8.58}$$

*which can be set in the form*

$$\begin{bmatrix} ^0\dot{\mathbf{d}}_2 \\ _0\boldsymbol{\omega}_2 \end{bmatrix} = \begin{bmatrix} {^0\hat{k}_0} \times {^0\mathbf{d}_2} & {^0\hat{k}_1} \times {_1^0\mathbf{d}_2} \\ {^0\hat{k}_0} & {^0\hat{k}_1} \end{bmatrix} \begin{bmatrix} \dot{\theta}_1 \\ \dot{\theta}_2 \end{bmatrix}$$

$$= \mathbf{J}\,\dot{\boldsymbol{\theta}}. \tag{8.59}$$

## 8.3   Jacobian Generating Vectors

The Jacobian matrix is a linear transformation, mapping joint velocities to Cartesian velocities

$$\dot{\mathbf{X}} = \mathbf{J}\,\dot{\mathbf{q}} \tag{8.60}$$

and is equal to

$$\mathbf{J} = \begin{bmatrix} {}^0\hat{k}_0 \times {}^0_0\mathbf{d}_n & {}^0\hat{k}_1 \times {}^0_1\mathbf{d}_n & \cdots & {}^0\hat{k}_{n-1} \times {}^0_{n-1}\mathbf{d}_n \\ {}^0\hat{k}_0 & {}^0\hat{k}_1 & \cdots & {}^0\hat{k}_{n-1} \end{bmatrix}. \tag{8.61}$$

$\mathbf{J}$ can be calculated column by column. The $i$th column of $\mathbf{J}$ is called *Jacobian generating vector* and is denoted by $\mathbf{c}_i(\mathbf{q})$.

$$\mathbf{c}_i(\mathbf{q}) = \begin{bmatrix} {}^0\hat{k}_{i-1} \times {}^0_{i-1}\mathbf{d}_n \\ {}^0\hat{k}_{i-1} \end{bmatrix}. \tag{8.62}$$

To calculate the $i$th column of the Jacobian matrix, we need to find two vectors ${}^0_{i-1}\mathbf{d}_n$ and ${}^0\hat{k}_{i-1}$. These vectors are position of origin and the joint axis unit vector of the frame attached to link $(i-1)$, both expressed in the base frame.

Calculating $\mathbf{J}$, based on the Jacobian generating vectors, shows that forward velocity kinematics is a consequence of the forward kinematics of robots.

**Proof.** Let ${}^0\mathbf{d}_i$ and ${}^0\mathbf{d}_{i-1}$ be the global position vector of the frames $B_i$ and $B_{i-1}$, while ${}^{i-1}\mathbf{d}_i$ is the position vector of the frame $B_i$ in $B_{i-1}$ as shown in Figure 8.3. These three position vectors are related according to a vector addition

$$\begin{aligned} {}^0\mathbf{d}_i &= {}^0\mathbf{d}_{i-1} + {}^0R_{i-1}\,{}^{i-1}\mathbf{d}_i \\ &= {}^0\mathbf{d}_{i-1} + d_i\,{}^0\hat{k}_{i-1} + a_i\,{}^0\hat{\imath}_i. \end{aligned} \tag{8.63}$$

in which we have used the Equation (8.6). Taking a time derivative,

$$\begin{aligned} {}^0\dot{\mathbf{d}}_i &= {}^0\dot{\mathbf{d}}_{i-1} + {}^0\dot{R}_{i-1}\,{}^{i-1}\mathbf{d}_i + {}^0R_{i-1}\,{}^{i-1}\dot{\mathbf{d}}_i \\ &= {}^0\dot{\mathbf{d}}_{i-1} + {}^0\dot{R}_{i-1}\left(d_i\,{}^{i-1}\hat{k}_{i-1} + a_i\,{}^{i-1}\hat{\imath}_i\right) + {}^0R_{i-1}\,\dot{d}_i\,{}^{i-1}\hat{k}_{i-1} \end{aligned} \tag{8.64}$$

shows that the global velocity of the origin of $B_i$ is a function of the translational and angular velocities of link $B_{i-1}$. However,

$$_{i-1}^{0}\dot{\mathbf{d}}_i = {}^0\dot{\mathbf{d}}_i - {}^0\dot{\mathbf{d}}_{i-1} \tag{8.65}$$

$$\begin{aligned} {}^0\dot{R}_{i-1}\,{}^{i-1}\mathbf{d}_i &= {}_0\omega_{i-1} \times {}^0R_{i-1}\,{}^{i-1}\mathbf{d}_i \\ &= {}_0\omega_{i-1} \times {}^0_{i-1}\mathbf{d}_i \\ &= \dot{\theta}_i\,{}^0\hat{k}_{i-1} \times {}^0_{i-1}\mathbf{d}_i \end{aligned} \tag{8.66}$$

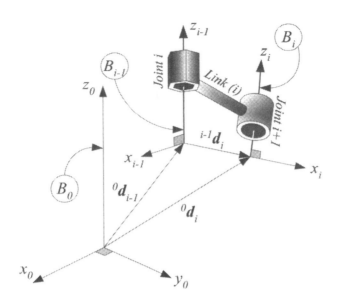

FIGURE 8.3. Link $(i)$ and associated coordinate frames.

and

$$
\begin{aligned}
{}^{0}R_{i-1}\,\dot{d}_{i}\,{}^{i-1}\hat{k}_{i-1} &= \dot{d}_{i}\,{}^{0}R_{i-1}\,{}^{i-1}\hat{k}_{i-1} \\
&= \dot{d}_{i}\,{}^{0}\hat{k}_{i-1}
\end{aligned}
\tag{8.67}
$$

therefore,

$$
{}^{0}_{i-1}\dot{\mathbf{d}}_{i} = \dot{\theta}_{i}\,{}^{0}\hat{k}_{i-1} \times {}^{0}_{i-1}\mathbf{d}_{i} + \dot{d}_{i}\,{}^{0}\hat{k}_{i-1}.
\tag{8.68}
$$

Since at each joint, either $\theta$ or $d$ is variable, we conclude that

$$
{}^{0}_{i-1}\dot{\mathbf{d}}_{i} = {}^{0}_{0}\boldsymbol{\omega}_{i} \times {}^{0}_{i-1}\mathbf{d}_{i} \qquad \text{if joint } i \text{ is R}
\tag{8.69}
$$

or

$$
{}^{0}_{i-1}\dot{\mathbf{d}}_{i} = \dot{d}_{i}\,{}^{0}\hat{k}_{i-1} + {}^{0}_{0}\boldsymbol{\omega}_{i} \times {}^{0}_{i-1}\mathbf{d}_{i} \qquad \text{if joint } i \text{ is P.}
\tag{8.70}
$$

The end-effector velocity is then expressed by

$$
{}^{0}_{0}\dot{\mathbf{d}}_{n} = \sum_{i=1}^{n} {}^{0}_{i-1}\dot{\mathbf{d}}_{i} = \sum_{i=1}^{n} \dot{\theta}_{i}\,{}^{0}\hat{k}_{i-1} \times {}^{0}_{i-1}\mathbf{d}_{n}
\tag{8.71}
$$

and

$$
{}^{0}_{0}\boldsymbol{\omega}_{n} = \sum_{i=1}^{n} {}^{0}_{i-1}\boldsymbol{\omega}_{i} = \sum_{i=1}^{n} \dot{\theta}_{i}\,{}^{0}\hat{k}_{i-1}.
\tag{8.72}
$$

They can be rearranged in a matrix form

$$\begin{bmatrix} {}^0_0\dot{\mathbf{d}}_n \\ {}^0_0\boldsymbol{\omega}_n \end{bmatrix} = \sum_{i=1}^{n} \dot{\theta}_i \begin{bmatrix} {}^0\hat{k}_{i-1} \times {}^0_{i-1}\mathbf{d}_n \\ {}^0\hat{k}_{i-1} \end{bmatrix}$$

$$= \begin{bmatrix} {}^0\hat{k}_0 \times {}^0_0\mathbf{d}_n & {}^0\hat{k}_1 \times {}^0_1\mathbf{d}_n & \cdots & {}^0\hat{k}_{n-1} \times {}^0_{n-1}\mathbf{d}_n \\ {}^0\hat{k}_0 & {}^0\hat{k}_1 & \cdots & {}^0\hat{k}_{n-1} \end{bmatrix} \begin{bmatrix} \dot{\theta}_1 \\ \dot{\theta}_2 \\ \vdots \\ \dot{\theta}_n \end{bmatrix}. \tag{8.73}$$

We usually show this equation by

$$\begin{bmatrix} {}^0_0\dot{\mathbf{d}}_n \\ {}^0_0\boldsymbol{\omega}_n \end{bmatrix} = \mathbf{J}\,\dot{\mathbf{q}} \tag{8.74}$$

where the vector $\dot{\mathbf{q}} = \begin{bmatrix} \dot{q}_1 & \dot{q}_2 & \cdots & \dot{q}_n \end{bmatrix}^T$ is the *joint velocities vector* and $\mathbf{J}$ is the *Jacobian* matrix.

$$\mathbf{J} = \begin{bmatrix} {}^0\hat{k}_0 \times {}^0_0\mathbf{d}_n & {}^0\hat{k}_1 \times {}^0_1\mathbf{d}_n & \cdots & {}^0\hat{k}_{n-1} \times {}^0_{n-1}\mathbf{d}_n \\ {}^0\hat{k}_0 & {}^0\hat{k}_1 & \cdots & {}^0\hat{k}_{n-1} \end{bmatrix} \tag{8.75}$$

Practically, we find the Jacobian matrix column by column. Each column is a *Jacobian generating vector*, $\mathbf{c}_i(\mathbf{q})$, and is associated to joint $i$. If joint $i$ is revolute, then

$$\mathbf{c}_i(\mathbf{q}) = \begin{bmatrix} {}^0\hat{k}_{i-1} \times {}^0_{i-1}\mathbf{d}_n \\ {}^0\hat{k}_{i-1} \end{bmatrix} \tag{8.76}$$

and if joint $i$ is prismatic then, $\mathbf{c}_i(\mathbf{q})$ simplifies to

$$\mathbf{c}_i(\mathbf{q}) = \begin{bmatrix} {}^0\hat{k}_{i-1} \\ 0 \end{bmatrix}. \tag{8.77}$$

Equation (8.35) provides a set of six equations, the first one the translational velocity of the end-effector along the $x_0$-axis. The rest of the equations are for translational and rotational velocities of the end-effector frame along and about the base frame axes. ∎

**Example 214** *Jacobian matrix for a spherical manipulator.*
    *Figure 8.4 depicts a spherical manipulator. To find its Jacobian, we start with determining the ${}^0\hat{k}_{i-1}$ axes for $i = 1, 2, 3$. It would be easier if we use the homogeneous definitions and write,*

$$ {}^0\hat{k}_0 = \begin{bmatrix} 0 \\ 0 \\ 1 \\ 0 \end{bmatrix} \tag{8.78}$$

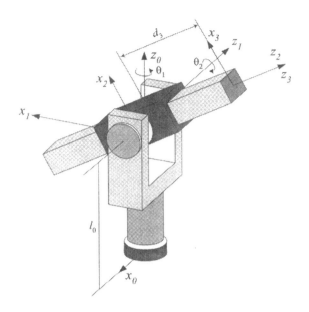

FIGURE 8.4. A spherical manipulator.

$$
{}^0\hat{k}_1 \;=\; {}^0T_1\,{}^1\hat{k}_1 \;=\;
\begin{bmatrix}
\cos\theta_1 & 0 & -\sin\theta_1 & 0 \\
\sin\theta_1 & 0 & \cos\theta_1 & 0 \\
0 & -1 & 0 & l_0 \\
0 & 0 & 0 & 1
\end{bmatrix}
\begin{bmatrix}
0 \\ 0 \\ 1 \\ 0
\end{bmatrix}
$$

$$
=\;
\begin{bmatrix}
-\sin\theta_1 \\
\cos\theta_1 \\
0 \\
0
\end{bmatrix}
\tag{8.79}
$$

$$
{}^0\hat{k}_2 \;=\; {}^0T_2\,{}^2\hat{k}_2 \;=\;
\begin{bmatrix}
c\theta_1 c\theta_2 & -s\theta_1 & c\theta_1 s\theta_2 & 0 \\
c\theta_2 s\theta_1 & c\theta_1 & s\theta_1 s\theta_2 & 0 \\
-s\theta_2 & 0 & c\theta_2 & l_0 \\
0 & 0 & 0 & 1
\end{bmatrix}
\begin{bmatrix}
0 \\ 0 \\ 1 \\ 0
\end{bmatrix}
$$

$$
=\;
\begin{bmatrix}
\cos\theta_1 \sin\theta_2 \\
\sin\theta_1 \sin\theta_2 \\
\cos\theta_2 \\
0
\end{bmatrix}.
\tag{8.80}
$$

*Then, the vectors ${}^0_{i-1}\mathbf{d}_n$ must be evaluated.*

$$
{}^0_0\mathbf{d}_2 = l_0\,{}^0\hat{k}_0 + d_3\,{}^0\hat{k}_2 =
\begin{bmatrix}
d_3 \cos\theta_1 \sin\theta_2 \\
d_3 \sin\theta_1 \sin\theta_2 \\
l_0 + d_3 \cos\theta_2 \\
0
\end{bmatrix}
\tag{8.81}
$$

$$
{}^0_1\mathbf{d}_2 = d_3\,{}^0\hat{k}_2 = \begin{bmatrix} d_3\cos\theta_1\sin\theta_2 \\ d_3\sin\theta_1\sin\theta_2 \\ d_3\cos\theta_2 \\ 0 \end{bmatrix}
\tag{8.82}
$$

Therefore, the Jacobian of the manipulator is

$$
\mathbf{J} = \begin{bmatrix} {}^0\hat{k}_0 \times {}^0_0\mathbf{d}_2 & {}^0\hat{k}_1 \times {}^0_1\mathbf{d}_2 & {}^0\hat{k}_2 \\ {}^0\hat{k}_0 & {}^0\hat{k}_1 & 0 \end{bmatrix}
$$

$$
= \begin{bmatrix} -d_3\sin\theta_1\sin\theta_2 & d_3\cos\theta_1\cos\theta_2 & \cos\theta_1\sin\theta_2 \\ d_3\cos\theta_1\sin\theta_2 & d_3\cos\theta_2\sin\theta_1 & \sin\theta_1\sin\theta_2 \\ 0 & -d_3\sin\theta_2 & \cos\theta_2 \\ 0 & -\sin\theta_1 & 0 \\ 0 & \cos\theta_1 & 0 \\ 1 & 0 & 0 \end{bmatrix}.
\tag{8.83}
$$

**Example 215** *Jacobian matrix for an articulated robot.*

The Jacobian matrix of a 6 DOF articulated robot is a $6 \times 6$ matrix. The robot was shown in Figure 6.1 and its transformation matrices are calculated in Example 163.

The $i$th column of the Jacobian, $\mathbf{c}_i(\mathbf{q})$, is

$$
\mathbf{c}_i(\mathbf{q}) = \begin{bmatrix} {}^0\hat{k}_{i-1} \times {}^0_{i-1}\mathbf{d}_6 \\ {}^0\hat{k}_{i-1} \end{bmatrix}.
\tag{8.84}
$$

For the first column of the Jacobian matrix, we need to find ${}^0\hat{k}_0$ and ${}^0\mathbf{d}_6$. The direction of the $z_0$-axis in the base coordinate frame is

$$
{}^0\hat{k}_0 = \begin{bmatrix} 0 \\ 0 \\ 1 \end{bmatrix}
\tag{8.85}
$$

and the position vector of the end-effector frame $B_6$ is ${}^0\mathbf{d}_6$, which can be directly determined from the fourth column of the transformation matrix ${}^0T_6$

$$
{}^0T_6 = {}^0T_1\,{}^1T_2\,{}^2T_3\,{}^3T_4\,{}^4T_5\,{}^5T_6
$$

$$
= \begin{bmatrix} {}^0R_6 & {}^0\mathbf{d}_6 \\ 0 & 1 \end{bmatrix} = \begin{bmatrix} t_{11} & t_{12} & t_{13} & t_{14} \\ t_{21} & t_{22} & t_{23} & t_{24} \\ t_{31} & t_{32} & t_{33} & t_{34} \\ 0 & 0 & 0 & 1 \end{bmatrix}
\tag{8.86}
$$

which is

$$
{}^0\mathbf{d}_6 = \begin{bmatrix} t_{14} \\ t_{24} \\ t_{34} \end{bmatrix}
\tag{8.87}
$$

*where,*

$$
\begin{aligned}
t_{14} &= d_6 \left( s\theta_1 s\theta_4 s\theta_5 + c\theta_1 \left( c\theta_4 s\theta_5 c \left( \theta_2 + \theta_3 \right) + c\theta_5 s \left( \theta_2 + \theta_3 \right) \right) \right) \\
&\quad + l_3 c\theta_1 s \left( \theta_2 + \theta_3 \right) + d_2 s\theta_1 + l_2 c\theta_1 c\theta_2 \quad (8.88)
\end{aligned}
$$

$$
\begin{aligned}
t_{24} &= d_6 \left( -c\theta_1 s\theta_4 s\theta_5 + s\theta_1 \left( c\theta_4 s\theta_5 c \left( \theta_2 + \theta_3 \right) + c\theta_5 s \left( \theta_2 + \theta_3 \right) \right) \right) \\
&\quad + s\theta_1 s \left( \theta_2 + \theta_3 \right) l_3 - d_2 c\theta_1 + l_2 c\theta_2 s\theta_1 \quad (8.89)
\end{aligned}
$$

$$
\begin{aligned}
t_{34} &= d_6 \left( c\theta_4 s\theta_5 s \left( \theta_2 + \theta_3 \right) - c\theta_5 c \left( \theta_2 + \theta_3 \right) \right) \\
&\quad + l_2 s\theta_2 + l_3 c \left( \theta_2 + \theta_3 \right). \quad (8.90)
\end{aligned}
$$

*Therefore,*

$$
{}^0\hat{k}_0 \times {}^0\mathbf{d}_6 = \begin{bmatrix} 0 \\ 0 \\ 1 \end{bmatrix} \times \begin{bmatrix} t_{14} \\ t_{24} \\ t_{34} \end{bmatrix} = \begin{bmatrix} -t_{24} \\ t_{14} \\ 0 \end{bmatrix} \quad (8.91)
$$

*and the first Jacobian generating vector is*

$$
\mathbf{c}_1 = \begin{bmatrix} -t_{24} \\ t_{14} \\ 0 \\ 0 \\ 0 \\ 1 \end{bmatrix}. \quad (8.92)
$$

*For the 2nd column we need to find ${}^0\hat{k}_1$ and ${}^0_1\mathbf{d}_6$. The $z_1$-axis in the base frame can be found by*

$$
\begin{aligned}
{}^0\hat{k}_1 &= {}^0R_1\, {}^1\hat{k}_1 = {}^0R_1 \begin{bmatrix} 0 \\ 0 \\ 1 \end{bmatrix} \\
&= \begin{bmatrix} c\theta_1 & 0 & s\theta_1 \\ s\theta_1 & 0 & -c\theta_1 \\ 0 & 1 & 0 \end{bmatrix} \begin{bmatrix} 0 \\ 0 \\ 1 \end{bmatrix} = \begin{bmatrix} \sin\theta_1 \\ -\cos\theta_1 \\ 0 \end{bmatrix}. \quad (8.93)
\end{aligned}
$$

*The first half of $\mathbf{c}_2$ is ${}^0\hat{k}_1 \times {}^0_1\mathbf{d}_6$. The vector ${}^0_1\mathbf{d}_6$ is the position of the end-effector in the coordinate frame $B_1$, however it must be expressed in the base frame to be able to perform the cross product. An easier method is to find ${}^1\hat{k}_1 \times {}^1\mathbf{d}_6$ and transform the resultant into the base frame. The vector ${}^1\mathbf{d}_6$ is the fourth column of ${}^1T_6 = {}^1T_2\,{}^2T_3\,{}^3T_4\,{}^4T_5\,{}^5T_6$, which, from Example 163, is equal to*

$$
{}^1\mathbf{d}_6 = \begin{bmatrix} l_2 \cos\theta_2 + l_3 \sin \left( \theta_2 + \theta_3 \right) \\ l_2 \sin\theta_2 - l_3 \cos \left( \theta_2 + \theta_3 \right) \\ d_2 \end{bmatrix}. \quad (8.94)
$$

*Therefore, the first half of* $c_2$ *is*

$$
\begin{aligned}
{}^0\hat{k}_1 \times {}^0_1d_6 &= {}^0R_1 \left( {}^1\hat{k}_1 \times {}^1d_6 \right) \\
&= {}^0R_1 \left( \begin{bmatrix} 0 \\ 0 \\ 1 \end{bmatrix} \times \begin{bmatrix} l_2 \cos \theta_2 + l_3 \sin (\theta_2 + \theta_3) \\ l_2 \sin \theta_2 - l_3 \cos (\theta_2 + \theta_3) \\ d_2 \end{bmatrix} \right) \\
&= \begin{bmatrix} \cos \theta_1 \left( -l_2 \sin \theta_2 + l_3 \cos (\theta_2 + \theta_3) \right) \\ \sin \theta_1 \left( -l_2 \sin \theta_2 + l_3 \cos (\theta_2 + \theta_3) \right) \\ l_2 \cos \theta_2 + l_3 \sin (\theta_2 + \theta_3) \end{bmatrix} \quad (8.95)
\end{aligned}
$$

*and* $c_2$ *is*

$$
c_2 = \begin{bmatrix} \cos \theta_1 \left( -l_2 \sin \theta_2 + l_3 \cos (\theta_2 + \theta_3) \right) \\ \sin \theta_1 \left( -l_2 \sin \theta_2 + l_3 \cos (\theta_2 + \theta_3) \right) \\ l_2 \cos \theta_2 + l_3 \sin (\theta_2 + \theta_3) \\ \sin \theta_1 \\ -\cos \theta_1 \\ 0 \end{bmatrix} . \quad (8.96)
$$

*The 3rd column is made by* ${}^0\hat{k}_2$ *and* ${}^0_2d_6$. *The vector* ${}^0_2d_6$ *is the position of the end-effector in the coordinate frame* $B_2$ *and is the fourth column of* ${}^2T_6 = {}^2T_3 \, {}^3T_4 \, {}^4T_5 \, {}^5T_6$. *The* $z_2$-*axis in the base frame can be found by*

$$
{}^0\hat{k}_2 = {}^0R_2 \, {}^2\hat{k}_2 = {}^0R_1 \, {}^1R_2 \begin{bmatrix} 0 \\ 0 \\ 1 \end{bmatrix} = \begin{bmatrix} \sin \theta_1 \\ -\cos \theta_1 \\ 0 \end{bmatrix} \quad (8.97)
$$

*and the cross product* ${}^0\hat{k}_2 \times {}^0_2d_6$ *can be found by transforming the resultant of* ${}^2\hat{k}_2 \times {}^2d_6$ *into the base coordinate frame.*

$$
{}^2\hat{k}_2 \times {}^2d_6 = \begin{bmatrix} l_3 \cos \theta_3 \\ l_3 \sin \theta_3 \\ 0 \end{bmatrix} \quad (8.98)
$$

$$
{}^0R_2 \left( {}^2\hat{k}_2 \times {}^2d_6 \right) = \begin{bmatrix} l_3 \cos \theta_1 \sin (\theta_2 + \theta_3) \\ l_3 \sin \theta_1 \sin (\theta_2 + \theta_3) \\ -l_3 \cos (\theta_2 + \theta_3) \end{bmatrix} \quad (8.99)
$$

*Therefore,* $c_3$ *is*

$$
c_3 = \begin{bmatrix} l_3 \cos \theta_1 \sin (\theta_2 + \theta_3) \\ l_3 \sin \theta_1 \sin (\theta_2 + \theta_3) \\ -l_3 \cos (\theta_2 + \theta_3) \\ \sin \theta_1 \\ -\cos \theta_1 \\ 0 \end{bmatrix} . \quad (8.100)
$$

*The 4th column needs $^0\hat{k}_3$ and $^0_3d_6$. The vector $^0\hat{k}_3$ can be found by transforming $^3\hat{k}_3$ to the base frame*

$$^0\hat{k}_3 = {}^0R_1\,{}^1R_2\,{}^2R_3 \begin{bmatrix} 0 \\ 0 \\ 1 \end{bmatrix} = \begin{bmatrix} \cos\theta_1\,(\cos\theta_2\sin\theta_3 + \cos\theta_3\sin\theta_2) \\ \sin\theta_1\,(\cos\theta_2\sin\theta_3 + \sin\theta_2\cos\theta_3) \\ -\cos\,(\theta_2 + \theta_3) \end{bmatrix}$$
(8.101)

*and the first half of $\mathbf{J}_4$ can be found by calculating $^3\hat{k}_3 \times {}^3d_6$ and transforming the resultant into the base coordinate frame.*

$$^0R_3\left({}^3\hat{k}_3 \times {}^3d_6\right) = {}^0R_3\left(\begin{bmatrix} 0 \\ 0 \\ 1 \end{bmatrix} \times \begin{bmatrix} 0 \\ 0 \\ l_3 \end{bmatrix}\right) = \begin{bmatrix} 0 \\ 0 \\ 0 \end{bmatrix}$$
(8.102)

*Therefore, $\mathbf{c}_4$ is*

$$\mathbf{c}_4 = \begin{bmatrix} 0 \\ 0 \\ 0 \\ \cos\theta_1\,(\cos\theta_2\sin\theta_3 + \cos\theta_3\sin\theta_2) \\ \sin\theta_1\,(\cos\theta_2\sin\theta_3 + \sin\theta_2\cos\theta_3) \\ -\cos\,(\theta_2 + \theta_3) \end{bmatrix}.$$
(8.103)

*The 5th column needs $^0\hat{k}_4$ and $^0_4d_6$. We can find the vector $^0\hat{k}_4$ by transforming $^4\hat{k}_4$ to the base frame*

$$^0\hat{k}_4 = {}^0R_4 \begin{bmatrix} 0 \\ 0 \\ 1 \end{bmatrix} = \begin{bmatrix} c\theta_4 s\theta_1 - c\theta_1 s\theta_4 c\,(\theta_2 + \theta_3) \\ -c\theta_1 c\theta_4 - s\theta_1 s\theta_4 c\,(\theta_2 + \theta_3) \\ -s\theta_4 s\,(\theta_2 + \theta_3) \end{bmatrix}.$$
(8.104)

*The first half of $\mathbf{c}_5$ is $^4\hat{k}_4 \times {}^4d_6$, expressed in the base coordinate frame.*

$$^0R_4\left({}^4\hat{k}_4 \times {}^4d_6\right) = {}^0R_4\left(\begin{bmatrix} 0 \\ 0 \\ 1 \end{bmatrix} \times \begin{bmatrix} 0 \\ 0 \\ 0 \end{bmatrix}\right) = \begin{bmatrix} 0 \\ 0 \\ 0 \end{bmatrix}$$
(8.105)

*Therefore, $\mathbf{c}_5$ is*

$$\mathbf{c}_5 = \begin{bmatrix} 0 \\ 0 \\ 0 \\ \cos\theta_4\sin\theta_1 - \cos\theta_1\sin\theta_4\cos\,(\theta_2 + \theta_3) \\ -\cos\theta_1\cos\theta_4 - \sin\theta_1\sin\theta_4\cos\,(\theta_2 + \theta_3) \\ -\sin\theta_4\sin\,(\theta_2 + \theta_3) \end{bmatrix}.$$
(8.106)

*The 6th column is found by calculating $^0\hat{k}_5$ and $^0\hat{k}_5 \times {}^0_5d_6$. The vector*

$^0\hat{k}_5$ is

$$
^0\hat{k}_5 \;=\; ^0R_5 \begin{bmatrix} 0 \\ 0 \\ 1 \end{bmatrix} \tag{8.107}
$$

$$
\;=\; \begin{bmatrix} -c\theta_1 c\theta_4 s\left(\theta_2+\theta_3\right) - s\theta_4\left(s\theta_1 s\theta_4 + c\theta_1 c\theta_4 c\left(\theta_2+\theta_3\right)\right) \\ -s\theta_1 c\theta_4 s\left(\theta_2+\theta_3\right) - s\theta_4\left(-c\theta_1 s\theta_4 + s\theta_1 c\theta_4 c\left(\theta_2+\theta_3\right)\right) \\ c\theta_4 c\left(\theta_2+\theta_3\right) - \frac{1}{2}s\left(\theta_2+\theta_3\right)s2\theta_4 \end{bmatrix}
$$

and the first half of $\mathbf{c}_6$ is $^5\hat{k}_5 \times\, ^5\mathbf{d}_6$, expressed in the base coordinate frame.

$$
^0R_5\left(^5\hat{k}_5 \times\, ^5\mathbf{d}_6\right) = \,^0R_5\left(\begin{bmatrix} 0 \\ 0 \\ 1 \end{bmatrix} \times \begin{bmatrix} 0 \\ 0 \\ 0 \end{bmatrix}\right) = \begin{bmatrix} 0 \\ 0 \\ 0 \end{bmatrix} \tag{8.108}
$$

Therefore, $\mathbf{c}_6$ is

$$
\mathbf{c}_6 = \begin{bmatrix} 0 \\ 0 \\ 0 \\ -c\theta_1 c\theta_4 s\left(\theta_2+\theta_3\right) - s\theta_4\left(s\theta_1 s\theta_4 + c\theta_1 c\theta_4 c\left(\theta_2+\theta_3\right)\right) \\ -s\theta_1 c\theta_4 s\left(\theta_2+\theta_3\right) - s\theta_4\left(-c\theta_1 s\theta_4 + s\theta_1 c\theta_4 c\left(\theta_2+\theta_3\right)\right) \\ c\theta_4 c\left(\theta_2+\theta_3\right) - \frac{1}{2}s\left(\theta_2+\theta_3\right)s2\theta_4 \end{bmatrix}
$$
$$\tag{8.109}$$

and the Jacobian matrix for the articulated robot is calculated.

$$
\mathbf{J} = \begin{bmatrix} \mathbf{c}_1 & \mathbf{c}_2 & \mathbf{c}_3 & \mathbf{c}_4 & \mathbf{c}_5 & \mathbf{c}_6 \end{bmatrix} \tag{8.110}
$$

**Example 216** *The effect of a spherical wrist on Jacobian matrix.*
    *The Jacobian matrix for a robot having a spherical wrist is always of the form*

$$
\mathbf{J} = \begin{bmatrix} ^0\hat{k}_0 \times\, ^0_0\mathbf{d}_6 & ^0\hat{k}_1 \times\, ^0_1\mathbf{d}_6 & ^0\hat{k}_2 \times\, ^0_2\mathbf{d}_6 & 0 & 0 & 0 \\ ^0\hat{k}_0 & ^0\hat{k}_1 & ^0\hat{k}_2 & ^0\hat{k}_3 & ^0\hat{k}_6 & ^0\hat{k}_5 \end{bmatrix} \tag{8.111}
$$

*which shows the upper $3 \times 3$ submatrix is zero. This is because of the spherical wrist structure and having a wrist point as the origin of the wrist coordinate frames $B_4$, $B_5$, and $B_6$.*

**Example 217** ★ *Jacobian matrix for an articulated robot using the direct differentiating method.*
    *Figure 6.1 illustrates an articulated robot with transformation matrices given in Example 163. Using the result of the forward kinematics*

$$
^0T_6 = \begin{bmatrix} ^0R_6 & ^0\mathbf{d}_6 \\ 0 & 1 \end{bmatrix} \tag{8.112}
$$

*we know that the position of the end-effector is at*

$$^0\mathbf{d}_6 = \begin{bmatrix} X_6 \\ Y_6 \\ Z_6 \end{bmatrix} = \begin{bmatrix} t_{14} \\ t_{24} \\ t_{34} \end{bmatrix} \qquad (8.113)$$

*where,*

$$
\begin{aligned}
t_{14} &= d_6 \left( s\theta_1 s\theta_4 s\theta_5 + c\theta_1 \left( c\theta_4 s\theta_5 c \left( \theta_2 + \theta_3 \right) + c\theta_5 s \left( \theta_2 + \theta_3 \right) \right) \right) \\
&\quad + l_3 c\theta_1 s \left( \theta_2 + \theta_3 \right) + d_2 s\theta_1 + l_2 c\theta_1 c\theta_2 & (8.114) \\
t_{24} &= d_6 \left( -c\theta_1 s\theta_4 s\theta_5 + s\theta_1 \left( c\theta_4 s\theta_5 c \left( \theta_2 + \theta_3 \right) + c\theta_5 s \left( \theta_2 + \theta_3 \right) \right) \right) \\
&\quad + s\theta_1 s \left( \theta_2 + \theta_3 \right) l_3 - d_2 c\theta_1 + l_2 c\theta_2 s\theta_1 & (8.115) \\
t_{34} &= d_6 \left( c\theta_4 s\theta_5 s \left( \theta_2 + \theta_3 \right) - c\theta_5 c \left( \theta_2 + \theta_3 \right) \right) \\
&\quad + l_2 s\theta_2 + l_3 c \left( \theta_2 + \theta_3 \right). & (8.116)
\end{aligned}
$$

*Taking the derivative of $X_6$ yields*

$$
\begin{aligned}
\dot{X}_6 &= \frac{\partial X_6}{\partial \theta_1} \dot{\theta}_1 + \frac{\partial X_6}{\partial \theta_2} \dot{\theta}_2 + \cdots + \frac{\partial X_6}{\partial \theta_6} \dot{\theta}_6 \\
&= J_{11} \dot{\theta}_1 + J_{12} \dot{\theta}_2 + \cdots + J_{16} \dot{\theta}_6 \\
&= -t_{24} \dot{\theta}_1 + c\theta_1 \left( -l_2 s\theta_2 + l_3 c \left( \theta_2 + \theta_3 \right) \right) \dot{\theta}_2 \\
&\quad + l_3 c\theta_1 s \left( \theta_2 + \theta_3 \right) \dot{\theta}_3 & (8.117)
\end{aligned}
$$

*that shows*

$$
\begin{aligned}
J_{11} &= -t_{24} \\
J_{12} &= \cos \theta_1 \left( -l_2 \sin \theta_2 + l_3 \cos \left( \theta_2 + \theta_3 \right) \right) \\
J_{13} &= l_3 \cos \theta_1 \sin \left( \theta_2 + \theta_3 \right) \\
J_{14} &= 0 \\
J_{15} &= 0 \\
J_{16} &= 0. & (8.118)
\end{aligned}
$$

*Similarly, the derivative of $Y_6$ and $Z_6$*

$$
\begin{aligned}
\dot{Y}_6 &= \frac{\partial Y_6}{\partial \theta_1} \dot{\theta}_1 + \frac{\partial Y_6}{\partial \theta_2} \dot{\theta}_2 + \cdots + \frac{\partial Y_6}{\partial \theta_6} \dot{\theta}_6 \\
&= J_{21} \dot{\theta}_1 + J_{22} \dot{\theta}_2 + \cdots + J_{26} \dot{\theta}_6 \\
&= t_{14} \dot{\theta}_1 + s\theta_1 \left( -l_2 s\theta_2 + l_3 c \left( \theta_2 + \theta_3 \right) \right) \dot{\theta}_2 \\
&\quad + l_3 s\theta_1 s \left( \theta_2 + \theta_3 \right) \dot{\theta}_3 & (8.119)
\end{aligned}
$$

$$
\begin{aligned}
\dot{Z}_6 &= \frac{\partial Z_6}{\partial \theta_1} \dot{\theta}_1 + \frac{\partial Z_6}{\partial \theta_2} \dot{\theta}_2 + \cdots + \frac{\partial Z_6}{\partial \theta_6} \dot{\theta}_6 \\
&= J_{31} \dot{\theta}_1 + J_{32} \dot{\theta}_2 + \cdots + J_{36} \dot{\theta}_6 \\
&= \left( l_2 \cos \theta_2 + l_3 \sin \left( \theta_2 + \theta_3 \right) \right) \dot{\theta}_2 - l_3 \cos \left( \theta_2 + \theta_3 \right) \dot{\theta}_3 & (8.120)
\end{aligned}
$$

*show that*

$$\begin{aligned}
J_{21} &= t_{14} \\
J_{22} &= \sin\theta_1 \left(-l_2 \sin\theta_2 + l_3 \cos(\theta_2 + \theta_3)\right) \\
J_{23} &= l_3 \sin\theta_1 \sin(\theta_2 + \theta_3) \\
J_{24} &= 0 \\
J_{25} &= 0 \\
J_{26} &= 0
\end{aligned}$$ (8.121)

$$\begin{aligned}
J_{31} &= 0 \\
J_{32} &= l_2 \cos\theta_2 + l_3 \sin(\theta_2 + \theta_3) \\
J_{33} &= -l_3 \cos(\theta_2 + \theta_3) \\
J_{34} &= 0 \\
J_{35} &= 0 \\
J_{36} &= 0.
\end{aligned}$$ (8.122)

Since there is no explicit equation for describing the rotations of the end-effector's frame about the axes, there is no equation to find differential rotations about the three axes. This is a reason for searching indirect or more systematic methods for evaluating the Jacobian matrix. However, the next three rows of the Jacobian matrix can be found by calculating the angular velocity vector based on the angular velocity matrix

$$_0\tilde{\omega}_6 = {}^0\dot{R}_0\,{}^0R_6^T = \begin{bmatrix} 0 & -\omega_Z & \omega_Y \\ \omega_Z & 0 & -\omega_X \\ -\omega_Y & \omega_X & 0 \end{bmatrix}$$ (8.123)

$$_0\boldsymbol{\omega}_6 = \begin{bmatrix} \omega_X \\ \omega_Y \\ \omega_Z \end{bmatrix}$$ (8.124)

and then rearranging the components to show the Jacobian elements.

$$\omega_X = \frac{\partial\omega_X}{\partial\theta_1}\dot{\theta}_1 + \frac{\partial\omega_X}{\partial\theta_2}\dot{\theta}_2 + \cdots + \frac{\partial\omega_X}{\partial\theta_6}\dot{\theta}_6$$ (8.125)

$$\omega_Y = \frac{\partial\omega_Y}{\partial\theta_1}\dot{\theta}_1 + \frac{\partial\omega_Y}{\partial\theta_2}\dot{\theta}_2 + \cdots + \frac{\partial\omega_Y}{\partial\theta_6}\dot{\theta}_6$$ (8.126)

$$\omega_Z = \frac{\partial\omega_Z}{\partial\theta_1}\dot{\theta}_1 + \frac{\partial\omega_Z}{\partial\theta_2}\dot{\theta}_2 + \cdots + \frac{\partial\omega_Z}{\partial\theta_6}\dot{\theta}_6$$ (8.127)

The angular velocity vector of the end-effector frame is

$$\begin{aligned}
\omega_X =\ & \sin\theta_1\dot{\theta}_2 + \sin\theta_1\dot{\theta}_3 + \cos\theta_1 \sin\theta_{23}\dot{\theta}_4 \\
& + \left(\cos\theta_4 \sin\theta_1 - \cos\theta_1 \sin\theta_4 \cos\theta_{23}\right)\dot{\theta}_5 \\
& - \left(c\theta_1 c\theta_4 s\theta_{23} + s\theta_4\left(s\theta_1 s\theta_4 + c\theta_1 c\theta_4 c\theta_{23}\right)\right)\dot{\theta}_6
\end{aligned}$$ (8.128)

$$\omega_Y = -\cos\theta_1\dot\theta_2 - \cos\theta_1\dot\theta_3 + \sin\theta_1\sin\theta_{23}\dot\theta_4$$
$$+ (-\cos\theta_1\cos\theta_4 - \sin\theta_1\sin\theta_4\cos\theta_{23})\dot\theta_5$$
$$+ (-s\theta_1 c\theta_4 s\theta_{23} - s\theta_4(-c\theta_1 s\theta_4 + s\theta_1 c\theta_4 c\theta_{23}))\dot\theta_6 \quad (8.129)$$

$$\omega_Z = \dot\theta_1 - \cos(\theta_2+\theta_3)\dot\theta_4 - \sin\theta_4\sin(\theta_2+\theta_3)\dot\theta_5$$
$$+ \left(\cos\theta_4\cos\theta_{23} - \frac{1}{2}\sin\theta_{23}\sin 2\theta_4\right)\dot\theta_6 \quad (8.130)$$

and therefore,

$$
\begin{aligned}
J_{41} &= 0 \\
J_{42} &= \sin\theta_1 \\
J_{43} &= \sin\theta_1 \\
J_{44} &= \cos\theta_1(\cos\theta_2\sin\theta_3 + \cos\theta_3\sin\theta_2) \\
J_{45} &= \cos\theta_4\sin\theta_1 - \cos\theta_1\sin\theta_4\cos(\theta_2+\theta_3) \\
J_{46} &= -c\theta_1 c\theta_4 s(\theta_2+\theta_3) \\
&\quad -s\theta_4(s\theta_1 s\theta_4 + c\theta_1 c\theta_4 c(\theta_2+\theta_3)) \quad (8.131)
\end{aligned}
$$

$$
\begin{aligned}
J_{51} &= 0 \\
J_{52} &= -\cos\theta_1 \\
J_{53} &= -\cos\theta_1 \\
J_{54} &= \sin\theta_1(\cos\theta_2\sin\theta_3 + \sin\theta_2\cos\theta_3) \\
J_{55} &= -\cos\theta_1\cos\theta_4 - \sin\theta_1\sin\theta_4\cos(\theta_2+\theta_3) \\
J_{56} &= -s\theta_1 c\theta_4 s(\theta_2+\theta_3) \\
&\quad -s\theta_4(-c\theta_1 s\theta_4 + s\theta_1 c\theta_4 c(\theta_2+\theta_3)) \quad (8.132)
\end{aligned}
$$

$$
\begin{aligned}
J_{61} &= 1 \\
J_{62} &= 0 \\
J_{63} &= 0 \\
J_{64} &= -\cos(\theta_2+\theta_3) \\
J_{65} &= -\sin\theta_4\sin(\theta_2+\theta_3) \\
J_{66} &= \cos\theta_4\cos(\theta_2+\theta_3) - \frac{1}{2}\sin(\theta_2+\theta_3)\sin 2\theta_4. \quad (8.133)
\end{aligned}
$$

**Example 218 ★** *Analytical Jacobian and geometrical Jacobian.*
   *Assume the global position and orientation of the end-effector frames are specified by a set of six parameters*

$$\mathbf{X} = \begin{bmatrix} {}^0\mathbf{r}_n \\ {}^0\phi_n \end{bmatrix} \quad (8.134)$$

*where*

$$^0\boldsymbol{\phi}_n = {}^0\boldsymbol{\phi}_n(\mathbf{q}) \tag{8.135}$$

*are three independent rotational parameters, and*

$$^0\mathbf{r}_n = {}^0\mathbf{r}_n(\mathbf{q}) \tag{8.136}$$

*is the Cartesian position of the end-effector frame, both functions of the joint variable vector, q.*

*The translational velocity of the end-effector frame can be expressed by*

$$^0\dot{\mathbf{r}}_n = \frac{\partial \mathbf{r}}{\partial \mathbf{q}}\dot{\mathbf{q}} = \mathbf{J}_D(\mathbf{q})\,\dot{\mathbf{q}} \tag{8.137}$$

*and the rotational velocity of the end-effector frame can be expressed by*

$$^0\dot{\boldsymbol{\phi}}_n = \frac{\partial \boldsymbol{\phi}}{\partial \mathbf{q}}\dot{\mathbf{q}} = \mathbf{J}_\phi(\mathbf{q})\,\dot{\mathbf{q}}. \tag{8.138}$$

*The rotational velocity vector $\dot{\boldsymbol{\phi}}$ in general differs from the angular velocity vector $\boldsymbol{\omega}$. The combination of the Jacobian matrices $\mathbf{J}_D$ and $\mathbf{J}_\phi$ in the form of*

$$\mathbf{J}_A = \begin{bmatrix} \mathbf{J}_D \\ \mathbf{J}_\phi \end{bmatrix} \tag{8.139}$$

*is called **analytical Jacobian** to indicate its difference with **geometrical Jacobian J.***

*Having a set of orientation angles, $\boldsymbol{\phi}$, it is possible to find the relationship between the angular velocity $\boldsymbol{\omega}$ and the rotational velocity $\dot{\boldsymbol{\phi}}$. As an example, consider the Euler angles $\varphi\theta\psi$ about $zxz$ axes defined in Section 6.3. The global angular velocity, in terms of Euler frequencies, is found in 2.81*

$$\begin{bmatrix} \omega_X \\ \omega_Y \\ \omega_Z \end{bmatrix} = \begin{bmatrix} 0 & \cos\varphi & \sin\theta\sin\varphi \\ 0 & \sin\varphi & -\cos\varphi\sin\theta \\ 1 & 0 & \cos\theta \end{bmatrix} \begin{bmatrix} \dot{\varphi} \\ \dot{\theta} \\ \dot{\psi} \end{bmatrix} \tag{8.140}$$

$$\boldsymbol{\omega} = T(\boldsymbol{\phi})\,\dot{\boldsymbol{\phi}}. \tag{8.141}$$

## 8.4   Inverse Velocity Kinematics

The *inverse velocity kinematics problem*, also known as the *resolved rates problem*, is searching for the joint velocity vector corresponding to the end-effector velocity vector. Six DOF are needed to be able to move the end-effector in an arbitrary direction with an arbitrary angular velocity. The velocity vector of the end-effector $\dot{\mathbf{X}}$ is related to the joint velocity vector $\dot{\mathbf{q}}$ by the Jacobian matrix

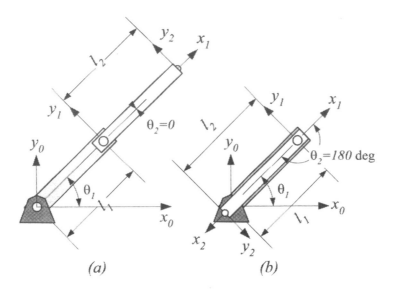

FIGURE 8.5. Singular configurations of a 2R planar manipulator.

$$\dot{\mathbf{X}} = \begin{bmatrix} {}^0\mathbf{v}_n \\ {}^0\omega_n \end{bmatrix} = \begin{bmatrix} \mathbf{J}_D \\ \mathbf{J}_R \end{bmatrix} \dot{\mathbf{q}} = \mathbf{J}\,\dot{\mathbf{q}}. \tag{8.142}$$

Consequently, for the inverse velocity kinematics, we require the differential change in joint coordinates expressed in terms of the Cartesian translation and angular velocities of the end-effector. If the Jacobian matrix is non-singular at the moment of calculation, the inverse Jacobian $\mathbf{J}^{-1}$ exists and we are able to find the required joint velocity vector as

$$\dot{\mathbf{q}} = \mathbf{J}^{-1}\dot{\mathbf{X}}. \tag{8.143}$$

Singular configuration is where the determinant of the Jacobian matrix is zero and therefore, $\mathbf{J}^{-1}$ is indeterminate. The Equation (8.143) determines the velocity required at the individual joints to produce a desired end-effector velocity $\dot{\mathbf{X}}$.

Since the inverse velocity kinematics is a consequence of the forward velocity and needs a matrix inversion, the problem is equivalent to the solution of a set of linear algebraic equations. To find $\mathbf{J}^{-1}$, every matrix inversion method may be applied.

**Example 219** *Inverse velocity of a 2R planar manipulator.*

*Forward and inverse kinematics of a 2R planar manipulator have been analyzed in Example 132 and Example 168. Its Jacobian and forward ve-*

*locity kinematics are also found in Example 212 as*

$$\dot{\mathbf{X}} = \mathbf{J}\dot{\mathbf{q}} \qquad (8.144)$$

$$\begin{bmatrix} \dot{X} \\ \dot{Y} \end{bmatrix} = \begin{bmatrix} -l_1 s\theta_1 - l_2 s(\theta_1 + \theta_2) & -l_2 s(\theta_1 + \theta_2) \\ l_1 c\theta_1 + l_2 c(\theta_1 + \theta_2) & l_2 c(\theta_1 + \theta_2) \end{bmatrix} \begin{bmatrix} \dot{\theta}_1 \\ \dot{\theta}_2 \end{bmatrix}. \qquad (8.145)$$

*The inverse velocity kinematics needs to find the inverse of the Jacobian. Therefore,*

$$\dot{\mathbf{q}} = \mathbf{J}^{-1}\dot{\mathbf{X}} \qquad (8.146)$$

$$\begin{bmatrix} \dot{\theta}_1 \\ \dot{\theta}_2 \end{bmatrix} = \begin{bmatrix} -l_1 s\theta_1 - l_2 s(\theta_1 + \theta_2) & -l_2 s(\theta_1 + \theta_2) \\ l_1 c\theta_1 + l_2 c(\theta_1 + \theta_2) & l_2 c(\theta_1 + \theta_2) \end{bmatrix}^{-1} \begin{bmatrix} \dot{X} \\ \dot{Y} \end{bmatrix} \qquad (8.147)$$

*where,*

$$\mathbf{J}^{-1} = \frac{-1}{l_1 l_2 s\theta_2} \begin{bmatrix} -l_2 c(\theta_1 + \theta_2) & -l_2 s(\theta_1 + \theta_2) \\ l_1 c\theta_1 + l_2 c(\theta_1 + \theta_2) & l_1 s\theta_1 + l_2 s(\theta_1 + \theta_2) \end{bmatrix} \qquad (8.148)$$

*and hence,*

$$\dot{\theta}_1 = \frac{\dot{X} c(\theta_1 + \theta_2) + \dot{Y} s(\theta_1 + \theta_2)}{l_1 s\theta_2} \qquad (8.149)$$

$$\dot{\theta}_2 = \frac{\dot{X}(l_1 c\theta_1 + l_2 c(\theta_1 + \theta_2)) + \dot{Y}(l_1 s\theta_1 + l_2 s(\theta_1 + \theta_2))}{-l_1 l_2 s\theta_2}. \qquad (8.150)$$

*Singularity occurs when the determinant of the Jacobian is zero.*

$$|\mathbf{J}| = l_1 l_2 \sin\theta_2 \qquad (8.151)$$

*Therefore, the singular configurations of the manipulator are*

$$\theta_2 = 0 \qquad \theta_2 = 180\,\text{deg} \qquad (8.152)$$

*corresponding to the fully extended or fully contracted configurations, as shown in Figure 8.5. At the singular configurations, the value of $\theta_1$ is indeterminate and may have any real value. The two columns of the Jacobian matrix become parallel because the Equation (8.145) becomes*

$$\begin{bmatrix} \dot{X} \\ \dot{Y} \end{bmatrix} = 2l_1 \begin{bmatrix} -s\theta_1 \\ c\theta_1 \end{bmatrix} \dot{\theta}_1 + l_2 \begin{bmatrix} -s\theta_1 \\ c\theta_1 \end{bmatrix} \dot{\theta}_2$$

$$= \cdot \left( 2l_1 \dot{\theta}_1 + l_2 \dot{\theta}_2 \right) \begin{bmatrix} -s\theta_1 \\ c\theta_1 \end{bmatrix}. \qquad (8.153)$$

*In this situation, the endpoint can only move in the direction perpendicular to the arm links.*

**Example 220** *Analytic method for inverse velocity kinematics.*

*Theoretically, we must be able to calculate the joint velocities from the equations describing the forward velocities, however, such a calculation is not easy in a general case.*

*As an example, consider a 2R planar manipulator shown in Figure 8.2. The endpoint velocity of the 2R manipulator was expressed in Equation (8.58) as*

$$^0\dot{\mathbf{d}}_2 = \dot{\theta}_1 \, ^0\hat{k}_0 \times \left(l_1 \, ^0\hat{\imath}_1 + l_2 \, ^0\hat{\imath}_2\right) + \dot{\theta}_2 \, ^0\hat{k}_1 \times l_2 \, ^0\hat{\imath}_2. \tag{8.154}$$

*A dot product of this equation with $^0\hat{\imath}_2$, gives*

$$
\begin{aligned}
^0\dot{\mathbf{d}}_2 \cdot {}^0\hat{\imath}_2 &= \dot{\theta}_1 \left({}^0\hat{k}_0 \times l_1 \, ^0\hat{\imath}_1\right) \cdot {}^0\hat{\imath}_2 \\
&= l_1\dot{\theta}_1 \, ^0\hat{k}_0 \cdot \left({}^0\hat{\imath}_1 \times {}^0\hat{\imath}_2\right) \\
&= l_1\dot{\theta}_1 \, ^0\hat{k}_0 \cdot {}^0\hat{\imath}_2 \sin\theta_2 \\
&= l_1\dot{\theta}_1 \sin\theta_2
\end{aligned}
\tag{8.155}
$$

*and therefore,*

$$\dot{\theta}_1 = \frac{^0\dot{\mathbf{d}}_2 \cdot {}^0\hat{\imath}_2}{l_1 \sin\theta_2}. \tag{8.156}$$

*Now a dot product of (8.154) with $^0\hat{\imath}_1$ reduces to*

$$
\begin{aligned}
^0\dot{\mathbf{d}}_2 \cdot {}^0\hat{\imath}_1 &= \dot{\theta}_1 \left({}^0\hat{k}_0 \times l_2 \, ^0\hat{\imath}_2\right) \cdot {}^0\hat{\imath}_1 + \dot{\theta}_2 \left({}^0\hat{k}_1 \times l_2 \, ^0\hat{\imath}_2\right) \cdot {}^0\hat{\imath}_1 \\
&= l_2 \left(\dot{\theta}_1 + \dot{\theta}_2\right) {}^0\hat{k}_0 \cdot \left({}^0\hat{\imath}_2 \times {}^0\hat{\imath}_1\right) \\
&= -l_2 \left(\dot{\theta}_1 + \dot{\theta}_2\right) \sin\theta_2
\end{aligned}
\tag{8.157}
$$

*and therefore,*

$$\dot{\theta}_2 = -\dot{\theta}_1 - \frac{^0\dot{\mathbf{d}}_2 \cdot {}^0\hat{\imath}_1}{l_2 \sin\theta_2}. \tag{8.158}$$

**Example 221** ★ *Inverse Jacobian matrix for a robot with spherical wrist.*

*The Jacobian matrix for an articulated robot with a spherical wrist is calculated in Example 215.*

$$\mathbf{J} = \begin{bmatrix} ^0\hat{k}_0 \times {}^0_0\mathbf{d}_6 & ^0\hat{k}_1 \times {}^0_1\mathbf{d}_6 & ^0\hat{k}_2 \times {}^0_2\mathbf{d}_6 & 0 & 0 & 0 \\ ^0\hat{k}_0 & ^0\hat{k}_1 & ^0\hat{k}_2 & ^0\hat{k}_3 & ^0\hat{k}_4 & ^0\hat{k}_5 \end{bmatrix} \tag{8.159}$$

*The upper right $3 \times 3$ submatrix of $\mathbf{J}$ is zero. This is because of spherical wrist structure and having the last three position vectors as zero.*

*Let us split the Jacobian matrix into four $3 \times 3$ submatrices and write it as*

$$\mathbf{J} = \begin{bmatrix} A & B \\ C & D \end{bmatrix} = \begin{bmatrix} A & 0 \\ C & D \end{bmatrix} \tag{8.160}$$

*where*

$$[A] = \begin{bmatrix} {}^0\hat{k}_0 \times {}^0_0\mathbf{d}_6 & {}^0\hat{k}_1 \times {}^0_1\mathbf{d}_6 & {}^0\hat{k}_2 \times {}^0_2\mathbf{d}_6 \end{bmatrix} \tag{8.161}$$

$$[C] = \begin{bmatrix} {}^0\hat{k}_0 & {}^0\hat{k}_1 & {}^0\hat{k}_2 \end{bmatrix} \tag{8.162}$$

$$[D] = \begin{bmatrix} {}^0\hat{k}_3 & {}^0\hat{k}_4 & {}^0\hat{k}_5 \end{bmatrix}. \tag{8.163}$$

*Inversion of such a Jacobian is simpler if we take advantage of $B = 0$. The forward velocity kinematics of the robot can be written as*

$$\dot{\mathbf{X}} = \mathbf{J}\dot{\mathbf{q}}$$

$$\begin{bmatrix} {}^0\dot{\mathbf{d}}_2 \\ {}_0\omega_2 \end{bmatrix} = \begin{bmatrix} A & 0 \\ C & D \end{bmatrix} \begin{bmatrix} \dot{\theta}_1 \\ \dot{\theta}_2 \\ \dot{\theta}_3 \\ \dot{\theta}_4 \\ \dot{\theta}_5 \\ \dot{\theta}_6 \end{bmatrix}. \tag{8.164}$$

*The upper half of the equation is*

$$ {}^0\dot{\mathbf{d}}_2 = [A] \begin{bmatrix} \dot{\theta}_1 \\ \dot{\theta}_2 \\ \dot{\theta}_3 \end{bmatrix} \tag{8.165}$$

*which can be inverted to*

$$\begin{bmatrix} \dot{\theta}_1 \\ \dot{\theta}_2 \\ \dot{\theta}_3 \end{bmatrix} = A^{-1}\, {}^0\dot{\mathbf{d}}_2. \tag{8.166}$$

*The lower half of the equation is*

$$ {}_0\omega_2 = \begin{bmatrix} C & D \end{bmatrix} \begin{bmatrix} \dot{\theta}_1 \\ \dot{\theta}_2 \\ \dot{\theta}_3 \\ \dot{\theta}_4 \\ \dot{\theta}_5 \\ \dot{\theta}_6 \end{bmatrix}$$

$$= [C] \begin{bmatrix} \dot{\theta}_1 \\ \dot{\theta}_2 \\ \dot{\theta}_3 \end{bmatrix} + [D] \begin{bmatrix} \dot{\theta}_4 \\ \dot{\theta}_5 \\ \dot{\theta}_6 \end{bmatrix} \tag{8.167}$$

*and therefore,*

$$\begin{bmatrix} \dot{\theta}_4 \\ \dot{\theta}_5 \\ \dot{\theta}_6 \end{bmatrix} = D^{-1} \left( {}_0\omega_2 - [C] \begin{bmatrix} \dot{\theta}_1 \\ \dot{\theta}_2 \\ \dot{\theta}_3 \end{bmatrix} \right). \tag{8.168}$$

## 8.5   Summary

Each link of a serial robot has an angular and a translational velocity. The angular velocity of link $(i)$ in the global coordinate frame can be found as a summation of the global angular velocities of its lower links

$$_0^0\omega_i = \sum_{j=1}^{i} {}_{j-1}^0\omega_j. \tag{8.169}$$

Utilizing DH parameters, the angular velocity of link $(j)$ with respect to link $(j-1)$ is

$$_{j-1}^0\omega_j = \begin{cases} \dot{\theta}_j \; {}^0\hat{k}_{j-1} & \text{if joint } i \text{ is R} \\ 0 & \text{if joint } i \text{ is P.} \end{cases} \tag{8.170}$$

The translational velocity of link $(i)$ is the global velocity of the origin of coordinate frame $B_i$ attached to link $(i)$

$$_{i-1}^0\dot{\mathbf{d}}_i = \begin{cases} _0^0\omega_i \times {}_{i-1}^0\mathbf{d}_i & \text{if joint } i \text{ is R} \\ \dot{d}_i \; {}^0\hat{k}_{i-1} + {}_0^0\omega_i \times {}_{i-1}^0\mathbf{d}_i & \text{if joint } i \text{ is P} \end{cases} \tag{8.171}$$

where $\theta$ and $d$ are DH parameters, and $\mathbf{d}$ is the frame's origin position vector.

The velocity kinematics of a robot is defined by the relationship between joint coordinate velocity

$$\dot{\mathbf{q}} = \begin{bmatrix} \dot{q}_n & \dot{q}_n & \dot{q}_n & \cdots & \dot{q}_n \end{bmatrix}^T \tag{8.172}$$

and global velocity of the end-effector

$$\dot{\mathbf{X}} = \begin{bmatrix} \dot{X}_n & \dot{Y}_n & \dot{Z}_n & \omega_{Xn} & \omega_{Yn} & \omega_{Zn} \end{bmatrix}^T. \tag{8.173}$$

Such a relationship introduces Jacobian matrix $[\mathbf{J}]$

$$\dot{\mathbf{X}} = \mathbf{J}\,\dot{\mathbf{q}}. \tag{8.174}$$

Having $[\mathbf{J}]$, we are able to find the velocity of the end-effector for a given joint coordinate velocity and vice versa. Jacobian is a function of joint coordinates and is the main tool in velocity kinematics of robots. Theoretically, $[\mathbf{J}]$ is defined by

$$\mathbf{J}(\mathbf{q}) = \frac{\partial T(\mathbf{q})}{\partial \mathbf{q}} \tag{8.175}$$

where $T(\mathbf{q}) = {}^0T_n(\mathbf{q})$ is the transformation matrix for the end-effector

$$^0T_n(\mathbf{q}) = \begin{bmatrix} {}^0R_n & {}^0\mathbf{d}_n \\ 0 & 1 \end{bmatrix}. \tag{8.176}$$

However practically, we calculate $[\mathbf{J}]$ by Jacobian generating vectors denoted by $\mathbf{c}_i(\mathbf{q})$

$$\mathbf{c}_i(\mathbf{q}) = \begin{bmatrix} {}^0\hat{k}_{i-1} \times {}^0_{i-1}\mathbf{d}_n \\ {}^0\hat{k}_{i-1} \end{bmatrix} \qquad (8.177)$$

where $\mathbf{c}_i(\mathbf{q})$ makes the column $i-1$ of the Jacobian matrix.

$$\mathbf{J} = \begin{bmatrix} \mathbf{c}_0 & \mathbf{c}_1 & \mathbf{c}_2 & \cdots & \mathbf{c}_{n-1} \end{bmatrix} \qquad (8.178)$$

# Exercises

1. Notation and symbols.

   Describe the meaning of

   a- ${}^{i-1}\mathbf{d}_i$    b- ${}^{0}_{i-1}\mathbf{d}_i$    c- ${}^{i}_{i-1}\mathbf{d}_i$    d- ${}^{i-1}_{i}\mathbf{d}_{i-1}$    e- ${}^{i-1}_{i-1}\mathbf{d}_i$    f- ${}^{i-1}_{0}\mathbf{d}_i$

   g- ${}^{i-1}\dot{\mathbf{d}}_i$    h- ${}^{0}_{i-1}\dot{\mathbf{d}}_i$    i- ${}^{i}_{i-1}\dot{\mathbf{d}}_i$    j- ${}^{i-1}_{i}\dot{\mathbf{d}}_{i-1}$    k- ${}^{i-1}_{i-1}\dot{\mathbf{d}}_i$    l- ${}^{i-1}_{0}\dot{\mathbf{d}}_i$.

   m- ${}^{0}\hat{k}_i$    n- ${}^{0}\hat{k}_{i-1}$    o- ${}^{0}_{i-1}\hat{k}_{i-2}$    p- ${}^{0}_{i-1}\mathbf{v}_i$    q- ${}^{i}_{i-1}\mathbf{v}_i$    r- ${}^{i}\mathbf{v}_i$

2. 3R planar manipulator velocity kinematics.

   Figure 5.21 illustrates an R‖R‖R planar manipulator. The forward kinematics of the manipulator provides the following transformation matrices:

   $$
   {}^{2}T_3 = \begin{bmatrix} \cos\theta_3 & -\sin\theta_3 & 0 & l_3\cos\theta_3 \\ \sin\theta_3 & \cos\theta_3 & 0 & l_3\sin\theta_3 \\ 0 & 0 & 1 & 0 \\ 0 & 0 & 0 & 1 \end{bmatrix}
   $$

   $$
   {}^{1}T_2 = \begin{bmatrix} \cos\theta_2 & -\sin\theta_2 & 0 & l_2\cos\theta_2 \\ \sin\theta_2 & \cos\theta_2 & 0 & l_2\sin\theta_2 \\ 0 & 0 & 1 & 0 \\ 0 & 0 & 0 & 1 \end{bmatrix}
   $$

   $$
   {}^{0}T_1 = \begin{bmatrix} \cos\theta_1 & -\sin\theta_1 & 0 & l_1\cos\theta_1 \\ \sin\theta_1 & \cos\theta_1 & 0 & l_1\sin\theta_1 \\ 0 & 0 & 1 & 0 \\ 0 & 0 & 0 & 1 \end{bmatrix}
   $$

   Calculate the Jacobian matrix, $J$, using direct differentiating, and find the Cartesian velocity vector of the endpoint for numerical values.

   $$
   \begin{aligned}
   \theta_1 &= 56\,\text{deg} \\
   \theta_2 &= -28\,\text{deg} \\
   \theta_3 &= -10\,\text{deg}
   \end{aligned}
   $$

   $$
   \begin{aligned}
   l_1 &= 100\,\text{cm} \\
   l_2 &= 55\,\text{cm} \\
   l_3 &= 30\,\text{cm}
   \end{aligned}
   $$

   $$
   \begin{aligned}
   \dot{\theta}_1 &= 30\,\text{deg}\,/\,\text{sec} \\
   \dot{\theta}_2 &= 10\,\text{deg}\,/\,\text{sec} \\
   \dot{\theta}_3 &= -10\,\text{deg}\,/\,\text{sec}
   \end{aligned}
   $$

3. Spherical wrist velocity kinematics.

Figure 5.22 illustrates a schematic of a spherical wrist. The associated transformation matrices are given below. Assume that the frame $B_3$ is the base frame. Find the angular velocity vector of the coordinate frame $B_6$.

$$^3T_4 = \begin{bmatrix} \cos\theta_4 & 0 & -\sin\theta_4 & 0 \\ \sin\theta_4 & 0 & \cos\theta_4 & 0 \\ 0 & -1 & 0 & 0 \\ 0 & 0 & 0 & 1 \end{bmatrix}$$

$$^4T_5 = \begin{bmatrix} \cos\theta_5 & 0 & \sin\theta_5 & 0 \\ \sin\theta_5 & 0 & -\cos\theta_5 & 0 \\ 0 & 1 & 0 & 0 \\ 0 & 0 & 0 & 1 \end{bmatrix}$$

$$^5T_6 = \begin{bmatrix} \cos\theta_6 & -\sin\theta_6 & 0 & 0 \\ \sin\theta_6 & \cos\theta_6 & 0 & 0 \\ 0 & 0 & 1 & 0 \\ 0 & 0 & 0 & 1 \end{bmatrix}$$

4. Spherical wrist and tool's frame velocity kinematics.

Assume that we attach a tools coordinate frame, with the following transformation matrix, to the last coordinate frame $B_6$ of a spherical wrist.

$$^6T_7 = \begin{bmatrix} 1 & 0 & 0 & 0 \\ 0 & 1 & 0 & 0 \\ 0 & 0 & 1 & d_6 \\ 0 & 0 & 0 & 1 \end{bmatrix}$$

The wrist transformation matrices are given in Exercise 3. Assume that the frame $B_3$ is the base frame and find the translational and angular velocities of the tools coordinate frame $B_7$.

5. SCARA manipulator velocity kinematics.

An R‖R‖R‖P SCARA manipulator is shown in Figure 5.27 with the following transformation matrices. Calculate the Jacobian matrix using the Jacobian-generating vector technique.

$$^0T_1 = \begin{bmatrix} \cos\theta_1 & -\sin\theta_1 & 0 & l_1\cos\theta_1 \\ \sin\theta_1 & \cos\theta_1 & 0 & l_1\sin\theta_1 \\ 0 & 0 & 1 & 0 \\ 0 & 0 & 0 & 1 \end{bmatrix} \tag{8.179}$$

$$^1T_2 = \begin{bmatrix} \cos\theta_2 & -\sin\theta_2 & 0 & l_2\cos\theta_2 \\ \sin\theta_2 & \cos\theta_2 & 0 & l_2\sin\theta_2 \\ 0 & 0 & 1 & 0 \\ 0 & 0 & 0 & 1 \end{bmatrix} \tag{8.180}$$

$$^2T_3 = \begin{bmatrix} \cos\theta_3 & -\sin\theta_3 & 0 & 0 \\ \sin\theta_3 & \cos\theta_3 & 0 & 0 \\ 0 & 0 & 1 & 0 \\ 0 & 0 & 0 & 1 \end{bmatrix} \qquad (8.181)$$

$$^3T_4 = \begin{bmatrix} 1 & 0 & 0 & 0 \\ 0 & 1 & 0 & 0 \\ 0 & 0 & 1 & d \\ 0 & 0 & 0 & 1 \end{bmatrix} \qquad (8.182)$$

6. R⊢R∥R articulated arm velocity kinematics.

   Figure 5.25 illustrates a 3 DOF R⊢R∥R manipulator with the following transformation matrices. Find the Jacobian matrix using direct differentiating, Jacobian-generating vector methods.

$$^0T_1 = \begin{bmatrix} \cos\theta_1 & 0 & -\sin\theta_1 & 0 \\ \sin\theta_1 & 0 & \cos\theta_1 & 0 \\ 0 & -1 & 0 & d_1 \\ 0 & 0 & 0 & 1 \end{bmatrix}$$

$$^1T_2 = \begin{bmatrix} \cos\theta_2 & -\sin\theta_2 & 0 & l_2\cos\theta_2 \\ \sin\theta_2 & \cos\theta_2 & 0 & l_2\sin\theta_2 \\ 0 & 0 & 1 & d_2 \\ 0 & 0 & 0 & 1 \end{bmatrix}$$

$$^2T_3 = \begin{bmatrix} \cos\theta_3 & 0 & \sin\theta_3 & 0 \\ \sin\theta_3 & 0 & -\cos\theta_3 & 0 \\ 0 & 1 & 0 & 0 \\ 0 & 0 & 0 & 1 \end{bmatrix}$$

7. Rigid link velocity.

   Figure 8.6 illustrates the coordinate frames and velocity vectors of a rigid link ($i$). Find

   (a) velocity $^0\mathbf{v}_i$ of the link at $C_i$ in terms of $\dot{\mathbf{d}}_i$ and $\dot{\mathbf{d}}_{i-1}$
   
   (b) angular velocity of the link $^0\boldsymbol{\omega}_i$ in terms of $\dot{\mathbf{d}}_i$ and $\dot{\mathbf{d}}_{i-1}$
   
   (c) velocity $^0\mathbf{v}_i$ of the link at $C_i$ in terms of proximal joint $i$ velocity
   
   (d) velocity $^0\mathbf{v}_i$ of the link at $C_i$ in terms of distal joint $i+1$ velocity
   
   (e) velocity of proximal joint $i$ in terms of distal joint $i+1$ velocity
   
   (f) velocity of distal joint $i+1$ in terms of proximal joint $i$ velocity

8. ★ Jacobian of a PRRR manipulator.

   Determine the Jacobian matrix for the manipulator shown in Figure 6.5.

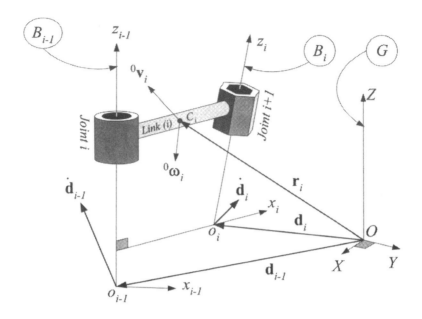

FIGURE 8.6. Rigid link velocity vectors.

9. ★ Spherical robot velocity kinematics.

A spherical manipulator R⊢R⊢P, equipped with a spherical wrist, is shown in Figure 5.26. The transformation matrices of the robot are given in Example 149. Solve the robot's forward and inverse velocity kinematics.

10. ★ Space station remote manipulator system velocity kinematics.

The transformation matrices for the shuttle remote manipulator system (SRMS), shown in Figure 5.28, are given in the Example 153. Solve the velocity kinematics of the SRMS by calculating the Jacobian matrix.

# 9

# Numerical Methods in Kinematics

## 9.1  Linear Algebraic Equations

In robotic analysis, there exist problems and situations, such as inverse velocity kinematics, that we need to solve a set of coupled linear or nonlinear algebraic equations. Every numerical method of solving nonlinear equations also works by iteratively solving a set of linear equations.

Consider a system of $n$ linear algebraic equations with real constant coefficients,

$$
\begin{aligned}
a_{11}x_1 + a_{12}x_2 + \cdots + a_{1n}x_n &= b_1 \\
a_{21}x_1 + a_{22}x_2 + \cdots + a_{2n}x_n &= b_2 \\
\cdots &= \cdots \\
a_{n1}x_1 + a_{n2}x_2 + \cdots + a_{nn}x_n &= b_n
\end{aligned}
\tag{9.1}
$$

which can also be written in matrix form

$$
[A]\,\mathbf{x} = \mathbf{b}.
\tag{9.2}
$$

There are numerous methods for solving this set of equations. Among the most efficient methods is the $LU$ factorization method.

For every nonsingular matrix $[A]$ there exists an upper triangular matrix $[U]$ with nonzero diagonal elements and a lower triangular matrix $[L]$ with unit diagonal elements, such that

$$
[A] = [L]\,[U]
\tag{9.3}
$$

$$
[A] =
\begin{bmatrix}
a_{11} & a_{12} & \cdots & a_{1n} \\
a_{21} & a_{22} & \cdots & a_{2n} \\
\cdots & \cdots & \cdots & \cdots \\
a_{n1} & a_{n2} & \cdots & a_{nn}
\end{bmatrix}
\tag{9.4}
$$

$$
[L] =
\begin{bmatrix}
1 & 0 & \cdots & 0 \\
l_{21} & 1 & \cdots & 0 \\
\cdots & \cdots & \cdots & \cdots \\
l_{n1} & l_{n2} & \cdots & 1
\end{bmatrix}
\tag{9.5}
$$

$$
[U] =
\begin{bmatrix}
u_{11} & u_{12} & \cdots & u_{1n} \\
0 & u_{22} & \cdots & u_{2n} \\
\cdots & \cdots & \cdots & \cdots \\
0 & 0 & \cdots & u_{nn}
\end{bmatrix}.
\tag{9.6}
$$

The process of factoring $[A]$ into the $[L][U]$ is called *LU* factorization. Once the $[L]$ and $[U]$ matrices are obtained, the equation

$$[L][U]\mathbf{x} = \mathbf{b} \tag{9.7}$$

can be solved by transforming into

$$[L]\mathbf{y} = \mathbf{b} \tag{9.8}$$

and

$$[U]\mathbf{x} = \mathbf{y}. \tag{9.9}$$

Since the Equations (9.8) and (9.9) are both a triangular set of equations, their solutions are easy to obtain by forward and backward substitution.
**Proof.** To show how $[A]$ can be transformed into $[L][U]$, we consider a $4 \times 4$ matrix.

$$\begin{bmatrix} a_{11} & a_{12} & a_{13} & a_{14} \\ a_{21} & a_{22} & a_{23} & a_{24} \\ a_{31} & a_{32} & a_{33} & a_{34} \\ a_{41} & a_{42} & a_{43} & a_{44} \end{bmatrix} = \begin{bmatrix} 1 & 0 & 0 & 0 \\ l_{21} & 1 & 0 & 0 \\ l_{31} & l_{32} & 1 & 0 \\ l_{41} & l_{42} & l_{43} & 1 \end{bmatrix} \begin{bmatrix} u_{11} & u_{12} & u_{13} & u_{14} \\ 0 & u_{22} & u_{23} & u_{24} \\ 0 & 0 & u_{33} & u_{34} \\ 0 & 0 & 0 & u_{44} \end{bmatrix} \tag{9.10}$$

Employing a dummy matrix $[B]$, we may combine the elements of $[L]$ and $[U]$ as

$$[B] = \begin{bmatrix} u_{11} & u_{12} & u_{13} & u_{14} \\ l_{21} & u_{22} & u_{23} & u_{24} \\ l_{31} & l_{32} & u_{33} & u_{34} \\ l_{41} & l_{42} & l_{43} & u_{44} \end{bmatrix}. \tag{9.11}$$

The elements of $[B]$ will be calculated one by one, in the following order:

$$[B] = \begin{bmatrix} (1) & (2) & (3) & (4) \\ (5) & (8) & (9) & (10) \\ (6) & (11) & (13) & (14) \\ (7) & (12) & (15) & (16) \end{bmatrix} \tag{9.12}$$

The process for generating a matrix $[B]$, associated to an $n \times n$ matrix $[A]$, is performed in $n - 1$ iterations. After $i - 1$ iterations, the matrix is in the following form:

$$[B] = \begin{bmatrix} u_{1,1} & u_{1,2} & \cdots & u_{1,i-1} & \cdots & \cdots & u_{1,n} \\ l_{2,1} & u_{2,2} & \cdots & \cdots & \cdots & \cdots & u_{2,n} \\ \cdots & \cdots & \cdots & \cdots & \cdots & \cdots & \cdots \\ \cdots & \cdots & \cdots & \cdots & \cdots & \cdots & u_{i-1,n} \\ \cdots & \cdots & \cdots & \cdots & & \begin{bmatrix} & \\ & D_i & \\ & \end{bmatrix} \\ \cdots & \cdots & \cdots & \cdots & & \\ l_{n,1} & l_{n,2} & \cdots & l_{n,i-1} & & \end{bmatrix} \tag{9.13}$$

The unprocessed $(n-i+1) \times (n-i+1)$ submatrix in the lower right corner is denoted by $[D_i]$ and has the same elements as $[A]$. In the $i$th step, the $LU$ factorization method converts $[D_i]$

$$[D_i] = \begin{bmatrix} d_{ii} & \mathbf{r}_i^T \\ \mathbf{s}_i & [H_{i+1}] \end{bmatrix} \tag{9.14}$$

to a new form

$$[D_i] = \begin{bmatrix} u_{ii} & \mathbf{u}_i^T \\ \mathbf{l}_i & [D_{i+1}] \end{bmatrix}. \tag{9.15}$$

Direct multiplication shows that

$$
\begin{aligned}
u_{11} &= a_{11} \\
u_{12} &= a_{12} \\
u_{13} &= a_{13} \\
u_{14} &= a_{14}
\end{aligned}
\tag{9.16}
$$

$$
\begin{aligned}
l_{21} &= \frac{a_{21}}{u_{11}} \\
l_{31} &= \frac{a_{31}}{u_{11}} \\
l_{41} &= \frac{a_{41}}{u_{11}}
\end{aligned}
\tag{9.17}
$$

$$
\begin{aligned}
u_{22} &= a_{22} - l_{21}u_{12} \\
u_{23} &= a_{23} - l_{21}u_{13} \\
u_{24} &= a_{24} - l_{21}u_{14}
\end{aligned}
\tag{9.18}
$$

$$
\begin{aligned}
l_{32} &= \frac{a_{32} - l_{31}u_{12}}{u_{22}} \\
l_{42} &= \frac{a_{42} - l_{41}u_{12}}{u_{22}}
\end{aligned}
\tag{9.19}
$$

$$
\begin{aligned}
u_{33} &= a_{33} - (l_{31}u_{13} + l_{32}u_{23}) \\
u_{34} &= a_{34} - (l_{31}u_{14} + l_{32}u_{24})
\end{aligned}
\tag{9.20}
$$

$$l_{43} = \frac{a_{43} - (l_{41}u_{13} + l_{42}u_{23})}{u_{33}} \tag{9.21}$$

$$u_{44} = a_{44} - (l_{41}u_{14} + l_{42}u_{24} + l_{43}u_{34}). \tag{9.22}$$

Therefore, the general formula for getting elements of $[L]$ and $[U]$ corresponding to an $n \times n$ coefficients matrix $[A]$ can be written as

$$u_{ij} = a_{ij} - \sum_{k=1}^{i-1} l_{ik}u_{kj} \qquad i \le j \qquad j = 1, \cdots n \tag{9.23}$$

$$l_{ij} = \frac{a_{ij} - \sum\limits_{k=1}^{j-1} l_{ik}u_{kj}}{u_{jj}} \qquad j \le i \qquad i = 1, \cdots n. \qquad (9.24)$$

For $i = 1$, the rule for $u$ reduces to

$$u_{1j} = a_{1j} \qquad (9.25)$$

and for $j = 1$, the rule for $l$ reduces to

$$l_{i1} = \frac{a_{i1}}{u_{11}}. \qquad (9.26)$$

The calculation of element $(k)$ of the dummy matrix $[B]$, which is an element of $[L]$ or $[U]$, involves only the elements of $[A]$ in the same position and some previously calculated elements of $[B]$.

The $LU$ factorization technique can be set up in an algorithm for easier numerical calculations.

**Algorithm 9.1.** *LU factorization technique for an $n \times n$ matrix $[A]$.*
*1- Set the initial counter $i = 1$.*
*2- Set $[D_1] = [A]$.*
*3- Calculate $[D_{i+1}]$ from $[D_i]$ according to*

$$
\begin{aligned}
u_{ii} &= d_{ii} & (9.27) \\
\mathbf{u}_i^T &= \mathbf{r}_i^T & (9.28) \\
\mathbf{l}_i &= \frac{1}{u_{ii}}\mathbf{s}_i & (9.29) \\
[D_{i+1}] &= [H_{i+1}] - \mathbf{l}_i\,\mathbf{u}_i^T. & (9.30)
\end{aligned}
$$

*4- Set $i = i + 1$. If $i = n$ then LU factorization is completed. Otherwise return to step 3.*

After decomposing the matrix $[A]$ into the matrices $[L]$ and $[U]$, the set of equations can be solved based on the following algorithm.

**Algorithm 9.2.** *LU solution technique.*
*1- Calculate $\mathbf{y}$ from $[L]\mathbf{y} = \mathbf{b}$ by*

$$
\begin{aligned}
y_1 &= b_1 \\
y_2 &= b_2 - y_1 l_{21} \\
y_3 &= b_3 - y_1 l_{31} - y_2 l_{32} \\
&\cdots \\
y_i &= b_i - \sum_{j=1}^{i-1} y_j l_{ij}. \qquad (9.31)
\end{aligned}
$$

2- *Calculate* x *from* $[U]\,x = y$ *by*

$$x_n = \frac{y_n}{u_{n,n}}$$

$$x_{n-1} = \frac{y_{n-1} - x_n u_{n-1,n}}{u_{n-1,n-1}}$$

$$\cdots$$

$$x_i = \frac{1}{u_{ii}}\left(y_i - \sum_{j=i+1}^{n} x_j u_{ij}\right). \tag{9.32}$$

∎

**Example 222** *Solution of a set of four equations.*
   *Consider a set of four linear algebraic equations*

$$[A]\,x = b \tag{9.33}$$

*where,*

$$[A] = \begin{bmatrix} 2 & 1 & 3 & -3 \\ 1 & 0 & -1 & -2 \\ 0 & 2 & 2 & 1 \\ 3 & 1 & 0 & -2 \end{bmatrix} \tag{9.34}$$

*and*

$$b = \begin{bmatrix} 1 \\ 2 \\ 0 \\ -2 \end{bmatrix}. \tag{9.35}$$

*Following the LU factorization algorithm we first set*

$$i = 1 \tag{9.36}$$
$$[D_1] = [A] \tag{9.37}$$

*to find*

$$d_{11} = 2 \tag{9.38}$$
$$r_1^T = \begin{bmatrix} 1 & 3 & -3 \end{bmatrix} \tag{9.39}$$

$$s_1 = \begin{bmatrix} 1 \\ 0 \\ 3 \end{bmatrix} \tag{9.40}$$

$$[H_2] = \begin{bmatrix} 0 & -1 & -2 \\ 2 & 2 & 1 \\ 1 & 0 & -2 \end{bmatrix} \tag{9.41}$$

*and calculate*

$$u_{11} = d_{11} = 2 \tag{9.42}$$
$$\mathbf{u}_1^T = \mathbf{r}_1^T = [\,1 \quad 3 \quad -3\,] \tag{9.43}$$

$$l_1 = \frac{1}{u_{11}}\mathbf{s}_1 = \begin{bmatrix} \frac{1}{2} \\ 0 \\ \frac{3}{2} \end{bmatrix} \tag{9.44}$$

$$[D_2] = [H_2] - l_1\,\mathbf{u}_1^T = \begin{bmatrix} -\frac{1}{2} & -\frac{5}{2} & -\frac{1}{2} \\ 2 & 2 & 1 \\ -\frac{1}{2} & -\frac{9}{2} & \frac{5}{2} \end{bmatrix}. \tag{9.45}$$

*In the second step we have*

$$i = 2 \tag{9.46}$$

*and*

$$d_{22} = -\frac{1}{2} \tag{9.47}$$
$$\mathbf{r}_2^T = [\,-\frac{5}{2} \quad -\frac{1}{2}\,] \tag{9.48}$$

$$\mathbf{s}_2 = \begin{bmatrix} 2 \\ -\frac{1}{2} \end{bmatrix} \tag{9.49}$$

$$[H_3] = \begin{bmatrix} 2 & 1 \\ -\frac{9}{2} & \frac{5}{2} \end{bmatrix} \tag{9.50}$$

*and calculate*

$$u_{22} = d_{22} = -\frac{1}{2} \tag{9.51}$$

$$\mathbf{u}_2^T = \mathbf{r}_2^T = [\,-\frac{5}{2} \quad -\frac{1}{2}\,] \tag{9.52}$$

$$l_2 = \frac{1}{u_{22}}\mathbf{s}_2 = \begin{bmatrix} -4 \\ 1 \end{bmatrix} \tag{9.53}$$

$$[D_3] = [H_3] - l_2\,\mathbf{u}_2^T = \begin{bmatrix} -8 & -1 \\ -2 & 3 \end{bmatrix}. \tag{9.54}$$

*In the third step we set*

$$i = 3 \tag{9.55}$$

*and find*

$$d_{33} = -8 \tag{9.56}$$
$$\mathbf{r}_3^T = [-1] \tag{9.57}$$

$$\mathbf{s}_3 = [-2] \tag{9.58}$$
$$[H_4] = [3] \tag{9.59}$$

*and therefore,*

$$u_{33} = d_{33} = -8 \tag{9.60}$$

$$\mathbf{u}_3^T = \mathbf{r}_3^T = [-1] \tag{9.61}$$

$$\mathbf{l}_3 = \frac{1}{u_{33}}\mathbf{s}_3 = \begin{bmatrix} 1 \\ 4 \end{bmatrix} \tag{9.62}$$

$$[D_4] = [H_4] - \mathbf{l}_3\,\mathbf{u}_3^T = \begin{bmatrix} \frac{13}{4} \end{bmatrix}. \tag{9.63}$$

*After these calculations, the matrix* $[B]$, $[L]$, *and* $[U]$ *become*

$$[B] = \begin{bmatrix} 2 & 1 & 3 & -3 \\ \frac{1}{2} & -\frac{1}{2} & -\frac{5}{2} & -\frac{1}{2} \\ 0 & 4 & -8 & -1 \\ \frac{3}{2} & 1 & \frac{1}{4} & \frac{13}{4} \end{bmatrix} \tag{9.64}$$

$$[L] = \begin{bmatrix} 1 & 0 & 0 & 0 \\ \frac{1}{2} & 1 & 0 & 0 \\ 0 & -4 & 1 & 0 \\ \frac{3}{2} & 1 & \frac{1}{4} & 1 \end{bmatrix} \tag{9.65}$$

$$[U] = \begin{bmatrix} 2 & 1 & 3 & -3 \\ 0 & -\frac{1}{2} & -\frac{5}{2} & -\frac{1}{2} \\ 0 & 0 & -8 & -1 \\ 0 & 0 & 0 & \frac{13}{4} \end{bmatrix}. \tag{9.66}$$

*Now a vector* $\mathbf{y}$ *can be found to satisfy*

$$[L]\,\mathbf{y} = \mathbf{b} \tag{9.67}$$

$$\mathbf{y} = \begin{bmatrix} 1 \\ \frac{3}{2} \\ 6 \\ -\frac{13}{2} \end{bmatrix} \tag{9.68}$$

*and finally the unknown vector* $\mathbf{x}$ *should be found to satisfy*

$$[U]\,\mathbf{x} = \mathbf{y} \tag{9.69}$$

$$\mathbf{x} = \begin{bmatrix} -\frac{5}{2} \\ \frac{3}{2} \\ -\frac{1}{2} \\ -2 \end{bmatrix}. \tag{9.70}$$

**Example 223** *LU factorization with* **pivoting**.

In the process of LU factorization, the situation $u_{ii} = 0$ generates a division by zero, which must be avoided. In this situation, pivoting must be applied. By pivoting, we change the order of equations to have a coefficient *matrix with the largest elements, in absolute value, as diagonal elements. The largest element is called the **pivot element**.*

*As an example, consider the following set of equations:*

$$[A]\mathbf{x} = \mathbf{b} \tag{9.71}$$

$$\begin{bmatrix} 2 & 1 & 3 & -3 \\ 1 & 0 & -1 & 2 \\ 0 & 2 & 0 & 1 \\ 3 & 1 & 4 & -2 \end{bmatrix} \begin{bmatrix} x_1 \\ x_2 \\ x_3 \\ x_4 \end{bmatrix} = \begin{bmatrix} 1 \\ 2 \\ 0 \\ -2 \end{bmatrix} \tag{9.72}$$

*however, we move the largest element to $d_{11}$ by interchanging row 1 with 4, and column 1 with 3.*

$$\begin{bmatrix} 3 & 1 & 4 & -2 \\ 1 & 0 & -1 & 2 \\ 0 & 2 & 0 & 1 \\ 2 & 1 & 3 & -3 \end{bmatrix} \begin{bmatrix} x_1 \\ x_2 \\ x_3 \\ x_4 \end{bmatrix} = \begin{bmatrix} -2 \\ 2 \\ 0 \\ 1 \end{bmatrix} \tag{9.73}$$

$$\begin{bmatrix} 4 & 1 & 3 & -2 \\ -1 & 0 & 1 & 2 \\ 0 & 2 & 0 & 1 \\ 3 & 1 & 2 & -3 \end{bmatrix} \begin{bmatrix} x_3 \\ x_2 \\ x_1 \\ x_4 \end{bmatrix} = \begin{bmatrix} -2 \\ 2 \\ 0 \\ 1 \end{bmatrix} \tag{9.74}$$

*Then the largest element in the $3 \times 3$ submatrix in the lower right corner will move to $d_{22}$*

$$\begin{bmatrix} 4 & -2 & 3 & 1 \\ -1 & 2 & 1 & 0 \\ 0 & 1 & 0 & 2 \\ 3 & -3 & 2 & 1 \end{bmatrix} \begin{bmatrix} x_3 \\ x_4 \\ x_1 \\ x_2 \end{bmatrix} = \begin{bmatrix} -2 \\ 2 \\ 0 \\ 1 \end{bmatrix} \tag{9.75}$$

$$\begin{bmatrix} 4 & -2 & 3 & 1 \\ 3 & -3 & 2 & 1 \\ 0 & 1 & 0 & 2 \\ -1 & 2 & 1 & 0 \end{bmatrix} \begin{bmatrix} x_3 \\ x_4 \\ x_1 \\ x_2 \end{bmatrix} = \begin{bmatrix} -2 \\ 1 \\ 0 \\ 2 \end{bmatrix} \tag{9.76}$$

*and finally the largest element in the $2 \times 2$ in the lower right corner will move to $d_{22}$.*

$$\begin{bmatrix} 4 & -2 & 1 & 3 \\ 3 & -3 & 1 & 2 \\ 0 & 1 & 2 & 0 \\ -1 & 2 & 0 & 1 \end{bmatrix} \begin{bmatrix} x_3 \\ x_4 \\ x_2 \\ x_1 \end{bmatrix} = \begin{bmatrix} -2 \\ 1 \\ 0 \\ 2 \end{bmatrix} \tag{9.77}$$

*To apply the LU factorization algorithm and LU solution algorithm, we define a new set of equations.*

$$[A']\mathbf{x}' = \mathbf{b}' \tag{9.78}$$

$$\begin{bmatrix} 4 & -2 & 1 & 3 \\ 3 & -3 & 1 & 2 \\ 0 & 1 & 2 & 0 \\ -1 & 2 & 0 & 1 \end{bmatrix} \begin{bmatrix} x_1' \\ x_2' \\ x_3' \\ x_4' \end{bmatrix} = \begin{bmatrix} -2 \\ 1 \\ 0 \\ 2 \end{bmatrix} \tag{9.79}$$

*Based on the LU factorization algorithm, in the first step we set*

$$i = 1 \tag{9.80}$$

*and find*

$$[D_1] = [A'] \tag{9.81}$$

$$d_{11} = 4 \tag{9.82}$$

$$\mathbf{r}_1^T = \begin{bmatrix} -2 & 1 & 3 \end{bmatrix} \tag{9.83}$$

$$\mathbf{s}_1 = \begin{bmatrix} 3 \\ 0 \\ -1 \end{bmatrix} \tag{9.84}$$

$$[H_2] = \begin{bmatrix} -3 & 1 & 2 \\ 1 & 2 & 0 \\ 2 & 0 & 1 \end{bmatrix} \tag{9.85}$$

*to calculate*

$$u_{11} = d_{11} = 4 \tag{9.86}$$

$$\mathbf{u}_1^T = \mathbf{r}_1^T = \begin{bmatrix} -2 & 1 & 3 \end{bmatrix} \tag{9.87}$$

$$\mathbf{l}_1 = \frac{1}{u_{11}}\mathbf{s}_1 = \begin{bmatrix} \frac{3}{4} \\ 0 \\ -\frac{1}{4} \end{bmatrix} \tag{9.88}$$

$$[D_2] = [H_2] - \mathbf{l}_1\mathbf{u}_1^T = \begin{bmatrix} -\frac{3}{2} & \frac{1}{4} & -\frac{1}{4} \\ 1 & 2 & 0 \\ \frac{3}{2} & \frac{1}{4} & \frac{7}{4} \end{bmatrix}. \tag{9.89}$$

*For the second step, we have*

$$i = 2 \tag{9.90}$$

*and*

$$d_{22} = -\frac{3}{2} \tag{9.91}$$

$$\mathbf{r}_2^T = \begin{bmatrix} \frac{1}{4} & -\frac{1}{4} \end{bmatrix} \tag{9.92}$$

$$s_2 = \begin{bmatrix} 1 \\ \frac{3}{2} \end{bmatrix} \tag{9.93}$$

$$[H_3] = \begin{bmatrix} 2 & 0 \\ \frac{1}{4} & \frac{7}{4} \end{bmatrix} \tag{9.94}$$

*and then*

$$u_{22} = d_{22} = -\frac{3}{2} \tag{9.95}$$

$$\mathbf{u}_2^T = \mathbf{r}_2^T = \begin{bmatrix} \frac{1}{4} & -\frac{1}{4} \end{bmatrix} \tag{9.96}$$

$$\mathbf{l}_2 = \frac{1}{u_{22}} s_2 = \begin{bmatrix} -\frac{2}{3} \\ -1 \end{bmatrix} \tag{9.97}$$

$$[D_3] = [H_3] - \mathbf{l}_2 \, \mathbf{u}_2^T = \begin{bmatrix} \frac{13}{6} & -\frac{1}{6} \\ \frac{1}{2} & \frac{3}{2} \end{bmatrix}. \tag{9.98}$$

*In the third step, we set*

$$i = 3 \tag{9.99}$$

*and find*

$$d_{33} = \frac{13}{6} \tag{9.100}$$

$$\mathbf{r}_3^T = \begin{bmatrix} -\frac{1}{6} \end{bmatrix} \tag{9.101}$$

$$s_3 = \begin{bmatrix} 1 \\ 2 \end{bmatrix} \tag{9.102}$$

$$[H_4] = \begin{bmatrix} 3 \\ 2 \end{bmatrix} \tag{9.103}$$

*and calculate*

$$u_{33} = d_{33} = \frac{13}{6} \tag{9.104}$$

$$\mathbf{u}_3^T = \mathbf{r}_3^T = \begin{bmatrix} -\frac{1}{6} \end{bmatrix} \tag{9.105}$$

$$\mathbf{l}_3 = \frac{1}{u_{33}} s_3 = \begin{bmatrix} \frac{3}{13} \end{bmatrix} \tag{9.106}$$

$$[D_4] = [H_4] - \mathbf{l}_3 \, \mathbf{u}_3^T = \begin{bmatrix} \frac{20}{13} \end{bmatrix}.$$

*Therefore, the matrices [L] and [U] are*

$$[L] = \begin{bmatrix} 1 & 0 & 0 & 0 \\ \frac{3}{4} & 1 & 0 & 0 \\ 0 & -\frac{2}{3} & 1 & 0 \\ -\frac{1}{4} & -1 & \frac{3}{13} & 1 \end{bmatrix} \tag{9.107}$$

$$[U] = \begin{bmatrix} 4 & -2 & 1 & 3 \\ 0 & -\frac{3}{2} & \frac{1}{4} & -\frac{1}{4} \\ 0 & 0 & \frac{13}{6} & -\frac{1}{6} \\ 0 & 0 & 0 & \frac{20}{13} \end{bmatrix} \tag{9.108}$$

and now we can find the vector **y**

$$[L]\,\mathbf{y} = \mathbf{b}' \tag{9.109}$$

$$\mathbf{y} = \begin{bmatrix} -2 \\ \frac{5}{2} \\ \frac{3}{3} \\ \frac{47}{13} \end{bmatrix}. \tag{9.110}$$

The unknown vector $\mathbf{x}'$ can then be calculated,

$$[U]\,\mathbf{x}' = \mathbf{y} \tag{9.111}$$

$$\mathbf{x}' = \begin{bmatrix} -\frac{69}{20} \\ -\frac{19}{10} \\ \frac{19}{20} \\ \frac{47}{20} \end{bmatrix} = \begin{bmatrix} x_3 \\ x_4 \\ x_2 \\ x_1 \end{bmatrix} \tag{9.112}$$

and therefore,

$$\mathbf{x} = \begin{bmatrix} \frac{47}{20} \\ \frac{19}{20} \\ -\frac{69}{20} \\ -\frac{19}{10} \end{bmatrix}. \tag{9.113}$$

**Example 224** ★ *Uniqueness of solution.*

Consider a set of $n$ linear equations, $[A]\,\mathbf{x} = \mathbf{b}$. If $[A]$ is square and non-singular, then there exists a unique solution $\mathbf{x} = [A]^{-1}\,\mathbf{b}$. However, if the linear system of equations involves $n$ variables and $m$ equations

$$
\begin{aligned}
a_{11}x_1 + a_{12}x_2 + \cdots + a_{1n}x_n &= b_1 \\
a_{21}x_1 + a_{22}x_2 + \cdots + a_{2n}x_n &= b_2 \\
\cdots &= \cdots \\
a_{m1}x_1 + a_{m2}x_2 + \cdots + a_{mn}x_n &= b_m
\end{aligned} \tag{9.114}
$$

then, three classes of solutions are possible.

1. A unique solution exists and the system is called consistent.

2. No solution exists and the system is called inconsistent.

3. Multiple solutions exist and the system is called undetermined.

**Example 225 ★** *Ill conditioned and well conditioned.*

*A system of equations, $[A]\mathbf{x} = \mathbf{b}$, is considered to be* **well conditioned**
*if a small change in $[A]$ or $\mathbf{b}$ results in a small change in the solution vector*
$\mathbf{x}$. *A system of equations, $[A]\mathbf{x} = \mathbf{b}$, is considered to be* **ill conditioned** *if*
*a small change in $[A]$ or $\mathbf{b}$ results in a big change in the solution vector $\mathbf{x}$.*
*The system of equations is ill conditioned when $[A]$ has rows or columns so*
*nearly dependent on each other.*

*Consider the following set of equations:*

$$[A]\mathbf{x} = \mathbf{b}$$
$$\begin{bmatrix} 2 & 3.99 \\ 1 & 2 \end{bmatrix}\begin{bmatrix} x_1 \\ x_2 \end{bmatrix} = \begin{bmatrix} 1.99 \\ 1 \end{bmatrix} \tag{9.115}$$

*The solution of this set of equations is*

$$\begin{bmatrix} x_1 \\ x_2 \end{bmatrix} = \begin{bmatrix} -1.0 \\ 1.0 \end{bmatrix}. \tag{9.116}$$

*Let's make a small change in the $\mathbf{b}$ vector*

$$\begin{bmatrix} 2 & 3.99 \\ 1 & 2 \end{bmatrix}\begin{bmatrix} x_1 \\ x_2 \end{bmatrix} = \begin{bmatrix} 1.98 \\ 1.01 \end{bmatrix} \tag{9.117}$$

*and see how the solution will change.*

$$\begin{bmatrix} x_1 \\ x_2 \end{bmatrix} = \begin{bmatrix} -6.99 \\ 4.0 \end{bmatrix} \tag{9.118}$$

*Now we make a small change in $[A]$ matrix*

$$\begin{bmatrix} 2.01 & 3.98 \\ 0.99 & 2.01 \end{bmatrix}\begin{bmatrix} x_1 \\ x_2 \end{bmatrix} = \begin{bmatrix} 1.99 \\ 1 \end{bmatrix} \tag{9.119}$$

*and solve the equations*

$$\begin{bmatrix} x_1 \\ x_2 \end{bmatrix} = \begin{bmatrix} 0.1988 \\ 0.3993 \end{bmatrix}. \tag{9.120}$$

*Therefore, the set of equations (9.115) is ill conditioned and is sensitive to*
*perturbation in $[A]$ and $\mathbf{b}$. However, the set of equations*

$$\begin{bmatrix} 2 & 3 \\ 1 & 2 \end{bmatrix}\begin{bmatrix} x_1 \\ x_2 \end{bmatrix} = \begin{bmatrix} 1 \\ 1 \end{bmatrix} \tag{9.121}$$

*is well conditioned because small changes in $[A]$ or $\mathbf{b}$ cannot change the*
*solution drastically.*

*The sensitivity of the solution $\mathbf{x}$ to small perturbations in $[A]$ and $\mathbf{b}$ is*
*measured in terms of the* **condition number** *of $[A]$ by*

$$\frac{\|\triangle\mathbf{x}\|}{\|\mathbf{x}\|} \leq con\,(A)\,\frac{\|\triangle A\|}{\|A\|} \tag{9.122}$$

*where*

$$\text{con}(A) = \left\|A^{-1}\right\| \|A\| . \tag{9.123}$$

*and $\|A\|$ is a norm of $[A]$. If $\text{con}(A) = 1$. Then $[A]$ is called perfectly conditioned. The matrix $[A]$ is well conditioned if $\text{con}(A) < 1$ and it is ill conditioned if $\text{con}(A) > 1$. In fact, the relative change in the norm of the coefficient matrix, $[A]$, can be amplified by $\text{con}(A)$ to make the upper limit of the relative change in the norm of the solution vector $\mathbf{x}$.*

**Proof.** *Start with a set of equations*

$$[A]\,\mathbf{x} = \mathbf{b} \tag{9.124}$$

*and change the matrix $[A]$ to $[A']$. Then the solution $\mathbf{x}$ will change to $\mathbf{x}'$ such that*

$$[A']\,\mathbf{x}' = \mathbf{b}. \tag{9.125}$$

*Therefore,*

$$\begin{aligned} [A]\,\mathbf{x} &= [A']\,\mathbf{x}' \\ &= ([A] + \triangle A)(\mathbf{x} + \triangle\mathbf{x}) \end{aligned} \tag{9.126}$$

*where*

$$\begin{aligned} \triangle A &= [A'] - [A] \\ \triangle\mathbf{x} &= \mathbf{x}' - \mathbf{x}. \end{aligned} \tag{9.127} \tag{9.128}$$

*Expanding (9.126)*

$$[A]\,\mathbf{x} = [A]\,\mathbf{x} + [A]\,\triangle\mathbf{x} + \triangle A\,(\mathbf{x} + \triangle\mathbf{x}) \tag{9.129}$$

*and simplifying*

$$\triangle\mathbf{x} = -A^{-1}\triangle A\,(\mathbf{x} + \triangle\mathbf{x}) \tag{9.130}$$

*shows that*

$$\|\triangle\mathbf{x}\| \leq \left\|A^{-1}\right\| \|\triangle A\| \|\mathbf{x} + \triangle\mathbf{x}\| . \tag{9.131}$$

*Multiplying both sides of (9.131) by the norm $\|A\|$ leads to Equation (9.122).*
■

**Example 226 ★** *Norm of a matrix.*

*The norm of a matrix is a scalar positive number, and is defined for every kind of matrices including square, rectangular, invertible, and non-invertible. There are several definitions for the norm of a matrix. The most important ones are*

$$\|A\|_1 = \underset{1 \leq j \leq n}{Max} \sum_{i=1}^{n} |a_{ij}| \tag{9.132}$$

$$\|A\|_2 = \lambda_{Max}\left(A^T A\right) \tag{9.133}$$

$$\|A\|_\infty = \underset{1\le i\le n}{Max} \sum_{j=1}^{n} |a_{ij}| \qquad (9.134)$$

$$\|A\|_F = \sum_{i=1}^{n}\sum_{j=1}^{n} a_{ij}^2. \qquad (9.135)$$

*The norm-infinity, $\|A\|_\infty$, is the one we accept to calculate the con $(A)$ in Equation (9.123). The norm-infinity, $\|A\|_\infty$, is also called the* **row sum norm** *and* **uniform norm**. *To calculate $\|A\|_\infty$, we find the sum of the absolute of the elements of each row of the matrix $[A]$ and pick the largest sum.*

*As an example, the norm of*

$$[A] = \begin{bmatrix} 1 & 3 & -3 \\ -1 & -1 & 2 \\ 2 & 4 & -2 \end{bmatrix} \qquad (9.136)$$

*is*

$$
\begin{aligned}
\|A\|_\infty &= \underset{1\le i\le n}{Max} \sum_{j=1}^{n} |a_{ij}| \\
&= Max\left\{ \left(|1| + |3| + |-3|\right), \left(|-1| + |-1| + |2|\right), \left(|2| + |4| + |-2|\right) \right\} \\
&= Max\left\{7, 4, 8\right\} \\
&= 8.
\end{aligned}
\qquad (9.137)
$$

*We may check the following relations between norms of matrices.*

$$
\begin{aligned}
\|[A] + [B]\| &\le \|[A]\| + \|[B]\| & (9.138) \\
\|[A]\,[B]\| &\le \|[A]\|\,\|[B]\| & (9.139)
\end{aligned}
$$

## 9.2   Matrix Inversion

There are numerous techniques for matrix inversion. However the method based on the $LU$ factorization can simplify our numerical calculations since we have already applied the method for solving a set of linear algebraic equations.

Assume a matrix $[A]$ could be decomposed into

$$[A] = [L]\,[U] \qquad (9.140)$$

where

$$[A] = \begin{bmatrix} a_{11} & a_{12} & \vdots & a_{1n} \\ a_{21} & a_{22} & \cdots & a_{2n} \\ \cdots & \cdots & \cdots & \cdots \\ a_{n1} & a_{n2} & \cdots & a_{nn} \end{bmatrix} \qquad (9.141)$$

$$[L] = \begin{bmatrix} 1 & 0 & \cdots & 0 \\ l_{21} & 1 & \cdots & 0 \\ \cdots & \cdots & \cdots & \cdots \\ l_{n1} & l_{n2} & \cdots & 1 \end{bmatrix} \tag{9.142}$$

$$[U] = \begin{bmatrix} u_{11} & u_{12} & \cdots & u_{1n} \\ 0 & u_{22} & \cdots & u_{2n} \\ \cdots & \cdots & \cdots & \cdots \\ 0 & 0 & \cdots & u_{nn} \end{bmatrix} \tag{9.143}$$

then its inverse would be

$$[A]^{-1} = [U]^{-1}\,[L]^{-1}. \tag{9.144}$$

Because $[L]$ and $[U]$ are triangular matrices, their inverses are also triangular. The elements of the matrix $[M]$

$$[M] = [L]^{-1} = \begin{bmatrix} 1 & 0 & \cdots & 0 \\ m_{21} & 1 & \cdots & 0 \\ \cdots & \cdots & \cdots & \cdots \\ m_{n1} & m_{n2} & \cdots & 1 \end{bmatrix} \tag{9.145}$$

are

$$m_{ij} = -l_{ij} - \sum_{k=j+1}^{i-1} l_{ik}\,m_{kj} \qquad j < i \qquad i = 2, 3, \cdots n - 1 \tag{9.146}$$

and the elements of the matrix $[V]$

$$[V] = [U]^{-1} = \begin{bmatrix} v_{11} & v_{12} & v_{13} & v_{14} \\ 0 & v_{22} & v_{23} & v_{24} \\ 0 & 0 & v_{33} & v_{34} \\ 0 & 0 & 0 & v_{44} \end{bmatrix} \tag{9.147}$$

are

$$v_{ij} = \begin{cases} \dfrac{1}{u_{ij}} & j = i \quad i = n, n-1, \cdots, 1 \\[2ex] \dfrac{-1}{u_{ii}} \sum_{k=i+1}^{j} u_{ik} v_{kj} & j \geq i \quad i = n-1, \cdots, 2. \end{cases} \tag{9.148}$$

**Example 227** *Solution of a set of equations by matrix inversion.*
  *Consider a set of four linear algebraic equations.*

$$[A]\,\mathbf{x} = \mathbf{b} \tag{9.149}$$

$$\begin{bmatrix} 2 & 1 & 3 & -3 \\ 1 & 0 & -1 & -2 \\ 0 & 2 & 2 & 1 \\ 3 & 1 & 0 & -2 \end{bmatrix} \begin{bmatrix} x_1 \\ x_2 \\ x_3 \\ x_4 \end{bmatrix} = \begin{bmatrix} 1 \\ 2 \\ 0 \\ -2 \end{bmatrix} \tag{9.150}$$

*Following the LU factorization algorithm, we can decompose the coefficient matrix to*

$$[A] = [L][U] \tag{9.151}$$

*where*

$$[L] = \begin{bmatrix} 1 & 0 & 0 & 0 \\ \frac{1}{2} & 1 & 0 & 0 \\ 0 & -4 & 1 & 0 \\ \frac{3}{2} & 1 & \frac{1}{4} & 1 \end{bmatrix} \tag{9.152}$$

$$[U] = \begin{bmatrix} 2 & 1 & 3 & -3 \\ 0 & -\frac{1}{2} & -\frac{5}{2} & -\frac{1}{2} \\ 0 & 0 & -8 & -1 \\ 0 & 0 & 0 & \frac{13}{4} \end{bmatrix}. \tag{9.153}$$

*The inverse of matrices $[L]$ and $[U]$ are*

$$[L]^{-1} = \begin{bmatrix} 1 & 0 & 0 & 0 \\ -\frac{1}{2} & 1 & 0 & 0 \\ -2 & 4 & 1 & 0 \\ -\frac{1}{2} & -2 & -\frac{1}{4} & 1 \end{bmatrix} \tag{9.154}$$

$$[U]^{-1} = \begin{bmatrix} \frac{1}{2} & 1 & -\frac{1}{8} & \frac{15}{26} \\ 0 & -2 & \frac{5}{8} & -\frac{3}{26} \\ 0 & 0 & -\frac{1}{8} & -\frac{1}{26} \\ 0 & 0 & 0 & \frac{4}{13} \end{bmatrix} \tag{9.155}$$

*and therefore the solution of the equations is*

$$\mathbf{x} = [U]^{-1}[L]^{-1}\mathbf{b} \tag{9.156}$$

$$= \begin{bmatrix} -\frac{5}{2} \\ \frac{3}{2} \\ -\frac{1}{2} \\ -2 \end{bmatrix}. \tag{9.157}$$

**Example 228 ★** *LU factorization method compared to other methods.*

*Every nonsingular matrix $[A]$ can be decomposed into lower and upper triangular matrices $[A] = [L][U]$. Then, the solution of a set of equations $[A]\mathbf{x} = \mathbf{b}$ is equivalent to*

$$[L][U]\mathbf{x} = \mathbf{b}. \tag{9.158}$$

*Multiplying both sides by $L^{-1}$ shows that*

$$[U]\mathbf{x} = [L]^{-1}\mathbf{b} \tag{9.159}$$

*and the problem is broken into two new sets of equations*

$$[L]\mathbf{y} = \mathbf{b} \tag{9.160}$$

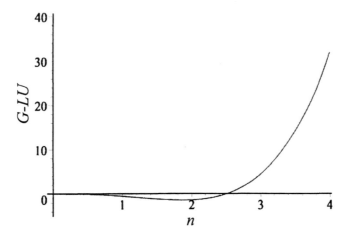

FIGURE 9.1. The number of calculations for the Gaussian elimination subtracted by the *LU* factorization methods, as a function of the size of the matrix.

*and*

$$[U]\, \mathbf{x} = \mathbf{y}. \tag{9.161}$$

*The computational time required to decompose $[A]$ into $[L]\,[U]$ is proportional to $n^3/3$, where $n$ is the number of equations. Then, the computational time for solving each set of $[L]\,\mathbf{y} = \mathbf{b}$ and $[U]\,\mathbf{x} = \mathbf{y}$ is proportional to $n^2/2$. Therefore, the total computational time for solving a set of equations by the LU factorization method is proportional to $n^2 + n^3/3$. However, the Gaussian elimination method takes a computational time proportional to $n^2/2 + n^3/3$, forward elimination takes a time proportional to $n^3/3$, and back substitution takes a time proportional to $n^2/2$.*

*On the other hand, the total computational time required to inverse a matrix using the LU factorization method is proportional to $4n^3/3$. However, the Gaussian elimination method needs $n^4/3 + n^3/2$, and*

$$\frac{n^4}{3} + \frac{n^3}{2} > \frac{4n^3}{3} \qquad n > 2. \tag{9.162}$$

*Figure 9.1 depicts a plot of the function $G-LU = \frac{n^4}{3} + \frac{n^3}{2} - \frac{4n^3}{3}$ and shows how fast the number of calculations for the Gaussian elimination, compared to the LU factorization methods, increases. As an example, for a $6 \times 6$ matrix inversion, we need 540 calculations for the Gaussian elimination method, compared to 288 calculations for the LU factorization method.*

**Example 229 ★** *Partitioning inverse method.*
*Assume that a matrix $[T]$ can be partitioned into*

$$[T] = \begin{bmatrix} A & B \\ C & D \end{bmatrix} \tag{9.163}$$

*then, $T^{-1}$ can be calculated by*

$$T^{-1} = \begin{bmatrix} E & F \\ G & H \end{bmatrix} \qquad (9.164)$$

*where*

$$[E] = \left[A - BD^{-1}\right]^{-1} \qquad (9.165)$$

$$[H] = \left[D - CA^{-1}B\right]^{-1} \qquad (9.166)$$

$$[F] = -A^{-1}BH \qquad (9.167)$$

$$[G] = -D^{-1}CE. \qquad (9.168)$$

*Sometimes, it is a shortcut inverse method.*

**Example 230 ★** *Analytic inversion method.*
   *If the $n \times n$ matrix $[A] = [a_{ij}]$ is non-singular, that is $\det(A) \neq 0$, we may compute the inverse, $A^{-1}$, by dividing the adjoint matrix $A^a$ by the determinant of $[A]$.*

$$A^{-1} = \frac{A^a}{\det(A)} \qquad (9.169)$$

*The **adjoint** or **adjugate matrix** of the matrix $[A]$, is the transpose of the cofactor matrix of $[A]$.*

$$A^a = A^{c^T} \qquad (9.170)$$

*The **cofactor matrix**, denoted by $A^c$, for a matrix $[A]$, is made of the matrix $[A]$ by replacing each of its elements by its cofactor. The **cofactor** associated with the element $a_{ij}$ is defined by*

$$A_{ij}^c = (-1)^{i+j} A_{ij} \qquad (9.171)$$

*where $A_{ij}$ is the $ij$-minor of $[A]$. Associated with each element $a_{ij}$ of the matrix $T$, there exists a **minor** $A_{ij}$ which is a number equal to the value of the determinant of the submatrix obtained by deleting row $i$ and column $j$ of the matrix $[A]$.*
   *The determinant of $[A]$ is calculated by*

$$\det(A) = \sum_{j=1}^{n} a_{ij} A_{ij}^c. \qquad (9.172)$$

*Therefore, if*

$$[A] = \begin{bmatrix} a_{11} & a_{12} & a_{13} \\ a_{21} & a_{22} & a_{23} \\ a_{31} & a_{32} & a_{33} \end{bmatrix} \qquad (9.173)$$

*then the elements of adjoint matrix $A^a$ are*

$$A_{11}^a = A_{11}^c = (-1)^2 \begin{vmatrix} a_{22} & a_{23} \\ a_{32} & a_{33} \end{vmatrix} \qquad (9.174)$$

$$A_{21}^a = A_{12}^c = (-1)^3 \begin{vmatrix} a_{21} & a_{23} \\ a_{31} & a_{33} \end{vmatrix} \qquad (9.175)$$

$$\vdots$$

$$A_{33}^a = A_{33}^c = (-1)^6 \begin{vmatrix} a_{11} & a_{12} \\ a_{21} & a_{22} \end{vmatrix}. \qquad (9.176)$$

and the determinant of $[A]$ is

$$\begin{aligned} \det(A) &= a_{11}a_{22}a_{33} - a_{11}a_{23}a_{32} - a_{12}a_{21}a_{33} \\ &\quad + a_{12}a_{31}a_{23} + a_{21}a_{13}a_{32} - a_{13}a_{22}a_{31}. \qquad (9.177) \end{aligned}$$

As an example, consider a $3 \times 3$ matrix as below

$$[A] = \begin{bmatrix} 3 & 4 & 8 \\ 7 & 2 & 5 \\ 9 & 6 & 1 \end{bmatrix}.$$

The associated adjoint matrix for $[A]$ is

$$\begin{aligned} A^a = A^{c^T} &= \begin{bmatrix} -28 & 38 & 24 \\ 44 & -69 & 18 \\ 4 & 41 & -22 \end{bmatrix}^T \\ &= \begin{bmatrix} -28 & 44 & 4 \\ 38 & -69 & 41 \\ 24 & 18 & -22 \end{bmatrix} \end{aligned}$$

and the determinant of $[A]$ is

$$\det[A] = 260$$

and therefore,

$$\begin{aligned} [A]^{-1} &= \frac{A^a}{\det(A)} \\ &= \begin{bmatrix} -\frac{7}{65} & \frac{11}{65} & \frac{1}{65} \\ \frac{19}{130} & -\frac{69}{260} & \frac{41}{260} \\ \frac{6}{65} & \frac{9}{130} & -\frac{11}{130} \end{bmatrix}. \end{aligned}$$

**Example 231 ★** *Cayley-Hamilton matrix inversion.*

The Cayley-Hamilton theorem says: *Every non-singular matrix satisfies its own characteristic equation.* The characteristic equation of an $n \times n$ matrix $[A] = [a_{ij}]$ is

$$\begin{aligned} \det(A - \lambda I) &= |A - \lambda I| \\ &= P(\lambda) = \lambda^n + a_{n-1}\lambda^{n-1} + \cdots + a_1\lambda + a_0 = 0. \quad (9.178) \end{aligned}$$

*Hence, the characteristic equation of an $n \times n$ matrix is a polynomial of degree $n$. Based on the Cayley-Hamilton theorem, we have*

$$P(A) = A^n + a_{n-1}A^{n-1} + \cdots + a_1 A + a_0 = 0. \tag{9.179}$$

*Multiplying both sides of this polynomial by $A^{-1}$ and solving for $A^{-1}$ provides*

$$A^{-1} = -\frac{1}{a_0} \left[ A^{n-1} + a_{n-1}A^{n-2} + \cdots + a_2 A + a_1 I \right]. \tag{9.180}$$

*Therefore, if*

$$[A] = \begin{bmatrix} a_{11} & a_{12} & a_{13} \\ a_{21} & a_{22} & a_{23} \\ a_{31} & a_{32} & a_{33} \end{bmatrix} \tag{9.181}$$

*and*

$$\det(A) = \sum_{j=1}^{n} a_{ij} A_{ij}^c \tag{9.182}$$

*then the characteristic equation of $[A]$ is*

$$\begin{aligned} P(\lambda) \quad = \quad & \det(A - \lambda I) = \lambda^3 + (-a_{11} - a_{22} - a_{33})\lambda^2 \\ & + (a_{11}a_{22} - a_{12}a_{21} + a_{11}a_{33} - a_{13}a_{31} + a_{22}a_{33} - a_{23}a_{32})\lambda \\ & + a_{11}a_{23}a_{32} + a_{12}a_{21}a_{33} + a_{13}a_{22}a_{31} \\ & - a_{11}a_{22}a_{33} - a_{12}a_{31}a_{23} - a_{21}a_{13}a_{32}. \end{aligned} \tag{9.183}$$

*As an example, consider a $3 \times 3$ matrix*

$$[A] = \begin{bmatrix} 1 & 2 & 3 \\ 4 & 6 & 7 \\ 5 & 8 & 9 \end{bmatrix} \tag{9.184}$$

*with following characteristic equation:*

$$\lambda^3 - 16\lambda^2 - 10\lambda - 2 = 0. \tag{9.185}$$

*Because $[A]$ satisfies its own characteristic equation, we have*

$$A^3 - 16A^2 - 10A - 2 = 0. \tag{9.186}$$

*Multiplying both sides by $A^{-1}$*

$$A^{-1}A^3 - 16A^{-1}A^2 - 10A^{-1}A = 2A^{-1} \tag{9.187}$$

*provides the inverse matrix.*

$$A^{-1} = \frac{1}{2}\left(A^2 - 16A - 10I\right)$$

$$= \frac{1}{2}\left[\begin{bmatrix} 1 & 2 & 3 \\ 4 & 6 & 7 \\ 5 & 8 & 9 \end{bmatrix}^2 - 16\begin{bmatrix} 1 & 2 & 3 \\ 4 & 6 & 7 \\ 5 & 8 & 9 \end{bmatrix} - 10\begin{bmatrix} 1 & 0 & 0 \\ 0 & 1 & 0 \\ 0 & 0 & 1 \end{bmatrix}\right]$$

$$= \begin{bmatrix} -1 & 3 & -2 \\ -\frac{1}{2} & -3 & \frac{5}{2} \\ 1 & 1 & -1 \end{bmatrix} \tag{9.188}$$

## 9.3   Nonlinear Algebraic Equations

Consider a set of nonlinear algebraic equations

$$\mathbf{f}(\mathbf{q}) = 0 \tag{9.189}$$

or

$$\begin{aligned} f_1(q_1, q_2, \cdots, q_n) &= 0 \\ f_2(q_1, q_2, \cdots, q_n) &= 0 \\ &\cdots \\ f_n(q_1, q_2, \cdots, q_n) &= 0 \end{aligned} \tag{9.190}$$

where the function and variable vectors are

$$\mathbf{f} = \begin{bmatrix} f_1(\mathbf{q}) \\ f_2(\mathbf{q}) \\ \cdots \\ f_n(\mathbf{q}) \end{bmatrix} \qquad \mathbf{q} = \begin{bmatrix} q_1 \\ q_2 \\ \cdots \\ q_n \end{bmatrix}. \tag{9.191}$$

To solve the set of equations (9.189), we begin with a guess solution vector $\mathbf{q}^{(0)}$, and employ the following *iteration formula* to search for a better solution

$$\mathbf{q}^{(i+1)} = \mathbf{q}^{(i)} - \mathbf{J}^{-1}(\mathbf{q}^{(i)})\,\mathbf{f}(\mathbf{q}^{(i)}) \tag{9.192}$$

where $\mathbf{J}^{-1}(\mathbf{q}^{(i)})$ is the Jacobian matrix of the system of equations evaluated at $\mathbf{q} = \mathbf{q}^{(i)}$.

$$[\mathbf{J}] = \left[\frac{\partial f_i}{\partial q_j}\right] \tag{9.193}$$

Utilizing the iteration formula (9.192), we can approach an exact solution as desired. The iteration method based on a guess solution is called *Newton-Raphson method*, which is the most common method for solving a set of nonlinear algebraic equations.

A set of nonlinear equations usually has multiple solutions and the main disadvantage of the Newton-Raphson method for solving a set of nonlinear equations is that the solution may not be the solution of interest. The solution that the method will provide depends highly on the initial estimation. Hence, having a correct estimate helps to detect the proper solution.

**Proof.** Let us define the increment $\delta^{(i)}$ as

$$\delta^{(i)} = q^{(i+1)} - q^{(i)} \tag{9.194}$$

and expand the set of equations around $q^{(i+1)}$

$$f(q^{(i+1)}) = f(q^{(i)}) + J(q^{(i)})\, \delta^{(i)}. \tag{9.195}$$

Assume that $q^{(i+1)}$ is the exact solution of Equation (9.189). Therefore, $f(q^{(i+1)}) = 0$ and we may use

$$J(q^{(i)})\, \delta^{(i)} = -f(q^{(i)}) \tag{9.196}$$

to find the increment $\delta^{(i)}$

$$\delta^{(i)} = -J^{-1}(q^{(i)})\, f(q^{(i)}) \tag{9.197}$$

and determine the solution

$$\begin{aligned} q^{(i+1)} &= q^{(i)} + \delta^{(i)} \\ &= q^{(i)} - J^{-1}(q^{(i)})\, f(q^{(i)}). \end{aligned} \tag{9.198}$$

The Newton-Raphson iteration method can be set up as an algorithm for better application.

**Algorithm 9.3.** *Newton-Raphson iteration method for* $f(q) = 0$.

1. *Set the initial counter* $i = 0$.

2. *Evaluate an estimate solution* $q = q^{(i)}$.

3. *Calculate the Jacobian matrix* $[J] = \left[\frac{\partial f_i}{\partial q_j}\right]$ *at* $q = q^{(i)}$.

4. *Solve for* $\delta^{(i)}$ *from the set of linear equations* $J(q^{(i)})\, \delta^{(i)} = -f(q^{(i)})$.

5. *If* $\left|\delta^{(i)}\right| < \epsilon$, *where* $\epsilon$ *is an arbitrary tolerance then,* $q^{(i)}$ *is the solution. Otherwise calculate* $q^{(i+1)} = q^{(i)} + \delta^{(i)}$.

6. *Set* $i = i + 1$ *and return to step 3.*

■

**Example 232** *Inverse kinematics problem for a 2R planar robot.*

*The endpoint of a 2R planar manipulator can be described by two non-linear algebraic equations.*

$$\begin{bmatrix} f_1(\theta_1, \theta_2) \\ f_2(\theta_1, \theta_2) \end{bmatrix} = \begin{bmatrix} l_1 \cos\theta_1 + l_2 \cos(\theta_1 + \theta_2) - X \\ l_1 \sin\theta_1 + l_2 \sin(\theta_1 + \theta_2) - Y \end{bmatrix} = 0 \qquad (9.199)$$

*Assuming*

$$l_1 = l_2 = 1 \qquad (9.200)$$

*and the endpoint is at*

$$\begin{bmatrix} X \\ Y \end{bmatrix} = \begin{bmatrix} 1 \\ 1 \end{bmatrix} \qquad (9.201)$$

*we are looking for the associated variables*

$$\boldsymbol{\theta} = \begin{bmatrix} \theta_1 \\ \theta_2 \end{bmatrix} \qquad (9.202)$$

*that provide the desired position of the endpoint. Due to simplicity of the system of equations, the Jacobian of the equations and its inverse can be found in closed form*

$$\begin{aligned} \mathbf{J}(\boldsymbol{\theta}) &= \begin{bmatrix} \dfrac{\partial f_i}{\partial \theta_j} \end{bmatrix} = \begin{bmatrix} \dfrac{\partial f_1}{\partial \theta_1} & \dfrac{\partial f_1}{\partial \theta_2} \\ \dfrac{\partial f_2}{\partial \theta_1} & \dfrac{\partial f_2}{\partial \theta_2} \end{bmatrix} \\ &= \begin{bmatrix} -l_1 \sin\theta_1 - l_2 \sin(\theta_1 + \theta_2) & -l_2 \sin(\theta_1 + \theta_2) \\ l_1 \cos\theta_1 + l_2 \cos(\theta_1 + \theta_2) & l_2 \cos(\theta_1 + \theta_2) \end{bmatrix} \end{aligned} \qquad (9.203)$$

$$\mathbf{J}^{-1} = \frac{-1}{l_1 l_2 s \theta_2} \begin{bmatrix} -l_2 c(\theta_1 + \theta_2) & -l_2 s(\theta_1 + \theta_2) \\ l_1 c\theta_1 + l_2 c(\theta_1 + \theta_2) & l_1 s\theta_1 + l_2 s(\theta_1 + \theta_2) \end{bmatrix}. \qquad (9.204)$$

*The Newton-Raphson iteration algorithm may now be started by setting $i = 0$ and evaluating an estimate solution*

$$\mathbf{q}^{(0)} = \begin{bmatrix} \theta_1 \\ \theta_2 \end{bmatrix}^{(0)} = \begin{bmatrix} \pi/3 \\ -\pi/3 \end{bmatrix}. \qquad (9.205)$$

*Therefore,*

$$\mathbf{J}(\frac{\pi}{3}, \frac{\pi}{3}) = \begin{bmatrix} -\sqrt{3} & -\frac{1}{2}\sqrt{3} \\ 0 & -\frac{1}{2} \end{bmatrix} \qquad (9.206)$$

$$\mathbf{f}(\frac{\pi}{3}, \frac{\pi}{3}) = \begin{bmatrix} -1 \\ \sqrt{3} - 1 \end{bmatrix} \qquad (9.207)$$

$$\boldsymbol{\delta}^{(0)} = -\mathbf{J}^{-1}(\boldsymbol{\theta}^{(0)})\, \mathbf{f}(\boldsymbol{\theta}^{(0)}) = \begin{bmatrix} -1.3094 \\ 1.4641 \end{bmatrix} \qquad (9.208)$$

*and a better solution is*

$$\begin{bmatrix} \theta_1 \\ \theta_2 \end{bmatrix}^{(1)} = \begin{bmatrix} \pi/3 \\ \pi/3 \end{bmatrix} + \begin{bmatrix} -1.3094 \\ 1.4641 \end{bmatrix} = \begin{bmatrix} -0.2622 \\ 2.5113 \end{bmatrix}. \qquad (9.209)$$

*In the next iterations we find*

$$\begin{bmatrix} \theta_1 \\ \theta_2 \end{bmatrix}^{(2)} = \begin{bmatrix} -0.2622 \\ 2.5113 \end{bmatrix} + \begin{bmatrix} -.06952 \\ -.80337 \end{bmatrix} = \begin{bmatrix} -0.3317 \\ 1.7079 \end{bmatrix}$$

$$\begin{bmatrix} \theta_1 \\ \theta_2 \end{bmatrix}^{(3)} = \begin{bmatrix} -0.3317 \\ 1.7079 \end{bmatrix} + \begin{bmatrix} .31414 \\ -.068348 \end{bmatrix} = \begin{bmatrix} -0.0176 \\ 1.63958 \end{bmatrix}$$

$$\begin{bmatrix} \theta_1 \\ \theta_2 \end{bmatrix}^{(4)} = \begin{bmatrix} -0.0176 \\ 1.63958 \end{bmatrix} + \begin{bmatrix} .016275 \\ -.06739 \end{bmatrix} = \begin{bmatrix} -.0013 \\ 1.5722 \end{bmatrix}$$

$$\begin{bmatrix} \theta_1 \\ \theta_2 \end{bmatrix}^{(5)} = \begin{bmatrix} -.0013 \\ 1.5722 \end{bmatrix} + \begin{bmatrix} .1304 \\ -.139 \end{bmatrix} = \begin{bmatrix} -.295 \times 10^{-8} \\ 1.571 \end{bmatrix}$$

$$\begin{bmatrix} \theta_1 \\ \theta_2 \end{bmatrix}^{(6)} = \begin{bmatrix} -.3 \times 10^{-8} \\ 1.571 \end{bmatrix} + \begin{bmatrix} .29 \times 10^{-8} \\ -.85 \times 10^{-6} \end{bmatrix} = \begin{bmatrix} -.49 \times 10^{-10} \\ 1.571 \end{bmatrix}$$

$$\begin{bmatrix} \theta_1 \\ \theta_2 \end{bmatrix}^{(7)} = \begin{bmatrix} -.49 \times 10^{-10} \\ 1.571 \end{bmatrix} + \begin{bmatrix} -.41 \times 10^{-19} \\ -.2 \times 10^{-9} \end{bmatrix} = \begin{bmatrix} -.49 \times 10^{-10} \\ 1.571 \end{bmatrix}$$

*and this answer is close enough to the exact elbow down answer*

$$\begin{bmatrix} \theta_1 \\ \theta_2 \end{bmatrix} = \begin{bmatrix} 0 \\ \pi/2 \end{bmatrix}. \tag{9.210}$$

**Example 233** ★ *Alternative and expanded proof for Newton-Raphson iteration method.*

*Consider the following set of equations in which we are searching for the exact solutions $q_j$*

$$y_i = f_i(q_j) \tag{9.211}$$
$$i = 1, \cdots, n$$
$$j = 1, \cdots, m$$

*where $j$ is the number of unknowns and $i$ is the number of equations.*

*Assume that, for a given $y_i$ an approximate solution $q_j^\star$ is available. The difference between the exact solution $q_j$ and the approximate solution $q_j^\star$ is*

$$\delta_j = q_j - q_j^\star \tag{9.212}$$

*where the value of equations for the approximate solution $q_j^\star$ is denoted by*

$$Y_i = f_i(q_j^\star). \tag{9.213}$$

*The iteration method is based on the minimization of $\delta_j$ to make the solution of*

$$y_i = f_i(q_j^\star + \delta_j) \tag{9.214}$$

*be as close as possible to the exact solution.*

*A first-order Taylor expansion of this equation is*

$$y_i = f_i(q_j^\star) + \sum_{i=1}^{m} \frac{\partial f_i}{\partial q_j} \delta_j + O(\delta_j^2). \tag{9.215}$$

*We may define*

$$Y_i = f_i(q_j^\star) \tag{9.216}$$

*and the residual quantity*

$$r_i = y_i - Y_i \tag{9.217}$$

*to write*

$$\mathbf{r} = \mathbf{J}\,\boldsymbol{\delta} + O(\delta^2) \tag{9.218}$$

*where $\mathbf{J}$ is the Jacobian matrix of the set of equations*

$$\mathbf{J} = \left[\frac{\partial f_i}{\partial q_j}\right]. \tag{9.219}$$

*The method of solution depends on the relative value of m and n. There-fore, three cases must be considered.*

**1- $m = n$**

*Provided that the Jacobian matrix remains non-singular, the linearized equation*

$$\mathbf{r} = \mathbf{J}\,\delta\mathbf{q} \tag{9.220}$$

*possesses a unique solution, and the Newton-Raphson technique may then be utilized to solve equation (9.211). The stepwise procedure is illustrated in figure 9.2.*

*The effectiveness of the procedure depends on the number of iterations to be performed, which depends on the initial estimate of $q_j^\star$ and on the dimension of the Jacobian matrix. Since the solution to nonlinear equations is not unique, it may generate different sets of solutions depending on the initial guess. Furthermore, convergence may not occur if the initial estimate of the solution falls outside the convergence domain of the algorithm. In this case, much effort is needed to attain a numerical solution.*

**2- $m < n$**

*This is the overdetermined case for which no solution exists in general, because the number of unknowns (such as the number of joints in robots) are not sufficient enough to generate a solution (such as an arbitrary con-figuration of the end-effector). A solution can, however, be generated that minimizes an error (such as position error).*

*Consider the problem*

$$\min\left(F = \frac{1}{2}\sum_{i=1}^{n} w_i\,[y_i - f_i(q_j)]^2\right) \tag{9.221}$$

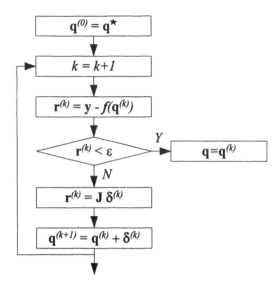

FIGURE 9.2. Newton-Raphson iteration method for solving a set of nonlinear algebraic equations.

*or, in matrix form,*

$$\min\left(F = \frac{1}{2}\sum_{i=1}^{n}[\mathbf{y}-\mathbf{f(q)}]^T\,\mathbf{W}\,[\mathbf{y}-\mathbf{f(q)}]\right) \tag{9.222}$$

*where*

$$\mathbf{W} = diag(w_1\cdots w_n) \tag{9.223}$$

*is a set of weighting factors giving a relative importance to each of the kinematic equations.*

*The error is minimum when*

$$\frac{\partial F}{\partial q_j} = -\sum_i \frac{\partial f_j}{\partial q_j}w_i\,[y_i - f_i(q_j)] = 0 \tag{9.224}$$

*or, in matrix form,*

$$\mathbf{J}^T\mathbf{W}\,[\mathbf{y}-\mathbf{f(q)}] = 0. \tag{9.225}$$

*A Taylor expansion of the third factor shows that the linear correction to an estimated solution* $\mathbf{q}^\star$ *is*

$$\mathbf{J}^T\mathbf{W}\,[\mathbf{y}-\mathbf{f(q^\star)}] - \mathbf{J}\,\delta = 0. \tag{9.226}$$

*The correction equation is*

$$\mathbf{J}^T\mathbf{W}\mathbf{J}\,\delta = \mathbf{J}^T\mathbf{W}\mathbf{r} \tag{9.227}$$

*where* $\mathbf{r}$ *is the residual vector defined by Equation (9.217).*

The weighting factor $\mathbf{W}$ is a positive definite diagonal matrix, and therefore, the matrix $\mathbf{J}^T\mathbf{W}\mathbf{J}$ is always symmetric and invertible. It provides the generalized inverse to the Jacobian matrix

$$\mathbf{J}^{-1} = \left[\mathbf{J}^T\mathbf{W}\mathbf{J}\right]^{-1}\mathbf{J}^T\mathbf{W} \tag{9.228}$$

which verifies the property

$$\mathbf{J}^{-1}\mathbf{J} = \mathbf{I}. \tag{9.229}$$

When the Jacobian is revertible, the solution for (9.222) is the solution to the nonlinear system (9.211).

**3- $m > n$**

This is the redundant case for which an infinity of solutions is generally available. Selection of an appropriate solution can be made under the condition that it is optimal in some sense. For example, let us find a solution for (9.211) which minimizes the deviation from a given reference configuration $q^{(0)}$. The problem may then be formulated as that of finding the minimum of a constrained function

$$\min\left(F = \frac{1}{2}\left[\mathbf{q} - \mathbf{q}^{(0)}\right]^T\mathbf{W}\left[\mathbf{q} - \mathbf{q}^{(0)}\right]\right) \tag{9.230}$$

subject to

$$\mathbf{y} - \mathbf{f}(\mathbf{q}) = 0. \tag{9.231}$$

Using the technique of Lagrangian multipliers, problem (9.230) and (9.231) may be replaced by an equivalent problem

$$\frac{\partial G}{\partial \mathbf{q}} = 0 \tag{9.232}$$

$$\frac{\partial G}{\partial \lambda} = 0 \tag{9.233}$$

with the definition of the functional

$$G(\mathbf{q}, \lambda) = \frac{1}{2}\left[\mathbf{q} - \mathbf{q}^{(0)}\right]^T\mathbf{W}\left[\mathbf{q} - \mathbf{q}^{(0)}\right] + \lambda^T\left[\mathbf{y} - \mathbf{f}(\mathbf{q})\right]. \tag{9.234}$$

It leads to a system of $m + n$ equations with $m + n$ unknowns

$$\begin{aligned}\mathbf{W}\left[\mathbf{q} - \mathbf{q}^{(0)}\right] - \mathbf{J}^T\lambda &= 0 \\ \mathbf{y} - \mathbf{f}(\mathbf{q}) &= 0.\end{aligned} \tag{9.235}$$

Linearization of equations (9.235) provides the system of equations for the displacement corrections and variations of Lagrangian multipliers

$$\mathbf{W}\,\delta - \mathbf{J}^T\,\lambda' = 0 \tag{9.236}$$

$$\mathbf{W}\,\delta = \mathbf{r} \tag{9.237}$$

*where $\lambda'$ is the increment of $\lambda$.*

*Substitution of the solution $\delta$ obtained from the first equation of (9.235) into the second one yields*

$$\mathbf{JW}^{-1}\mathbf{J}^T \delta\lambda = \mathbf{r} \tag{9.238}$$

*or, in terms of the displacement correction*

$$\delta\mathbf{q} = \mathbf{W}^{-1}\mathbf{J}^T \left( \mathbf{JW}^{-1}\mathbf{J}^T \right). \tag{9.239}$$

*The matrix*

$$\mathbf{J}^+ = \mathbf{W}^{-1}\mathbf{J}^T \left( \mathbf{JW}^{-1}\mathbf{J}^T \right)^{-1} \tag{9.240}$$

*has the meaning of a pseudo-inverse to the singular Jacobian matrix $\mathbf{J}$. It verifies the identity*

$$\mathbf{JJ}^+ = \mathbf{I} \tag{9.241}$$

*and, whenever $\mathbf{J}$ is invertible,*

$$\mathbf{J}^+ = \mathbf{J}^{-1}. \tag{9.242}$$

# 9.4  ★ Jacobian Matrix From Link Transformation Matrices

In robot motion, we need to calculate the Jacobian matrix in a very short time for every configuration of the robot. The Jacobian matrix of a robot can be found easier and in an algorithmic way by evaluating columns of the Jacobian

$$\begin{aligned} \mathbf{J} &= \begin{bmatrix} \mathbf{c}_1 & \mathbf{c}_2 & \cdots & \mathbf{c}_n \end{bmatrix} \\ &= \begin{bmatrix} {}^0\tilde{k}_0\, {}^0_0\mathbf{d}_n & {}^0\tilde{k}_1\, {}^0_1\mathbf{d}_n & \cdots & {}^0\tilde{k}_{n-1}\, {}^0_{n-1}\mathbf{d}_n \\ {}^0\hat{k}_0 & {}^0\hat{k}_1 & \cdots & {}^0\hat{k}_{n-1} \end{bmatrix} & (9.243) \\ &= \begin{bmatrix} {}^0\hat{k}_0 \times {}^0_0\mathbf{d}_n & {}^0\hat{k}_1 \times {}^0_1\mathbf{d}_n & \cdots & {}^0\hat{k}_{n-1} \times {}^0_{n-1}\mathbf{d}_n \\ {}^0\hat{k}_0 & {}^0\hat{k}_1 & \cdots & {}^0\hat{k}_{n-1} \end{bmatrix} & (9.244) \end{aligned}$$

where $\mathbf{c}_i$ is called the *Jacobian generating vector*

$$\mathbf{c}_i = \begin{bmatrix} {}^0\tilde{k}_{i-1}\, {}^0_{i-1}\mathbf{d}_n \\ {}^0\hat{k}_{i-1} \end{bmatrix} = \begin{bmatrix} {}^0\hat{k}_{i-1} \times {}^0_{i-1}\mathbf{d}_n \\ {}^0\hat{k}_{i-1} \end{bmatrix} \tag{9.245}$$

and ${}^0\hat{k}_{i-1}$ is the vector associated to the skew matrix ${}^0\tilde{k}_{i-1}$. This method is solely based on link transformation matrices found in forward kinematics and does not involve differentiation.

The matrix ${}^0\tilde{k}_{i-1}$ is

$$ {}^0\tilde{k}_{i-1} = {}^0R_{i-1}\, {}^{i-1}\tilde{k}_{i-1}\, {}^0R_{i-1}^T \tag{9.246}$$

which means $^0\hat{k}_{i-1}$ is a unit vector in the direction of joint axis $i$ in the global coordinate frame. For a revolute joint, we have

$$\mathbf{c}_i = \begin{bmatrix} ^0\tilde{k}_{i-1}\,^0_{i-1}\mathbf{d}_n \\ ^0\hat{k}_{i-1} \end{bmatrix} \tag{9.247}$$

and for a prismatic joint we have

$$\mathbf{c}_i = \begin{bmatrix} ^0\hat{k}_{i-1} \\ 0 \end{bmatrix}. \tag{9.248}$$

**Proof.** Transformation between two coordinate frames

$$^G\mathbf{r} = {}^GT_B\,^B\mathbf{r} \tag{9.249}$$

is based on a transformation matrix that is a combination of rotation matrix $R$ and the position vector $\mathbf{d}$.

$$T = \begin{bmatrix} R & \mathbf{d} \\ 0 & 1 \end{bmatrix}. \tag{9.250}$$

Introducing the infinitesimal transformation matrix

$$\delta T = \begin{bmatrix} \delta R & \delta \mathbf{d} \\ 0 & 0 \end{bmatrix} \tag{9.251}$$

leads to

$$\delta T\, T^{-1} = \begin{bmatrix} \tilde{\delta\theta} & \delta \mathbf{v} \\ 0 & 0 \end{bmatrix} \tag{9.252}$$

where

$$T^{-1} = \begin{bmatrix} R^T & -R^T\mathbf{d} \\ 0 & 1 \end{bmatrix} \tag{9.253}$$

and therefore, $\tilde{\delta\theta}$ is the matrix of infinitesimal rotations,

$$\tilde{\delta\theta} = \delta R\, R^T = \begin{bmatrix} 0 & -\delta\theta_z & \delta\theta_y \\ \delta\theta_z & 0 & -\delta\theta_x \\ -\delta\theta_y & \delta\theta_x & 0 \end{bmatrix} \tag{9.254}$$

and $\delta\mathbf{v}$ is a vector related to infinitesimal displacements,

$$\delta\mathbf{v} = \delta\mathbf{d} - \tilde{\delta\theta}\,\mathbf{d}. \tag{9.255}$$

Let's define a $6 \times 1$ coordinate vector describing the rotational and translational coordinates of the end-effector

$$\mathbf{X} = \begin{bmatrix} \mathbf{d} \\ \theta \end{bmatrix} \tag{9.256}$$

which its variation is

$$\delta \mathbf{X} = \begin{bmatrix} \delta \mathbf{d} \\ \delta \theta \end{bmatrix}. \tag{9.257}$$

The Jacobian matrix $\mathbf{J}$ is then a matrix that maps differential joint variables to differential end-effector motion.

$$\delta \mathbf{X} = \frac{\partial \mathbf{T}(\mathbf{q})}{\partial \mathbf{q}} \delta \mathbf{q} = \mathbf{J} \, \delta \mathbf{q} \tag{9.258}$$

The transformation matrix $T$, generated in forward kinematics, is a function of joint coordinates

$$\begin{aligned} {}^0T_n &= \mathbf{T}(\mathbf{q}) \\ &= {}^0T_1(q_1)\,{}^1T_2(q_2)\,{}^2T_3(q_3)\,{}^3T_4(q_4) \cdots {}^{n-1}T_n(q_n) \end{aligned} \tag{9.259}$$

therefore, the infinitesimal transformation matrix is

$$\delta T = \sum_{i=1}^{n} {}^0T_1(q_1)\,{}^1T_2(q_2) \cdots \frac{\delta\left({}^{i-1}T_i\right)}{\delta q_i} \cdots {}^{n-1}T_n(q_n) \cdot \delta q_i. \tag{9.260}$$

Interestingly, the partial derivative of the transformation matrix can be arranged in the form

$$\frac{\delta\left({}^{i-1}T_i\right)}{\delta q_i} = {}^{i-1}\Delta_{i-1}\,{}^{i-1}T_i \tag{9.261}$$

where according to DH transformation matrix (5.11)

$$\begin{aligned} {}^{i-1}T_i &= \begin{bmatrix} \cos\theta_i & -\sin\theta_i\cos\alpha_i & \sin\theta_i\sin\alpha_i & a_i\cos\theta_i \\ \sin\theta_i & \cos\theta_i\cos\alpha_i & -\cos\theta_i\sin\alpha_i & a_i\sin\theta_i \\ 0 & \sin\alpha_i & \cos\alpha_i & d_i \\ 0 & 0 & 0 & 1 \end{bmatrix} \\ &= \begin{bmatrix} {}^{i-1}R_i & {}^{i-1}\mathbf{d}_i \\ 0 & 1 \end{bmatrix} \end{aligned} \tag{9.262}$$

we can find the *velocity coefficient matrices matrix* $\Delta_i$ for a revolute joint to be

$$ {}^{i-1}\Delta_{i-1} = \Delta_R = \begin{bmatrix} {}^{i-1}\tilde{k}_{i-1} & 0 \\ 0 & 0 \end{bmatrix} = \begin{bmatrix} 0 & -1 & 0 & 0 \\ 1 & 0 & 0 & 0 \\ 0 & 0 & 0 & 0 \\ 0 & 0 & 0 & 0 \end{bmatrix} \tag{9.263}$$

and for a prismatic joint to be

$$ {}^{i-1}\Delta_{i-1} = \Delta_P = \begin{bmatrix} 0 & {}^{i-1}\hat{k}_{i-1} \\ 0 & 0 \end{bmatrix} = \begin{bmatrix} 0 & 0 & 0 & 0 \\ 0 & 0 & 0 & 0 \\ 0 & 0 & 0 & 1 \\ 0 & 0 & 0 & 0 \end{bmatrix}. \tag{9.264}$$

We may now express each term of (9.260) in the form

$$
{}^0T_1(q_1)\,{}^1T_2(q_2)\,\cdots\,\frac{\delta\left({}^{i-1}T_i\right)}{\delta q_i}\,\cdots\,{}^{n-1}T_n(q_n) = C_i\,T \tag{9.265}
$$

where

$$
\begin{aligned}
C_i &= \left[{}^0T_1\,{}^1T_2\,\cdots\,{}^{i-2}T_{i-1}\right]\frac{\delta\left({}^{i-1}T_i\right)}{\delta q_i}\left[{}^0T_1\,{}^1T_2\,\cdots\,{}^{i-1}T_i\right]^{-1}\\[2mm]
&= \left[{}^0T_1\,{}^1T_2\,\cdots\,{}^{i-2}T_{i-1}\right]{}^{i-1}\Delta_{i-1}\,{}^{i-1}T_i\left[{}^0T_1\,{}^1T_2\,\cdots\,{}^{i-1}T_i\right]^{-1}\\[2mm]
&= \left[{}^0T_1\,{}^1T_2\,\cdots\,{}^{i-2}T_{i-1}\right]{}^{i-1}\Delta_{i-1}\left[{}^0T_1\,{}^1T_2\,\cdots\,{}^{i-2}T_{i-1}\right]^{-1}\\[2mm]
&= {}^0T_1\,{}^1T_2\,\cdots\,{}^{i-2}T_{i-1}\,{}^{i-1}\Delta_{i-1}\,{}^{i-2}T_{i-1}^{-1}\,\cdots\,{}^1T_2^{-1}\,{}^0T_1^{-1}. \tag{9.266}
\end{aligned}
$$

The matrix $C_i$ can be rearranged for a revolute joint, in the form

$$
C_i = \left[\begin{array}{cc} {}^0\tilde{k}_{i-1} & {}^0\tilde{k}_{i-1}\,{}^0d_n - {}^0\tilde{k}_{i-1}\,{}^0d_{i-1} \\ 0 & 0 \end{array}\right] \tag{9.267}
$$

and for a prismatic joint in the form

$$
C_i = \left[\begin{array}{cc} {}^0\tilde{k}_{i-1} & 0 \\ 0 & 0 \end{array}\right] \tag{9.268}
$$

$C_i$ has six independent terms that can be combined in a $6 \times 1$ vector. This vector makes the $i$th column of the Jacobian matrix and is called the generating vector $\mathbf{c}_i$. The Jacobian generating vector for a revolute joint is

$$
\mathbf{c}_i = \left[\begin{array}{c} {}^0\tilde{k}_{i-1}\,{}^0_{i-1}\mathbf{d}_n \\ {}^0\hat{k}_{i-1} \end{array}\right] \tag{9.269}
$$

and for a revolute joint is

$$
\mathbf{c}_i = \left[\begin{array}{c} 0 \\ {}^0\hat{k}_{i-1} \end{array}\right]. \tag{9.270}
$$

The position vector ${}^0\mathbf{d}_i$ indicated the origin of the coordinate frame $B_i$ in the base frame $B_0$. Hence, ${}^0_{i-1}\mathbf{d}_n$ indicated the origin of the end-effector coordinate frame $B_n$ with respect to coordinate frame $B_{i-1}$ and expressed in the base frame $B_0$.

Therefore, the Jacobian matrix describing the instantaneous kinematics of the robot can be obtained from

$$
\mathbf{J} = \left[\begin{array}{cccc} {}^0\tilde{k}_0\,{}^0_0\mathbf{d}_n & {}^0\tilde{k}_1\,{}^0_1\mathbf{d}_n & \cdots & {}^0\tilde{k}_{n-1}\,{}^0_{n-1}\mathbf{d}_n \\ {}^0\hat{k}_0 & {}^0\hat{k}_1 & \cdots & {}^0\hat{k}_{n-1} \end{array}\right]. \tag{9.271}
$$

■

**Example 234** *Jacobian matrix for articulated robots.*

*The forward and inverse kinematics of the articulated robot has been analyzed in Example 163 with the following individual transformation matrices:*

$$
{}^0T_1 = \begin{bmatrix} c\theta_1 & 0 & s\theta_1 & 0 \\ s\theta_1 & 0 & -c\theta_1 & 0 \\ 0 & 1 & 0 & 0 \\ 0 & 0 & 0 & 1 \end{bmatrix} \qquad {}^1T_2 = \begin{bmatrix} c\theta_2 & -s\theta_2 & 0 & l_2c\theta_2 \\ s\theta_2 & c\theta_2 & 0 & l_2s\theta_2 \\ 0 & 0 & 1 & d_2 \\ 0 & 0 & 0 & 1 \end{bmatrix}
$$

$$
{}^2T_3 = \begin{bmatrix} c\theta_3 & 0 & s\theta_3 & 0 \\ s\theta_3 & 0 & -c\theta_3 & 0 \\ 0 & 1 & 0 & 0 \\ 0 & 0 & 0 & 1 \end{bmatrix} \qquad {}^3T_4 = \begin{bmatrix} c\theta_4 & 0 & -s\theta_4 & 0 \\ s\theta_4 & 0 & c\theta_4 & 0 \\ 0 & -1 & 0 & l_3 \\ 0 & 0 & 0 & 1 \end{bmatrix}
$$

$$
{}^4T_5 = \begin{bmatrix} c\theta_5 & 0 & s\theta_5 & 0 \\ s\theta_5 & 0 & -c\theta_5 & 0 \\ 0 & 1 & 0 & 0 \\ 0 & 0 & 0 & 1 \end{bmatrix} \qquad {}^5T_6 = \begin{bmatrix} c\theta_6 & -s\theta_6 & 0 & 0 \\ s\theta_6 & c\theta_6 & 0 & 0 \\ 0 & 0 & 1 & 0 \\ 0 & 0 & 0 & 1 \end{bmatrix} \qquad (9.272)
$$

*The articulated robot has 6 DOF and therefore, its Jacobian matrix is a $6 \times 6$ matrix*

$$
\mathbf{J}(\mathbf{q}) = \begin{bmatrix} \mathbf{c}_1(\mathbf{q}) & \mathbf{c}_2(\mathbf{q}) & \cdots & \mathbf{c}_6(\mathbf{q}) \end{bmatrix} \qquad (9.273)
$$

*that relates the translational and angular velocities of the end-effector to the joints' velocities $\dot{\mathbf{q}}$.*

$$
\begin{bmatrix} \mathbf{v} \\ \boldsymbol{\omega} \end{bmatrix} = \mathbf{J}(\mathbf{q})\,\dot{\mathbf{q}} \qquad (9.274)
$$

*The $i$th column vector $\mathbf{c}_i(\mathbf{q})$ for a revolute joint is given by*

$$
\mathbf{c}_i(\mathbf{q}) = \begin{bmatrix} {}^0\hat{k}_{i-1} \times {}^{i-1}\mathbf{d}_6 \\ {}^0\hat{k}_{i-1} \end{bmatrix} \qquad (9.275)
$$

*and for a prismatic joint is given by*

$$
\mathbf{c}_i(\mathbf{q}) = \begin{bmatrix} {}^0\hat{k}_{i-1} \\ 0 \end{bmatrix}. \qquad (9.276)
$$

**Column 1.** *The first column of the Jacobian matrix has the simplest calculation, since it is based on the contribution of the $z_0$-axis and the position of the end-effector frame ${}^0\mathbf{d}_6$. The direction of the $z_0$-axis in the base coordinate frame is*

$$
{}^0\hat{k}_0 = \begin{bmatrix} 0 \\ 0 \\ 1 \end{bmatrix} \qquad (9.277)
$$

*and the position vector of the end-effector frame $B_6$ is given by $^0\mathbf{d}_6$ directly determined from $^0T_6$*

$$^0T_6 = {}^0T_1\,{}^1T_2\,{}^2T_3\,{}^3T_4\,{}^4T_5\,{}^5T_6$$

$$= \begin{bmatrix} {}^0R_6 & {}^0\mathbf{d}_6 \\ 0 & 1 \end{bmatrix} = \begin{bmatrix} t_{11} & t_{12} & t_{13} & t_{14} \\ t_{21} & t_{22} & t_{23} & t_{24} \\ t_{31} & t_{32} & t_{33} & t_{34} \\ 0 & 0 & 0 & 1 \end{bmatrix} \tag{9.278}$$

$$^0\mathbf{d}_6 = \begin{bmatrix} t_{14} \\ t_{24} \\ t_{34} \end{bmatrix} \tag{9.279}$$

*where,*

$$
\begin{aligned}
t_{14} &= d_6\left(s\theta_1 s\theta_4 s\theta_5 + c\theta_1\left(c\theta_4 s\theta_5 c\left(\theta_2+\theta_3\right)+c\theta_5 s\left(\theta_2+\theta_3\right)\right)\right) \\
&\quad +l_3 c\theta_1 s\left(\theta_2+\theta_3\right)+d_2 s\theta_1+l_2 c\theta_1 c\theta_2 \tag{9.280}
\end{aligned}
$$

$$
\begin{aligned}
t_{24} &= d_6\left(-c\theta_1 s\theta_4 s\theta_5 + s\theta_1\left(c\theta_4 s\theta_5 c\left(\theta_2+\theta_3\right)+c\theta_5 s\left(\theta_2+\theta_3\right)\right)\right) \\
&\quad +s\theta_1 s\left(\theta_2+\theta_3\right)l_3 - d_2 c\theta_1 + l_2 c\theta_2 s\theta_1 \tag{9.281}
\end{aligned}
$$

$$
\begin{aligned}
t_{34} &= d_6\left(c\theta_4 s\theta_5 s\left(\theta_2+\theta_3\right)-c\theta_5 c\left(\theta_2+\theta_3\right)\right) \\
&\quad +l_2 s\theta_2 + l_3 c\left(\theta_2+\theta_3\right). \tag{9.282}
\end{aligned}
$$

*Therefore,*

$$
\begin{aligned}
{}^0\hat{k}_0 \times {}^0\mathbf{d}_6 &= {}^0\tilde{k}_0\,{}^0\mathbf{d}_6 \\
&= \begin{bmatrix} 0 & -1 & 0 \\ 1 & 0 & 0 \\ 0 & 0 & 0 \end{bmatrix} \begin{bmatrix} t_{14} \\ t_{24} \\ t_{34} \end{bmatrix} \\
&= \begin{bmatrix} -t_{24} \\ t_{14} \\ 0 \end{bmatrix} \tag{9.283}
\end{aligned}
$$

*and the first Jacobian generating vector is*

$$\mathbf{c}_1 = \begin{bmatrix} -t_{24} \\ t_{14} \\ 0 \\ 0 \\ 0 \\ 1 \end{bmatrix}. \tag{9.284}$$

**Column 2.** *The $z_1$-axis in the base frame can be found by*

$$
\begin{aligned}
{}^0\hat{k}_1 &= {}^0R_1 \begin{bmatrix} 0 \\ 0 \\ 1 \end{bmatrix} = \begin{bmatrix} c\theta_1 & 0 & s\theta_1 \\ s\theta_1 & 0 & -c\theta_1 \\ 0 & 1 & 0 \end{bmatrix} \begin{bmatrix} 0 \\ 0 \\ 1 \end{bmatrix} \\
&= \begin{bmatrix} \sin\theta_1 \\ -\cos\theta_1 \\ 0 \end{bmatrix}. \tag{9.285}
\end{aligned}
$$

The second half of $c_2$ needs the cross product of $^0\hat{k}_1$ and position vector $^1d_6$. The vector $^1d_6$ is the position of the end-effector in the coordinate frame $B_1$, however it must be described in the base frame to be able to perform the cross product. An easier method is to find $^1\hat{k}_1 \times {}^1d_6$ and transform the resultant into the base frame.

$$
\begin{aligned}
^0\hat{k}_1 \times {}^1d_6 &= {}^0R_1 \left( {}^1\hat{k}_1 \times {}^1d_6 \right) \\
&= {}^0R_1 \left( \begin{bmatrix} 0 \\ 0 \\ 1 \end{bmatrix} \times \begin{bmatrix} l_2 \cos\theta_2 + l_3 \sin(\theta_2 + \theta_3) \\ l_2 \sin\theta_2 - l_3 \cos(\theta_2 + \theta_3) \\ d_2 \end{bmatrix} \right) \\
&= \begin{bmatrix} \cos\theta_1 \left( -l_2 \sin\theta_2 + l_3 \cos(\theta_2 + \theta_3) \right) \\ \sin\theta_1 \left( -l_2 \sin\theta_2 + l_3 \cos(\theta_2 + \theta_3) \right) \\ l_2 \cos\theta_2 + l_3 \sin(\theta_2 + \theta_3) \end{bmatrix}
\end{aligned}
\tag{9.286}
$$

Therefore, $c_2$ is found as a $6 \times 1$ vector,

$$
c_2 = \begin{bmatrix} \cos\theta_1 \left( -l_2 \sin\theta_2 + l_3 \cos(\theta_2 + \theta_3) \right) \\ \sin\theta_1 \left( -l_2 \sin\theta_2 + l_3 \cos(\theta_2 + \theta_3) \right) \\ l_2 \cos\theta_2 + l_3 \sin(\theta_2 + \theta_3) \\ \sin\theta_1 \\ -\cos\theta_1 \\ 0 \end{bmatrix} .
\tag{9.287}
$$

**Column 3.** The $z_2$-axis in the base frame can be found using the same method:

$$
^0\hat{k}_2 = {}^0R_2 {}^2\hat{k}_2 = {}^0R_1 {}^1R_2 \begin{bmatrix} 0 \\ 0 \\ 1 \end{bmatrix} = \begin{bmatrix} \sin\theta_1 \\ -\cos\theta_1 \\ 0 \end{bmatrix} .
\tag{9.288}
$$

The second half of $c_3$ can be found by finding $^2\hat{k}_2 \times {}^2d_6$ and transforming the resultant into the base coordinate frame.

$$
^2\hat{k}_2 \times {}^2d_6 = \begin{bmatrix} l_3 \cos\theta_3 \\ l_3 \sin\theta_3 \\ 0 \end{bmatrix}
\tag{9.289}
$$

$$
^0R_2 \left( {}^2\hat{k}_2 \times {}^2d_6 \right) = \begin{bmatrix} l_3 \cos\theta_1 \sin(\theta_2 + \theta_3) \\ l_3 \sin\theta_1 \sin(\theta_2 + \theta_3) \\ -l_3 \cos(\theta_2 + \theta_3) \end{bmatrix}
\tag{9.290}
$$

Therefore, $c_3$ is

$$
c_3 = \begin{bmatrix} l_3 \cos\theta_1 \sin(\theta_2 + \theta_3) \\ l_3 \sin\theta_1 \sin(\theta_2 + \theta_3) \\ -l_3 \cos(\theta_2 + \theta_3) \\ \sin\theta_1 \\ -\cos\theta_1 \\ 0 \end{bmatrix} .
\tag{9.291}
$$

*Column 4. The $z_3$-axis in the base frame is*

$$
{}^0\hat{k}_3 \;=\; {}^0R_1\,{}^1R_2\,{}^2R_3 \begin{bmatrix} 0 \\ 0 \\ 1 \end{bmatrix}
$$

$$
= \begin{bmatrix} \cos\theta_1\,(\cos\theta_2\sin\theta_3 + \cos\theta_3\sin\theta_2) \\ \sin\theta_1\,(\cos\theta_2\sin\theta_3 + \sin\theta_2\cos\theta_3) \\ -\cos(\theta_2+\theta_3) \end{bmatrix} \tag{9.292}
$$

*and the second half of $c_4$ can be found by finding ${}^3\hat{k}_3 \times {}^3d_6$ and transforming the resultant into the base coordinate frame.*

$$
{}^0R_3\left({}^3\hat{k}_3 \times {}^3d_6\right) = {}^0R_3\left(\begin{bmatrix} 0 \\ 0 \\ 1 \end{bmatrix} \times \begin{bmatrix} 0 \\ 0 \\ l_3 \end{bmatrix}\right) = \begin{bmatrix} 0 \\ 0 \\ 0 \end{bmatrix} \tag{9.293}
$$

*Therefore, $c_4$ is*

$$
c_4 = \begin{bmatrix} 0 \\ 0 \\ 0 \\ \cos\theta_1\,(\cos\theta_2\sin\theta_3 + \cos\theta_3\sin\theta_2) \\ \sin\theta_1\,(\cos\theta_2\sin\theta_3 + \sin\theta_2\cos\theta_3) \\ -\cos(\theta_2+\theta_3) \end{bmatrix}. \tag{9.294}
$$

*Column 5. The $z_4$-axis in the base frame is*

$$
{}^0\hat{k}_4 \;=\; {}^0R_4 \begin{bmatrix} 0 \\ 0 \\ 1 \end{bmatrix}
$$

$$
= \begin{bmatrix} c\theta_4 s\theta_1 - c\theta_1 s\theta_4 c\,(\theta_2+\theta_3) \\ -c\theta_1 c\theta_4 - s\theta_1 s\theta_4 c\,(\theta_2+\theta_3) \\ -s\theta_4 s\,(\theta_2+\theta_3) \end{bmatrix} \tag{9.295}
$$

*and the second half of $c_5$ can be found by finding ${}^4\hat{k}_4 \times {}^4d_6$ and transforming the resultant into the base coordinate frame.*

$$
{}^0R_4\left({}^4\hat{k}_4 \times {}^4d_6\right) = {}^0R_4\left(\begin{bmatrix} 0 \\ 0 \\ 1 \end{bmatrix} \times \begin{bmatrix} 0 \\ 0 \\ 0 \end{bmatrix}\right) = \begin{bmatrix} 0 \\ 0 \\ 0 \end{bmatrix} \tag{9.296}
$$

*Therefore, $c_5$ is*

$$
c_5 = \begin{bmatrix} 0 \\ 0 \\ 0 \\ \cos\theta_4\sin\theta_1 - \cos\theta_1\sin\theta_4\cos(\theta_2+\theta_3) \\ -\cos\theta_1\cos\theta_4 - \sin\theta_1\sin\theta_4\cos(\theta_2+\theta_3) \\ -\sin\theta_4\sin(\theta_2+\theta_3) \end{bmatrix}. \tag{9.297}
$$

**Column 6.** *The $z_5$-axis in the base frame is*

$$
{}^0\hat{k}_5 = {}^0R_5 \begin{bmatrix} 0 \\ 0 \\ 1 \end{bmatrix} \tag{9.298}
$$

$$
= \begin{bmatrix} -c\theta_1 c\theta_4 s\,(\theta_2 + \theta_3) - s\theta_4\,(s\theta_1 s\theta_4 + c\theta_1 c\theta_4 c\,(\theta_2 + \theta_3)) \\ -s\theta_1 c\theta_4 s\,(\theta_2 + \theta_3) - s\theta_4\,(-c\theta_1 s\theta_4 + s\theta_1 c\theta_4 c\,(\theta_2 + \theta_3)) \\ c\theta_4 c\,(\theta_2 + \theta_3) - \tfrac{1}{2}s\,(\theta_2 + \theta_3)\,s2\theta_4 \end{bmatrix}
$$

*and the second half of* $c_6$ *can be found by finding* ${}^5\hat{k}_5 \times {}^5d_6$ *and transforming the resultant into the base coordinate frame.*

$$
{}^0R_5 \left({}^5\hat{k}_5 \times {}^5d_6\right) = {}^0R_5 \left( \begin{bmatrix} 0 \\ 0 \\ 1 \end{bmatrix} \times \begin{bmatrix} 0 \\ 0 \\ 0 \end{bmatrix} \right) = \begin{bmatrix} 0 \\ 0 \\ 0 \end{bmatrix} \tag{9.299}
$$

*Therefore,* $c_6$ *is*

$$
c_6 = \begin{bmatrix} 0 \\ 0 \\ 0 \\ -c\theta_1 c\theta_4 s\,(\theta_2 + \theta_3) - s\theta_4\,(s\theta_1 s\theta_4 + c\theta_1 c\theta_4 c\,(\theta_2 + \theta_3)) \\ -s\theta_1 c\theta_4 s\,(\theta_2 + \theta_3) - s\theta_4\,(-c\theta_1 s\theta_4 + s\theta_1 c\theta_4 c\,(\theta_2 + \theta_3)) \\ c\theta_4 c\,(\theta_2 + \theta_3) - \tfrac{1}{2}s\,(\theta_2 + \theta_3)\,s2\theta_4 \end{bmatrix}. \tag{9.300}
$$

# 9.5 ★ Kinematics Recursive Equations

## 9.5.1 ★ Recursive Velocity in Base Frame

The time derivative of a homogeneous transformation matrix $[T]$ can be arranged in the form

$$
\dot{T} = [V]\,[T] \tag{9.301}
$$

where $[T]$ may be a link transformation matrix or the result of the forward kinematics of a robot as

$$
{}^0T_6 = {}^0T_1\,{}^1T_2\,{}^2T_3\,{}^3T_4\,{}^4T_5\,{}^5T_6. \tag{9.302}
$$

Therefore, the time derivative of a transformation matrix can be computed when the velocity transformation matrix $[V]$ is calculated.

The transformation matrix ${}^0T_i$ is

$$
{}^0T_i = {}^0T_1\,{}^1T_2\,{}^2T_3 \cdots {}^{i-1}T_i \tag{9.303}
$$

and the matrices ${}^0V_i$ and ${}^0V_{i+1}$ are

$$
{}^0V_i = {}^0\dot{T}_i\,{}^0T_i^{-1} \tag{9.304}
$$

$$^0V_{i+1} = {^0\dot{T}_{i+1}} \, ^0T_{i+1}^{-1}$$

$$= \frac{d}{dt} \left( ^0T_i \, ^iT_{i+1} \right) {^0T_{i+1}^{-1}}$$

$$= \left( ^0T_i \, ^i\dot{T}_{i+1} + {^0\dot{T}_i} \, ^iT_{i+1} \right) {^0T_{i+1}^{-1}}. \tag{9.305}$$

These two equations can be combined as a recursive formula

$$^0V_{i+1} = {^0V_i} + {^0T_i} \, ^iV_{i+1} \, ^0T_i^{-1}. \tag{9.306}$$

The recursive velocity transformation matrix formula can be simplified according to the type of joints connecting two links.

For a revolute joint, the velocity transformation matrix is

$$^iV_{i+1} = \dot{q}_{i+1} \Delta_R = \dot{\theta}_{i+1} \Delta_R$$

$$= \dot{\theta}_{i+1} \begin{bmatrix} 0 & -1 & 0 & 0 \\ 1 & 0 & 0 & 0 \\ 0 & 0 & 0 & 0 \\ 0 & 0 & 0 & 0 \end{bmatrix} = \dot{\theta}_{i+1} \begin{bmatrix} ^{i-1}\tilde{k}_{i-1} & 0 \\ 0 & 0 \end{bmatrix} \tag{9.307}$$

then we have

$$_0\omega_{i+1} = {_0\omega_i} + {_i^0\omega_{i+1}}$$

$$= {_0\omega_i} + \dot{\theta}_{i+1} \, {_i^0\hat{k}_{i+1}} \tag{9.308}$$

which shows that the angular velocity of frame $B_{i+1}$ is the angular velocity of frame $B_i$ plus the relative angular velocity produced by joint variable $q_{i+1}$. Furthermore,

$$^0\dot{\mathbf{d}}_{i+1} = {^0\dot{\mathbf{d}}_i} + {_i^0\dot{\mathbf{d}}_{i+1}}$$

$$= {^0\dot{\mathbf{d}}_i} + {_0\tilde{\omega}_{i+1}} \left( ^0\mathbf{d}_{i+1} - {^0\mathbf{d}_i} \right) \tag{9.309}$$

which shows the translational velocity is obtained by adding the translational velocity of frame $B_i$ to the contribution of rotation of link $B_{i+1}$.

For a prismatic joint, the velocity coefficient matrix formula is

$$^iV_{i+1} = \dot{q}_{i+1} \Delta_P = \dot{d}_{i+1} \Delta_R$$

$$= \dot{d}_{i+1} \begin{bmatrix} 0 & 0 & 0 & 0 \\ 0 & 0 & 0 & 0 \\ 0 & 0 & 0 & 1 \\ 0 & 0 & 0 & 0 \end{bmatrix} = \dot{d}_{i+1} \begin{bmatrix} 0 & ^{i-1}\hat{k}_{i-1} \\ 0 & 0 \end{bmatrix} \tag{9.310}$$

and therefore, we have

$$\omega_{i+1} = \omega_i \tag{9.311}$$

and

$$^0\dot{\mathbf{d}}_{i+1} = {^0\dot{\mathbf{d}}_i} + {_0\tilde{\omega}_{i+1}} \left( ^0\mathbf{d}_{i+1} - {^0\mathbf{d}_i} \right) + \dot{d}_{i+1} \, {_i^0\hat{k}_{i+1}} \tag{9.312}$$

which shows that the angular velocity of frame $B_{i+1}$ is the same as the angular velocity of frame $B_i$. Furthermore, the translational velocity is obtained by adding the translational velocity due to $\dot{d}_{i+1}$ to the relative velocity due to rotation of link $B_{i+1}$.

## 9.5.2 ★ Recursive Acceleration in Base Frame

The acceleration of connected links can also be calculated recursively. Let us start with computing

$$^0\dot{V}_i = {}^0\ddot{T}_i \, {}^0T_i^{-1} \tag{9.313}$$

for the absolute accelerations of link $(i)$, and then calculating the acceleration matrix for link $(i+1)$

$$
\begin{aligned}
^0\dot{V}_{i+1} &= {}^0\ddot{T}_{i+1} \, {}^0T_{i+1}^{-1} \\
&= \left( {}^0\ddot{T}_i \, {}^iT_{i+1} + 2\,{}^0\dot{T}_i \, {}^i\dot{T}_{i+1} + {}^0T_i \, {}^i\ddot{T}_{i+1} \right) {}^0T_{i+1}^{-1} \\
&= {}^0\ddot{T}_i \, {}^0T_i^{-1} + 2\,{}^0\dot{T}_i \, {}^0T_i^{-1} \, {}^0T_i \, {}^i\dot{T}_{i+1} \, {}^iT_{i+1}^{-1} \, {}^0T_i^{-1} \\
&\quad + {}^0T_i \, {}^i\ddot{T}_{i+1} \, {}^iT_{i+1}^{-1} \, {}^0T_i^{-1}.
\end{aligned} \tag{9.314}
$$

These two equations can be put in a recursive form

$$^0\dot{V}_{i+1} = {}^0\dot{V}_i + 2\,{}^0V_i \, {}^0T_i \, {}^i\dot{V}_{i+1} \, {}^0T_i^{-1} + {}^0T_i \, {}^i\dot{V}_{i+1} \, {}^0T_i^{-1}. \tag{9.315}$$

For a revolute joint, we have

$$^i\dot{V}_{i+1} = \ddot{q}_{i+1}\,\Delta_R = \ddot{\theta}_{i+1}\,\Delta_R = \ddot{\theta}_{i+1}
\begin{bmatrix}
0 & -1 & 0 & 0 \\
1 & 0 & 0 & 0 \\
0 & 0 & 0 & 0 \\
0 & 0 & 0 & 0
\end{bmatrix} \tag{9.316}$$

and therefore,

$$
\begin{aligned}
^i\dot{V}_{i+1} &= \ddot{\theta}_{i+1}\,\Delta_R - \dot{\theta}_{i+1}^2\,\Delta_R\,\Delta_R^T \\
&= \ddot{\theta}_{i+1}
\begin{bmatrix}
0 & -1 & 0 & 0 \\
1 & 0 & 0 & 0 \\
0 & 0 & 0 & 0 \\
0 & 0 & 0 & 0
\end{bmatrix}
- \dot{\theta}_{i+1}^2
\begin{bmatrix}
1 & 0 & 0 & 0 \\
0 & 1 & 0 & 0 \\
0 & 0 & 0 & 0 \\
0 & 0 & 0 & 0
\end{bmatrix}. \tag{9.317}
\end{aligned}
$$

## 9.6 Summary

There are some general numerical calculations needed in robot kinematics. Solutions to a set of linear and nonlinear algebraic equations are the most important ones for calculating a matrix inversion and a Jacobian matrix. An applied solution for a set of linear equations is *LU* factorization, and a practical method for a set of nonlinear equations is the Newton-Raphson method. Both of these methods are cast in applied algorithms.

# Exercises

1. Notation and symbols.

   Describe the meaning of

   a- $[L]$     b- $[U]$     c- $[B]$     d- $[D_i]$     e- $u_{ii}$   f- $d_{ii}$

   g- con $(A)$   h- $\|A\|_\infty$   i- $\|A\|_1$     j- $\|A\|_2$   k- $c_i$   l- $X$

   m- $J$     n- $\dot{q}$     o- $^{i-1}\Delta_{i-1}$   p- $\dot{T}$     q- $\dot{V}$   r- $\theta$.

2. *LU* factorization method.

   Use the *LU* factorization method and find the associated $[L]$ and $[U]$ for the following matrices.

   (a)
   $$[A] = \begin{bmatrix} 1 & 4 & 8 \\ 5 & 2 & 7 \\ 9 & 6 & 3 \end{bmatrix}$$

   (b)
   $$[A] = \begin{bmatrix} 2 & -1 & 3 & -3 \\ 1 & 3 & -1 & -2 \\ 0 & 2 & 2 & 4 \\ 3 & 1 & 5 & -2 \end{bmatrix}$$

   (c)
   $$[A] = \begin{bmatrix} -2 & -1 & 3 & -3 & 6 \\ 1 & 3 & -1 & -2 & 0 \\ 1 & 2 & 2 & 4 & -2 \\ 3 & 1 & 5 & -2 & -1 \\ 7 & -5 & 2 & 1 & 1 \end{bmatrix}$$

3. Gaussian elimination method.

   There are two situations where the Gaussian elimination method fails: division by zero and round-off errors.

   Examine the *LU* factorization method for the possibility of division by zero.

4. Number of subtractions as a source of round-off error.

   Round-off error is common in numerical techniques, however it increases by increasing the number of subtractions. Apply the Gaussian elimination and *LU* factorization methods for solving a set of four equations

$$
\begin{bmatrix}
2 & 1 & 3 & -3 \\
1 & 0 & -1 & -2 \\
0 & 2 & 2 & 1 \\
3 & 1 & 0 & -2
\end{bmatrix}
\begin{bmatrix}
x_1 \\
x_2 \\
x_3 \\
x_4
\end{bmatrix}
=
\begin{bmatrix}
1 \\
2 \\
0 \\
-2
\end{bmatrix}
$$

and count the number of subtractions in each method.

5. Jacobian matrix from transformation matrices.

   Use the Jacobian matrix technique from links' transformation matrices and find the Jacobian matrix of the R‖R‖R planar manipulator shown in Figure 5.21. Choose a set of sample data for the dimensions and kinematics of the manipulator and find the inverse of the Jacobian matrix.

6. Jacobian matrix for a spherical wrist.

   Use the Jacobian matrix technique from links' transformation matrices and find the Jacobian matrix of the spherical wrist shown in Figure 5.22. Assume that the frame $B_3$ is the base frame.

7. Jacobian matrix for a SCARA manipulator.

   Use the Jacobian matrix technique from links' transformation matrices and find the Jacobian matrix of the R‖R‖R‖P robot shown in Figure 5.27.

8. Jacobian matrix for an R⊢R‖R articulated manipulator.

   Figure 5.25 illustrates a 3 DOF R⊢R‖R manipulator. Use the Jacobian matrix technique from links' transformation matrices and find the Jacobian matrix for the manipulator.

9. ★ Partitioning inverse method.

   Calculate the matrix inversion for the matrices in Exercise 2 using the partitioning inverse method.

10. ★ Analytic matrix inversion.

    Use the analytic and *LU* factorization methods and find the inverse of

$$
[A] =
\begin{bmatrix}
1 & 4 & 8 \\
5 & 2 & 7 \\
9 & 6 & 3
\end{bmatrix}
$$

    or an arbitrary $3 \times 3$ matrix. Count and compare the number of arithmetic operations.

11. ★ Cayley-Hamilton matrix inversion.

Use the Cayley-Hamilton and $LU$ factorization methods and find the inverse of

$$[A] = \begin{bmatrix} 1 & 4 & 8 \\ 5 & 2 & 7 \\ 9 & 6 & 3 \end{bmatrix}$$

or an arbitrary $3 \times 3$ matrix. Count and compare the number of arithmetic operations.

12. ★ Norms of matrices.

Calculate the following norms of the matrices in Exercise 2.

$$\|A\|_1 = \underset{1 \leq j \leq n}{Max} \sum_{i=1}^{n} |a_{ij}|$$

$$\|A\|_2 = \lambda_{Max}\left(A^T A\right)$$

$$\|A\|_\infty = \underset{1 \leq i \leq n}{Max} \sum_{j=1}^{n} |a_{ij}|$$

$$\|A\|_F = \sum_{i=1}^{n} \sum_{j=1}^{n} a_{ij}^2$$

# Part II

# Dynamics

Dynamics is the science of *motion*. It describes why and how a motion occurs when forces and moments are applied on massive bodies. The motion can be considered as evolution of the position, orientation, and their time derivatives. In robotics, the dynamic equation of motion for manipulators is utilized to set up the fundamental equations for control.

The links and arms in a robotic system are modeled as rigid bodies. Therefore, the dynamic properties of the rigid body takes a central place in robot dynamics. Since the arms of a robot may rotate or translate with respect to each other, translational and rotational equations of motion must be developed and described in body-attached coordinate frames $B_1, B_2, B_3, \cdots$ or in the global reference frame $G$.

There are basically two problems in robot dynamics.

**Problem 1**. We want the links of a robot to move in a specified manner. What forces and moments are required to achieve the motion?

Problem 1 is called *direct dynamics* and is easier to solve when the equations of motion are in hand because it needs differentiating of kinematics equations. The first problem includes robots statics because the specified motion can be the rest of a robot. In this condition, the problem reduces to finding forces such that no motion takes place when they act. However, there are many meaningful problems of the first type that involve robot motion rather than rest. An important example is that of finding the required forces that must act on a robot such that its end-effector moves on a given path and with a prescribed time history from the start configuration to the final configuration.

**Problem 2**. The applied forces and moments on a robot are completely specified. How will the robot move?

The second problem is called *inverse dynamics* and is more difficult to solve since it needs integration of equations of motion. However, the variety of the applied problems of the second type is interesting. Problem 2 is essentially a prediction since we wish to find the robot motion for all future times when the initial state of each link is given.

In this Part we develop techniques to derive the equations of motion for a robot.

# 10

# Acceleration Kinematics

## 10.1 Angular Acceleration Vector and Matrix

Consider a rotating rigid body $B(Oxyz)$ with a fixed point $O$ in a reference frame $G(OXYZ)$ as shown in Figure 10.1.

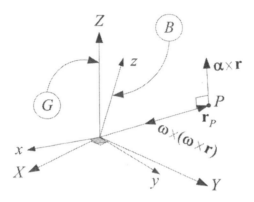

FIGURE 10.1. A rotating rigid body $B(Oxyz)$ with a fixed point $O$ in a reference frame $G(OXYZ)$.

Equation (7.5), for the velocity vector of a point in a fixed origin body frame,

$$
\begin{aligned}
{}^G\dot{\mathbf{r}}(t) &= {}^G\mathbf{v}(t) \\
&= {}_G\tilde{\omega}_B \, {}^G\mathbf{r}(t) \\
&= {}_G\boldsymbol{\omega}_B \times {}^G\mathbf{r}(t)
\end{aligned}
\tag{10.1}
$$

can be utilized to find the acceleration vector of the body point

$$
\begin{aligned}
{}^G\ddot{\mathbf{r}} &= \frac{{}^Gd}{dt} {}^G\dot{\mathbf{r}}(t) \\
&= {}_G\boldsymbol{\alpha}_B \times {}^G\mathbf{r} + {}_G\boldsymbol{\omega}_B \times \left( {}_G\boldsymbol{\omega}_B \times {}^G\mathbf{r} \right) \tag{10.2} \\
&= \left( \ddot{\phi}\hat{u} + \dot{\phi}\dot{\hat{u}} \right) \times {}^G\mathbf{r} + \dot{\phi}^2 \hat{u} \times \left( \hat{u} \times {}^G\mathbf{r} \right). \tag{10.3}
\end{aligned}
$$

${}_G\boldsymbol{\alpha}_B$ is the *angular acceleration vector* of the body with respect to the $G$ frame.

$$
{}_G\boldsymbol{\alpha}_B = \frac{{}^Gd}{dt} {}_G\boldsymbol{\omega}_B
\tag{10.4}
$$

The acceleration ${}^G\ddot{\mathbf{r}}$ can also be defined using the *angular acceleration matrix* ${}_G\tilde{\alpha}_B = {}_G\dot{\tilde{\omega}}_B$

$$
{}^G\ddot{\mathbf{r}} = {}_G\tilde{\alpha}_B \, {}^G\mathbf{r} \tag{10.5}
$$

where

$$
{}_G\tilde{\alpha}_B = {}_G\dot{\tilde{\omega}}_B = \left[\ddot{\phi}\tilde{u} + \dot{\phi}\dot{\tilde{u}} + \dot{\phi}^2\tilde{u}^2\right]. \tag{10.6}
$$

**Proof.** Differentiating Equation (10.1) gives

$$
\begin{aligned}
{}^G\ddot{\mathbf{r}} &= {}_G\dot{\omega}_B \times {}^G\mathbf{r} + {}_G\omega_B \times {}^G\dot{\mathbf{r}} \\
&= {}_G\alpha_B \times {}^G\mathbf{r} + {}_G\omega_B \times \left({}_G\omega_B \times {}^G\mathbf{r}\right)
\end{aligned} \tag{10.7}
$$

and since

$$
\omega = \dot{\phi}\hat{u} \tag{10.8}
$$
$$
\alpha = \ddot{\phi}\hat{u} + \dot{\phi}\dot{\hat{u}} \tag{10.9}
$$

we derive the Equation (10.3), which can also be written in a matrix form

$$
{}^G\ddot{\mathbf{r}} = \left[\ddot{\phi}\tilde{u} + \dot{\phi}\dot{\tilde{u}} + \dot{\phi}^2\tilde{u}^2\right]{}^G\mathbf{r} = {}_G\tilde{\alpha}_B \, {}^G\mathbf{r} \tag{10.10}
$$

and the angular acceleration matrix $\tilde{\alpha}$ can be found.

$$
\tilde{\alpha} = \begin{bmatrix}
-(u_1^2 - 1)\dot{\phi}^2 & u_1u_2\dot{\phi}^2 - \dot{u}_3\dot{\phi} - u_3\ddot{\phi} & u_1u_3\dot{\phi}^2 + \dot{u}_2\dot{\phi} + u_2\ddot{\phi} \\
u_1u_2\dot{\phi}^2 + \dot{u}_3\dot{\phi} + u_3\ddot{\phi} & -(u_2^2 - 1)\dot{\phi}^2 & u_2u_3\dot{\phi}^2 - \dot{u}_1\dot{\phi} - u_1\ddot{\phi} \\
u_1u_3\dot{\phi}^2 - \dot{u}_2\dot{\phi} - u_2\ddot{\phi} & u_2u_3\dot{\phi}^2 - \dot{u}_1\dot{\phi} + u_1\ddot{\phi} & -(u_3^2 - 1)\dot{\phi}^2
\end{bmatrix} \tag{10.11}
$$

Therefore, the position, velocity, and acceleration vectors of a body point are

$$
{}^B\mathbf{r}_P = x\hat{i} + y\hat{j} + z\hat{k} \tag{10.12}
$$

$$
\begin{aligned}
{}^G\mathbf{v}_P &= {}^G\dot{\mathbf{r}}_P = \frac{{}^Gd}{dt}{}^B\mathbf{r}_P \\
&= {}_G\omega_B \times {}^G\mathbf{r}
\end{aligned} \tag{10.13}
$$

$$
\begin{aligned}
{}^G\mathbf{a}_P &= {}^G\dot{\mathbf{v}}_P = {}^G\ddot{\mathbf{r}}_P = \frac{{}^Gd^2}{dt^2}{}^B\mathbf{r}_P \\
&= {}_G\alpha_B \times {}^G\mathbf{r} + {}_G\omega_B \times {}^G\dot{\mathbf{r}} \\
&= {}_G\alpha_B \times {}^G\mathbf{r} + {}_G\omega_B \times \left({}_G\omega_B \times {}^G\mathbf{r}\right).
\end{aligned} \tag{10.14}
$$

The angular acceleration expressed in the body frame is the body derivative of the angular velocity vector. To show this, we use the derivative transport formula (7.154)

$$
\begin{aligned}
{}_G^B\boldsymbol{\alpha}_B &= \frac{{}^Gd}{dt}\,{}_G^B\boldsymbol{\omega}_B \\
&= \frac{{}^Bd}{dt}\,{}_G^B\boldsymbol{\omega}_B + {}_G^B\boldsymbol{\omega}_B \times {}_G^B\boldsymbol{\omega}_B \\
&= \frac{{}^Bd}{dt}\,{}_G^B\boldsymbol{\omega}_B \\
&= {}_G^B\dot{\boldsymbol{\omega}}_B .
\end{aligned}
\tag{10.15}
$$

The angular acceleration of $B$ in $G$ can always be expressed in the form

$$
{}_G\boldsymbol{\alpha}_B = {}_G\alpha_B\,\hat{u}_\alpha
\tag{10.16}
$$

where $\hat{u}_\alpha$ is a unit vector parallel to ${}_G\boldsymbol{\alpha}_B$. The angular velocity and angular acceleration vectors are not parallel in general, and therefore,

$$
\hat{u}_\alpha \neq \hat{u}_\omega
\tag{10.17}
$$

$$
{}_G\alpha_B \neq {}_G\dot{\omega}_B .
\tag{10.18}
$$

However, the only special case is when the axis of rotation is fixed in both $G$ and $B$ frames. In this case

$$
{}_G\boldsymbol{\alpha}_B = \alpha\,\hat{u} = \dot{\omega}\,\hat{u} = \ddot{\phi}\,\hat{u} .
\tag{10.19}
$$

Using the Rodriguez rotation formula, it can also be shown that

$$
\begin{aligned}
{}_G\tilde{\alpha}_B &= {}_G\dot{\tilde{\omega}}_B = \lim_{\phi \to 0} \frac{{}^Gd^2}{dt^2} R_{\hat{u},\phi} \\
&= \lim_{\phi \to 0} \frac{{}^Gd^2}{dt^2}\left(-\tilde{u}^2\cos\phi + \tilde{u}\sin\phi + \tilde{u}^2 + \mathbf{I}\right) \\
&= \ddot{\phi}\tilde{u} + \dot{\phi}\dot{\tilde{u}} + \dot{\phi}^2\tilde{u}\tilde{u} .
\end{aligned}
\tag{10.20}
$$

∎

**Example 235** *Velocity and acceleration of a simple pendulum.*

*A point mass attached to a massless rod and hanging from a revolute joint is called a simple pendulum. Figure 10.2 illustrates a simple pendulum. A local coordinate frame $B$ is attached to the pendulum that rotates in a global frame $G$. The position vector of the bob and the angular velocity vector ${}_G\boldsymbol{\omega}_B$ are*

$$
{}^B\mathbf{r} = l\hat{i}
\tag{10.21}
$$

$$
{}^G\mathbf{r} = {}^GR_B\,{}^B\mathbf{r} = \begin{bmatrix} l\sin\phi \\ -l\cos\phi \\ 0 \end{bmatrix}
\tag{10.22}
$$

$$
{}_G^B\boldsymbol{\omega}_B = \dot{\phi}\hat{k}
\tag{10.23}
$$

$$
{}_G\boldsymbol{\omega}_B = {}^GR_B^T\,{}_G^B\boldsymbol{\omega}_B = \dot{\phi}\hat{K} .
\tag{10.24}
$$

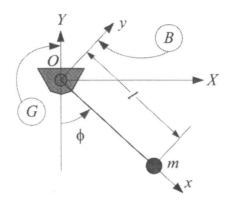

FIGURE 10.2. Illustration of a simple pendulum.

$$
{}^G R_B = \begin{bmatrix} \cos\left(\frac{3}{2}\pi + \phi\right) & -\sin\left(\frac{3}{2}\pi + \phi\right) & 0 \\ \sin\left(\frac{3}{2}\pi + \phi\right) & \cos\left(\frac{3}{2}\pi + \phi\right) & 0 \\ 0 & 0 & 1 \end{bmatrix}
$$

$$
= \begin{bmatrix} \sin\phi & \cos\phi & 0 \\ -\cos\phi & \sin\phi & 0 \\ 0 & 0 & 1 \end{bmatrix} \tag{10.25}
$$

*Its velocity is therefore given by*

$$
\begin{aligned}
{}^B_G \mathbf{v} &= {}^B \dot{\mathbf{r}} + {}^B_G \boldsymbol{\omega}_B \times {}^B_G \mathbf{r} \\
&= 0 + \dot{\phi}\hat{k} \times l\hat{i} \\
&= l\dot{\phi}\hat{j}
\end{aligned} \tag{10.26}
$$

$$
{}^G \mathbf{v} = {}^G R_B \, {}^B \mathbf{v} = \begin{bmatrix} l\dot{\phi}\cos\phi \\ l\dot{\phi}\sin\phi \\ 0 \end{bmatrix}. \tag{10.27}
$$

*The acceleration of the bob is then equal to*

$$
\begin{aligned}
{}^B_G \mathbf{a} &= {}^B_G \dot{\mathbf{v}} + {}^B_G \boldsymbol{\omega}_B \times {}^B_G \mathbf{v} \\
&= l\ddot{\phi}\hat{j} + \dot{\phi}\hat{k} \times l\dot{\phi}\hat{j} \\
&= l\ddot{\phi}\hat{j} - l\dot{\phi}^2\hat{i}
\end{aligned} \tag{10.28}
$$

$$
{}^G \mathbf{a} = {}^G R_B \, {}^B \mathbf{a} = \begin{bmatrix} l\ddot{\phi}\cos\phi - l\dot{\phi}^2\sin\phi \\ l\ddot{\phi}\sin\phi + l\dot{\phi}^2\cos\phi \\ 0 \end{bmatrix}. \tag{10.29}
$$

**Example 236** *Motion of a vehicle on the Earth.*

*Consider the motion of a vehicle on the Earth at latitude 30 deg and heading north, as shown in Figure 10.3. The vehicle has the velocity v =*

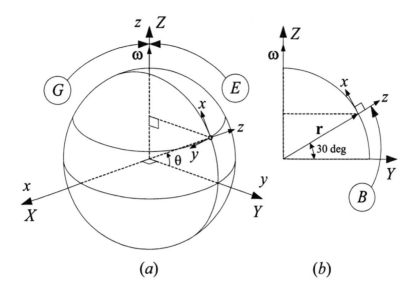

(a)                                (b)

FIGURE 10.3. Motion of a vehicle at latitude 30 deg and heading north on the Earth.

$^B_E\dot{\mathbf{r}} = 80$ km/h $= 22.22$ m/s *and acceleration* $a = ^B_E\ddot{\mathbf{r}} = 0.1$ m/s², *both with respect to the road. Radius of the Earth is* $R$, *and hence, the vehicle's kinematics are*

$$^B_E\mathbf{r} = R\hat{k} \text{ m} \tag{10.30}$$

$$^B_E\dot{\mathbf{r}} = 22.22\hat{\imath} \text{ m/s} \tag{10.31}$$

$$^B_E\ddot{\mathbf{r}} = 0.1\hat{\imath} \text{ m/s}^2 \tag{10.32}$$

$$\dot{\theta} = \frac{v}{R} \text{ rad/s} \tag{10.33}$$

$$\ddot{\theta} = \frac{a}{R} \text{ rad/s}^2. \tag{10.34}$$

*There are three coordinate frames involved. A body coordinate frame B is attached to the vehicle as shown in the Figure. A global coordinate G is set up at the center of the Earth. Another local coordinate frame E that is rigidly attached to the Earth and turns with the Earth. The frames E and G are assumed coincident at the moment. The angular velocity of B is*

$$
\begin{aligned}
^B_G\boldsymbol{\omega}_B &= \ _G\boldsymbol{\omega}_E + ^G_E\boldsymbol{\omega}_B \\
&= \ ^B R_G \left( \omega_E \hat{K} + \dot{\theta}\hat{I} \right) \\
&= (\omega_E \cos\theta)\,\hat{\imath} + (\omega_E \sin\theta)\,\hat{k} + \dot{\theta}\hat{\jmath} \\
&= (\omega_E \cos\theta)\,\hat{\imath} + (\omega_E \sin\theta)\,\hat{k} + \frac{v}{R}\hat{\jmath}. \tag{10.35}
\end{aligned}
$$

*Therefore, the velocity and acceleration of the vehicle are*

$$
\begin{aligned}
{}^B_G\mathbf{v} &= {}^B\dot{\mathbf{r}} + {}^B_G\boldsymbol{\omega}_B \times {}^B_G\mathbf{r} \\
&= 0 + {}^B_G\boldsymbol{\omega}_B \times R\hat{k} \\
&= v\hat{\imath} - (R\omega_E \cos\theta)\,\hat{\jmath}
\end{aligned}
\tag{10.36}
$$

$$
\begin{aligned}
{}^B_G\mathbf{a} &= {}^B_G\dot{\mathbf{v}} + {}^B_G\boldsymbol{\omega}_B \times {}^B_G\mathbf{v} \\
&= a\hat{\imath} + \left(R\omega_E\,\dot\theta\sin\theta\right)\hat{\jmath} + \begin{bmatrix} \omega_E\cos\theta \\ \frac{v}{R} \\ \omega_E\sin\theta \end{bmatrix} \times \begin{bmatrix} v \\ -R\omega_E\cos\theta \\ 0 \end{bmatrix} \\
&= a\hat{\imath} + \left(R\omega_E\,\dot\theta\sin\theta\right)\hat{\jmath} + \begin{bmatrix} R\omega_E^2\,\cos\theta\sin\theta \\ v\omega_E\sin\theta \\ -\frac{1}{R}v^2 - R\omega_E^2\cos^2\theta \end{bmatrix} \\
&= \begin{bmatrix} a + R\omega_E^2\,\cos\theta\sin\theta \\ 2R\omega_E\,\dot\theta\sin\theta \\ -\frac{1}{R}v^2 - R\omega_E^2\cos^2\theta \end{bmatrix}.
\end{aligned}
\tag{10.37}
$$

*The term $a\hat{\imath}$ is the acceleration relative to Earth, $(2R\omega_E\,\dot\theta\sin\theta)\hat{\jmath}$ is the Coriolis acceleration, $-\frac{v^2}{R}\hat{k}$ is the centrifugal acceleration due to traveling, and $-(R\omega_E^2\cos^2\theta)$ is the centrifugal acceleration due to Earth's rotation.*
   *Substituting the numerical values and accepting $R = 6.3677 \times 10^6$ m provide*

$$
\begin{aligned}
{}^B_G\mathbf{v} &= 22.22\hat{\imath} - 6.3677 \times 10^6 \left( \frac{2\pi}{24 \times 3600}\frac{366.25}{365.25} \right)\cos\frac{\pi}{6}\,\hat{\jmath} \\
&= 22.22\hat{\imath} - 402.13\hat{\jmath}\ \text{m/s}
\end{aligned}
\tag{10.38}
$$

$$
{}^B_G\mathbf{a} = 1.5662 \times 10^{-2}\hat{\imath} + 1.6203 \times 10^{-3}\hat{\jmath} - 2.5473 \times 10^{-2}\hat{k}\ \text{m/s}^2. \tag{10.39}
$$

**Example 237** *Combination of angular accelerations.*
   *It is shown that the angular velocity of several bodies rotating relative to each other can be related according to (7.68)*

$$
{}_0\boldsymbol{\omega}_n = {}_0\boldsymbol{\omega}_1 + {}^0_1\boldsymbol{\omega}_2 + {}^0_2\boldsymbol{\omega}_3 + \cdots + {}^0_{n-1}\boldsymbol{\omega}_n.
$$

*However, in general, there is no such equation for angular accelerations*

$$
{}_0\boldsymbol{\alpha}_n \neq {}_0\boldsymbol{\alpha}_1 + {}^0_1\boldsymbol{\alpha}_2 + {}^0_2\boldsymbol{\alpha}_3 + \cdots + {}^0_{n-1}\boldsymbol{\alpha}_n. \tag{10.40}
$$

*To show this, consider a pair of rigid links connected by revolute joints. The angular velocities of the links are*

$$
{}_0\boldsymbol{\omega}_1 = \dot\theta_1\,{}^0\hat{k}_0 \tag{10.41}
$$

$$
{}^0_1\boldsymbol{\omega}_2 = \dot\theta_2\,{}^0\hat{k}_1 \tag{10.42}
$$

$$_0\omega_2 = {}_0\omega_1 + {}_1^0\omega_2$$
$$= \dot{\theta}_1\,{}^0\hat{k}_0 + \dot{\theta}_2\,{}^0\hat{k}_1$$
$$= \dot{\theta}_1\,{}^0\hat{k}_0 + \dot{\theta}_2\,{}^0R_1\,{}^1\hat{k}_1 \qquad (10.43)$$

however, the angular accelerations show that

$$_0\alpha_1 = \frac{{}^0d}{dt}\,{}_0\omega_1 = {}_0\dot{\omega}_1\,{}^0\hat{k}_0 \qquad (10.44)$$

$$_0\alpha_2 = \frac{{}^0d}{dt}\,{}_0\omega_2 = \frac{{}^0d}{dt}\left(\dot{\theta}_1\,{}^0\hat{k}_0 + \dot{\theta}_2\,{}^0R_1\,{}^1\hat{k}_1\right)$$
$$= \ddot{\theta}_1\,{}^0\hat{k}_0 + \ddot{\theta}_2\,{}^0R_1\,{}^1\hat{k}_1 + {}_0\omega_1 \times \dot{\theta}_2\,{}^0R_1\,{}^1\hat{k}_1$$
$$= \ddot{\theta}_1\,{}^0\hat{k}_0 + \ddot{\theta}_2\,{}^0\hat{k}_1 + \dot{\theta}_1\dot{\theta}_2\,{}^0\hat{k}_0 \times {}^0\hat{k}_1$$
$$= {}_0\alpha_1 + {}_1^0\alpha_2 + {}_0\omega_1 \times {}_1^0\omega_2 \qquad (10.45)$$

and therefore,

$$_0\alpha_2 \neq {}_0\alpha_1 + {}_1^0\alpha_2. \qquad (10.46)$$

Equation (10.45) is the relative acceleration equation. It expresses the relative accelerations for a multi-link robot. As an example, consider a 6R articulated robot with six revolute joints. The angular acceleration of the end-effector frame in the base frame would be

$$_0\alpha_6 = {}_0\alpha_1 + {}_1^0\alpha_2 + {}_2^0\alpha_3 + {}_3^0\alpha_4 + {}_4^0\alpha_5 + {}_5^0\alpha_6$$
$$+ {}_0\omega_1 \times \left({}_1^0\omega_2 + {}_2^0\omega_3 + {}_3^0\omega_4 + {}_4^0\omega_5 + {}_5^0\omega_6\right)$$
$$+ {}_1^0\omega_2 \times \left({}_2^0\omega_3 + {}_3^0\omega_4 + {}_4^0\omega_5 + {}_5^0\omega_6\right)$$
$$\vdots$$
$$+ {}_4^0\omega_5 \times \left({}_5^0\omega_6\right). \qquad (10.47)$$

**Example 238** *Angular acceleration and Euler angles.*
The angular velocity $_G^B\omega_B$ in terms of Euler angles is

$$_G^G\omega_B = \begin{bmatrix} \omega_X \\ \omega_Y \\ \omega_Z \end{bmatrix} = \begin{bmatrix} 0 & \cos\varphi & \sin\theta\sin\varphi \\ 0 & \sin\varphi & -\cos\varphi\sin\theta \\ 1 & 0 & \cos\theta \end{bmatrix} \begin{bmatrix} \dot{\varphi} \\ \dot{\theta} \\ \dot{\psi} \end{bmatrix}$$
$$= \begin{bmatrix} \dot{\theta}\cos\varphi + \dot{\psi}\sin\theta\sin\varphi \\ \dot{\theta}\sin\varphi - \dot{\psi}\cos\varphi\sin\theta \\ \dot{\varphi} + \dot{\psi}\cos\theta \end{bmatrix}. \qquad (10.48)$$

*The angular acceleration is then equal to*

$$
{}^{G}_{G}\boldsymbol{\alpha}_B \;=\; \frac{{}^{G}d}{dt}{}^{G}_{G}\boldsymbol{\omega}_B \tag{10.49}
$$

$$
= \begin{bmatrix}
\cos\varphi\left(\ddot{\theta}+\dot{\varphi}\dot{\psi}\sin\theta\right)+\sin\varphi\left(\ddot{\psi}\sin\theta+\dot{\theta}\dot{\psi}\cos\theta-\dot{\theta}\dot{\varphi}\right)\\
\sin\varphi\left(\ddot{\theta}+\dot{\varphi}\dot{\psi}\sin\theta\right)+\cos\varphi\left(\dot{\theta}\dot{\varphi}-\ddot{\psi}\sin\theta-\dot{\theta}\dot{\psi}\cos\theta\right)\\
\ddot{\varphi}+\ddot{\psi}\cos\theta-\dot{\theta}\dot{\psi}\sin\theta
\end{bmatrix}.
$$

*The angular acceleration vector in the body coordinate frame is then equal to*

$$
{}^{B}_{G}\boldsymbol{\alpha}_B \;=\; {}^{G}R_B^{T}\,{}^{G}_{G}\boldsymbol{\alpha}_B \tag{10.50}
$$

$$
= \begin{bmatrix}
c\varphi c\psi - c\theta s\varphi s\psi & c\psi s\varphi + c\theta c\varphi s\psi & s\theta s\psi\\
-c\varphi s\psi - c\theta c\psi s\varphi & -s\varphi s\psi + c\theta c\varphi c\psi & s\theta c\psi\\
s\theta s\varphi & -c\varphi s\theta & c\theta
\end{bmatrix}{}^{G}_{G}\boldsymbol{\alpha}_B
$$

$$
= \begin{bmatrix}
\cos\psi\left(\ddot{\theta}+\dot{\varphi}\dot{\psi}\sin\theta\right)+\sin\psi\left(\varphi''\sin\theta+\dot{\theta}\dot{\varphi}\cos\theta-\dot{\theta}\dot{\psi}\right)\\
\cos\psi\left(\ddot{\varphi}\sin\theta+\dot{\theta}\dot{\varphi}\cos\theta-\dot{\theta}\dot{\psi}\right)-\sin\psi\left(\ddot{\theta}+\dot{\varphi}\dot{\psi}\sin\theta\right)\\
\ddot{\varphi}\cos\theta-\ddot{\psi}-\dot{\theta}\dot{\varphi}\sin\theta
\end{bmatrix}.
$$

**Example 239 ★** *Angular jerk.*

*Angular second acceleration matrix* $\tilde{\chi}=\dot{\tilde{\alpha}}=\ddot{\tilde{\omega}}$ *is a skew symmetric matrix associated to the angular jerk* $\chi=\dot{\alpha}=\ddot{\omega}$. *The angular jerk can be found by differentiating either (10.10) or (10.11).*

$$
{}^{G}\dddot{\mathbf{r}} = \ddot{\tilde{\omega}}\,{}^{G}\mathbf{r}
$$
$$
= \left[\dot{\phi}\tilde{u}+2\dot{\phi}\dot{\tilde{u}}+\dot{\phi}\ddot{\tilde{u}}+3\dot{\phi}\ddot{\phi}\tilde{u}\tilde{u}+2\dot{\phi}^{2}\dot{\tilde{u}}\tilde{u}+\dot{\phi}^{2}\tilde{u}\dot{\tilde{u}}+\dot{\phi}^{3}\tilde{u}\tilde{u}\tilde{u}\right]{}^{G}\mathbf{r}
$$
$$
= \dot{\tilde{\alpha}}\,{}^{G}\mathbf{r}=\tilde{\chi}\,{}^{G}\mathbf{r} \tag{10.51}
$$

*Hence, the angular jerk matrix would be*

$$
\tilde{\chi}=\dot{\tilde{\alpha}}=\ddot{\tilde{\omega}}=\begin{bmatrix}
j_{11} & j_{12} & j_{13}\\
j_{21} & j_{22} & j_{23}\\
j_{31} & j_{32} & j_{33}
\end{bmatrix} \tag{10.52}
$$

*where*

$$
j_{11} = 3u_1\dot{u}_1\dot{\phi}^{2}+3\left(u_1^{2}-1\right)\dot{\phi}\ddot{\phi}
$$
$$
j_{21} = \left(2u_2\dot{u}_1+\dot{u}_2u_1\right)\dot{\phi}^{2}+3u_2u_1\dot{\phi}\ddot{\phi}+\left(\ddot{u}_3\dot{\phi}+2\dot{u}_3\ddot{\phi}+u_3\dddot{\phi}-u_3\dot{\phi}^{3}\right)
$$
$$
j_{31} = \left(2u_3\dot{u}_1+\dot{u}_3u_1\right)\dot{\phi}^{2}+3u_3u_1\dot{\phi}\ddot{\phi}+\left(\ddot{u}_2\dot{\phi}+2\dot{u}_2\ddot{\phi}+u_2\dddot{\phi}-u_2\dot{\phi}^{3}\right)
$$

$$j_{12} = (2u_1\dot{u}_2 + \dot{u}_1 u_2)\dot{\phi}^2 + 3u_1 u_2 \dot{\phi}\ddot{\phi} + \left(\ddot{u}_3\dot{\phi} + 2\dot{u}_3\ddot{\phi} + u_3\dddot{\phi} - u_3\dot{\phi}^3\right)$$

$$j_{22} = 3u_2\dot{u}_2\dot{\phi}^2 + 3\left(u_2^2 - 1\right)\dot{\phi}\ddot{\phi}$$

$$j_{32} = (2u_3\dot{u}_2 + \dot{u}_3 u_2)\dot{\phi}^2 + 3u_3 u_2 \dot{\phi}\ddot{\phi} + \left(\ddot{u}_1\dot{\phi} + 2\dot{u}_1\ddot{\phi} + u_1\dddot{\phi} - u_1\dot{\phi}^3\right)$$

$$j_{13} = (2u_1\dot{u}_3 + \dot{u}_1 u_3)\dot{\phi}^2 + 3u_1 u_3 \dot{\phi}\ddot{\phi} + \left(\ddot{u}_2\dot{\phi} + 2\dot{u}_2\ddot{\phi} + u_2\dddot{\phi} - u_2\dot{\phi}^3\right)$$

$$j_{23} = (2u_2\dot{u}_3 + \dot{u}_2 u_3)\dot{\phi}^2 + 3u_2 u_3 \dot{\phi}\ddot{\phi} + \left(\ddot{u}_1\dot{\phi} + 2\dot{u}_1\ddot{\phi} + u_1\dddot{\phi} - u_1\dot{\phi}^3\right)$$

$$j_{33} = 3u_3\dot{u}_3\dot{\phi}^2 + 3\left(u_3^2 - 1\right)\dot{\phi}\ddot{\phi}. \tag{10.53}$$

**Example 240 ★** *Angular acceleration in terms of quaternion and Euler parameters.*

Utilizing the definition of the angular velocity based on rotational quaternion

$$\overleftrightarrow{_G\omega_B} = 2\,\overleftrightarrow{\dot{e}}\,\overleftrightarrow{e^*} \tag{10.54}$$

$$\overleftrightarrow{_G^B\omega_B} = 2\overleftrightarrow{e^*}\,\overleftrightarrow{\dot{e}} \tag{10.55}$$

we are able to define the angular acceleration quaternion

$$\overleftrightarrow{_G\alpha_B} = 2\,\overleftrightarrow{\ddot{e}}\,\overleftrightarrow{e^*} + 2\,\overleftrightarrow{\dot{e}}\,\overleftrightarrow{\dot{e}^*} \tag{10.56}$$

$$\overleftrightarrow{_G^B\alpha_B} = 2\overleftrightarrow{e^*}\,\overleftrightarrow{\ddot{e}} + 2\overleftrightarrow{\dot{e}^*}\,\overleftrightarrow{\dot{e}}. \tag{10.57}$$

## 10.2  Rigid Body Acceleration

Consider a rigid body with an attached local coordinate frame $B\,(oxyz)$ moving freely in a fixed global coordinate frame $G(OXYZ)$. The rigid body can rotate in the global frame, while the origin of the body frame $B$ can translate relative to the origin of $G$. The coordinates of a body point $P$ in local and global frames, as shown in Figure 10.4, are related by the following equation

$$^G\mathbf{r}_P = {}^G R_B\, {}^B\mathbf{r}_P + {}^G\mathbf{d}_B \tag{10.58}$$

where $^G\mathbf{d}_B$ indicates the position of the moving origin $o$ relative to the fixed origin $O$.

The acceleration of point $P$ in $G$ is

$$\begin{aligned}
^G\mathbf{a}_P &= {}^G\dot{\mathbf{v}}_P = {}^G\ddot{\mathbf{r}}_P \\
&= {}_G\alpha_B \times \left({}^G\mathbf{r}_P - {}^G\mathbf{d}_B\right) \\
&\quad + {}_G\omega_B \times \left({}_G\omega_B \times \left({}^G\mathbf{r}_P - {}^G\mathbf{d}_B\right)\right) + {}^G\ddot{\mathbf{d}}_B. \tag{10.59}
\end{aligned}$$

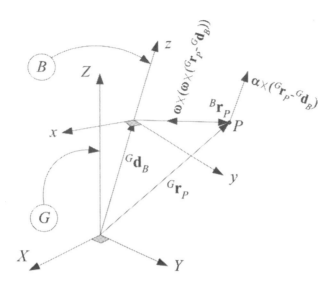

FIGURE 10.4. A rigid body with coordinate frame $B(oxyz)$ moving freely in a fixed global coordinate frame $G(OXYZ)$.

**Proof.** The acceleration of point $P$ is a consequence of differentiating the velocity equation (7.191) or (7.192).

$$
\begin{aligned}
{}^G\mathbf{a}_P &= \frac{{}^Gd}{dt}\,{}^G\mathbf{v}_P \\
&= {}_G\boldsymbol{\alpha}_B \times {}^G_B\mathbf{r}_P + {}_G\boldsymbol{\omega}_B \times {}^G_B\dot{\mathbf{r}}_P + {}^G\ddot{\mathbf{d}}_B \\
&= {}_G\boldsymbol{\alpha}_B \times {}^G_B\mathbf{r}_P + {}_G\boldsymbol{\omega}_B \times \left({}_G\boldsymbol{\omega}_B \times {}^G_B\mathbf{r}_P\right) + {}^G\ddot{\mathbf{d}}_B \\
&= {}_G\boldsymbol{\alpha}_B \times \left({}^G\mathbf{r}_P - {}^G\mathbf{d}_B\right) \\
&\quad + {}_G\boldsymbol{\omega}_B \times \left({}_G\boldsymbol{\omega}_B \times \left({}^G\mathbf{r}_P - {}^G\mathbf{d}_B\right)\right) + {}^G\ddot{\mathbf{d}}_B. \qquad (10.60)
\end{aligned}
$$

The term ${}_G\boldsymbol{\omega}_B \times \left({}_G\boldsymbol{\omega}_B \times {}^G_B\mathbf{r}_P\right)$ is called *centripetal acceleration* and is independent of the angular acceleration. The term ${}_G\boldsymbol{\alpha}_B \times {}^G_B\mathbf{r}_P$ is called *tangential acceleration* and is perpendicular to ${}^G_B\mathbf{r}_P$. ∎

**Example 241** *Acceleration of a body point.*

*Consider a rigid body is moving and rotating in a global frame. The acceleration of a body point can be found by taking twice the time derivative of its position vector*

$$
{}^G\mathbf{r}_P = {}^G R_B \,{}^B\mathbf{r}_P + {}^G\mathbf{d}_B \qquad (10.61)
$$

$$
{}^G\dot{\mathbf{r}}_P = {}^G\dot{R}_B \,{}^B\mathbf{r}_P + {}^G\dot{\mathbf{d}}_B \qquad (10.62)
$$

$$
\begin{aligned}
{}^G\ddot{\mathbf{r}}_P &= {}^G\ddot{R}_B \,{}^B\mathbf{r}_P + {}^G\ddot{\mathbf{d}}_B \\
&= {}^G\ddot{R}_B \,{}^G R_B^T \left({}^G\mathbf{r}_P - {}^G\mathbf{d}_B\right) + {}^G\ddot{\mathbf{d}}_B. \qquad (10.63)
\end{aligned}
$$

*Differentiating the angular velocity matrix*

$$_G\tilde{\omega}_B = {}^G\dot{R}_B \, {}^G R_B^T \tag{10.64}$$

*shows that*

$$
\begin{aligned}
_G\dot{\tilde{\omega}}_B &= \frac{{}^G d}{dt}\, {}_G\tilde{\omega}_B = {}^G\ddot{R}_B \, {}^G R_B^T + {}^G\dot{R}_B \, {}^G\dot{R}_B^T \\
&= {}^G\ddot{R}_B \, {}^G R_B^T + {}_G\tilde{\omega}_B \, {}_G\tilde{\omega}_B^T
\end{aligned} \tag{10.65}
$$

*and therefore,*

$$^G\ddot{R}_B \, {}^G R_B^T = {}_G\dot{\tilde{\omega}}_B - {}_G\tilde{\omega}_B \, {}_G\tilde{\omega}_B^T. \tag{10.66}$$

*Hence, the acceleration vector of the body point becomes*

$$^G\ddot{\mathbf{r}}_P = \left( {}_G\dot{\tilde{\omega}}_B - {}_G\tilde{\omega}_B \, {}_G\tilde{\omega}_B^T \right) \left( {}^G\mathbf{r}_P - {}^G\mathbf{d}_B \right) + {}^G\ddot{\mathbf{d}}_B \tag{10.67}$$

*where*

$$_G\dot{\tilde{\omega}}_B = {}_G\tilde{\alpha}_B = \begin{bmatrix} 0 & -\dot{\omega}_3 & \dot{\omega}_2 \\ \dot{\omega}_3 & 0 & -\dot{\omega}_1 \\ -\dot{\omega}_2 & \dot{\omega}_1 & 0 \end{bmatrix} \tag{10.68}$$

*and*

$$_G\tilde{\omega}_B \, {}_G\tilde{\omega}_B^T = \begin{bmatrix} \omega_2^2 + \omega_3^2 & -\omega_1\omega_2 & -\omega_1\omega_3 \\ -\omega_1\omega_2 & \omega_1^2 + \omega_3^2 & -\omega_2\omega_3 \\ -\omega_1\omega_3 & -\omega_2\omega_3 & \omega_1^2 + \omega_2^2 \end{bmatrix}. \tag{10.69}$$

**Example 242** *Acceleration of joint 2 of a 2R planar manipulator.*

*A 2R planar manipulator is illustrated in Figure 5.9. The elbow joint has a circular motion about the base joint. Knowing that*

$$_0\boldsymbol{\omega}_1 = \dot{\theta}_1 \, {}^0\hat{k}_0 \tag{10.70}$$

*we can write*

$$
\begin{aligned}
_0\boldsymbol{\alpha}_1 &= {}_0\dot{\boldsymbol{\omega}}_1 = \ddot{\theta}_1 \, {}^0\hat{k}_0 \tag{10.71} \\
_0\dot{\boldsymbol{\omega}}_1 \times {}^0\mathbf{r}_1 &= \ddot{\theta}_1 \, {}^0\hat{k}_0 \times {}^0\mathbf{r}_1 \\
&= \ddot{\theta}_1 \, R_{Z,\theta+90} \, {}^0\mathbf{r}_1 \tag{10.72} \\
_0\boldsymbol{\omega}_1 \times \left( {}_0\boldsymbol{\omega}_1 \times {}^0\mathbf{r}_1 \right) &= -\dot{\theta}_1^2 \, {}^0\mathbf{r}_1 \tag{10.73}
\end{aligned}
$$

*and calculate the acceleration of the elbow joint*

$$^0\ddot{\mathbf{r}}_1 = \ddot{\theta}_1 \, R_{Z,\theta+90} \, {}^0\mathbf{r}_1 - \dot{\theta}_1^2 \, {}^0\mathbf{r}_1. \tag{10.74}$$

**Example 243** *Acceleration of a moving point in a moving body frame.*

Assume the point $P$ in Figure 10.4 is indicated by a time varying local position vector ${}^B\mathbf{r}_P(t)$. Then, the velocity and acceleration of $P$ can be found by applying the derivative transformation formula (7.154).

$$
\begin{aligned}
{}^G\mathbf{v}_P &= {}^G\dot{\mathbf{d}}_B + {}^B\dot{\mathbf{r}}_P + {}^B_G\boldsymbol{\omega}_B \times {}^B\mathbf{r}_P \\
&= {}^G\dot{\mathbf{d}}_B + {}^B\mathbf{v}_P + {}^B_G\boldsymbol{\omega}_B \times {}^B\mathbf{r}_P
\end{aligned}
\tag{10.75}
$$

$$
\begin{aligned}
{}^G\mathbf{a}_P &= {}^G\ddot{\mathbf{d}}_B + {}^B\ddot{\mathbf{r}}_P + {}^B_G\boldsymbol{\omega}_B \times {}^B\dot{\mathbf{r}}_P + {}^B_G\dot{\boldsymbol{\omega}}_B \times {}^B\mathbf{r}_P \\
&\quad + {}^B_G\boldsymbol{\omega}_B \times \left( {}^B\dot{\mathbf{r}}_P + {}^B_G\boldsymbol{\omega}_B \times {}^B\mathbf{r}_P \right) \\
&= {}^G\ddot{\mathbf{d}}_B + {}^B\mathbf{a}_P + 2{}^B_G\boldsymbol{\omega}_B \times {}^B\mathbf{v}_P + {}^B_G\dot{\boldsymbol{\omega}}_B \times {}^B\mathbf{r}_P \\
&\quad + {}^B_G\boldsymbol{\omega}_B \times \left( {}^B_G\boldsymbol{\omega}_B \times {}^B\mathbf{r}_P \right).
\end{aligned}
\tag{10.76}
$$

It is also possible to take the derivative from Equation (7.188) with the assumption ${}^B\dot{\mathbf{r}}_P \neq 0$ and find the acceleration of $P$.

$$
{}^G\mathbf{r}_P = {}^G R_B \, {}^B\mathbf{r}_P + {}^G\mathbf{d}_B
\tag{10.77}
$$

$$
\begin{aligned}
{}^G\dot{\mathbf{r}}_P &= {}^G\dot{R}_B \, {}^B\mathbf{r}_P + {}^G R_B \, {}^B\dot{\mathbf{r}}_P + {}^G\dot{\mathbf{d}}_B \\
&= {}_G\boldsymbol{\omega}_B \times {}^G R_B \, {}^B\mathbf{r}_P + {}^G R_B \, {}^B\dot{\mathbf{r}}_P + {}^G\dot{\mathbf{d}}_B
\end{aligned}
\tag{10.78}
$$

$$
\begin{aligned}
{}^G\ddot{\mathbf{r}}_P &= {}_G\dot{\boldsymbol{\omega}}_B \times {}^G R_B \, {}^B\mathbf{r}_P + {}_G\boldsymbol{\omega}_B \times {}^G\dot{R}_B \, {}^B\mathbf{r}_P + {}_G\boldsymbol{\omega}_B \times {}^G R_B \, {}^B\dot{\mathbf{r}}_P \\
&\quad + {}^G\dot{R}_B \, {}^B\dot{\mathbf{r}}_P + {}^G R_B \, {}^B\ddot{\mathbf{r}}_P + {}^G\ddot{\mathbf{d}}_B \\
&= {}_G\dot{\boldsymbol{\omega}}_B \times {}^G_B\mathbf{r}_P + {}_G\boldsymbol{\omega}_B \times \left( {}_G\boldsymbol{\omega}_B \times {}^G\mathbf{r}_P \right) + 2\,{}_G\boldsymbol{\omega}_B \times {}^G_B\dot{\mathbf{r}}_P \\
&\quad + {}^G_B\ddot{\mathbf{r}}_P + {}^G\ddot{\mathbf{d}}_B
\end{aligned}
\tag{10.79}
$$

The third term on the right-hand side is called the **Coriolis acceleration**. The Coriolis acceleration is perpendicular to both ${}_G\boldsymbol{\omega}_B$ and ${}^B\dot{\mathbf{r}}_P$.

## 10.3   Acceleration Transformation Matrix

Consider the motion of a rigid body $B$ in the global coordinate frame $G$, as shown in Figure 10.4. Assume the body fixed frame $B(oxyz)$ is coincident at some initial time $t_0$ with the global frame $G(OXYZ)$. At any time $t \neq t_0$, $B$ is not necessarily coincident with $G$, and therefore, the homogeneous transformation matrix ${}^G T_B(t)$ is time varying.

The acceleration of a body point in the global coordinate frame can be found by applying a homogeneous acceleration transformation matrix

$$
{}^G\mathbf{a}_P(t) = {}^G A_B \, {}^G\mathbf{r}_P(t)
\tag{10.80}
$$

where, hereafter $^GA_B$ is the *acceleration transformation matrix*

$$^GA_B = \begin{bmatrix} _G\tilde{\alpha}_B - \tilde{\omega}\tilde{\omega}^T & ^G\ddot{\mathbf{d}}_B - \left( _G\tilde{\alpha}_B - \tilde{\omega}\tilde{\omega}^T \right){}^G\mathbf{d}_B \\ 0 & 0 \end{bmatrix}. \qquad (10.81)$$

**Proof.** Based on homogeneous coordinate transformation we have,

$$^G\mathbf{r}_P(t) = {}^GT_B{}^B\mathbf{r}_P = {}^GT_B{}^G\mathbf{r}_P(t_0) \qquad (10.82)$$

$$\begin{aligned}
^G\mathbf{v}_P &= {}^G\dot{T}_B{}^GT_B^{-1}{}^G\mathbf{r}_P(t) \\
&= \begin{bmatrix} ^G\dot{R}_B{}^GR_B^T & ^G\dot{\mathbf{d}}_B - {}^G\dot{R}_B{}^GR_B^T{}^G\mathbf{d}_B \\ 0 & 0 \end{bmatrix}{}^G\mathbf{r}_P(t) \\
&= \begin{bmatrix} _G\tilde{\omega}_B & ^G\dot{\mathbf{d}}_B - {}_G\tilde{\omega}_B{}^G\mathbf{d}_B \\ 0 & 0 \end{bmatrix}{}^G\mathbf{r}_P(t) \\
&= {}^GV_B{}^G\mathbf{r}_P(t).
\end{aligned} \qquad (10.83)$$

To find the acceleration of a body point in the global frame, we take twice the time derivative from $^G\mathbf{r}_P(t) = {}^GT_B{}^B\mathbf{r}_P$

$$\begin{aligned}
^G\mathbf{a}_P(t) &= \frac{d^2}{dt^2}{}^GT_B{}^B\mathbf{r}_P \\
&= {}^G\ddot{T}_B{}^B\mathbf{r}_P
\end{aligned} \qquad (10.84)$$

and substitute for $^B\mathbf{r}_P$

$$^G\mathbf{a}_P(t) = {}^G\ddot{T}_B{}^GT_B^{-1}{}^G\mathbf{r}_P(t). \qquad (10.85)$$

Substituting for $^G\ddot{T}_B$ and $^GT_B^{-1}$ provides

$$\begin{aligned}
^G\mathbf{a}_P(t) &= \begin{bmatrix} ^G\ddot{R}_B & ^G\ddot{\mathbf{d}}_B \\ 0 & 0 \end{bmatrix}\begin{bmatrix} ^GR_B^T & -{}^GR_B^T{}^G\mathbf{d}_B \\ 0 & 1 \end{bmatrix}{}^G\mathbf{r}_P(t) \\
&= \begin{bmatrix} ^G\ddot{R}_B{}^GR_B^T & ^G\ddot{\mathbf{d}}_B - {}^G\ddot{R}_B{}^GR_B^T{}^G\mathbf{d}_B \\ 0 & 0 \end{bmatrix}{}^G\mathbf{r}_P(t) \\
&= \begin{bmatrix} _G\tilde{\alpha}_B - \tilde{\omega}\tilde{\omega}^T & ^G\ddot{\mathbf{d}}_B - \left( _G\tilde{\alpha}_B - \tilde{\omega}\tilde{\omega}^T \right){}^G\dot{\mathbf{d}}_B \\ 0 & 0 \end{bmatrix}{}^G\mathbf{r}_P(t) \\
&= {}^GA_B{}^G\mathbf{r}_P(t)
\end{aligned} \qquad (10.86)$$

where

$$\begin{aligned}
^G\ddot{R}_B{}^GR_B^T &= {}_G\dot{\tilde{\omega}}_B - \tilde{\omega}\tilde{\omega}^T \\
&= {}_G\tilde{\alpha}_B - \tilde{\omega}\tilde{\omega}^T.
\end{aligned} \qquad (10.87)$$

Following the same pattern we may define a jerk transformation as

$$^G\mathbf{j}_P(t) = {}_G\tilde{\chi}_B{}^G\mathbf{r}_P(t) \qquad (10.88)$$

where,

$$
{}^G J_B = \begin{bmatrix} {}^G\ddot{R}_B & {}^G\ddot{d}_B \\ 0 & 0 \end{bmatrix} \begin{bmatrix} {}^G R_B^T & -{}^G R_B^T \,{}^G d_B \\ 0 & 1 \end{bmatrix}
$$

$$
= \begin{bmatrix} {}^G\ddot{R}_B\,{}^G R_B^T & {}^G\ddot{d}_B - {}^G\ddot{R}_B\,{}^G R_B^T\,{}^G d_B \\ 0 & 0 \end{bmatrix} \tag{10.89}
$$

and

$$
{}^G\ddot{R}_B\,{}^G R_B^T = {}_G\tilde{\dddot{\omega}}_B - 2\left({}_G\dot{\tilde{\omega}}_B - \tilde{\omega}\tilde{\omega}^T\right)\tilde{\omega}^T - \tilde{\omega}\left({}_G\dot{\tilde{\omega}}_B - \tilde{\omega}\tilde{\omega}^T\right)^T
$$

$$
= {}_G\tilde{\chi}_B - 2\left({}_G\tilde{\alpha}_B - \tilde{\omega}\tilde{\omega}^T\right)\tilde{\omega}^T - \tilde{\omega}\left({}_G\tilde{\alpha}_B - \tilde{\omega}\tilde{\omega}^T\right)^T. \tag{10.90}
$$

∎

**Example 244** *Velocity, acceleration, and jerk transformation matrices.*
*The velocity transformation matrix is a matrix to map a position vector to its velocity vector. Assume* **p** *and* **q** *denote the position of two body points P and Q. Then,*

$$
{}^G\dot{\mathbf{q}} - {}^G\dot{\mathbf{p}} = {}_G\tilde{\omega}_B \left({}^G\mathbf{q} - {}^G\mathbf{p}\right) \tag{10.91}
$$

*which can be converted to*

$$
\begin{bmatrix} {}^G\dot{\mathbf{q}} \\ 0 \end{bmatrix} = \begin{bmatrix} {}_G\tilde{\omega}_B & {}^G\dot{\mathbf{p}} - {}_G\tilde{\omega}_B\,{}^G\mathbf{p} \\ 0 & 0 \end{bmatrix} \begin{bmatrix} {}^G\mathbf{q} \\ 1 \end{bmatrix}
$$

$$
= {}^G V_B \begin{bmatrix} {}^G\mathbf{q} \\ 1 \end{bmatrix} \tag{10.92}
$$

*and* [V] *is the velocity transformation matrix. Similarly, we obtain the acceleration equation*

$$
{}^G\ddot{\mathbf{q}} - {}^G\ddot{\mathbf{p}} = {}_G\tilde{\alpha}_B \left({}^G\mathbf{q} - {}^G\mathbf{p}\right) + {}_G\tilde{\omega}_B \left({}^G\dot{\mathbf{q}} - {}^G\dot{\mathbf{p}}\right)
$$

$$
= {}_G\tilde{\alpha}_B \left({}^G\mathbf{q} - {}^G\mathbf{p}\right) + {}_G\tilde{\omega}_B\,{}_G\tilde{\omega}_B \left({}^G\mathbf{q} - {}^G\mathbf{p}\right)
$$

$$
= {}_G\tilde{\alpha}_B \left({}^G\mathbf{q} - {}^G\mathbf{p}\right) - {}_G\tilde{\omega}_B\,{}_G\tilde{\omega}_B^T \left({}^G\mathbf{q} - {}^G\mathbf{p}\right)
$$

$$
= \left({}_G\tilde{\alpha}_B - {}_G\tilde{\omega}_B\,{}_G\tilde{\omega}_B^T\right)\left({}^G\mathbf{q} - {}^G\mathbf{p}\right) \tag{10.93}
$$

*which can be converted to*

$$
\begin{bmatrix} {}^G\ddot{\mathbf{q}} \\ 0 \end{bmatrix} = [A] \begin{bmatrix} {}^G\mathbf{q} \\ 1 \end{bmatrix} \tag{10.94}
$$

$$
[A] = \begin{bmatrix} {}_G\tilde{\alpha}_B - {}_G\tilde{\omega}_B\,{}_G\tilde{\omega}_B^T & {}^G\ddot{\mathbf{p}} - \left({}_G\tilde{\alpha}_B - {}_G\tilde{\omega}_B\,{}_G\tilde{\omega}_B^T\right){}^G\mathbf{p} \\ 0 & 0 \end{bmatrix} \begin{bmatrix} {}^G\mathbf{q} \\ 1 \end{bmatrix}
$$

$$
\tag{10.95}
$$

*where* [A] *is the acceleration transformation matrix for rigid motion.*

*The jerk or second acceleration matrix can be found after another differentiation*

$$
\begin{aligned}
{}^{G}\dddot{\mathbf{q}} - {}^{G}\dddot{\mathbf{p}} \;=\;& {}_{G}\dot{\tilde{\alpha}}_{B}\left({}^{G}\mathbf{q} - {}^{G}\mathbf{p}\right) + 2{}_{G}\tilde{\alpha}_{B}\left({}^{G}\dot{\mathbf{q}} - {}^{G}\dot{\mathbf{p}}\right) + {}_{G}\tilde{\omega}_{B}\left({}^{G}\ddot{\mathbf{q}} - {}^{G}\ddot{\mathbf{p}}\right) \\
=\;& {}_{G}\dot{\tilde{\alpha}}_{B}\left({}^{G}\mathbf{q} - {}^{G}\mathbf{p}\right) - \left(2{}_{G}\tilde{\alpha}_{B} + {}_{G}\tilde{\omega}_{B}\right){}_{G}\tilde{\omega}_{B}^{T}\left({}^{G}\mathbf{q} - {}^{G}\mathbf{p}\right) \\
=\;& \left({}_{G}\dot{\tilde{\alpha}}_{B} - \left(2{}_{G}\tilde{\alpha}_{B} + {}_{G}\tilde{\omega}_{B}\right){}_{G}\tilde{\omega}_{B}^{T}\right)\left({}^{G}\mathbf{q} - {}^{G}\mathbf{p}\right) \qquad (10.96)
\end{aligned}
$$

*which can be converted to*

$$
\begin{bmatrix} {}^{G}\dddot{\mathbf{q}} \\ 0 \end{bmatrix} = [J] \begin{bmatrix} \mathbf{q} \\ 1 \end{bmatrix} \qquad (10.97)
$$

*where $[J]$ is the second acceleration or jerk transformation matrix.*

$$
[J] = \begin{bmatrix} J_{11} & J_{12} \\ 0 & 0 \end{bmatrix} \qquad (10.98)
$$

$$
\begin{aligned}
J_{11} &= {}_{G}\dot{\tilde{\alpha}}_{B} - \left(2{}_{G}\tilde{\alpha}_{B} + {}_{G}\tilde{\omega}_{B}\right){}_{G}\tilde{\omega}_{B}^{T} \qquad (10.99) \\
J_{12} &= {}^{G}\dddot{\mathbf{p}} - \left({}_{G}\dot{\tilde{\alpha}}_{B} - \left(2{}_{G}\tilde{\alpha}_{B} + {}_{G}\tilde{\omega}_{B}\right){}_{G}\tilde{\omega}_{B}^{T}\right){}^{G}\mathbf{p} \qquad (10.100)
\end{aligned}
$$

## 10.4   Forward Acceleration Kinematics

The forward acceleration kinematics problem is the method of relating the joint accelerations, $\ddot{\mathbf{q}}$, to the end-effector accelerations $\ddot{\mathbf{X}}$. It is

$$
\ddot{\mathbf{X}} = \mathbf{J}\ddot{\mathbf{q}} + \dot{\mathbf{J}}\dot{\mathbf{q}} \qquad (10.101)
$$

where $\mathbf{J}$ is the Jacobian matrix, $\mathbf{q}$ is the joint variable vector, $\dot{\mathbf{q}}$ is the joint velocity vector, and $\ddot{\mathbf{q}}$ is the *joint acceleration vector*

$$
\mathbf{q} = \begin{bmatrix} q_n & q_n & q_n & \cdots & q_n \end{bmatrix}^{T} \qquad (10.102)
$$

$$
\dot{\mathbf{q}} = \begin{bmatrix} \dot{q}_n & \dot{q}_n & \dot{q}_n & \cdots & \dot{q}_n \end{bmatrix}^{T} \qquad (10.103)
$$

$$
\ddot{\mathbf{q}} = \begin{bmatrix} \ddot{q}_n & \ddot{q}_n & \ddot{q}_n & \cdots & \ddot{q}_n \end{bmatrix}^{T}. \qquad (10.104)
$$

However, $\mathbf{X}$ is the *end-effector configuration vector* and $\dot{\mathbf{X}}$ is the *end-effector configuration velocity vector.*

$$
\mathbf{X} = \begin{bmatrix} X_n & Y_n & Z_n & \varphi_n & \theta_n & \psi_n \end{bmatrix}^{T} \qquad (10.105)
$$

$$\dot{\mathbf{X}} = \begin{bmatrix} {}^0\mathbf{v}_n \\ {}^0\boldsymbol{\omega}_n \end{bmatrix} = \begin{bmatrix} {}^0\dot{\mathbf{d}}_n \\ {}^0\boldsymbol{\omega}_n \end{bmatrix} \tag{10.106}$$

$$= \begin{bmatrix} \dot{X}_n & \dot{Y}_n & \dot{Z}_n & \omega_{Xn} & \omega_{Yn} & \omega_{Zn} \end{bmatrix}^T \tag{10.107}$$

$$\ddot{\mathbf{X}} = \begin{bmatrix} {}^0\mathbf{a}_n \\ {}^0\boldsymbol{\alpha}_n \end{bmatrix} = \begin{bmatrix} {}^0\ddot{\mathbf{d}}_n \\ {}^0\dot{\boldsymbol{\omega}}_n \end{bmatrix}$$

$$= \begin{bmatrix} \ddot{X}_n & \ddot{Y}_n & \ddot{Z}_n & \dot{\omega}_{Xn} & \dot{\omega}_{Yn} & \dot{\omega}_{Zn} \end{bmatrix}^T \tag{10.108}$$

Since calculating the time derivative of the Jacobian matrix, $\begin{bmatrix} \dot{\mathbf{J}} \end{bmatrix}$, is not simple in general, to find the forward acceleration kinematics it is better to use Equations (8.69) to (8.72)

$$_{i-1}^{0}\dot{\mathbf{d}}_i = \begin{cases} {}_0^0\boldsymbol{\omega}_i \times {}_{i-1}^{0}\mathbf{d}_i & \text{if joint } i \text{ is R} \\ \dot{d}_i\, {}^0\hat{k}_{i-1} + {}_0^0\boldsymbol{\omega}_i \times {}_{i-1}^{0}\mathbf{d}_i & \text{if joint } i \text{ is P} \end{cases} \tag{10.109}$$

$$_{i-1}^{0}\boldsymbol{\omega}_i = \begin{cases} \dot{\theta}_i\, {}^0\hat{k}_{i-1} & \text{if joint } i \text{ is R} \\ 0 & \text{if joint } i \text{ is P} \end{cases} \tag{10.110}$$

and take a derivative to find the acceleration of link $(i)$ with respect to its previous link $(i-1)$.

$$_{i-1}^{0}\ddot{\mathbf{d}}_i = \begin{cases} {}_0^0\dot{\boldsymbol{\omega}}_i \times {}_{i-1}^{0}\mathbf{d}_i + {}_0^0\boldsymbol{\omega}_i \times \left( {}_0^0\boldsymbol{\omega}_i \times {}_{i-1}^{0}\mathbf{d}_i \right) & \text{if joint } i \text{ is R} \\[1em] {}_0^0\dot{\boldsymbol{\omega}}_i \times {}_{i-1}^{0}\mathbf{d}_i + {}_0^0\boldsymbol{\omega}_i \times \left( {}_0^0\boldsymbol{\omega}_i \times {}_{i-1}^{0}\mathbf{d}_i \right) \\ \quad + \ddot{d}_i\, {}^0\hat{k}_{i-1} + 2\dot{d}_i\, {}_0^0\boldsymbol{\omega}_{i-1} \times {}^0\hat{k}_{i-1} & \text{if joint } i \text{ is P} \end{cases} \tag{10.111}$$

$$_{i-1}^{0}\dot{\boldsymbol{\omega}}_i = \begin{cases} \ddot{\theta}_i\, {}^0\hat{k}_{i-1} + \dot{\theta}_i\, {}_0^0\boldsymbol{\omega}_{i-1} \times {}^0\hat{k}_{i-1} & \text{if joint } i \text{ is R} \\ 0 & \text{if joint } i \text{ is P} \end{cases} \tag{10.112}$$

Therefore, the acceleration vectors of the end-effector frame are

$$_0^0\ddot{\mathbf{d}}_n = \sum_{i=1}^{n} {}_{i-1}^{0}\ddot{\mathbf{d}}_i \tag{10.113}$$

and

$$_0^0\dot{\boldsymbol{\omega}}_n = \sum_{i=1}^{n} {}_{i-1}^{0}\dot{\boldsymbol{\omega}}_i. \tag{10.114}$$

The acceleration relationships can also be rearranged in a recursive form.

$$_{i-1}^{0}\dot{\boldsymbol{\omega}}_i = \begin{cases} \ddot{\theta}_i\, {}^0\hat{k}_{i-1} + \dot{\theta}_i\, {}_0^0\boldsymbol{\omega}_{i-1} \times {}^0\hat{k}_{i-1} & \text{if joint } i \text{ is R} \\ 0 & \text{if joint } i \text{ is P} \end{cases} \tag{10.115}$$

**Example 245** *Forward acceleration of the 2R planar manipulator.*
*The forward velocity of the 2R planar manipulator is found as*

$$\dot{\mathbf{X}} = \mathbf{J}\dot{\mathbf{q}} \qquad (10.116)$$

$$\left[ \begin{array}{c} \dot{X} \\ \dot{Y} \end{array} \right] = \left[ \begin{array}{cc} -l_1 s\theta_1 - l_2 s\,(\theta_1 + \theta_2) & -l_2 s\,(\theta_1 + \theta_2) \\ l_1 c\theta_1 + l_2 c\,(\theta_1 + \theta_2) & l_2 c\,(\theta_1 + \theta_2) \end{array} \right] \left[ \begin{array}{c} \dot{\theta}_1 \\ \dot{\theta}_2 \end{array} \right].$$

*The differential of the Jacobian matrix is*

$$\dot{\mathbf{J}} = \left[ \begin{array}{cc} \dot{J}_{11} & \dot{J}_{12} \\ \dot{J}_{21} & \dot{J}_{22} \end{array} \right] \qquad (10.117)$$

*where*

$$\begin{array}{rcl} \dot{J}_{11} & = & (-l_1 \cos\theta_1 - l_2 \cos(\theta_1 + \theta_2))\,\dot{\theta}_1 - l_2 \cos(\theta_1 + \theta_2)\,\dot{\theta}_2 \\ \dot{J}_{12} & = & -l_2 \cos(\theta_1 + \theta_2)\,\dot{\theta}_1 - l_2 \cos(\theta_1 + \theta_2)\,\dot{\theta}_2 \\ \dot{J}_{21} & = & (-l_1 \sin\theta_1 - l_2 \sin(\theta_1 + \theta_2))\dot{\theta}_1 - l_2 \sin(\theta_1 + \theta_2)\,\dot{\theta}_2 \\ \dot{J}_{22} & = & -l_2 \sin(\theta_1 + \theta_2)\,\dot{\theta}_1 - l_2 \sin(\theta_1 + \theta_2)\,\dot{\theta}_2 \qquad (10.118) \end{array}$$

*and therefore, the forward acceleration kinematics of the manipulator can be rearranged in the form*

$$\ddot{\mathbf{X}} = \mathbf{J}\ddot{\mathbf{q}} + \dot{\mathbf{J}}\dot{\mathbf{q}}. \qquad (10.119)$$

*However, for the 2R manipulator it is easier to show the acceleration in the form*

$$\begin{array}{rcl} \left[ \begin{array}{c} \ddot{X} \\ \ddot{Y} \end{array} \right] & = & \left[ \begin{array}{cc} -l_1 \sin\theta_1 & -l_2 \sin(\theta_1 + \theta_2) \\ l_1 \cos\theta_1 & l_2 \cos(\theta_1 + \theta_2) \end{array} \right] \left[ \begin{array}{c} \ddot{\theta}_1 \\ \ddot{\theta}_1 + \ddot{\theta}_2 \end{array} \right] \\[4mm] & & - \left[ \begin{array}{cc} l_1 \cos\theta_1 & l_2 \cos(\theta_1 + \theta_2) \\ l_1 \sin\theta_1 & l_2 \sin(\theta_1 + \theta_2) \end{array} \right] \left[ \begin{array}{c} \dot{\theta}_1^2 \\ \left(\dot{\theta}_1 + \dot{\theta}_2\right)^2 \end{array} \right]. \qquad (10.120) \end{array}$$

## 10.5  ★ Inverse Acceleration Kinematics

The forward acceleration kinematics shows that

$$\ddot{\mathbf{X}} = \left[ \begin{array}{c} {}^0\mathbf{a}_n \\ {}^0\boldsymbol{\alpha}_n \end{array} \right] = \mathbf{J}\ddot{\mathbf{q}} + \dot{\mathbf{J}}\dot{\mathbf{q}}. \qquad (10.121)$$

Assuming that the Jacobian matrix, $\mathbf{J}$, is square and nonsingular, the joint acceleration vector $\ddot{\mathbf{q}}$ can be found by matrix inversion.

$$\ddot{\mathbf{q}} = \mathbf{J}^{-1}\left(\ddot{\mathbf{X}} - \dot{\mathbf{J}}\dot{\mathbf{q}}\right) \qquad (10.122)$$

However, calculating $\dot{\mathbf{J}}$ and $\mathbf{J}^{-1}$ becomes more tedious by increasing the DOF of robots. The alternative technique is to write the equation in a new form

$$\ddot{\mathbf{X}} - \dot{\mathbf{J}}\dot{\mathbf{q}} = \mathbf{J}\ddot{\mathbf{q}} \tag{10.123}$$

which is similar to Equation (8.160) for a robot with a spherical wrist

$$\ddot{\mathbf{X}} - \dot{\mathbf{J}}\dot{\mathbf{q}} = \begin{bmatrix} A & 0 \\ C & D \end{bmatrix} \ddot{\mathbf{q}} \tag{10.124}$$

or

$$\begin{bmatrix} \mathbf{m} \\ \mathbf{n} \end{bmatrix} = \begin{bmatrix} A & 0 \\ C & D \end{bmatrix} \begin{bmatrix} \ddot{q}_1 \\ \ddot{q}_2 \\ \ddot{q}_3 \\ \ddot{q}_4 \\ \ddot{q}_5 \\ \ddot{q}_6 \end{bmatrix} \tag{10.125}$$

where

$$\begin{bmatrix} \mathbf{m} \\ \mathbf{n} \end{bmatrix} = \ddot{\mathbf{X}} - \dot{\mathbf{J}}\dot{\mathbf{q}}. \tag{10.126}$$

Therefore, the inverse acceleration kinematics problem can be solved as

$$\begin{bmatrix} \ddot{\theta}_1 \\ \ddot{\theta}_2 \\ \ddot{\theta}_3 \end{bmatrix} = A^{-1}\,[\mathbf{m}] \tag{10.127}$$

and

$$\begin{bmatrix} \ddot{q}_4 \\ \ddot{q}_5 \\ \ddot{q}_6 \end{bmatrix} = D^{-1}\left([\mathbf{n}] - [C]\begin{bmatrix} \ddot{q}_1 \\ \ddot{q}_2 \\ \ddot{q}_3 \end{bmatrix}\right). \tag{10.128}$$

The matrix $\dot{\mathbf{J}}\dot{\mathbf{q}} = \ddot{\mathbf{X}} - \mathbf{J}\ddot{\mathbf{q}}$ is called the *acceleration bias vector* and can be calculated by differentiating from

$$^0_0\dot{\mathbf{d}}_6 = \sum_{i=1}^{3} {}^0_0\boldsymbol{\omega}_i \times {}^{\,0}_{i-1}\mathbf{d}_i \tag{10.129}$$

$$^0_0\boldsymbol{\omega}_6 = \sum_{i=1}^{6} \dot{\theta}_i\,{}^0\hat{k}_{i-1} \tag{10.130}$$

to find

$$\begin{aligned} ^0_0\mathbf{a}_6 &= {}^0_0\ddot{\mathbf{d}}_6 \\ &= \sum_{i=1}^{3} \left({}^0_0\dot{\boldsymbol{\omega}}_i \times {}^{\,0}_{i-1}\mathbf{d}_i + {}^0_0\boldsymbol{\omega}_i \times \left({}^0_0\boldsymbol{\omega}_i \times {}^{\,0}_{i-1}\mathbf{d}_i\right)\right) \end{aligned} \tag{10.131}$$

and

$$
\begin{aligned}
{}^0_0\alpha_6 &= {}^0_0\dot{\omega}_6 \\
&= \sum_{i=1}^{6} \left( \ddot{\theta}_i \, {}^0\hat{k}_{i-1} + {}^0_0\omega_i \times \dot{\theta}_i \, {}^0\hat{k}_{i-1} \right).
\end{aligned}
\tag{10.132}
$$

The angular acceleration vector ${}^0_0\alpha_6$ is the second half of $\ddot{\mathbf{X}}$. Then, subtracting the second half of $\mathbf{J}\ddot{\mathbf{q}}$ from ${}^0_0\alpha_6$ provides the second half of the bias vector.

$$
\sum_{i=1}^{6} {}^0_0\omega_i \times \dot{\theta}_i \, {}^0\hat{k}_{i-1}
\tag{10.133}
$$

We substitute ${}^0_0\omega_i$ and ${}^0_0\dot{\omega}_i$

$$
{}^0_0\omega_i = \sum_{j=1}^{i} \dot{\theta}_j \, {}^0\hat{k}_{j-1}
\tag{10.134}
$$

$$
{}^0_0\dot{\omega}_i = \sum_{j=1}^{i} \left( \ddot{\theta}_j \, {}^0\hat{k}_{j-1} + {}^0_0\omega_{j-1} \times \dot{\theta}_j \, {}^0\hat{k}_{j-1} \right)
\tag{10.135}
$$

in Equation (10.131)

$$
\begin{aligned}
{}^0_0\ddot{d}_6 &= \sum_{i=1}^{3}\sum_{j=1}^{i} \left( \ddot{\theta}_j \, {}^0\hat{k}_{j-1} + {}^0_0\omega_{j-1} \times \dot{\theta}_j \, {}^0\hat{k}_{j-1} \right) \times {}^0_{i-1}\mathbf{d}_i \\
&\quad + \sum_{i=1}^{3}\sum_{j=1}^{i} \dot{\theta}_j \, {}^0\hat{k}_{j-1} \times \left( {}^0_0\omega_i \times {}^0_{i-1}\mathbf{d}_i \right) \\
&= \sum_{i=1}^{3}\sum_{j=1}^{i} \ddot{\theta}_j \, {}^0\hat{k}_{j-1} \times {}^0_{i-1}\mathbf{d}_i + \sum_{i=1}^{3}\sum_{j=1}^{i} \left( {}^0_0\omega_{j-1} \times \dot{\theta}_j \, {}^0\hat{k}_{j-1} \right) \times {}^0_{i-1}\mathbf{d}_i \\
&\quad + \sum_{i=1}^{3}\sum_{j=1}^{i} \dot{\theta}_j \, {}^0\hat{k}_{j-1} \times \left( {}^0_0\omega_i \times {}^0_{i-1}\mathbf{d}_i \right)
\end{aligned}
\tag{10.136}
$$

to find the first half of the bias vector

$$
\sum_{i=1}^{3}\sum_{j=1}^{i} \left( {}^0_0\omega_{j-1} \times \dot{\theta}_j \, {}^0\hat{k}_{j-1} \right) \times {}^0_{i-1}\mathbf{d}_i + \sum_{i=1}^{3}\sum_{j=1}^{i} \dot{\theta}_j \, {}^0\hat{k}_{j-1} \times \left( {}^0_0\omega_i \times {}^0_{i-1}\mathbf{d}_i \right).
\tag{10.137}
$$

**Example 246** *Inverse acceleration of a 2R planar manipulator.*
*The forward velocity and acceleration of the 2R planar manipulator is found as*

$$
\dot{\mathbf{X}} = \mathbf{J}\dot{\mathbf{q}}
\tag{10.138}
$$

$$
\begin{bmatrix} \dot{X} \\ \dot{Y} \end{bmatrix} = \begin{bmatrix} -l_1 s\theta_1 - l_2 s(\theta_1 + \theta_2) & -l_2 s(\theta_1 + \theta_2) \\ l_1 c\theta_1 + l_2 c(\theta_1 + \theta_2) & l_2 c(\theta_1 + \theta_2) \end{bmatrix} \begin{bmatrix} \dot{\theta}_1 \\ \dot{\theta}_2 \end{bmatrix}
$$

$$\ddot{X} = J\ddot{q} + \dot{J}\dot{q} \tag{10.139}$$

$$\begin{bmatrix} \ddot{X} \\ \ddot{Y} \end{bmatrix} = \begin{bmatrix} -l_1 \sin\theta_1 & -l_2 \sin(\theta_1 + \theta_2) \\ l_1 \cos\theta_1 & l_2 \cos(\theta_1 + \theta_2) \end{bmatrix} \begin{bmatrix} \ddot{\theta}_1 \\ \ddot{\theta}_1 + \ddot{\theta}_2 \end{bmatrix}$$

$$- \begin{bmatrix} l_1 \cos\theta_1 & l_2 \cos(\theta_1 + \theta_2) \\ l_1 \sin\theta_1 & l_2 \sin(\theta_1 + \theta_2) \end{bmatrix} \begin{bmatrix} \dot{\theta}_1^2 \\ \left(\dot{\theta}_1 + \dot{\theta}_2\right)^2 \end{bmatrix}$$

*along with the following derivative and inverse Jacobian matrices.*

$$\dot{J} = \begin{bmatrix} -l_1\dot{\theta}_1 c\theta_1 - l_2\left(\dot{\theta}_1 + \dot{\theta}_2\right) c(\theta_1 + \theta_2) & -l_2\left(\dot{\theta}_1 + \dot{\theta}_2\right) c(\theta_1 + \theta_2) \\ -l_1\dot{\theta}_1 s\theta_1 - l_2\left(\dot{\theta}_1 + \dot{\theta}_2\right) s(\theta_1 + \theta_2) & -l_2\left(\dot{\theta}_1 + \dot{\theta}_2\right) s(\theta_1 + \theta_2) \end{bmatrix} \tag{10.140}$$

$$J^{-1} = \frac{-1}{l_1 l_2 s\theta_2} \begin{bmatrix} -l_2 c(\theta_1 + \theta_2) & -l_2 s(\theta_1 + \theta_2) \\ l_1 c\theta_1 + l_2 c(\theta_1 + \theta_2) & l_1 s\theta_1 + l_2 s(\theta_1 + \theta_2) \end{bmatrix} \tag{10.141}$$

*The inverse acceleration kinematics of the manipulator*

$$\ddot{q} = J^{-1}\left(\ddot{X} - \dot{J}\dot{q}\right) \tag{10.142}$$

*can be arranged to*

$$\begin{bmatrix} \ddot{\theta}_1 \\ \ddot{\theta}_1 + \ddot{\theta}_2 \end{bmatrix} =$$

$$\frac{1}{l_1 l_2 s\theta_2} \begin{bmatrix} l_2 \cos(\theta_1 + \theta_2) & l_2 \sin(\theta_1 + \theta_2) \\ -l_1 \cos\theta_1 & -l_1 \sin\theta_1 \end{bmatrix} \begin{bmatrix} \ddot{X} \\ \ddot{Y} \end{bmatrix}$$

$$+ \frac{1}{l_1 l_2 s\theta_2} \begin{bmatrix} l_1 l_2 \cos\theta_1 & l_2^2 \\ -l_1^2 & -l_1 l_2 \cos\theta_1 \end{bmatrix} \begin{bmatrix} \dot{\theta}_1^2 \\ \left(\dot{\theta}_1 + \dot{\theta}_2\right)^2 \end{bmatrix}. \tag{10.143}$$

## 10.6   Summary

When a body coordinate frame $B$ and a global frame $G$ have a common origin, the global acceleration of a point $P$ in frame $B$ is

$$^G\ddot{r} = \frac{^G d}{dt}\, ^G v_P = {}_G\alpha_B \times {}^G r + {}_G\omega_B \times \left({}_G\omega_B \times {}^G r\right) \tag{10.144}$$

where, $_G\alpha_B$ is the angular acceleration of $B$ with respect to $G$

$$_G\alpha_B = \frac{^G d}{dt}\, _G\omega_B. \tag{10.145}$$

However, when the body coordinate frame $B$ has a rigid motion with respect to $G$, then

$$
\begin{aligned}
{}^{G}\mathbf{a}_P &= \frac{{}^{G}d}{dt}{}^{G}\mathbf{v}_P \\
&= {}_{G}\boldsymbol{\alpha}_B \times \left({}^{G}\mathbf{r}_P - {}^{G}\mathbf{d}_B\right) \\
&\quad + {}_{G}\boldsymbol{\omega}_B \times \left({}_{G}\boldsymbol{\omega}_B \times \left({}^{G}\mathbf{r}_P - {}^{G}\mathbf{d}_B\right)\right) + {}^{G}\ddot{\mathbf{d}}_B \qquad (10.146)
\end{aligned}
$$

where ${}^{G}\mathbf{d}_B$ indicates the position of the origin of $B$ with respect to the origin of $G$.

Angular accelerations of two links are related according to

$$
{}_0\boldsymbol{\alpha}_2 = {}_0\boldsymbol{\alpha}_1 + {}_1^0\boldsymbol{\alpha}_2 + {}_0\boldsymbol{\omega}_1 \times {}_1^0\boldsymbol{\omega}_2. \qquad (10.147)
$$

The acceleration relationship for a body $B$ having a rigid motion in $G$ may also be expressed by a homogeneous acceleration transformation matrix ${}^{G}A_B$

$$
{}^{G}\mathbf{a}_P(t) = {}^{G}A_B\,{}^{G}\mathbf{r}_P(t) \qquad (10.148)
$$

where,

$$
\begin{aligned}
{}^{G}A_B &= {}^{G}\ddot{T}_B\,{}^{G}T_B^{-1} \\
&= \begin{bmatrix} {}_G\tilde{\alpha}_B - \tilde{\omega}\tilde{\omega}^T & {}^{G}\ddot{\mathbf{d}}_B - \left({}_G\tilde{\alpha}_B - \tilde{\omega}\tilde{\omega}^T\right){}^{G}\mathbf{d}_B \\ 0 & 0 \end{bmatrix}. \qquad (10.149)
\end{aligned}
$$

The forward acceleration kinematics of a robot is defined by

$$
\ddot{\mathbf{X}} = \mathbf{J}\,\ddot{\mathbf{q}} + \dot{\mathbf{J}}\,\dot{\mathbf{q}} \qquad (10.150)
$$

that is a relationship between joint coordinate acceleration

$$
\ddot{\mathbf{q}} = \begin{bmatrix} \ddot{q}_n & \ddot{q}_n & \ddot{q}_n & \cdots & \ddot{q}_n \end{bmatrix}^T \qquad (10.151)
$$

and global acceleration of the end-effector

$$
\ddot{\mathbf{X}} = \begin{bmatrix} \ddot{X}_n & \ddot{Y}_n & \ddot{Z}_n & \dot{\omega}_{Xn} & \dot{\omega}_{Yn} & \dot{\omega}_{Zn} \end{bmatrix}^T. \qquad (10.152)
$$

Such a relationship introduces the time derivative of Jacobian matrix $\begin{bmatrix}\dot{\mathbf{J}}\end{bmatrix}$.

# Exercises

1. Notation and symbols.

   Describe their meaning.

   a- $_G\alpha_B$   b- $_B\alpha_G$   c- $_G^G\alpha_B$   d- $_G^B\alpha_B$   e- $_B^B\alpha_G$   f- $_B^G\alpha_G$

   g- $_2^0\alpha_1$   h- $_1^0\alpha_2$   i- $_2^1\alpha_1$   j- $_2^0\alpha_1$   k- $_2^3\alpha_1$   l- $_j^k\alpha_i$

   m- $_2^0\tilde{\alpha}_1$   n- $_1^2\dot{\tilde{\omega}}_2$   o- $^G A_B$   p- $^G\dot{V}_B$   q- $\dot{J}$   r- $\ddot{X}$

2. Local position, global acceleration.

   A body is turning about the $Z$-axis at a constant angular acceleration $\ddot{\alpha} = 2\,\mathrm{rad/sec^2}$. Find the global velocity of a point, when $\dot{\alpha} = 2\,\mathrm{rad/sec}$, $\alpha = \pi/3\,\mathrm{rad}$ and

   $$^B\mathbf{r} = \begin{bmatrix} 5 \\ 30 \\ 10 \end{bmatrix}.$$

3. Global position, constant angular acceleration.

   A body is turning about the $Z$-axis at a constant angular acceleration $\ddot{\alpha} = 2\,\mathrm{rad/sec^2}$. Find the global position of a point at

   $$^B\mathbf{r} = \begin{bmatrix} 5 \\ 30 \\ 10 \end{bmatrix}$$

   after $t = 3\,\mathrm{sec}$ if the body and global coordinate frames were coincident at $t = 0\,\mathrm{sec}$.

4. Turning about $x$-axis.

   Find the angular acceleration matrix when the body coordinate frame is turning $-5\,\mathrm{deg/sec^2}$, $35\,\mathrm{deg/sec}$ at $45\,\mathrm{deg}$ about the $x$-axis.

5. Angular acceleration and Euler angles.

   Calculate the angular velocity and acceleration vectors in body and global coordinate frames if the Euler angles and their derivatives are:

   $$\begin{array}{lll}
   \varphi = .25\,\mathrm{rad} & \dot{\varphi} = 2.5\,\mathrm{rad/sec} & \ddot{\varphi} = 25\,\mathrm{rad/sec^2} \\
   \theta = -.25\,\mathrm{rad} & \dot{\theta} = -4.5\,\mathrm{rad/sec} & \ddot{\theta} = 35\,\mathrm{rad/sec^2} \\
   \psi = .5\,\mathrm{rad} & \dot{\psi} = 3\,\mathrm{rad/sec} & \ddot{\psi} = 25\,\mathrm{rad/sec^2}
   \end{array}$$

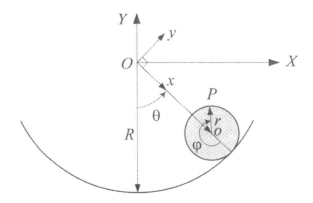

FIGURE 10.5. A roller in a circular path.

6. Combined rotation and angular acceleration.

   Find the rotation matrix for a body frame after 30 deg rotation about the $Z$-axis, followed by 30 deg about the $X$-axis, and then 90 deg about the $Y$-axis. Then calculate the angular velocity of the body if it is turning with $\dot{\alpha} = 20\,\mathrm{deg/sec}$, $\dot{\beta} = -40\,\mathrm{deg/sec}$, and $\dot{\gamma} = 55\,\mathrm{deg/sec}$ about the $Z$, $Y$, and $X$ axes respectively. Finally, calculate the angular acceleration of the body if it is turning with $\ddot{\alpha} = 2\,\mathrm{deg/sec^2}$, $\ddot{\beta} = 4\,\mathrm{deg/sec^2}$, and $\ddot{\gamma} = -6\,\mathrm{deg/sec^2}$ about the $Z$, $Y$, and $X$ axes.

7. ★ Differentiation and coordinate frame.

   How can we define these?

   $$\frac{^G d}{dt}\frac{^G d}{dt}\,{}^G\mathbf{r}$$

   $$\frac{^G d}{dt}\frac{^G d}{dt}\,{}^B\mathbf{r}\,, \qquad \frac{^G d}{dt}\frac{^B d}{dt}\,{}^G\mathbf{r}\,, \qquad \frac{^B d}{dt}\frac{^G d}{dt}\,{}^G\mathbf{r}$$

   $$\frac{^G d}{dt}\frac{^B d}{dt}\,{}^B\mathbf{r}\,, \qquad \frac{^B d}{dt}\frac{^B d}{dt}\,{}^G\mathbf{r}\,, \qquad \frac{^B d}{dt}\frac{^G d}{dt}\,{}^B\mathbf{r}$$

   $$\frac{^B d}{dt}\frac{^B d}{dt}\,{}^B\mathbf{r}$$

8. A roller in a circle.

   Figure 10.5 shows a roller in a circular path. Find the velocity and acceleration of a point $P$ on the circumference of the roller.

9. An RPR manipulator.

   Label the coordinate frames and find the velocity and acceleration of point $P$ at the endpoint of the manipulator shown in Figure 10.6.

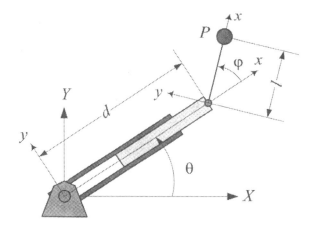

FIGURE 10.6. A planar RPR manipulator.

10. Rigid link acceleration.

Figure 10.7 illustrates the coordinate frames and kinematics of a rigid link ($i$). Assume that the angular velocity of the link, as well as the velocity and acceleration at proximal and distal joints, are given. Find

(a) velocity and acceleration of the link at $C_i$ in terms of proximal joint $i$ velocity and acceleration

(b) velocity and acceleration of the link at $C_i$ in terms of distal joint $i + 1$ velocity and acceleration

(c) velocity and acceleration of the of proximal joint $i$ in terms of distal joint $i + 1$ velocity and acceleration

(d) velocity and acceleration of the of distal joint $i + 1$ in terms of proximal joint $i$ velocity and acceleration.

11. ★ Jerk of a point in a roller.

Figure 10.8 shows a rolling disc on an inclined flat surface. Find the velocity, acceleration, and jerk of point $P$ in local and global coordinate frames.

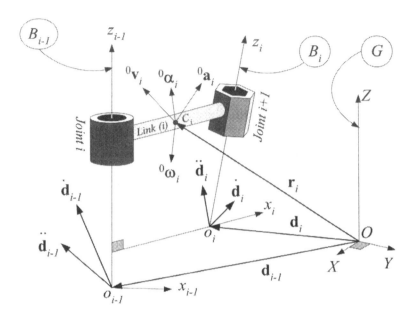

FIGURE 10.7. Kinematics of a rigid link.

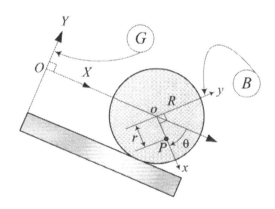

FIGURE 10.8. A rolling disc on an inclined flat.

# 11

# Motion Dynamics

## 11.1 Force and Moment

From a Newtonian viewpoint, the forces acting on a rigid body are divided into *internal* and *external forces*. Internal forces are acting between particles of the body, and external forces are acting from outside the body. An external force is either *contact force*, such as actuating force at a joint of a robot, or *body force*, such as gravitational force on the links of a robot. External forces and moments are called *load*, and a set of forces and moments acting on a rigid body is called a *force system*. The *resultant* or *total force* **F** is the sum of all the external forces acting on a body, and the *resultant* or *total moment* **M** is the sum of all the moments of the external forces about an origin.

$$\mathbf{F} = \sum_i \mathbf{F}_i \tag{11.1}$$

$$\mathbf{M} = \sum_i \mathbf{M}_i \tag{11.2}$$

Consider a force **F** acting at a point $P$ indicated by position vector $\mathbf{r}_P$. The *moment of the force* about a directional line $l$ passing through the origin is

$$\mathbf{M}_l = l\hat{u} \cdot (\mathbf{r}_P \times \mathbf{F}) \tag{11.3}$$

where $\hat{u}$ is a unit vector indicating the direction of $l$. The moment of a force may also be called *torque* or *moment*.

The moment of a force **F**, acting at a point $P$, about a point $Q$ at $\mathbf{r}_Q$ is

$$\mathbf{M}_Q = (\mathbf{r}_P - \mathbf{r}_Q) \times \mathbf{F} \tag{11.4}$$

and therefore, the moment of **F** about the origin is

$$\mathbf{M} = \mathbf{r}_P \times \mathbf{F}. \tag{11.5}$$

The effect of a force system, acting on a rigid body, is equivalent to the effect of the resultant force and resultant moment of the force system. Any two force systems are equivalent if their resultant forces and resultant moments are equal. It is possible that the resultant force of a force system be zero. In this condition the resultant moment of the force system is independent of the origin of the coordinate frame. Such a resultant moment is called *torque*.

When a force system is reduced to a resultant $\mathbf{F}_P$ and $\mathbf{M}_P$ with respect to a reference point $P$, we may change the reference point to another point $Q$ and find the new resultants as

$$\mathbf{F}_Q = \mathbf{F}_P \qquad (11.6)$$
$$\mathbf{M}_Q = \mathbf{M}_P + (\mathbf{r}_P - \mathbf{r}_Q) \times \mathbf{F}_P$$
$$= \mathbf{M}_P + {}_Q\mathbf{r}_P \times \mathbf{F}_P. \qquad (11.7)$$

The momentum of a rigid body is a vector quantity proportional to the total mass of the body times the translational velocity of the mass center of the body.

$$\mathbf{p} = m\mathbf{v} \qquad (11.8)$$

The momentum is also called *linear momentum* or *translational momentum*. Consider a rigid body with momentum $\mathbf{p}$. The *moment of momentum*, $\mathbf{L}$, about a directional line $l$ passing through the origin is

$$\mathbf{L}_l = l\hat{u} \cdot (\mathbf{r}_C \times \mathbf{p}) \qquad (11.9)$$

where $\hat{u}$ is a unit vector indicating the direction of the line, $\mathbf{r}_C$ is the position vector of the mass center $(CM)$, and $\mathbf{L} = \mathbf{r}_C \times \mathbf{p}$ is the moment of momentum about the origin. The moment of momentum is also called *angular momentum* to justify the word linear momentum.

A *bounded vector* is a vector fixed at a point in space. A *sliding* or *line vector* is a vector free to slide on its *line of action*. A *free vector* can move to any point as long as it keeps its direction. Force is a sliding vector and moment is a free vector. However, the moment of a force is dependent on the distance between the origin of the coordinate frame and the line of action.

The application of a force system is emphasized by *Newton's second and third laws of motion*. The second law of motion, also called *Newton equation of motion*, says that the global rate of change of *linear momentum* is proportional to the global *applied force*.

$$ {}^G\mathbf{F} = \frac{{}^Gd}{dt}{}^G\mathbf{p} = \frac{{}^Gd}{dt}\left(m\,{}^G\mathbf{v}\right) \qquad (11.10) $$

The third law of motion says that the action and reaction forces acting between two bodies are equal and opposite.

The second law of motion can be expanded to include rotational motions. Hence, the second law of motion also says that the global rate of change of *angular momentum* is proportional to the global *applied moment*.

$$ {}^G\mathbf{M} = \frac{{}^Gd}{dt}{}^G\mathbf{L} \qquad (11.11) $$

**Proof.** Differentiating from angular momentum shows that

$$\frac{^Gd}{dt}{}^G\mathbf{L} = \frac{^Gd}{dt}(\mathbf{r}_C \times \mathbf{p})$$

$$= \left(\frac{^Gd\mathbf{r}_C}{dt} \times \mathbf{p} + \mathbf{r}_C \times \frac{^Gd\mathbf{p}}{dt}\right)$$

$$= {}^G\mathbf{r}_C \times \frac{^Gd\mathbf{p}}{dt}$$

$$= {}^G\mathbf{r}_C \times {}^G\mathbf{F}$$

$$= {}^G\mathbf{M}. \tag{11.12}$$

■

*Kinetic energy* of a point $P$ from a moving frame $B$, with mass $m$ at a position pointed by ${}^G\mathbf{r}_P$ and having a velocity ${}^G\mathbf{v}_P$, is

$$K = \frac{1}{2}m\,{}^G\mathbf{v}_P^2$$

$$= \frac{1}{2}m\left({}^G\dot{\mathbf{d}}_B + {}^B\mathbf{v}_P + {}^B_G\boldsymbol{\omega}_B \times {}^B\mathbf{r}_P\right)^2. \tag{11.13}$$

The work done by the applied force ${}^G\mathbf{F}$ on $m$ in going from point 1 to point 2 on a path indicated by a vector $\mathbf{r}$ is

$$_1W_2 = \int_1^2 {}^G\mathbf{F} \cdot d\mathbf{r}. \tag{11.14}$$

However,

$$\int_1^2 {}^G\mathbf{F} \cdot d\mathbf{r} = m\int_1^2 \frac{^Gd}{dt}{}^G\mathbf{v} \cdot {}^G\mathbf{v}\,dt$$

$$= \frac{1}{2}m\int_1^2 \frac{d}{dt}v^2\,dt$$

$$= \frac{1}{2}m\left(v_2^2 - v_1^2\right)$$

$$= K_2 - K_1 \tag{11.15}$$

that shows $_1W_2$ is equal to the difference of the kinetic energy between terminal and initial points.

$$_1W_2 = K_2 - K_1 \tag{11.16}$$

**Example 247** *Position of center of mass.*

*The position of the mass center of a rigid body in a coordinate frame is indicated by $\mathbf{r}_C$ and is usually measured in the body coordinate frame.*

$$^B\mathbf{r}_C = \frac{1}{m}\int_B \mathbf{r}\,dm \tag{11.17}$$

$$\begin{bmatrix} x_C \\ y_C \\ z_C \end{bmatrix} = \begin{bmatrix} \frac{1}{m}\int_B x\,dm \\ \frac{1}{m}\int_B y\,dm \\ \frac{1}{m}\int_B z\,dm \end{bmatrix} \tag{11.18}$$

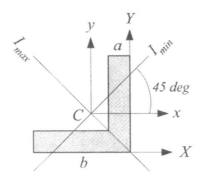

FIGURE 11.1. Principal coordinate frame for a symmetric L-section.

*Applying the mass center integral on the symmetric L-section rigid body with $\rho = 1$ shown in Figure 11.1 provides the CM of the section. The $x$ position of $C$ is*

$$
\begin{aligned}
x_C &= \frac{1}{m} \int_B x \, dm \\
&= \frac{1}{A} \int_B x \, dA \\
&= -\frac{b^2 + ab - a^2}{4ab + 2a^2}
\end{aligned}
\tag{11.19}
$$

*and because of symmetry, we have*

$$
y_C = -x_C = \frac{b^2 + ab - a^2}{4ab + 2a^2}.
\tag{11.20}
$$

**Example 248** *Every force system is equivalent to a wrench.*

*The Poinsot theorem says: Every force system is equivalent to a single force, plus a moment parallel to the force. Let $\mathbf{F}$ and $\mathbf{M}$ be the resultant force and moment of the force system. We decompose the moment into parallel and perpendicular components, $\mathbf{M}_\parallel$ and $\mathbf{M}_\perp$, to the force axis. The force $\mathbf{F}$ and the perpendicular moment $\mathbf{M}_\perp$ can be replaced by a single force $\mathbf{F}'$ parallel to $\mathbf{F}$. Therefore, the force system is reduced to a force $\mathbf{F}'$ and a moment $\mathbf{M}_\parallel$ parallel to each other. A force and a moment about the force axis is a **wrench**.*

*Poinsot theorem is similar to the Chasles theorem that says every rigid body motion is equivalent to a screw, which is a translation plus a rotation about the axis of translation.*

**Example 249** *Motion of a moving point in a moving body frame.*

*The velocity and acceleration of a moving point $P$ as shown in Figure 7.5 are found in Example 243.*

$$
{}^G\mathbf{v}_P = {}^G\dot{\mathbf{d}}_B + {}^G R_B \left( {}^B\mathbf{v}_P + {}^B_G\boldsymbol{\omega}_B \times {}^B\mathbf{r}_P \right)
\tag{11.21}
$$

$$^G\mathbf{a}_P = {}^G\ddot{\mathbf{d}}_B + {}^G R_B \left({}^B\mathbf{a}_P + 2\,{}^B_G\boldsymbol{\omega}_B \times {}^B\mathbf{v}_P + {}^B_G\dot{\boldsymbol{\omega}}_B \times {}^B\mathbf{r}_P\right)$$
$$+ {}^G R_B \left({}^B_G\boldsymbol{\omega}_B \times \left({}^B_G\boldsymbol{\omega}_B \times {}^B\mathbf{r}_P\right)\right) \tag{11.22}$$

*Therefore, the equation of motion for the point mass $P$ is*

$$^G\mathbf{F} = m\,{}^G\mathbf{a}_P$$
$$= m\left({}^G\ddot{\mathbf{d}}_B + {}^G R_B \left({}^B\mathbf{a}_P + 2\,{}^B_G\boldsymbol{\omega}_B \times {}^B\mathbf{v}_P + {}^B_G\dot{\boldsymbol{\omega}}_B \times {}^B\mathbf{r}_P\right)\right)$$
$$+ m\,{}^G R_B \left({}^B_G\boldsymbol{\omega}_B \times \left({}^B_G\boldsymbol{\omega}_B \times {}^B\mathbf{r}_P\right)\right). \tag{11.23}$$

**Example 250** *Newton equation in a rotating frame.*
*Consider a spherical rigid body (such as Earth) with a fixed point that is rotating with a constant angular velocity. The equation of motion for a moving point $P$ on the rigid body is found by setting ${}^G\ddot{\mathbf{d}}_B = {}^B_G\dot{\boldsymbol{\omega}}_B = 0$ in the equation of motion of a moving point in a moving body frame (11.23)*

$$^B\mathbf{F} = m\,{}^B\mathbf{a}_P + m\,{}^B_G\boldsymbol{\omega}_B \times \left({}^B_G\boldsymbol{\omega}_B \times {}^B\mathbf{r}_P\right) + 2m\,{}^B_G\boldsymbol{\omega}_B \times {}^B\dot{\mathbf{r}}_P \tag{11.24}$$
$$\neq m\,{}^B\mathbf{a}_P$$

*which shows that the Newton equation of motion $\mathbf{F} = m\,\mathbf{a}$ is not correct in a rotating frame.*

**Example 251** *Coriolis force.*
*The equation of motion of a moving point on the surface of the Earth is*

$$^B\mathbf{F} = m\,{}^B\mathbf{a}_P + m\,{}^B_G\boldsymbol{\omega}_B \times \left({}^B_G\boldsymbol{\omega}_B \times {}^B\mathbf{r}_P\right) + 2m\,{}^B_G\boldsymbol{\omega}_B \times {}^B\mathbf{v}_P \tag{11.25}$$

*which can be rearranged to*

$$^B\mathbf{F} - m\,{}^B_G\boldsymbol{\omega}_B \times \left({}^B_G\boldsymbol{\omega}_B \times {}^B\mathbf{r}_P\right) - 2m\,{}^B_G\boldsymbol{\omega}_B \times {}^B\mathbf{v}_P = m\,{}^B\mathbf{a}_P. \tag{11.26}$$

*Equation (11.26) is the equation of motion to an observer in the rotating frame, which in this case is an observer on the Earth. The left-hand side of this equation is called the effective force,*

$$\mathbf{F}_{eff} = {}^B\mathbf{F} - m\,{}^B_G\boldsymbol{\omega}_B \times \left({}^B_G\boldsymbol{\omega}_B \times {}^B\mathbf{r}_P\right) - 2m\,{}^B_G\boldsymbol{\omega}_B \times {}^B\mathbf{v}_P \tag{11.27}$$

*because it seems that the particle is moving under the influence of this force.*
*The second term is negative of the centrifugal force and pointing outward. The maximum value of this force on the Earth is on the equator*

$$r\omega^2 = 6378.388 \times 10^3 \times \left(\frac{2\pi}{24 \times 3600}\frac{366.25}{365.25}\right)^2$$
$$= 3.3917 \times 10^{-2}\,\mathrm{m/s^2}$$

*which is about 0.3% of the acceleration of gravity. If we add the variation of the gravitational acceleration because of a change of radius from*

$R = 6356912$ m *at the pole to* $R = 6378388$ m *on the equator, then the variation of the acceleration of gravity becomes 0.53%. So, generally speaking, a sportsman such as a pole-vaulter practicing in the north pole can show a better record in a competition held on the equator.*

*The third term is called the **Coriolis effect**,* $F_C$, *which is perpendicular to both* $\omega$ *and* ${}^B\mathbf{v}_P$. *For a mass m moving on the north hemisphere at a latitude* $\theta$ *towards the equator, we should provide a lateral eastward force equal to the Coriolis effect to force the mass, keeping its direction relative to the ground.*

$$
\begin{aligned}
F_C &= 2m\, {}^B_G\omega_B \times {}^B\mathbf{v}_m \\
&= 1.4584 \times 10^{-4}\, {}^B p_m \cos\theta \ \ \mathrm{kg\,m/s^2}
\end{aligned}
$$

*The Coriolis effect is the reason of wearing the west side of railways, roads, and rivers. The lack of providing Coriolis force is the reason for turning the direction of winds, projectiles, and falling objects westward.*

**Example 252** *Work, force, and kinetic energy in a unidirectional motion.*

*A mass* $m = 2$ kg *has an initial kinetic energy* $K = 12$ J. *The mass is under a constant force* $\mathbf{F} = F\hat{I} = 4\hat{I}$ *and moves from* $X(0) = 1$ *to* $X(t_f) = 22$ m *at a terminal time* $t_f$. *The work done by the force during this motion is*

$$
W = \int_{\mathbf{r}(0)}^{\mathbf{r}(t_f)} \mathbf{F} \cdot d\mathbf{r} = \int_1^{22} 4\, dX = 21\,\mathrm{N\,m} = 21\,\mathrm{J}
$$

*The kinetic energy at the terminal time is*

$$
K(t_f) = W - K(0) = 9\,\mathrm{J}
$$

*which shows that the terminal speed of the mass is*

$$
v_2 = \sqrt{\frac{2K(t_f)}{m}} = 3\,\mathrm{m/s}.
$$

**Example 253** *Time varying force.*

*When the applied force is time varying,*

$$
\mathbf{F}(t) = m\ddot{\mathbf{r}} \tag{11.28}
$$

*then there is a general solution for the equation of motion because the characteristic and category of the motion differs for different* $\mathbf{F}(t)$.

$$
\dot{\mathbf{r}}(t) = \dot{\mathbf{r}}(t_0) + \frac{1}{m}\int_{t_0}^t \mathbf{F}(t)dt \tag{11.29}
$$

$$
\mathbf{r}(t) = \mathbf{r}(t_0) + \dot{\mathbf{r}}(t_0)(t - t_0) + \frac{1}{m}\int_{t_0}^t \int_{t_0}^t \mathbf{F}(t)dt\,dt \tag{11.30}
$$

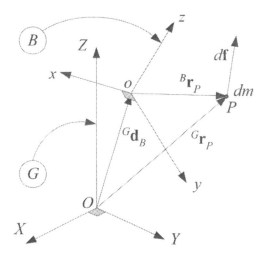

FIGURE 11.2. A body point mass moving with velocity $^G\mathbf{v}_P$ and acted on by force $d\mathbf{f}$.

## 11.2   Rigid Body Translational Kinetics

Figure 11.2 depicts a moving body $B$ in a global frame $G$. Assume that the body frame is attached at the center of mass $(CM)$ of the body. Point $P$ indicates an infinitesimal sphere of the body which has a very small mass $dm$. The point mass $dm$ is acted on by an infinitesimal force $d\mathbf{f}$ and has a global velocity $^G\mathbf{v}_P$.

According to Newton law of motion we have

$$d\mathbf{f} = {}^G\mathbf{a}_P \, dm. \tag{11.31}$$

however, the equation of motion for the whole body in global coordinate frame is

$$^G\mathbf{F} = m \, {}^G\mathbf{a}_B \tag{11.32}$$

which can be expressed in the body coordinate frame as

$$^B\mathbf{F} = m \, {}^B_G\mathbf{a}_B + m \, {}^B_G\boldsymbol{\omega}_B \times {}^B\mathbf{v}_B. \tag{11.33}$$

In these equations, $^G\mathbf{a}_B$ is the acceleration vector of the body $CM$ in global frame, $m$ is the total mass of the body, and $\mathbf{F}$ is the resultant of the external forces acted on the body at $CM$.

**Proof.** A body coordinate frame at center of mass is called a *central frame*. If the frame $B$ is a central frame, then the *center of mass*, $CM$, is defined such that

$$\int_B {}^B\mathbf{r}_{dm} \, dm = 0. \tag{11.34}$$

The global position vector of $dm$ is related to its local position vector by

$$^G\mathbf{r}_{dm} = {}^G\mathbf{d}_B + {}^GR_B \, {}^B\mathbf{r}_{dm} \tag{11.35}$$

where $^G\mathbf{d}_B$ is the global position vector of the central body frame, and therefore,

$$
\begin{aligned}
\int_B {}^G\mathbf{r}_{dm}\, dm &= \int_B {}^G\mathbf{d}_B\, dm + {}^GR_B \int_m {}^B\mathbf{r}_{dm}\, dm \\
&= \int_B {}^G\mathbf{d}_B\, dm \\
&= {}^G\mathbf{d}_B \int_B dm \\
&= m\, {}^G\mathbf{d}_B. \tag{11.36}
\end{aligned}
$$

The time derivative of both sides shows that

$$m\, {}^G\dot{\mathbf{d}}_B = m\, {}^G\mathbf{v}_B = \int_B {}^G\dot{\mathbf{r}}_{dm}\, dm = \int_B {}^G\mathbf{v}_{dm}\, dm \tag{11.37}$$

and another derivative is

$$m\, {}^G\dot{\mathbf{v}}_B = m\, {}^G\mathbf{a}_B = \int_B {}^G\dot{\mathbf{v}}_{dm}\, dm. \tag{11.38}$$

However, we have $d\mathbf{f} = {}^G\dot{\mathbf{v}}_P\, dm$ and,

$$m\, {}^G\mathbf{a}_B = \int_B d\mathbf{f}. \tag{11.39}$$

The integral on the right-hand side accounts for all the forces acting on particles of mass in the body. The internal forces cancel one another out, so the net result is the vector sum of all the externally applied forces, $\mathbf{F}$, and therefore,

$$^G\mathbf{F} = m\, {}^G\mathbf{a}_B = m\, {}^G\dot{\mathbf{v}}_B. \tag{11.40}$$

In the body coordinate frame we have

$$
\begin{aligned}
{}^B\mathbf{F} &= {}^BR_G\, {}^G\mathbf{F} \\
&= m\, {}^BR_G\, {}^G\mathbf{a}_B \\
&= m\, {}^B_G\mathbf{a}_B \\
&= m\, {}^B\mathbf{a}_B + m\, {}^B_G\boldsymbol{\omega}_B \times {}^B\mathbf{v}_B. \tag{11.41}
\end{aligned}
$$

■

## 11.3   Rigid Body Rotational Kinetics

The rigid body rotational equation of motion is the *Euler equation*

$$
\begin{aligned}
{}^{B}\mathbf{M} &= \frac{{}^{G}d}{dt}\,{}^{B}\mathbf{L} \\
&= {}^{B}\dot{\mathbf{L}} + {}^{B}_{G}\boldsymbol{\omega}_{B} \times {}^{B}\mathbf{L} \\
&= {}^{B}I\,{}^{B}_{G}\dot{\boldsymbol{\omega}}_{B} + {}^{B}_{G}\boldsymbol{\omega}_{B} \times \left({}^{B}I\,{}^{B}_{G}\boldsymbol{\omega}_{B}\right)
\end{aligned}
\tag{11.42}
$$

where $\mathbf{L}$ is the *angular momentum*

$$
{}^{B}\mathbf{L} = {}^{B}I\,{}^{B}_{G}\boldsymbol{\omega}_{B}
\tag{11.43}
$$

and $I$ is the *moment of inertia* of the rigid body.

$$
I = \begin{bmatrix}
I_{xx} & I_{xy} & I_{xz} \\
I_{yx} & I_{yy} & I_{yz} \\
I_{zx} & I_{zy} & I_{zz}
\end{bmatrix}
\tag{11.44}
$$

The elements of $I$ are only functions of the mass distribution of the rigid body and may be defined by

$$
I_{ij} = \int_{B} \left( r_i^2 \delta_{mn} - x_{im} x_{jn} \right) dm \quad , \quad i,j = 1,2,3
\tag{11.45}
$$

where $\delta_{ij}$ is Kronecker's delta.

The expanded form of the Euler equation is

$$
\begin{aligned}
M_x &= I_{xx}\dot{\omega}_x + I_{xy}\dot{\omega}_y + I_{xz}\dot{\omega}_z - (I_{yy} - I_{zz})\,\omega_y\omega_z \\
&\quad - I_{yz}\left(\omega_z^2 - \omega_y^2\right) - \omega_x\left(\omega_z I_{xy} - \omega_y I_{xz}\right)
\end{aligned}
\tag{11.46}
$$

$$
\begin{aligned}
M_y &= I_{yx}\dot{\omega}_x + I_{yy}\dot{\omega}_y + I_{yz}\dot{\omega}_z - (I_{zz} - I_{xx})\,\omega_z\omega_x \\
&\quad - I_{xz}\left(\omega_x^2 - \omega_z^2\right) - \omega_y\left(\omega_x I_{yz} - \omega_z I_{xy}\right)
\end{aligned}
\tag{11.47}
$$

$$
\begin{aligned}
M_z &= I_{zx}\dot{\omega}_x + I_{zy}\dot{\omega}_y + I_{zz}\dot{\omega}_z - (I_{xx} - I_{yy})\,\omega_x\omega_y \\
&\quad - I_{xy}\left(\omega_y^2 - \omega_x^2\right) - \omega_z\left(\omega_y I_{xz} - \omega_x I_{yz}\right).
\end{aligned}
\tag{11.48}
$$

which can be reduced to

$$
\begin{aligned}
M_1 &= I_1\dot{\omega}_1 - (I_2 - I_2)\,\omega_2\omega_3 \\
M_2 &= I_2\dot{\omega}_2 - (I_3 - I_1)\,\omega_3\omega_1 \\
M_3 &= I_3\dot{\omega}_3 - (I_1 - I_2)\,\omega_1\omega_2
\end{aligned}
\tag{11.49}
$$

in a special Cartesian coordinate frame called the *principal coordinate frame*. The principal coordinate frame is denoted by numbers 123 to indicate the first, second, and third *principal axes*. The parameters $I_{ij}$, $i \neq j$

are zero in the principal frame. The body and principal coordinate frame are assumed to sit at $CM$ of the body.

Kinetic energy of a rotating rigid body is

$$
\begin{aligned}
K &= \frac{1}{2}\left(I_{xx}\omega_x^2 + I_{yy}\omega_y^2 + I_{zz}\omega_z^2\right) \\
&\quad - I_{xy}\omega_x\omega_y - I_{yz}\omega_y\omega_z - I_{zx}\omega_z\omega_x && (11.50) \\
&= \frac{1}{2}\boldsymbol{\omega}\cdot\mathbf{L} && (11.51) \\
&= \frac{1}{2}\boldsymbol{\omega}^T I\,\boldsymbol{\omega} && (11.52)
\end{aligned}
$$

that in the principal coordinate frame reduces to

$$
K = \frac{1}{2}\left(I_1\omega_1^2 + I_2\omega_2^2 + I_3\omega_3^2\right). \tag{11.53}
$$

**Proof.** Let $m_i$ be the mass of the $i$th particle of a rigid body $B$, which is made of $n$ particles and let $\mathbf{r}_i = {}^B\mathbf{r}_i = \begin{bmatrix} x_i & y_i & z_i \end{bmatrix}^T$ be the Cartesian position vector of $m_i$ in a central body fixed coordinate frame $Oxyz$. Assume that $\boldsymbol{\omega} = {}^B_G\boldsymbol{\omega}_B = \begin{bmatrix} \omega_x & \omega_y & \omega_z \end{bmatrix}^T$ is the angular velocity of the rigid body with respect to the ground, expressed in the body coordinate frame.

The angular momentum of $m_i$ is

$$
\begin{aligned}
\mathbf{L}_i &= \mathbf{r}_i \times m_i\dot{\mathbf{r}}_i \\
&= m_i\left[\mathbf{r}_i \times (\boldsymbol{\omega} \times \mathbf{r}_i)\right] \\
&= m_i\left[(\mathbf{r}_i \cdot \mathbf{r}_i)\boldsymbol{\omega} - (\mathbf{r}_i \cdot \boldsymbol{\omega})\mathbf{r}_i\right] \\
&= m_i r_i^2 \boldsymbol{\omega} - m_i(\mathbf{r}_i \cdot \boldsymbol{\omega})\mathbf{r}_i. && (11.54)
\end{aligned}
$$

Hence, the angular momentum of the rigid body would be

$$
\mathbf{L} = \boldsymbol{\omega}\sum_{i=1}^n m_i r_i^2 - \sum_{i=1}^n m_i(\mathbf{r}_i \cdot \boldsymbol{\omega})\mathbf{r}_i. \tag{11.55}
$$

Substitution $\mathbf{r}_i$ and $\boldsymbol{\omega}$ gives us

$$
\begin{aligned}
\mathbf{L} &= \left(\omega_x\hat{i} + \omega_y\hat{j} + \omega_z\hat{k}\right)\sum_{i=1}^n m_i\left(x_i^2 + y_i^2 + z_i^2\right) \\
&\quad - \sum_{i=1}^n m_i\left(x_i\omega_x + y_i\omega_y + z_i\omega_z\right)\cdot\left(x_i\hat{i} + y_i\hat{j} + z_i\hat{k}\right) && (11.56)
\end{aligned}
$$

and therefore,

$$\mathbf{L} = \sum_{i=1}^{n} m_i \left(x_i^2 + y_i^2 + z_i^2\right) \omega_x \hat{\imath} + \sum_{i=1}^{n} m_i \left(x_i^2 + y_i^2 + z_i^2\right) \omega_y \hat{\jmath}$$

$$+ \sum_{i=1}^{n} m_i \left(x_i^2 + y_i^2 + z_i^2\right) \omega_z \hat{k} - \sum_{i=1}^{n} m_i \left(x_i \omega_x + y_i \omega_y + z_i \omega_z\right) x_i \hat{\imath}$$

$$- \sum_{i=1}^{n} m_i \left(x_i \omega_x + y_i \omega_y + z_i \omega_z\right) y_i \hat{\jmath}$$

$$- \sum_{i=1}^{n} m_i \left(x_i \omega_x + y_i \omega_y + z_i \omega_z\right) z_i \hat{k} \tag{11.57}$$

or

$$\mathbf{L} = \sum_{i=1}^{n} m_i \left[ \left(x_i^2 + y_i^2 + z_i^2\right) \omega_x - \left(x_i \omega_x + y_i \omega_y + z_i \omega_z\right) x_i \right] \hat{\imath}$$

$$+ \sum_{i=1}^{n} m_i \left[ \left(x_i^2 + y_i^2 + z_i^2\right) \omega_y - \left(x_i \omega_x + y_i \omega_y + z_i \omega_z\right) y_i \right] \hat{\jmath}$$

$$+ \sum_{i=1}^{n} m_i \left[ \left(x_i^2 + y_i^2 + z_i^2\right) \omega_z - \left(x_i \omega_x + y_i \omega_y + z_i \omega_z\right) z_i \right] \hat{k} \tag{11.58}$$

which can be rearranged as

$$\mathbf{L} = \sum_{i=1}^{n} \left[m_i \left(y_i^2 + z_i^2\right)\right] \omega_x \hat{\imath} - \left( \sum_{i=1}^{n} (m_i x_i y_i) \omega_y + \sum_{i=1}^{n} (m_i x_i z_i) \omega_z \right) \hat{\imath}$$

$$- \sum_{i=1}^{n} (m_i y_i x_i) \omega_x \hat{\jmath} + \sum_{i=1}^{n} \left[m_i \left(z_i^2 + x_i^2\right)\right] \omega_y \hat{\jmath}$$

$$- \sum_{i=1}^{n} (m_i y_i z_i) \omega_z \hat{\jmath} - \left( \sum_{i=1}^{n} (m_i z_i x_i) \omega_x + \sum_{i=1}^{n} (m_i z_i y_i) \omega_y \right) \hat{k}$$

$$+ \sum_{i=1}^{n} \left[m_i \left(x_i^2 + y_i^2\right)\right] \omega_z \hat{k}. \tag{11.59}$$

The angular momentum can be written in a concise form

$$L_x = I_{xx}\omega_x + I_{xy}\omega_y + I_{xz}\omega_z \tag{11.60}$$

$$L_y = I_{yx}\omega_x + I_{yy}\omega_y + I_{yz}\omega_z \tag{11.61}$$

$$L_z = I_{zx}\omega_x + I_{zy}\omega_y + I_{zz}\omega_z \tag{11.62}$$

or in a matrix form

$$\mathbf{L} = I \cdot \boldsymbol{\omega} \tag{11.63}$$

$$\begin{bmatrix} L_x \\ L_y \\ L_z \end{bmatrix} = \begin{bmatrix} I_{xx} & I_{xy} & I_{xz} \\ I_{yx} & I_{yy} & I_{yz} \\ I_{zx} & I_{zy} & I_{zz} \end{bmatrix} \begin{bmatrix} \omega_x \\ \omega_y \\ \omega_z \end{bmatrix} \tag{11.64}$$

by introducing the moment of inertia matrix $I$ where

$$I_{xx} = \sum_{i=1}^{n} \left[ m_i \left( y_i^2 + z_i^2 \right) \right] \tag{11.65}$$

$$I_{yy} = \sum_{i=1}^{n} \left[ m_i \left( z_i^2 + x_i^2 \right) \right] \tag{11.66}$$

$$I_{zz} = \sum_{i=1}^{n} \left[ m_i \left( x_i^2 + y_i^2 \right) \right] \tag{11.67}$$

$$I_{xy} = I_{yx} = -\sum_{i=1}^{n} \left( m_i x_i y_i \right) \tag{11.68}$$

$$I_{yz} = I_{zy} = -\sum_{i=1}^{n} \left( m_i y_i z_i \right) \tag{11.69}$$

$$I_{zx} = I_{xz} = -\sum_{i=1}^{n} \left( m_i z_i x_i \right). \tag{11.70}$$

For a rigid body that is a continuous solid, the summations must be replaced by integrations over the volume of the body as in Equation (11.45).

The Euler equation of motion for a rigid body is

$$^{B}\mathbf{M} = \frac{^{G}d}{dt} \, ^{B}\mathbf{L} \tag{11.71}$$

where $^{B}\mathbf{M}$ is the resultant of the external moments applied on the rigid body. The angular momentum $^{B}\mathbf{L}$ is a vector defined in the body coordinate frame. Hence, its time derivative in the global coordinate frame is

$$\frac{^{G}d \, ^{B}\mathbf{L}}{dt} = {}^{B}\dot{\mathbf{L}} + {}_{G}^{B}\boldsymbol{\omega}_B \times {}^{B}\mathbf{L}. \tag{11.72}$$

Therefore,

$$\begin{aligned} {}^{B}\mathbf{M} &= \frac{d\mathbf{L}}{dt} = \dot{\mathbf{L}} + \boldsymbol{\omega} \times \mathbf{L} \\ &= I\dot{\boldsymbol{\omega}} + \boldsymbol{\omega} \times (I\boldsymbol{\omega}) \end{aligned} \tag{11.73}$$

$$
\begin{aligned}
{}^{B}\mathbf{M} = & \left(I_{xx}\dot{\omega}_x + I_{xy}\dot{\omega}_y + I_{xz}\dot{\omega}_z\right)\hat{\imath} \\
& + \left(I_{yx}\dot{\omega}_x + I_{yy}\dot{\omega}_y + I_{yz}\dot{\omega}_z\right)\hat{\jmath} \\
& + \left(I_{zx}\dot{\omega}_x + I_{zy}\dot{\omega}_y + I_{zz}\dot{\omega}_z\right)\hat{k} \\
& + \omega_y \left(I_{xz}\omega_x + I_{yz}\omega_y + I_{zz}\omega_z\right)\hat{\imath} \\
& - \omega_z \left(I_{xy}\omega_x + I_{yy}\omega_y + I_{yz}\omega_z\right)\hat{\imath} \\
& + \omega_z \left(I_{xx}\omega_x + I_{xy}\omega_y + I_{xz}\omega_z\right)\hat{\jmath} \\
& - \omega_x \left(I_{xz}\omega_x + I_{yz}\omega_y + I_{zz}\omega_z\right)\hat{\jmath} \\
& + \omega_x \left(I_{xy}\omega_x + I_{yy}\omega_y + I_{yz}\omega_z\right)\hat{k} \\
& - \omega_y \left(I_{xx}\omega_x + I_{xy}\omega_y + I_{xz}\omega_z\right)\hat{k} \qquad (11.74)
\end{aligned}
$$

and the most general form of the Euler equations of motion for a rigid body in a body frame attached to $CM$ of the body are

$$
\begin{aligned}
M_x = & \; I_{xx}\dot{\omega}_x + I_{xy}\dot{\omega}_y + I_{xz}\dot{\omega}_z - (I_{yy} - I_{zz})\,\omega_y\omega_z \\
& - I_{yz}\left(\omega_z^2 - \omega_y^2\right) - \omega_x\left(\omega_z I_{xy} - \omega_y I_{xz}\right) \qquad (11.75)
\end{aligned}
$$

$$
\begin{aligned}
M_y = & \; I_{yx}\dot{\omega}_x + I_{yy}\dot{\omega}_y + I_{yz}\dot{\omega}_z - (I_{zz} - I_{xx})\,\omega_z\omega_x \\
& - I_{xz}\left(\omega_x^2 - \omega_z^2\right) - \omega_y\left(\omega_x I_{yz} - \omega_z I_{xy}\right) \qquad (11.76)
\end{aligned}
$$

$$
\begin{aligned}
M_z = & \; I_{zx}\dot{\omega}_x + I_{zy}\dot{\omega}_y + I_{zz}\dot{\omega}_z - (I_{xx} - I_{yy})\,\omega_x\omega_y \\
& - I_{xy}\left(\omega_y^2 - \omega_x^2\right) - \omega_z\left(\omega_y I_{xz} - \omega_x I_{yz}\right). \qquad (11.77)
\end{aligned}
$$

Assume that we can rotate the body frame about its origin to find an orientation that makes $I_{ij} = 0$, for $i \neq j$. In such a coordinate frame, which is called a principal frame, the Euler equations reduce to

$$
\begin{aligned}
M_1 & = I_1\dot{\omega}_1 - (I_2 - I_2)\,\omega_2\omega_3 & (11.78) \\
M_2 & = I_2\dot{\omega}_2 - (I_3 - I_1)\,\omega_3\omega_1 & (11.79) \\
M_3 & = I_3\dot{\omega}_3 - (I_1 - I_2)\,\omega_1\omega_2. & (11.80)
\end{aligned}
$$

The kinetic energy of a rigid body may be found by the integral of the kinetic energy of a mass element $dm$, over the whole body.

$$
\begin{aligned}
K = & \; \frac{1}{2}\int_B \dot{\mathbf{v}}^2\,dm \\
= & \; \frac{1}{2}\int_B (\boldsymbol{\omega}\times\mathbf{r})\cdot(\boldsymbol{\omega}\times\mathbf{r})\,dm \\
= & \; \frac{\omega_x^2}{2}\int_B \left(y^2 + z^2\right)dm + \frac{\omega_y^2}{2}\int_B \left(z^2 + x^2\right)dm + \frac{\omega_z^2}{2}\int_B \left(x^2 + y^2\right)dm \\
& - \omega_x\omega_y\int_B xy\,dm - \omega_y\omega_z\int_B yz\,dm - \omega_z\omega_x\int_B zx\,dm \\
= & \; \frac{1}{2}\left(I_{xx}\omega_x^2 + I_{yy}\omega_y^2 + I_{zz}\omega_z^2\right) \\
& - I_{xy}\omega_x\omega_y - I_{yz}\omega_y\omega_z - I_{zx}\omega_z\omega_x \qquad (11.81)
\end{aligned}
$$

The kinetic energy can be rearranged to a matrix multiplication form

$$K = \frac{1}{2}\omega^T I \omega \tag{11.82}$$

$$= \frac{1}{2}\omega \cdot \mathbf{L}. \tag{11.83}$$

When the body frame is principal, the kinetic energy will simplify to

$$K = \frac{1}{2}\left(I_1\omega_1^2 + I_2\omega_2^2 + I_3\omega_3^2\right). \tag{11.84}$$

■

**Example 254** *Steady rotation of a freely rotating rigid body.*
  *The Newton-Euler equations of motion for a rigid body are*

$$^G\mathbf{F} = m\,^G\dot{\mathbf{v}} \tag{11.85}$$

$$^B\mathbf{M} = I\,_G^B\dot{\omega}_B + _G^B\omega_B \times\,^B\mathbf{L}. \tag{11.86}$$

*Consider a situation where the resultant applied force and moment on the body are zero.*

$$\mathbf{F} = 0 \tag{11.87}$$

$$\mathbf{M} = 0 \tag{11.88}$$

*Based on the Newton equation, the velocity of the mass center will be constant in the global coordinate frame. However, the Euler equation reduces to*

$$\dot{\omega}_1 = \frac{I_2 - I_3}{I_1}\omega_2\omega_3 \tag{11.89}$$

$$\dot{\omega}_2 = \frac{I_3 - I_1}{I_{22}}\omega_3\omega_1 \tag{11.90}$$

$$\dot{\omega}_3 = \frac{I_1 - I_2}{I_3}\omega_1\omega_2 \tag{11.91}$$

*showing that the angular velocity can be constant if*

$$I_1 = I_2 = I_3 \tag{11.92}$$

*or if two principal moments of inertia, say $I_1$ and $I_2$, are zero and the third angular velocity, in this case $\omega_3$, is initially zero, or if the angular velocity vector is initially parallel to a principal axis.*

**Example 255** *Angular momentum of a two-link manipulator.*
  *A two-link manipulator is shown in Figure 11.3. Link A rotates with angular velocity $\dot{\varphi}$ about the z-axis of its local coordinate frame. Link B is attached to link A and has angular velocity $\dot{\psi}$ with respect to A about the*

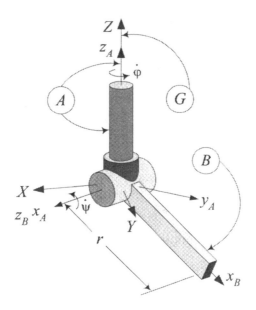

FIGURE 11.3. A two-link manipulator.

$x_A$-axis. We assume that $A$ and $G$ were coincident at $\varphi = 0$, therefore, the rotation matrix between $A$ and $G$ is

$$
{}^G R_A = \begin{bmatrix} \cos\varphi(t) & -\sin\varphi(t) & 0 \\ \sin\varphi(t) & \cos\varphi(t) & 0 \\ 0 & 0 & 1 \end{bmatrix}.
\tag{11.93}
$$

The frame $B$ is related to the frame $A$ by Euler angles $\varphi = 90\deg$, $\theta = 90\deg$, and $\psi = \psi$, hence,

$$
\begin{aligned}
{}^A R_B &= \begin{bmatrix} c\pi c\psi - c\pi s\pi s\psi & -c\pi s\psi - c\pi c\psi s\pi & s\pi s\pi \\ c\psi s\pi + c\pi c\pi s\psi & -s\pi s\psi + c\pi c\pi c\psi & -c\pi s\pi \\ s\pi s\psi & s\pi c\psi & c\pi \end{bmatrix} \\[4pt]
&\phantom{=}\begin{bmatrix} -\cos\psi & \sin\psi & 0 \\ \sin\psi & \cos\psi & 0 \\ 0 & 0 & -1 \end{bmatrix}
\end{aligned}
\tag{11.94}
$$

and therefore,

$$
\begin{aligned}
{}^G R_B &= {}^G R_A \, {}^A R_B \tag{11.95} \\[4pt]
&= \begin{bmatrix} -\cos\varphi\cos\psi - \sin\varphi\sin\psi & \cos\varphi\sin\psi - \cos\psi\sin\varphi & 0 \\ \cos\varphi\sin\psi - \cos\psi\sin\varphi & \cos\varphi\cos\psi + \sin\varphi\sin\psi & 0 \\ 0 & 0 & -1 \end{bmatrix}.
\end{aligned}
$$

*The angular velocity of A in G, and B in A is*

$$_G\boldsymbol{\omega}_A = \dot{\varphi}\hat{K} \tag{11.96}$$

$$_A\boldsymbol{\omega}_B = \dot{\psi}\hat{\imath}_A. \tag{11.97}$$

*Moment of inertia matrices for the arms A and B can be defined as*

$$^A I_A = \begin{bmatrix} I_{A1} & 0 & 0 \\ 0 & I_{A2} & 0 \\ 0 & 0 & I_{A3} \end{bmatrix} \tag{11.98}$$

$$^B I_B = \begin{bmatrix} I_{B1} & 0 & 0 \\ 0 & I_{B2} & 0 \\ 0 & 0 & I_{B3} \end{bmatrix}. \tag{11.99}$$

*These moments of inertia must be transformed to the global frame*

$$^G I_A = {}^G R_B \, {}^A I_A \, {}^G R_A^T \tag{11.100}$$

$$^G I_B = {}^G R_B \, {}^B I_B \, {}^G R_B^T. \tag{11.101}$$

*The total angular momentum of the manipulator is*

$$^G\mathbf{L} = {}^G\mathbf{L}_A + {}^G\mathbf{L}_B \tag{11.102}$$

*where*

$$^G\mathbf{L}_A = {}^G I_A \, {}_G\boldsymbol{\omega}_A \tag{11.103}$$

$$^G\mathbf{L}_B = {}^G I_B \, {}_G\boldsymbol{\omega}_B$$

$$= {}^G I_B \left( {}_A^G\boldsymbol{\omega}_B + {}_G\boldsymbol{\omega}_A \right). \tag{11.104}$$

**Example 256** *Poinsot's construction.*
*Consider a freely rotating rigid body with an attached principal coordinate frame. Having* $\mathbf{M} = 0$ *provides a motion under constant angular momentum and kinetic energy*

$$\mathbf{L} = I\boldsymbol{\omega} = cte \tag{11.105}$$

$$K = \frac{1}{2}\boldsymbol{\omega}^T I \boldsymbol{\omega} = cte. \tag{11.106}$$

*Because the length of the angular momentum* $\mathbf{L}$ *is constant, the equation*

$$L^2 = \mathbf{L} \cdot \mathbf{L}$$

$$= L_x^2 + L_y^2 + L_z^2$$

$$= I_1^2\omega_1^2 + I_2^2\omega_2^2 + I_3^2\omega_3^2 \tag{11.107}$$

*introduces an ellipsoid in the* $(\omega_1, \omega_2, \omega_3)$ *coordinate frame, called the* **momentum ellipsoid***. The tip of all possible angular velocity vectors must*

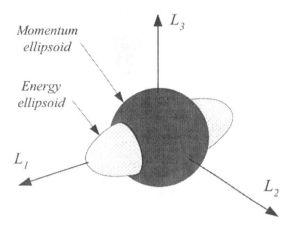

FIGURE 11.4. Intersection of the momentum and energy ellipsoids.

*lie on the surface of the momentum ellipsoid. The kinetic energy also de-
fines an **energy ellipsoid** in the same coordinate frame so that the tip of
angular velocity vectors must also lie on its surface.*

$$K = \frac{1}{2}\left(I_1\omega_1^2 + I_2\omega_2^2 + I_3\omega_3^2\right) \tag{11.108}$$

*In other words, the dynamics of moment-free motion of a rigid body requires
that the corresponding angular velocity $\omega(t)$ satisfy both Equations (11.107)
and (11.108) and therefore lie on the intersection of the momentum and
energy ellipsoids.*

*For a better visualization, we can define the ellipsoids in the $(L_x, L_y, L_z)$
coordinate system as*

$$L_x^2 + L_y^2 + L_z^2 = L^2 \tag{11.109}$$

$$\frac{L_x^2}{2I_1K} + \frac{L_y^2}{2I_2K} + \frac{L_z^2}{2I_3K} = 1. \tag{11.110}$$

*Equation (11.109) is a sphere and Equation (11.110) defines an ellipsoid
with $\sqrt{2I_iK}$ as semi-axes. To have a meaningful motion, these two shapes
must intersect. The intersection may form a trajectory, as shown in Figure
11.4.*

*It can be deduced that for a certain value of angular momentum there are
maximum and minimum limit values for acceptable kinetic energy. Assuming*

$$I_1 > I_3 > I_3 \tag{11.111}$$

*the limits of possible kinetic energy are*

$$K_{\min} = \frac{L^2}{2I_1} \tag{11.112}$$

$$K_{\max} = \frac{L^2}{2I_3} \tag{11.113}$$

*and the corresponding motions are turning about the axes $I_1$ and $I_3$ respectively.*

**Example 257** ★ *Alternative derivation of Euler equations of motion.*
   *The moment of the small force* $d\mathbf{f}$ *is* $d\mathbf{m}$

$$\begin{aligned} d\mathbf{m} &= {}^G\mathbf{r}_{dm} \times d\mathbf{f} \\ &= {}^G\mathbf{r}_{dm} \times {}^G\dot{\mathbf{v}}_{dm}\, dm. \end{aligned} \tag{11.114}$$

*The inertial angular momentum* $d\mathbf{l}$ *of* $dm$ *is equal to*

$$d\mathbf{l} = {}^G\mathbf{r}_{dm} \times {}^G\mathbf{v}_{dm}\, dm \tag{11.115}$$

*and according to (11.11) we have*

$$d\mathbf{m} = \frac{{}^G d}{dt} d\mathbf{l} \tag{11.116}$$

$$^G\mathbf{r}_{dm} \times d\mathbf{f} = \frac{{}^G d}{dt}\left({}^G\mathbf{r}_{dm} \times {}^G\mathbf{v}_{dm}\, dm\right). \tag{11.117}$$

*Integrating over the body is*

$$\begin{aligned} \int_B {}^G\mathbf{r}_{dm} \times d\mathbf{f} &= \int_B \frac{{}^G d}{dt}\left({}^G\mathbf{r}_{dm} \times {}^G\mathbf{v}_{dm}\, dm\right) \\ &= \frac{{}^G d}{dt}\int_B \left({}^G\mathbf{r}_{dm} \times {}^G\mathbf{v}_{dm}\, dm\right). \end{aligned} \tag{11.118}$$

*However, utilizing*

$$^G\mathbf{r}_{dm} = {}^G\mathbf{d}_B + {}^G R_B\, {}^B\mathbf{r}_{dm} \tag{11.119}$$

*where* ${}^G\mathbf{d}_B$ *is the global position vector of the central body frame, may simplify the left-hand side of the integral to*

$$\begin{aligned} \int_B {}^G\mathbf{r}_{dm} \times d\mathbf{f} &= \int_B \left({}^G\mathbf{d}_B + {}^G R_B\, {}^B\mathbf{r}_{dm}\right) \times d\mathbf{f} \\ &= \int_B {}^G\mathbf{d}_B \times d\mathbf{f} + \int_B {}^G_B\mathbf{r}_{dm} \times d\mathbf{f} \\ &= {}^G\mathbf{d}_B \times {}^G\mathbf{F} + {}^G\mathbf{M}_C \end{aligned} \tag{11.120}$$

*where $\mathbf{M}_C$ is the resultant external moment about the body mass center $C$.
The right-hand side of the Equation (11.118) is*

$$\frac{{}^{G}d}{dt}\int_B \left({}^{G}\mathbf{r}_{dm} \times {}^{G}\mathbf{v}_{dm}\, dm\right)$$

$$= \frac{{}^{G}d}{dt}\int_B \left(\left({}^{G}\mathbf{d}_B + {}^{G}R_B\, {}^{B}\mathbf{r}_{dm}\right) \times {}^{G}\mathbf{v}_{dm}\, dm\right)$$

$$= \frac{{}^{G}d}{dt}\int_B \left({}^{G}\mathbf{d}_B \times {}^{G}\mathbf{v}_{dm}\right) dm + \frac{{}^{G}d}{dt}\int_B \left({}^{G}_{B}\mathbf{r}_{dm} \times {}^{G}\mathbf{v}_{dm}\right) dm$$

$$= \frac{{}^{G}d}{dt}\left({}^{G}\mathbf{d}_B \times \int_B {}^{G}\mathbf{v}_{dm}dm\right) + \frac{{}^{G}d}{dt}\mathbf{L}_C$$

$$= {}^{G}\dot{\mathbf{d}}_B \times \int_B {}^{G}\mathbf{v}_{dm}dm + {}^{G}\mathbf{d}_B \times \int_B {}^{G}\dot{\mathbf{v}}_{dm}dm + \frac{d}{dt}\mathbf{L}_C. \quad (11.121)$$

*We use $\mathbf{L}_C$ for angular momentum about the body mass center. Since the
body frame is at center of mass, we have*

$$\int_B {}^{G}\mathbf{r}_{dm}\, dm = m\,{}^{G}\mathbf{d}_B = m\,{}^{G}\mathbf{r}_C \quad (11.122)$$

$$\int_B {}^{G}\mathbf{v}_{dm}\, dm = m\,{}^{G}\dot{\mathbf{d}}_B = m\,{}^{G}\mathbf{v}_C \quad (11.123)$$

$$\int_B {}^{G}\dot{\mathbf{v}}_{dm}\, dm = m\,{}^{G}\ddot{\mathbf{d}}_B = m\,{}^{G}\mathbf{a}_C \quad (11.124)$$

*and therefore,*

$$\frac{{}^{G}d}{dt}\int_B \left({}^{G}\mathbf{r}_{dm} \times {}^{G}\mathbf{v}_{dm}\, dm\right) = {}^{G}\mathbf{d}_B \times {}^{G}\mathbf{F} + \frac{{}^{G}d}{dt}\,{}^{G}\mathbf{L}_C. \quad (11.125)$$

*Substituting (11.120) and (11.125) in (11.118) provides the Euler equation
of motion in the global frame, indicating that the resultant of externally
applied moments about $CM$ is equal to the global derivative of angular
momentum about $CM$.*

$$^{G}\mathbf{M}_C = \frac{{}^{G}d}{dt}\,{}^{G}\mathbf{L}_C. \quad (11.126)$$

*The Euler equation in the body coordinate can be found by transforming
(11.126)*

$$^{B}\mathbf{M}_C = {}^{G}R_B^{T}\,{}^{G}\mathbf{M}_C$$

$$= {}^{G}R_B^{T}\,\frac{{}^{G}d}{dt}\mathbf{L}_C$$

$$= \frac{{}^{G}d}{dt}\,{}^{G}R_B^{T}\,\mathbf{L}_C$$

$$= \frac{{}^{G}d}{dt}\,{}^{B}\mathbf{L}_C$$

$$= {}^{B}\dot{\mathbf{L}}_C + {}^{B}_{G}\boldsymbol{\omega}_B \times {}^{B}\mathbf{L}_C. \quad (11.127)$$

## 11.4   Mass Moment of Inertia Matrix

In analyzing the motion of rigid bodies, two types of integrals arise that belong to the geometry of the body. The first type defines the center of mass and arises when the translation motion of the body is considered. The second is the *moment of inertia* that appears when the rotational motion of the body is considered. The moment of inertia is also called *centrifugal moments*, or *deviation moments*. Every rigid body has a $3 \times 3$ moment of inertia matrix $I$, which is denoted by

$$I = \begin{bmatrix} I_{xx} & I_{xy} & I_{xz} \\ I_{yx} & I_{yy} & I_{yz} \\ I_{zx} & I_{zy} & I_{zz} \end{bmatrix}. \tag{11.128}$$

The diagonal elements $I_{ij}$, $i = j$ are called *polar moments of inertia*

$$I_{xx} = I_x = \int_B \left( y^2 + z^2 \right) dm \tag{11.129}$$

$$I_{yy} = I_y = \int_B \left( z^2 + x^2 \right) dm \tag{11.130}$$

$$I_{zz} = I_z = \int_B \left( x^2 + y^2 \right) dm \tag{11.131}$$

and the off-diagonal elements $I_{ij}$, $i \neq j$ are called *products of inertia*

$$I_{xy} = I_{yx} = -\int_B xy \, dm \tag{11.132}$$

$$I_{yz} = I_{zy} = -\int_B yz \, dm \tag{11.133}$$

$$I_{zx} = I_{xz} = -\int_B zx \, dm. \tag{11.134}$$

The elements of $I$ for a rigid body, made of discrete point masses, are defined in Equation (11.45).

The elements of $I$ are moments of inertia about a body coordinate frame attached to the $CM$ of the body. Therefore, $I$ is a frame-dependent quantity and must be written like $^B I$ to show the frame it is computed in.

$$
\begin{aligned}
^B I &= \int_B \begin{bmatrix} y^2 + z^2 & -xy & -zx \\ -xy & z^2 + x^2 & -yz \\ -zx & -yz & x^2 + y^2 \end{bmatrix} dm & (11.135) \\
&= \int_B \left( r^2 \mathbf{I} - \mathbf{r}\,\mathbf{r}^T \right) dm & (11.136) \\
&= \int_B -\tilde{r}\,\tilde{r} \, dm. & (11.137)
\end{aligned}
$$

Moments of inertia can be transformed from a coordinate frame $B_1$ to another coordinate frame $B_2$, both installed at the mass center of the body, according to the rule of the *rotated-axes theorem*

$$^{B_2}I = {}^{B_2}R_{B_1} \; {}^{B_1}I \; {}^{B_2}R_{B_1}^T . \tag{11.138}$$

Transformation of the moment of inertia from a central frame $B_1$ located at $^{B_2}\mathbf{r}_C$ to another frame $B_2$, which is parallel to $B_1$, is, according to the rule of *parallel-axes theorem*,

$$^{B_2}I = {}^{B_1}I + m\tilde{r}_C \tilde{r}_C^T . \tag{11.139}$$

If the local coordinate frame $Oxyz$ is located such that the products of inertia vanish, the local coordinate frame is called the *principal coordinate frame* and the associated moments of inertia are called *principal moments of inertia*. Principal axes and principal moments of inertia can be found by solving the following equation for $I$

$$\begin{vmatrix} I_{xx} - I & I_{xy} & I_{xz} \\ I_{yx} & I_{yy} - I & I_{yz} \\ I_{zx} & I_{zy} & I_{zz} - I \end{vmatrix} = 0 \tag{11.140}$$

$$\det\left([I_{ij}] - I\left[\delta_{ij}\right]\right) = 0. \tag{11.141}$$

Since Equation (11.141) is a cubic equation in $I$, we obtain three eigenvalues

$$I_1 = I_x \qquad I_2 = I_y \qquad I_3 = I_z \tag{11.142}$$

which are the principal moments of inertia.

We may utilize the homogeneous position vectors and define a more general moment of inertia, called *pseudo inertia matrix*

$$^{B}\bar{I} = \int_B \mathbf{r}\,\mathbf{r}^T \, dm \tag{11.143}$$

$$= \begin{bmatrix} \int_B x^2 \, dm & \int_B xy \, dm & \int_B xz \, dm & \int_B x \, dm \\ \int_B xy \, dm & \int_B y^2 \, dm & \int_B yz \, dm & \int_B y \, dm \\ \int_B xz \, dm & \int_B yz \, dm & \int_B z^2 \, dm & \int_B z \, dm \\ \int_B x \, dm & \int_B y \, dm & \int_B z \, dm & \int_B dm \end{bmatrix}.$$

This pseudo inertia matrix can be expanded to show that

$$^{B}\bar{I} = \begin{bmatrix} \dfrac{-I_{xx} + I_{yy} + I_{zz}}{2} & I_{xy} & I_{xz} & mx_C \\ I_{yx} & \dfrac{I_{xx} - I_{yy} + I_{zz}}{2} & I_{yz} & my_C \\ I_{zx} & I_{zy} & \dfrac{I_{xx} + I_{yy} - I_{zz}}{2} & mz_C \\ mx_C & mz_C & my_C & m \end{bmatrix} \tag{11.144}$$

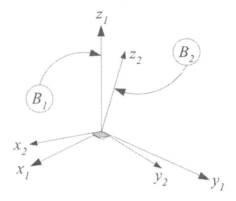

FIGURE 11.5. Two coordinate frames with a common origin at mass center of a rigid body.

where,

$$
{}^{B}\mathbf{r}_C = \left[ \begin{array}{c} x_C \\ y_C \\ z_C \end{array} \right] = \left[ \begin{array}{c} \frac{1}{m} \int_B x\, dm \\ \frac{1}{m} \int_B y\, dm \\ \frac{1}{m} \int_B z\, dm \end{array} \right] \tag{11.145}
$$

is the position of the mass center in the body frame. This vector is zero if the body frame is central.

**Proof.** Two coordinate frames with a common origin at the mass center of a rigid body are shown in Figure 11.5. The angular velocity and angular momentum of a rigid body transform from the frame $B_1$ to a the frame $B_2$ by vector transformation rule

$$
{}^{B_2}\boldsymbol{\omega} = {}^{B_2}R_{B_1}\, {}^{B_1}\boldsymbol{\omega} \tag{11.146}
$$

$$
{}^{B_2}\mathbf{L} = {}^{B_2}R_{B_1}\, {}^{B_1}\mathbf{L}. \tag{11.147}
$$

However, $\mathbf{L}$ and $\boldsymbol{\omega}$ are related according to Equation (11.43)

$$
{}^{B_1}\mathbf{L} = {}^{B_1}I \, {}^{B_1}\boldsymbol{\omega} \tag{11.148}
$$

and therefore,

$$
\begin{aligned}
{}^{B_2}\mathbf{L} &= {}^{B_2}R_{B_1}\, {}^{B_1}I \, {}^{B_2}R_{B_1}^{T}\, {}^{B_2}\boldsymbol{\omega} \\
&= {}^{B_2}I \, {}^{B_2}\boldsymbol{\omega}
\end{aligned} \tag{11.149}
$$

which shows how to transfer the moment of inertia from the coordinate frame $B_1$ to a rotated frame $B_2$

$$
{}^{B_2}I = {}^{B_2}R_{B_1}\, {}^{B_1}I \, {}^{B_2}R_{B_1}^{T}. \tag{11.150}
$$

Now consider a central frame $B_1$, shown in Figure 11.6, at ${}^{B_2}\mathbf{r}_C$, which rotates about the origin of a fixed frame $B_2$ such that their axes remain

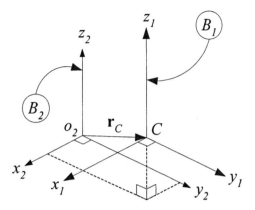

FIGURE 11.6. A central coordinate frame $B_1$ and a translated frame $B_2$.

parallel. The angular velocity and angular momentum of the rigid body transform from the frame $B_1$ to the frame $B_2$ by

$$^{B_2}\boldsymbol{\omega} = {}^{B_1}\boldsymbol{\omega} \tag{11.151}$$

$$^{B_2}\mathbf{L} = {}^{B_1}\mathbf{L} + (\mathbf{r}_C \times m\mathbf{v}_C). \tag{11.152}$$

Therefore,

$$
\begin{aligned}
^{B_2}\mathbf{L} &= {}^{B_1}\mathbf{L} + m\,{}^{B_2}\mathbf{r}_C \times \left({}^{B_2}\boldsymbol{\omega} \times {}^{B_2}\mathbf{r}_C\right) \\
&= {}^{B_1}\mathbf{L} + \left(m\,{}^{B_2}\tilde{r}_C\,{}^{B_2}\tilde{r}_C^T\right){}^{B_2}\boldsymbol{\omega} \\
&= \left({}^{B_1}I + m\,{}^{B_2}\tilde{r}_C\,{}^{B_2}\tilde{r}_C^T\right){}^{B_2}\boldsymbol{\omega}
\end{aligned}
\tag{11.153}
$$

which shows how to transfer the moment of inertia from frame $B_1$ to a parallel frame $B_2$

$$^{B_2}I = {}^{B_1}I + m\tilde{r}_C\,\tilde{r}_C^T. \tag{11.154}$$

The parallel-axes theorem is also called the *Huygens-Steiner theorem*.

Referring to Equation (11.150) for transformation of the moment of inertia to a rotated frame, we can always find a frame in which $^{B_2}I$ is diagonal. In such a frame, we have

$$^{B_2}R_{B_1}\,{}^{B_1}I = {}^{B_2}I\,{}^{B_2}R_{B_1} \tag{11.155}$$

or

$$
\begin{bmatrix}
I_{xx} & I_{xy} & I_{xz} \\
I_{yx} & I_{yy} & I_{yz} \\
I_{zx} & I_{zy} & I_{zz}
\end{bmatrix}
\begin{bmatrix}
r_{11} & r_{12} & r_{13} \\
r_{21} & r_{22} & r_{23} \\
r_{31} & r_{32} & r_{33}
\end{bmatrix}
$$
$$
=
\begin{bmatrix}
I_1 & 0 & 0 \\
0 & I_2 & 0 \\
0 & 0 & I_3
\end{bmatrix}
\begin{bmatrix}
r_{11} & r_{12} & r_{13} \\
r_{21} & r_{22} & r_{23} \\
r_{31} & r_{32} & r_{33}
\end{bmatrix}
\tag{11.156}
$$

which shows that $I_1$, $I_2$, and $I_3$ are eigenvalues of $^{B_1}I$. These eigenvalues can be found by solving the following equation for $\lambda$.

$$
\begin{vmatrix}
I_{xx} - \lambda & I_{xy} & I_{xz} \\
I_{yx} & I_{yy} - \lambda & I_{yz} \\
I_{zx} & I_{zy} & I_{zz} - \lambda
\end{vmatrix} = 0.
\tag{11.157}
$$

The eigenvalues $I_1$, $I_2$, and $I_3$ are *principal moments of inertia*, and their associated eigenvectors are called *principal directions*. The coordinate frame made by the eigenvectors is the *principal body coordinate frame*. In the principal coordinate frame, the rigid body angular momentum is

$$
\begin{bmatrix} L_1 \\ L_2 \\ L_3 \end{bmatrix} =
\begin{bmatrix} I_1 & 0 & 0 \\ 0 & I_2 & 0 \\ 0 & 0 & I_3 \end{bmatrix}
\begin{bmatrix} \omega_1 \\ \omega_2 \\ \omega_3 \end{bmatrix}.
\tag{11.158}
$$

■

**Example 258** *Principal moments of inertia.*
  *Consider the inertia matrix*

$$
I = \begin{bmatrix} 20 & -2 & 0 \\ -2 & 30 & 0 \\ 0 & 0 & 40 \end{bmatrix}
\tag{11.159}
$$

*we set up the determinant (11.141)*

$$
\begin{vmatrix}
20 - \lambda & -2 & 0 \\
-2 & 30 - \lambda & 0 \\
0 & 0 & 40 - \lambda
\end{vmatrix} = 0
\tag{11.160}
$$

*which leads to the characteristic equation*

$$
(20 - \lambda)(30 - \lambda)(40 - \lambda) - 4(40 - \lambda) = 0.
\tag{11.161}
$$

  *Three roots of Equation (11.161) are*

$$
I_1 = 30.385, \qquad I_2 = 19.615, \qquad I_3 = 40
\tag{11.162}
$$

*and therefore, the principal moment of inertia matrix is*

$$
I = \begin{bmatrix} 30.385 & 0 & 0 \\ 0 & 19.615 & 0 \\ 0 & 0 & 40 \end{bmatrix}.
\tag{11.163}
$$

**Example 259** *Principal coordinate frame.*
  *Consider the inertia matrix*

$$
I = \begin{bmatrix} 20 & -2 & 0 \\ -2 & 30 & 0 \\ 0 & 0 & 40 \end{bmatrix}
\tag{11.164}
$$

*the direction of a principal axis $x_i$ is established by solving*

$$\begin{bmatrix} I_{xx} - I_i & I_{xy} & I_{xz} \\ I_{yx} & I_{yy} - I_i & I_{yz} \\ I_{zx} & I_{zy} & I_{zz} - I_i \end{bmatrix} \begin{bmatrix} \cos \alpha_i \\ \cos \beta_i \\ \cos \gamma_i \end{bmatrix} = \begin{bmatrix} 0 \\ 0 \\ 0 \end{bmatrix} \tag{11.165}$$

*for direction cosines, which must also satisfy*

$$\cos^2 \alpha_i + \cos^2 \beta_i + \cos^2 \gamma_i = 1. \tag{11.166}$$

*For the first principal moment of inertia $I_1 = 30.385$ we have*

$$\begin{bmatrix} 20 - 30.385 & -2 & 0 \\ -2 & 30 - 30.385 & 0 \\ 0 & 0 & 40 - 30.385 \end{bmatrix} \begin{bmatrix} \cos \alpha_1 \\ \cos \beta_1 \\ \cos \gamma_1 \end{bmatrix} = \begin{bmatrix} 0 \\ 0 \\ 0 \end{bmatrix} \tag{11.167}$$

*or*

$$-10.385 \cos \alpha_1 - 2 \cos \beta_1 + 0 = 0 \tag{11.168}$$

$$-2 \cos \alpha_1 - 0.385 \cos \beta_1 + 0 = 0 \tag{11.169}$$

$$0 + 0 + 9.615 \cos \gamma_1 = 0 \tag{11.170}$$

*and we obtain*

$$\alpha_1 = 79.1 \deg \tag{11.171}$$

$$\beta_1 = 169.1 \deg \tag{11.172}$$

$$\gamma_1 = 90.0 \deg. \tag{11.173}$$

*Using $I_2 = 19.615$ for the second principal axis*

$$\begin{bmatrix} 20 - 19.62 & -2 & 0 \\ -2 & 30 - 19.62 & 0 \\ 0 & 0 & 40 - 19.62 \end{bmatrix} \begin{bmatrix} \cos \alpha_2 \\ \cos \beta_2 \\ \cos \gamma_2 \end{bmatrix} = \begin{bmatrix} 0 \\ 0 \\ 0 \end{bmatrix} \tag{11.174}$$

*we obtain*

$$\alpha_2 = 10.9 \deg \tag{11.175}$$

$$\beta_2 = 79.1 \deg \tag{11.176}$$

$$\gamma_2 = 90.0 \deg. \tag{11.177}$$

*The third principal axis is for $I_3 = 40$*

$$\begin{bmatrix} 20 - 40 & -2 & 0 \\ -2 & 30 - 40 & 0 \\ 0 & 0 & 40 - 40 \end{bmatrix} \begin{bmatrix} \cos \alpha_3 \\ \cos \beta_3 \\ \cos \gamma_3 \end{bmatrix} = \begin{bmatrix} 0 \\ 0 \\ 0 \end{bmatrix} \tag{11.178}$$

*which leads to*

$$\alpha_3 = 90.0 \deg \tag{11.179}$$

$$\beta_3 = 90.0 \deg \tag{11.180}$$

$$\gamma_3 = 0.0 \deg. \tag{11.181}$$

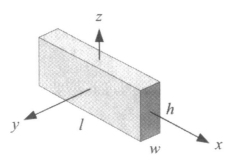

FIGURE 11.7. A homogeneous rectangular link.

**Example 260** *Moment of inertia of a rigid rectangular link.*

*Consider a homogeneous rectangular link with mass m, length l, width w, and height h, as shown in Figure 11.7.*

*The local central coordinate frame is attached to the link at its mass center. The moments of inertia matrix of the link can be found by integral method. We begin with calculating $I_{xx}$*

$$
\begin{aligned}
I_{xx} &= \int_B \left(y^2 + z^2\right) dm \\
&= \int_v \left(y^2 + z^2\right) \rho dv \\
&= \frac{m}{lwh} \int_v \left(y^2 + z^2\right) dv \\
&= \frac{m}{lwh} \int_{-h/2}^{h/2} \int_{-w/2}^{w/2} \int_{-l/2}^{l/2} \left(y^2 + z^2\right) dx\, dy\, dz \\
&= \frac{m}{12} \left(w^2 + h^2\right)
\end{aligned}
\tag{11.182}
$$

*which shows $I_{yy}$ and $I_{zz}$ can be calculated similarly*

$$
I_{yy} = \frac{m}{12} \left(h^2 + l^2\right) \tag{11.183}
$$

$$
I_{zz} = \frac{m}{12} \left(l^2 + w^2\right). \tag{11.184}
$$

*Since the coordinate frame is central, the products of inertia must be zero.*

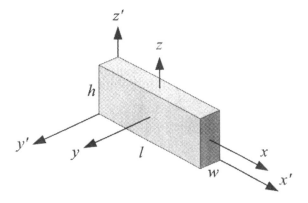

FIGURE 11.8. A rigid rectangular link in the principal and non principal frames.

*To show this, we examine* $I_{xy}$.

$$
\begin{aligned}
I_{xy} &= I_{yx} = -\int_B xy\, dm \\
&= \int_v xy\rho\, dv \\
&= \frac{m}{lwh} \int_{-h/2}^{h/2} \int_{-w/2}^{w/2} \int_{-l/2}^{l/2} xy\, dx\, dy\, dz \\
&= 0
\end{aligned}
\tag{11.185}
$$

*Therefore, the moment of inertia matrix for the rigid rectangular link in its central frame is*

$$
I = \begin{bmatrix}
\frac{m}{12}\left(w^2 + h^2\right) & 0 & 0 \\
0 & \frac{m}{12}\left(h^2 + l^2\right) & 0 \\
0 & 0 & \frac{m}{12}\left(l^2 + w^2\right)
\end{bmatrix}.
\tag{11.186}
$$

**Example 261** *Translation of the inertia matrix.*
*The moment of inertia matrix of the rigid link shown in Figure 11.8, in the principal frame $B(oxyz)$ is give in Equation (11.186). The moment of inertia matrix in the non principal frame $B'(ox'y'z')$ can be found by applying the parallel-axes transformation formula (11.154).*

$$
{}^{B'}I = {}^{B}I + m\, {}^{B'}\tilde{r}_C\, {}^{B'}\tilde{r}_C^T
\tag{11.187}
$$

*The center of mass position vector is*

$$
{}^{B'}\mathbf{r}_C = \frac{1}{2}\begin{bmatrix} l \\ w \\ h \end{bmatrix}
\tag{11.188}
$$

*and therefore,*

$$
{}^{B'}\tilde{r}_C = \frac{1}{2}
\begin{bmatrix}
0 & -h & w \\
h & 0 & -l \\
-w & l & 0
\end{bmatrix}
\tag{11.189}
$$

*that provides*

$$
{}^{B'}I =
\begin{bmatrix}
\frac{1}{3}h^2m + \frac{1}{3}mw^2 & -\frac{1}{4}lmw & -\frac{1}{4}hlm \\
-\frac{1}{4}lmw & \frac{1}{3}h^2m + \frac{1}{3}l^2m & -\frac{1}{4}hmw \\
-\frac{1}{4}hlm & -\frac{1}{4}hmw & \frac{1}{3}l^2m + \frac{1}{3}mw^2
\end{bmatrix}.
\tag{11.190}
$$

**Example 262** *Principal rotation matrix.*
  *Consider a body inertia matrix as*

$$
I =
\begin{bmatrix}
2/3 & -1/2 & -1/2 \\
-1/2 & 5/3 & -1/4 \\
-1/2 & -1/4 & 5/3
\end{bmatrix}.
$$

*The eigenvalues and eigenvectors of $I$ are*

$$
I_1 = 0.2413 ,
\begin{bmatrix}
2.351 \\
1 \\
1
\end{bmatrix}
$$

$$
I_2 = 1.8421 ,
\begin{bmatrix}
-0.851 \\
1 \\
1
\end{bmatrix}
$$

$$
I_3 = 1.9167 ,
\begin{bmatrix}
0 \\
-1 \\
1
\end{bmatrix}.
$$

*The normalized eigenvector matrix $W$ is equal to the transpose of the required transformation matrix to make the inertia matrix diagonal*

$$
W =
\begin{bmatrix}
| & | & | \\
\mathbf{w}_1 & \mathbf{w}_2 & \mathbf{w}_3 \\
| & | & |
\end{bmatrix}
= {}^2R_1^T
$$

$$
=
\begin{bmatrix}
0.8569 & -0.5156 & 0.0 \\
0.36448 & 0.60588 & -0.70711 \\
0.36448 & 0.60588 & 0.70711
\end{bmatrix}.
$$

*We may verify that*

$$
{}^2I \approx {}^2R_1\,{}^1I\,{}^2R_1^T = W^T\,{}^1I\,W
$$

$$
=
\begin{bmatrix}
0.2413 & -1 \times 10^{-4} & 0.0 \\
-1 \times 10^{-4} & 1.8421 & -1 \times 10^{-19} \\
0.0 & 0.0 & 1.9167
\end{bmatrix}.
$$

**Example 263 ★** *Relative diagonal moments of inertia.*
 *Using the definitions for moments of inertia 11.129, 11.130, and 11.131
it is seen that the inertia matrix is symmetric, and*

$$\int_B \left(x^2 + y^2 + z^2\right) dm = \frac{1}{2}\left(I_{xx} + I_{yy} + I_{zz}\right) \tag{11.191}$$

*and also*

$$I_{xx} + I_{yy} \geq I_{zz} \tag{11.192}$$
$$I_{yy} + I_{zz} \geq I_{xx} \tag{11.193}$$
$$I_{zz} + I_{xx} \geq I_{yy}. \tag{11.194}$$

 *Noting that*

$$(y - z)^2 \geq 0$$

*it is evident that*

$$\left(y^2 + z^2\right) \geq 2yz$$

*and therefore*

$$I_{xx} \geq 2I_{yz} \tag{11.195}$$

*and similarly*

$$I_{yy} \geq 2I_{zx} \tag{11.196}$$
$$I_{zz} \geq 2I_{xy}. \tag{11.197}$$

**Example 264 ★** *Coefficients of the characteristic equation.*
 *The determinant (11.157)*

$$\begin{vmatrix} I_{xx} - \lambda & I_{xy} & I_{xz} \\ I_{yx} & I_{yy} - \lambda & I_{yz} \\ I_{zx} & I_{zy} & I_{zz} - \lambda \end{vmatrix} = 0 \tag{11.198}$$

*for calculating the principal moments of inertia, leads to a third degree
equation of $\lambda$, called the **characteristic equation**.*

$$\lambda^3 - a_1 \lambda^2 + a_2 \lambda - a_3 = 0 \tag{11.199}$$

*The coefficients of the characteristic equation are called the **principal invariants** of $[I]$. The coefficients can directly be found from the following
equations:*

$$\begin{aligned} a_1 &= I_{xx} + I_{yy} + I_{zz} \\ &= \operatorname{tr}[I] \end{aligned} \tag{11.200}$$

$$\begin{aligned} a_2 &= I_{xx}I_{yy} + I_{yy}I_{zz} + I_{zz}I_{xx} - I_{xy}^2 - I_{yz}^2 - I_{zx}^2 \\ &= \begin{vmatrix} I_{xx} & I_{xy} \\ I_{yx} & I_{yy} \end{vmatrix} + \begin{vmatrix} I_{yy} & I_{yz} \\ I_{zy} & I_{zz} \end{vmatrix} + \begin{vmatrix} I_{xx} & I_{xz} \\ I_{zx} & I_{zz} \end{vmatrix} \\ &= \frac{1}{2}\left(a_1^2 - \operatorname{tr}\left[I^2\right]\right) \end{aligned} \tag{11.201}$$

$$
\begin{aligned}
a_3 &= I_{xx}I_{yy}I_{zz} + I_{xy}I_{yz}I_{zx} + I_{zy}I_{yx}I_{xz} \\
&\quad - (I_{xx}I_{yz}I_{zy} + I_{yy}I_{zx}I_{xz} + I_{zz}I_{xy}I_{yx}) \\
&= I_{xx}I_{yy}I_{zz} + 2I_{xy}I_{yz}I_{zx} - \left(I_{xx}I_{yz}^2 + I_{yy}I_{zx}^2 + I_{zz}I_{xy}^2\right) \\
&= \det[I] \tag{11.202}
\end{aligned}
$$

**Example 265** ★ *The principal moments of inertia are coordinate invariants.*

*The roots of the inertia characteristic equation are the principal moments of inertia and all real but not necessarily different. The principal moments of inertia are extreme. That is, the principal moments of inertia determine the smallest and the largest values of $I_{ii}$ for a rigid body. Since the smallest and largest values of $I_{ii}$ do not depend on the choice of the body coordinate frame, the solution of the characteristic equation is not dependent of the coordinate frame.*

*In other words, if $I_1$, $I_2$, and $I_3$ are the principal moments of inertia for $^{B_1}I$, the principal moments of inertia for $^{B_2}I$ are also $I_1$, $I_2$, and $I_3$ when*

$$
^{B_2}I = {}^{B_2}R_{B_1} \, {}^{B_1}I \, {}^{B_2}R_{B_1}^T.
$$

*We conclude that $I_1$, $I_2$, and $I_3$ are coordinate invariants of the matrix $[I]$, and therefore any quantity that depends on $I_1$, $I_2$, and $I_3$ is also coordinate invariant. The matrix $[I]$ has only three independent invariants and every other invariant can be expressed in terms of $I_1$, $I_2$, and $I_3$.*

*Since $I_1$, $I_2$, and $I_3$ are the solutions of the characteristic equation of $[I]$ given in (11.199), we may write the determinant (11.157) in the form*

$$
(\lambda - I_1)(\lambda - I_2)(\lambda - I_3) = 0. \tag{11.203}
$$

*Expanded form of this equation is*

$$
\lambda^3 - (I_1 + I_2 + I_3)\lambda^2 + (I_1 I_2 + I_2 I_3 + I_3 I_1) a_2 \lambda - I_1 I_2 I_3 = 0. \tag{11.204}
$$

*By comparing (11.204) and (11.199) we conclude that*

$$
\begin{aligned}
a_1 &= I_{xx} + I_{yy} + I_{zz} = I_1 + I_2 + I_3 \tag{11.205} \\
a_2 &= I_{xx}I_{yy} + I_{yy}I_{zz} + I_{zz}I_{xx} - I_{xy}^2 - I_{yz}^2 - I_{zx}^2 \\
&= I_1 I_2 + I_2 I_3 + I_3 I_1 \tag{11.206} \\
a_3 &= I_{xx}I_{yy}I_{zz} + 2I_{xy}I_{yz}I_{zx} - \left(I_{xx}I_{yz}^2 + I_{yy}I_{zx}^2 + I_{zz}I_{xy}^2\right) \\
&= I_1 I_2 I_3. \tag{11.207}
\end{aligned}
$$

*Being able to express the coefficients $a_1$, $a_2$, and $a_3$ as functions of $I_1$, $I_2$, and $I_3$ determines that the coefficients of the characteristic equation are coordinate invariant.*

**Example 266 ★** *Short notation of the elements of inertia matrix.*

Taking advantage of the Kronecker's delta (2.133) we may write the elements of the moment of inertia matrix $I_{ij}$ in short notation forms.

$$I_{ij} = \int_B \left( (x_1^2 + x_2^2 + x_3^2) \delta_{ij} - x_i x_j \right) dm \qquad (11.208)$$

$$I_{ij} = \int_B \left( r^2 \delta_{ij} - x_i x_j \right) dm \qquad (11.209)$$

$$I_{ij} = \int_B \left( \sum_{k=1}^{3} x_k x_k \delta_{ij} - x_i x_j \right) dm \qquad (11.210)$$

where we utilized the following notations:

$$x_1 = x \qquad x_2 = y \qquad x_3 = z. \qquad (11.211)$$

**Example 267 ★** *Moment of inertia with respect to a plane, a line, and a point.*

The moment of inertia of a system of particles may be defined with respect to a plane, a line, or a point as the sum of the products of the mass of the particles into the square of the perpendicular distance from the particle to the plane, line, or point. For a continuous body, the sum would be definite integral over the volume of the body.

The moments of inertia with respect to the $xy$, $yz$, and $zx$-plane are

$$I_{z^2} = \int_B z^2 dm \qquad (11.212)$$

$$I_{y^2} = \int_B y^2 dm \qquad (11.213)$$

$$I_{x^2} = \int_B x^2 dm. \qquad (11.214)$$

The moments of inertia with respect to the $x$, $y$, and $z$-axis are

$$I_x = \int_B \left( y^2 + z^2 \right) dm \qquad (11.215)$$

$$I_y = \int_B \left( z^2 + x^2 \right) dm \qquad (11.216)$$

$$I_z = \int_B \left( x^2 + y^2 \right) dm \qquad (11.217)$$

and therefore,

$$I_x = I_{y^2} + I_{z^2} \qquad (11.218)$$

$$I_y = I_{z^2} + I_{x^2} \qquad (11.219)$$

$$I_z = I_{x^2} + I_{y^2}. \qquad (11.220)$$

*The moment of inertia with respect to the origin is*

$$I_o = \int_B \left(x^2 + y^2 + z^2\right) dm$$

$$= I_{x^2} + I_{y^2} + I_{z^2}$$

$$= \frac{1}{2}\left(I_x + I_y + I_z\right). \tag{11.221}$$

*Because the choice of the coordinate frame is arbitrary, we can say that the moment of inertia with respect to a line is the sum of the moments of inertia with respect to any two mutually orthogonal planes that pass through the line. The moment of inertia with respect to a point has similar meaning for three mutually orthogonal planes intersecting at the point.*

## 11.5  Lagrange's Form of Newton's Equations of Motion

Newton's equation of motion can be transformed to

$$\frac{d}{dt}\left(\frac{\partial K}{\partial \dot{q}_r}\right) - \frac{\partial K}{\partial q_r} = F_r \qquad r = 1, 2, \cdots n \tag{11.222}$$

where

$$F_r = \sum_{i=1}^{n}\left(F_{ix}\frac{\partial f_i}{\partial q_1} + F_{iy}\frac{\partial g_i}{\partial q_2} + F_{iz}\frac{\partial h_i}{\partial q_n}\right). \tag{11.223}$$

Equation (11.222) is called the *Lagrange equation of motion*, where $K$ is the kinetic energy of the $n$ DOF system, $q_r$, $r = 1, 2, \cdots, n$ are the generalized coordinates of the system, $\mathbf{F} = \begin{bmatrix} F_{ix} & F_{iy} & F_{iz} \end{bmatrix}^T$ is the external force acting on the $i$th particle of the system, and $F_r$ is the generalized force associated to $q_r$.

**Proof.** Let $m_i$ be the mass of one of the particles of a system and let $(x_i, y_i, z_i)$ be its Cartesian coordinates in a globally fixed coordinate frame. Assume that the coordinates of every individual particle are functions of another set of coordinates $q_1, q_2, q_3, \cdots, q_n$ and possibly time $t$.

$$x_i = f_i(q_1, q_2, q_3, \cdots, q_n, t) \tag{11.224}$$

$$y_i = g_i(q_1, q_2, q_3, \cdots, q_n, t) \tag{11.225}$$

$$z_i = h_i(q_1, q_2, q_3, \cdots, q_n, t) \tag{11.226}$$

If $F_{xi}, F_{yi}, F_{zi}$ are components of the total force acting on the particle $m_i$ then, the Newton equations of motion for the particle would be

$$F_{xi} = m_i \ddot{x}_i \tag{11.227}$$

$$F_{yi} = m_i \ddot{y}_i \tag{11.228}$$

$$F_{zi} = m_i \ddot{z}_i. \tag{11.229}$$

We multiply both sides of these equations by

$$\frac{\partial f_i}{\partial q_r}$$
$$\frac{\partial g_i}{\partial q_r}$$
$$\frac{\partial h_i}{\partial q_r}$$

respectively, and add them up for all the particles to have

$$\sum_{i=1}^{n} m_i \left( \ddot{x}_i \frac{\partial f_i}{\partial q_r} + \ddot{y}_i \frac{\partial g_i}{\partial q_r} + \ddot{z}_i \frac{\partial h_i}{\partial q_r} \right) = \sum_{i=1}^{n} \left( F_{xi} \frac{\partial f_i}{\partial q_r} + F_{yi} \frac{\partial g_i}{\partial q_r} + F_{zi} \frac{\partial h_i}{\partial q_r} \right) \tag{11.230}$$

where $n$ is the total number of particles.

Taking a time derivative of Equation (11.224),

$$\dot{x}_i = \frac{\partial f_i}{\partial q_1} \dot{q}_1 + \frac{\partial f_i}{\partial q_2} \dot{q}_2 + \frac{\partial f_i}{\partial q_3} \dot{q}_3 + \cdots + \frac{\partial f_i}{\partial q_n} \dot{q}_n + \frac{\partial f_i}{\partial t} \tag{11.231}$$

we find

$$\begin{aligned} \frac{\partial \dot{x}_i}{\partial \dot{q}_r} &= \frac{\partial}{\partial \dot{q}_r} \left( \frac{\partial f_i}{\partial q_1} \dot{q}_1 + \frac{\partial f_i}{\partial q_2} \dot{q}_2 + \cdots + \frac{\partial f_i}{\partial q_n} \dot{q}_n + \frac{\partial f_i}{\partial t} \right) \\ &= \frac{\partial f_i}{\partial q_r}. \end{aligned} \tag{11.232}$$

and therefore,

$$\begin{aligned} \ddot{x}_i \frac{\partial f_i}{\partial q_r} &= \ddot{x}_i \frac{\partial \dot{x}_i}{\partial \dot{q}_r} \\ &= \frac{d}{dt} \left( \dot{x}_i \frac{\partial \dot{x}_i}{\partial \dot{q}_r} \right) - \dot{x}_i \frac{d}{dt} \left( \frac{\partial \dot{x}_i}{\partial \dot{q}_r} \right). \end{aligned} \tag{11.233}$$

However,

$$\begin{aligned} \dot{x}_i \frac{d}{dt} \left( \frac{\partial \dot{x}_i}{\partial \dot{q}_r} \right) &= \dot{x}_i \frac{d}{dt} \left( \frac{\partial f_i}{\partial q_r} \right) \\ &= \dot{x}_i \left( \frac{\partial^2 f_i}{\partial q_1 \partial q_r} \dot{q}_1 + \cdots + \frac{\partial^2 f_i}{\partial q_n \partial q_r} \dot{q}_n + \frac{\partial^2 f_i}{\partial t \partial q_r} \right) \\ &= \dot{x}_i \frac{\partial}{\partial q_r} \left( \frac{\partial f_i}{\partial q_1} \dot{q}_1 + \frac{\partial f_i}{\partial q_2} \dot{q}_2 + \cdots + \frac{\partial f_i}{\partial q_n} \dot{q}_n + \frac{\partial f_i}{\partial t} \right) \\ &= \dot{x}_i \frac{\partial \dot{x}_i}{\partial q_r} \end{aligned} \tag{11.234}$$

and we have

$$\ddot{x}_i \frac{\partial \dot{x}_i}{\partial \dot{q}_r} = \frac{d}{dt} \left( \dot{x}_i \frac{\partial \dot{x}_i}{\partial \dot{q}_r} \right) - \dot{x}_i \frac{\partial \dot{x}_i}{\partial q_r} \tag{11.235}$$

which is equal to

$$\ddot{x}_i \frac{\dot{x}_i}{\dot{q}_r} = \frac{d}{dt}\left[\frac{\partial}{\partial \dot{q}_r}\left(\frac{1}{2}\dot{x}_i^2\right)\right] - \frac{\partial}{\partial q_r}\left(\frac{1}{2}\dot{x}_i^2\right). \tag{11.236}$$

Now substituting (11.233) and (11.236) in the left-hand side of (11.230) leads to

$$\sum_{i=1}^{n} m_i\left(\ddot{x}_i \frac{\partial f_i}{\partial q_r} + \ddot{y}_i \frac{\partial g_i}{\partial q_r} + \ddot{z}_i \frac{\partial h_i}{\partial q_r}\right)$$

$$= \sum_{i=1}^{n} m_i \frac{d}{dt}\left[\frac{\partial}{\partial \dot{q}_r}\left(\frac{1}{2}\dot{x}_i^2 + \frac{1}{2}\dot{y}_i^2 + \frac{1}{2}\dot{z}_i^2\right)\right]$$

$$- \sum_{i=1}^{n} m_i \frac{\partial}{\partial q_r}\left(\frac{1}{2}\dot{x}_i^2 + \frac{1}{2}\dot{y}_i^2 + \frac{1}{2}\dot{z}_i^2\right)$$

$$= \frac{1}{2}\sum_{i=1}^{n} m_i \frac{d}{dt}\left[\frac{\partial}{\partial \dot{q}_r}\left(\dot{x}_i^2 + \dot{y}_i^2 + \dot{z}_i^2\right)\right]$$

$$- \frac{1}{2}\sum_{i=1}^{n} m_i \frac{\partial}{\partial q_r}\left(\dot{x}_i^2 + \dot{y}_i^2 + \dot{z}_i^2\right). \tag{11.237}$$

where

$$\frac{1}{2}\sum_{i=1}^{n} m_i\left(\dot{x}_i^2 + \dot{y}_i^2 + \dot{z}_i^2\right) = K \tag{11.238}$$

is the *kinetic energy* of the system. Therefore, the Newton equations of motion (11.227), (11.228), and (11.229) are converted to

$$\frac{d}{dt}\left(\frac{\partial K}{\partial \dot{q}_r}\right) - \frac{\partial K}{\partial q_r} = \sum_{i=1}^{n}\left(F_{xi}\frac{\partial f_i}{\partial q_r} + F_{yi}\frac{\partial g_i}{\partial q_r} + F_{zi}\frac{\partial h_i}{\partial q_r}\right). \tag{11.239}$$

Because of (11.224), (11.225), and (11.226), the kinetic energy is a function of $q_1, q_2, q_3, \cdots, q_n$ and time $t$. The left-hand side of Equation (11.239) includes the kinetic energy of the whole system and the right-hand side is a generalized force and shows the effect of changing coordinates from $x_i$ to $q_j$ on the external forces. Let us assume that the coordinate $q_r$ alters to $q_r + \delta q_r$ while the other coordinates $q_1, q_2, q_3, \cdots, q_{r-1}, q_{r+1}, \cdots, q_n$ and time $t$ are unaltered. So, the coordinates of $m_i$ are changed to

$$x_i + \frac{\partial f_i}{\partial q_r}\delta q_r \tag{11.240}$$

$$y_i + \frac{\partial g_i}{\partial q_r}\delta q_r \tag{11.241}$$

$$z_i + \frac{\partial h_i}{\partial q_r}\delta q_r \tag{11.242}$$

and the work done in this virtual displacement by all forces acting on the particles of the system is

$$\delta W = \sum_{i=1}^{n} \left( F_{xi} \frac{\partial f_i}{\partial q_r} + F_{yi} \frac{\partial g_i}{\partial q_r} + F_{zi} \frac{\partial h_i}{\partial q_r} \right) \delta q_r. \tag{11.243}$$

Since the work done by internal forces appears in opposite pairs, only the work done by external forces remains in Equation (11.243). Let's denote the virtual work by

$$\delta W = F_r \left( q_1, q_2, q_3, \cdots, q_n, t \right) \delta q_r. \tag{11.244}$$

Then we have

$$\frac{d}{dt} \left( \frac{\partial K}{\partial \dot{q}_r} \right) - \frac{\partial K}{\partial q_r} = F_r \tag{11.245}$$

where

$$F_r = \sum_{i=1}^{n} \left( F_{xi} \frac{\partial f_i}{\partial q_r} + F_{yi} \frac{\partial g_i}{\partial q_r} + F_{zi} \frac{\partial h_i}{\partial q_r} \right). \tag{11.246}$$

Equation (11.245) is the Lagrange form of equations of motion. This equation is true for all values of $r$ from 1 to $n$. We thus have $n$ second order ordinary differential equations in which $q_1, q_2, q_3, \cdots, q_n$ are the dependent variables and $t$ is the independent variable. The coordinates $q_1, q_2, q_3, \cdots, q_n$ are called *generalized coordinates* and can be any measurable parameters to provide the configuration of the system. Since the number of equations and the number of dependent variables are equal, the equations are theoretically sufficient to determine the motion of all $m_i$. ∎

**Example 268** *A simple pendulum.*
*A pendulum is shown in Figure 11.9. Using $x$ and $y$ for Cartesian position of $m$, and using $\theta = q$ as the generalized coordinate, we have*

$$x = f(\theta) = l \sin \theta \tag{11.247}$$
$$y = g(\theta) = l \cos \theta \tag{11.248}$$
$$K = \frac{1}{2} m \left( \dot{x}^2 + \dot{y}^2 \right) = \frac{1}{2} m l^2 \dot{\theta}^2 \tag{11.249}$$

*and therefore,*

$$\frac{d}{dt} \left( \frac{\partial K}{\partial \dot{\theta}} \right) - \frac{\partial K}{\partial \theta} = \frac{d}{dt} (ml^2 \dot{\theta}) = ml^2 \ddot{\theta}. \tag{11.250}$$

*The external force components, acting on $m$, are*

$$F_x = 0 \tag{11.251}$$
$$F_y = mg \tag{11.252}$$

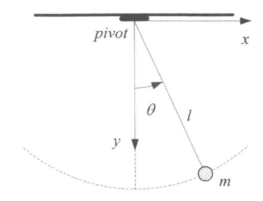

FIGURE 11.9. A simple pendulum.

and therefore,

$$F_\theta = F_x \frac{\partial f}{\partial \theta} + F_y \frac{\partial g}{\partial \theta} = -mgl\sin\theta. \qquad (11.253)$$

Hence, the equation of motion for the pendulum is

$$ml^2\ddot\theta = -mgl\sin\theta. \qquad (11.254)$$

**Example 269** *A pendulum attached to an oscillating mass.*
   *Figure 11.10 illustrates a vibrating mass with a hanging pendulum. The pendulum can act as a vibration absorber if designed properly.*
   *Starting with coordinate relationships*

$$
\begin{aligned}
x_M &= f_M = x & (11.255)\\
y_M &= g_M = 0 & (11.256)\\
x_m &= f_m = x + l\sin\theta & (11.257)\\
y_m &= g_m = l\cos\theta & (11.258)
\end{aligned}
$$

*we may find the kinetic energy in terms of the generalized coordinates $x$ and $\theta$.*

$$
\begin{aligned}
K &= \frac{1}{2}M\left(\dot x_M^2 + \dot y_M^2\right) + \frac{1}{2}m\left(\dot x_m^2 + \dot y_m^2\right)\\
&= \frac{1}{2}M\dot x^2 + \frac{1}{2}m\left(\dot x^2 + l^2\dot\theta^2 + 2l\dot x\dot\theta\cos\theta\right) \qquad (11.259)
\end{aligned}
$$

*Then, the left-hand side of Lagrange equations are*

$$\frac{d}{dt}\left(\frac{\partial K}{\partial \dot x}\right) - \frac{\partial K}{\partial x} = (M+m)\ddot x + ml\ddot\theta\cos\theta - ml\dot\theta^2\sin\theta \quad (11.260)$$

$$\frac{d}{dt}\left(\frac{\partial K}{\partial \dot\theta}\right) - \frac{\partial K}{\partial \theta} = ml^2\ddot\theta + ml\ddot x\cos\theta. \qquad (11.261)$$

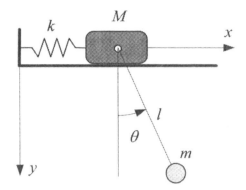

FIGURE 11.10. A vibration absorber.

*The external forces acting on $M$ and $m$ are*

$$F_{x_M} = -kx \tag{11.262}$$
$$F_{y_M} = 0 \tag{11.263}$$
$$F_{x_m} = 0 \tag{11.264}$$
$$F_{y_m} = mg. \tag{11.265}$$

*Therefore, the generalized forces are*

$$
\begin{aligned}
F_x &= F_{x_M}\frac{\partial f_M}{\partial x} + F_{y_M}\frac{\partial g_M}{\partial x} + F_{x_m}\frac{\partial f_m}{\partial x} + F_{y_m}\frac{\partial g_m}{\partial x} \\
&= -kx \tag{11.266}
\end{aligned}
$$

$$
\begin{aligned}
F_\theta &= F_{x_M}\frac{\partial f_M}{\partial \theta} + F_{y_M}\frac{\partial g_M}{\partial \theta} + F_{x_m}\frac{\partial f_m}{\partial \theta} + F_{y_m}\frac{\partial g_m}{\partial \theta} \\
&= -mgl\sin\theta \tag{11.267}
\end{aligned}
$$

*and finally the Lagrange equations of motion are*

$$(M+m)\ddot{x} + ml\ddot{\theta}\cos\theta - ml\dot{\theta}^2\sin\theta = -kx \tag{11.268}$$
$$ml^2\ddot{\theta} + ml\ddot{x}\cos\theta = -mgl\sin\theta. \tag{11.269}$$

**Example 270 ★** *Potential force field.*
  *If a system of masses $m_i$ are moving in a potential force field*

$$\mathbf{F}_{m_i} = -\nabla_i V \tag{11.270}$$

*their Newton equations of motion will be*

$$m_i\ddot{\mathbf{r}}_i = -\nabla_i V \qquad i = 1, 2, \cdots n. \tag{11.271}$$

*Inner product of equations of motion with $\dot{\mathbf{r}}_i$ and adding the equations*

$$\sum_{i=1}^{n} m_i \dot{\mathbf{r}}_i \cdot \ddot{\mathbf{r}}_i = -\sum_{i=1}^{n} \dot{\mathbf{r}}_i \cdot \nabla_i V \qquad (11.272)$$

*and then, integrating over time*

$$\frac{1}{2}\sum_{i=1}^{n} m_i \dot{\mathbf{r}}_i \cdot \dot{\mathbf{r}}_i = -\int \sum_{i=1}^{n} \mathbf{r}_i \cdot \nabla_i V \qquad (11.273)$$

*shows that*

$$
\begin{aligned}
K &= -\int \sum_{i=1}^{n} \left( \frac{\partial V}{\partial x_i} x_i + \frac{\partial V}{\partial y_i} y_i + \frac{\partial V}{\partial z_i} z_i \right) \\
&= -V + E \qquad (11.274)
\end{aligned}
$$

*where $E$ is the constant of integration. $E$ is called **mechanical energy** of the system and is equal to kinetic plus potential energies.*

**Example 271** *Kinetic energy of the Earth.*

*Earth is approximately a rotating rigid body about a fixed axis. The two motions of the Earth are called **revolution** about the sun, and **rotation** about an axis approximately fixed in the Earth. The kinetic energy of the Earth due to its rotation is*

$$
\begin{aligned}
K_1 &= \frac{1}{2} I \omega_1^2 \\
&= \frac{1}{2}\frac{2}{5} (5.9742 \times 10^{24}) \left( \frac{6356912 + 6378388}{2} \right)^2 \left( \frac{2\pi}{24 \times 3600} \frac{366.25}{365.25} \right)^2 \\
&= 2.5762 \times 10^{29} \text{ J}
\end{aligned}
$$

*and the kinetic energy of the Earth due to its revolution is*

$$
\begin{aligned}
K_2 &= \frac{1}{2} M r^2 \omega_2^2 \\
&= \frac{1}{2} (5.9742 \times 10^{24}) (1.49475 \times 10^{11})^2 \left( \frac{2\pi}{24 \times 3600} \frac{1}{365.25} \right)^2 \\
&= 2.6457 \times 10^{33} \text{ J}
\end{aligned}
$$

*where $r$ is the distance from the sun and $\omega_2$ is the angular speed about the sun. The total kinetic energy of the Earth is $K = K_1 + K_2$. However, the ratio of the revolutionary to rotational kinetic energies is*

$$\frac{K_2}{K_1} = \frac{2.6457 \times 10^{33}}{2.5762 \times 10^{29}} \approx 10000.$$

**Example 272 ★** *Non Cartesian coordinate system.*
*The parabolic coordinate system and Cartesian coordinate systems are related according to*

$$x = \eta\xi\cos\varphi \tag{11.275}$$

$$y = \eta\xi\sin\varphi \tag{11.276}$$

$$z = \frac{(\xi^2 - \eta^2)}{2} \tag{11.277}$$

$$\xi^2 = \sqrt{x^2 + y^2 + z^2} + z \tag{11.278}$$

$$\eta^2 = \sqrt{x^2 + y^2 + z^2} - z \tag{11.279}$$

$$\varphi = \tan^{-1}\frac{y}{x}. \tag{11.280}$$

*An electron in a uniform electric field along the positive z-axis is also under the action of an attractive central force field due to the nuclei of the atom.*

$$\mathbf{F} = -\frac{k}{r^2}\hat{e}_r = -\nabla\left(-\frac{k}{r}\right). \tag{11.281}$$

*The influence of a uniform electric field on the motion of the electrons in atoms is called the Stark effect and it is easier to analyze its motion in a parabolic coordinate system.*
*The kinetic energy in a parabolic coordinate system is*

$$
\begin{aligned}
K &= \frac{1}{2}m\left(\dot{x}^2 + \dot{y}^2 + \dot{z}^2\right) \\
&= \frac{1}{2}m\left[\left(\eta^2 + \xi^2\right)\left(\dot{\eta}^2 + \dot{\xi}^2\right) + \eta^2\xi^2\dot{\varphi}^2\right]
\end{aligned} \tag{11.282}
$$

*and the force acting on the electron is*

$$
\begin{aligned}
\mathbf{F} &= -\nabla\left(-\frac{k}{r} + eEz\right) \\
&= -\nabla\left(-\frac{2k}{\xi^2 + \eta^2} + \frac{eE}{2}\left(\xi^2 - \eta^2\right)\right)
\end{aligned} \tag{11.283}
$$

*which leads to the following generalized forces:*

$$F_\eta = \mathbf{F}\cdot\mathbf{b}_\eta = -\frac{4k\eta}{\left(\xi^2 + \eta^2\right)^2} + eE\eta \tag{11.284}$$

$$F_\xi = \mathbf{F}\cdot\mathbf{b}_\xi = -\frac{4k\xi}{\left(\xi^2 + \eta^2\right)^2} - eE\eta \tag{11.285}$$

$$F_\varphi = 0 \tag{11.286}$$

*where, $\mathbf{b}_\xi$, $\mathbf{b}_\eta$, and $\mathbf{b}_\varphi$ are base vectors of the coordinate system*

$$\mathbf{b}_\xi \;=\; \frac{\partial \mathbf{r}}{\partial \xi} = \eta \cos\varphi \,\hat{\imath} + \eta \sin\varphi \,\hat{\jmath} + \xi \hat{k} \qquad (11.287)$$

$$\mathbf{b}_\eta \;=\; \frac{\partial \mathbf{r}}{\partial \eta} = \xi \cos\varphi \,\hat{\imath} + \xi \sin\varphi \,\hat{\jmath} - \eta \hat{k} \qquad (11.288)$$

$$\mathbf{b}_\varphi \;=\; \frac{\partial \mathbf{r}}{\partial \varphi} = -\eta\xi \sin\varphi \,\hat{\imath} + \eta\xi \cos\varphi \,\hat{\jmath}. \qquad (11.289)$$

*Therefore, following the Lagrange method, the equations of motion of the electron are*

$$F_\eta \;=\; \frac{d}{dt}\left[ m\dot{\eta}\left(\xi^2 + \eta^2\right)\right] - mn\left(\dot{\eta}^2 + \dot{\xi}^2\right) - mn\xi^2\dot{\varphi}^2 \quad (11.290)$$

$$F_\xi \;=\; \frac{d}{dt}\left[ m\dot{\xi}\left(\xi^2 + \eta^2\right)\right] - mn\left(\dot{\eta}^2 + \dot{\xi}^2\right) - m\xi\eta^2\dot{\varphi}^2 \quad (11.291)$$

$$F_\varphi \;=\; \frac{d}{dt}\left( mn^2\xi^2\dot{\varphi}^2\right). \qquad (11.292)$$

**Example 273 ★** *Explicit form of Lagrange equations.*

   *Assume that the coordinates of every particle are functions of the coordinates $q_1, q_2, q_3, \cdots, q_n$ but not the time $t$. The kinetic energy of the system made of $n$ massive particles can be written as*

$$K \;=\; \frac{1}{2}\sum_{i=1}^{n} m_i\left(\dot{x}_i^2 + \dot{y}_i^2 + \dot{z}_i^2\right)$$

$$=\; \frac{1}{2}\sum_{j=1}^{n}\sum_{k=1}^{n} a_{jk}\dot{q}_j\dot{q}_k \qquad (11.293)$$

*where the coefficients $a_{jk}$ are functions of $q_1, q_2, q_3, \cdots, q_n$ and*

$$a_{jk} = a_{kj}. \qquad (11.294)$$

*The Lagrange equations of motion*

$$\frac{d}{dt}\left(\frac{\partial K}{\partial \dot{q}_r}\right) - \frac{\partial K}{\partial q_r} = F_r \qquad r = 1, 2, \cdots n \qquad (11.295)$$

*are then equal to*

$$\frac{d}{dt}\sum_{m=1}^{n} a_{mr}\dot{q}_m - \frac{1}{2}\sum_{j=1}^{n}\sum_{k=1}^{n}\frac{\partial a_{jk}}{\partial q_r}\dot{q}_j\dot{q}_k = F_r \qquad (11.296)$$

*or*

$$\sum_{m=1}^{n} a_{mr}\ddot{q}_m + \sum_{k=1}^{n}\sum_{n=1}^{n}\Gamma^r_{k,n}\dot{q}_k\dot{q}_n = F_r \qquad (11.297)$$

*where $\Gamma^i_{j,k}$ is called the* **Christoffel operator**

$$\Gamma^i_{j,k} = \frac{1}{2}\left(\frac{\partial a_{ij}}{\partial q_k} + \frac{\partial a_{ik}}{\partial q_j} - \frac{\partial a_{kj}}{\partial q_i}\right). \qquad (11.298)$$

## 11.6   Lagrangian Mechanics

Assume for some forces $\mathbf{F} = \begin{bmatrix} F_{ix} & F_{iy} & F_{iz} \end{bmatrix}^T$ there is a function $V$, called *potential energy*, such that the force is derivable from $V$

$$\mathbf{F} = -\nabla V. \tag{11.299}$$

Such a force is called *potential* or *conservative force*. Then, the Lagrange equation of motion can be written as

$$\frac{d}{dt}\left(\frac{\partial \mathcal{L}}{\partial \dot{q}_r}\right) - \frac{\partial \mathcal{L}}{\partial q_r} = Q_r \qquad r = 1, 2, \cdots n \tag{11.300}$$

where

$$\mathcal{L} = K - V \tag{11.301}$$

is the *Lagrangean* of the system and $Q_r$ is the nonpotential generalized force.

**Proof.** Assume the external forces $\mathbf{F} = \begin{bmatrix} F_{xi} & F_{yi} & F_{zi} \end{bmatrix}^T$ acting on the system are conservative.

$$\mathbf{F} = -\nabla V \tag{11.302}$$

The work done by these forces in an arbitrary virtual displacement $\delta q_1$, $\delta q_2$, $\delta q_3$, $\cdots$, $\delta q_n$ is

$$\partial W = -\frac{\partial V}{\partial q_1}\delta q_1 - \frac{\partial V}{\partial q_2}\delta q_2 - \cdots \frac{\partial V}{\partial q_n}\delta q_n \tag{11.303}$$

then the Lagrange equation becomes

$$\frac{d}{dt}\left(\frac{\partial K}{\partial \dot{q}_r}\right) - \frac{\partial K}{\partial q_r} = -\frac{\partial V}{\partial q_1} \qquad r = 1, 2, \cdots n. \tag{11.304}$$

Introducing the Lagrangean function $\mathcal{L} = K - V$ converts the Lagrange equation to

$$\frac{d}{dt}\left(\frac{\partial \mathcal{L}}{\partial \dot{q}_r}\right) - \frac{\partial \mathcal{L}}{\partial q_r} = 0 \qquad r = 1, 2, \cdots n \tag{11.305}$$

for a conservative system. The Lagrangean is also called *kinetic potential*.

If a force is not conservative, then the work done by the force is

$$\delta W = \sum_{i=1}^{n}\left(F_{xi}\frac{\partial f_i}{\partial q_r} + F_{yi}\frac{\partial g_i}{\partial q_r} + F_{zi}\frac{\partial h_i}{\partial q_r}\right)\delta q_r$$

$$= Q_r\,\delta q_r \tag{11.306}$$

and the equation of motion would be

$$\frac{d}{dt}\left(\frac{\partial \mathcal{L}}{\partial \dot{q}_r}\right) - \frac{\partial \mathcal{L}}{\partial q_r} = Q_r \qquad r = 1, 2, \cdots n \tag{11.307}$$

where $Q_r$ is the nonpotential generalized force doing work in a virtual displacement of the $r$th generalized coordinate $q_r$. ∎

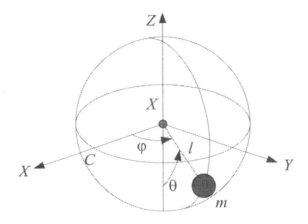

FIGURE 11.11. A spherical pendulum.

**Example 274** *Spherical pendulum.*

*A pendulum analogy is utilized in modeling of many dynamical problems. Figure 11.11 illustrates a spherical pendulum with mass m and length l. The angles $\varphi$ and $\theta$ may be used as describing coordinates of the system.*

*The Cartesian coordinates of the mass as a function of the generalized coordinates are*

$$\begin{bmatrix} X \\ Y \\ Z \end{bmatrix} = \begin{bmatrix} r\cos\varphi\sin\theta \\ r\sin\theta\sin\varphi \\ -r\cos\theta \end{bmatrix} \tag{11.308}$$

*and therefore, the kinetic and potential energies of the pendulum are*

$$K = \frac{1}{2}m\left(l^2\dot{\theta}^2 + l^2\dot{\varphi}^2\sin^2\theta\right) \tag{11.309}$$

$$V = -mgl\cos\theta. \tag{11.310}$$

*The kinetic potential function of this system is then equal to*

$$\mathcal{L} = \frac{1}{2}m\left(l^2\dot{\theta}^2 + l^2\dot{\varphi}^2\sin^2\theta\right) + mgl\cos\theta \tag{11.311}$$

*that leads to the following equations of motion:*

$$\ddot{\theta} - \dot{\varphi}^2\sin\theta\cos\theta + \frac{g}{l}\sin\theta = 0 \tag{11.312}$$

$$\ddot{\varphi}\sin^2\theta + 2\dot{\varphi}\dot{\theta}\sin\theta\cos\theta = 0. \tag{11.313}$$

**Example 275** *A one-link manipulator.*

*A one-link manipulator is illustrated in Figure 11.12. Assume that there is viscous friction in the joint where an ideal motor can apply the torque Q to move the arm. The rotor of an ideal motor has no moment of inertia by assumption.*

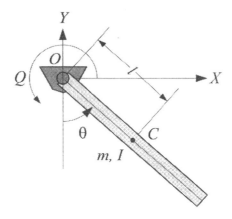

FIGURE 11.12. A one link manipulator.

*The kinetic and potential energies of the manipulator are*

$$K = \frac{1}{2}I\dot{\theta}^2$$
$$= \frac{1}{2}\left(I_C + ml^2\right)\dot{\theta}^2 \tag{11.314}$$

$$V = -mg\cos\theta \tag{11.315}$$

*where $m$ is the mass and $I$ is the moment of inertia of the pendulum about $O$. The Lagrangean of the manipulator is*

$$\mathcal{L} = K - V$$
$$= \frac{1}{2}I\dot{\theta}^2 + mg\cos\theta \tag{11.316}$$

*and therefore, the equation of motion of the manipulator is*

$$M = \frac{d}{dt}\left(\frac{\partial\mathcal{L}}{\partial\dot{\theta}}\right) - \frac{\partial\mathcal{L}}{\partial\theta}$$
$$= I\ddot{\theta} + mgl\sin\theta. \tag{11.317}$$

*The generalized force $M$ is the contribution of the motor torque $Q$ and the viscous friction torque $-c\dot{\theta}$. Hence, the equation of motion of the manipulator is*

$$Q = I\ddot{\theta} + c\dot{\theta} + mgl\sin\theta. \tag{11.318}$$

**Example 276** *The ideal 2R planar manipulator dynamics.*
*An ideal model of a 2R planar manipulator is illustrated in Figure 11.13. It is called ideal because we assumed that the links are massless and there is no friction. The masses $m_1$ and $m_2$ are the mass of the second motor*

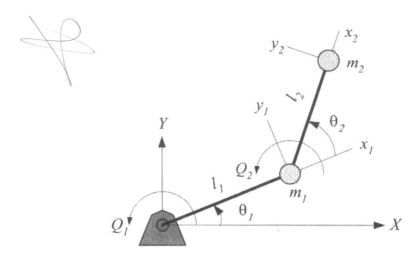

FIGURE 11.13. An ideal model of a 2R planar manipulator.

*to run the second link and the load at the endpoint. We take the absolute angle $\theta_1$ and the relative angle $\theta_2$ as the generalized coordinates to express the configuration of the manipulator.*

*The global position of $m_1$ and $m_2$ are*

$$\begin{bmatrix} X_1 \\ Y_2 \end{bmatrix} = \begin{bmatrix} l_1 \cos \theta_1 \\ l_1 \sin \theta_1 \end{bmatrix} \tag{11.319}$$

$$\begin{bmatrix} X_2 \\ Y_2 \end{bmatrix} = \begin{bmatrix} l_1 \cos \theta_1 + l_2 \cos (\theta_1 + \theta_2) \\ l_1 \sin \theta_1 + l_2 \sin (\theta_1 + \theta_2) \end{bmatrix} \tag{11.320}$$

*and therefore, the global velocity of the masses are*

$$\begin{bmatrix} \dot{X}_1 \\ \dot{Y}_1 \end{bmatrix} = \begin{bmatrix} -l_1 \dot{\theta}_1 \sin \theta_1 \\ l_1 \dot{\theta}_1 \cos \theta_1 \end{bmatrix} \tag{11.321}$$

$$\begin{bmatrix} \dot{X}_2 \\ \dot{Y}_2 \end{bmatrix} = \begin{bmatrix} -l_1 \dot{\theta}_1 \sin \theta_1 - l_2 \left( \dot{\theta}_1 + \dot{\theta}_2 \right) \sin (\theta_1 + \theta_2) \\ l_1 \dot{\theta}_1 \cos \theta_1 + l_2 \left( \dot{\theta}_1 + \dot{\theta}_2 \right) \cos (\theta_1 + \theta_2) \end{bmatrix} . \tag{11.322}$$

*The kinetic energy of this manipulator is made of kinetic energy of the masses and is equal to*

$$\begin{aligned} K &= K_1 + K_2 \\ &= \frac{1}{2} m_1 \left( \dot{X}_1^2 + \dot{Y}_1^2 \right) + \frac{1}{2} m_2 \left( \dot{X}_2^2 + \dot{Y}_2^2 \right) \\ &= \frac{1}{2} m_1 l_1^2 \dot{\theta}_1^2 \\ &\quad + \frac{1}{2} m_2 \left( l_1^2 \dot{\theta}_1^2 + l_2^2 \left( \dot{\theta}_1 + \dot{\theta}_2 \right)^2 + 2 l_1 l_2 \dot{\theta}_1 \left( \dot{\theta}_1 + \dot{\theta}_2 \right) \cos \theta_2 \right) . \end{aligned} \tag{11.323}$$

*The potential energy of the manipulator is*

$$V = V_1 + V_2$$
$$= m_1 g Y_1 + m_2 g Y_2$$
$$= m_1 g l_1 \sin \theta_1 + m_2 g (l_1 \sin \theta_1 + l_2 \sin (\theta_1 + \theta_2)). \quad (11.324)$$

*The Lagrangean is then obtained from Equations (11.323) and (11.324)*

$$\mathcal{L} = K - V \qquad (11.325)$$
$$= \frac{1}{2} m_1 l_1^2 \dot{\theta}_1^2$$
$$+ \frac{1}{2} m_2 \left( l_1^2 \dot{\theta}_1^2 + l_2^2 \left( \dot{\theta}_1 + \dot{\theta}_2 \right)^2 + 2 l_1 l_2 \dot{\theta}_1 \left( \dot{\theta}_1 + \dot{\theta}_2 \right) \cos \theta_2 \right)$$
$$- \left( m_1 g l_1 \sin \theta_1 + m_2 g (l_1 \sin \theta_1 + l_2 \sin (\theta_1 + \theta_2)) \right).$$

*which provides the required partial derivatives as follows:*

$$\frac{\partial \mathcal{L}}{\partial \theta_1} = - (m_1 + m_2) g l_1 \cos \theta_1 - m_2 g l_2 \cos (\theta_1 + \theta_2) \qquad (11.326)$$

$$\frac{\partial \mathcal{L}}{\partial \dot{\theta}_1} = (m_1 + m_2) l_1^2 \dot{\theta}_1 + m_2 l_2^2 \left( \dot{\theta}_1 + \dot{\theta}_2 \right)$$
$$+ m_2 l_1 l_2 \left( 2 \dot{\theta}_1 + \dot{\theta}_2 \right) \cos \theta_2 \qquad (11.327)$$

$$\frac{d}{dt} \left( \frac{\partial \mathcal{L}}{\partial \dot{\theta}_1} \right) = (m_1 + m_2) l_1^2 \ddot{\theta}_1 + m_2 l_2^2 \left( \ddot{\theta}_1 + \ddot{\theta}_2 \right)$$
$$+ m_2 l_1 l_2 \left( 2 \ddot{\theta}_1 + \ddot{\theta}_2 \right) \cos \theta_2$$
$$- m_2 l_1 l_2 \dot{\theta}_2 \left( 2 \dot{\theta}_1 + \dot{\theta}_2 \right) \sin \theta_2 \qquad (11.328)$$

$$\frac{\partial \mathcal{L}}{\partial \theta_2} = - m_2 l_1 l_2 \dot{\theta}_1 \left( \dot{\theta}_1 + \dot{\theta}_2 \right) \sin \theta_2 - m_2 g l_2 \cos (\theta_1 + \theta_2) \qquad (11.329)$$

$$\frac{\partial \mathcal{L}}{\partial \dot{\theta}_2} = m_2 l_2^2 \left( \dot{\theta}_1 + \dot{\theta}_2 \right) + m_2 l_1 l_2 \dot{\theta}_1 \cos \theta_2 \qquad (11.330)$$

$$\frac{d}{dt} \left( \frac{\partial \mathcal{L}}{\partial \dot{\theta}_2} \right) = m_2 l_2^2 \left( \ddot{\theta}_1 + \ddot{\theta}_2 \right) + m_2 l_1 l_2 \ddot{\theta}_1 \cos \theta_2 - m_2 l_1 l_2 \dot{\theta}_1 \dot{\theta}_2 \sin \theta_2 \quad (11.331)$$

*Therefore, the equations of motion for the 2R manipulator are*

$$Q_1 = \frac{d}{dt} \left( \frac{\partial \mathcal{L}}{\partial \dot{\theta}_1} \right) - \frac{\partial \mathcal{L}}{\partial \theta_1}$$
$$= (m_1 + m_2) l_1^2 \ddot{\theta}_1 + m_2 l_2^2 \left( \ddot{\theta}_1 + \ddot{\theta}_2 \right)$$
$$+ m_2 l_1 l_2 \left( 2 \ddot{\theta}_1 + \ddot{\theta}_2 \right) \cos \theta_2 - m_2 l_1 l_2 \dot{\theta}_2 \left( 2 \dot{\theta}_1 + \dot{\theta}_2 \right) \sin \theta_2$$
$$+ (m_1 + m_2) g l_1 \cos \theta_1 + m_2 g l_2 \cos (\theta_1 + \theta_2) \qquad (11.332)$$

$$Q_2 = \frac{d}{dt}\left(\frac{\partial \mathcal{L}}{\partial \dot{\theta}_2}\right) - \frac{\partial \mathcal{L}}{\partial \theta_2}$$

$$= m_2 l_2^2 \left(\ddot{\theta}_1 + \ddot{\theta}_2\right) + m_2 l_1 l_2 \ddot{\theta}_1 \cos\theta_2 - m_2 l_1 l_2 \dot{\theta}_1 \dot{\theta}_2 \sin\theta_2$$

$$+ m_2 l_1 l_2 \dot{\theta}_1 \left(\dot{\theta}_1 + \dot{\theta}_2\right) \sin\theta_2 + m_2 g l_2 \cos\left(\theta_1 + \theta_2\right) \quad (11.333)$$

*The generalized forces $Q_1$ and $Q_2$ are the required forces to drive the generalized coordinates. In this case, $Q_1$ is the torque at the base motor and $Q_2$ is the torque of the motor at $m_1$.*

*The equations of motion can be rearranged to have a more systematic form*

$$Q_1 = \left((m_1 + m_2) l_1^2 + m_2 l_2 (l_2 + 2l_1 \cos\theta_2)\right) \ddot{\theta}_1$$

$$+ m_2 l_2 (l_2 + l_1 \cos\theta_2) \ddot{\theta}_2$$

$$- 2m_2 l_1 l_2 \sin\theta_2 \, \dot{\theta}_1 \dot{\theta}_2 - m_2 l_1 l_2 \sin\theta_2 \, \dot{\theta}_2^2$$

$$+ (m_1 + m_2) g l_1 \cos\theta_1 + m_2 g l_2 \cos\left(\theta_1 + \theta_2\right) \quad (11.334)$$

$$Q_2 = m_2 l_2 (l_2 + l_1 \cos\theta_2) \ddot{\theta}_1 + m_2 l_2^2 \ddot{\theta}_2$$

$$+ m_2 l_1 l_2 \sin\theta_2 \, \dot{\theta}_1^2 + m_2 g l_2 \cos\left(\theta_1 + \theta_2\right). \quad (11.335)$$

## 11.7  Summary

The translational and rotational equations of motion for a rigid body, expressed in the global coordinate frame, are

$$^G\mathbf{F} = \frac{^Gd}{dt}\,^G\mathbf{p} \quad (11.336)$$

$$^G\mathbf{M} = \frac{^Gd}{dt}\,^G\mathbf{L} \quad (11.337)$$

where $^G\mathbf{F}$ and $^G\mathbf{M}$ indicate the resultant of the external forces and moments applied on the rigid body and measured at $CM$. The vector $^G\mathbf{p}$ is the momentum and $^G\mathbf{L}$ is the moment of momentum for the rigid body at $CM$

$$\mathbf{p} = m\,\mathbf{v} \quad (11.338)$$

$$\mathbf{L} = \mathbf{r}_C \times \mathbf{p}. \quad (11.339)$$

The expression of the equations of motion in the body coordinate frame are

$$^B\mathbf{F} = {}^G\dot{\mathbf{p}} + {}^B_G\omega_B \times {}^B\mathbf{p}$$

$$= m\,{}^B\mathbf{a}_B + m\,{}^B_G\omega_B \times {}^B\mathbf{v}_B \quad (11.340)$$

$$
\begin{aligned}
{}^B\mathbf{M} &= {}^B\dot{\mathbf{L}} + {}^B_G\omega_B \times {}^B\mathbf{L} \\
&= {}^BI\,{}^B_G\dot{\omega}_B + {}^B_G\omega_B \times \left({}^BI\,{}^B_G\omega_B\right)
\end{aligned}
\tag{11.341}
$$

where $I$ is the moment of inertia for the rigid body.

$$
I = \begin{bmatrix} I_{xx} & I_{xy} & I_{xz} \\ I_{yx} & I_{yy} & I_{yz} \\ I_{zx} & I_{zy} & I_{zz} \end{bmatrix}.
\tag{11.342}
$$

The elements of $I$ are only functions of the mass distribution of the rigid body and are defined by

$$
I_{ij} = \int_B \left(r_i^2 \delta_{mn} - x_{im} x_{jn}\right) dm \quad , \quad i, j = 1, 2, 3
\tag{11.343}
$$

where $\delta_{ij}$ is Kronecker's delta.

Every rigid body has a principal body coordinate frame in which the moment of inertia is in the form

$$
{}^BI = \begin{bmatrix} I_1 & 0 & 0 \\ 0 & I_2 & 0 \\ 0 & 0 & I_3 \end{bmatrix}.
\tag{11.344}
$$

The rotational equation of motion in the principal coordinate frame simplifies to

$$
\begin{aligned}
M_1 &= I_1 \dot{\omega}_1 - (I_2 - I_2)\,\omega_2 \omega_3 \\
M_2 &= I_2 \dot{\omega}_2 - (I_3 - I_1)\,\omega_3 \omega_1 \\
M_3 &= I_3 \dot{\omega}_3 - (I_1 - I_2)\,\omega_1 \omega_2.
\end{aligned}
\tag{11.345}
$$

Utilizing homogeneous position vectors we also define pseudo inertia matrix $\bar{I}$ with application in robot dynamics as

$$
\begin{aligned}
{}^B\bar{I} &= \int_B \mathbf{r}\,\mathbf{r}^T\, dm \\[2mm]
&= \begin{bmatrix}
\frac{-I_{xx}+I_{yy}+I_{zz}}{2} & I_{xy} & I_{xz} & mx_C \\
I_{yx} & \frac{I_{xx}-I_{yy}+I_{zz}}{2} & I_{yz} & my_C \\
I_{zx} & I_{zy} & \frac{I_{xx}+I_{yy}-I_{zz}}{2} & mz_C \\
mx_C & mz_C & my_C & m
\end{bmatrix}
\end{aligned}
\tag{11.346}
$$

where, ${}^B\mathbf{r}_C$ is the position of the mass center in the body frame.

$$
{}^B\mathbf{r}_C = \begin{bmatrix} x_C \\ y_C \\ z_C \end{bmatrix} = \begin{bmatrix} \frac{1}{m}\int_B x\,dm \\ \frac{1}{m}\int_B y\,dm \\ \frac{1}{m}\int_B z\,dm \end{bmatrix}
\tag{11.347}
$$

The equations of motion for a mechanical system having $n$ DOF can also be found by the Lagrange equation

$$\frac{d}{dt}\left(\frac{\partial \mathcal{L}}{\partial \dot{q}_r}\right) - \frac{\partial \mathcal{L}}{\partial q_r} = Q_r \qquad r = 1, 2, \cdots n \qquad (11.348)$$

$$\mathcal{L} = K - V \qquad (11.349)$$

where $\mathcal{L}$ is the *Lagrangean* of the system, $K$ is the kinetic energy, $V$ is the potential energy, and $Q_r$ is the nonpotential generalized force.

$$Q_r = \sum_{i=1}^{n}\left(Q_{ix}\frac{\partial f_i}{\partial q_1} + Q_{iy}\frac{\partial g_i}{\partial q_2} + Q_{iz}\frac{\partial h_i}{\partial q_n}\right) \qquad (11.350)$$

The parameters $q_r$, $r = 1, 2, \cdots, n$ are the generalized coordinates of the system, $\mathbf{Q} = \begin{bmatrix} Q_{ix} & Q_{iy} & Q_{iz} \end{bmatrix}^T$ is the external force acting on the $i$th particle of the system, and $Q_r$ is the generalized force associated to $q_r$. When $(x_i, y_i, z_i)$ is the Cartesian coordinates in a globally fixed coordinate frame for the particle $m_i$, then its coordinates may be functions of another set of coordinates $q_1, q_2, q_3, \cdots, q_n$ and possibly time $t$.

$$\begin{aligned} x_i &= f_i(q_1, q_2, q_3, \cdots, q_n, t) & (11.351) \\ y_i &= g_i(q_1, q_2, q_3, \cdots, q_n, t) & (11.352) \\ z_i &= h_i(q_1, q_2, q_3, \cdots, q_n, t) & (11.353) \end{aligned}$$

# Exercises

1. Notation and symbols.

   Describe their meaning.

   a- $^G\mathbf{p}_P$   b- $\mathbf{L}_l$   c- $^G\mathbf{F}$   d- $_1W_2$   e- $K$   f- $V$

   g- $dm$   h- $I$   i- $I_{ij}$   j- $\bar{I}$   k- $^G\mathbf{M}$   l- $I_1$

   m- $^0_2\tilde{\alpha}_1$   n- $^2_1\tilde{\omega}_2$   o- $^G A_B$   p- $^G\dot{V}_B$   q- $\mathbf{J}$   r- $\ddot{\mathbf{X}}$

2. Kinetic energy of a rigid link.

   Consider a straight and uniform bar as a rigid link of a manipulator. The link has a mass $m$. Show that the kinetic energy of the bar can be expressed as

   $$K = \frac{1}{6}m\left(\mathbf{v}_1 \cdot \mathbf{v}_1 + \mathbf{v}_1 \cdot \mathbf{v}_2 + \mathbf{v}_2 \cdot \mathbf{v}_2\right)$$

   where $\mathbf{v}_1$ and $\mathbf{v}_2$ are the velocity vectors of the endpoints of the link.

3. Discrete particles.

   There are three particles $m_1 = 1\,\mathrm{kg}$, $m_2 = 2\,\mathrm{kg}$, $m_3 = 3\,\mathrm{kg}$, at

   $$\mathbf{r}_1 = \begin{bmatrix} 1 \\ -1 \\ 1 \end{bmatrix} \quad \mathbf{r}_1 = \begin{bmatrix} -1 \\ -3 \\ 2 \end{bmatrix} \quad \mathbf{r}_1 = \begin{bmatrix} 2 \\ -1 \\ -3 \end{bmatrix}.$$

   Their velocities are

   $$\mathbf{v}_1 = \begin{bmatrix} 2 \\ 1 \\ 1 \end{bmatrix} \quad \mathbf{v}_1 = \begin{bmatrix} -1 \\ 0 \\ 2 \end{bmatrix} \quad \mathbf{v}_1 = \begin{bmatrix} 3 \\ -2 \\ -1 \end{bmatrix}.$$

   Find the position and velocity of the system at $CM$. Calculate the system's momentum and moment of momentum. Calculate the system's kinetic energy and determine the rotational and translational parts of the kinetic energy.

4. Newton's equation of motion in the body frame.

   Show that Newton's equation of motion in the body frame is

   $$\begin{bmatrix} F_x \\ F_y \\ F_z \end{bmatrix} = m\begin{bmatrix} a_x \\ a_y \\ a_z \end{bmatrix} + \begin{bmatrix} 0 & -\omega_z & \omega_y \\ \omega_z & 0 & -\omega_x \\ -\omega_y & \omega_x & 0 \end{bmatrix}\begin{bmatrix} v_x \\ v_y \\ v_z \end{bmatrix}.$$

5. Work on a curved path.

A particle of mass $m$ is moving on a circular path given by

$$^G\mathbf{r}_P = \cos\theta\,\hat{I} + \sin\theta\,\hat{J} + 4\,\hat{K}.$$

Calculate the work done by a force $^G\mathbf{F}$ when the particle moves from $\theta = 0$ to $\theta = \frac{\pi}{2}$.

(a)

$$^G\mathbf{F} = \frac{z^2 - y^2}{(x+y)^2}\,\hat{I} + \frac{y^2 - x^2}{(x+y)^2}\,\hat{J} + \frac{x^2 - y^2}{(x+z)^2}\,\hat{K}$$

(b)

$$^G\mathbf{F} = \frac{z^2 - y^2}{(x+y)^2}\,\hat{I} + \frac{2y}{x+y}\,\hat{J} + \frac{x^2 - y^2}{(x+z)^2}\,\hat{K}$$

6. Newton's equation of motion.

Find the equations of motion for the system shown in Figure 11.10 based on Newton method.

7. Acceleration in the body frame.

Find the acceleration vector for the endpoint of the two-link manipulator shown in Figure 11.3, expressed in frame $A$.

8. Principal moments of inertia.

Find the principal moments of inertia and directions for the following inertia matrices:

(a)

$$[I] = \begin{bmatrix} 3 & 2 & 2 \\ 2 & 2 & 0 \\ 2 & 0 & 4 \end{bmatrix}$$

(b)

$$[I] = \begin{bmatrix} 3 & 2 & 4 \\ 2 & 0 & 2 \\ 4 & 2 & 3 \end{bmatrix}$$

(c)

$$[I] = \begin{bmatrix} 100 & 20\sqrt{3} & 0 \\ 20\sqrt{3} & 60 & 0 \\ 0 & 0 & 10 \end{bmatrix}$$

9. Moment of inertia.

Calculate the moment of inertia for the following objects in a principal Cartesian coordinate frame at $CM$.

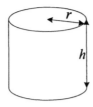

FIGURE 11.14. A cylinder.

(a) A cylinder similar to a uniform arm with a circular cross section.

(b) A rectangular box similar to a uniform arm with a rectangular cross section.

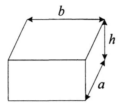

FIGURE 11.15. A box.

(c) A house similar to a prismatic bar with a nonsymmetric polygon cross section.

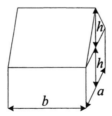

FIGURE 11.16. A solid house.

10. Rotated moment of inertia matrix.

    A principal moment of inertia matrix $^{B_2}I$ is given as

    $$[I] = \begin{bmatrix} 3 & 0 & 0 \\ 0 & 5 & 0 \\ 0 & 0 & 4 \end{bmatrix}.$$

    The principal frame was achieved by rotating the initial body coordinate frame 30 deg about the $x$-axis, followed by 45 deg about the $z$-axis. Find the initial moment of inertia matrix $^{B_1}I$.

11. Rotation of moment of inertia matrix.

Find the required rotation matrix that transforms the moment of inertia matrix $[I]$ to an diagonal matrix.

$$[I] = \begin{bmatrix} 3 & 2 & 2 \\ 2 & 2 & 0.1 \\ 2 & 0.1 & 4 \end{bmatrix}$$

12. Pseudo inertia matrix.

Calculate the pseudo inertia matrix for the following objects in a principal Cartesian coordinate frame at $CM$.

(a) A uniform arm with a rectangular cross section.

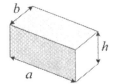

FIGURE 11.17. A prismatic bar.

(b) A compound shape as a housing for wrist joints.

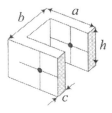

FIGURE 11.18. A housing for wrist joints.

(c) ★ A gripper, which is a compound shape made of some 3D geometrical shapes.

13. Cubic equations.

The solution of a cubic equation

$$ax^3 + bx^2 + cx + d = 0$$

where $a \neq 0$, can be found in a systematic way.

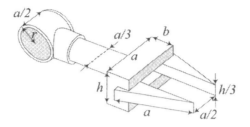

FIGURE 11.19. A gripper.

Transform the equation to a new form with discriminant $4p^3 + q^2$,

$$y^3 + 3py + q = 0$$

using the transformation $x = y - \frac{b}{3a}$, where,

$$p = \frac{3ac - b^2}{9a^2}$$

$$q = \frac{2b^3 - 9abc + 27a^2 d}{27a^3}.$$

The solutions are then

$$y_1 = \sqrt[3]{\alpha} - \sqrt[3]{\beta}$$

$$y_2 = e^{\frac{2\pi i}{3}} \sqrt[3]{\alpha} - e^{\frac{4\pi i}{3}} \sqrt[3]{\beta}$$

$$y_3 = e^{\frac{4\pi i}{3}} \sqrt[3]{\alpha} - e^{\frac{2\pi i}{3}} \sqrt[3]{\beta}$$

where,

$$\alpha = \frac{-q + \sqrt{q^2 + 4p^3}}{2}$$

$$\beta = \frac{-q + \sqrt{q^2 + 4p^3}}{2}.$$

For real values of $p$ and $q$, if the discriminant is positive, then one root is real, and two roots are complex conjugates. If the discriminant is zero, then there are three real roots, of which at least two are equal. If the discriminant is negative, then there are three unequal real roots.

Apply this theory for the characteristic equation of the matrix $[I]$ and show that the principal moments of inertia are real.

14. Kinematics of a moving car on the Earth.

The location of a vehicle on the Earth is described by its longitude $\varphi$ from a fixed meridian, say, the Greenwich meridian, and its latitude $\theta$ from the equator, as shown in Figure 11.20. We attach a coordinate

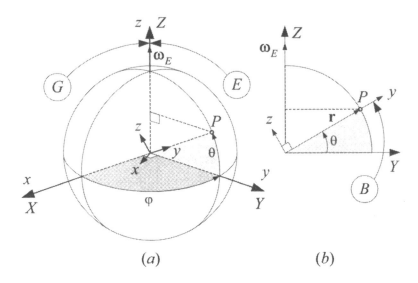

$(a)$ $(b)$

FIGURE 11.20. The location on the Earth is defined by longitude $\varphi$ and latitude $\theta$.

frame $B$ at the center of the Earth with $x$-axis on the equator's plane and $y$-axis pointing the vehicle. There are also two coordinate frames $E$ and $G$ where $E$ is attached to the Earth and $G$ is the global coordinate frame. Show that the angular velocity of $B$ and the velocity of the vehicle are

$$
\begin{aligned}
{}^{B}_{G}\boldsymbol{\omega}_B &= \dot\theta\,\hat\imath_B + (\omega_E + \dot\varphi)\sin\theta\,\hat\jmath_B + (\omega_E + \dot\varphi)\cos\theta\,\hat k \\
{}^{B}_{G}\mathbf{v}_P &= -r\,(\omega_E + \dot\varphi)\cos\theta\,\hat\imath_B + r\dot\theta\,\hat k .
\end{aligned}
$$

Calculate the acceleration of the vehicle.

15. Global differentiating of angular momentum.

Convert the moment of inertia ${}^{B}I$ and the angular velocity ${}^{B}_{G}\boldsymbol{\omega}_B$ to the global coordinate frame and then find the differential of angular momentum. It is an alternative method to show that

$$
\begin{aligned}
\frac{{}^{G}d}{dt}\,{}^{B}\mathbf{L} &= \frac{{}^{G}d}{dt}\left({}^{B}I\;{}^{B}_{G}\boldsymbol{\omega}_B\right) \\
&= {}^{B}\dot{\mathbf{L}} + {}^{B}_{G}\boldsymbol{\omega}_B \times {}^{B}\mathbf{L} \\
&= I\dot{\boldsymbol\omega} + \boldsymbol\omega \times (I\boldsymbol\omega).
\end{aligned}
$$

16. Equations of motion for a rotating arm.

Find the equations of motion for the rotating link shown in Figure 11.21 based on the Lagrange method.

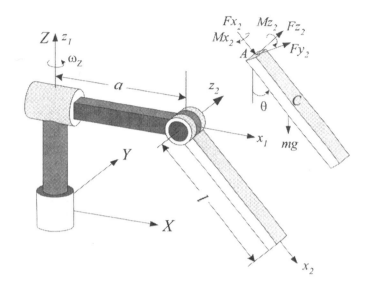

FIGURE 11.21. A rotating link.

17. Lagrange method and nonlinear vibrating system.

Use the Lagrange method and find the equation of motion for the link shown in Figure 11.22. The stiffness of the linear spring is $k$.

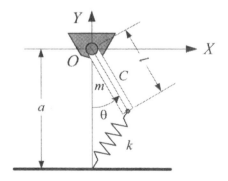

FIGURE 11.22. A compound pendulum attached with a linear spring at the tip point.

18. Forced vibration of a pendulum.

Figure 11.23 illustrates a simple pendulum having a length $l$ and a bob with mass $m$. Find the equation of motion if

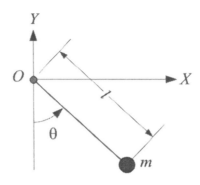

FIGURE 11.23. A pendulum with a vibrating pivot.

(a) the pivot $O$ has a dictated motion in $X$ direction

$$X_O = a \sin \omega t$$

(b) the pivot $O$ has a dictated motion in $Y$ direction

$$Y_O = b \sin \omega t$$

(c) the pivot $O$ has a uniform motion on a circle

$$\mathbf{r}_O = R \cos \omega t \; \hat{I} + R \sin \omega t \; \hat{J}.$$

19. **Equations of motion from Lagrangean.**

   Consider a physical system with a Lagrangean as

   $$\mathcal{L} = \frac{1}{2} m \left( a\dot{x} + b\dot{y} \right)^2 - \frac{1}{2} k \left( ax + by \right)^2 .$$

   The coefficients $m$, $k$, $a$, and $b$ are constant.

20. **Lagrangean from equation of motion.**

   Find the Lagrangean associated to the following equations of motions:

   (a)

   $$mr^2\ddot{\theta} + k_1 l_1 \theta + k_2 l_2 \theta + mgl = 0$$

   (b)

   $$\ddot{r} - r\dot{\theta}^2 = 0$$
   $$r^2 \ddot{\theta} + 2r \dot{r} \dot{\theta} = 0$$

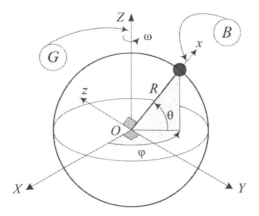

FIGURE 11.24. A mass on a rotating ring.

21. A mass on a rotating ring.

A particle of mass $m$ is free to slide on a rotating vertical ring as shown in Figure 11.24. The ring is turning with a constant angular velocity $\omega = \dot{\varphi}$. Determine the equation of motion of the particle. The local coordinate frame is set up with the $x$-axis pointing the particle, and the $y$-axis in the plane of the ring and parallel with the tangent to the ring at the position of the mass.

22. ★ Particle in electromagnetic field.

Show that equations of motion of a particle with mass $m$ with a Lagrangian $\mathcal{L}$

$$\mathcal{L} = \frac{1}{2}m\dot{\mathbf{r}}^2 - e\Phi + e\dot{\mathbf{r}} \cdot \mathbf{A}$$

are

$$m\ddot{q}_i = e\left(-\frac{\partial \Phi}{\partial q_i} - \frac{\partial A_i}{\partial t}\right) + e\sum_{j=1}^{3} \dot{q}_j \left(\frac{\partial A_j}{\partial q_i} - \frac{\partial A_i}{\partial q_j}\right)$$

where

$$\mathbf{q} = \begin{bmatrix} q_1 \\ q_2 \\ q_3 \end{bmatrix} = \begin{bmatrix} x \\ y \\ z \end{bmatrix}.$$

Then convert the equations of motion to a vectorial form

$$m\ddot{\mathbf{r}} = e\mathbf{E}(\mathbf{r}, t) + e\dot{\mathbf{r}} \times \mathbf{B}(\mathbf{r}, t)$$

where $\mathbf{E}$ and $\mathbf{B}$ are electric and magnetic fields.

$$\begin{aligned} \mathbf{E} &= -\nabla\Phi - \frac{\partial \mathbf{A}}{\partial t} \\ \mathbf{B} &= \nabla \times \mathbf{A}. \end{aligned}$$

# 12

# Robot Dynamics

## 12.1 Rigid Link Recursive Acceleration

Figure 12.1 illustrates a link $(i)$ of a manipulator and shows its velocity and acceleration vectorial characteristics. Based on velocity and acceleration kinematics of rigid links we can write a set of recursive equations to relate the kinematics information of each link in the coordinate frame attached to the previous link. We may then generalize the method of analysis to be suitable for a robot with any number of links.

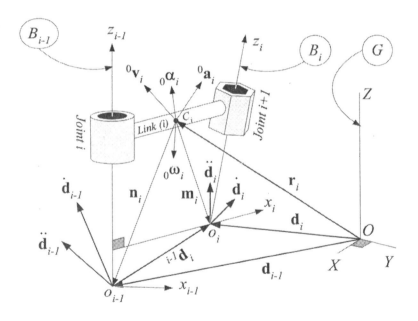

FIGURE 12.1. Illustration of vectorial kinematics information of a link $(i)$.

Translational acceleration of the link $(i)$ is denoted by ${}^0\mathbf{a}_i$ and is measured at the mass center $C_i$. Angular acceleration of the link $(i)$ is denoted by ${}^0\boldsymbol{\alpha}_i$ and is usually shown at the mass center $C_i$, although ${}^0\boldsymbol{\alpha}_i$ is the same for all points of a rigid link. We calculate the translational acceleration of the link $(i)$ at $C_i$ by referring to the acceleration at the distal end of the link. Hence,

$$
\begin{aligned}
{}^0\mathbf{a}_i \;=\;& {}^0\ddot{\mathbf{d}}_i + {}_0\boldsymbol{\alpha}_i \times \left({}^0\mathbf{r}_i - {}^0\mathbf{d}_i\right) \\
& + {}_0\boldsymbol{\omega}_i \times \left({}_0\boldsymbol{\omega}_i \times \left({}^0\mathbf{r}_i - {}^0\mathbf{d}_i\right)\right)
\end{aligned}
\tag{12.1}
$$

which can also be written in the form

$$
\begin{aligned}
{}^i_0\mathbf{a}_i \;=\;& {}^i\ddot{\mathbf{d}}_i + {}^i_0\boldsymbol{\alpha}_i \times \left({}^i\mathbf{r}_i - {}^i\mathbf{d}_i\right) \\
& + {}^i_0\boldsymbol{\omega}_i \times \left({}^i_0\boldsymbol{\omega}_i \times \left({}^i\mathbf{r}_i - {}^i\mathbf{d}_i\right)\right).
\end{aligned}
\tag{12.2}
$$

The angular acceleration of the link $(i)$ is

$$
{}_0\boldsymbol{\alpha}_i = \begin{cases} {}_0\boldsymbol{\alpha}_{i-1} + \ddot{\theta}_i\,{}^0\hat{k}_{i-1} + {}_0\boldsymbol{\omega}_{i-1} \times \dot{\theta}_i\,{}^0\hat{k}_{i-1} & \text{if joint } i \text{ is R} \\ {}_0\boldsymbol{\alpha}_{i-1} & \text{if joint } i \text{ is P} \end{cases}
\tag{12.3}
$$

that can also be written in the the local frame

$$
{}^i_0\boldsymbol{\alpha}_i = \begin{cases} {}^iT_{i-1}\left({}^{i-1}_0\boldsymbol{\alpha}_{i-1} + \ddot{\theta}_i\,{}^{i-1}\hat{k}_{i-1}\right) \\ \quad + {}^iT_{i-1}\left({}^{i-1}_0\boldsymbol{\omega}_{i-1} \times \dot{\theta}_i\,{}^{i-1}\hat{k}_{i-1}\right) & \text{if joint } i \text{ is R} \\[2mm] {}^iT_{i-1}\,{}^{i-1}_0\boldsymbol{\alpha}_{i-1} & \text{if joint } i \text{ is P.} \end{cases}
\tag{12.4}
$$

**Proof.** According to rigid body acceleration in Equation (10.59), the acceleration of the point $C_i$ is

$$
\begin{aligned}
{}^0\mathbf{a}_i \;=\;& \frac{{}^0d}{dt}\,{}^0\mathbf{v}_i \\
=\;& {}_0\boldsymbol{\alpha}_i \times \left({}^0\mathbf{r}_i - {}^0\mathbf{d}_i\right) \\
& + {}_0\boldsymbol{\omega}_i \times \left({}_0\boldsymbol{\omega}_i \times \left({}^0\mathbf{r}_i - {}^0\mathbf{d}_i\right)\right) + {}^0\ddot{\mathbf{d}}_i.
\end{aligned}
\tag{12.5}
$$

The acceleration of $C_i$ may be transformed to the body frame $B_i$ by

$$
\begin{aligned}
{}^i\mathbf{a}_i \;=\;& {}^0T_i^{-1}\,{}^0\mathbf{a}_i \tag{12.6} \\
=\;& {}^i_0\boldsymbol{\alpha}_i \times \left({}^i\mathbf{r}_i - {}^i\mathbf{d}_i\right) + {}^i_0\boldsymbol{\omega}_i \times \left({}^i_0\boldsymbol{\omega}_i \times \left({}^i\mathbf{r}_i - {}^i\mathbf{d}_i\right)\right) + {}^i\ddot{\mathbf{d}}_i.
\end{aligned}
$$

We define the position of $o_{i-1}$ and $o_i$ with respect to $C_i$ by

$$
{}^0\mathbf{n}_i = {}^0\mathbf{d}_{i-1} - {}^0\mathbf{r}_i \tag{12.7}
$$

$$
{}^0\mathbf{m}_i = {}^0\mathbf{d}_i - {}^0\mathbf{r}_i \tag{12.8}
$$

to simplify the dynamic equation of robots.

Using Equation (10.45) or (10.47) as a rule for adding relative angular accelerations, we find

$$
{}_0\boldsymbol{\alpha}_i = {}_0\boldsymbol{\alpha}_{i-1} + {}^0_{i-1}\boldsymbol{\alpha}_i + {}_0\boldsymbol{\omega}_{i-1} \times {}^0_{i-1}\boldsymbol{\omega}_i
\tag{12.9}
$$

however, the angular velocity $_{i-1}^{\quad 0}\boldsymbol{\omega}_i$ and angular acceleration $_{i-1}^{\quad 0}\boldsymbol{\alpha}_i$ are

$$_{i-1}^{\quad 0}\boldsymbol{\omega}_i = \dot{\theta}_i\,{}^0\hat{k}_{i-1} \qquad \text{if joint } i \text{ is R} \qquad (12.10)$$

$$_{i-1}^{\quad 0}\boldsymbol{\alpha}_i = \ddot{\theta}_i\,{}^0\hat{k}_{i-1} \qquad \text{if joint } i \text{ is R} \qquad (12.11)$$

or

$$_{i-1}^{\quad 0}\boldsymbol{\omega}_i = 0 \qquad \text{if joint } i \text{ is P} \qquad (12.12)$$

$$_{i-1}^{\quad 0}\boldsymbol{\alpha}_i = 0 \qquad \text{if joint } i \text{ is P.} \qquad (12.13)$$

Therefore,

$$_0\boldsymbol{\alpha}_i = \begin{cases} _0\boldsymbol{\alpha}_{i-1} + \ddot{\theta}_i\,{}^0\hat{k}_{i-1} + {}_0\boldsymbol{\omega}_{i-1} \times \dot{\theta}_i\,{}^0\hat{k}_{i-1} & \text{if joint } i \text{ is R} \\ _0\boldsymbol{\alpha}_{i-1} & \text{if joint } i \text{ is P} \end{cases} \qquad (12.14)$$

and it can be transformed to any coordinate frame, including $B_i$, to find the Equation (12.4).

$$\begin{aligned} _0^i\boldsymbol{\alpha}_i &= {}^0T_i^{-1}\,_0\boldsymbol{\alpha}_i \\ &= {}_0^i\boldsymbol{\alpha}_{i-1} + \ddot{\theta}_i\,_0^i\hat{k}_{i-1} + {}_0^i\boldsymbol{\omega}_{i-1} \times \dot{\theta}_i\,_0^i\hat{k}_{i-1} \\ &= {}^iT_{i-1}\left( {}_0^{i-1}\boldsymbol{\alpha}_{i-1} + \ddot{\theta}_i\,{}_0^{i-1}\hat{k}_{i-1} + {}_0^{i-1}\boldsymbol{\omega}_{i-1} \times \dot{\theta}_i\,{}_0^{i-1}\hat{k}_{i-1}\right) \end{aligned} \qquad (12.15)$$

■

**Example 277** *Recursive angular velocity equation for link (i).*
*The recursive global angular velocity equation for a link (i) is*

$$_0\boldsymbol{\omega}_i = \begin{cases} _0\boldsymbol{\omega}_{i-1} + \dot{\theta}_i\,{}^0\hat{k}_{i-1} & \text{if joint } i \text{ is R} \\ _0\boldsymbol{\omega}_{i-1} & \text{if joint } i \text{ is P.} \end{cases} \qquad (12.16)$$

*We may transform this equation to the local frame $B_i$*

$$\begin{aligned} _0^i\boldsymbol{\omega}_i &= {}^0T_i^{-1}\,_0\boldsymbol{\omega}_i \\ &= {}^0T_i^{-1}\left( {}_0\boldsymbol{\omega}_{i-1} + \dot{\theta}_i\,{}^0\hat{k}_{i-1}\right) \\ &= {}_0^i\boldsymbol{\omega}_{i-1} + \dot{\theta}_i\,{}^i\hat{k}_{i-1} \\ &= {}^iT_{i-1}\left( {}_0^{i-1}\boldsymbol{\omega}_{i-1} + \dot{\theta}_i\,{}^{i-1}\hat{k}_{i-1}\right). \end{aligned} \qquad (12.17)$$

*Therefore, we can use a recursive equation to find the locally expressed angular velocity of link (i) by having the angular velocity of its lower link (i − 1). In this equation, every vector is expressed in its own coordinate frame.*

$$_0^i\boldsymbol{\omega}_i = \begin{cases} {}^iT_{i-1}\left( {}_0^{i-1}\boldsymbol{\omega}_{i-1} + \dot{\theta}_i\,{}^{i-1}\hat{k}_{i-1}\right) & \text{if joint } i \text{ is R} \\ {}^iT_{i-1}\,{}_0^{i-1}\boldsymbol{\omega}_{i-1} & \text{if joint } i \text{ is P.} \end{cases} \qquad (12.18)$$

**Example 278** *Recursive translational velocity equation for link (i).*
*The recursive global translational velocity equation for a link (i) is*

$$
{}^0\dot{\mathbf{d}}_i = \begin{cases} {}^0\dot{\mathbf{d}}_{i-1} + {}_0\boldsymbol{\omega}_i \times {}^0_{i-1}\mathbf{d}_i & \text{if joint } i \text{ is } R \\ {}^0\dot{\mathbf{d}}_{i-1} + {}_0\boldsymbol{\omega}_i \times {}^0_{i-1}\mathbf{d}_i + \dot{d}_i\, {}^0\hat{k}_{i-1} & \text{if joint } i \text{ is } P. \end{cases} \tag{12.19}
$$

*We may transform this equation to the local frame $B_i$*

$$
\begin{aligned}
{}^i_0\dot{\mathbf{d}}_i &= {}^0T_i^{-1}\, {}_0\dot{\mathbf{d}}_i \\
&= {}^0T_i^{-1}\left( {}^0\dot{\mathbf{d}}_{i-1} + {}_0\boldsymbol{\omega}_i \times {}^0_{i-1}\mathbf{d}_i + \dot{d}_i\, {}^0\hat{k}_{i-1} \right) \\
&= {}^i_0\dot{\mathbf{d}}_{i-1} + {}^i_0\boldsymbol{\omega}_i \times {}^i_{i-1}\mathbf{d}_i + \dot{d}_i\, {}^i\hat{k}_{i-1} \\
&= {}^iT_{i-1}\left( {}^{i-1}_0\dot{\mathbf{d}}_{i-1} + \dot{d}_i\, {}^{i-1}\hat{k}_{i-1} \right) + {}^i_0\boldsymbol{\omega}_i \times {}^i_{i-1}\mathbf{d}_i. \tag{12.20}
\end{aligned}
$$

*Therefore, we may use a recursive equation to find the locally expressed translational velocity of link (i) by having the translational velocity of its lower link (i − 1). In this equation, every vector is expressed in its own coordinate frame.*

$$
{}^i\dot{\mathbf{d}}_i = \begin{cases} {}^iT_{i-1}\left( {}^{i-1}_0\dot{\mathbf{d}}_{i-1} + \dot{d}_i\, {}^{i-1}\hat{k}_{i-1} \right) \\ \quad + {}^i_0\boldsymbol{\omega}_i \times {}^i_{i-1}\mathbf{d}_i & \text{if joint } i \text{ is } R \\[6pt] {}^iT_{i-1}\, {}^{i-1}_0\dot{\mathbf{d}}_{i-1} + {}^i_0\boldsymbol{\omega}_i \times {}^i_{i-1}\mathbf{d}_i & \text{if joint } i \text{ is } P. \end{cases} \tag{12.21}
$$

*The angular velocity equation is a consequence of the relative velocity equation (7.67) and the rigid link's angular velocity based on DH parameters (8.2). The translational velocity equation also comes from rigid link velocity analysis (8.3).*

**Example 279** *Recursive translational acceleration equations for link (i).*
*The recursive translational acceleration equations for a link (i) are*

$$
{}^0\ddot{\mathbf{d}}_i = \begin{cases} {}^0\ddot{\mathbf{d}}_{i-1} + {}_0\dot{\boldsymbol{\omega}}_i \times {}^0_{i-1}\mathbf{d}_i \\ \quad + {}_0\boldsymbol{\omega}_i \times \left( {}_0\boldsymbol{\omega}_i \times {}^0_{i-1}\mathbf{d}_i \right) & \text{if joint } i \text{ is } R \\[6pt] {}^0\ddot{\mathbf{d}}_{i-1} + {}_0\dot{\boldsymbol{\omega}}_i \times {}^0_{i-1}\mathbf{d}_i \\ \quad + {}_0\boldsymbol{\omega}_i \times \left( {}_0\boldsymbol{\omega}_i \times {}^0_{i-1}\mathbf{d}_i \right) \\ \quad + \ddot{d}_i\, {}^0\hat{k}_{i-1} + 2\, {}_0\boldsymbol{\omega}_i \times \dot{d}_i\, {}^0\hat{k}_{i-1} & \text{if joint } i \text{ is } P. \end{cases} \tag{12.22}
$$

*where*

$$
{}^0_{i-1}\mathbf{d}_i = {}^0\mathbf{d}_i - {}^0\mathbf{d}_{i-1}. \tag{12.23}
$$

*We may transform this equation to the local frame $B_i$*

$$
\begin{aligned}
{}^i_0\ddot{\mathbf{d}}_i &= {}^0T_i^{-1}\,{}^0\ddot{\mathbf{d}}_i \\
&= {}^i\ddot{\mathbf{d}}_{i-1} + {}^i_0\dot{\boldsymbol{\omega}}_i \times {}^i_{i-1}\mathbf{d}_i + {}^i_0\boldsymbol{\omega}_i \times \left({}^i_0\boldsymbol{\omega}_i \times {}^i_{i-1}\mathbf{d}_i\right) \\
&\quad + \ddot{d}_i\,{}^i\hat{k}_{i-1} + 2\,{}^i_0\boldsymbol{\omega}_i \times \dot{d}_i\,{}^i\hat{k}_{i-1} \\
&= {}^iT_{i-1}\left({}^{i-1}_0\ddot{\mathbf{d}}_{i-1} + \ddot{d}_i\,{}^{i-1}\hat{k}_{i-1}\right) + 2\,{}^i_0\boldsymbol{\omega}_i \times \dot{d}_i\,{}^iT_{i-1}\,{}^{i-1}\hat{k}_{i-1} \\
&\quad + {}^i_0\boldsymbol{\omega}_i \times \left({}^i_0\boldsymbol{\omega}_i \times {}^i_{i-1}\mathbf{d}_i\right) + {}^i_0\dot{\boldsymbol{\omega}}_i \times {}^i_{i-1}\mathbf{d}_i. \qquad (12.24)
\end{aligned}
$$

*Therefore, we may use a recursive equation to find the locally expressed translational acceleration of link $(i)$ by having the translational acceleration of its lower link $(i-1)$. In this equation, every vector is expressed in its own coordinate frame.*

$$
{}^i_0\ddot{\mathbf{d}}_i = \begin{cases}
\begin{aligned}
&{}^iT_{i-1}\,{}^{i-1}_0\ddot{\mathbf{d}}_{i-1} \\
&+ {}^i_0\boldsymbol{\omega}_i \times \left({}^i_0\boldsymbol{\omega}_i \times {}^i_{i-1}\mathbf{d}_i\right) \\
&+ {}^i_0\dot{\boldsymbol{\omega}}_i \times {}^i_{i-1}\mathbf{d}_i \qquad\qquad\text{if joint } i \text{ is } R
\end{aligned} \\[2em]
\begin{aligned}
&{}^iT_{i-1}\left({}^{i-1}_0\ddot{\mathbf{d}}_{i-1} + \ddot{d}_i\,{}^{i-1}\hat{k}_{i-1}\right) \\
&+ 2\,{}^i_0\boldsymbol{\omega}_i \times \dot{d}_i\,{}^iT_{i-1}\,{}^{i-1}\hat{k}_{i-1} \\
&+ {}^i_0\boldsymbol{\omega}_i \times \left({}^i_0\boldsymbol{\omega}_i \times {}^i_{i-1}\mathbf{d}_i\right) \\
&+ {}^i_0\dot{\boldsymbol{\omega}}_i \times {}^i_{i-1}\mathbf{d}_i \qquad\qquad\text{if joint } i \text{ is } P.
\end{aligned}
\end{cases} \qquad (12.25)
$$

*Equation (12.22) is the result of differentiating (12.19), and Equation (12.25) is the result of transforming (12.22) to the local frame $B_i$.*

## 12.2 Rigid Link Newton-Euler Dynamics

Figure 12.2 illustrates free body diagram of a link $(i)$. The force $\mathbf{F}_{i-1}$ and moment $\mathbf{M}_{i-1}$ are the resultant force and moment that link $(i-1)$ applies to link $(i)$ at joint $i$. Similarly, $\mathbf{F}_i$ and $\mathbf{M}_i$ are the resultant force and moment that link $(i)$ applies to link $(i+1)$ at joint $i+1$. We measure and show the force systems $(\mathbf{F}_{i-1}, \mathbf{M}_{i-1})$ and $(\mathbf{F}_i, \mathbf{M}_i)$ at the origin of the coordinate frames $B_{i-1}$ and $B_i$ respectively. The sum of the external loads acting on the link $(i)$ are shown by $\sum \mathbf{F}_{e_i}$ and $\sum \mathbf{M}_{e_i}$.

The *Newton-Euler equations of motion* for the link $(i)$ in the global coordinate frame are

$$
{}^0\mathbf{F}_{i-1} - {}^0\mathbf{F}_i + \sum {}^0\mathbf{F}_{e_i} = m_i\,{}^0\mathbf{a}_i \qquad (12.26)
$$

$$
\begin{aligned}
&{}^0\mathbf{M}_{i-1} - {}^0\mathbf{M}_i + \sum {}^0\mathbf{M}_{e_i} \\
&\quad + \left({}^0\mathbf{d}_{i-1} - {}^0\mathbf{r}_i\right) \times {}^0\mathbf{F}_{i-1} - \left({}^0\mathbf{d}_i - {}^0\mathbf{r}_i\right) \times {}^0\mathbf{F}_i = {}^0I_i\,{}_0\boldsymbol{\alpha}_i. \qquad (12.27)
\end{aligned}
$$

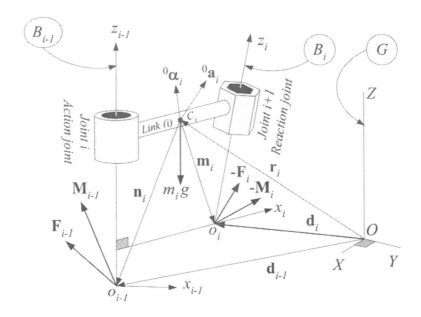

FIGURE 12.2. Force system on link $(i)$.

**Proof.** The force system at the distal end of a link $(i)$ is made of a force $\mathbf{F}_i$ and a moment $\mathbf{M}_i$ measured at the origin of $B_i$. The right subscript on $\mathbf{F}_i$ and $\mathbf{M}_i$ is a number indicating the number of coordinate frame $B_i$.

At joint $i+1$ there is always an *action force* $\mathbf{F}_i$, that link $(i)$ applies to link $(i+1)$, and a *reaction force* $-\mathbf{F}_i$, that the link $(i+1)$ applies to the link $(i)$. Therefore, on link $(i)$ there is always an action force $\mathbf{F}_{i-1}$ coming from link $(i-1)$, and a reaction force $-\mathbf{F}_i$ coming from link $(i+1)$. Action force is called *driving force*, and reaction force is called *driven force*.

Similarly, at joint $i+1$ there is always an *action moment* $\mathbf{M}_i$, that link $(i)$ applies to the link $(i+1)$, and a *reaction moment* $-\mathbf{M}_i$, that link $(i+1)$ applies to the link $(i)$. Hence, on link $(i)$ there is always an action moment $\mathbf{M}_{i-1}$ coming from link $(i-1)$, and a reaction moment $-\mathbf{M}_i$ coming from link $(i+1)$. The action moment is called the *driving moment*, and the reaction moment is called the *driven moment*.

Therefore, there is a driving force system $(\mathbf{F}_{i-1}$ , $\mathbf{M}_{i-1})$ at the origin of the coordinate frame $B_{i-1}$, and a driven force system $(\mathbf{F}_i$ , $\mathbf{M}_i)$ at the origin of the coordinate frame $B_i$. The driving force system $(\mathbf{F}_{i-1}$ , $\mathbf{M}_{i-1})$ gives motion to link $(i)$ and the driven force system $(\mathbf{F}_i$ , $\mathbf{M}_i)$ gives motion to link $(i+1)$.

In addition to the action and reaction force systems, there might be some external forces acting on the link $(i)$ that their resultant makes a force system $(\sum \mathbf{F}_{e_i}$ , $\sum \mathbf{M}_{e_i})$ at the mass center $C_i$. In robotic application, weight is usually the only external load on middle links, and reactions from the environment are extra external force systems on the base and end-effector links. The force and moment that the base actuator applies to the first link are $\mathbf{F}_0$ and $\mathbf{M}_0$, and the force and moment that the end-effector applies to the environment are $\mathbf{F}_n$ and $\mathbf{M}_n$. If weight is the only external load on link $(i)$ and it is in $-{}^0\hat{k}_0$ direction, then we have

$$\sum {}^0\mathbf{F}_{e_i} = m_i\,{}^0\mathbf{g} = -m_i\,g\,{}^0\hat{k}_0 \tag{12.28}$$

$$\sum {}^0\mathbf{M}_{e_i} = {}^0\mathbf{r}_i \times m_i\,{}^0\mathbf{g} = -{}^0\mathbf{r}_i \times m_i\,g\,{}^0\hat{k}_0 \tag{12.29}$$

where $\mathbf{g}$ is the gravitational acceleration vector.

As shown in Figure 12.2, we indicate the global position of the mass center of the link by ${}^0\mathbf{r}_i$, and the global position of the origin of body frames $B_i$ and $B_{i-1}$ by ${}^0\mathbf{d}_i$ and ${}^0\mathbf{d}_{i-1}$ respectively. The link's velocities ${}^0\mathbf{v}_i$, ${}_0\boldsymbol{\omega}_i$ and accelerations ${}^0\mathbf{a}_i$, ${}_0\boldsymbol{\alpha}_i$ are measured and shown at $C_i$. The physical properties of the link $(i)$ are specified by its moment of inertia ${}^0I_i$ about the link's mass center $C_i$, and its mass $m_i$.

The Newton's equation of motion determines that the sum of forces applied to the link $(i)$ is equal to the mass of the link times its acceleration at $C_i$.

$$ {}^0\mathbf{F}_{i-1} - {}^0\mathbf{F}_i + \sum {}^0\mathbf{F}_{e_i} = m_i\,{}^0\mathbf{a}_i \tag{12.30}$$

For the Euler equation, in addition to the action and reaction moments, we must add the moments of the action and reaction forces about $C_i$. The moment of $-\mathbf{F}_i$ and $\mathbf{F}_{i-1}$ are equal to $-\mathbf{m}_i \times \mathbf{F}_i$ and $\mathbf{n}_i \times \mathbf{F}_{i-1}$ where $\mathbf{m}_i$ is the position vector of $o_i$ from $C_i$ and $\mathbf{n}_i$ is the position vector of $o_{i-1}$ from $C_i$. Therefore, the link's Euler equation of motion is

$$\begin{aligned} {}^0\mathbf{M}_{i-1} - {}^0\mathbf{M}_i + \sum {}^0\mathbf{M}_{e_i} \\ + {}^0\mathbf{n}_i \times {}^0\mathbf{F}_{i-1} - {}^0\mathbf{m}_i \times {}^0\mathbf{F}_i = {}^0I_i\,{}_0\boldsymbol{\alpha}_i \end{aligned} \tag{12.31}$$

however, $\mathbf{n}_i$ and $\mathbf{m}_i$ can be expressed by

$$ {}^0\mathbf{n}_i = {}^0\mathbf{d}_{i-1} - {}^0\mathbf{r}_i \tag{12.32}$$

$$ {}^0\mathbf{m}_i = {}^0\mathbf{d}_i - {}^0\mathbf{r}_i \tag{12.33}$$

to derive Equation (12.27).

Since there is one translational and one rotational equation of motion for each link of a robot, there are $2n$ vectorial equations of motion for an $n$ link robot. However, there are $2(n+1)$ forces and moments involved. Therefore, one set of force systems (usually $\mathbf{F}_n$ and $\mathbf{M}_n$) must be specified to solve the equations and find the joints' force and moment. ∎

**Example 280** *One-link manipulator.*

*Figure 12.3 depicts a link attached to the ground via a joint at O. The free body diagram of the link is made of an external force and moment at*

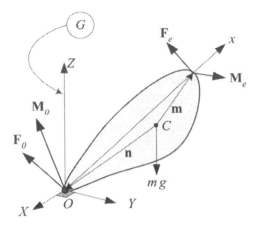

FIGURE 12.3. One link manipulator.

*the endpoint, gravity, and the driving force and moment at the joint. The Newton-Euler equations for the link are*

$$\mathbf{F}_0 + \mathbf{F}_e + mg\,\hat{K} \;=\; m\mathbf{a} \tag{12.34}$$
$$\mathbf{M}_0 + \mathbf{M}_e + \mathbf{n} \times \mathbf{F}_0 + \mathbf{m} \times \mathbf{F}_e \;=\; I\boldsymbol{\alpha}. \tag{12.35}$$

**Example 281** *A four-bar linkage dynamics.*

*Figure 12.4(a) illustrates a closed loop four-bar linkage along with the free body diagrams of the links, shown in Figure 12.4(b). The position of the mass centers are given, and therefore the vectors $^0\mathbf{n}_i$ and $^0\mathbf{m}_i$ for each link are also known. The Newton-Euler equations for the link (i) are*

$$^0\mathbf{F}_{i-1} - {}^0\mathbf{F}_i + m_i g\,\hat{J} \;=\; m_i\,{}^0\mathbf{a}_i \tag{12.36}$$
$$^0\mathbf{M}_{i-1} - {}^0\mathbf{M}_i + {}^0\mathbf{n}_i \times {}^0\mathbf{F}_{i-1} - {}^0\mathbf{m}_i \times \mathbf{F}_i \;=\; I_{i\,0}\boldsymbol{\alpha}_i \tag{12.37}$$

*and therefore, we have three sets of equations.*

$$^0\mathbf{F}_0 - {}^0\mathbf{F}_1 + m_1 g\,\hat{J} \;=\; m_1\,{}^0\mathbf{a}_1 \tag{12.38}$$
$$^0\mathbf{M}_0 - {}^0\mathbf{M}_1 + {}^0\mathbf{n}_1 \times {}^0\mathbf{F}_0 - {}^0\mathbf{m}_1 \times \mathbf{F}_1 \;=\; I_{1\,0}\boldsymbol{\alpha}_1 \tag{12.39}$$

$$^0\mathbf{F}_1 - {}^0\mathbf{F}_2 + m_2 g\,\hat{J} \;=\; m_2\,{}^0\mathbf{a}_2 \tag{12.40}$$
$$^0\mathbf{M}_1 - {}^0\mathbf{M}_2 + {}^0\mathbf{n}_2 \times {}^0\mathbf{F}_1 - {}^0\mathbf{m}_2 \times \mathbf{F}_2 \;=\; I_{2\,0}\boldsymbol{\alpha}_2 \tag{12.41}$$

$$^0\mathbf{F}_2 - {}^0\mathbf{F}_3 + m_3 g\,\hat{J} \;=\; m_2\,{}^0\mathbf{a}_2 \tag{12.42}$$
$$^0\mathbf{M}_2 - {}^0\mathbf{M}_3 + {}^0\mathbf{n}_3 \times {}^0\mathbf{F}_2 - {}^0\mathbf{m}_3 \times \mathbf{F}_3 \;=\; I_{3\,0}\boldsymbol{\alpha}_3 \tag{12.43}$$

*We assume that there is no friction in joints and the mechanism is planar. Therefore, the force vectors are in the XY plane, and the moments are*

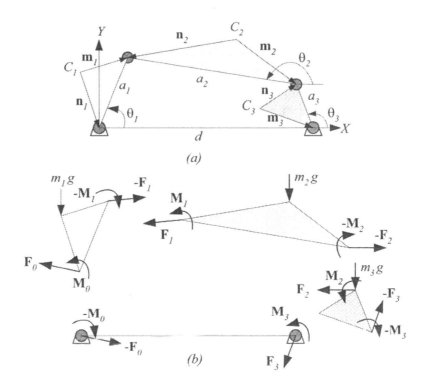

FIGURE 12.4. A four-bar linkage, and free body diagram of each link.

parallel to the Z-axis. In this condition, the equations of motion simplify to,

$$^0\mathbf{F}_0 - {}^0\mathbf{F}_1 + m_1 g\,\hat{J} = m_1\,{}^0\mathbf{a}_1 \qquad (12.44)$$

$$^0\mathbf{M}_0 + {}^0\mathbf{n}_1 \times {}^0\mathbf{F}_0 - {}^0\mathbf{m}_1 \times \mathbf{F}_1 = I_1\,_0\alpha_1 \qquad (12.45)$$

$$^0\mathbf{F}_1 - {}^0\mathbf{F}_2 + m_2 g\,\hat{J} = m_2\,{}^0\mathbf{a}_2 \qquad (12.46)$$

$$^0\mathbf{n}_2 \times {}^0\mathbf{F}_1 - {}^0\mathbf{m}_2 \times \mathbf{F}_2 = I_2\,_0\alpha_2 \qquad (12.47)$$

$$^0\mathbf{F}_2 - {}^0\mathbf{F}_3 + m_3 g\,\hat{J} = m_2\,{}^0\mathbf{a}_2 \qquad (12.48)$$

$$^0\mathbf{n}_3 \times {}^0\mathbf{F}_2 - {}^0\mathbf{m}_3 \times \mathbf{F}_3 = I_3\,_0\alpha_3 \qquad (12.49)$$

where $^0\mathbf{M}_0$ is the driving torque of the mechanism. The number of equations reduces to 9 and the unknowns of the mechanism are

$$F_{0x}, F_{0y}, F_{1x}, F_{1y}, F_{2x}, F_{2y}, F_{3x}, F_{3y}, M_0.$$

The set of equations can be arranged in a matrix form

$$[A]\,\mathbf{x} = \mathbf{b} \qquad (12.50)$$

*where*

$$[A] = \begin{bmatrix} 1 & 0 & -1 & 0 & 0 & 0 & 0 & 0 & 0 \\ 0 & 1 & 0 & -1 & 0 & 0 & 0 & 0 & 0 \\ -n_{1y} & n_{1x} & m_{1y} & -m_{1x} & 0 & 0 & 0 & 0 & 1 \\ 0 & 0 & 1 & 0 & -1 & 0 & 0 & 0 & 0 \\ 0 & 0 & 0 & 1 & 0 & -1 & 0 & 0 & 0 \\ 0 & 0 & -n_{2y} & n_{2x} & m_{2y} & -m_{2x} & 0 & 0 & 0 \\ 0 & 0 & 0 & 0 & 1 & 0 & -1 & 0 & 0 \\ 0 & 0 & 0 & 0 & 0 & 1 & 0 & -1 & 0 \\ 0 & 0 & 0 & 0 & -n_{3y} & n_{3x} & m_{3y} & -m_{3x} & 0 \end{bmatrix} \tag{12.51}$$

$$\mathbf{x} = \begin{bmatrix} F_{0x} \\ F_{0y} \\ F_{1x} \\ F_{1y} \\ F_{2x} \\ F_{2y} \\ F_{3x} \\ F_{3y} \\ M_0 \end{bmatrix} \tag{12.52}$$

$$\mathbf{b} = \begin{bmatrix} m_1 a_{1x} \\ m_1 a_{1y} - m_1 g \\ I_1 \alpha_1 \\ m_2 a_{2x} \\ m_2 a_{2y} - m_2 g \\ I_2 \alpha_2 \\ m_3 a_{3x} \\ m_3 a_{3y} - m_3 g \\ I_3 \alpha_3 \end{bmatrix} \tag{12.53}$$

The matrix $[A]$ describes the geometry of the mechanism, the vector $\mathbf{x}$ is the unknown forces, and the vector $\mathbf{b}$ indicates the dynamic terms. To solve the dynamics of the four-bar mechanism, we must calculate the accelerations $^0\mathbf{a}_i$ and $^0\boldsymbol{\alpha}_i$ and then find the required driving moment $^0\mathbf{M}_0$ and the joints' forces.

The force

$$\mathbf{F}_s = \mathbf{F}_3 - \mathbf{F}_0 \tag{12.54}$$

is called the **shaking force** and shows the reaction of the mechanism on the ground.

**Example 282** *2R planar manipulator Newton-Euler dynamics.*
A 2R planar manipulator and its free body diagram are shown in Figure 12.5. The torques of actuators are parallel to the Z-axis and are shown by

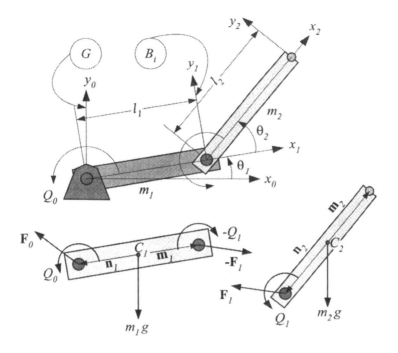

FIGURE 12.5. Free body diagram of a 2R palanar manipulator.

$Q_0$ and $Q_1$. The Newton-Euler equations of motion for the first link are

$$\mathbf{F}_0 - {}^0\mathbf{F}_1 + m_1 g\,\hat{\jmath} = m_1\,{}^0\mathbf{a}_1 \qquad (12.55)$$
$$\mathbf{Q}_0 - \mathbf{Q}_1 + {}^0\mathbf{n}_1 \times \mathbf{F}_0 - {}^0\mathbf{m}_1 \times {}^0\mathbf{F}_1 = {}^0I_1\,{}_0\boldsymbol{\alpha}_1 \qquad (12.56)$$

and the equations of motion for the second link are

$${}^0\mathbf{F}_1 + m_2 g\,\hat{\jmath} = m_2\,{}^0\mathbf{a}_2 \qquad (12.57)$$
$$\mathbf{Q}_1 + {}^0\mathbf{n}_2 \times {}^0\mathbf{F}_1 = {}^0I_2\,{}_0\boldsymbol{\alpha}_2. \qquad (12.58)$$

There are four equations for four unknowns $\mathbf{F}_0$, $\mathbf{F}_1$, $\mathbf{Q}_0$, and $\mathbf{Q}_1$. These equations can be set in a matrix form

$$[A]\,\mathbf{x} = \mathbf{b} \qquad (12.59)$$

where

$$[A] = \begin{bmatrix} 1 & 0 & 0 & -1 & 0 & 0 \\ 0 & 1 & 0 & 0 & -1 & 0 \\ n_{1y} & -n_{1x} & 1 & -m_{1y} & m_{1x} & -1 \\ 0 & 0 & 0 & 1 & 0 & 0 \\ 0 & 0 & 0 & 0 & 1 & 0 \\ 0 & 0 & 0 & n_{2y} & -n_{2x} & 1 \end{bmatrix} \qquad (12.60)$$

$$\mathbf{x} = \begin{bmatrix} F_{0x} \\ F_{0y} \\ Q_0 \\ F_{1x} \\ F_{1y} \\ Q_1 \end{bmatrix} \tag{12.61}$$

$$\mathbf{b} = \begin{bmatrix} m_1 a_{1x} \\ m_1 a_{1y} - m_1 g \\ {}^0 I_1 \alpha_1 \\ m_2 a_{2x} \\ m_2 a_{2y} - m_2 g \\ {}^0 I_2 \alpha_2 \end{bmatrix}. \tag{12.62}$$

**Example 283** *Equations for joint actuators.*

*In robot dynamics, we usually do not need to find the joint forces. Actuators' torque are much more important since we use them to control a robot. In Example 282 we found four equations*

$$
\begin{align}
{}^0\mathbf{F}_0 - {}^0\mathbf{F}_1 + m_1 g\,\hat{J} &= m_1\,{}^0\mathbf{a}_1 \tag{12.63} \\
{}^0\mathbf{Q}_0 - {}^0\mathbf{Q}_1 + {}^0\mathbf{n}_1 \times {}^0\mathbf{F}_0 - {}^0\mathbf{m}_1 \times {}^0\mathbf{F}_1 &= {}^0 I_1 {}_0\alpha_1 \tag{12.64} \\
{}^0\mathbf{F}_1 + m_2 g\,\hat{J} &= m_2\,{}^0\mathbf{a}_2 \tag{12.65} \\
{}^0\mathbf{Q}_1 + {}^0\mathbf{n}_2 \times {}^0\mathbf{F}_1 &= {}^0 I_2 {}_0\alpha_2. \tag{12.66}
\end{align}
$$

*for the joints' force system of a 2R manipulator. However, we may eliminate the joint forces $\mathbf{F}_0$, $\mathbf{F}_1$, and reduce the number of equations to two for the two torques $\mathbf{Q}_0$ and $\mathbf{Q}_1$. Eliminating $\mathbf{F}_1$ between (12.65) and (12.66) provides*

$$
{}^0\mathbf{Q}_1 = {}^0 I_2 {}_0\alpha_2 - {}^0\mathbf{n}_2 \times \left( m_2\,{}^0\mathbf{a}_2 - m_2 g\,\hat{J} \right) \tag{12.67}
$$

*and eliminating $\mathbf{F}_0$ and $\mathbf{F}_1$ between (12.63) and (12.66) gives*

$$
\begin{align}
{}^0\mathbf{Q}_0 = {}&^0\mathbf{Q}_1 + {}^0 I_1 {}_0\alpha_1 + {}^0\mathbf{m}_1 \times \left( m_2\,{}^0\mathbf{a}_2 - m_2 g\,\hat{J} \right) \\
&- {}^0\mathbf{n}_1 \times \left( m_1\,{}^0\mathbf{a}_1 - m_1 g\,\hat{J} + m_2\,{}^0\mathbf{a}_2 - m_2 g\,\hat{J} \right). \tag{12.68}
\end{align}
$$

*The forces $\mathbf{F}_0$ and $\mathbf{F}_1$, if we are interested, are equal to*

$$
\begin{align}
{}^0\mathbf{F}_1 &= m_2\,{}^0\mathbf{a}_2 - m_2 g\,\hat{J} \tag{12.69} \\
{}^0\mathbf{F}_0 &= m_1\,{}^0\mathbf{a}_1 + m_2\,{}^0\mathbf{a}_2 - (m_1 + m_2)\,g\,\hat{J}. \tag{12.70}
\end{align}
$$

*The position of $C_i$ and the links' angular velocity and acceleration are*

$$
\begin{align}
{}^0\mathbf{r}_1 &= -{}^0\mathbf{n}_1 \tag{12.71} \\
{}^0\mathbf{r}_2 &= -{}^0\mathbf{n}_1 + {}^0\mathbf{m}_1 - {}^0\mathbf{n}_2 \tag{12.72} \\
{}^0\mathbf{d}_1 &= -{}^0\mathbf{n}_1 + {}^0\mathbf{m}_1 \tag{12.73} \\
{}^0_1\mathbf{d}_2 &= -{}^0\mathbf{n}_2 + {}^0\mathbf{m}_2 \tag{12.74} \\
{}^0\mathbf{d}_2 &= -{}^0\mathbf{n}_1 + {}^0\mathbf{m}_1 - {}^0\mathbf{n}_2 + {}^0\mathbf{m}_2 \tag{12.75}
\end{align}
$$

$$_0\omega_1 = \dot{\theta}_1\,\hat{K} \tag{12.76}$$

$$_0\alpha_1 = {}_0\dot{\omega}_1 = \ddot{\theta}_1\,\hat{K} \tag{12.77}$$

$$_0\omega_2 = \left(\dot{\theta}_1 + \dot{\theta}_2\right)\hat{K} \tag{12.78}$$

$$_0\alpha_2 = {}_0\dot{\omega}_2 = \left(\ddot{\theta}_1 + \ddot{\theta}_2\right)\hat{K} \tag{12.79}$$

*and the local position vectors of $C_i$ are*

$$
\begin{aligned}
{}^0\mathbf{n}_1 &= {}^0R_1\,{}^1\mathbf{n}_1 \\
&= R_{Z,\theta_1}\,{}^1\mathbf{n}_1 \\
&= R_{Z,\theta_1}\,\frac{-l_1}{2}\,{}^1\hat{\imath}_1 \\
&= \begin{bmatrix} -\frac{1}{2}l_1\cos\theta_1 \\ -\frac{1}{2}l_1\sin\theta_1 \\ 0 \end{bmatrix}
\end{aligned} \tag{12.80}
$$

$$
\begin{aligned}
{}^0\mathbf{n}_2 &= {}^0R_2\,{}^2\mathbf{n}_2 \\
&= R_{Z,\theta_1}\,R_{Z,\theta_2}\,{}^2\mathbf{n}_2 \\
&= R_{Z,\theta_1}\,R_{Z,\theta_2}\,\frac{-l_2}{2}\,{}^2\hat{\imath}_2 \\
&= \begin{bmatrix} -\frac{1}{2}l_2\cos(\theta_1+\theta_2) \\ -\frac{1}{2}l_2\sin(\theta_1+\theta_2) \\ 0 \end{bmatrix}
\end{aligned} \tag{12.81}
$$

$$
\begin{aligned}
{}^0\mathbf{m}_1 &= {}^0R_1\,{}^1\mathbf{m}_1 \\
&= R_{Z,\theta_1}\,{}^1\mathbf{m}_1 \\
&= R_{Z,\theta_1}\,\frac{l_1}{2}\,{}^1\hat{\imath}_1 \\
&= \begin{bmatrix} \frac{1}{2}l_1\cos\theta_1 \\ \frac{1}{2}l_1\sin\theta_1 \\ 0 \end{bmatrix}
\end{aligned} \tag{12.82}
$$

$$
\begin{aligned}
{}^0\mathbf{m}_2 &= {}^0R_2\,{}^2\mathbf{m}_2 \\
&= R_{Z,\theta_1}\,R_{Z,\theta_2}\,\frac{l_2}{2}\,{}^2\hat{\imath}_2 \\
&= \begin{bmatrix} \frac{1}{2}l_2\cos(\theta_1+\theta_2) \\ \frac{1}{2}l_2\sin(\theta_1+\theta_2) \\ 0 \end{bmatrix}.
\end{aligned} \tag{12.83}
$$

*The translational acceleration of $C_1$ is*

$$
\begin{aligned}
{}^0\mathbf{a}_1 &= -{}_0\boldsymbol{\alpha}_1 \times {}^0\mathbf{m}_1 - {}_0\boldsymbol{\omega}_1 \times \left({}_0\boldsymbol{\omega}_1 \times {}^0\mathbf{m}_1\right) + {}^0\ddot{\mathbf{d}}_1 \\
&= {}_0\boldsymbol{\alpha}_1 \times {}^0\mathbf{m}_1 + {}_0\boldsymbol{\omega}_1 \times \left({}_0\boldsymbol{\omega}_1 \times {}^0\mathbf{m}_1\right) \\
&= \begin{bmatrix} -\frac{1}{2}l_1\ddot{\theta}_1 \sin\theta_1 - \frac{1}{2}l_1\dot{\theta}_1^2 \cos\theta_1 \\ \frac{1}{2}l_1\ddot{\theta}_1 \cos\theta_1 - \frac{1}{2}l_1\dot{\theta}_1^2 \sin\theta_1 \\ 0 \end{bmatrix}
\end{aligned}
\tag{12.84}
$$

*because*

$$
{}^0\ddot{\mathbf{d}}_1 = 2\,{}^0\mathbf{a}_1.
\tag{12.85}
$$

*The translational acceleration of $C_2$ is*

$$
\begin{aligned}
{}^0\mathbf{a}_2 &= {}_0\boldsymbol{\alpha}_2 \times \left({}^0\mathbf{r}_2 - {}^0\mathbf{d}_2\right) \\
&\quad + {}_0\boldsymbol{\omega}_2 \times \left({}_0\boldsymbol{\omega}_2 \times \left({}^0\mathbf{r}_2 - {}^0\mathbf{d}_2\right)\right) + {}^0\ddot{\mathbf{d}}_2 \\
&= -{}_0\boldsymbol{\alpha}_2 \times {}^0\mathbf{m}_2 - {}_0\boldsymbol{\omega}_2 \times \left({}_0\boldsymbol{\omega}_2 \times {}^0\mathbf{m}_2\right) + {}^0\ddot{\mathbf{d}}_2 \\
&= \begin{bmatrix} {}^0a_{2x} \\ {}^0a_{2y} \\ 0 \end{bmatrix}
\end{aligned}
\tag{12.86}
$$

$$
\begin{aligned}
{}^0a_{2x} &= -l_1\ddot{\theta}_1 \sin\theta_1 - \frac{1}{2}l_2\left(\ddot{\theta}_1 + \ddot{\theta}_2\right)\sin\left(\theta_1 + \theta_2\right) \\
&\quad - l_1\dot{\theta}_1^2 \cos\theta_1 - \frac{1}{2}l_2\left(\ddot{\theta}_1 + \ddot{\theta}_2\right)^2 \cos\left(\theta_1 + \theta_2\right)
\end{aligned}
\tag{12.87}
$$

$$
\begin{aligned}
{}^0a_{2y} &= l_1\ddot{\theta}_1 \cos\theta_1 + \frac{1}{2}l_2\left(\ddot{\theta}_1 + \ddot{\theta}_2\right)\cos\left(\theta_1 + \theta_2\right) \\
&\quad - l_1\dot{\theta}_1^2 \sin\theta_1 + \frac{1}{2}l_2\left(\ddot{\theta}_1 + \ddot{\theta}_2\right)^2 \sin\left(\theta_1 + \theta_2\right)
\end{aligned}
\tag{12.88}
$$

*because*

$$
\begin{aligned}
{}^0\ddot{\mathbf{d}}_2 &= {}^0\ddot{\mathbf{d}}_1 + {}_0\dot{\boldsymbol{\omega}}_2 \times {}^0_1\mathbf{d}_2 + {}_0\boldsymbol{\omega}_2 \times \left({}_0\boldsymbol{\omega}_2 \times {}^0_1\mathbf{d}_2\right) \\
&= \begin{bmatrix} {}^0\ddot{d}_{2x} \\ {}^0\ddot{d}_{2y} \\ 0 \end{bmatrix}
\end{aligned}
\tag{12.89}
$$

$$
\begin{aligned}
{}^0\ddot{d}_{2x} &= -l_1\ddot{\theta}_1 \sin\theta_1 - l_2\alpha_2 \sin\left(\theta_1 + \theta_2\right) \\
&\quad - l_1\dot{\theta}_1^2 \cos\theta_1 - l_2\omega_2^2 \cos\left(\theta_1 + \theta_2\right)
\end{aligned}
\tag{12.90}
$$

$$
\begin{aligned}
{}^0\ddot{d}_{2y} &= l_1\ddot{\theta}_1 \cos\theta_1 + l_2\alpha_2 \cos\left(\theta_1 + \theta_2\right) \\
&\quad - l_1\dot{\theta}_1^2 \sin\theta_1 - l_2\omega_2^2 \sin\left(\theta_1 + \theta_2\right)
\end{aligned}
\tag{12.91}
$$

*and the moment of inertia matrices in the global coordinate frame are*

$$
\begin{aligned}
^0I_1 &= R_{Z,\theta_1}\,^1I_1\,R_{Z,\theta_1}^T \\
&= {}^0R_1 \begin{bmatrix} 0 & 0 & 0 \\ 0 & I_1 & 0 \\ 0 & 0 & I_1 \end{bmatrix} {}^0R_1^T \\
&\quad \begin{bmatrix} I_1 \sin^2\theta_1 & -I_1\cos\theta_1\sin\theta_1 & 0 \\ -I_1\cos\theta_1\sin\theta_1 & I_1\cos^2\theta_1 & 0 \\ 0 & 0 & I_1 \end{bmatrix}
\end{aligned}
\tag{12.92}
$$

$$
\begin{aligned}
^0I_2 &= {}^0R_2\,^2I_2\,{}^0R_2^T \\
&= {}^0R_2 \begin{bmatrix} 0 & 0 & 0 \\ 0 & I_2 & 0 \\ 0 & 0 & I_2 \end{bmatrix} {}^0R_2^T \\
&= \begin{bmatrix} I_2\sin^2\theta_{12} & -I_2\cos\theta_{12}\sin\theta_{12} & 0 \\ -I_2\cos\theta_{12}\sin\theta_{12} & I_2\cos^2\theta_{12} & 0 \\ 0 & 0 & I_2 \end{bmatrix}.
\end{aligned}
\tag{12.93}
$$

*Substituting these results in Equations (12.67) and (12.68), and solving for $Q_0$ and $Q_1$, provides the dynamic equations for the 2R manipulator in an appropriate form.*

$$
^0\mathbf{Q}_1 = \begin{bmatrix} 0 \\ 0 \\ ^0Q_{1z} \end{bmatrix}
\tag{12.94}
$$

$$
\begin{aligned}
^0Q_{1z} &= I_2\left(\ddot\theta_1 + \ddot\theta_2\right) + \frac{1}{4}l_2^2 m_2\left(\ddot\theta_1 + \ddot\theta_2\right) + \frac{1}{2}l_1 l_2 m_2 \ddot\theta_1 \cos\theta_2 \\
&\quad + \frac{1}{2}l_1 l_2 m_2 \dot\theta_1^2 \sin\theta_2 - \frac{1}{2}g l_2 m_2 \cos\left(\theta_1 + \theta_2\right)
\end{aligned}
\tag{12.95}
$$

$$
^0\mathbf{Q}_0 = {}^0\mathbf{Q}_1 + \begin{bmatrix} 0 \\ 0 \\ ^0Q_{0z} \end{bmatrix}
\tag{12.96}
$$

$$
\begin{aligned}
^0Q_{0z} &= \left(I_1 + \frac{1}{4}l_1^2 m_1 + l_1^2 m_2\right)\ddot\theta_1 + \frac{1}{2}l_1 l_2 m_2\left(\ddot\theta_1 + \ddot\theta_2\right)\cos\theta_2 \\
&\quad - \frac{1}{2}l_1 l_2 m_2\left(\dot\theta_1 + \dot\theta_2\right)^2 \sin\theta_2 \\
&\quad - \left(\frac{1}{2}m_1 + m_2\right)g l_1 \cos\theta_1.
\end{aligned}
\tag{12.97}
$$

*We may also find the joint forces.*

$$
\begin{aligned}
^0\mathbf{F}_1 &= m_2\,^0\mathbf{a}_2 - m_2 g\,\hat{J} \\
&= \begin{bmatrix} ^0F_{1x} \\ ^0F_{1y} \\ 0 \end{bmatrix}
\end{aligned}
\tag{12.98}
$$

$$^0F_{1x} = m_2 {}^0a_{2x} \tag{12.99}$$
$$^0F_{1y} = m_2 {}^0a_{2y} - m_2 g \tag{12.100}$$

$$^0\mathbf{F}_0 = m_1 {}^0\mathbf{a}_1 + m_2 {}^0\mathbf{a}_2 - (m_1 + m_2) g \, \hat{J}$$
$$= \begin{bmatrix} ^0F_{0x} \\ ^0F_{0y} \\ 0 \end{bmatrix} \tag{12.101}$$

$$^0F_{0x} = -\frac{1}{2} l_2 m_2 \left( \ddot{\theta}_1 + \ddot{\theta}_2 \right) \sin (\theta_1 + \theta_2) - \left( \frac{1}{2} m_1 + m_2 \right) l_1 \dot{\theta}_1^2 \cos \theta_1$$
$$- \frac{1}{2} l_2 m_2 \left( \dot{\theta}_1 + \dot{\theta}_2 \right)^2 \cos (\theta_1 + \theta_2)$$
$$- \left( \frac{1}{2} m_1 + m_2 \right) l_1 \ddot{\theta}_1 \sin \theta_1 \tag{12.102}$$

$$^0F_{0y} = -\frac{1}{2} l_2 m_2 \left( \ddot{\theta}_1 + \ddot{\theta}_2 \right) \cos (\theta_1 + \theta_2) - \left( \frac{1}{2} m_1 + m_2 \right) l_1 \dot{\theta}_1^2 \sin \theta_1$$
$$- \frac{1}{2} l_2 m_2 \left( \dot{\theta}_1 + \dot{\theta}_2 \right)^2 \sin (\theta_1 + \theta_2)$$
$$+ \left( \frac{1}{2} m_1 + m_2 \right) l_1 \ddot{\theta}_1 \cos \theta_1 - (m_1 + m_2) g \tag{12.103}$$

## 12.3   Recursive Newton-Euler Dynamics

An advantage of the Newton-Euler equations of motion in robotic application is that we can calculate the joint forces one link at a time. Therefore, starting from the end-effector link, we can analyze the links one by one and end up at the base link. For such an analysis, we need to reform the Newton-Euler equations of motion to work in the interested link's frame.

The *backward Newton-Euler equations of motion* for the link $(i)$ in the the local coordinate frame $B_i$ are

$$^i\mathbf{F}_{i-1} = {}^i\mathbf{F}_i - \sum {}^i\mathbf{F}_{e_i} + m_i \, {}^i_0\mathbf{a}_i \tag{12.104}$$

$$^i\mathbf{M}_{i-1} = {}^i\mathbf{M}_i - \sum {}^i\mathbf{M}_{e_i} - \left( {}^i\mathbf{d}_{i-1} - {}^i\mathbf{r}_i \right) \times {}^i\mathbf{F}_{i-1}$$
$$+ \left( {}^i\mathbf{d}_i - {}^i\mathbf{r}_i \right) \times {}^i\mathbf{F}_i + {}^iI_i \, {}^i_0\boldsymbol{\alpha}_i + {}^i_0\boldsymbol{\omega}_i \times {}^iI_i \, {}^i_0\boldsymbol{\omega}_i \tag{12.105}$$

$$^i\mathbf{n}_i = {}^i\mathbf{d}_{i-1} - {}^i\mathbf{r}_i \tag{12.106}$$
$$^i\mathbf{m}_i = {}^i\mathbf{d}_i - {}^i\mathbf{r}_i. \tag{12.107}$$

When the driving force system $(^i\mathbf{F}_{i-1}, {}^i\mathbf{M}_{i-1})$ is found in frame $B_i$, we can transform them to the frame $B_{i-1}$ and apply the Newton-Euler equation for link $(i-1)$.

$$
\begin{aligned}
{}^{i-1}\mathbf{F}_{i-1} &= {}^{i-1}T_i\, {}^i\mathbf{F}_{i-1} & (12.108) \\
{}^{i-1}\mathbf{M}_{i-1} &= {}^{i-1}T_i\, {}^i\mathbf{M}_{i-1} & (12.109)
\end{aligned}
$$

The negative of the converted force system acts as the driven force system $(-\,{}^{i-1}\mathbf{F}_{i-1}, -\,{}^{i-1}\mathbf{M}_{i-1})$ for the link $(i-1)$.

**Proof.** The Euler equation for a rigid link in body coordinate frame is

$$
\begin{aligned}
{}^B\mathbf{M} &= \frac{{}^Gd}{dt}\, {}^B\mathbf{L} \\
&= {}^B\dot{\mathbf{L}} + {}^B_G\boldsymbol{\omega}_B \times {}^B\mathbf{L} & (12.110) \\
&= {}^iI_i\, {}_i\boldsymbol{\alpha}_i + {}^B_G\boldsymbol{\omega}_B \times {}^iI_i\, {}_i\boldsymbol{\omega}_i & (12.111)
\end{aligned}
$$

where $\mathbf{L}$ is the angular momentum of the link

$$
{}^B\mathbf{L} = {}^BI\, {}^B_G\boldsymbol{\omega}_B. \qquad (12.112)
$$

We may solve the Newton-Euler equations of motion (12.26) and (12.27) for the action force system

$$
\begin{aligned}
{}^0\mathbf{F}_{i-1} &= {}^0\mathbf{F}_i - \sum {}^0\mathbf{F}_{e_i} + m_i\, {}^0\mathbf{a}_i & (12.113) \\
{}^0\mathbf{M}_{i-1} &= {}^0\mathbf{M}_i - \sum {}^0\mathbf{M}_{e_i} - \left({}^0\mathbf{d}_{i-1} - {}^0\mathbf{r}_i\right) \times {}^0\mathbf{F}_{i-1} \\
&\quad + \left({}^0\mathbf{d}_i - {}^0\mathbf{r}_i\right) \times {}^0\mathbf{F}_i + \frac{{}^0d}{dt}\, {}^0\mathbf{L}_i & (12.114)
\end{aligned}
$$

and then, transform the equations to the coordinate frame $B_i$ attached to the link's $(i)$ to make the recursive form of the Newton-Euler equations of motion.

$$
\begin{aligned}
{}^i\mathbf{F}_{i-1} &= {}^0T_i^{-1}\, {}^0\mathbf{F}_{i-1} \\
&= {}^i\mathbf{F}_i - \sum {}^i\mathbf{F}_{e_i} + m_i\, {}^i_0\mathbf{a}_i & (12.115)
\end{aligned}
$$

$$
\begin{aligned}
{}^i\mathbf{M}_{i-1} &= {}^0T_i^{-1}\, {}^0\mathbf{M}_{i-1} \\
&= {}^i\mathbf{M}_i - \sum {}^i\mathbf{M}_{e_i} - \left({}^i\mathbf{d}_{i-1} - {}^i\mathbf{r}_i\right) \times {}^i\mathbf{F}_{i-1} \\
&\quad + \left({}^i\mathbf{d}_i - {}^i\mathbf{r}_i\right) \times {}^i\mathbf{F}_i + \frac{{}^0d}{dt}\, {}^i\mathbf{L}_i \\
&= {}^i\mathbf{M}_i - \sum {}^i\mathbf{M}_{e_i} - \left({}^i\mathbf{d}_{i-1} - {}^i\mathbf{r}_i\right) \times {}^i\mathbf{F}_{i-1} \\
&\quad + \left({}^i\mathbf{d}_i - {}^i\mathbf{r}_i\right) \times {}^i\mathbf{F}_i + {}^iI_i\, {}^i_0\boldsymbol{\alpha}_i + {}^i_0\boldsymbol{\omega}_i \times {}^iI_i\, {}^i_0\boldsymbol{\omega}_i. \quad (12.116)
\end{aligned}
$$

∎

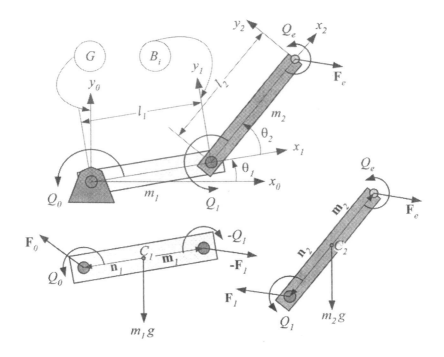

FIGURE 12.6. A 2R planar manipulator carrying a load at the endpoint.

**Example 284** *Recursive dynamics of a 2R planar manipulator.*

*Consider the 2R planar manipulator shown in Figure 12.6. The manipulator is carrying a force system at the endpoint. We use this manipulator to show how we can, step by step, develop the dynamic equations for a robot.*

*The backward recursive Newton-Euler equations of motion for the first link are*

$$
\begin{aligned}
^1\mathbf{F}_0 &= {}^1\mathbf{F}_1 - \sum {}^1\mathbf{F}_{e_1} + m_1 \, {}^1_0\mathbf{a}_1 \\
&= {}^1\mathbf{F}_1 - m_1 \, {}^1\mathbf{g} + m_1 \, {}^1_0\mathbf{a}_1
\end{aligned}
\tag{12.117}
$$

$$
\begin{aligned}
^1\mathbf{M}_0 &= {}^1\mathbf{M}_1 - \sum {}^1\mathbf{M}_{e_1} - \left({}^1\mathbf{d}_0 - {}^1\mathbf{r}_1\right) \times {}^1\mathbf{F}_0 \\
&\quad + \left({}^1\mathbf{d}_1 - {}^1\mathbf{r}_1\right) \times {}^1\mathbf{F}_1 + {}^1I_1 \, {}^1_0\boldsymbol{\alpha}_1 + {}^1_0\boldsymbol{\omega}_1 \times {}^1I_1 \, {}^1_0\boldsymbol{\omega}_1 \\
&= {}^1\mathbf{M}_1 - {}^1\mathbf{n}_1 \times {}^1\mathbf{F}_0 + {}^1\mathbf{m}_1 \times {}^1\mathbf{F}_1 \\
&\quad + {}^1I_1 \, {}^1_0\boldsymbol{\alpha}_1 + {}^1_0\boldsymbol{\omega}_1 \times {}^1I_1 \, {}^1_0\boldsymbol{\omega}_1
\end{aligned}
\tag{12.118}
$$

*and the backward recursive equations of motion for the second link are*

$$
\begin{aligned}
^2\mathbf{F}_1 &= {}^2\mathbf{F}_2 - \sum {}^2\mathbf{F}_{e_2} + m_2 \, {}^2_0\mathbf{a}_2 \\
&= -m_2 \, {}^2\mathbf{g} - {}^2\mathbf{F}_e + m_2 \, {}^2_0\mathbf{a}_2
\end{aligned}
\tag{12.119}
$$

$$
{}^{2}\mathbf{M}_{1} = {}^{2}\mathbf{M}_{2} - \sum {}^{2}\mathbf{M}_{e_{2}} - \left({}^{2}\mathbf{d}_{1} - {}^{2}\mathbf{r}_{2}\right) \times {}^{2}\mathbf{F}_{1}
$$
$$
+ \left({}^{2}\mathbf{d}_{2} - {}^{2}\mathbf{r}_{2}\right) \times {}^{2}\mathbf{F}_{2} + {}^{2}I_{2} {}_{0}^{2}\boldsymbol{\alpha}_{2} + {}_{0}^{2}\boldsymbol{\omega}_{2} \times {}^{2}I_{2} {}_{0}^{2}\boldsymbol{\omega}_{2}
$$
$$
{}^{2}\mathbf{M}_{1} = - {}^{2}\mathbf{M}_{e} - {}^{2}\mathbf{m}_{2} \times {}^{2}\mathbf{F}_{e} - {}^{2}\mathbf{n}_{2} \times {}^{2}\mathbf{F}_{1}
$$
$$
+ {}^{2}I_{2} {}_{0}^{2}\boldsymbol{\alpha}_{2} + {}_{0}^{2}\boldsymbol{\omega}_{2} \times {}^{2}I_{2} {}_{0}^{2}\boldsymbol{\omega}_{2} \tag{12.120}
$$

*The manipulator consists of two $R\|R(0)$ links, therefore their transformation matrices ${}^{i-1}T_{i}$ are of class (5.32). Substituting $d_{i} = 0$ and $a_{i} = l_{i}$, produces the following transformation matrices*

$$
{}^{0}T_{1} = \begin{bmatrix} \cos\theta_{1} & -\sin\theta_{1} & 0 & l_{1}\cos\theta_{1} \\ \sin\theta_{1} & \cos\theta_{1} & 0 & l_{1}\sin\theta_{1} \\ 0 & 0 & 1 & 0 \\ 0 & 0 & 0 & 1 \end{bmatrix} \tag{12.121}
$$

$$
{}^{1}T_{2} = \begin{bmatrix} \cos\theta_{2} & -\sin\theta_{2} & 0 & l_{2}\cos\theta_{2} \\ \sin\theta_{2} & \cos\theta_{2} & 0 & l_{2}\sin\theta_{2} \\ 0 & 0 & 1 & 0 \\ 0 & 0 & 0 & 1 \end{bmatrix}. \tag{12.122}
$$

*The homogeneous moments of inertia matrices are*

$$
{}^{1}I_{1} = \frac{m_{1}l_{1}^{2}}{12} \begin{bmatrix} 0 & 0 & 0 & 0 \\ 0 & 1 & 0 & 0 \\ 0 & 0 & 1 & 0 \\ 0 & 0 & 0 & 0 \end{bmatrix} \qquad {}^{2}I_{2} = \frac{m_{2}l_{2}^{2}}{12} \begin{bmatrix} 0 & 0 & 0 & 0 \\ 0 & 1 & 0 & 0 \\ 0 & 0 & 1 & 0 \\ 0 & 0 & 0 & 0 \end{bmatrix}. \tag{12.123}
$$

*The homogeneous moment of inertia matrix is obtained by appending a zero row and column to the I matrix.*

*The position vectors involved are*

$$
{}^{1}\mathbf{n}_{1} = \begin{bmatrix} -l_{1}/2 \\ 0 \\ 0 \\ 0 \end{bmatrix} \qquad {}^{2}\mathbf{n}_{2} = \begin{bmatrix} -l_{2}/2 \\ 0 \\ 0 \\ 0 \end{bmatrix} \tag{12.124}
$$

$$
{}^{1}\mathbf{m}_{1} = \begin{bmatrix} l_{1}/2 \\ 0 \\ 0 \\ 0 \end{bmatrix} \qquad {}^{2}\mathbf{m}_{2} = \begin{bmatrix} l_{2}/2 \\ 0 \\ 0 \\ 0 \end{bmatrix} \tag{12.125}
$$

$$
{}^{1}\mathbf{r}_{1} = - {}^{1}\mathbf{n}_{1} \qquad {}^{2}\mathbf{r}_{2} = - {}^{2}\mathbf{n}_{1} + {}^{2}\mathbf{m}_{2} - {}^{2}\mathbf{n}_{2}. \tag{12.126}
$$

*The angular velocities and accelerations are*

$$
{}_{0}^{1}\boldsymbol{\omega}_{1} = \begin{bmatrix} 0 \\ 0 \\ \dot{\theta}_{1} \\ 0 \end{bmatrix} \qquad {}_{0}^{2}\boldsymbol{\omega}_{2} = \begin{bmatrix} 0 \\ 0 \\ \dot{\theta}_{1} + \dot{\theta}_{2} \\ 0 \end{bmatrix} \tag{12.127}
$$

$$\begin{array}{cc} {}^1_0\alpha_1 = \begin{bmatrix} 0 \\ 0 \\ \ddot{\theta}_1 \\ 0 \end{bmatrix} & {}^2_0\alpha_2 = \begin{bmatrix} 0 \\ 0 \\ \ddot{\theta}_1 + \ddot{\theta}_2 \\ 0 \end{bmatrix}. \end{array} \quad (12.128)$$

The translational acceleration of $C_1$ is

$$\begin{aligned} {}^1_0\mathbf{a}_1 &= {}^1_0\alpha_1 \times (-{}^1\mathbf{m}_1) + {}^1_0\omega_1 \times \left({}^1_0\omega_1 \times (-{}^1\mathbf{m}_1)\right) + {}^1_0\ddot{\mathbf{d}}_1 \\ &= \begin{bmatrix} -\frac{1}{2}l_1\dot{\theta}_1^2 \\ \frac{1}{2}l_1\ddot{\theta}_1 \\ 0 \\ 0 \end{bmatrix} \end{aligned} \quad (12.129)$$

because

$$ {}^1\ddot{\mathbf{d}}_1 = 2\,{}^1\mathbf{a}_1. $$

The translational acceleration of $C_2$ is

$$\begin{aligned} {}^2_0\mathbf{a}_2 &= {}^2_0\alpha_2 \times (-{}^2\mathbf{m}_2) + {}^2_0\omega_2 \times \left({}^2_0\omega_2 \times (-{}^2\mathbf{m}_2)\right) + {}^2_0\ddot{\mathbf{d}}_2 \\ &= \begin{bmatrix} -\frac{1}{2}l_2\left(\dot{\theta}_1 + \dot{\theta}_2\right)^2 \\ \frac{1}{2}l_2\left(\ddot{\theta}_1 + \ddot{\theta}_2\right) \\ 0 \\ 0 \end{bmatrix} \end{aligned} \quad (12.130)$$

because

$$ {}^2\ddot{\mathbf{d}}_2 = 2\,{}^2\mathbf{a}_2. \quad (12.131)$$

The gravitational acceleration vector in the links' frame is

$$\begin{aligned} {}^1\mathbf{g} &= {}^0T_1^{-1}\,{}^0\mathbf{g} \\ &= \begin{bmatrix} -g\sin\theta_2 \\ g\cos\theta_2 \\ 0 \\ 0 \end{bmatrix} \end{aligned} \quad (12.132)$$

$$\begin{aligned} {}^2\mathbf{g} &= {}^0T_2^{-1}\,{}^0\mathbf{g} \\ &= \begin{bmatrix} -g\sin(\theta_1 + \theta_2) \\ g\cos(\theta_1 + \theta_2) \\ 0 \\ 0 \end{bmatrix}. \end{aligned} \quad (12.133)$$

The external load is usually given in the global coordinate frame. We must transform them to the interested link's frame to apply the recursive equa-

*tions of motion. Therefore, the external force system expressed in $B_2$ is*

$$
\begin{aligned}
^2\mathbf{F}_e &= {}^0T_2^{-1}\,{}^0\mathbf{F}_e \\
&= \begin{bmatrix} F_{ex}\cos(\theta_1+\theta_2) + F_{ey}\sin(\theta_1+\theta_2) \\ F_{ey}\cos(\theta_1+\theta_2) - F_{ex}\sin(\theta_1+\theta_2) \\ 0 \\ 0 \end{bmatrix}
\end{aligned} \tag{12.134}
$$

$$
\begin{aligned}
^2\mathbf{M}_e &= {}^0T_2^{-1}\,{}^0\mathbf{M}_e \\
&= \begin{bmatrix} 0 \\ 0 \\ M_e \\ 0 \end{bmatrix}.
\end{aligned} \tag{12.135}
$$

*Now, we start from the final link and calculate its action force system. The backward Newton equation for link (2) is*

$$
\begin{aligned}
^2\mathbf{F}_1 &= -m_2\,{}^2\mathbf{g} - {}^2\mathbf{F}_e + m_2\,{}_0^2\mathbf{a}_2 \\
&= \begin{bmatrix} {}^2F_{1x} \\ {}^2F_{1y} \\ 0 \\ 0 \end{bmatrix}
\end{aligned} \tag{12.136}
$$

$$
\begin{aligned}
^2F_{1x} &= -\frac{1}{2}l_2 m_2\left(\dot\theta_1+\dot\theta_2\right)^2 - F_{ex}\cos(\theta_1+\theta_2) \\
&\quad - (F_{ey} - gm_2)\sin(\theta_1+\theta_2)
\end{aligned} \tag{12.137}
$$

$$
\begin{aligned}
^2F_{1y} &= \frac{1}{2}l_2 m_2\left(\ddot\theta_1+\ddot\theta_2\right) + F_{ex}\sin(\theta_1+\theta_2) \\
&\quad - (F_{ey} + gm_2)\cos(\theta_1+\theta_2)
\end{aligned} \tag{12.138}
$$

*and the backward Euler equation for link (2) is*

$$
\begin{aligned}
^2\mathbf{M}_1 &= -{}^2\mathbf{M}_e - {}^2\mathbf{m}_2 \times {}^2\mathbf{F}_e - {}^2\mathbf{n}_2 \times {}^2\mathbf{F}_1 \\
&\quad + {}^2I_2\,{}_0^2\boldsymbol{\alpha}_2 + {}_0^2\boldsymbol{\omega}_2 \times {}^2I_2\,{}_0^2\boldsymbol{\omega}_2 \\
&= \begin{bmatrix} 0 \\ 0 \\ {}^2M_{1z} \\ 0 \end{bmatrix}
\end{aligned} \tag{12.139}
$$

*where*

$$
\begin{aligned}
^2M_{1z} &= -M_e + l_2 F_{ex}\sin(\theta_1+\theta_2) - l_2 F_{ey}\cos(\theta_1+\theta_2) \\
&\quad + \frac{1}{3}l_2^2 m_2\left(\ddot\theta_1+\ddot\theta_2\right) - \frac{1}{2}gl_2 m_2\cos(\theta_1+\theta_2).
\end{aligned} \tag{12.140}
$$

*Finally the action force on link (1) is*

$$
\begin{aligned}
{}^{1}\mathbf{F}_0 &= {}^{1}\mathbf{F}_1 - m_1\,{}^{1}\mathbf{g} + m_1\,{}^{1}_{0}\mathbf{a}_1 \\
&= {}^{1}T_2\,{}^{2}\mathbf{F}_1 - m_1\,{}^{1}\mathbf{g} + m_1\,{}^{1}_{0}\mathbf{a}_1 \\
&= \begin{bmatrix} {}^{1}F_{0x} \\ {}^{1}F_{0y} \\ 0 \\ 0 \end{bmatrix}
\end{aligned}
\tag{12.141}
$$

*where*

$$
\begin{aligned}
{}^{1}F_{0x} &= -F_{ex}\cos\theta_1 - (F_{ey} - gm_1)\sin\theta_1 \\
&\quad - \frac{1}{2}l_2 m_2\left(\ddot{\theta}_1 + \ddot{\theta}_2\right)\sin\theta_2 - \frac{1}{2}l_2 m_2\left(\dot{\theta}_1 + \dot{\theta}_2\right)^2\cos\theta_2 \\
&\quad + gm_2\sin(2\theta_2 + \theta_1) - \frac{1}{2}l_1 m_1 \dot{\theta}_1^2
\end{aligned}
\tag{12.142}
$$

$$
\begin{aligned}
{}^{1}F_{0y} &= F_{ex}\sin\theta_1 - (F_{ey} + gm_1)\cos\theta_1 \\
&\quad + \frac{1}{2}l_2 m_2\left(\ddot{\theta}_1 + \ddot{\theta}_2\right)\cos\theta_2 - \frac{1}{2}l_2 m_2\left(\dot{\theta}_1 + \dot{\theta}_2\right)^2\sin\theta_2 \\
&\quad - gm_2\cos(2\theta_2 + \theta_1) + \frac{1}{2}l_1 m_1 \ddot{\theta}_1
\end{aligned}
\tag{12.143}
$$

*and the action moment on link (1) is*

$$
\begin{aligned}
{}^{1}\mathbf{M}_0 &= {}^{1}\mathbf{M}_1 - {}^{1}\mathbf{n}_1 \times {}^{1}\mathbf{F}_0 + {}^{1}\mathbf{m}_1 \times {}^{1}\mathbf{F}_1 \\
&\quad + {}^{1}I_1\,{}^{1}_{0}\boldsymbol{\alpha}_1 + {}^{1}_{0}\boldsymbol{\omega}_1 \times {}^{1}I_1\,{}^{1}_{0}\boldsymbol{\omega}_1 \\
&= {}^{1}T_2\,{}^{2}\mathbf{M}_1 - {}^{1}\mathbf{n}_1 \times {}^{1}\mathbf{F}_0 + {}^{1}\mathbf{m}_1 \times {}^{1}T_2\,{}^{2}\mathbf{F}_1 \\
&\quad + {}^{1}I_1\,{}^{1}_{0}\boldsymbol{\alpha}_1 + {}^{1}_{0}\boldsymbol{\omega}_1 \times {}^{1}I_1\,{}^{1}_{0}\boldsymbol{\omega}_1 \\
&= \begin{bmatrix} 0 \\ 0 \\ {}^{1}M_{0z} \\ 0 \end{bmatrix}
\end{aligned}
\tag{12.144}
$$

*where*

$$
\begin{aligned}
{}^{1}M_{0z} &= -M_e + \frac{1}{3}l_2^2 m_2\left(\ddot{\theta}_1 + \ddot{\theta}_2\right) + \frac{1}{3}l_1^2 m_1 \ddot{\theta}_1 \\
&\quad - \left(F_{ey}l_2 + \frac{1}{2}gl_2 m_2\right)\cos(\theta_1 + \theta_2) \\
&\quad - \frac{1}{2}l_1 m_1 g\cos\theta_1 + F_{ex}l_2\sin(\theta_1 + \theta_2).
\end{aligned}
\tag{12.145}
$$

**Example 285** *Actuator's force and torque.*

*Applying a backward recursive force analysis ends up with a set of known force systems at joints. Each joint is driven by a motor known as an actuator that applies a force in a P joint, or a torque in an R joint. When the joint i is prismatic, the force of the driving actuator is along the $z_{i-1}$-axis*

$$F_m = {}^0\hat{k}_{i-1}^T \, {}^0\mathbf{F}_i \qquad (12.146)$$

*showing that the $\hat{k}_{i-1}$ component of the joint force $\mathbf{F}_i$ is supported by the actuator. The $\hat{i}_{i-1}$ and $\hat{j}_{i-1}$ components of $\mathbf{F}_i$ must be supported by the bearings of the joint. Similarly, when the joint i is revolute, the torque of the driving actuator is along the $z_{i-1}$-axis*

$$M_m = {}^0\hat{k}_{i-1}^T \, {}^0\mathbf{M}_i \qquad (12.147)$$

*showing that the $\hat{k}_{i-1}$ component of the joint torque $\mathbf{M}_i$ is supported by the actuator. The $\hat{i}_{i-1}$ and $\hat{j}_{i-1}$ components of $\mathbf{M}_i$ must be supported by the bearings of the joint.*

**Example 286 ★** *Forward Newton-Euler equations of motion.*

*The Newton-Euler equations of motion (12.104) and (12.105) can be written in a forward recursive form in coordinate frame $B_i$ attached to the link (i).*

$$^i\mathbf{F}_i = {}^i\mathbf{F}_{i-1} + \sum {}^i\mathbf{F}_{e_i} - m_i \, {}^i_0\mathbf{a}_i \qquad (12.148)$$

$$
\begin{aligned}
^i\mathbf{M}_i &= {}^i\mathbf{M}_{i-1} + \sum {}^i\mathbf{M}_{e_i} + \left({}^i\mathbf{d}_{i-1} - {}^i\mathbf{r}_i\right) \times {}^i\mathbf{F}_{i-1} \\
&\quad - \left({}^i\mathbf{d}_i - {}^i\mathbf{r}_i\right) \times {}^i\mathbf{F}_i - {}^iI_i \, {}^i_0\boldsymbol{\alpha}_i - {}^i_0\boldsymbol{\omega}_i \times {}^iI_i \, {}^i_0\boldsymbol{\omega}_i.
\end{aligned} \qquad (12.149)
$$

$$^i\mathbf{n}_i = {}^i\mathbf{d}_{i-1} - {}^i\mathbf{r}_i \qquad (12.150)$$

$$^i\mathbf{m}_i = {}^i\mathbf{d}_i - {}^i\mathbf{r}_i \qquad (12.151)$$

*Using the forward Newton-Euler equations of motion (12.148) and (12.149), we can calculate the reaction force system $({}^i\mathbf{F}_i, {}^i\mathbf{M}_i)$ by having the action force system $({}^i\mathbf{F}_{i-1}, {}^i\mathbf{M}_{i-1})$. When the reaction force system $({}^i\mathbf{F}_i, {}^i\mathbf{M}_i)$ is found in frame $B_i$, we can transform them to the frame $B_{i+1}$*

$$^{i+1}\mathbf{F}_i = {}^iT_{i+1}^{-1} \, {}^i\mathbf{F}_i \qquad (12.152)$$

$$^{i+1}\mathbf{M}_i = {}^iT_{i+1}^{-1} \, {}^i\mathbf{M}_i. \qquad (12.153)$$

*The negative of the converted force system acts as the action force system $(-{}^{i+1}\mathbf{F}_i, -{}^{i+1}\mathbf{M}_i)$ for the link $(i+1)$ and we can apply the Newton-Euler equation to the link $(i+1)$.*

*The forward Newton-Euler equations of motion allows us to start from a known action force system $({}^1\mathbf{F}_0, {}^1\mathbf{M}_0)$, that the base link applies to the* *link (1), and calculate the action force of the next link. Therefore, analyzing the links of a robot, one by one, we end up with the force system that the end-effector applies to the environment.*

*Using the forward or backward recursive Newton-Euler equations of motion depends on the measurement and sensory system of the robot.*

## 12.4   Robot Lagrange Dynamics

The Lagrange equation of motion provides a systematic approach to obtain the dynamics equations for robots. The Lagrangean is defined as the difference between the kinetic and potential energies

$$\mathcal{L} = K - V. \tag{12.154}$$

The Lagrange equation of motion for a robotic system can be found by applying the Lagrange equation

$$\frac{d}{dt}\left(\frac{\partial \mathcal{L}}{\partial \dot{q}_i}\right) - \frac{\partial \mathcal{L}}{\partial q_i} = Q_i \qquad i = 1, 2, \cdots n \tag{12.155}$$

where $q_i$ is the coordinates by which the energies are expressed, and $Q_i$ is the corresponding generalized nonpotential force that drives $q_i$.

The equations of motion for an $n$ link serial manipulator can be set in a matrix form

$$\mathbf{D(q)}\,\ddot{\mathbf{q}} + \mathbf{H(q,\dot{q})} + \mathbf{G(q)} = \mathbf{Q} \tag{12.156}$$

or in a summation form

$$\sum_{j=1}^{n} D_{ij}(q)\,\ddot{q}_j + \sum_{k=1}^{n}\sum_{m=1}^{n} H_{ikm}\dot{q}_k\dot{q}_m + G_i = Q_i. \tag{12.157}$$

$\mathbf{D(q)}$ is an $n \times n$ inertial-type symmetric matrix

$$D = \sum_{i=1}^{n}\left(\mathbf{J}_{Di}^{T}\,m_i\,\mathbf{J}_{Di} + \frac{1}{2}\mathbf{J}_{Ri}^{T}\,{}^{0}I_i\,\mathbf{J}_{Ri}\right) \tag{12.158}$$

$H_{ikm}$ is the velocity coupling vector

$$H_{ijk} = \sum_{j=1}^{n}\sum_{k=1}^{n}\left(\frac{\partial D_{ij}}{\partial q_k} - \frac{1}{2}\frac{\partial D_{jk}}{\partial q_i}\right) \tag{12.159}$$

and $G_i$ is the gravitational vector

$$G_i = \sum_{j=1}^{n} m_j \mathbf{g}^{T}\,\mathbf{J}_{Dj}^{(i)}. \tag{12.160}$$

**Proof.** Kinetic energy of link $(i)$ may be written as

$$K_i = \frac{1}{2}\,{}^0\mathbf{v}_i^T\, m_i\,{}^0\mathbf{v}_i + \frac{1}{2}\,{}_0\omega_i^T\,{}^iI_i\,{}_0\omega_i \tag{12.161}$$

where $m_i$ is the mass of the link, ${}^iI_i$ is the moment of inertia matrix of the link in the link's frame $B_i$, ${}^0\mathbf{v}_i$ is the global velocity of the $CM$ of the link, and ${}_0\omega_i$ is the global angular velocity of the link.

The translational and angular velocity vectors can be expressed based on the joint coordinate velocities, utilizing the *Jacobian of the link* $\mathbf{J}_i$

$$\dot{\mathbf{X}}_i = \begin{bmatrix} {}^0\mathbf{v}_i \\ {}_0\omega_i \end{bmatrix} = \begin{bmatrix} \mathbf{J}_{Di} \\ \mathbf{J}_{Ri} \end{bmatrix}\dot{\mathbf{q}} = \mathbf{J}_i\,\dot{\mathbf{q}}. \tag{12.162}$$

The link's Jacobian $\mathbf{J}_i$ is a $6 \times n$ matrix that transforms the instantaneous joint coordinate velocities into the instantaneous link's $CM$ translational and angular velocities. The $j$th column of $\mathbf{J}_i$ is made of $\mathbf{c}_{Di}^{(j)}$ and $\mathbf{c}_{Ri}^{(j)}$, where for $j \le i$

$$\begin{aligned} \mathbf{c}_{Di}^{(j)} &= \hat{k}_{j-1} \times {}_{j-1}^{\;0}\mathbf{r}_i && \text{for a R joint} \\ &= \hat{k}_{j-1} && \text{for a P joint} \end{aligned} \tag{12.163}$$

and

$$\begin{aligned} \mathbf{c}_{Ri}^{(j)} &= \hat{k}_{j-1} && \text{for a R joint} \\ &= 0 && \text{for a P joint} \end{aligned} \tag{12.164}$$

and ${}_{j-1}^{\;0}\mathbf{r}_i$ is the position of $CM$ of the link $(i)$ in the coordinate frame $B_{j-1}$ expressed in the base frame. The columns of $\mathbf{J}_i$ are zero for $j > i$.

The kinetic energy of the whole robot is then

$$\begin{aligned} K &= \sum_{i=1}^{n} K_i \\ &= \frac{1}{2}\sum_{i=1}^{n}\left({}^0\mathbf{v}_i^T\, m_i\,{}^0\mathbf{v}_i + \frac{1}{2}\,{}_0\omega_i^T\,{}^0I_i\,{}_0\omega_i\right) \\ &= \frac{1}{2}\sum_{i=1}^{n}\left((\mathbf{J}_{Di}\,\dot{\mathbf{q}}_i)^T\, m_i\,(\mathbf{J}_{Di}\,\dot{\mathbf{q}}_i) + \frac{1}{2}\,(\mathbf{J}_{Ri}\,\dot{\mathbf{q}}_i)^T\,{}^0I_i\,(\mathbf{J}_{Ri}\,\dot{\mathbf{q}}_i)\right) \\ &= \frac{1}{2}\dot{\mathbf{q}}_i^T\left(\sum_{i=1}^{n}\left(\mathbf{J}_{Di}^T\, m_i\,\mathbf{J}_{Di} + \frac{1}{2}\,\mathbf{J}_{Ri}^T\,{}^0I_i\,\mathbf{J}_{Ri}\right)\right)\dot{\mathbf{q}}_i. \end{aligned} \tag{12.165}$$

where $I_i$ is the inertia matrix of the link $(i)$ about its $CM$ and expressed in the base frame.

$$^0I_i = {}^0R_i\,{}^iI_i\,{}^0R_i^T \tag{12.166}$$

The kinetic energy may be written in a more convenient form as

$$K = \frac{1}{2}\dot{\mathbf{q}}_i^T\, D\,\dot{\mathbf{q}}_i \tag{12.167}$$

where $D$ is an $n \times n$ matrix called the *manipulator inertia matrix*

$$D = \sum_{i=1}^{n} \left( \mathbf{J}_{Di}^T \, m_i \, \mathbf{J}_{Di} + \frac{1}{2} \mathbf{J}_{Ri}^T \, {}^0 I_i \, \mathbf{J}_{Ri} \right). \qquad (12.168)$$

The potential energy of the link $(i)$ is due to gravity

$$V_i = -m_i \, {}^0\mathbf{g} \cdot {}^0\mathbf{r}_i \qquad (12.169)$$

and therefore, the total potential energy of the manipulator is

$$V = \sum_{i=1}^{n} V_i = -\sum_{i=1}^{n} m_i \, {}^0\mathbf{g}^T \, {}^0\mathbf{r}_i. \qquad (12.170)$$

where ${}^0\mathbf{g}$ is the gravitational acceleration vector expressed in the base frame.

The Lagrangean of the manipulator is

$$
\begin{aligned}
\mathcal{L} &= K - V \\
&= \frac{1}{2}\dot{\mathbf{q}}_i^T \, D \, \dot{\mathbf{q}}_i + \sum_{i=1}^{n} m_i \, {}^0\mathbf{g}^T \, {}^0\mathbf{r}_i \\
&= \frac{1}{2}\sum_{i=1}^{n}\sum_{j=1}^{n} D_{ij} \, \dot{q}_i \dot{q}_j + \sum_{i=1}^{n} m_i \, {}^0\mathbf{g}^T \, {}^0\mathbf{r}_i. \qquad (12.171)
\end{aligned}
$$

Based on the Lagrangean $\mathcal{L}$ we can find

$$
\begin{aligned}
\frac{\partial \mathcal{L}}{\partial q_i} &= \frac{1}{2}\frac{\partial}{\partial q_i}\left( \sum_{j=1}^{n}\sum_{k=1}^{n} D_{jk} \, \dot{q}_j \dot{q}_k \right) + \sum_{j=1}^{n} m_j \, {}^0\mathbf{g}^T \frac{\partial \, {}^0\mathbf{r}_j}{\partial q_i} \\
&= \frac{1}{2}\sum_{j=1}^{n}\sum_{k=1}^{n} \frac{\partial D_{jk}}{\partial q_i} \, \dot{q}_j \dot{q}_k + \sum_{j=1}^{n} m_j \, {}^0\mathbf{g}^T \, \mathbf{J}_{Dj}^{(i)} \qquad (12.172)
\end{aligned}
$$

$$\frac{\partial \mathcal{L}}{\partial \dot{q}_i} = \sum_{j=1}^{n} D_{ij} \, \dot{q}_j \qquad (12.173)$$

and

$$
\begin{aligned}
\frac{d}{dt}\frac{\partial \mathcal{L}}{\partial \dot{q}_i} &= \sum_{j=1}^{n} D_{ij} \, \ddot{q}_j + \sum_{j=1}^{n} \frac{dD_{ij}}{dt} \, \dot{q}_j \\
&= \sum_{j=1}^{n} D_{ij} \, \ddot{q}_j + \sum_{j=1}^{n}\sum_{k=1}^{n} \frac{\partial D_{ij}}{\partial q_k} \, \dot{q}_k \, \dot{q}_j. \qquad (12.174)
\end{aligned}
$$

The generalized force of the Lagrange equations are

$$Q_i = Q_i + \mathbf{J}^T \, \mathbf{F}_e \qquad (12.175)$$

where $Q_i$ is the $i$th actuator force at joint $i$, and $\mathbf{F}_e = \begin{bmatrix} -\mathbf{F}_{en}^T & -\mathbf{M}_{en}^T \end{bmatrix}^T$ is the external force system applied on the end-effector.

Finally, the Lagrange equations of motion for an $n$-link manipulator are

$$\sum_{j=1}^{n} D_{ij}(q)\,\ddot{q}_j + H_{ikm}\dot{q}_k\dot{q}_m + G_i = Q_i \qquad (12.176)$$

where

$$H_{ijk} = \sum_{j=1}^{n}\sum_{k=1}^{n}\left(\frac{\partial D_{ij}}{\partial q_k} - \frac{1}{2}\frac{\partial D_{jk}}{\partial q_i}\right) \qquad (12.177)$$

$$G_i = \sum_{j=1}^{n} m_j\mathbf{g}^T \mathbf{J}_{Dj}^{(i)} \qquad (12.178)$$

The equations of motion for a manipulator may be shown in a more concise form to simplify matrix calculations.

$$\mathbf{D}(\mathbf{q})\,\ddot{\mathbf{q}} + \mathbf{H}(\mathbf{q},\dot{\mathbf{q}}) + \mathbf{G}(\mathbf{q}) = \mathbf{Q} \qquad (12.179)$$

The term $\mathbf{G}(\mathbf{q})$ is called the *gravitational force vector* and the term $\mathbf{H}(\mathbf{q},\dot{\mathbf{q}})$ is called the *velocity coupling vector*. The velocity coupling vector may sometimes be written in the form

$$\mathbf{H}(\mathbf{q},\dot{\mathbf{q}}) = \mathbf{C}(\mathbf{q},\dot{\mathbf{q}})\dot{\mathbf{q}}. \qquad (12.180)$$

■

**Example 287** *2R planar manipulator Lagrangian dynamics.*

*The 2R planar manipulator is shown in Figure 5.9 and its homogeneous transformation matrices are given in Equations (5.29) and (5.30).*

$$^0R_1 = \begin{bmatrix} \cos\theta_1 & -\sin\theta_1 & 0 \\ \sin\theta_1 & \cos\theta_1 & 0 \\ 0 & 0 & 1 \end{bmatrix} \qquad (12.181)$$

$$^1R_2 = \begin{bmatrix} \cos\theta_2 & -\sin\theta_2 & 0 \\ \sin\theta_2 & \cos\theta_2 & 0 \\ 0 & 0 & 1 \end{bmatrix} \qquad (12.182)$$

*Assuming that the links are made of homogeneous material in a bar shape, the CM position vectors are at*

$$^i\mathbf{r}_i = \begin{bmatrix} -l_i/2 \\ 0 \\ 0 \end{bmatrix} \qquad i = 1,2 \qquad (12.183)$$

*and the inertia matrices are*

$$^iI_i = \frac{1}{12}m_i l_i^2 \begin{bmatrix} 0 & 0 & 0 \\ 0 & 1 & 0 \\ 0 & 0 & 1 \end{bmatrix}. \tag{12.184}$$

*Therefore,*

$$\begin{aligned} ^0I_1 &= {}^0R_1 \, {}^1I_1 \, {}^0R_1^T \\ &= \frac{1}{12}m_1 l_1^2 \begin{bmatrix} \sin^2\theta_1 & -\cos\theta_1\sin\theta_1 & 0 \\ -\cos\theta_1\sin\theta_1 & \cos^2\theta_1 & 0 \\ 0 & 0 & 1 \end{bmatrix} \end{aligned} \tag{12.185}$$

$$\begin{aligned} ^0I_2 &= {}^0R_2 \, {}^2I_2 \, {}^0R_2^T \\ &= \frac{1}{12}m_2 l_2^2 \begin{bmatrix} \sin^2\theta_{12} & -\cos\theta_{12}\sin\theta_{12} & 0 \\ -\cos\theta_{12}\sin\theta_{12} & \cos^2\theta_{12} & 0 \\ 0 & 0 & 1 \end{bmatrix}. \tag{12.186} \end{aligned}$$

*The gravity is assumed to be in $-\hat{j}_0$ direction*

$$\mathbf{g} = \begin{bmatrix} 0 \\ g \\ 0 \end{bmatrix} \tag{12.187}$$

*and the link Jacobian matrices are*

$$\mathbf{J}_{D1} = \begin{bmatrix} -\frac{1}{2}l_1\sin\theta_1 & 0 \\ \frac{1}{2}l_1\cos\theta_1 & 0 \\ 0 & 0 \end{bmatrix} \tag{12.188}$$

$$\mathbf{J}_{R1} = \begin{bmatrix} 0 & 0 \\ 0 & 0 \\ 1 & 0 \end{bmatrix} \tag{12.189}$$

$$\mathbf{J}_{D2} = \begin{bmatrix} -l_1\sin\theta_1 - \frac{1}{2}l_2\sin\theta_{12} & -\frac{1}{2}l_2\sin\theta_{12} \\ l_1\cos\theta_1 + \frac{1}{2}l_2\cos\theta_{12} & \frac{1}{2}l_2\cos\theta_{12} \\ 0 & 0 \end{bmatrix} \tag{12.190}$$

$$\mathbf{J}_{R2} = \begin{bmatrix} 0 & 0 \\ 0 & 0 \\ 1 & 0 \end{bmatrix}. \tag{12.191}$$

*We can calculate the manipulator inertia matrix by substituting $^0I_i$, $\mathbf{J}_{Di}$, and $\mathbf{J}_{Ri}$ in Equation (12.168)*

$$\begin{aligned} D &= \sum_{i=1}^{2} \left( \mathbf{J}_{Di}^T \, m_i \, \mathbf{J}_{Di} + \frac{1}{2}\mathbf{J}_{Ri}^T \, {}^0I_i \, \mathbf{J}_{Ri} \right) \tag{12.192} \\ &= \mathbf{J}_{D1}^T \, m_1 \, \mathbf{J}_{D1} + \frac{1}{2}\mathbf{J}_{R1}^T \, {}^0I_1 \, \mathbf{J}_{R1} + \mathbf{J}_{D2}^T \, m_2 \, \mathbf{J}_{D2} + \frac{1}{2}\mathbf{J}_{R2}^T \, {}^0I_2 \, \mathbf{J}_{R2} \\ &= \begin{bmatrix} \frac{1}{3}m_1 l_1^2 + m_2\left(l_1^2 + l_1 l_2 c\theta_2 + \frac{1}{3}l_2^2\right) & m_2\left(\frac{1}{2}l_1 l_2 c\theta_2 + \frac{1}{3}l_2^2\right) \\ m_2\left(\frac{1}{2}l_1 l_2 c\theta_2 + \frac{1}{3}l_2^2\right) & \frac{1}{3}m_2 l_2^2 \end{bmatrix} \end{aligned}$$

*The velocity coupling vector* **H** *has two elements that are*

$$H_1 = \sum_{j=1}^{1}\sum_{k=1}^{1}\left(\frac{\partial D_{1j}}{\partial q_k} - \frac{1}{2}\frac{\partial D_{jk}}{\partial q_1}\right)\dot{q}_j\dot{q}_k$$

$$= -m_2 l_1 l_2 \left(\dot{\theta}_1 + \frac{1}{2}\dot{\theta}_2\right)\dot{\theta}_2 \sin\theta_2 \qquad (12.193)$$

$$H_2 = \sum_{j=1}^{2}\sum_{k=1}^{2}\left(\frac{\partial D_{2j}}{\partial q_k} - \frac{1}{2}\frac{\partial D_{jk}}{\partial q_2}\right)\dot{q}_j\dot{q}_k$$

$$= \frac{1}{2}m_2 l_1 l_2 \dot{\theta}_1^2 \sin\theta_2. \qquad (12.194)$$

*The elements of the gravitational force vector* **G** *are*

$$G_1 = \frac{1}{2}m_1 g l_1 \cos\theta_1 + m_2 g l_1 \cos\theta_1 + \frac{1}{2}m_2 g l_2 \cos\theta_{12} \quad (12.195)$$

$$G_2 = \frac{1}{2}m_2 g l_2 \cos\theta_{12}. \qquad (12.196)$$

*Now we can assemble the equations of motion for the 2R planar manipulator utilizing the matrix and vectors we found above. Assuming no external force on the end-effector, the equations of motion are*

$$Q_1 = \left(\frac{1}{3}m_1 l_1^2 + m_2\left(l_1^2 + l_1 l_2 c\theta_2 + \frac{1}{3}l_2^2\right)\right)\ddot{\theta}_1$$

$$+ m_2 l_2\left(\frac{1}{2}l_1 c\theta_2 + \frac{1}{3}l_2\right)\ddot{\theta}_2 - m_2 l_1 l_2\left(\dot{\theta}_1 + \frac{1}{2}\dot{\theta}_2\right)\dot{\theta}_2 \sin\theta_2$$

$$+ \left(\frac{1}{2}m_1 + m_2\right)g l_1 \cos\theta_1 + \frac{1}{2}m_2 g l_2 \cos\theta_{12} \qquad (12.197)$$

$$Q_2 = m_2\left(\frac{1}{2}l_1 l_2 c\theta_2 + \frac{1}{3}l_2^2\right)\ddot{\theta}_1 + \frac{1}{3}m_2 l_2^2 \ddot{\theta}_2$$

$$+ \frac{1}{2}m_2 l_1 l_2 \dot{\theta}_1^2 \sin\theta_2 + \frac{1}{2}m_2 g l_2 \cos\theta_{12}. \qquad (12.198)$$

**Example 288 ★** *Christoffel operator.*
   *The symbol* $\Gamma^i_{j,k}$ *is called the Christoffel symbol or Christoffel operator with the following definition:*

$$\Gamma^i_{j,k} = \frac{1}{2}\left(\frac{\partial a_{ij}}{\partial q_k} + \frac{\partial a_{ik}}{\partial q_j} - \frac{\partial a_{jk}}{\partial q_i}\right) \qquad (12.199)$$

*The velocity coupling vector $H_{ijk}$ is a Christoffel operator.*

$$H_{ijk} = \sum_{j=1}^{n}\sum_{k=1}^{n}\left(\frac{\partial D_{ij}}{\partial q_k} - \frac{1}{2}\frac{\partial D_{jk}}{\partial q_i}\right)$$

$$= \frac{1}{2}\sum_{j=1}^{n}\sum_{k=1}^{n}\left(\frac{\partial D_{ij}}{\partial q_k} + \frac{\partial D_{ik}}{\partial q_j} - \frac{\partial D_{jk}}{\partial q_i}\right) \qquad (12.200)$$

**Example 289** *No gravity and no external force.*

*Assume there is no gravity and there is no external force applied on the end-effector of a robot. In these conditions, the Lagrangean of the manipulator simplifies to*

$$\mathcal{L} = \frac{1}{2}\sum_{i=1}^{n}\sum_{j=1}^{n} D_{ij}\,\dot{q}_i\dot{q}_j \qquad (12.201)$$

*and the equations of motion reduce to*

$$\frac{d}{dt}\left(\frac{\partial \mathcal{L}}{\partial \dot{q}_i}\right) - \frac{\partial \mathcal{L}}{\partial q_i} = \sum_{i=1}^{n}\sum_{j=1}^{n} D_{ij}\left(\ddot{q}_i + \Gamma^j_{l,m}\dot{q}_l\dot{q}_m\right). \qquad (12.202)$$

## 12.5  ★ Lagrange Equations and Link Transformation Matrices

The matrix form of the equations of motion for a robotic manipulator, based on the Lagrange equations, is

$$\mathbf{D(q)}\,\ddot{\mathbf{q}} + \mathbf{H(q,\dot{q})} + \mathbf{G(q)} = \mathbf{Q} \qquad (12.203)$$

which can be written in a summation form as

$$\sum_{j=1}^{n} D_{ij}(q)\,\ddot{q}_j + H_{ikm}\dot{q}_k\dot{q}_m + G_i = Q_i. \qquad (12.204)$$

The matrix $\mathbf{D(q)}$ is an $n \times n$ inertial-type symmetric matrix

$$D_{ij} = \sum_{r=\max i,j}^{n} \mathrm{tr}\left(\frac{\partial\,^{0}T_r}{\partial q_i}\,^{r}\bar{I}_r\,\frac{\partial\,^{0}T_r}{\partial q_j}^{T}\right) \qquad (12.205)$$

and $H_{ikm}$ is the velocity coupling term

$$H_{ijk} = \sum_{r=\max i,j,k}^{n} \mathrm{tr}\left(\frac{\partial^2\,^{0}T_r}{\partial q_j\partial q_k}\,^{r}\bar{I}_r\,\frac{\partial\,^{0}T_r}{\partial q_i}^{T}\right) \qquad (12.206)$$

and $G_i$ is the gravitational vector

$$G_i = -\sum_{r=i}^{n} m_r \mathbf{g}^T \frac{\partial {}^0 T_r}{\partial q_i} {}^r \mathbf{r}_r. \tag{12.207}$$

**Proof.** Position vector of a point $P$ of the link $(i)$ at ${}^i \mathbf{r}_P$ in the body coordinate $B_i$, can be transformed to the base frame by

$$ {}^0 \mathbf{r}_P = {}^0 T_i \, {}^i \mathbf{r}_P. \tag{12.208}$$

Therefore, its velocity and square of velocity in the base frame are

$$ {}^0 \dot{\mathbf{r}}_P = \sum_{j=1}^{i} \frac{\partial {}^0 T_i}{\partial q_j} \dot{q}_j \, {}^i \mathbf{r}_P \tag{12.209}$$

and

$$
\begin{aligned}
{}^0 \dot{\mathbf{r}}_P^2 &= {}^0 \dot{\mathbf{r}}_P \cdot {}^0 \dot{\mathbf{r}}_P \\
&= \mathrm{tr} \left( {}^0 \dot{\mathbf{r}}_P \, {}^0 \dot{\mathbf{r}}_P^T \right) \\
&= \mathrm{tr} \left( \sum_{j=1}^{i} \frac{\partial {}^0 T_i}{\partial q_j} \dot{q}_j \, {}^i \mathbf{r}_P \sum_{k=1}^{i} \left[ \frac{\partial {}^0 T_i}{\partial q_k} \dot{q}_k \, {}^i \mathbf{r}_P \right]^T \right) \\
&= \mathrm{tr} \left( \sum_{j=1}^{i} \sum_{k=1}^{i} \frac{\partial {}^0 T_i}{\partial q_j} \, {}^i \mathbf{r}_P \, {}^i \mathbf{r}_P^T \left[ \frac{\partial {}^0 T_i}{\partial q_k} \right]^T \dot{q}_j \, \dot{q}_k \right). \tag{12.210}
\end{aligned}
$$

The kinetic energy of point $P$ having a small mass $dm$ is then equal to

$$
\begin{aligned}
dK_P &= \frac{1}{2} \mathrm{tr} \left( \sum_{j=1}^{i} \sum_{k=1}^{i} \frac{\partial {}^0 T_i}{\partial q_j} \, {}^i \mathbf{r}_P \, {}^i \mathbf{r}_P^T \frac{\partial {}^0 T_i}{\partial q_k}^T \dot{q}_j \, \dot{q}_k \right) dm \\
&= \frac{1}{2} \mathrm{tr} \left( \sum_{j=1}^{i} \sum_{k=1}^{i} \frac{\partial {}^0 T_i}{\partial q_j} \left( {}^i \mathbf{r}_P \, dm \, {}^i \mathbf{r}_P^T \right) \frac{\partial {}^0 T_i}{\partial q_k}^T \dot{q}_j \, \dot{q}_k \right) \tag{12.211}
\end{aligned}
$$

and the kinetic energy of the link $(i)$ is

$$
\begin{aligned}
K_i &= \int_{B_i} dK_P \\
&= \frac{1}{2} \mathrm{tr} \left( \sum_{j=1}^{i} \sum_{k=1}^{i} \frac{\partial {}^0 T_i}{\partial q_j} \left( \int_{B_i} {}^i \mathbf{r}_P \, {}^i \mathbf{r}_P^T \, dm \right) \frac{\partial {}^0 T_i}{\partial q_k}^T \dot{q}_j \, \dot{q}_k \right). \tag{12.212}
\end{aligned}
$$

The integral in Equation (12.212) is the pseudo inertia matrix (11.143) for the link $(i)$

$$ {}^i \bar{I}_i = \int_{B_i} {}^i \mathbf{r}_P \, {}^i \mathbf{r}_P^T \, dm. \tag{12.213}$$

Hence, the kinetic energy of link $(i)$ becomes

$$K_i = \frac{1}{2} \operatorname{tr} \left( \sum_{j=1}^{i} \sum_{k=1}^{i} \frac{\partial^0 T_i}{\partial q_j} {}^i \bar{I}_i \frac{\partial^0 T_i}{\partial q_k}^T \dot{q}_j \dot{q}_k \right). \tag{12.214}$$

The kinetic energy of a robot having $n$ links is a summation of the kinetic energies of each link

$$K = \sum_{i=1}^{n} K_i = \frac{1}{2} \operatorname{tr} \sum_{i=1}^{n} \left( \sum_{j=1}^{i} \sum_{k=1}^{i} \frac{\partial^0 T_i}{\partial q_j} {}^i \bar{I}_i \frac{\partial^0 T_i}{\partial q_k}^T \dot{q}_j \dot{q}_k \right). \tag{12.215}$$

We may also add the kinetic energy due to the actuating motors that are installed at the joints of the robot. However, we may assume that the motors are concentrated masses at joints and add the mass of the motor at joint $i$ to the mass of the link $(i-1)$ and adjust the inertial parameters of the link. The motor at joint $i$ will drive the link $(i)$.

For the potential energy we assume the gravity is the only source of potential energy. Therefore, the potential energy of the link $(i)$ with respect to the base coordinate frame is

$$\begin{aligned} V_i &= -m_i \,{}^0\mathbf{g} \cdot {}^0\mathbf{r}_i \\ &= -m_i \,{}^0\mathbf{g}^T \,{}^0 T_i \,{}^i\mathbf{r}_i \end{aligned} \tag{12.216}$$

where ${}^0\mathbf{g} = \begin{bmatrix} g_x & g_y & g_z & 0 \end{bmatrix}^T$ is the gravitational acceleration usually in the direction $-z_0$ and ${}^0\mathbf{r}_i$ is the position vector of the $CM$ of link $(i)$ in the base frame. The potential energy of the whole robot is then equal to

$$V = \sum_{i=1}^{n} V_i = -\sum_{i=1}^{n} m_i \mathbf{g}^T \,{}^0 T_i \,{}^i\mathbf{r}_i. \tag{12.217}$$

The Lagrangean of a robot is found by substituting (12.215) and (12.217) in the Lagrange equation (12.154).

$$\begin{aligned} \mathcal{L} &= K - V \\ &= \frac{1}{2} \sum_{i=1}^{n} \sum_{j=1}^{i} \sum_{k=1}^{i} \operatorname{tr} \left( \frac{\partial^0 T_i}{\partial q_j} {}^i \bar{I}_i \frac{\partial^0 T_i}{\partial q_k}^T \right) \dot{q}_j \dot{q}_k \\ &\quad + \sum_{i=1}^{n} m_i \,{}^0\mathbf{g}^T \,{}^0 T_i \,{}^i\mathbf{r}_i. \end{aligned} \tag{12.218}$$

The dynamic equations of motion of a robot can now be found by applying the Lagrange equations (12.155) to Equation (12.218). We develop the equations of motion term by term. Differentiating the $\mathcal{L}$ with respect to $\dot{q}_r$ is

$$\frac{\partial \mathcal{L}}{\partial \dot{q}_r} = \frac{1}{2} \sum_{i=1}^{n} \sum_{k=1}^{i} \mathrm{tr} \left( \frac{\partial\, {}^0T_i}{\partial q_r} \, {}^i\bar{I}_i \, \frac{\partial\, {}^0T_i}{\partial q_k}^T \right) \dot{q}_k$$

$$+ \frac{1}{2} \sum_{i=1}^{n} \sum_{j=1}^{i} \mathrm{tr} \left( \frac{\partial\, {}^0T_i}{\partial q_j} \, {}^i\bar{I}_i \, \frac{\partial\, {}^0T_i}{\partial q_r}^T \right) \dot{q}_j$$

$$= \sum_{i=r}^{n} \sum_{j=1}^{i} \mathrm{tr} \left( \frac{\partial\, {}^0T_i}{\partial q_j} \, {}^i\bar{I}_i \, \frac{\partial\, {}^0T_i}{\partial q_r}^T \right) \dot{q}_j \qquad (12.219)$$

because

$$\frac{\partial\, {}^0T_i}{\partial q_r} = 0 \qquad \text{for } r > i \qquad (12.220)$$

and

$$\mathrm{tr} \left( \frac{\partial\, {}^0T_i}{\partial q_j} \, {}^i\bar{I}_i \, \frac{\partial\, {}^0T_i}{\partial q_k}^T \right) = \mathrm{tr} \left( \frac{\partial\, {}^0T_i}{\partial q_k} \, {}^i\bar{I}_i \, \frac{\partial\, {}^0T_i}{\partial q_j}^T \right). \qquad (12.221)$$

Time derivative of $\partial \mathcal{L}/\partial \dot{q}_r$ is

$$\frac{d}{dt}\frac{\partial \mathcal{L}}{\partial \dot{q}_r} = \sum_{i=r}^{n} \sum_{j=1}^{i} \mathrm{tr} \left( \frac{\partial\, {}^0T_i}{\partial q_j} \, {}^i\bar{I}_i \, \frac{\partial\, {}^0T_i}{\partial q_r}^T \right) \ddot{q}_j$$

$$+ \sum_{i=r}^{n} \sum_{j=1}^{i} \sum_{k=1}^{i} \mathrm{tr} \left( \frac{\partial^2\, {}^0T_i}{\partial q_j \partial q_k} \, {}^i\bar{I}_i \, \frac{\partial\, {}^0T_i}{\partial q_r}^T \right) \dot{q}_j \dot{q}_k$$

$$+ \sum_{i=r}^{n} \sum_{j=1}^{i} \sum_{k=1}^{i} \mathrm{tr} \left( \frac{\partial^2\, {}^0T_i}{\partial q_r \partial q_k} \, {}^i\bar{I}_i \, \frac{\partial\, {}^0T_i}{\partial q_j}^T \right) \dot{q}_j \dot{q}_k. \qquad (12.222)$$

The last term of the Lagrange equation is

$$\frac{\partial \mathcal{L}}{\partial q_r} = \frac{1}{2} \sum_{i=r}^{n} \sum_{j=1}^{i} \sum_{k=1}^{i} \mathrm{tr} \left( \frac{\partial^2\, {}^0T_i}{\partial q_j \partial q_r} \, {}^i\bar{I}_i \, \frac{\partial\, {}^0T_i}{\partial q_k}^T \right) \dot{q}_j \dot{q}_k$$

$$+ \frac{1}{2} \sum_{i=r}^{n} \sum_{j=1}^{i} \sum_{k=1}^{i} \mathrm{tr} \left( \frac{\partial^2\, {}^0T_i}{\partial q_k \partial q_r} \, {}^i\bar{I}_i \, \frac{\partial\, {}^0T_i}{\partial q_j}^T \right) \dot{q}_j \dot{q}_k$$

$$+ \sum_{i=r}^{n} m_i \mathbf{g}^T \frac{\partial\, {}^0T_i}{\partial q_r} \, {}^i\mathbf{r}_i \qquad (12.223)$$

which can be simplified to

$$\frac{\partial \mathcal{L}}{\partial q_r} = \sum_{i=r}^{n} \sum_{j=1}^{i} \sum_{k=1}^{i} \mathrm{tr} \left( \frac{\partial^2\, {}^0T_i}{\partial q_r \partial q_j} \, {}^i\bar{I}_i \, \frac{\partial\, {}^0T_i}{\partial q_k}^T \right) \dot{q}_j \dot{q}_k$$

$$+ \sum_{i=r}^{n} m_i \mathbf{g}^T \frac{\partial\, {}^0T_i}{\partial q_r} \, {}^i\mathbf{r}_i. \qquad (12.224)$$

Interestingly, the third term in Equation (12.222) is equal to the first term in (12.224). So, substituting these equations in the Lagrange equation can be simplified to

$$\frac{d}{dt}\left(\frac{\partial \mathcal{L}}{\partial \dot{q}_i}\right) - \frac{\partial \mathcal{L}}{\partial q_i} = \sum_{j=i}^{n}\sum_{k=1}^{j} \mathrm{tr}\left(\frac{\partial\, ^0T_j}{\partial q_k}\, ^j\bar{I}_j\, \frac{\partial\, ^0T_j}{\partial q_i}^T\right)\ddot{q}_k$$

$$+ \sum_{j=i}^{n}\sum_{k=1}^{j}\sum_{m=1}^{j} \mathrm{tr}\left(\frac{\partial^2\, ^0T_j}{\partial q_k \partial q_m}\, ^j\bar{I}_j\, \frac{\partial\, ^0T_j}{\partial q_i}^T\right)\dot{q}_k\dot{q}_m$$

$$- \sum_{j=i}^{n} m_j \mathbf{g}^T \frac{\partial\, ^0T_j}{\partial q_i}\, ^j\mathbf{r}_j. \tag{12.225}$$

Finally, the equations of motion for an $n$ link robot are

$$Q_i = \sum_{i=i}^{n}\sum_{k=1}^{j} \mathrm{tr}\left(\frac{\partial\, ^0T_j}{\partial q_k}\, ^j\bar{I}_j\, \frac{\partial\, ^0T_j}{\partial q_i}^T\right)\ddot{q}_k$$

$$+ \sum_{j=i}^{n}\sum_{k=1}^{j}\sum_{m=1}^{j} \mathrm{tr}\left(\frac{\partial^2\, ^0T_j}{\partial q_k \partial q_m}\, ^j\bar{I}_j\, \frac{\partial\, ^0T_j}{\partial q_i}^T\right)\dot{q}_k\dot{q}_m$$

$$- \sum_{j=i}^{n} m_j \mathbf{g}^T \frac{\partial\, ^0T_j}{\partial q_i}\, ^j\mathbf{r}_j. \tag{12.226}$$

The equations of motion can be written in a more concise form

$$Q_i = \sum_{j=1}^{n} D_{ij}\ddot{q}_j + H_{ijk}\dot{q}_j\,\dot{q}_k + G_i \tag{12.227}$$

where

$$D_{ij} = \sum_{r=\max i,j}^{n} \mathrm{tr}\left(\frac{\partial\, ^0T_r}{\partial q_j}\, ^r\bar{I}_r\, \frac{\partial\, ^0T_r}{\partial q_i}^T\right) \tag{12.228}$$

$$H_{ijk} = \sum_{r=\max i,j,k}^{n} \mathrm{tr}\left(\frac{\partial^2\, ^0T_r}{\partial q_j \partial q_k}\, ^r\bar{I}_r\, \frac{\partial\, ^0T_r}{\partial q_i}^T\right) \tag{12.229}$$

$$G_i = -\sum_{r=i}^{n} m_r \mathbf{g}^T \frac{\partial\, ^0T_r}{\partial q_i}\, ^r\mathbf{r}_r. \tag{12.230}$$

■

**Example 290** *2R manipulator mounted on ceiling.*

*Figure 12.7 depicts an ideal 2R planar manipulator mounted on a ceiling. Ceiling mounting is an applied method in some robotic operated assembly lines.*

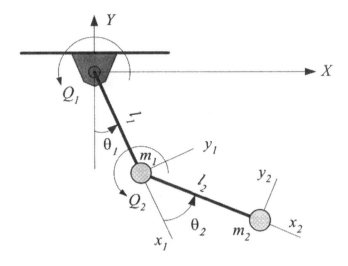

FIGURE 12.7. 2R manipulator mounted on a ceiling.

*The Lagrangean of the manipulator is*

$$
\begin{aligned}
\mathcal{L} &= K - V \\
&= \frac{1}{2} m_1 l_1^2 \dot{\theta}_1^2 \\
&\quad + \frac{1}{2} m_2 \left( l_1^2 \dot{\theta}_1^2 + l_2^2 \left( \dot{\theta}_1 + \dot{\theta}_2 \right)^2 + 2 l_1 l_2 \dot{\theta}_1 \left( \dot{\theta}_1 + \dot{\theta}_2 \right) \cos \theta_2 \right) \\
&\quad + m_1 g l_1 \cos \theta_1 + m_2 g \left( l_1 \cos \theta_1 + l_2 \cos \left( \theta_1 + \theta_2 \right) \right) \qquad (12.231)
\end{aligned}
$$

*which leads to the following equations of motion:*

$$
\begin{aligned}
Q_1 &= \left( (m_1 + m_2) l_1^2 + m_2 l_2^2 + 2 m_2 l_1 l_2 \cos \theta_2 \right) \ddot{\theta}_1 \\
&\quad + m_2 l_2 (l_2 + l_1 \cos \theta_2) \ddot{\theta}_2 \\
&\quad - 2 m_2 l_1 l_2 \sin \theta_2 \dot{\theta}_1 \dot{\theta}_2 - m_2 l_1 l_2 \sin \theta_2 \dot{\theta}_2^2 \\
&\quad + (m_1 + m_2) g l_1 \sin \theta_1 + m_2 g l_2 \sin \left( \theta_1 + \theta_2 \right) \qquad (12.232)
\end{aligned}
$$

$$
\begin{aligned}
Q_2 &= m_2 l_2 (l_2 + l_1 \cos \theta_2) \ddot{\theta}_1 + m_2 l_2^2 \ddot{\theta}_2 \\
&\quad - 2 m_2 l_1 l_2 \sin \theta_2 \dot{\theta}_1 \left( \dot{\theta}_1 + \dot{\theta}_2 \right) - m_2 g l_2 \sin \left( \theta_1 + \theta_2 \right). \quad (12.233)
\end{aligned}
$$

*The equations of motion can be rearranged to*

$$
\begin{aligned}
Q_1 &= D_{11} \ddot{\theta}_1 + D_{12} \ddot{\theta}_2 + H_{111} \dot{\theta}_1^2 + H_{122} \dot{\theta}_2^2 \\
&\quad + D_{112} \dot{\theta}_1 \dot{\theta}_2 + D_{121} \dot{\theta}_2 \dot{\theta}_1 + G_1 \qquad (12.234)
\end{aligned}
$$

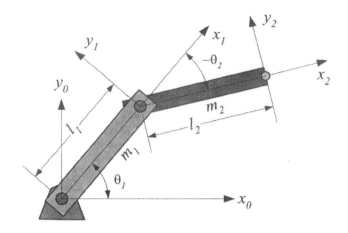

FIGURE 12.8. A 2R planar manipulator with massive links.

$$Q_2 = D_{12}\ddot{\theta}_1 + D_{22}\ddot{\theta}_2 + H_{211}\dot{\theta}_1^2 + H_{222}\dot{\theta}_2^2$$
$$+ D_{212}\dot{\theta}_1\dot{\theta}_2 + D_{221}\dot{\theta}_2\dot{\theta}_1 + G_2 \qquad (12.235)$$

*where,*

$$D_{11} = (m_1 + m_2)\, l_1^2 + m_2 l_2^2 + 2m_2 l_1 l_2 \cos\theta_2 \qquad (12.236)$$

$$D_{12} = m_2 l_2 \left(l_2 + l_1 \cos\theta_2\right) \qquad (12.237)$$

$$D_{21} = D_{12} = m_2 l_2 \left(l_2 + l_1 \cos\theta_2\right) \qquad (12.238)$$

$$D_{22} = m_2 l_2^2 \qquad (12.239)$$

$$H_{111} = 0 \qquad (12.240)$$

$$H_{122} = -m_2 l_1 l_2 \sin\theta_2 \qquad (12.241)$$

$$H_{211} = -m_2 l_1 l_2 \sin\theta_2 \qquad (12.242)$$

$$H_{222} = 0 \qquad (12.243)$$

$$H_{112} = H_{121} = -m_2 l_1 l_2 \sin\theta_2 \qquad (12.244)$$

$$H_{212} = H_{221} = -m_2 l_1 l_2 \sin\theta_2 \qquad (12.245)$$

$$G_1 = (m_1 + m_2)\, g l_1 \sin\theta_1 + m_2 g l_2 \sin(\theta_1 + \theta_2) \qquad (12.246)$$

$$G_2 = m_2 g l_2 \sin(\theta_1 + \theta_2)\,. \qquad (12.247)$$

**Example 291** *2R manipulator with massive links.*

*A 2R planar manipulator with massive links is shown in Figure 12.8. We assume the CM of each link is in the middle of the link and the motors at each joint is massless. The links' transformation matrices are*

$$
{}^{0}T_{1} = \begin{bmatrix} \cos\theta_{1} & -\sin\theta_{1} & 0 & l_{1}\cos\theta_{1} \\ \sin\theta_{1} & \cos\theta_{1} & 0 & l_{1}\sin\theta_{1} \\ 0 & 0 & 1 & 0 \\ 0 & 0 & 0 & 1 \end{bmatrix} \tag{12.248}
$$

$$
{}^{1}T_{2} = \begin{bmatrix} \cos\theta_{2} & -\sin\theta_{2} & 0 & l_{2}\cos\theta_{2} \\ \sin\theta_{2} & \cos\theta_{2} & 0 & l_{2}\sin\theta_{2} \\ 0 & 0 & 1 & 0 \\ 0 & 0 & 0 & 1 \end{bmatrix} \tag{12.249}
$$

$$
{}^{0}T_{2} = {}^{0}T_{1}\,{}^{1}T_{2} \tag{12.250}
$$

$$
= \begin{bmatrix} c(\theta_{1}+\theta_{2}) & -s(\theta_{1}+\theta_{2}) & 0 & l_{1}c\theta_{1}+l_{2}c(\theta_{1}+\theta_{2}) \\ s(\theta_{1}+\theta_{2}) & c(\theta_{1}+\theta_{2}) & 0 & l_{1}s\theta_{1}+l_{2}s(\theta_{1}+\theta_{2}) \\ 0 & 0 & 1 & 0 \\ 0 & 0 & 0 & 1 \end{bmatrix}.
$$

*Employing the velocity coefficient matrix $\Delta_{R}$ for revolute joints, we can write*

$$
\frac{\partial {}^{0}T_{1}}{\partial\theta_{1}} = \Delta_{R}\,{}^{0}T_{1} \tag{12.251}
$$

$$
= \begin{bmatrix} 0 & -1 & 0 & 0 \\ 1 & 0 & 0 & 0 \\ 0 & 0 & 0 & 0 \\ 0 & 0 & 0 & 0 \end{bmatrix} \begin{bmatrix} \cos\theta_{1} & -\sin\theta_{1} & 0 & l_{1}\cos\theta_{1} \\ \sin\theta_{1} & \cos\theta_{1} & 0 & l_{1}\sin\theta_{1} \\ 0 & 0 & 1 & 0 \\ 0 & 0 & 0 & 1 \end{bmatrix}
$$

$$
= \begin{bmatrix} -\sin\theta_{1} & -\cos\theta_{1} & 0 & -l_{1}\sin\theta_{1} \\ \cos\theta_{1} & -\sin\theta_{1} & 0 & l_{1}\cos\theta_{1} \\ 0 & 0 & 0 & 0 \\ 0 & 0 & 0 & 0 \end{bmatrix}
$$

*and similarly,*

$$
\frac{\partial {}^{0}T_{2}}{\partial\theta_{1}} = \Delta_{R}\,{}^{0}T_{2} \tag{12.252}
$$

$$
= \begin{bmatrix} 0 & -1 & 0 & 0 \\ 1 & 0 & 0 & 0 \\ 0 & 0 & 0 & 0 \\ 0 & 0 & 0 & 0 \end{bmatrix} \begin{bmatrix} c\theta_{12} & -s\theta_{12} & 0 & l_{1}c\theta_{1}+l_{2}c\theta_{12} \\ s\theta_{12} & c\theta_{12} & 0 & l_{1}s\theta_{1}+l_{2}s\theta_{12} \\ 0 & 0 & 1 & 0 \\ 0 & 0 & 0 & 1 \end{bmatrix}
$$

$$
= \begin{bmatrix} -s(\theta_{1}+\theta_{2}) & -c(\theta_{1}+\theta_{2}) & 0 & -l_{1}s\theta_{1}-l_{2}s(\theta_{1}+\theta_{2}) \\ c(\theta_{1}+\theta_{2}) & -s(\theta_{1}+\theta_{2}) & 0 & l_{1}c\theta_{1}+l_{2}c(\theta_{1}+\theta_{2}) \\ 0 & 0 & 0 & 0 \\ 0 & 0 & 0 & 0 \end{bmatrix}
$$

$$\frac{\partial \,^0T_2}{\partial \theta_2} = \,^0T_1 \Delta_R \,^1T_2 \tag{12.253}$$

$$= \begin{bmatrix} -s\,(\theta_1 + \theta_2) & -c\,(\theta_1 + \theta_2) & 0 & -l_2 s\,(\theta_1 + \theta_2) \\ c\,(\theta_1 + \theta_2) & -s\,(\theta_1 + \theta_2) & 0 & l_2 c\,(\theta_1 + \theta_2) \\ 0 & 0 & 0 & 0 \\ 0 & 0 & 0 & 0 \end{bmatrix}.$$

*Assuming all the product of inertias are zero, we find*

$$\,^1\bar{I}_1 = \begin{bmatrix} \frac{1}{3}m_1 l_1^2 & 0 & 0 & -\frac{1}{2}m_1 l_1 \\ 0 & 0 & 0 & 0 \\ 0 & 0 & 0 & 0 \\ -\frac{1}{2}m_1 l_1 & 0 & 0 & m_1 \end{bmatrix} \tag{12.254}$$

$$\,^2\bar{I}_2 = \begin{bmatrix} \frac{1}{3}m_2 l_2^2 & 0 & 0 & -\frac{1}{2}m_2 l_2 \\ 0 & 0 & 0 & 0 \\ 0 & 0 & 0 & 0 \\ -\frac{1}{2}m_2 l_2 & 0 & 0 & m_2 \end{bmatrix}. \tag{12.255}$$

*Utilizing inertia and derivative of transformation matrices we can calculate the inertial-type symmetric matrix* $\mathbf{D}(\mathbf{q})$

$$\begin{aligned} D_{11} &= \mathrm{tr}\left(\frac{\partial \,^0T_1}{\partial q_1}\,^1\bar{I}_1\, \frac{\partial \,^0T_1}{\partial q_1}^T\right) + \mathrm{tr}\left(\frac{\partial \,^0T_2}{\partial q_1}\,^2\bar{I}_2\, \frac{\partial \,^0T_2}{\partial q_1}^T\right) \\ &= \mathrm{tr}\begin{bmatrix} \frac{1}{3}m_1 l_1^2 \sin^2 \theta_1 & -\frac{1}{3}m_1 l_1^2 \cos \theta_1 \sin \theta_1 & 0 & 0 \\ -\frac{1}{3}m_1 l_1^2 \cos \theta_1 \sin \theta_1 & \frac{1}{3}m_1 l_1^2 \cos^2 \theta_1 & 0 & 0 \\ 0 & 0 & 0 & 0 \\ 0 & 0 & 0 & 0 \end{bmatrix} \\ &\quad + \mathrm{tr}\left(\frac{\partial \,^0T_2}{\partial q_1}\,^2\bar{I}_2\, \frac{\partial \,^0T_2}{\partial q_1}^T\right) \\ &= \frac{1}{3}m_1 l_1^2 + \frac{4}{3}m_2 l_2^2 + m_2 l_2^2 \cos \theta_2 \tag{12.256} \end{aligned}$$

$$\begin{aligned} D_{12} &= D_{21} = \mathrm{tr}\left(\frac{\partial \,^0T_2}{\partial q_1}\,^2\bar{I}_2\, \frac{\partial \,^0T_2}{\partial q_1}^T\right) \\ &= \frac{1}{3}m_2 l_2^2 + \frac{1}{2}m_2 l_2^2 \cos \theta_2 \tag{12.257} \end{aligned}$$

$$D_{22} = \text{tr}\left(\frac{\partial^0 T_2}{\partial q_2}\,^2\bar{I}_2\,\frac{\partial^0 T_2}{\partial q_2}^T\right)$$

$$= \frac{1}{3}l_2^2 m_2\,\text{tr}\begin{bmatrix} \sin^2\theta_{12} & -\cos\theta_{12}\sin\theta_{12} & 0 & 0 \\ -\cos\theta_{12}\sin\theta_{12} & \cos^2\theta_{12} & 0 & 0 \\ 0 & 0 & 0 & 0 \\ 0 & 0 & 0 & 0 \end{bmatrix}$$

$$= \frac{1}{3}l_2^2 m_2. \tag{12.258}$$

The coupling terms $\mathbf{H}(\mathbf{q},\dot{\mathbf{q}})$ are calculated as below

$$H_1 = \sum_{k=1}^{2}\sum_{m=1}^{2} H_{1km}\dot{q}_k\dot{q}_m$$

$$= H_{111}\dot{q}_1\dot{q}_1 + H_{112}\dot{q}_1\dot{q}_2 + H_{121}\dot{q}_2\dot{q}_1 + H_{122}\dot{q}_2\dot{q}_2 \tag{12.259}$$

$$H_2 = \sum_{k=1}^{2}\sum_{m=1}^{2} H_{2km}\dot{q}_k\dot{q}_m$$

$$= H_{211}\dot{q}_1\dot{q}_1 + H_{212}\dot{q}_1\dot{q}_2 + H_{221}\dot{q}_2\dot{q}_1 + H_{222}\dot{q}_2\dot{q}_2 \tag{12.260}$$

where

$$H_{ijk} = \sum_{r=\max i,j,k}^{n} \text{tr}\left(\frac{\partial^2\,^0T_r}{\partial q_j\partial q_k}\,^r\bar{I}_r\,\frac{\partial^0 T_r}{\partial q_i}^T\right). \tag{12.261}$$

These calculations lead to

$$\mathbf{H} = \begin{bmatrix} -\frac{1}{2}m_2 l_2^2\dot{\theta}_2^2\sin\theta_2 - m_2 l_2^2\dot{\theta}_1\dot{\theta}_2\sin\theta_2 \\ \frac{1}{2}m_2 l_2^2\dot{\theta}_1^2\sin\theta_2 \end{bmatrix}. \tag{12.262}$$

The last terms are the gravitational vector $\mathbf{G}(\mathbf{q})$

$$G_1 = -m_1\mathbf{g}^T\frac{\partial^0 T_1}{\partial q_1}\,^1\mathbf{r}_1 - m_2\mathbf{g}^T\frac{\partial^0 T_2}{\partial q_1}\,^2\mathbf{r}_2$$

$$= -m_1\begin{bmatrix} 0 \\ -g \\ 0 \\ 0 \end{bmatrix}^T\begin{bmatrix} -\sin\theta_1 & -\cos\theta_1 & 0 & -l_1\sin\theta_1 \\ \cos\theta_1 & -\sin\theta_1 & 0 & l_1\cos\theta_1 \\ 0 & 0 & 0 & 0 \\ 0 & 0 & 0 & 0 \end{bmatrix}\begin{bmatrix} -\frac{l_1}{2} \\ 0 \\ 0 \\ 1 \end{bmatrix}$$

$$-m_2\begin{bmatrix} 0 \\ -g \\ 0 \\ 0 \end{bmatrix}^T\begin{bmatrix} -s\theta_{12} & -c\theta_{12} & 0 & -l_1 s\theta_1 - l_2 s\theta_{12} \\ c\theta_{12} & -s\theta_{12} & 0 & l_1 c\theta_1 + l_2 c\theta_{12} \\ 0 & 0 & 0 & 0 \\ 0 & 0 & 0 & 0 \end{bmatrix}\begin{bmatrix} -\frac{l_1}{2} \\ 0 \\ 0 \\ 1 \end{bmatrix}$$

$$= \frac{1}{2}m_1 g l_1\cos\theta_1$$

$$+\frac{1}{2}m_2 g l_1\cos(\theta_1 + \theta_2) + m_2 g l_1\cos\theta_1 \tag{12.263}$$

$$G_2 = -m_2\mathbf{g}^T \frac{\partial\, {}^0T_2}{\partial q_2}\, {}^2\mathbf{r}_2$$

$$= -m_2 \begin{bmatrix} 0 \\ -g \\ 0 \\ 0 \end{bmatrix}^T \begin{bmatrix} -s\theta_{12} & -c\theta_{12} & 0 & -l_2 s\theta_{12} \\ c\theta_{12} & -s\theta_{12} & 0 & l_2 c\theta_{12} \\ 0 & 0 & 0 & 0 \\ 0 & 0 & 0 & 0 \end{bmatrix} \begin{bmatrix} -\frac{l_1}{2} \\ 0 \\ 0 \\ 1 \end{bmatrix}$$

$$= \frac{1}{2} m_2 g l_2 \cos(\theta_1 + \theta_2). \tag{12.264}$$

*Finally the equations of motion for the 2R planar manipulator are*

$$\begin{bmatrix} Q_1 \\ Q_2 \end{bmatrix} = \begin{bmatrix} \frac{1}{3} m_1 l_1^2 + \frac{4}{3} m_2 l_2^2 + m_2 l_2^2 c\theta_2 & \frac{1}{3} m_2 l_2^2 + \frac{1}{2} m_2 l_2^2 c\theta_2 \\ \frac{1}{3} m_2 l_2^2 + \frac{1}{2} m_2 l_2^2 c\theta_2 & \frac{1}{3} l_2^2 m_2 \end{bmatrix} \begin{bmatrix} \ddot{\theta}_1 \\ \ddot{\theta}_2 \end{bmatrix}$$

$$+ \begin{bmatrix} -\frac{1}{2} m_2 l_2^2 \dot{\theta}_2^2 \sin\theta_2 - m_2 l_2^2 \dot{\theta}_1 \dot{\theta}_2 \sin\theta_2 \\ \frac{1}{2} m_2 l_2^2 \dot{\theta}_1^2 \sin\theta_2 \end{bmatrix}$$

$$+ \begin{bmatrix} \frac{1}{2} m_1 g l_1 \cos\theta_1 + \frac{1}{2} m_2 g l_1 \cos\theta_{12} + m_2 g l_1 \cos\theta_1 \\ \frac{1}{2} m_2 g l_2 \cos\theta_{12} \end{bmatrix} \tag{12.265}$$

## 12.6 Robot Statics

At the beginning and at the end of a rest-to-rest mission, a robot must keep the specified configurations. To hold the position and orientation, the actuators must apply some required forces to balance the external loads applied to the robot. Calculating the required actuators' force to hold a robot in a specific configuration is called *robot statics*.

In a static condition, the globally expressed Newton-Euler equations for the link $(i)$ can be written in a recursive form

$$^0\mathbf{F}_{i-1} = {}^0\mathbf{F}_i - \sum {}^0\mathbf{F}_{e_i} \tag{12.266}$$

$$^0\mathbf{M}_{i-1} = {}^0\mathbf{M}_i - \sum {}^0\mathbf{M}_{e_i} + {}_{i-1}^0\mathbf{d}_i \times {}^0\mathbf{F}_i \tag{12.267}$$

where

$$_{i-1}^0\mathbf{d}_i = {}^0\mathbf{d}_i - {}^0\mathbf{d}_{i-1}. \tag{12.268}$$

Therefore, we are able to calculate the action force system $(\mathbf{F}_{i-1}, \mathbf{M}_{i-1})$ when the reaction force system $(-\mathbf{F}_i, -\mathbf{M}_i)$ is given. The position vectors and force systems on link $(i)$ are shown in Figure 12.9.

**Proof.** In a static condition, the Newton-Euler equations of motion (12.26) and (12.27) for the link $(i)$ reduce to force and moment balance equations

$$^0\mathbf{F}_{i-1} - {}^0\mathbf{F}_i + \sum {}^0\mathbf{F}_{e_i} = 0 \tag{12.269}$$

$$^0\mathbf{M}_{i-1} - {}^0\mathbf{M}_i + \sum {}^0\mathbf{M}_{e_i} + {}^0\mathbf{n}_i \times {}^0\mathbf{F}_{i-1} - {}^0\mathbf{m}_i \times {}^0\mathbf{F}_i = 0. \tag{12.270}$$

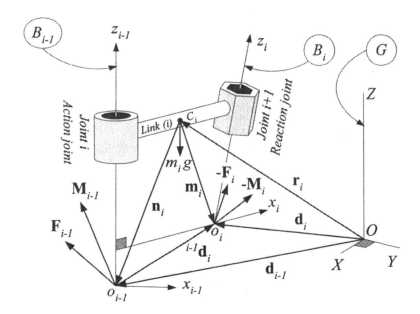

FIGURE 12.9. Illustration of position vectors and force system on link $(i)$.

These equations can be rearranged into a *backward recursive* form

$$^0\mathbf{F}_{i-1} = {}^0\mathbf{F}_i - \sum {}^0\mathbf{F}_{e_i} \tag{12.271}$$

$$^0\mathbf{M}_{i-1} = {}^0\mathbf{M}_i - \sum {}^0\mathbf{M}_{e_i} - {}^0\mathbf{n}_i \times {}^0\mathbf{F}_{i-1} + {}^0\mathbf{m}_i \times {}^0\mathbf{F}_i. \tag{12.272}$$

However, we may transform the Euler equation from $C_i$ to $O_{i-1}$ and find the Equation (12.267).

Practically, we measure the position of mass center $\mathbf{r}_i$ and the relative position of $B_i$ and $B_{i-1}$ in the coordinate frame $B_i$ attached to the link $(i)$. Hence, we must transform ${}^i\mathbf{r}_i$ and ${}_{i-1}^i\mathbf{d}_i$ to the base frame.

$$^0\mathbf{r}_i = {}^0T_i \, {}^i\mathbf{r}_i \tag{12.273}$$

$$_{i-1}^0\mathbf{d}_i = {}^0T_i \, {}_{i-1}^i\mathbf{d}_i \tag{12.274}$$

The external load is usually the gravitational force $m_i\mathbf{g}$ and hence,

$$\sum {}^0\mathbf{F}_{e_i} = m_i \, {}^0\mathbf{g} \tag{12.275}$$

$$\sum {}^0\mathbf{M}_{e_i} = {}^0\mathbf{r}_i \times m_i \, {}^0\mathbf{g}. \tag{12.276}$$

Using the DH parameters, we may express the relative position vector $_{i-1}^i\mathbf{d}_i$

by

$$_{i-1}^{i}\mathbf{d}_i = \begin{bmatrix} a_i \\ d_i \sin \alpha_i \\ d_i \cos \alpha_i \\ 1 \end{bmatrix}. \qquad (12.277)$$

The backward recursive equations (12.266) and (12.267) allow us to start with a known force system $(\mathbf{F}_n, \mathbf{M}_n)$ at $B_n$, applied from the end-effector to the environment, and calculate the force system at $B_{n-1}$.

$$^0\mathbf{F}_{n-1} = {}^0\mathbf{F}_n - \sum {}^0\mathbf{F}_{e_n} \qquad (12.278)$$

$$^0\mathbf{M}_{n-1} = {}^0\mathbf{M}_n - \sum {}^0\mathbf{M}_{e_n} + {}_{n-1}^{0}\mathbf{d}_n \times {}^0\mathbf{F}_n \qquad (12.279)$$

Following the same procedure and calculating force system at proximal end by having the force system at distal end of each link, ends up to the force system at the base. In this procedure, the force system applied by the end-effector to the environment is assumed to be known.

It is also possible to rearrange the static Equations (12.269) and (12.270) into a *forward recursive* form

$$^0\mathbf{F}_i = {}^0\mathbf{F}_{i-1} + \sum {}^0\mathbf{F}_{e_i} \qquad (12.280)$$

$$^0\mathbf{M}_i = {}^0\mathbf{M}_{i-1} + \sum {}^0\mathbf{M}_{e_i} + {}^0\mathbf{n}_i \times {}^0\mathbf{F}_{i-1} - {}^0\mathbf{m}_i \times {}^0\mathbf{F}_i. \qquad (12.281)$$

Transforming the Euler equation from $C_i$ to $O_i$ simplifies the forward recursive equations into the more practical equations

$$^0\mathbf{F}_i = {}^0\mathbf{F}_{i-1} + \sum {}^0\mathbf{F}_{e_i} \qquad (12.282)$$

$$^0\mathbf{M}_i = {}^0\mathbf{M}_{i-1} + \sum {}^0\mathbf{M}_{e_i} - {}_{i-1}^{0}\mathbf{d}_i \times {}^0\mathbf{F}_{i-1}. \qquad (12.283)$$

Using the forward recursive Equations (12.282) and (12.283) we can start with a known force system $(\mathbf{F}_0, \mathbf{M}_0)$ at $B_0$, applied from the base to the link (1), and calculate the force system at $B_1$.

$$^0\mathbf{F}_1 = {}^0\mathbf{F}_0 + \sum {}^0\mathbf{F}_{e_1} \qquad (12.284)$$

$$^0\mathbf{M}_1 = {}^0\mathbf{M}_0 + \sum {}^0\mathbf{M}_{e_1} - {}^0\mathbf{d}_1 \times {}^0\mathbf{F}_0 \qquad (12.285)$$

Following this procedure and calculating force system at the distal end by having the force system at the proximal end of each link, ends up at the force system applied to the environment by the end-effector. In this procedure, the force system applied by the base actuators to the first link is assumed to be known. ∎

**Example 292** *Statics of a 4R planar manipulator.*

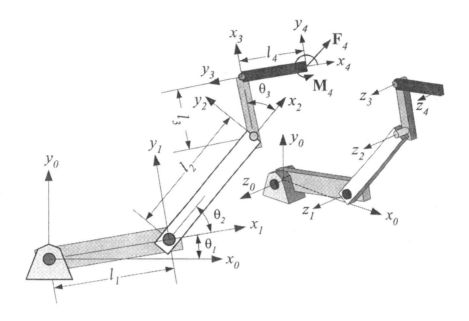

FIGURE 12.10. A 4R planar manipulator.

*Figure 12.10 illustrates a 4R planar manipulator with the DH coordinate frames set up for each link. Assume the end-effector force system applied to the environment is measured as*

$$^4\mathbf{F}_4 = \begin{bmatrix} F_x \\ F_y \\ 0 \end{bmatrix} \qquad ^4\mathbf{M}_4 = \begin{bmatrix} 0 \\ 0 \\ M_z \end{bmatrix}. \qquad (12.286)$$

*In addition, we assume that the links are uniform such that their CM are located at the midpoint of each link, and the gravitational acceleration is*

$$\mathbf{g} = -g\,\hat{\jmath}_0. \qquad (12.287)$$

*The manipulator consists of four $R\|R(0)$ links, therefore their transformation matrices $^{i-1}T_i$ are of class (5.32) that because $d_i = 0$ and $a_i = l_i$, simplifies to*

$$^{i-1}T_i = \begin{bmatrix} \cos\theta_i & -\sin\theta_i & 0 & l_i\cos\theta_i \\ \sin\theta_i & \cos\theta_i & 0 & l_i\sin\theta_i \\ 0 & 0 & 1 & 0 \\ 0 & 0 & 0 & 1 \end{bmatrix}. \qquad (12.288)$$

The CM position vectors ${}^i\mathbf{r}_i$ and the relative position vectors ${}_{i-1}^{0}\mathbf{d}_i$ are

$$
{}^i\mathbf{r}_i = \begin{bmatrix} l_i/2 \\ 0 \\ 0 \\ 0 \end{bmatrix} \tag{12.289}
$$

$$
{}_{i-1}^{i}\mathbf{d}_i = \begin{bmatrix} l_i \\ 0 \\ 0 \\ 0 \end{bmatrix} \tag{12.290}
$$

and therefore,

$$
\begin{aligned}
{}^0\mathbf{r}_i &= {}^0T_i\,{}^i\mathbf{r}_i & (12.291) \\
{}_{i-1}^{0}\mathbf{d}_i &= {}^0T_i\,{}_{i-1}^{i}\mathbf{d}_i & (12.292)
\end{aligned}
$$

where

$$
{}^0T_i = {}^0T_1 \cdots {}^{i-1}T_i. \tag{12.293}
$$

The static force at joints 3, 2, and 1 are

$$
\begin{aligned}
{}^0\mathbf{F}_3 &= {}^0\mathbf{F}_4 - \sum {}^0\mathbf{F}_{e_4} \\
&= {}^0\mathbf{F}_4 + m_4 g\,{}^0\hat{\jmath}_0 \\
&= \begin{bmatrix} F_x \\ F_y \\ 0 \\ 0 \end{bmatrix} + m_4 g \begin{bmatrix} 0 \\ 1 \\ 0 \\ 0 \end{bmatrix} \\
&= \begin{bmatrix} F_x \\ F_y + m_4 g \\ 0 \\ 0 \end{bmatrix} \tag{12.294}
\end{aligned}
$$

$$
\begin{aligned}
{}^0\mathbf{F}_2 &= {}^0\mathbf{F}_3 - m_3 g\,{}^0\hat{\jmath}_0 \\
&= \begin{bmatrix} F_x \\ F_y + m_4 g \\ 0 \\ 0 \end{bmatrix} + m_3 g \begin{bmatrix} 0 \\ 1 \\ 0 \\ 0 \end{bmatrix} \\
&= \begin{bmatrix} F_x \\ F_y + (m_3 + m_4)\,g \\ 0 \\ 0 \end{bmatrix} \tag{12.295}
\end{aligned}
$$

$$
\begin{aligned}
{}^0\mathbf{F}_1 &= {}^0\mathbf{F}_2 - m_2 g \, {}^0\hat{\jmath}_0 \\
&= \begin{bmatrix} F_x \\ F_y + (m_3 + m_4)\, g \\ 0 \\ 0 \end{bmatrix} + m_2 g \begin{bmatrix} 0 \\ 1 \\ 0 \\ 0 \end{bmatrix} \\
&= \begin{bmatrix} F_x \\ F_y + g\, (m_2 + m_3 + m_4) \\ 0 \\ 0 \end{bmatrix}
\end{aligned}
\tag{12.296}
$$

$$
\begin{aligned}
{}^0\mathbf{F}_0 &= {}^0\mathbf{F}_1 - m_1 g \, {}^0\hat{\jmath}_0 \\
&= \begin{bmatrix} F_x \\ F_y + g\, (m_2 + m_3 + m_4) \\ 0 \\ 0 \end{bmatrix} + m_1 g \begin{bmatrix} 0 \\ 1 \\ 0 \\ 0 \end{bmatrix} \\
&= \begin{bmatrix} F_x \\ F_y + g\, (m_1 + m_2 + m_3 + m_4) \\ 0 \\ 0 \end{bmatrix}.
\end{aligned}
\tag{12.297}
$$

The static moment at joints 3, 2, and 1 are

$$
\begin{aligned}
{}^0\mathbf{M}_3 &= {}^0\mathbf{M}_4 - \sum {}^0\mathbf{M}_{e_i} + {}^0_3\mathbf{d}_4 \times {}^0\mathbf{F}_4 \\
&= {}^0\mathbf{M}_4 + {}^0\mathbf{r}_4 \times m_4 g \, {}^0\hat{\jmath}_0 + {}^0_3\mathbf{d}_4 \times {}^0\mathbf{F}_4 \\
&= {}^0\mathbf{M}_4 + {}^0\mathbf{r}_4 \times m_4 g \, {}^0\hat{\jmath}_0 + {}^0_3\mathbf{d}_4 \times {}^0\mathbf{F}_4 \\
&= {}^0\mathbf{M}_4 + m_4 g \left( {}^0T_4 \, {}^4\mathbf{r}_4 \times {}^0\hat{\jmath}_0 \right) + \left( {}^0T_4 \, {}^4_3\mathbf{d}_4 \right) \times {}^0\mathbf{F}_4 \\
&= \begin{bmatrix} 0 \\ 0 \\ M_{3z} \\ 0 \end{bmatrix}
\end{aligned}
\tag{12.298}
$$

$$
M_{3z} = M_z + l_4 F_y \cos\theta_{1234} - l_4 F_x \sin\theta_{1234} + \frac{1}{2} g l_4 m_4 \cos\theta_{1234}
\tag{12.299}
$$

$$
\begin{aligned}
{}^0\mathbf{M}_2 &= {}^0\mathbf{M}_3 + m_3 g \left( {}^0T_3 \, {}^3\mathbf{r}_3 \times {}^0\hat{\jmath}_0 \right) + \left( {}^0T_3 \, {}^3_2\mathbf{d}_3 \right) \times {}^0\mathbf{F}_3 \\
&= \begin{bmatrix} 0 \\ 0 \\ M_{2z} \\ 0 \end{bmatrix}
\end{aligned}
\tag{12.300}
$$

$$M_{2z} = M_z + l_4 F_y \cos\theta_{1234} - l_4 F_x \sin\theta_{1234} + \frac{1}{2} g l_4 m_4 \cos\theta_{1234}$$
$$+ \frac{1}{2} g l_3 m_3 \cos\theta_{123} - l_3 F_x \sin\theta_{123}$$
$$+ l_3 \left(F_y + g m_4\right) \cos\theta_{123} \tag{12.301}$$

$$^0\mathbf{M}_1 = {}^0\mathbf{M}_2 + m_2 g \left({}^0T_2\, {}^2\mathbf{r}_2 \times {}^0\hat{\jmath}_0\right) + \left({}^0T_2\, {}^2_1\mathbf{d}_2\right) \times {}^0\mathbf{F}_2$$
$$= \begin{bmatrix} 0 \\ 0 \\ M_{1z} \\ 0 \end{bmatrix} \tag{12.302}$$

$$M_{1z} = M_z + l_4 F_y \cos\theta_{1234} - l_4 F_x \sin\theta_{1234} + \frac{1}{2} g l_4 m_4 \cos\theta_{1234}$$
$$+ \frac{1}{2} g l_3 m_3 \cos\theta_{123} - l_3 F_x \sin\theta_{123} + l_3 \cos\theta_{123} \left(F_y + g m_4\right)$$
$$+ \frac{1}{2} g l_2 m_2 \cos\theta_{12} - l_2 F_x \sin\theta_{12}$$
$$+ l_2 \cos\theta_{12} \left(F_y + g\left(m_3 + m_4\right)\right) \tag{12.303}$$

$$^0\mathbf{M}_0 = {}^0\mathbf{M}_1 + m_1 g \left({}^0T_1\, {}^1\mathbf{r}_1 \times {}^0\hat{\jmath}_0\right) + \left({}^0T_1\, {}^1_0\mathbf{d}_1\right) \times {}^0\mathbf{F}_1$$
$$= \begin{bmatrix} 0 \\ 0 \\ M_{0z} \\ 0 \end{bmatrix} \tag{12.304}$$

$$M_{0z} = M_z + l_4 F_y \cos\theta_{1234} - l_4 F_x \sin\theta_{1234} + \frac{1}{2} g l_4 m_4 \cos\theta_{1234}$$
$$+ \frac{1}{2} g l_3 m_3 \cos\theta_{123} - l_3 F_x \sin\theta_{123} + l_3 \cos\theta_{123} \left(F_y + g m_4\right)$$
$$- l_2 F_x \sin\theta_{12} + \frac{1}{2} g l_2 m_2 \cos\theta_{12} + l_2 \cos\theta_{12} \left(F_y + g\left(m_3 + m_4\right)\right)$$
$$+ \frac{1}{2} g l_1 m_1 \cos\theta_1 - l_1 F_x \sin\theta_1$$
$$+ l_1 \cos\theta_1 \left(F_y + g\left(m_2 + m_3 + m_4\right)\right) \tag{12.305}$$

*where*

$$\theta_{1234} = \theta_1 + \theta_2 + \theta_3 + \theta_4 \tag{12.306}$$
$$\theta_{123} = \theta_1 + \theta_2 + \theta_3 \tag{12.307}$$
$$\theta_{12} = \theta_1 + \theta_2. \tag{12.308}$$

**Example 293** *Recursive force equation in link's frame.*

Practically, it is easier to measure and calculate the force systems in the kink's frame. Therefore, we may write the backward recursive Equations (12.266) and (12.267) in the following form:

$$^{i}\mathbf{F}_{i-1} = {}^{i}\mathbf{F}_i - \sum {}^{i}\mathbf{F}_{e_i} \tag{12.309}$$

$$^{i}\mathbf{M}_{i-1} = {}^{i}\mathbf{M}_i - \sum {}^{i}\mathbf{M}_{e_i} + {}_{i-1}^{i}\mathbf{d}_i \times {}^{i}\mathbf{F}_i \tag{12.310}$$

and calculate the proximal force system from the distal force system in the link's frame. The calculated force system, then, may be transformed to the previous link's coordinate frame by a transformation

$$^{i-1}\mathbf{F}_{i-1} = {}^{i-1}T_i \, {}^{i}\mathbf{F}_{i-1} \tag{12.311}$$

$$^{i-1}\mathbf{M}_{i-1} = {}^{i-1}T_i \, {}^{i}\mathbf{M}_{i-1} \tag{12.312}$$

or they may be transformed to any other coordinate frame including the base frame.

$$^{0}\mathbf{F}_{i-1} = {}^{0}T_i \, {}^{i}\mathbf{F}_{i-1} \tag{12.313}$$

$$^{0}\mathbf{M}_{i-1} = {}^{0}T_i \, {}^{i}\mathbf{M}_{i-1} \tag{12.314}$$

**Example 294** *Actuator's force and torque.*

Applying a backward or forward recursive static force analysis ends up with a set of known force systems at joints. Each joint is driven by a motor or generally an actuator that applies a force in a $P$ joint, or a torque in an $R$ joint. When the joint $i$ is prismatic, the actuator force is applied along the axis of the joint $i$. Therefore, the force of the driving motor is along the $z_{i-1}$-axis

$$F_m = {}^{0}\hat{k}_{i-1}^T \, {}^{0}\mathbf{F}_i \tag{12.315}$$

showing that the $\hat{k}_{i-1}$ component of the joint force $\mathbf{F}_i$ is supported by the actuator, while the $\hat{i}_{i-1}$ and $\hat{j}_{i-1}$ components of $\mathbf{F}_i$ must be supported by the bearings of the joint.

Similarly, when joint $i$ is revolute, the actuator torque is applied about the axis of joint $i$. Therefore, the torque of the driving motor is along the $z_{i-1}$-axis

$$M_m = {}^{0}\hat{k}_{i-1}^T \, {}^{0}\mathbf{M}_i \tag{12.316}$$

showing that the $\hat{k}_{i-1}$ component of the joint torque $\mathbf{M}_i$ is supported by the actuator, while the $\hat{i}_{i-1}$ and $\hat{j}_{i-1}$ components of $\mathbf{M}_i$ must be supported by the bearings of the joint.

## 12.7  Summary

Dynamics equations of motion for a robot can be found by both Newton-Euler and Lagrange methods. In the Newton-Euler method, each link $(i)$

is a rigid body and therefore, its translational and rotational equations of
motion in the base coordinate frame are

$$
{}^0\mathbf{F}_{i-1} - {}^0\mathbf{F}_i + \sum {}^0\mathbf{F}_{e_i} = m_i\,{}^0\mathbf{a}_i \tag{12.317}
$$

$$
\begin{aligned}
{}^0\mathbf{M}_{i-1} - {}^0\mathbf{M}_i + \sum {}^0\mathbf{M}_{e_i} \\
+ \left({}^0\mathbf{d}_{i-1} - {}^0\mathbf{r}_i\right) \times {}^0\mathbf{F}_{i-1} \\
- \left({}^0\mathbf{d}_i - {}^0\mathbf{r}_i\right) \times {}^0\mathbf{F}_i = {}^0I_i\,{}_0\boldsymbol{\alpha}_i.
\end{aligned} \tag{12.318}
$$

The force $\mathbf{F}_{i-1}$ and moment $\mathbf{M}_{i-1}$ are the resultant force and moment
that link $(i-1)$ applies to link $(i)$ at joint $i$. Similarly, $\mathbf{F}_i$ and $\mathbf{M}_i$ are the
resultant force and moment that link $(i)$ applies to link $(i+1)$ at joint
$i+1$. We measure the force systems $(\mathbf{F}_{i-1}$ , $\mathbf{M}_{i-1})$ and $(\mathbf{F}_i$ , $\mathbf{M}_i)$ at the
origin of the coordinate frames $B_{i-1}$ and $B_i$ respectively. The sum of the
external loads acting on the link $(i)$ are $\sum \mathbf{F}_{e_i}$ and $\sum \mathbf{M}_{e_i}$. The vector ${}^0\mathbf{r}_i$
is the global position vector of $C_i$ and ${}^0\mathbf{d}_i$ is the global position vector of
the origin of $B_i$. The vector ${}^0\boldsymbol{\alpha}_i$ is the angular acceleration and ${}^0\mathbf{a}_i$ is the
translational acceleration of the link $(i)$ measured at the mass center $C_i$.

$$
{}^0\mathbf{a}_i = {}^0\ddot{\mathbf{d}}_i + {}_0\boldsymbol{\alpha}_i \times \left({}^0\mathbf{r}_i - {}^0\mathbf{d}_i\right) + {}_0\boldsymbol{\omega}_i \times \left({}_0\boldsymbol{\omega}_i \times \left({}^0\mathbf{r}_i - {}^0\mathbf{d}_i\right)\right) \tag{12.319}
$$

$$
{}_0\boldsymbol{\alpha}_i = \begin{cases} {}_0\boldsymbol{\alpha}_{i-1} + \ddot{\theta}_i\,{}^0\hat{k}_{i-1} + {}_0\boldsymbol{\omega}_{i-1} \times \dot{\theta}_i\,{}^0\hat{k}_{i-1} & \text{if joint } i \text{ is R} \\ {}_0\boldsymbol{\alpha}_{i-1} & \text{if joint } i \text{ is P} \end{cases} \tag{12.320}
$$

Weight is usually the only external load on middle links of a robot, and
reactions from the environment are extra external force systems on the
base and end-effector links. The force and moment that the base actuator
applies to the first link are $\mathbf{F}_0$ and $\mathbf{M}_0$, and the force and moment that the
end-effector applies to the environment are $\mathbf{F}_n$ and $\mathbf{M}_n$. If weight is the
only external load on link $(i)$ and it is in $-{}^0\hat{k}_0$ direction, then we have

$$
\sum {}^0\mathbf{F}_{e_i} = m_i\,{}^0\mathbf{g} = -m_i\,g\,{}^0\hat{k}_0 \tag{12.321}
$$

$$
\sum {}^0\mathbf{M}_{e_i} = {}^0\mathbf{r}_i \times m_i\,{}^0\mathbf{g} = -{}^0\mathbf{r}_i \times m_i\,g\,{}^0\hat{k}_0 \tag{12.322}
$$

where $\mathbf{g}$ is the gravitational acceleration vector.

The Newton-Euler equation of motion can also be written in link's coordinate frame in a forward or backward method. The backward Newton-Euler
equations of motion for link $(i)$ in the the local coordinate frame $B_i$ are

$$
{}^i\mathbf{F}_{i-1} = {}^i\mathbf{F}_i - \sum {}^i\mathbf{F}_{e_i} + m_i\,{}^i_0\mathbf{a}_i \tag{12.323}
$$

$$
\begin{aligned}
{}^i\mathbf{M}_{i-1} = {}^i\mathbf{M}_i - \sum {}^i\mathbf{M}_{e_i} - \left({}^i\mathbf{d}_{i-1} - {}^i\mathbf{r}_i\right) \times {}^i\mathbf{F}_{i-1} \\
+ \left({}^i\mathbf{d}_i - {}^i\mathbf{r}_i\right) \times {}^i\mathbf{F}_i + {}^iI_i\,{}^i_0\boldsymbol{\alpha}_i + {}^i_0\boldsymbol{\omega}_i \times {}^iI_i\,{}^i_0\boldsymbol{\omega}_i
\end{aligned} \tag{12.324}
$$

where

$$^i\mathbf{n}_i = {}^i\mathbf{d}_{i-1} - {}^i\mathbf{r}_i \tag{12.325}$$

$$^i\mathbf{m}_i = {}^i\mathbf{d}_i - {}^i\mathbf{r}_i. \tag{12.326}$$

and

$$_0^i\mathbf{a}_i = {}^i\ddot{\mathbf{d}}_i + {}_0^i\boldsymbol{\alpha}_i \times \left({}^i\mathbf{r}_i - {}^i\mathbf{d}_i\right) + {}_0^i\boldsymbol{\omega}_i \times \left({}_0^i\boldsymbol{\omega}_i \times \left({}^i\mathbf{r}_i - {}^i\mathbf{d}_i\right)\right) \tag{12.327}$$

$$_0^i\boldsymbol{\alpha}_i = \begin{cases} {}^iT_{i-1}\left({}_0^{i-1}\boldsymbol{\alpha}_{i-1} + \ddot{\theta}_i{}^{i-1}\hat{k}_{i-1}\right) \\ \quad + {}^iT_{i-1}\left({}_0^{i-1}\boldsymbol{\omega}_{i-1} \times \dot{\theta}_i{}^{i-1}\hat{k}_{i-1}\right) & \text{if joint } i \text{ is R} \\ {}^iT_{i-1}{}_0^{i-1}\boldsymbol{\alpha}_{i-1} & \text{if joint } i \text{ is P.} \end{cases} \tag{12.328}$$

In this method, we search for the driving force system $({}^i\mathbf{F}_{i-1}, {}^i\mathbf{M}_{i-1})$ by having the driven force system $({}^i\mathbf{F}_i, {}^i\mathbf{M}_i)$ and the resultant external force system $({}^i\mathbf{F}_{e_i}, {}^i\mathbf{M}_{e_i})$. When the driving force system $({}^i\mathbf{F}_{i-1}, {}^i\mathbf{M}_{i-1})$ is found in frame $B_i$, we can transform them to the frame $B_{i-1}$ and apply the Newton-Euler equation for link $(i-1)$.

$$^{i-1}\mathbf{F}_{i-1} = {}^{i-1}T_i{}^i\mathbf{F}_{i-1} \tag{12.329}$$

$$^{i-1}\mathbf{M}_{i-1} = {}^{i-1}T_i{}^i\mathbf{M}_{i-1} \tag{12.330}$$

The negative of the converted force system acts as the driven force system $(-{}^{i-1}\mathbf{F}_{i-1}, -{}^{i-1}\mathbf{M}_{i-1})$ for the link $(i-1)$.

The forward Newton-Euler equations of motion for link $(i)$ in the the local coordinate frame $B_i$ are

$$^i\mathbf{F}_i = {}^i\mathbf{F}_{i-1} + \sum {}^i\mathbf{F}_{e_i} - m_i{}_0^i\mathbf{a}_i \tag{12.331}$$

$$^i\mathbf{M}_i = {}^i\mathbf{M}_{i-1} + \sum {}^i\mathbf{M}_{e_i} + \left({}^i\mathbf{d}_{i-1} - {}^i\mathbf{r}_i\right) \times {}^i\mathbf{F}_{i-1} - \left({}^i\mathbf{d}_i - {}^i\mathbf{r}_i\right) \times {}^i\mathbf{F}_i - {}^iI_i{}_0^i\boldsymbol{\alpha}_i - {}_0^i\boldsymbol{\omega}_i \times {}^iI_i{}_0^i\boldsymbol{\omega}_i. \tag{12.332}$$

$$^i\mathbf{n}_i = {}^i\mathbf{d}_{i-1} - {}^i\mathbf{r}_i \tag{12.333}$$

$$^i\mathbf{m}_i = {}^i\mathbf{d}_i - {}^i\mathbf{r}_i \tag{12.334}$$

Using the forward Newton-Euler equations of motion, we can calculate the reaction force system $({}^i\mathbf{F}_i, {}^i\mathbf{M}_i)$ by having the action force system $({}^i\mathbf{F}_{i-1}, {}^i\mathbf{M}_{i-1})$. When the reaction force system $({}^i\mathbf{F}_i, {}^i\mathbf{M}_i)$ is found in frame $B_i$, we can transform them to the frame $B_{i+1}$

$$^{i+1}\mathbf{F}_i = {}^iT_{i+1}^{-1}{}^i\mathbf{F}_i \tag{12.335}$$

$$^{i+1}\mathbf{M}_i = {}^iT_{i+1}^{-1}{}^i\mathbf{M}_i. \tag{12.336}$$

The negative of the converted force system acts as the action force system $(-\,^{i+1}\mathbf{F}_i, -\,^{i+1}\mathbf{M}_i)$ for the link $(i+1)$ and we can apply the Newton-Euler equation to the link $(i+1)$. The forward Newton-Euler equations of motion allows us to start from a known action force system $(^1\mathbf{F}_0, \,^1\mathbf{M}_0)$, that the base link applies to the link $(1)$, and calculate the action force of the next link. Therefore, analyzing the links of a robot, one by one, we end up with the force system that the end-effector applies to the environment.

The Lagrange equation of motion

$$\frac{d}{dt}\left(\frac{\partial \mathcal{L}}{\partial \dot{q}_i}\right) - \frac{\partial \mathcal{L}}{\partial q_i} \;=\; Q_i \qquad i = 1, 2, \cdots n \qquad (12.337)$$

$$\mathcal{L} \;=\; K - V \qquad\qquad (12.338)$$

provides a systematic approach to obtain the dynamics equations for robots. The variables $q_i$ are the coordinates by which the energies are expressed and the $Q_i$ is the corresponding generalized nonpotential force.

The equations of motion for an $n$ link serial manipulator, based on Newton-Euler or Lagrangian, can always be set in a matrix form

$$\mathbf{D}(\mathbf{q})\,\ddot{\mathbf{q}} + \mathbf{H}(\mathbf{q}, \dot{\mathbf{q}}) + \mathbf{G}(\mathbf{q}) = \mathbf{Q} \qquad (12.339)$$

or in a summation form

$$\sum_{j=1}^{n} D_{ij}(q)\,\ddot{q}_j + \sum_{k=1}^{n}\sum_{m=1}^{n} H_{ikm}\dot{q}_k\dot{q}_m + G_i = Q_i \qquad (12.340)$$

where, $\mathbf{D}(\mathbf{q})$ is an $n \times n$ inertial-type symmetric matrix

$$D = \sum_{i=1}^{n}\left(\mathbf{J}_{Di}^T\, m_i\, \mathbf{J}_{Di} + \frac{1}{2}\mathbf{J}_{Ri}^T\, {}^0 I_i\, \mathbf{J}_{Ri}\right) \qquad (12.341)$$

$H_{ikm}$ is the velocity coupling vector

$$H_{ijk} = \sum_{j=1}^{n}\sum_{k=1}^{n}\left(\frac{\partial D_{ij}}{\partial q_k} - \frac{1}{2}\frac{\partial D_{jk}}{\partial q_i}\right) \qquad (12.342)$$

$G_i$ is the gravitational vector

$$G_i = \sum_{j=1}^{n} m_j \mathbf{g}^T\, \mathbf{J}_{Dj}^{(i)} \qquad (12.343)$$

and $\mathbf{J}_i$ is the Jacobian matrix of the robot

$$\dot{\mathbf{X}}_i = \begin{bmatrix} {}^0\mathbf{v}_i \\ {}^0\boldsymbol{\omega}_i \end{bmatrix} = \begin{bmatrix} \mathbf{J}_{Di} \\ \mathbf{J}_{Ri} \end{bmatrix}\dot{\mathbf{q}} = \mathbf{J}_i\,\dot{\mathbf{q}}. \qquad (12.344)$$

To hold a robot in a stationary configuration, the actuators must apply some required forces to balance the external loads applied to the robot. In the static condition, the globally expressed Newton-Euler equations for the link $(i)$, can be written in a recursive form

$${}^0\mathbf{F}_{i-1} \; = \; {}^0\mathbf{F}_i - \sum {}^0\mathbf{F}_{e_i} \tag{12.345}$$

$${}^0\mathbf{M}_{i-1} \; = \; {}^0\mathbf{M}_i - \sum {}^0\mathbf{M}_{e_i} + {}_{i-1}^{\;\;\;0}\mathbf{d}_i \times {}^0\mathbf{F}_i. \tag{12.346}$$

Now we are able to calculate the action force system $(\mathbf{F}_{i-1}, \mathbf{M}_{i-1})$ when the reaction force system $(-\mathbf{F}_i, -\mathbf{M}_i)$ is given.

# Exercises

1. Notation and symbols.

   Describe their meaning.

   a- $\mathbf{F}_2$     b- $^0\mathbf{F}_1$     c- $^1\mathbf{F}_1$     d- $^2\mathbf{M}_1$     e- $^2\mathbf{M}_{e1}$     f- $^B\mathbf{M}$

   g- $\mathbf{m}_2$     h- $^0\mathbf{n}_2$     i- $^0_{i-1}\mathbf{d}_i$     j- $^0\mathbf{d}_i$     k- $^0\mathbf{r}_i$     l- $^{i-1}\mathbf{d}_i$

   m- $^0\mathbf{L}_2$     n- $^0\mathbf{I}_2$     o- $^0_{i-1}\mathbf{L}_i$     p- $\mathbf{K}_i$     q- $\mathbf{V}_i$     r- $^{i-1}\mathbf{I}_i$

2. Even order recursive translational velocity.

   Find an equation to relate the velocity of link $(i)$ to the velocity of link $(i-2)$, and the velocity of link $(i)$ to the velocity of link $(i+2)$.

3. Even order recursive angular velocity.

   Find an equation to relate the angular velocity of link $(i)$ to the angular velocity of link $(i-2)$, and the angular velocity of link $(i)$ to the angular velocity of link $(i+2)$.

4. Even order recursive translational acceleration.

   Find an equation to relate the acceleration of link $(i)$ to the acceleration of link $(i-2)$, and the acceleration of link $(i)$ to the acceleration of link $(i+2)$.

5. Even order recursive angular acceleration.

   Find an equation to relate the angular acceleration of link $(i)$ to the angular acceleration of link $(i-2)$, and the angular acceleration of link $(i)$ to the angular acceleration of link $(i+2)$.

6. Acceleration in different frames.

   For the 2R planar manipulator shown in Figure 12.5, find $^0_1\mathbf{a}_2$, $^1_0\mathbf{a}_2$, $^0_2\mathbf{a}_1$, $^2_0\mathbf{a}_1$, $^2_0\mathbf{a}_2$, and $^0_1\mathbf{a}_1$.

7. Slider-crank mechanism dynamics.

   A planar slider-crank mechanism is shown in Figure 12.11. Set up the link coordinate frames, develop the Newton-Euler equations of motion, and find the driving moment at the base revolute joint.

8. PR manipulator dynamics.

   Find the equations of motion for the planar polar manipulator shown in Figure 5.37. Eliminate the joints' constraint force and moment to derive the equations for the actuators' force or moment.

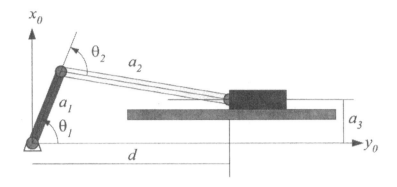

FIGURE 12.11. A planar slider-crank machanism.

9. Global differential of a link momentum.

In recursive Newton-Euler equations of motion, why we do not use the following Newton equation?

$$^i\mathbf{F} = \frac{^G d}{dt} {}^i\mathbf{F} = \frac{^G d}{dt} m {}^i\mathbf{v} = m {}^i\dot{\mathbf{v}} + {}_0^i\omega_i \times m {}^i\mathbf{v}$$

10. 3R planar manipulator dynamics.

A 3R planar manipulator is shown in Figure 12.13. The manipulator is attached to a wall and therefore, $\mathbf{g} = g {}^0\hat{\imath}_0$.

   (a) Find the Newton-Euler equations of motion for the manipulator. Do your calculations in the global frame and derive the dynamic force and moment at each joint.

   (b) Reduce the number of equations to three for moments at joints.

   (c) Substitute the vectorial quantities and calculate the moments in terms of geometry and angular variables of the manipulator.

11. A planar Cartesian manipulator dynamics.

Determine the Newton-Euler equations of motion for the planar Cartesian manipulator shown in Figure 5.38.

12. Polar planar manipulator dynamics.

A polar planar manipulator with 2 DOF is shown in Figure 5.37.

   (a) Determine the Newton-Euler equations of motion for the manipulator.

   (b) Reduce the number of equations to two, for moments at the base joint and force at the P joint.

(c) Substitute the vectorial quantities and calculate the action force and moment in terms of geometry and angular variables of the manipulator.

13. ★ Dynamics of a spherical manipulator.

Figure 5.26 illustrates a spherical manipulator attached with a spherical wrist. Analyze the robot and derive the equations of motion for joints action force and moment. Assume $\mathbf{g} = g\,^0\hat{k}_0$ and the end-effector is carrying a mass $m$.

14. ★ Dynamics of an articulated manipulator.

Figure 5.25 illustrates an articulated manipulator R⊢R‖R. Use $\mathbf{g} = g\,^0\hat{k}_0$ and find the manipulator's equations of motion.

15. ★ Dynamics of a SCARA robot.

Calculate the dynamic joints' force system for the SCARA robot R‖R‖R‖P shown in Figure 5.27 if $\mathbf{g} = g\,^0\hat{k}_0$.

16. ★ Dynamics of an SRMS manipulator.

Figure 5.28 shows a model of the Shuttle remote manipulator system (SRMS).

(a) Derive the equations of motion for the SRMS and calculate the joints' force system for $\mathbf{g} = 0$.

(b) Derive the equations of motion for the SRMS and calculate the joints' force system for $\mathbf{g} = g\,^0\hat{k}_0$.

(c) Eliminate the constraint forces and reduce the number of equations equal to the number of action moments.

(d) Assume the links are made of a uniform cylinder with radius $r = .25\,\mathrm{m}$ and $m = 12\,\mathrm{kg/m}$. Use the characteristics indicated in Table 5.11 and find the equations of motion when the end-effector is holding a 24 kg mass.

17. 3R planar manipulator recursive dynamics.

The manipulator shown in Figure 12.13 is a 3R planar manipulator attached to a wall and therefore, $\mathbf{g} = g\,^0\hat{i}_0$.

(a) Find the equations of motion for the manipulator utilizing the backward recursive Newton-Euler technique.

(b) ★ Find the equations of motion for the manipulator utilizing the forward recursive Newton-Euler technique.

18. Polar planar manipulator recursive dynamics.

Figure 5.37 depicts a polar planar manipulator with 2 DOF.

(a) Find the equations of motion for the manipulator utilizing the backward recursive Newton-Euler technique.

(b) ★ Find the equations of motion for the manipulator utilizing the forward recursive Newton-Euler technique.

19. ★ Recursive dynamics of an articulated manipulator.

Figure 5.25 illustrates an articulated manipulator R⊢R‖R. Use $\mathbf{g} = g\,^0\hat{k}_0$ and find the manipulator's equations of motion

(a) utilizing the backward recursive Newton-Euler technique.

(b) utilizing the forward recursive Newton-Euler technique.

20. ★ Recursive dynamics of a SCARA robot.

A SCARA robot R‖R‖R‖P is shown in Figure 5.27. If $\mathbf{g} = g\,^0\hat{k}_0$ determine the dynamic equations of motion by

(a) utilizing the backward recursive Newton-Euler technique.

(b) utilizing the forward recursive Newton-Euler technique.

21. ★ Recursive dynamics of an SRMS manipulator.

Figure 5.28 shows a model of the Shuttle remote manipulator system (SRMS).

(a) Derive the equations of motion for the SRMS utilizing the backward recursive Newton-Euler technique for $\mathbf{g} = 0$.

(b) Derive the equations of motion for the SRMS utilizing the forward recursive Newton-Euler technique for $\mathbf{g} = 0$.

22. 3R planar manipulator Lagrange dynamics.

Find the equations of motion for the 3R planar manipulator shown in Figure 12.13 utilizing the Lagrange technique. The manipulator is attached to a wall and therefore, $\mathbf{g} = g\,^0\hat{\imath}_0$.

23. Polar planar manipulator Lagrange dynamics.

Find the equations of motion for the polar planar manipulator, shown in Figure 5.37, utilizing the Lagrange technique.

24. ★ Lagrange dynamics of an articulated manipulator.

Figure 5.25 illustrates an articulated manipulator R⊢R‖R. Use $\mathbf{g} = g\,^0\hat{k}_0$ and find the manipulator's equations of motion utilizing the Lagrange technique.

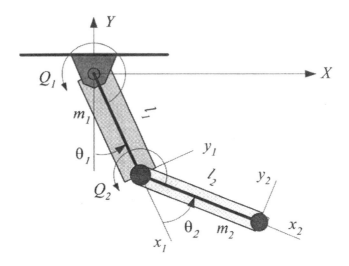

FIGURE 12.12. A 2R planar manipulator attached to a ceiling in static condition.

25. ★ Lagrange dynamics of a SCARA robot.

A SCARA robot R‖R‖R‖P is shown in Figure 5.27. If $\mathbf{g} = g\,{}^0\hat{k}_0$ determine the dynamic equations of motion by applying the Lagrange technique.

26. ★ Lagrange dynamics of an SRMS manipulator.

Figure 5.28 shows a model of the Shuttle remote manipulator system (SRMS). Derive the equations of motion for the SRMS utilizing the Lagrange technique for

(a) $\mathbf{g} = 0$

(b) $\mathbf{g} = g\,{}^0\hat{k}_0$.

27. ★ Work done by actuators.

Consider a 2R planar manipulator moving on a given path. Assume that the endpoint of a 2R manipulator moves with constant speed $v = 1\,\mathrm{m/sec}$ from $P_1$ to $P_2$, on a path made of two semi-circles as shown in Figure 13.16. Calculate the work done by the actuators if $l_1 = l_2 = 1\,\mathrm{m}$ and the manipulator is carrying a 12 kg mass. The center of the circles are at $(0.75\,\mathrm{m},\,0.5\,\mathrm{m})$ and $(-0.75\,\mathrm{m},\,0.5\,\mathrm{m})$.

28. Statics of a 2R planar manipulator.

Figure 12.12 illustrates a 2R planar manipulator attached to a ceiling.

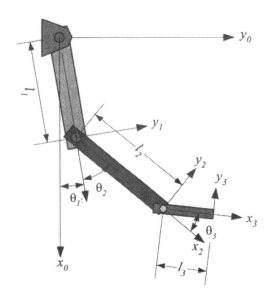

FIGURE 12.13. A 3R planar manipulator attached to a wall.

The links are uniform with

$$
\begin{aligned}
m_1 &= 24\,\text{kg} \\
m_2 &= 18\,\text{kg} \\
l_1 &= 1\,\text{m} \\
l_2 &= 1\,\text{m} \\
\mathbf{g} &= -g\,{}^0\hat{\jmath}_0.
\end{aligned}
$$

There is a load $\mathbf{F}_e = -14g\,{}^0\hat{\jmath}_0\,\text{N}$ at the endpoint. Calculate the static moments $Q_1$ and $Q_2$ for $\theta_1 = 30\,\text{deg}$ and $\theta_2 = 45\,\text{deg}$.

29. **Statics of a 2R planar manipulator at a different base angle.**

   In Exercise 28 keep $\theta_2 = 45\,\text{deg}$ and calculate the static moments $Q_1$ and $Q_2$ as functions of $\theta_1$. Plot $Q_1$ and $Q_2$ versus $\theta_1$ and find the configuration that minimizes $Q_1$, $Q_2$, $Q_1 + Q_2$, and the potential energy $V$.

30. **Statics of a 3R planar manipulator.**

   Figure 12.13 illustrates a 3R planar manipulator attached to a wall. Derive the static force and moment at each joint to keep the configuration of the manipulator if $\mathbf{g} = g\,{}^0\hat{\imath}_0$.

31. ★ Statics of an articulated manipulator.

An articulated manipulator R⊢R∥R is shown in Figure 5.25. Find the static force and moment at joints for $\mathbf{g} = g^0\hat{k}_0$. The end-effector is carrying a 20 kg mass. Calculate the maximum base force moment.

32. ★ Statics of a SCARA robot.

Calculate the static joints' force system for the SCARA robot R∥R∥R∥P shown in Figure 5.27 if $\mathbf{g} = g^0\hat{k}_0$ and the end-effector is carrying a 10 kg mass.

33. ★ Statics of a spherical manipulator.

Figure 5.26 illustrates a spherical manipulator attached with a spherical wrist. Analyze the robot and calculate the static force system in joints for $\mathbf{g} = g^0\hat{k}_0$ if the end-effector is carrying a 12 kg mass.

34. ★ Statics of an SRMS manipulator.

A model of the Shuttle remote manipulator system (SRMS) is shown in Figure 5.28. Analyze the static configuration of the SRMS and calculate the joints' force system for $\mathbf{g} = g^0\hat{k}_0$.

Assume the links are made of a uniform cylinder with radius $r = .25$ m and $m = 12$ kg/m. Use the characteristics indicated in Table 5.11 and find the maximum value of the base force system for a 24 kg mass held by the end-effector. The SRMS is supposed to work in a no-gravity field.

# Part III

# Control

Control is the science of *desired motion*. It relates the dynamics and kinematics of a robot to a prescribed motion. It includes optimization problems to determine forces so that the system will behave optimally. A typical example is the situation in which the initial and terminal configurations of the robot are given and the forces acting on the robot must be found to have the motion in minimum time.

Path or trajectory planning is a part of control, in which we plan a path followed by the manipulator in a planned time profile. Paths can be planned in joint or Cartesian space. Joint path planning directly specifies the time evolution of the joint variables. However, Cartesian path specifies the position and orientation of the end frame. So, path includes attaining a desired target from an initial configuration. It may include avoiding obstacles. Joint path planning is simple because it does not involve inverse kinematics, but it is hard to digest the motion of the manipulator in Cartesian space. However, Cartesian coordinates make sense but need inverse kinematics calculation.

In this Part, we develop techniques to derive the required commands to control the robot's task.

# 13

# Path Planning

## 13.1  Joint Cubic Path

A cubic path in joint space for the joint variable $q_i(t)$ between two points $q_i(t_0)$ and $q_i(t_f)$ is

$$q_i(t) = a_0 + a_1 t + a_2 t^2 + a_3 t^3 \tag{13.1}$$

where

$$
\begin{aligned}
a_0 &= -\frac{q_1 \left(t_0^3 - 3t_0^2 t_f\right) + q_0 \left(3t_0 t_f^2 - t_f^3\right)}{(t_f - t_0)^3} \\
&\quad - \frac{q_0' \left(t_0 t_f^3 - t_0^2 t_f^2\right) + q_1' \left(t_0^2 t_f^2 - t_0^3 t_f\right)}{(t_f - t_0)^3}
\end{aligned} \tag{13.2}
$$

$$
\begin{aligned}
a_1 &= \frac{6 q_0 t_0 t_f - 6 q_1 t_0 t_f}{(t_f - t_0)^3} \\
&\quad + \frac{q_0' \left(t_f^3 + t_0 t_f^2 - 2t_0^2 t_f\right) + q_1' \left(2t_0 t_f^2 - t_0^3 - t_0^2 t_f\right)}{(t_f - t_0)^3}
\end{aligned} \tag{13.3}
$$

$$
\begin{aligned}
a_2 &= -\frac{q_0 \left(3t_0 + 3t_f\right) + q_1 \left(-3t_0 - 3t_f\right)}{(t_f - t_0)^3} \\
&\quad - \frac{q_1' \left(t_0 t_f - 2t_0^2 + t_f^2\right) + q_0' \left(2t_f^2 - t_0^2 - t_0 t_f\right)}{(t_f - t_0)^3}
\end{aligned} \tag{13.4}
$$

$$
a_3 = \frac{2q_0 - 2q_1 + q_0' \left(t_f - t_0\right) + q_1' \left(t_f - t_0\right)}{(t_f - t_0)^3} \tag{13.5}
$$

and

$$
\begin{aligned}
q_i(t_0) &= q_0 \\
\dot{q}_i(t_0) &= q_0' \\
q_i(t_f) &= q_f \\
\dot{q}_i(t_f) &= q_f'.
\end{aligned} \tag{13.6}
$$

**Proof.** A cubic polynomial has four coefficients. Therefore, it can satisfy the position and velocity constraints at the initial and final points. For simplicity, we call the value of the joint variable, the *position*, and the rate of the joint variable, the *velocity*. Assume that the position and velocity of a joint variable at the initial time $t_0$ and at the final time $t_f$ are given as (13.6).

Substituting the boundary conditions in the position and velocity equations of the joint variable

$$q_i(t) = a_0 + a_1 t + a_2 t^2 + a_3 t^3 \tag{13.7}$$
$$\dot{q}_i(t) = a_1 + 2a_2 t + 3a_3 t^2 \tag{13.8}$$

generates four equations for the coefficients of the path

$$
\begin{bmatrix}
1 & t_0 & t_0^2 & t_0^3 \\
0 & 1 & 2t_0 & 3t_0^2 \\
1 & t_f & t_f^2 & t_f^3 \\
0 & 1 & 2t_f & 3t_f^2
\end{bmatrix}
\begin{bmatrix}
a_0 \\ a_1 \\ a_2 \\ a_3
\end{bmatrix}
=
\begin{bmatrix}
q_0 \\ q_0' \\ q_f \\ q_f'
\end{bmatrix}. \tag{13.9}
$$

Their solutions are given in (13.2) to (13.5).

In case that $t_0 = 0$, the coefficients simplify to

$$a_0 = q_0 \tag{13.10}$$
$$a_1 = q_0' \tag{13.11}$$
$$a_2 = \frac{3(q_f - q_0) - \left(2q_0' + q_f'\right)t_f}{t_f^2} \tag{13.12}$$
$$a_3 = \frac{-2(q_f - q_0) + \left(q_0' + q_f'\right)t_f}{t_f^3}. \tag{13.13}$$

It is also possible to employ a time shift and search for a cubic polynomial of the form

$$q_i(t) = a_0 + a_1(t - t_0) + a_2 t(t - t_0)^2 + a_3(t - t_0)^3. \tag{13.14}$$

Now, the boundary conditions (13.6) generate a set of equations

$$
\begin{bmatrix}
1 & 0 & 0 & 0 \\
0 & 1 & 0 & 0 \\
1 & (t_f - t_0) & (t_f - t_0)^2 & (t_f - t_0)^3 \\
0 & 1 & 2(t_f - t_0) & 3(t_f - t_0)^2
\end{bmatrix}
\begin{bmatrix}
a_0 \\ a_1 \\ a_2 \\ a_3
\end{bmatrix}
=
\begin{bmatrix}
q_0 \\ q_0' \\ q_f \\ q_f'
\end{bmatrix} \tag{13.15}
$$

with the following solutions:

$$
\begin{bmatrix}
a_0 \\ a_1 \\ a_2 \\ a_3
\end{bmatrix}
=
\begin{bmatrix}
q_0 \\
q_0' \\
-(t_f - t_0)^{-2}\left(3q_0 - 3q_f - 2t_0 q_0' - t_0 q_f' + 2t_f q_0' + t_f q_f'\right) \\
(t_f - t_0)^{-3}\left(2q_0 - 2q_f - t_0 q_0' - t_0 q_f' + t_f q_0' + t_f q_f'\right)
\end{bmatrix} \tag{13.16}
$$

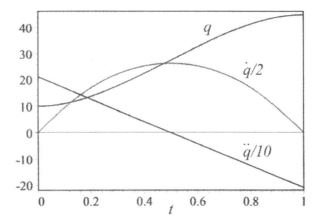

FIGURE 13.1. Kinematics of a rest-to-rest cubic path in joint space.

A disadvantage of cubic paths is the acceleration jump at boundaries that introduces infinite jerks. ∎

**Example 295** *Rest-to-rest cubic path.*
*Assume $q(0) = 10\,\mathrm{deg}$, $q(1) = 45\,\mathrm{deg}$, and $\dot{q}(0) = \dot{q}(1) = 0$. The coefficients of the cubic path are*

$$
\begin{aligned}
a_0 &= 10 \\
a_1 &= 0 \\
a_2 &= 105 \\
a_3 &= -70
\end{aligned}
\tag{13.17}
$$

*that generate a path for the joint variable as*

$$
q(t) = 10 + 105t^2 - 70t^3 \ \mathrm{deg}.
\tag{13.18}
$$

*The path information is shown in Figure 13.1.*

**Example 296** *To-rest cubic path.*
*Assume the angle of a joint starts from $\theta(0) = 10\,\mathrm{deg}$, $\dot{\theta}(0) = 12\,\mathrm{deg}/s$ and ends at $\theta(2) = 45\,\mathrm{deg}$, $\dot{q}_i(0) = 0$. The coefficients of a cubic path for this motion are*

$$
\begin{aligned}
a_0 &= 10 \\
a_1 &= 12 \\
a_2 &= \frac{81}{2} \\
a_3 &= \frac{-29}{2}.
\end{aligned}
\tag{13.19}
$$

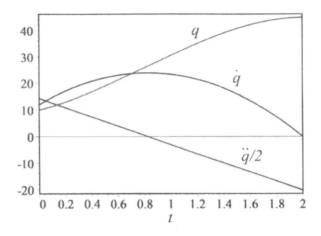

FIGURE 13.2. Kinematics of a to-rest cubic path in joint space.

*The kinematics of this path are*

$$\theta(t) \quad = \quad 10 + 12t + 40.5t^2 - 14.5t^3 \text{ deg} \qquad (13.20)$$
$$\dot{\theta}(t) \quad = \quad 81t - 43.5t^2 + 12 \text{ deg / s} \qquad (13.21)$$
$$\ddot{\theta}(t) \quad = \quad 81 - 87t \text{ deg / s}^2 \qquad (13.22)$$

*and are shown graphically in Figure 13.2.*

**Example 297** *Rest-to-rest path with a constant velocity in the middle.*
   *Assume we need a rest-to-rest path with a constant given velocity $\dot{q} = \dot{q}_c$ for $t_1 < t < t_2$ where $t_0 < t_1 < t_2 < t_f$. We show the boundary conditions to be satisfied as*

$$q(t_0) \quad = \quad q_0$$
$$\dot{q}(t_0) \quad = \quad q_0'$$
$$\dot{q}(t) \quad = \quad q_c' \qquad t_1 < t < t_2$$
$$q(t_f) \quad = \quad q_f$$
$$\dot{q}(t_f) \quad = \quad q_f'. \qquad (13.23)$$

   *The path has three parts: rest-to, constant-velocity, and to-rest. We need an equation for the rest-to part of the motion to achieve the given velocity. A quadratic path has three coefficients and can be utilized to satisfy three conditions.*

$$q_1(t) \quad = \quad a_0 + a_1 t + a_2 t^2 \qquad (13.24)$$
$$\dot{q}_1(t) \quad = \quad a_1 + 2a_2 t \qquad (13.25)$$

*The conditions are the initial position and velocity, and the final constant* *velocity. Assuming $t_0 = 0$ the conditions for the rest-to path are*

$$q_1(0) = q_0$$
$$\dot{q}_1(0) = 0$$
$$\dot{q}_1(t_1) = q_c' \qquad (13.26)$$

*that generate the following equations:*

$$q_0 = a_0 \qquad (13.27)$$
$$0 = a_1 \qquad (13.28)$$
$$q_c' = 2a_2 t_1. \qquad (13.29)$$

*Therefore, the rest-to path is*

$$q_1(t) = q_0 + \frac{q_c'}{2t_1} t^2 \qquad 0 < t < t_1. \qquad (13.30)$$

*Given the specific constant velocity $q_c'$ shows that the path in the middle part is*

$$\dot{q}_2(t) = q_c' \qquad (13.31)$$
$$q_2(t) = q_c' t + C_1 \qquad t_1 < t < t_2. \qquad (13.32)$$

*The constant of integration can be found by utilizing the position condition at $t = t_1$*

$$q_0 + \frac{q_c'}{2t_1} t_1^2 = q_c' t_1 + C_1 \qquad (13.33)$$

$$C_1 = q_0 - \frac{1}{2} t_1 q_c'. \qquad (13.34)$$

*There are four conditions for the to-rest part of the path. Therefore, it can be calculated utilizing a cubic equation*

$$q_3(t) = b_0 + b_1 t + b_2 t^2 + b_3 t^3 \qquad (13.35)$$
$$\dot{q}_3(t) = b_1 + 2b_2 t + 3b_3 t^2 \qquad (13.36)$$

*and the following boundary conditions:*

$$q_3(t_f) = q_f \qquad (13.37)$$
$$\dot{q}_3(t_f) = 0 \qquad (13.38)$$
$$q_3(t_2) = q_2(t_2) = q_2 = q_c' t_2 + q_0 - \frac{1}{2} t_1 q_c' \qquad (13.39)$$
$$\dot{q}(t_2) = \dot{q}_2(t_2) = q_c'. \qquad (13.40)$$

*These conditions generate three equations*

$$
\begin{bmatrix}
1 & t_f & t_f^2 & t_f^3 \\
0 & 1 & 2t_f & 3t_f^2 \\
1 & t_2 & t_2^2 & t_2^3 \\
0 & 1 & 2t_2 & 3t_2^2
\end{bmatrix}
\begin{bmatrix}
b_0 \\
b_1 \\
b_2 \\
b_3
\end{bmatrix}
=
\begin{bmatrix}
q_f \\
0 \\
q_c' t_2 + q_0 - \frac{1}{2} t_1 q_c' \\
q_c'
\end{bmatrix}
\tag{13.41}
$$

*with the following solutions:*

$$
b_0 = -t_2 t_f^2 \frac{q_c'}{-2t_2 t_f + t_2^2 + t_f^2} + q_2 \frac{t_f^3 - 3t_2 t_f^2}{-t_2^3 + t_f^3 - 3t_2 t_f^2 + 3t_2^2 t_f}
$$

$$
+ q_f \frac{-t_2^3 + 3t_2^2 t_f}{-t_2^3 + t_f^3 - 3t_2 t_f^2 + 3t_2^2 t_f}
$$

$$
b_1 = q_c' \frac{2t_2 t_f + t_f^2}{-2t_2 t_f + t_2^2 + t_f^2} + 6q_2 t_2 \frac{t_f}{-t_2^3 + t_f^3 - 3t_2 t_f^2 + 3t_2^2 t_f}
$$

$$
- 6t_2 q_f \frac{t_f}{-t_2^3 + t_f^3 - 3t_2 t_f^2 + 3t_2^2 t_f}
$$

$$
b_2 = q_c' \frac{-t_2 - 2t_f}{-2t_2 t_f + t_2^2 + t_f^2} + q_2 \frac{-3t_2 - 3t_f}{-t_2^3 + t_f^3 - 3t_2 t_f^2 + 3t_2^2 t_f}
$$

$$
+ q_f \frac{3t_2 + 3t_f}{-t_2^3 + t_f^3 - 3t_2 t_f^2 + 3t_2^2 t_f}
$$

$$
b_3 = \frac{q_c'}{-2t_2 t_f + t_2^2 + t_f^2} + 2\frac{q_2}{-t_2^3 + t_f^3 - 3t_2 t_f^2 + 3t_2^2 t_f}
$$

$$
- 2\frac{q_f}{-t_2^3 + t_f^3 - 3t_2 t_f^2 + 3t_2^2 t_f}
\tag{13.42}
$$

*for*

$$
t_2 < t < t_f.
$$

*A graph of the path for the following values is illustrated in Figure 13.3.*

$$
\begin{aligned}
t_1 &= 0.4\,\text{s} \\
t_2 &= 0.7\,\text{s} \\
t_f &= 1\,\text{s} \\
q_0 &= 0 \\
q_f &= 60\,\text{deg} \\
q_c' &= 50\,\text{deg}\,/\,\text{s}
\end{aligned}
$$

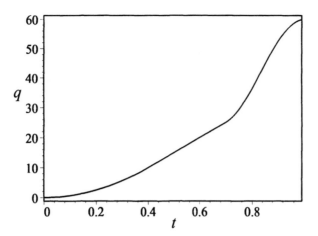

FIGURE 13.3. A piecewise rest-to-rest path with a constant velocity in the middle.

**Example 298** *A quadratic path through three points.*
*A quadratic path passing through three points* $(q_1, t_1)$, $(q_2, t_2)$, *and* $(q_3, t_3)$
*is*

$$q(t) = \frac{(t - t_2)(t - t_3)}{(t_1 - t_2)(t_1 - t_3)} q_1$$
$$+ \frac{(t - t_3)(t - t_1)}{(t_2 - t_3)(t_2 - t_1)} q_2$$
$$+ \frac{(t - t_1)(t - t_2)}{(t_3 - t_1)(t_3 - t_1)} q_3. \qquad (13.43)$$

*As an example, the path passing through* $(10 \deg, 0)$, $(25 \deg, 0.5)$, *and* $(45 \deg, 1)$ *is*

$$q(t) = \frac{(t - 0.5)(t - 1)}{-0.5(-1)} 10 + \frac{(t - 1)t}{(-0.5)(0.5)} 25 + \frac{t(t - 0.5)}{1} 45$$
$$= -\frac{5}{2} (14.0t^2 - 19.0t - 4.0). \qquad (13.44)$$

*The velocity of the path at both ends are*

$$\dot{q}(0) = 47.5 \deg / s$$
$$\dot{q}(1) = -22.5 \deg / s.$$

## 13.2  Higher Polynomial Path

The number of required conditions determines the degree of the polynomial for $q = q(t)$. In general, a polynomial path of degree $n$,

$$q(t) = a_0 + a_1 t + a_2 t^2 + \cdots + a_n t^n \tag{13.45}$$

needs $n + 1$ conditions. The conditions may be of two types: positions at a series of points, so that the trajectory will pass through all specified points; or position, velocity, acceleration, and jerk at two points, so that the smoothness of the path can be controlled.

The problem of searching for the coefficients of a polynomial reduces to a set of linear algebraic equations and may be solved numerically. However, the path planning can be simplified by splitting the whole path into a series of segments and utilizing combinations of lower order polynomials for different segments of the path. The polynomials must then be joined together to satisfy all the required boundary conditions.

**Example 299** *Quintic path.*
*Forcing a joint variable to have specific position, velocity, and accelera-
tion at boundaries introduces six conditions:*

$$
\begin{aligned}
q(t_0) &= q_0 \\
\dot{q}(t_0) &= q_0' \\
\ddot{q}(t_0) &= q_0'' \\
q(t_f) &= q_f \\
\dot{q}_i(t_f) &= q_f' \\
\ddot{q}(t_f) &= q_f''
\end{aligned}
\tag{13.46}
$$

*A five degree polynomial can satisfy these conditions*

$$q(t) = a_0 + a_1 t + a_2 t^2 + a_3 t^3 + a_4 t^4 + a_5 t^5 \tag{13.47}$$

*and generates a set of six equations:*

$$
\begin{bmatrix}
1 & t_0 & t_0^2 & t_0^3 & t_0^4 & t_0^5 \\
0 & 1 & 2t_0 & 3t_0^2 & 4t_0^3 & 5t_0^4 \\
0 & 0 & 2 & 6t_0 & 12t_0^2 & 20t_0^3 \\
1 & t_f & t_f^2 & t_f^3 & t_f^4 & t_f^5 \\
0 & 1 & 2t_f & 3t_f^2 & 4t_f^3 & 5t_f^4 \\
0 & 0 & 2 & 6t_f & 12t_f^2 & 20t_f^3
\end{bmatrix}
\begin{bmatrix}
a_0 \\
a_1 \\
a_2 \\
a_3 \\
a_4 \\
a_5
\end{bmatrix}
=
\begin{bmatrix}
q_0 \\
q_0' \\
q_0'' \\
q_f \\
q_f' \\
q_f''
\end{bmatrix}
\tag{13.48}
$$

*A rest-to-rest path with no acceleration at the rest positions with the following conditions:*

$$
\begin{aligned}
q(0) &= 10\,\mathrm{deg} \\
\dot{q}(0) &= 0 \\
\ddot{q}(0) &= 0 \\
q(1) &= 45\,\mathrm{deg} \\
\dot{q}_i(1) &= 0 \\
\ddot{q}(1) &= 0
\end{aligned}
\tag{13.49}
$$

*can be found by solving a set of equations for the coefficients of the polynomial*

$$
\begin{bmatrix}
1 & 0 & 0 & 0 & 0 & 0 \\
0 & 1 & 0 & 0 & 0 & 0 \\
0 & 0 & 2 & 0 & 0 & 0 \\
1 & 1 & 1 & 1 & 1 & 1 \\
0 & 1 & 2 & 3 & 4 & 5 \\
0 & 0 & 2 & 6 & 12 & 20
\end{bmatrix}
\begin{bmatrix}
a_0 \\ a_1 \\ a_2 \\ a_3 \\ a_4 \\ a_5
\end{bmatrix}
=
\begin{bmatrix}
10 \\ 0 \\ 0 \\ 45 \\ 0 \\ 0
\end{bmatrix}
\tag{13.50}
$$

*which shows*

$$
\begin{bmatrix}
a_0 \\ a_1 \\ a_2 \\ a_3 \\ a_4 \\ a_5
\end{bmatrix}
=
\begin{bmatrix}
10 \\ 0 \\ 0 \\ 350 \\ -525 \\ 210
\end{bmatrix}.
\tag{13.51}
$$

*The path equation is then equal to*

$$
q(t) = 10 + 350t^3 - 525t^4 + 210t^5
\tag{13.52}
$$

*which is shown in Figure 13.4.*

**Example 300** *A jerk zero at a start-stop path.*

To make a path start and stop with zero jerk, a seven degree polynomial

$$
q(t) = a_0 + a_1 t + a_2 t^2 + a_3 t^3 + a_4 t^4 + a_5 t^5 + a_6 t^6 + a_7 t^7
\tag{13.53}
$$

*and eight boundary conditions must be employed.*

$$
\begin{aligned}
q(0) &= q_0 \\
\dot{q}(0) &= 0 \\
\ddot{q}(0) &= 0 \\
\dddot{q}(0) &= 0 \\
q(1) &= q_f \\
\dot{q}_i(1) &= 0 \\
\ddot{q}(1) &= 0 \\
\dddot{q}(1) &= 0
\end{aligned}
\tag{13.54}
$$

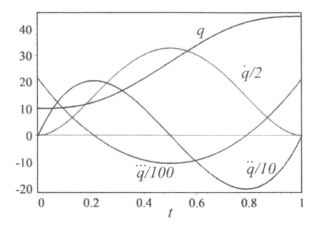

FIGURE 13.4. A quintic rest-to-rest path.

*Such a zero jerk start-stop path for $q(0) = 10\,\text{deg}$ and $q(1) = 45\,\text{deg}$, can be found by solving the following set of equations for the unknown coefficients $a_0, a_1, \cdots, a_7$*

$$
\begin{bmatrix}
1 & 0 & 0 & 0 & 0 & 0 & 0 & 0 \\
0 & 1 & 0 & 0 & 0 & 0 & 0 & 0 \\
0 & 0 & 2 & 0 & 0 & 0 & 0 & 0 \\
0 & 0 & 0 & 6 & 0 & 0 & 0 & 0 \\
1 & 1 & 1 & 1 & 1 & 1 & 1 & 1 \\
0 & 1 & 2 & 3 & 4 & 5 & 6 & 7 \\
0 & 0 & 2 & 6 & 12 & 20 & 30 & 42 \\
0 & 0 & 0 & 6 & 24 & 60 & 120 & 210
\end{bmatrix}
\begin{bmatrix}
a_0 \\ a_1 \\ a_2 \\ a_3 \\ a_4 \\ a_5 \\ a_6 \\ a_7
\end{bmatrix}
=
\begin{bmatrix}
10 \\ 0 \\ 0 \\ 0 \\ 45 \\ 0 \\ 0 \\ 0
\end{bmatrix}
\qquad (13.55)
$$

*which provides*

$$q(t) = 10 + 1225t^4 - 2940t^5 + 2450t^6 - 700t^7. \qquad (13.56)$$

*A graph of this path is illustrated in Figure 13.5.*

 *Figures 13.1, depicts the path of a rest-to-rest motion with no condition on the acceleration and jerk. Figure 13.4 shows an improvement by forcing the motion to have zero accelerations at start and stop. In Figure 13.5, the motion is forced to have zero acceleration and zero jerk at start and stop. Hence, it shows the smoothest start and stop. However, increasing the smoothness of the start and stop increases the peak value of acceleration.*

**Example 301** *Constant acceleration path.*

 *A constant acceleration path has two segments with positive and negative accelerations. Let's assume the absolute value of the positive and negative accelerations are given.*

$$|\ddot{q}(t_0)| = a_c \qquad (13.57)$$

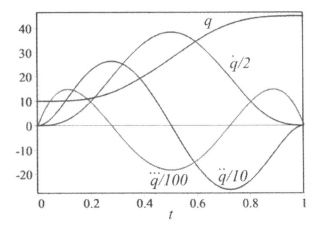

FIGURE 13.5. A jerk zero at start-stop path.

The first half of the motion has a positive acceleration that needs a second degree polynomial

$$\dot{q}_1(t_0) = a_c t \tag{13.58}$$

$$q_1(t_0) = \frac{1}{2} a_c t^2 + q_0 \tag{13.59}$$

for

$$0 < t < \frac{1}{2} t_f. \tag{13.60}$$

The constants of integration are found based on the initial conditions.

$$q_1(0) = q_0$$
$$\dot{q}_1(0) = 0 \tag{13.61}$$

For the second half of the path, we may start with a second degree polynomial

$$q_2(t) = a_0 + a_1 t + a_2 t^2 \qquad \frac{1}{2} t_f < t < t_f \tag{13.62}$$

and impose the following boundary conditions:

$$q_2(t_f) = q_f$$
$$\dot{q}_2(t_f) = 0$$
$$q_1(\frac{t_f}{2}) = q_2(\frac{t_f}{2}) = \frac{1}{8} a_c t_f^2 + q_0 \tag{13.63}$$

These conditions generate three equations for the unknown coefficients

$$\begin{bmatrix} 1 & t_f & t_f^2 \\ 0 & 1 & 2t_f \\ 0 & 1 & t_f \end{bmatrix} \begin{bmatrix} a_0 \\ a_1 \\ a_2 \end{bmatrix} = \begin{bmatrix} q_f \\ 0 \\ \frac{1}{8} a_c t_f^2 + q_0 \end{bmatrix} \tag{13.64}$$

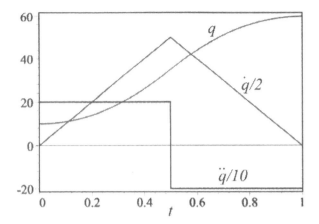

FIGURE 13.6. A constant acceleration path.

*with the following solution:*

$$
\begin{bmatrix} a_0 \\ a_1 \\ a_2 \end{bmatrix} = \begin{bmatrix} q_f - t_f \left( q_0 + \frac{1}{8} a_c t_f^2 \right) \\ 2q_0 + \frac{1}{4} a_c t_f^2 \\ -\frac{1}{t_f} \left( q_0 + \frac{1}{8} a_c t_f^2 \right) \end{bmatrix}
\tag{13.65}
$$

*A constant acceleration path is shown in Figure 13.6 for the conditions $q_0 = 10\,\mathrm{deg}$, $q_f = 45\,\mathrm{deg}$, $t_f = 1$, and $a_c = 200\,\mathrm{deg}/\mathrm{s}^2$.*

**Example 302** *Point sequence path.*
*A path can be assigned via a series of points to approximate a trajectory. Consider an example path specified by four points $q_0$, $q_1$, $q_2$, and $q_3$, such that the points are reached at times $t_0$, $t_1$, $t_2$, and $t_3$ respectively. In addition to positions, we usually impose constraint on initial and final velocities and accelerations. The conditions for such a sequence of points can be*

$$
\begin{aligned}
q(t_0) &= q_0 \\
\dot{q}(t_0) &= 0 \\
\ddot{q}(t_0) &= 0 \\
q(t_1) &= q_1 \\
q(t_2) &= q_2 \\
q(t_3) &= q_3 \\
\dot{q}_i(t_3) &= 0 \\
\ddot{q}(t_3) &= 0.
\end{aligned}
\tag{13.66}
$$

*A seven degree polynomial can be utilized to satisfy these eight conditions.*

$$q(t) = a_0 + a_1 t + a_2 t^2 + a_3 t^3 + a_4 t^4$$
$$+ a_5 t^5 + a_6 t^6 + a_7 t^7 \tag{13.67}$$

*The set of equations for the unknown coefficients is*

$$
\begin{bmatrix}
1 & t_0 & t_0^2 & t_0^3 & t_0^4 & t_0^5 & t_0^6 & t_0^7 \\
0 & 1 & 2t_0 & 3t_0^2 & 4t_0^3 & 5t_0^4 & 6t_0^5 & 7t_0^6 \\
0 & 0 & 2 & 6t_0 & 12t_0^2 & 20t_0^3 & 30t_0^4 & 42t_0^5 \\
1 & t_1 & t_1^2 & t_1^3 & t_1^4 & t_1^5 & t_1^6 & t_1^7 \\
1 & t_2 & t_2^2 & t_2^3 & t_2^4 & t_2^5 & t_2^6 & t_2^7 \\
1 & t_3 & t_3^2 & t_3^3 & t_3^4 & t_3^5 & t_3^6 & t_3^7 \\
0 & 1 & 2t_3 & 3t_3^2 & 4t_3^3 & 5t_3^4 & 6t_3^5 & 7t_3^6 \\
0 & 0 & 2 & 6t_3 & 12t_3^2 & 20t_3^3 & 30t_3^4 & 42t_3^5
\end{bmatrix}
\begin{bmatrix}
a_0 \\ a_1 \\ a_2 \\ a_3 \\ a_4 \\ a_5 \\ a_6 \\ a_7
\end{bmatrix}
=
\begin{bmatrix}
q_0 \\ 0 \\ 0 \\ q_1 \\ q_2 \\ q_3 \\ 0 \\ 0
\end{bmatrix}
\tag{13.68}
$$

*that can be simplified to*

$$
\begin{bmatrix}
1 & 0 & 0 & 0 & 0 & 0 & 0 & 0 \\
0 & 1 & 0 & 0 & 0 & 0 & 0 & 0 \\
0 & 0 & 2 & 0 & 0 & 0 & 0 & 0 \\
1 & 0.4 & 0.4^2 & 0.4^3 & 0.4^4 & 0.4^5 & 0.4^6 & 0.4^7 \\
1 & 0.7 & 0.7^2 & 0.7^3 & 0.7^4 & 0.7^5 & 0.7^6 & 0.7^7 \\
1 & 1 & 1 & 1 & 1 & 1 & 1 & 1 \\
0 & 1 & 2 & 3 & 4 & 5 & 6 & 7 \\
0 & 0 & 2 & 6 & 12 & 20 & 30 & 42
\end{bmatrix}
\begin{bmatrix}
a_0 \\ a_1 \\ a_2 \\ a_3 \\ a_4 \\ a_5 \\ a_6 \\ a_7
\end{bmatrix}
=
\begin{bmatrix}
10 \\ 0 \\ 0 \\ 20 \\ 30 \\ 45 \\ 0 \\ 0
\end{bmatrix}
\tag{13.69}
$$

*for the example data given below.*

$$
\begin{aligned}
q(0) &= 10 \, \text{deg} \\
\dot{q}(0) &= 0 \\
\ddot{q}(0) &= 0 \\
q(0.4) &= 20 \, \text{deg} \\
q(0.7) &= 30 \, \text{deg} \\
q(1) &= 45 \, \text{deg} \\
\dot{q}_i(1) &= 0 \\
\ddot{q}(1) &= 0.
\end{aligned}
\tag{13.70}
$$

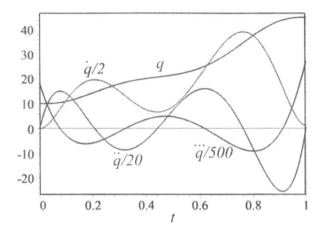

FIGURE 13.7. A point sequence path.

*The solution for the coefficients is*

$$
\begin{bmatrix} a_0 \\ a_1 \\ a_2 \\ a_3 \\ a_4 \\ a_5 \\ a_6 \\ a_7 \end{bmatrix} = \begin{bmatrix} 10 \\ 0 \\ 0 \\ 1500.5 \\ -7053 \\ 12891 \\ -10380 \\ 3076.9 \end{bmatrix}. \tag{13.71}
$$

*These coefficients generate a path as shown in Figure 13.7.*

*This method provides a continuous and differentiable function for joint variables. Continuity and differentiability of $q = q(t)$ is an advantage that provides a continuous velocity, acceleration, and jerk. However, the number of equations increases by increasing the number of points, which needs larger data storage and increases the calculating time.*

**Example 303** *Splitting a path into a series of segments.*

*Instead of using a single high degree polynomial for the entire trajectory, we may prefer to split the trajectory into some segments and use a series of low order polynomials.*

*Consider a path for the following boundary conditions:*

$$
\begin{aligned}
q(t_0) &= q_0 \\
\dot{q}(t_0) &= 0 \\
\ddot{q}(t_0) &= 0 \\
q(t_4) &= q_3 \\
\dot{q}(t_4) &= 0 \\
\ddot{q}(t_4) &= 0.
\end{aligned} \tag{13.72}
$$

*which must also pass through three middle points given below.*

$$
\begin{aligned}
q(t_1) &= q_1 \\
q(t_2) &= q_2 \\
q(t_3) &= q_3
\end{aligned} \tag{13.73}
$$

*Let's split the entire path into four segments, namely $q_1(t)$, $q_2(t)$, $q_3(t)$, and $q_4(t)$.*

$$
\begin{array}{llll}
q_1(t) & for & q(t_0) < q_1(t) < q(t_1) & and & t_0 < t < t_1 \\
q_2(t) & for & q(t_1) < q_2(t) < q(t_2) & and & t_1 < t < t_2 \\
q_3(t) & for & q(t_2) < q_3(t) < q(t_3) & and & t_2 < t < t_3 \\
q_4(t) & for & q(t_3) < q_4(t) < q(t_4) & and & t_3 < t < t_4
\end{array}
$$

*The boundary conditions for the first segment are*

$$
\begin{aligned}
q_1(t_0) &= q_0 \\
\dot{q}_1(t_0) &= 0 \\
\ddot{q}_1(t_0) &= 0 \\
q_1(t_1) &= q_1
\end{aligned} \tag{13.74}
$$

*which can be satisfied by a cubic function.*

$$
q_1(t) = a_0 + a_1 (t - t_0) + a_2 (t - t_0)^2 + a_3 (t - t_0)^3 \tag{13.75}
$$

*The coefficients can be calculated by solving a set of equations*

$$
\begin{bmatrix}
1 & 0 & 0 & 0 \\
0 & 1 & 0 & 0 \\
0 & 0 & 2 & 0 \\
1 & (t_1 - t_0) & (t_1 - t_0)^2 & (t_1 - t_0)^3
\end{bmatrix}
\begin{bmatrix}
a_0 \\ a_1 \\ a_2 \\ a_3
\end{bmatrix}
=
\begin{bmatrix}
q_0 \\ 0 \\ 0 \\ q_1
\end{bmatrix} \tag{13.76}
$$

*that provides*

$$
\begin{aligned}
a_0 &= q_0 \\
a_1 &= 0 \\
a_2 &= 0 \\
a_3 &= \frac{q_1 - q_0}{(t_1 - t_0)^3}.
\end{aligned} \tag{13.77}
$$

The path in the second segment must satisfy the following boundary conditions:

$$
\begin{aligned}
q_2(t_1) &= q_1 \\
\dot{q}_2(t_1) &= \dot{q}_1(t_1) \\
&= a_1 + 2a_2(t_1 - t_0)^2 + 3a_3(t_1 - t_0)^2 \\
&= q_0 + 3\frac{q_1 - q_0}{t_1 - t_0} \\
q_2(t_2) &= q_2.
\end{aligned}
\tag{13.78}
$$

A quadratic polynomial will satisfy these conditions:

$$
q_2(t) = b_0 + b_1 t + b_2 t^2.
\tag{13.79}
$$

The coefficients are the solutions of

$$
\begin{bmatrix}
1 & t_1 & t_1^2 \\
0 & 1 & 2t_1 \\
1 & t_2 & t_2^2
\end{bmatrix}
\begin{bmatrix}
b_0 \\
b_1 \\
b_2
\end{bmatrix}
=
\begin{bmatrix}
q_1 \\
q_0 + 3\frac{q_1 - q_0}{t_1 - t_0} \\
q_2
\end{bmatrix}
\tag{13.80}
$$

that provide

$$
\begin{aligned}
b_0 &= q_2 \frac{t_1^2}{-2t_1 t_2 + t_1^2 + t_2^2} + q_1 \frac{-2t_1 t_2 + t_2^2}{-2t_1 t_2 + t_1^2 + t_2^2} \\
&\quad - t_1 \frac{t_2}{-t_1 + t_2}\left(q_0 + 3\frac{-q_0 + q_1}{-t_0 + t_1}\right) \\
b_1 &= 2q_1 \frac{t_1}{-2t_1 t_2 + t_1^2 + t_2^2} - 2q_2 \frac{t_1}{-2t_1 t_2 + t_1^2 + t_2^2} \\
&\quad + \frac{t_1 + t_2}{-t_1 + t_2}\left(q_0 + 3\frac{-q_0 + q_1}{-t_0 + t_1}\right) \\
b_2 &= -\frac{q_1}{-2t_1 t_2 + t_1^2 + t_2^2} + \frac{q_2}{-2t_1 t_2 + t_1^2 + t_2^2} \\
&\quad - \frac{1}{-t_1 + t_2}\left(q_0 + 3\frac{-q_0 + q_1}{-t_0 + t_1}\right).
\end{aligned}
\tag{13.81}
$$

The boundary conditions in the third segment are

$$
\begin{aligned}
q_3(t_2) &= q_2 \\
\dot{q}_3(t_2) &= \dot{q}_2(t_2) \\
&= b_1 + 2b_2 t_2 \\
q_3(t_3) &= q_3.
\end{aligned}
\tag{13.82}
$$

We can satisfy these conditions with a quadratic equation

$$
q_3(t) = c_0 + c_1 t + c_2 t^2
\tag{13.83}
$$

*that provides three equations for the unknown coefficients*

$$\begin{bmatrix} 1 & t_2 & t_2^2 \\ 0 & 1 & 2t_2 \\ 1 & t_3 & t_3^2 \end{bmatrix} \begin{bmatrix} c_0 \\ c_1 \\ c_2 \end{bmatrix} = \begin{bmatrix} q_2 \\ b_1 + 2b_2t_2 \\ q_3 \end{bmatrix}. \tag{13.84}$$

*The coefficients are*

$$\begin{aligned}
c_0 &= -t_2\frac{t_3}{-t_2+t_3}(b_1+2b_2t_2)+q_3\frac{t_2^2}{-2t_2t_3+t_2^2+t_3^2} \\
&\quad +q_2\frac{-2t_2t_3+t_3^2}{-2t_2t_3+t_2^2+t_3^2} \\
c_1 &= \frac{t_2+t_3}{-t_2+t_3}(b_1+2b_2t_2)+2q_2\frac{t_2}{-2t_2t_3+t_2^2+t_3^2} \\
&\quad -2q_3\frac{t_2}{-2t_2t_3+t_2^2+t_3^2} \\
c_2 &= -\frac{1}{-t_2+t_3}(b_1+2b_2t_2)-\frac{q_2}{-2t_2t_3+t_2^2+t_3^2} \\
&\quad +\frac{q_3}{-2t_2t_3+t_2^2+t_3^2}. \tag{13.85}
\end{aligned}$$

*The boundary conditions for the fourth segment are*

$$\begin{aligned}
q_4(t_3) &= q_3 \\
\dot{q}_4(t_3) &= \dot{q}_3(t_3) = c_1 + 2c_2t_3 \\
q_4(t_4) &= q_4 \\
\dot{q}_4(t_4) &= 0 \\
\ddot{q}_4(t_4) &= 0 \tag{13.86}
\end{aligned}$$

*which needs a fourth degree polynomial to be satisfied.*

$$q_4(t) = d_0 + d_1t + d_2t^2 + d_3t^3 + d_4t^4 \tag{13.87}$$

*Substituting the boundary conditions generates a set of four equations for the coefficient*

$$\begin{bmatrix} 1 & t_3 & t_3^2 & t_3^3 & t_3^4 \\ 0 & 1 & 2t_2 & 3t_2^2 & 4t_2^3 \\ 1 & t_4 & t_4^2 & t_4^3 & t_4^4 \\ 0 & 1 & 2t_4 & 3t_4^2 & 4t_4^3 \\ 0 & 0 & 2 & 6t_4 & 12t_4^2 \end{bmatrix} \begin{bmatrix} d_0 \\ d_1 \\ d_2 \\ d_3 \\ d_4 \end{bmatrix} = \begin{bmatrix} q_3 \\ c_1 + 2c_2t_3 \\ q_4 \\ 0 \\ 0 \end{bmatrix}. \tag{13.88}$$

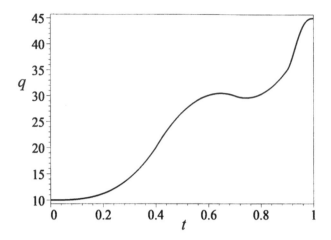

FIGURE 13.8. Spliting a path into a series of segments.

*As an example, a set of conditions given by*

$$
\begin{aligned}
q(0) &= 10\,\text{deg} \\
\dot{q}(0) &= 0 \\
\ddot{q}(0) &= 0 \\
q(0.4) &= 20\,\text{deg} \\
q(0.7) &= 30\,\text{deg} \\
q(0.9) &= 35\,\text{deg} \\
q(1) &= 45\,\text{deg} \\
\dot{q}_i(1) &= 0 \\
\ddot{q}(1) &= 0
\end{aligned}
\tag{13.89}
$$

*provides*

$$
\begin{aligned}
q_1(t) &= 10 + 156.25t^3 & (13.90) \\
q_2(t) &= -41.56 + 222.78t - 172.2t^2 & (13.91) \\
q_3(t) &= 148.99 - 321.67t + 216.67t^2 & (13.92) \\
q_4(t) &= 198545 - 827166.6672t \\
&\quad + 1290500t^2 - 893500t^3 + 231666.67t^4 & (13.93)
\end{aligned}
$$

*which is shown in Figure 13.8 graphically.*

*The disadvantage of the segment method is the lack of a smooth overall path and having a discontinuous acceleration. To increase the smoothness of the path, we need to use higher degree polynomials and put constraints on acceleration and possibly jerk. A piecewise path with continuous acceleration is called* **spline***.*

## 13.3   Non-Polynomial Path Planning

A path of motion in either joint or Cartesian spaces may be defined based on different mathematical functions. Harmonic and cycloid functions are the most common paths,

$$q(t) = a_0 + a_1 \cos a_2 t + a_3 \sin a_2 t \qquad (13.94)$$
$$q(t) = a_0 + a_1 t - a_2 \sin a_3 t \qquad (13.95)$$

however, Fourier, Legendre, Chebyshev, and other series methods may also be utilized for path planning.

**Example 304** *Harmonic path.*
   *Consider a harmonic path between two points $q(t_0)$ and $q(t_f)$*

$$q(t) = a_0 + a_1 \cos a_2 t + a_3 \sin a_2 t \qquad (13.96)$$

*with the rest-to-rest boundary conditions*

$$
\begin{aligned}
q(t_0) &= q_0 \\
\dot{q}(t_0) &= 0 \\
q(t_f) &= q_f \\
\dot{q}(t_f) &= 0.
\end{aligned}
\qquad (13.97)
$$

*Applying the conditions to the harmonic equation (13.96) provides the following solution:*

$$q(t) = \frac{1}{2}\left( q_f + q_0 - (q_f - q_0)\cos\frac{\pi(t - t_0)}{t_f - t_0} \right). \qquad (13.98)$$

*A plot of the solution is depicted in Figure 13.9 for the following numerical values:*

$$
\begin{aligned}
t_0 &= 0 \\
t_f &= 0 \\
q_0 &= 10\,\text{deg} \\
q_0' &= 0 \\
q_f &= 45\,\text{deg} \\
q_f' &= 0.
\end{aligned}
\qquad (13.99)
$$

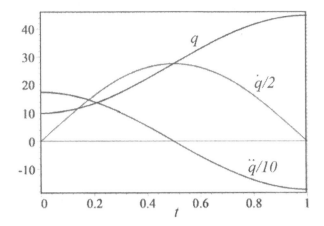

FIGURE 13.9. A harmonic path.

**Example 305** *A cycloid path.*

A cycloid path between two points $q(t_0)$ and $q(t_f)$ with rest-to-rest boundary conditions

$$
\begin{aligned}
q(t_0) &= q_0 \\
\dot{q}(t_0) &= 0 \\
q(t_f) &= q_f \\
\dot{q}(t_f) &= 0
\end{aligned}
\tag{13.100}
$$

*is*

$$
q(t) = q_0 + \frac{q_f - q_0}{\pi} \left( \frac{\pi (t - t_0)}{t_f - t_0} - \frac{1}{2} \sin \frac{2\pi (t - t_0)}{t_f - t_0} \right).
\tag{13.101}
$$

*A plot of the cycloid path is illustrated in Figure 13.10 for the following numerical values:*

$$
\begin{aligned}
t_0 &= 0 \\
t_f &= 0 \\
q_0 &= 10\,\mathrm{deg} \\
q_0' &= 0 \\
q_f &= 45\,\mathrm{deg} \\
q_f' &= 0.
\end{aligned}
\tag{13.102}
$$

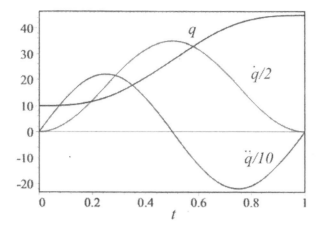

FIGURE 13.10. A cycloid path.

## 13.4  Manipulator Motion by Joint Path

Having the joint variables as functions of time, and employing the forward kinematics of manipulators, allows us to calculate the path of motion for the end-effector.

**Example 306** *2R manipulator motion based on joints' path.*

*Assume that we have calculated the paths of the two joints of a 2R planar manipulator according to cubic functions, and they are*

$$\theta_1(t) = 10 + 105t^2 - 70t^3 \deg \tag{13.103}$$
$$\theta_2(t) = 10 + 350t^3 - 525t^4 + 210t^5 \deg. \tag{13.104}$$

*The joints' paths satisfy the following conditions:*

$$\theta_1(0) = 10 \deg$$
$$\theta_1(1) = 45 \deg$$
$$\dot{\theta}_1(0) = 0$$
$$\dot{\theta}_1(1) = 0 \tag{13.105}$$

$$\theta_2(0) = 10 \deg$$
$$\dot{\theta}_1(0) = 0$$
$$\ddot{\theta}_1(0) = 0$$
$$\theta_1(1) = 45 \deg$$
$$\dot{\theta}_1(1) = 0$$
$$\ddot{\theta}_1(1) = 0 \tag{13.106}$$

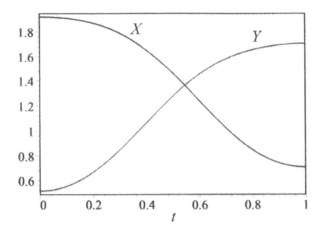

FIGURE 13.11. $X$ and $Y$ components of the tip point position of a 2R planar manipulator.

*The forward kinematics of a 2R manipulator are found in Example 132 as below.*

$$
\begin{aligned}
{}^{0}T_2 &= {}^{0}T_1\,{}^{1}T_2 \\[4pt]
&= \begin{bmatrix}
c\,(\theta_1 + \theta_2) & -s\,(\theta_1 + \theta_2) & 0 & l_1 c\theta_1 + l_2 c\,(\theta_1 + \theta_2) \\
s\,(\theta_1 + \theta_2) & c\,(\theta_1 + \theta_2) & 0 & l_1 s\theta_1 + l_2 s\,(\theta_1 + \theta_2) \\
0 & 0 & 1 & 0 \\
0 & 0 & 0 & 1
\end{bmatrix}
\end{aligned}
\quad (13.107)
$$

*The fourth column of ${}^{0}T_2$ indicates the Cartesian position of the tip point of the manipulator in the base frame. Therefore, the $X$ and $Y$ components of the tip point are*

$$
\begin{aligned}
X &= l_1 \cos\theta_1 + l_2 \cos(\theta_1 + \theta_2) & (13.108) \\
Y &= l_1 \sin\theta_1 + l_2 \sin(\theta_1 + \theta_2). & (13.109)
\end{aligned}
$$

*Substituting $\theta_1$ and $\theta_2$ from (13.103) and (13.104) provides the time variation of the tip position. These variations, for $l_1 = l_2 = 1\,\mathrm{m}$ are shown in Figure 13.11, while the configurations of the manipulator at initial and final positions are shown in Figure 13.12.*

## 13.5   Cartesian Path

Cartesian path planning is similar to joint space path planning. The point-to-point path can be planned by connecting the points, or designing a path to pass close to but not necessarily through the points. A practical method

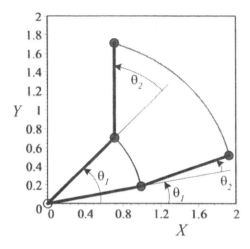

FIGURE 13.12. Configuration of a 2R manipulator at initial and final positions.

is to design a path utilizing straight lines with constant velocity, and deform the corners to have a smooth transition.

The path connecting points $\mathbf{r}_0$ to $\mathbf{r}_2$, and passing close to the corner $\mathbf{r}_1$ on a transition curve, can be designed by a piecewise motion.

$$
\begin{aligned}
\mathbf{r}(t) &= \mathbf{r}_1 - \frac{t_1 - t}{t_1 - t_0} (\mathbf{r}_1 - \mathbf{r}_0) && t_0 \leq t \leq t_1 - t' \\
\mathbf{r}(t) &= \mathbf{r}_1 - \frac{(t - t' - t_1)^2}{4t'(t_1 - t_0)} (\mathbf{r}_1 - \mathbf{r}_0) \\
&\quad + \frac{(t + t' - t_1)^2}{4t'(t_2 - t_1)} (\mathbf{r}_2 - \mathbf{r}_1) && t_1 - t' \leq t \leq t_1 + t' \\
\mathbf{r}(t) &= \mathbf{r}_1 - \frac{t_1 - t}{t_2 - t_1} (\mathbf{r}_2 - \mathbf{r}_1) && t_1 + t' \leq t \leq t_2
\end{aligned}
\tag{13.110}
$$

The path starts from $\mathbf{r}_0$ at time $t_0$ and moves with constant velocity along a line until a point at switching time $t_1 - t'$. At this time, the path switches to a constant acceleration parabola. At another switching point at time $t_1 + t'$, the path switches to the second line and moves with constant velocity toward the destination at point $\mathbf{r}_2$. The time $t_1 - t_0$ is the required time to move from $\mathbf{r}_0$ to $\mathbf{r}_1$ and $t_2 - t_1$ is the required time to move from $\mathbf{r}_1$ to $\mathbf{r}_2$, if there were no transition path. The path is shown in Figure 13.13 schematically.

**Proof.** The first line segment starts from a point $\mathbf{r}_0$ at time $t_0$ and, without any deformation, it arrives at point $\mathbf{r}_1$ at time $t_1$ via a constant velocity. The second line ends with a constant velocity at point $\mathbf{r}_2$ at time $t_2$ and,

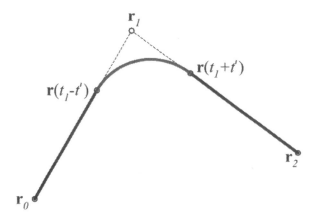

FIGURE 13.13. Transition parabola between two line segments as a path in Cartesian space.

without deformation, it would start from point $\mathbf{r}_1$ at time $t_1$.

$$\mathbf{r}(t) = \mathbf{r}_1 - \frac{t_1 - t}{t_1 - t_0}(\mathbf{r}_1 - \mathbf{r}_0) \qquad t_0 \le t \le t_1$$

$$\mathbf{r}(t) = \mathbf{r}_1 - \frac{t_1 - t}{t_2 - t_1}(\mathbf{r}_2 - \mathbf{r}_1) \qquad t_1 \le t \le t_2 \qquad (13.111)$$

We introduce an interval time $t'$ before arriving at $\mathbf{r}_1$ to switch from the line to a transition curve. The transition curve is then between times $t_1 - t'$ and $t_1 + t'$. The simplest transition curve is a parabola, which, at the end points, has the same speed as the lines.

The boundary positions of the transition curve on the first and second lines are respectively at

$$\mathbf{r}(t_1 - t') = \mathbf{r}_1 - \frac{t'}{t_1 - t_0}\delta_1 \qquad (13.112)$$

$$\mathbf{r}(t_1 + t') = \mathbf{r}_1 + \frac{t'}{t_2 - t_1}\delta_2 \qquad (13.113)$$

where

$$\delta_1 = \mathbf{r}_1 - \mathbf{r}_0 \qquad (13.114)$$

$$\delta_2 = \mathbf{r}_2 - \mathbf{r}_1. \qquad (13.115)$$

The velocity at the beginning and final points of the transition curve are respectively equal to

$$\dot{\mathbf{r}}(t_1 - t') = \frac{1}{t_1 - t_0}\delta_1 \qquad (13.116)$$

$$\dot{\mathbf{r}}(t_1 + t') = \frac{1}{t_2 - t_1}\delta_2. \qquad (13.117)$$

Assume the acceleration of motion along the transition curve is constant

$$\ddot{\mathbf{r}}(t) = \ddot{\mathbf{r}}_c = cte \tag{13.118}$$

and therefore, the transition curve after integration is equal to

$$\mathbf{r}(t) = \mathbf{r}(t_1 - t') + (t - t_1 + t')\dot{\mathbf{r}}(t_1 - t') + \frac{1}{2}(t - t_1 + t')^2\ddot{\mathbf{r}}_c. \tag{13.119}$$

Substituting (13.112) and (13.116) provides

$$\mathbf{r}(t) = \mathbf{r}_1 + \frac{t - t_1}{t_1 - t_0}\boldsymbol{\delta}_1 + \frac{1}{2}\ddot{\mathbf{r}}_c(t - t_1 + t')^2. \tag{13.120}$$

The transition curve $\mathbf{r}(t)$ must be at the end point when $t = t_1 + t'$

$$\begin{aligned}
\mathbf{r}(t_1 + t') &= \mathbf{r}_1 + \frac{t'}{t_2 - t_1}\boldsymbol{\delta}_2 \\
&= \mathbf{r}_1 + \frac{t'}{t_1 - t_0}\boldsymbol{\delta}_1 + 2\ddot{\mathbf{r}}_c t_1^2
\end{aligned}$$

therefore, the acceleration on the curve must be

$$\ddot{\mathbf{r}}_c = \frac{1}{2t'}\left(\frac{\boldsymbol{\delta}_2}{t_2 - t_1} - \frac{\boldsymbol{\delta}_1}{t_1 - t_0}\right). \tag{13.121}$$

Hence, the curve equation becomes

$$\mathbf{r}(t) = \mathbf{r}_1 - \boldsymbol{\delta}_1\frac{(t - t' - t_1)^2}{4t'(t_1 - t_0)} + \boldsymbol{\delta}_2\frac{(t + t' - t_1)^2}{4t'(t_2 - t_1)} \tag{13.122}$$

showing that the path between $\mathbf{r}_0$ and $\mathbf{r}_2$ has a piecewise character given in (13.110).

A Cartesian path followed by the manipulator, plus the time profile along the path, specify the position and orientation of the end frame. Issues in Cartesian path planning include attaining a specific target from an initial starting point, avoiding obstacles, and staying within manipulator capabilities. A path is modeled by $n$ points called *control points*. The control points are connected via straight lines and the transient parabolas will be implemented to exclude the sharp corners.

An alternative method is applying an interpolating method to design a continuous path over the control points, or close to them. ∎

**Example 307** *A path in 2D Cartesian space.*

*Consider a line in the XY plane connecting $(1,0)$ and $(1,1)$, and another line connecting $(1,1)$ and $(0,1)$. Assume that the time is zero at $(1,0)$, is $t = 1$ at $(1,1)$, and is $t = 2$ at $(0,1)$. For an interval time $t' = 0.1$, the position vector at control points are*

$$\begin{aligned}
\mathbf{r}_0 &= \hat{\imath} & (13.123) \\
\mathbf{r}_1 &= \hat{\imath} + \hat{\jmath} & (13.124) \\
\mathbf{r}_2 &= \hat{\jmath} & (13.125)
\end{aligned}$$

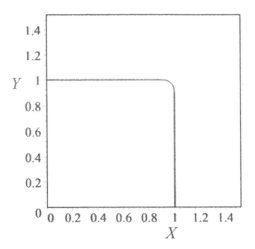

FIGURE 13.14. A transition parabola connecting two lines.

$$\mathbf{r}(t_1 - t') = \mathbf{r}_1 - \frac{t'}{t_1}\delta_1$$

$$= \hat{\imath} + \left(1 - \frac{t'}{t_1}\right)\hat{\jmath} \qquad (13.126)$$

$$\mathbf{r}(t_1 + t') = \mathbf{r}_1 + \frac{t'}{t_2}\delta_2$$

$$= \left(1 - \frac{t'}{t_2}\right)\hat{\imath} + \hat{\jmath} \qquad (13.127)$$

*where*

$$\delta_1 = \mathbf{r}_1 - \mathbf{r}_0 = \hat{\jmath} \qquad (13.128)$$
$$\delta_2 = \mathbf{r}_2 - \mathbf{r}_1 = -\hat{\imath}. \qquad (13.129)$$

*The path of motion is then expressed by the following piecewise function as shown in Figure 13.14:*

$$\begin{array}{ll}
\mathbf{r}(t) = \hat{\imath} + t\hat{\jmath} & 0 \le t \le 0.9 \\
\mathbf{r}(t) = \left(1 - \frac{(t-0.9)^2}{0.4}\right)\hat{\imath} + \left(1 - \frac{(t-1.1)^2}{0.4}\right)\hat{\jmath} & 0.9 \le t \le 1.1 \\
\mathbf{r}(t) = (2 - t)\hat{\imath} + \hat{\jmath} & 1.1 \le t \le 2
\end{array} \qquad (13.130)$$

*The velocity of motion along the path is also a piecewise function given below.*

$$\begin{array}{ll}
\dot{\mathbf{r}}(t) = \hat{\jmath} & 0 \le t \le 0.9 \\
\mathbf{r}(t) = \frac{t-0.9}{0.2}\hat{\imath} - \frac{t-1.1}{0.2}\hat{\jmath} & 0.9 \le t \le 1.1 \\
\mathbf{r}(t) = -\hat{\imath} & 1.1 \le t \le 2
\end{array} \qquad (13.131)$$

## 13.6 ★ Rotational Path

Consider an end-effector frame to have a rotation matrix $^G R_0$ at an initial orientation at time $t_0$. The end-effector must be at a final orientation $^G R_f$ at time $t_f$. The *rotational path* is defined by the angle-axis rotation matrix $R_{\hat{u},\phi}$

$$R_{0\hat{u},\phi} = {}^0 R_f = {}^G R_0^T {}^G R_f \tag{13.132}$$

that transforms the end-effector frame from the final orientation $^G R_f$ to the initial orientation $^G R_0$. The axis of rotation $^0 \hat{u}$ is defined by a unit vector expressed in the initial frame. Therefore, the desired rotation matrix for going from initial to the final orientation, would be

$$R_{0\hat{u},\phi}^T = {}^G R_f^T {}^G R_0. \tag{13.133}$$

To control a rotation, we may define a series of control orientations $^G R_1$, $^G R_2$, $\cdots$, $^G R_n$ between the initial and final orientations, and rotate the end-effector frame through the control orientations. When there is a control orientation $^G R_1$ between the initial and final orientations, then the initial orientation $^G R_0$ transforms to the control orientation $^G R_1$ using an angle-axis rotation $R_{0\hat{u},\phi_0}$, and then it transforms from the control orientation $^G R_1$ to the final orientation using a second-angle axis rotation $R_{1\hat{u},\phi_1}$.

$$R_{0\hat{u},\phi_0} = {}^G R_0^T {}^G R_1 \tag{13.134}$$

$$R_{1\hat{u},\phi_1} = {}^G R_1^T {}^G R_f \tag{13.135}$$

**Proof.** According to the Rodriguez rotation formula (3.4),

$$^0 R_f = R_{0\hat{u},\phi} = \mathbf{I} \cos\phi + {}^0\hat{u} \, {}^0\hat{u}^T \, \text{vers}\,\phi + {}^0\tilde{u} \sin\phi \tag{13.136}$$

the angle and axis that transforms a frame $B_f$ to another frame $B_0$ are found from

$$\cos\phi = \frac{1}{2}\left(\text{tr}\left({}^0 R_f\right) - 1\right) \tag{13.137}$$

$$^0\tilde{u} = \frac{1}{2\sin\phi}\left({}^0 R_f - {}^0 R_f^T\right). \tag{13.138}$$

If $^G R_0$ is the rotation matrix from $B_0$ to the global frame $G$, and $^G R_f$ is the rotation matrix from $B_f$ to $G$, then

$$^G R_f = {}^G R_0 \, {}^0 R_f \tag{13.139}$$

and therefore,

$$^0 R_f = R_{0\hat{u},\phi} = {}^G R_0^T {}^G R_f. \tag{13.140}$$

We define a linearly time dependent rotation matrix by varying the angle
of rotation about the axis of rotation

$$^0R_f(t) \;=\; R_{0\hat{u},(\frac{t-t_0}{t_f-t_0})\phi} \tag{13.141}$$

$$= \begin{bmatrix} r_{11}(t) & r_{12}(t) & r_{13}(t) \\ r_{21}(t) & r_{22}(t) & r_{23}(t) \\ r_{31}(t) & r_{32}(t) & r_{33}(t) \end{bmatrix} \qquad t_0 \le t \le t_f$$

where, $t_0$ is the time when the end-effector frame is at orientation $^GR_0$ and
$t_f$ is the time at which the end-effector frame is at orientation $^GR_f$, and

$$r_{11}(t) \;=\; u_1^2 \operatorname{vers}\left(\frac{t-t_0}{t_f-t_0}\right)\phi + \cos\left(\frac{t-t_0}{t_f-t_0}\right)\phi$$

$$r_{21}(t) \;=\; u_1 u_2 \operatorname{vers}\left(\frac{t-t_0}{t_f-t_0}\right)\phi + u_3 \sin\left(\frac{t-t_0}{t_f-t_0}\right)\phi$$

$$r_{31}(t) \;=\; u_1 u_3 \operatorname{vers}\left(\frac{t-t_0}{t_f-t_0}\right)\phi - u_2 \sin\left(\frac{t-t_0}{t_f-t_0}\right)\phi$$

$$r_{12}(t) \;=\; u_1 u_2 \operatorname{vers}\left(\frac{t-t_0}{t_f-t_0}\right)\phi - u_3 \sin\left(\frac{t-t_0}{t_f-t_0}\right)\phi$$

$$r_{22}(t) \;=\; u_2^2 \operatorname{vers}\left(\frac{t-t_0}{t_f-t_0}\right)\phi + \cos\left(\frac{t-t_0}{t_f-t_0}\right)\phi$$

$$r_{32}(t) \;=\; u_2 u_3 \operatorname{vers}\left(\frac{t-t_0}{t_f-t_0}\right)\phi + u_1 \sin\left(\frac{t-t_0}{t_f-t_0}\right)\phi$$

$$r_{13}(t) \;=\; u_1 u_3 \operatorname{vers}\left(\frac{t-t_0}{t_f-t_0}\right)\phi + u_2 \sin\left(\frac{t-t_0}{t_f-t_0}\right)\phi$$

$$r_{23}(t) \;=\; u_2 u_3 \operatorname{vers}\left(\frac{t-t_0}{t_f-t_0}\right)\phi - u_1 \sin\left(\frac{t-t_0}{t_f-t_0}\right)\phi$$

$$r_{33}(t) \;=\; u_3^2 \operatorname{vers}\left(\frac{t-t_0}{t_f-t_0}\right)\phi + \cos\left(\frac{t-t_0}{t_f-t_0}\right)\phi.$$

The matrix $^0R_f(t)$ can turn the final frame about the axis of rotation $^0\hat{u}$
onto the initial frame, and therefore,

$$^GR_f = {}^GR_0\, {}^0R_f(t). \tag{13.142}$$

If there is a control orientation frame $^GR_1$ between the initial and final
orientations, then

$$^GR_1 \;=\; {}^GR_0\, {}^0R_1 \tag{13.143}$$

$$^GR_f \;=\; {}^GR_1\, {}^1R_f \tag{13.144}$$

and therefore,

$$R_{0\hat{u},\phi_0} = {}^0R_1 = {}^GR_0^T {}^GR_1 \tag{13.145}$$

$$R_{1\hat{u},\phi_1} = {}^1R_f = {}^GR_1^T {}^GR_f. \tag{13.146}$$

The rotation matrices ${}^0R_1$ and ${}^1R_f$ may be defined as linearly time varying rotation matrices by

$$\begin{aligned} {}^0R_1(t) &= R_{0\hat{u},(\frac{t-t_0}{t_1-t_0})\phi_0} & t_0 \leq t \leq t_1 \\ {}^1R_f(t) &= R_{1\hat{u},(\frac{t-t_1}{t_f-t_1})\phi_1} & t_1 \leq t \leq t_f. \end{aligned} \tag{13.147}$$

Utilizing these variable matrices, we can turn the end-effector frame from the initial orientation ${}^GR_0$ about ${}^0\hat{u}$ to achieve the control orientation ${}^GR_1$, and then turn the end-effector frame about ${}^1\hat{u}$ to achieve the final orientation ${}^GR_f$.

Following the parabola transition technique, we may define an orientation path connecting ${}^GR_0$ and ${}^GR_f$, and passing close to the corner orientation ${}^GR_1$ on a transient rotation path. The path starts from ${}^GR_0$ at time $t_0$ and turns with constant angular velocity along an axis until $t = t_1 - t'$. At this time, the path switches to a parabolic path with constant angular acceleration. At another switching orientation at time $t = t_1 + t'$, the path switches to the second path and turns with constant velocity toward the destination orientation ${}^GR_f$. The time $t_1 - t_0$ is the required time to move from ${}^GR_0$ to ${}^GR_1$, and $t_2 - t_1$ is the required time to move from ${}^GR_1$ to ${}^GR_f$ if there were no transition path.

We introduce an interval time $t'$ before arriving at orientation ${}^GR_1$ to switch from the first path segment to a transition path. The transition path is then between times $t_1 - t'$ and $t_1 + t'$. At the second switching orientation, the transition path ends at the same angular velocity as the third path segment.

The boundary positions of the transition path between the first and third segments are respectively

$$\begin{aligned} {}^GR_1(t_1 - t') &= {}^GR_0 \, {}^0R_1(t_1 - t') \\ &= {}^GR_0 \, R_{0\hat{u},(1-\frac{t'}{t_1-t_0})\phi_0} & t = t_1 - t' \quad (13.148) \end{aligned}$$

$$\begin{aligned} {}^GR_f(t_1 + t') &= {}^GR_1 \, {}^1R_f(t_1 + t') \\ &= {}^GR_1 \, R_{1\hat{u},(\frac{t'}{t_f-t_1})\phi_1} & t = t_1 + t'. \quad (13.149) \end{aligned}$$

The transition path is then equal to

$$\begin{aligned} R_t(t) &= {}^GR_0 \, {}^0R_1 \left( \frac{t_1 - t' - t}{2t'} - \frac{(t - t' - t_1)^2}{4t'(t_1 - t_0)} \right) {}^1R_f \left( \frac{(t + t' - t_1)^2}{4t'(t_f - t_1)} \right) \\ &= {}^GR_0 \, R_{0\hat{u},(\frac{t_1-t'-t}{2t'} - \frac{(t-t'-t_1)^2}{4t'(t_1-t_0)})\phi_0} \, R_{1\hat{u},(\frac{(t+t'-t_1)^2}{4t'(t_f-t_1)})\phi_1} \end{aligned} \tag{13.150}$$

$$t_1 - t' \leq t \leq t_1 + t'$$

and the entire path is

$$
\begin{aligned}
R(t) &= {}^0R_1(t) = R_{0\hat{u},(\frac{t-t_0}{t_1-t_0})\phi_0} && t_0 \le t \le t_1 - t' \\
R(t) &= R_t(t) && t_1 - t' \le t \le t_1 + t' \qquad (13.151) \\
R(t) &= {}^1R_f(t) = R_{1\hat{u},(\frac{t-t_1}{t_f-t_1})\phi_1} && t_1 + t' \le t \le t_2.
\end{aligned}
$$

■

# 13.7  Manipulator Motion by End-Effector Path

Cartesian path planning has more application in robotics, because it can control the level of force and jerk inserted by the hand of a robot to the carrying object. Path planning in Cartesian space also determines the geometric constraints of the external world. However, a Cartesian path needs inverse kinematics to determine the time history of the joint variables.

**Example 308** *Joint path for a designed Cartesian path.*
*Consider a rest-to-rest Cartesian path from point $(1, 1.5)$ to point $(-1, 1.5)$ on a straight line $Y = 1.5$. A cubic polynomial can satisfy the position and velocity constraints at initial and final points.*

$$
\begin{aligned}
X(0) &= X_0 = 1 \\
X(1) &= X_f = -1 \\
\dot{X}(0) &= \dot{X}_0 = 0 \\
\dot{X}(1) &= \dot{X}_f = 0 \qquad (13.152)
\end{aligned}
$$

*The coefficients of the polynomial are*

$$
\begin{aligned}
a_0 &= 1 \\
a_1 &= 0 \\
a_2 &= -6 \\
a_3 &= 4 \qquad (13.153)
\end{aligned}
$$

*and the Cartesian path is*

$$
\begin{aligned}
X &= 1 - 6t^2 + 4t^3 \qquad (13.154) \\
Y &= 1.5. \qquad (13.155)
\end{aligned}
$$

*The inverse kinematics of a 2R planar manipulator is calculated in Example 168 as*

$$
\theta_2 = \pm 2\operatorname{atan2}\sqrt{\frac{(l_1 + l_2)^2 - (X + Y)^2}{(X^2 + Y^2) - (l_1^2 + l_2^2)}} \qquad (13.156)
$$

$$
\theta_1 = \operatorname{atan2}\frac{X(l_1 + l_2 \cos\theta_2) + Yl_2 \sin\theta_2}{Y(l_1 + l_2 \cos\theta_2) - Xl_2 \sin\theta_2} \qquad (13.157)
$$

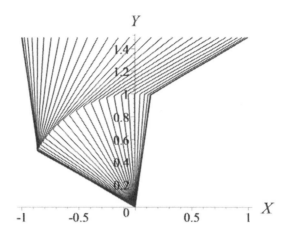

FIGURE 13.15. Illustration of a 2R panipulator when the tip point moves on a straight line $y = 1.5$.

*where the sign ($\pm$) indicates the elbow-up and elbow-down configurations of the manipulator. Depending on the initial configuration at point $(1, 1.5)$, the manipulator is supposed to stay in that configuration. Let's consider an elbow-up configuration. Therefore, we accept only those values of the joint valuables that belong to the elbow-up configuration. Substituting (13.154) and (13.155) in (13.156) and (13.157) provides the path in joint space*

$$\theta_2 = \pm 2 \operatorname{atan2} \sqrt{\frac{(l_1 + l_2)^2 - (4l^3 - 6t^2 + 2.5)^2}{(4t^3 - 6t^2 + 1)^2 - l_2^2 - l_1^2 + 2.25}} \qquad (13.158)$$

$$\theta_1 = \operatorname{atan2} \frac{(1 - 6t^2 + 4t^3)(l_1 + l_2 \cos\theta_2) + 1.5 l_2 \sin\theta_2}{1.5 (l_1 + l_2 \cos\theta_2) - (1 - 6t^2 + 4t^3) l_2 \sin\theta_2}. \qquad (13.159)$$

*A graphical illustration of the manipulator at every 1/30 th of the total time is shown in Figure 13.15.*

## 13.8  Summary

A serial robot may be assumed as a variable geometrical chain of links that relates the configuration of its end-effector to the Cartesian coordinate frame in which the base frame is attached. Forward kinematics are mathematical-geometrical relations that provide the end-effector configuration by having the joint coordinates. On the other hand, the inverse kinematics are mathematical-geometrical relations that provide joint coordinates for a given end-effector configuration.

The Cartesian path of motion for the end-effector must be expressed as a function of time to find the links' velocity and acceleration. The first applied path function that can provide a rest-to-rest motion is a cubic path for the joint variable $q_i(t)$ between two points $q_i(t_0)$ and $q_i(t_f)$

$$q_i(t) = a_0 + a_1 t + a_2 t^2 + a_3 t^3. \tag{13.160}$$

By increasing the requirements, such as zero acceleration or jerk at some points on the path, we need to employ higher polynomials to satisfy the conditions. An $n$ degree polynomial can satisfy $n + 1$ conditions. It is also possible to split a multiple conditional path into some intervals with fewer conditions. The interval paths must then be connected to satisfy their boundary conditions.

A path of motion may also be defined based on different mathematical functions. Harmonic and cycloid functions are the most common paths. Non-polynomial equations introduce some advantages, due to simpler expression, and some disadvantages due to nonlinearities.

When a path of motion either in joint or Cartesian coordinates space is defined, forward and inverse kinematics must be utilized to find the path of motion in the other space.

Rotational maneuver of the end-effector about the wrist point needs a rotational path. A rotational path may mathematically be defined similar to a Cartesian path utilizing the Rodriguez formula and rotation matrices.

# Exercises

1. Notation and symbols.

   Describe their meaning.

$$\text{a- } t_0 \quad \text{b- } t_f \quad \text{c- } q_i(t) \quad \text{d- } t'$$

2. Rest-to-rest cubic path.

   Find a cubic path for a joint coordinate to satisfy the following conditions:

$$q(0) = -10 \deg, \ q(1) = 45 \deg, \ \dot{q}(0) = \dot{q}(1) = 0.$$

3. To-rest path.

   Find a quadratic path to satisfy the following conditions:

$$q(0) = -10 \deg, \ q(1) = 45 \deg, \ \dot{q}(1) = 0.$$

   Calculate the initial velocity of the path using the quadratic path. Then, find a cubic path to satisfy the same boundary conditions as the quadratic path. Compare the maximum accelerations of the two paths.

4. Constant velocity path.

   Calculate a path to satisfy the following conditions:

$$q(0) = -10 \deg, \ q(1) = 45 \deg, \ \dot{q}(0) = \dot{q}(1) = 0$$

   and move with constant velocity $\dot{q} = 25 \deg / \sec$ between $12 \deg$ and $35 \deg$.

5. Constant acceleration path.

   Calculate a path with constant acceleration $\ddot{q} = 25 \deg / \sec^2$ between $12 \deg$ and $35 \deg$, and satisfy the following conditions:

$$q(0) = -10 \deg, \ q(1) = 45 \deg, \ \dot{q}(0) = \dot{q}(1) = 0.$$

6. Zero jerk path.

   Find a path to satisfy the following boundary conditions:

$$q(0) = 0, \ q(1) = 66 \deg, \ \dot{q}(0) = \dot{q}(1) = 0$$

   and have zero jerk at the beginning, middle, and end points.

7. Control points.

Find a path to satisfy the conditions

$$q(0) = 10 \deg, \ q(1) = 95 \deg, \ \dot{q}(0) = \dot{q}(1) = 0$$

and pass through the following control points:

$$q(0.25) = 30 \deg, \ q(0.5) = 65 \deg$$

8. A jerk zero at start-middle-stop path.

To make a path have jerk as close to zero as possible, an eight degree polynomial

$$q(t) = a_0 + a_1 t + a_2 t^2 + a_3 t^3 + a_4 t^4 + a_5 t^5 + a_6 t^6 + a_7 t^7 + a_7 t^8$$

and nine boundary conditions can be employed. Find the path.

$$q(0) = 0 \qquad \dot{q}(0) = 0 \quad \ddot{q}(0) = 0 \quad \dddot{q}(0) = 0$$
$$\dddot{q}(0.5) = 0$$
$$q(1) = 120 \deg \quad \dot{q}_i(1) = 0 \quad \ddot{q}(1) = 0 \quad \dddot{q}(1) = 0$$

9. Point sequence path.

The conditions for a sequence of points are given here. Find a path to satisfy the conditions given below.

(a)
$$\begin{aligned} q(0) &= 5 \deg & \dot{q}(0) &= 0 & \ddot{q}(0) &= 0 \\ q(0.4) &= 35 \deg \\ q(0.75) &= 65 \deg \\ q(1) &= 100 \deg & \dot{q}_i(1) &= 0 & \ddot{q}(1) &= 0. \end{aligned}$$

(b)
$$\begin{aligned} q(0) &= 5 \deg & \dot{q}(0) &= 0 & \ddot{q}(0) &= 0 \\ q(2) &= 15 \deg \\ q(4) &= 35 \deg \\ q(7.5) &= 65 \deg \\ q(10) &= 100 \deg & \dot{q}_i(10) &= 0 & \ddot{q}(10) &= 0. \end{aligned}$$

10. Splitting a path into a series of segments.

Find a path for the following conditions:

$$\begin{aligned} q(0) &= 5 \deg & \dot{q}(0) &= 0 & \ddot{q}(0) &= 0 \\ q(2) &= 15 \deg \\ q(4) &= 35 \deg \\ q(7.5) &= 65 \deg \\ q(10) &= 100 \deg & \dot{q}_i(10) &= 0 & \ddot{q}(10) &= 0. \end{aligned}$$

by the splitting method and breaking the entire path into four segments.

$$
\begin{array}{llll}
q_1(t) & \text{for} & q(0) < q_1(t) < q(2) & \text{and} & 0 < t < 2 \\
q_2(t) & \text{for} & q(2) < q_2(t) < q(4) & \text{and} & 2 < t < 4 \\
q_3(t) & \text{for} & q(4) < q_3(t) < q(7.5) & \text{and} & 4 < t < 7.5 \\
q_4(t) & \text{for} & q(7.5) < q_4(t) < q(10) & \text{and} & 7.5 < t < 10
\end{array}
$$

11. ★ Extra conditions.

To have a smooth overall path in the splitting method, we may add extra conditions to match the segments. Solve Exercise 10 having a zero jerk transition between segments.

12. 2R manipulator motion to follow a joint path.

Find the path of the endpoint of a 2R manipulator, with $l_1 = l_2 = 1\,\text{m}$, if the joints' variable follow the given paths:

$$
\begin{aligned}
\theta_1(t) &= 10 + 156.25t^3 \\
\theta_2(t) &= -41.56 + 222.78t - 172.2t^2.
\end{aligned}
$$

13. 3R planar manipulator motion to follow a joint path.

Find the path of the endpoint of a 3R manipulator, with

$$
\begin{aligned}
l_1 &= 1\,\text{m} \\
l_2 &= 0.65\,\text{m} \\
l_3 &= 0.35\,\text{m}
\end{aligned}
$$

if the joints' variable follow these given paths

$$
\begin{aligned}
\theta_1(t) &= -41.56 + 222.78t - 172.2t^2 \\
\theta_2(t) &= 148.99 - 321.67t + 216.67t^2 \\
\theta_3(t) &= 198545 - 827166.6672t \\
&\quad + 1290500t^2 - 893500t^3 + 231666.67t^4.
\end{aligned}
$$

Calculate the maximum acceleration and jerk of the endpoint.

14. R⊢R∥R articulated arm motion.

Find the Cartesian trajectory of the endpoint of an articulated manipulator, shown in Figure 5.25, if the geometric parameters are

$$
\begin{aligned}
d_1 &= 1\,\text{m} \\
d_2 &= 0 \\
l_2 &= 1\,\text{m} \\
l_3 &= 1\,\text{m}
\end{aligned}
$$

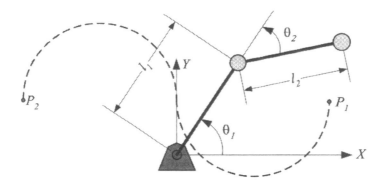

FIGURE 13.16. A 2R manipulator moves on a path made of two semi-circles.

and the joints' paths are

$$\begin{aligned}
\theta_1(t) &= -41.56 + 222.78t - 172.2t^2 \\
\theta_2(t) &= 148.99 - 321.67t + 216.67t^2 \\
\theta_3(t) &= 198545 - 827166.6672t \\
&\quad +1290500t^2 - 893500t^3 + 231666.67t^4.
\end{aligned}$$

15. Cartesian path for a 2R manipulator.

Calculate a cubic rest-to-rest path in Cartesian space to join the following points with a straight line.

$$\begin{aligned}
P_1 &= (1.5, 1) \\
P_2 &= (-0.5, 1.5).
\end{aligned}$$

Then calculate and plot the joint coordinates of a 2R manipulator, with $l_1 = l_2 = 1\,\mathrm{m}$, that follows the Cartesian path. What is the maximum angular acceleration of the joints' variable?

16. ★ Joint path for a given Cartesian path.

Assume that the endpoint of a 2R manipulator moves with constant speed $v = 1\,\mathrm{m/sec}$ from $P_1$ to $P_2$, on a path made of two semi-circles, as shown in Figure 13.16. Calculate and plot the joints' path if $l_1 = l_2 = 1\,\mathrm{m}$. Calculate the value and position of the maximum angular acceleration and jerk in joint variables. The center of the circles are at $(0.75\,\mathrm{m}, 0.5\,\mathrm{m})$ and $(-0.75\,\mathrm{m}, 0.5\,\mathrm{m})$.

17. ★ Obstacle avoidance and path planning.

Find a path between $P_1 = (1.5, 1)$ and $P_2 = (-1, 1)$ to avoid the obstacle shown in Figure 13.17. The path may be made of two straight lines with a transition circular path in the middle. The radius of the

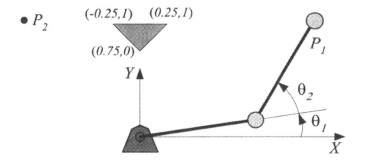

$P_2$

$(-0.25,1)$    $(0.25,1)$

$(0.75,0)$

$Y$

$P_1$

$\theta_2$

$\theta_1$

$X$

FIGURE 13.17. An obstacle in the Cartesian space of motion for a 2R manipulator.

circle is $r = 0.5\,\text{m}$ and the center of the circle is at a point that makes the total length of the path minimum. The lines connect to the circle smoothly.

The endpoint of a 2R manipulator, with $l_1 = l_2 = 1\,\text{m}$, starts at rest from $P_1$ and moves along the first line with constant acceleration. The endpoint keeps its speed constant $v = 1\,\text{m}/\sec$ on the circular path and then moves with constant acceleration on the final line segment to stop at $P_2$.

Calculate and plot the joints' paths. Find the value and position of the maximum angular velocity, acceleration, and jerk for both joints' variable.

18. ★ Joint path for a given Cartesian path.

Connect the points $P_1 = (1.1, 0.8, 0.5)$ and $P_2 = (-1, 1, 0.35)$ with a straight line. Find a rest-to-rest cubic path and plot the Cartesian coordinates $X$, $Y$, and $Z$ as functions of time. Then, calculate the joints' path for an articulated manipulator, shown in Figure 5.25, if the geometric parameters are:

$$d_1 = 1\,\text{m}$$
$$d_2 = 0$$
$$l_2 = 1\,\text{m}$$
$$l_3 = 1\,\text{m}$$

Find the value and position of the maximum angular velocity, acceleration, and jerk for the joints' variable.

19. ★ Transition parabola.

In Exercise 17, connect the points $P_1 = (1.5, 1)$ and $P_2 = (-1, 1)$ with two straight lines, using $P_0 = (0, 0.6)$ as a corner. Design a parabolic transition path to avoid the corner if the interval time is $t' = 5\sec$ and the total time of motion is $12\sec$.

20. ★ Euler angles rotational path.

Assume that the spherical wrist of a 6 DOF robot starts from rest position and turns about the axes of the final coordinate frame $B_6$ in order $z$-$x$-$z$ for $\varphi = 15\deg$, $\theta = 38\deg$, and $\psi = 77\deg$. The frame $B_6$ is installed at the wrist point. Design a rest-to-rest cubic rotational path for the angles $\varphi$, $\theta$, and $\psi$, if each rotation takes $1\sec$. Then, find the axis and angle of rotation, $(\hat{u}, \phi)$, that moves the wrist from the initial to the final orientation. Design a cubic rotational path for the axis-angle rotation if it takes $3\sec$. Calculate the Euler angles path $\varphi(t)$, $\theta(t)$, and $\psi(t)$ for this motion. Calculate and compare the maximum angular velocity, acceleration, and jerk for $\varphi$, $\theta$, and $\psi$ in the first and second motions. What is the maximum angular velocity, acceleration, and jerk of $\phi$ in the second motions?

# 14

# ★ Time Optimal Control

## 14.1 ★ Minimum Time and Bang-Bang Control

The most important job of industrial robots is moving between two points rest-to-rest. Minimum time control is what we need to increase industrial robots productivity. The objective of time-optimal control is to transfer the end-effector of a robot from an initial position to a desired destination in minimum time. Consider a system with the following equation of motion:

$$\dot{\mathbf{x}} = \mathbf{f}\left(\mathbf{x}(t), \mathbf{Q}(t)\right) \tag{14.1}$$

where $\mathbf{Q}$ is the control input, and $\mathbf{x}$ is the state vector of the system

$$\mathbf{x} = \left[\begin{array}{c} \mathbf{q} \\ \dot{\mathbf{q}} \end{array}\right]. \tag{14.2}$$

The minimum time problem is always subject to bounded input such as

$$|\mathbf{Q}(t)| \leq \mathbf{Q}_{Max}. \tag{14.3}$$

The solution of the time-optimal control problem subject to bounded input is *bang-bang control*. The control in which the input variable takes either the maximum or minimum values is called bang-bang control.

**Proof.** The goal of minimum time control is to find the trajectory $\mathbf{x}(t)$ and input $\mathbf{Q}(t)$ starting from an initial state $\mathbf{x}_0(t)$ and arriving at the final state $\mathbf{x}_f(t)$ under the condition that the whole trajectory minimizes the time integral

$$J = \int_{t_0}^{t_f} dt. \tag{14.4}$$

The input command vector $\mathbf{Q}(t)$ usually has the constraint (14.3).

We define a scalar function $H$, and a vector $\mathbf{p}$

$$H(\mathbf{x}, \mathbf{Q}, \mathbf{p}) = \mathbf{p}^T \mathbf{f}\left(\mathbf{x}(t), \mathbf{Q}(t)\right) \tag{14.5}$$

that provide the following two equations:

$$\dot{\mathbf{x}} = \frac{\partial H^T}{\partial \mathbf{p}} \tag{14.6}$$

$$\dot{\mathbf{p}} = -\frac{\partial H^T}{\partial \mathbf{x}}. \tag{14.7}$$

Based on the *Pontryagin principle*, the optimal input $Q(t)$ is the one that minimizes the function $H$. Such an optimal input is to apply the maximum effort, $Q_{Max}$ or $-Q_{Max}$, over the entire time interval. When the control command takes a value at the boundary of its admissible region, it is said to be *saturated*. The function $H$ is called *Hamiltonian*, and the vector $\mathbf{p}$ is called a *co-state*. ∎

**Example 309** ★ *A linear dynamical system.*
  *Consider a linear dynamical system given by*

$$Q = \ddot{x}$$

*or*

$$\dot{\mathbf{x}} = [A]\,\mathbf{x} + \mathbf{b}Q \tag{14.8}$$

*where*

$$\mathbf{x} = \begin{bmatrix} x_1 \\ x_2 \end{bmatrix} \qquad [A] = \begin{bmatrix} 0 & 1 \\ 0 & 0 \end{bmatrix} \qquad \mathbf{b} = \begin{bmatrix} 0 \\ 1 \end{bmatrix} \tag{14.9}$$

*along with a constraint on the input variable*

$$Q \le 1. \tag{14.10}$$

*By defining a co-state vector*

$$\mathbf{p} = \begin{bmatrix} p_1 \\ p_2 \end{bmatrix} \tag{14.11}$$

*the Hamiltonian (14.5) becomes*

$$H(\mathbf{x}, Q, \mathbf{p}) = \mathbf{p}^T\,([A]\,\mathbf{x} + \mathbf{b}Q) \tag{14.12}$$

*that provides two first-order differential equations*

$$\dot{\mathbf{x}} = \frac{\partial H}{\partial \mathbf{p}}^T = [A]\,\mathbf{x} + \mathbf{b}Q \tag{14.13}$$

$$\dot{\mathbf{p}} = -\frac{\partial H}{\partial \mathbf{x}}^T = -[A]\,\mathbf{p}. \tag{14.14}$$

*Equation (14.14) is*

$$\begin{bmatrix} \dot{p}_1 \\ \dot{p}_2 \end{bmatrix} = \begin{bmatrix} 0 \\ -p_2 \end{bmatrix} \tag{14.15}$$

*which can be integrated to find* $\mathbf{p}$

$$\mathbf{p} = \begin{bmatrix} p_1 \\ p_2 \end{bmatrix} = \begin{bmatrix} C_1 \\ -C_1 t + C_2 \end{bmatrix}. \tag{14.16}$$

*The Hamiltonian is then equal to*

$$\begin{aligned} H &= Q p_2 + p_1 x_2 \\ &= (-C_1 t + C_2)\,Q + p_1 x_2 \end{aligned} \tag{14.17}$$

*The control command $Q$ only appears in*

$$\mathbf{p}^T \mathbf{b} Q = (-C_1 t + C_2) Q \qquad (14.18)$$

*which can be maximized by*

$$Q(t) \quad \begin{array}{ll} = 1 & \text{if} \quad -C_1 t + C_2 \geq 0 \\ = -1 & \text{if} \quad -C_1 t + C_2 < 0 \end{array} \qquad (14.19)$$

*This solution implies that $Q(t)$ has a jump point at $t = \frac{C_2}{C_1}$. The jump point, at which the control command suddenly changes from maximum to minimum or from minimum to maximum, is called the **switching point**.*

*Substituting the control input (14.19) into (14.8) gives us two first-order differential equations*

$$\begin{bmatrix} \dot{x}_1 \\ \dot{x}_2 \end{bmatrix} = \begin{bmatrix} x_2 \\ Q \end{bmatrix}. \qquad (14.20)$$

*Equation (14.20) can be integrated to find the path $\mathbf{x}(t)$.*

$$\begin{bmatrix} x_1 \\ x_2 \end{bmatrix} = \begin{bmatrix} \frac{1}{2}(t + C_3)^2 + C_4 \\ t + C_3 \end{bmatrix} \quad \text{if} \quad Q = 1$$

$$\begin{bmatrix} x_1 \\ x_2 \end{bmatrix} = \begin{bmatrix} -\frac{1}{2}(t - C_3)^2 + C_4 \\ -t + C_3 \end{bmatrix} \quad \text{if} \quad Q = -1 \qquad (14.21)$$

*The constants of integration, $C_1$, $C_2$, $C_3$, and $C_4$, must be calculated based on the following boundary conditions:*

$$\mathbf{x}_0 = \mathbf{x}(t_0) \qquad \mathbf{x}_f = \mathbf{x}(t_f). \qquad (14.22)$$

*Eliminating $t$ between equations in (14.21) provides the relationship between the state variables $x_1$ and $x_2$.*

$$\begin{array}{ll} x_1 = \frac{1}{2} x_2^2 + C_4 & \text{if} \quad Q = 1 \\ x_1 = -\frac{1}{2} x_2^2 + C_4 & \text{if} \quad Q = -1 \end{array} \qquad (14.23)$$

*These equations show a series of parabolic curves in the $x_1 x_2$-plane with $C_4$ as a parameter. The parabolas are shown in Figure 14.1(a) and (b) with the arrows indicating the direction of motion on the paths. The $x_1 x_2$-plane is called the **phase plane**.*

*Considering that there is one switching point in this system, the overall optimal paths are shown in Figure 14.1(c). As an example, assume the state of the system at initial and final times are $\mathbf{x}(t_0)$ and $\mathbf{x}(t_f)$ respectively. The motion starts with $Q = 1$, which forces the system to move on the control path $x_1 = \frac{1}{2} x_2^2 + \left(x_{10} - \frac{1}{2} x_{20}^2\right)$ up to the intersection point with $x_1 = -\frac{1}{2} x_2^2 + \left(x_{1f} + \frac{1}{2} x_{2f}^2\right)$. The intersection is the switching point at which the control input changes to $Q = -1$. The switching point is at*

$$x_1 = \frac{1}{4}\left(2x_{10} + 2x_f - x_{20}^2 + x_{2f}^2\right) \qquad (14.24)$$

$$x_2 = \sqrt{\left(x_{1f} + \frac{1}{2} x_{2f}^2\right) - \left(x_{10} - \frac{1}{2} x_{20}^2\right)}. \qquad (14.25)$$

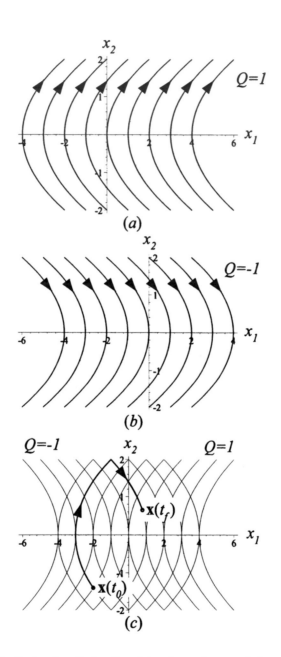

FIGURE 14.1. Optimal path for $Q = \ddot{x}$ in phase plane and the mesh of optimal paths in phase plane.

**Example 310** ★ *Robot equations in state equations.*
*The vector form of the equations of motion of a robot is*

$$\mathbf{D}(\mathbf{q})\,\ddot{\mathbf{q}} + \mathbf{H}(\mathbf{q}, \dot{\mathbf{q}}) + \mathbf{G}(\mathbf{q}) = \mathbf{Q}. \tag{14.26}$$

*We can define a state vector*

$$\mathbf{x} = \begin{bmatrix} \mathbf{q} \\ \dot{\mathbf{q}} \end{bmatrix} \tag{14.27}$$

*and transform the equations of motion to an equation in state space*

$$\dot{\mathbf{x}} = \mathbf{f}\left(\mathbf{x}(t), \mathbf{Q}(t)\right) \tag{14.28}$$

*where*

$$\mathbf{f}\left(\mathbf{x}(t), \mathbf{Q}(t)\right) = \begin{bmatrix} \dot{\mathbf{q}} \\ \mathbf{D}^{-1}\left(\mathbf{Q} - \mathbf{H} - \mathbf{G}\right) \end{bmatrix}. \tag{14.29}$$

**Example 311** ★ *Time-optimal control for robots.*
*Assume that a robot is initially at*

$$\mathbf{x}(\mathbf{t_0}) = \mathbf{x}_0 = \begin{bmatrix} \mathbf{q}_0 \\ \dot{\mathbf{q}}_0 \end{bmatrix} \tag{14.30}$$

*and it is supposed to be finally at*

$$\mathbf{x}(\mathbf{t}_f) = \mathbf{x}_f = \begin{bmatrix} \mathbf{q}_f \\ \dot{\mathbf{q}}_f \end{bmatrix} \tag{14.31}$$

*in the shortest possible time. The torques of the actuators at each joint is
assumed to be bounded*

$$|Q_i| \leq Q_{iMax}. \tag{14.32}$$

*The optimal control problem is to minimize the time performance index*

$$J = \int_{t_0}^{t_f} dt = t_f - t_0. \tag{14.33}$$

*The Hamiltonian H is defined as*

$$H(\mathbf{x}, \mathbf{Q}, \mathbf{p}) = \mathbf{p}^T \mathbf{f}\left(\mathbf{x}(t), \mathbf{Q}(t)\right) \tag{14.34}$$

*which provides the following two sets of equations:*

$$\dot{\mathbf{x}} = \frac{\partial H^T}{\partial \mathbf{p}} \tag{14.35}$$

$$\dot{\mathbf{p}} = -\frac{\partial H^T}{\partial \mathbf{x}}. \tag{14.36}$$

*The optimal control input $\mathbf{Q}_t(t)$ is the one that minimizes the function $H$. Hamiltonian minimization reduces the time-optimal control problem to a two-point boundary value problem. The boundary conditions are the states of the robot at times $t_0$ and $t_f$. Due to nonlinearity of the robots' equations of motion, there is no analytic solution for the boundary value problem. Hence, a numerical technique must be developed.*

**Example 312 ★** *Euler-Lagrange equation.*
    *To show that a path $x = x^\star(t)$ is a minimizing path for the functional $J$*

$$J(x) = \int_{t_0}^{t_f} f(x, \dot{x}, t)dt \tag{14.37}$$

*with boundary conditions $x(t_0) = x_0$, $x(t_f) = x_f$, we need to show that*

$$J(x) \geq J(x^\star) \tag{14.38}$$

*for all continuous paths $x(t)$ satisfying the boundary conditions. Any path $x(t)$ satisfying the boundary conditions $x(t_0) = x_0$, $x(t_f) = x_f$, is called admissible. To see that $x^\star(t)$ is the optimal path, we may examine the integral $J$ for every admissible path. An admissible path may be defined by*

$$x(t) = x^\star + \epsilon y(t) \tag{14.39}$$

*where*
$$y(t_0) = y(t_f) = 0 \tag{14.40}$$

*and*
$$\epsilon \ll 1 \tag{14.41}$$

*is a small number. Substituting $x(t)$ in $J$ and subtracting from (14.37) provides $\Delta J$*

$$
\begin{aligned}
\Delta J &= J\left(x^\star + \epsilon y(t)\right) - J\left(x^\star\right) \tag{14.42}\\
&= \int_{t_0}^{t_f} f(x^\star + \epsilon y, \dot{x}^\star + \epsilon \dot{y}, t)dt - \int_{t_0}^{t_f} f(x^\star, x^\star, t)dt.
\end{aligned}
$$

*Let us expand $f(x^\star + \epsilon y, \dot{x}^\star + \epsilon \dot{y}, t)$ about $(x^\star, \dot{x}^\star)$*

$$
\begin{aligned}
f(x^\star + \epsilon y, \dot{x}^\star + \epsilon \dot{y}, t) = {}& f(x^\star, \dot{x}^\star, t) + \epsilon \left(y\frac{\partial f}{\partial x} + \dot{y}\frac{\partial f}{\partial \dot{x}}\right)\\
&+ \epsilon^2 \left(y^2\frac{\partial^2 f}{\partial x^2} + 2y\dot{y}\frac{\partial^2 f}{\partial x\,\partial \dot{x}} + \dot{y}^2\frac{\partial^2 f}{\partial \dot{x}^2}\right) dt\\
&+ O\left(\epsilon^3\right) \tag{14.43}
\end{aligned}
$$

*and find*
$$\Delta J = \epsilon V_1 + \epsilon^2 V_2 + O\left(\epsilon^3\right). \tag{14.44}$$

*where*

$$V_1 = \int_{t_0}^{t_f} \left( y\frac{\partial f}{\partial x} + \dot{y}\frac{\partial f}{\partial \dot{x}} \right) dt \tag{14.45}$$

$$V_2 = \int_{t_0}^{t_f} \left( y^2\frac{\partial^2 f}{\partial x^2} + 2y\dot{y}\frac{\partial^2 f}{\partial x \partial \dot{x}} + \dot{y}^2\frac{\partial^2 f}{\partial \dot{x}^2} \right) \tag{14.46}$$

*The first integral, $V_1$ is called the **first variation** of J, and the second integral, $V_2$ is called the **second variation** of J. All the higher variations are combined and shown as $O\left(\epsilon^3\right)$. If $x^\star$ is the minimizing curve, then it is necessary that $\Delta J \geq 0$ for every admissible $y(t)$. If we divide $\Delta J$ by $\epsilon$ and make $\epsilon \rightarrow 0$ then we find a necessary condition for $x^\star$ to be the optimal path as $V_1 = 0$. This condition is equivalent to*

$$\int_{t_0}^{t_f} \left( y\frac{\partial f}{\partial x} + \dot{y}\frac{\partial f}{\partial \dot{x}} \right) dt = 0. \tag{14.47}$$

*By integrating by parts we may write*

$$\int_{t_0}^{t_f} \dot{y}\frac{\partial f}{\partial \dot{x}} dt = \left( y\frac{\partial f}{\partial \dot{x}} \right)_{t_0}^{t_f} - \int_{t_0}^{t_f} y\frac{d}{dt}\left( \frac{\partial f}{\partial \dot{x}} \right) dt. \tag{14.48}$$

*Since $y(t_0) = y(t_f) = 0$, the first term on the right-hand side is zero. Therefore, the minimization integral condition (14.47), for every admissible $y(t)$, reduces to*

$$\int_{t_0}^{t_f} y\left( \frac{\partial f}{\partial x} - \frac{d}{dt}\frac{\partial f}{\partial \dot{x}} \right) dt = 0. \tag{14.49}$$

*The terms in the parentheses are continuous functions of t, evaluated on the optimal path $x^\star$, and they do not involve $y(t)$. So, the only way that the bounded integral of the parentheses $\left( \frac{\partial f}{\partial x} - \frac{d}{dt}\frac{\partial f}{\partial \dot{x}} \right)$, multiplied by a nonzero function $y(t)$, from $t_0$ and $t_f$ to be zero, is that*

$$\frac{\partial f}{\partial x} - \frac{d}{dt}\frac{\partial f}{\partial \dot{x}} = 0. \tag{14.50}$$

*Equation (14.50) is a necessary condition for $x = x^\star(t)$ to be a solution of the minimization problem (14.37). This differential equation is called the **Euler-Lagrange** equation. It is the same Lagrange equation that we utilized to derive the equations of motion of a robot. The second necessary condition to have $x = x^\star(t)$ as a minimizing solution is that the second variation, evaluated on $x^\star(t)$, must be negative.*

**Example 313 ★** *The Lagrange equation for extremizing $J = \int_1^2 \dot{x}^2 dt$.*
  *The Lagrange equation for extremizing the functional*

$$J = \int_1^2 \dot{x}^2 dt \tag{14.51}$$

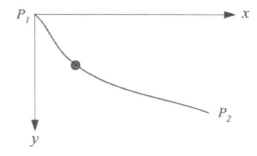

FIGURE 14.2. A curve joining points $P_1$ and $P_2$, and a frictionless sliding point.

is

$$\frac{\partial f}{\partial x} - \frac{d}{dt}\frac{\partial f}{\partial \dot{x}} = -\ddot{x} = 0 \tag{14.52}$$

that shows the optimal path is

$$x = C_1 t + C_2. \tag{14.53}$$

Considering the boundary conditions $x(1) = 0$, $x(2) = 3$ provides

$$x = 3t - 3.$$

**Example 314** ★ *Brachistochrone problem.*
  *We may utilize the Lagrange equation and find the frictionless curve joining two points as shown if Figure 14.2, along which a particle falling from rest due to gravity, travels from the higher to the lower point in the minimum time. This is the well-known* **brachistochrone** *problem.*
  *If v is the velocity of the falling point along the curve, then the time required to fall an arc length ds is ds/v. Then, the objective function to find the curve of minimum time is*

$$J = \int_1^2 \frac{ds}{v}. \tag{14.54}$$

*However,*

$$ds = \sqrt{1 + y'^2}dx \tag{14.55}$$

*and according to the conservation of energy*

$$v = \sqrt{2gy}. \tag{14.56}$$

*Therefore, the objective function simplifies to*

$$J = \int_1^2 \sqrt{\frac{1 + y'^2}{2gy}}dx. \tag{14.57}$$

*Applying the Lagrange equations we find*

$$y \left(1 + y'^2\right) = 2r \tag{14.58}$$

*where r is a constant. The optimal curve starting from $y(0) = 0$ can be expressed by two parametric equations*

$$\begin{aligned} x &= r \left(\beta - \sin \beta\right) & (14.59) \\ y &= r \left(1 - \cos \beta\right). & (14.60) \end{aligned}$$

*The optimal curve is a **cycloid**.*

*The name of the problem is derived from the Greek word "βραχιστος," meaning "shortest," and "χρονος," meaning "time." The brachistochrone problem was originally discussed by Galilei in 1630 and later solved by Johann and Jacob Bernoulli in 1696.*

**Example 315 ★** *Lagrange multiplier.*

*Assume $f(x)$ is defined on an open interval $(a, b)$ and has continuous first and second order derivatives in some neighborhood of $x_0 \in (a, b)$. The point $x_0$ is a local extremum of $f(x)$ if*

$$\frac{df(x_0)}{dx} = 0. \tag{14.61}$$

*Assume $f(\mathbf{x}) = 0$, $\mathbf{x} \in \mathbb{R}^n$ and $g_i(\mathbf{x}) = 0$, $i = 1, 2, \cdots, m$ are functions defined on an open region $\mathbb{R}^n$ and have continuous first and second order derivatives in $\mathbb{R}^n$. The necessary condition that $\mathbf{x}_0$ be an extremum of $f(\mathbf{x})$ subject to the constraints $g_i(\mathbf{x}) = 0$ is that there exist m **Lagrange multipliers** $\lambda_i$, $i = 1, 2, \cdots, m$ such that*

$$\nabla \left(s + \sum \lambda_i g_i\right) = 0. \tag{14.62}$$

*As an example, we can find the minimum of*

$$f = 1 - x_1^2 - x_2^2 \tag{14.63}$$

*subject to*

$$g = x_1^2 + x_2 - 1 = 0 \tag{14.64}$$

*by finding the gradient of $f + \lambda g$*

$$\nabla \left(1 - x_1^2 - x_2^2 + \lambda \left(x_1^2 + x_2 - 1\right)\right) = 0.$$

*That leads to*

$$\begin{aligned} \frac{\partial f}{\partial x_1} &= -2x_1 + 2\lambda x_1 = 0 & (14.65) \\ \frac{\partial f}{\partial x_2} &= -2x_2 + \lambda = 0. & (14.66) \end{aligned}$$

To find the three unknowns, $x_1$, $x_2$, and $\lambda$, we employ Equations (14.65), (14.66), and (14.64). There are two sets of solutions as follows:

$$
\begin{array}{lll}
x_1 = 0 & x_2 = 1 & \lambda = 2 \\
x_1 = \pm 1/\sqrt{2} & x_2 = 1/2 & \lambda = 1
\end{array}
\tag{14.67}
$$

**Example 316 ★** *Admissible control function.*

*The components of the control command* $\mathbf{Q}(t)$ *are allowed to be piecewise continuous and the values they can take may be any number within the bounded region of the control space. As an example, consider a 2 DOF system* $\mathbf{Q}(t) = \begin{bmatrix} Q_1 & Q_2 \end{bmatrix}^T$ *with the restriction* $|Q_i| < 1$, $i = 1, 2$. *The control space is a circle in the plane* $Q_1 Q_2$. *The control components may have any piecewise continuous value within the circle. Such controls are called admissible.*

**Example 317** *Description of the time optimal control problem.*

*The aim of minimum time control is to guide the robot on a path in minimum time to increase the robot's productivity. Except for low order, autonomous, and linear problems, there is no general analytic solution for the time optimal control problems of dynamical systems. The problem of time optimal control is always a bounded input problem. If there exists an admissible time optimal control for a given initial condition and final target, then, at any time, at least one of the control variables attains its maximum or minimum value. Based on Pontryagin's principle, the solution of minimum time problems with bounded inputs is a bang-bang control, indicating that at least one of the input actuators must be saturated at any time. However, finding the switching points at which the saturated input signal is replaced with another saturated signal is not straightforward, and is the main concern of numerical solution methods.*

*In a general case, the problem reduces to a two-points boundary value problem that is difficult to solve. The corresponding nonsingular, nonlinear two-point boundary value problem must be solved to determine the switching times. A successful approach is to assume that the configuration trajectory of the dynamical system is preplanned, and then reduce the problem to a minimum time motion along the trajectory.*

## 14.2    ★ Floating Time Method

Consider a particle with mass $m$, as shown in Figure 14.3, is moving according to the following equation of motion:

$$
m\ddot{x} = g(x, \dot{x}) + f(t)
\tag{14.68}
$$

where $g(x, \dot{x})$ is a general nonlinear external force function, and $f(t)$ is the unknown input control force function. The control command $f(t)$ is

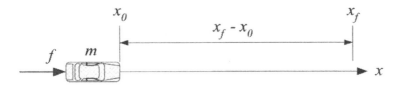

FIGURE 14.3. Rest-to-rest motion of a mass on a straight line time-optimally.

bounded to

$$|f(t)| \leq F. \tag{14.69}$$

The particle starts from rest at position $x(0) = x_0$ and moves on a straight line to the destination point $x(t_f) = x_f$ at which it stops.

We can solve this rest-to-rest control problem and find the required $f(t)$ to move $m$ from $\dot{x}_0$ to $\dot{x}_f$ in minimum time utilizing the *floating time algorithm*.

**Algorithm 14.1.** Floating time technique.

1. *Divide the preplanned path of motion $x(t)$ into $s + 1$ intervals and specify all coordinate values $x_i$, $(i = 0, 1, 2, 3, ..., s + 1)$*

2. *Set $f_0 = +F$ and calculate*

$$\tau_0 = \sqrt{\frac{2m(x_1 - x_0)}{F}} \tag{14.70}$$

3. *Set $f_{s+1} = -F$ and calculate*

$$\tau_s = \sqrt{\frac{2m(x_s - x_{s-1})}{-F}} \tag{14.71}$$

4. *For $i$ from $1$ to $s - 1$, calculate $\tau_i$ such that $f_i = +F$ and*

$$
\begin{aligned}
f_i &= m\ddot{x}_i - g(x_i, \dot{x}_i) \\
&= \frac{4m}{\tau_i^2 + \tau_{i-1}^2}\left(\frac{\tau_{i-1}}{\tau_i + \tau_{i-1}}x_{i+1} + \frac{\tau_i}{\tau_i + \tau_{i-1}}x_{i-1} + x_i\right) \\
&\quad -g(x_i, \dot{x}_i)
\end{aligned} \tag{14.72}
$$

5. *If $|f_i| \leq F$, then stop, otherwise set $j = s$,*

6. *Calculate $\tau_{j-1}$ such that $f_j = -F$*

7. *If $|f_{j-1}| \leq F$, then stop, otherwise set $j = j - 1$ and return to step 6*

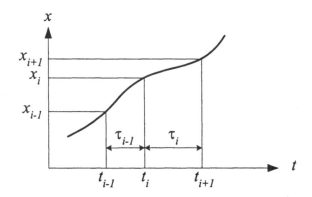

FIGURE 14.4. Time history of motion for the point mass $m$.

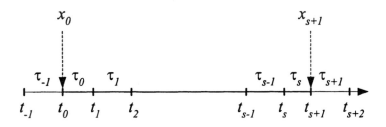

FIGURE 14.5. Introducing two extra points, $x_{-1}$ and $x_{s+2}$, before the initial and after the final points.

**Proof.** Assume $g(x_i, \dot{x}_i) = 0$ and $x(t)$, as shown in Figure 14.4, is the time history of motion for the point mass $m$. We divide the path of motion into $s+1$ arbitrary, and not necessarily equal, segments. Hence, the coordinates $x_i$, $(i = 0, 1, 2, ..., s+1)$ are known. The *floating-time* $\tau_i = t_i - t_{i-1}$ is defined as the required time to move $m$ from $x_i$ to $x_{i+1}$.

Utilizing the central difference method, we may define the first and second derivatives at point $i$ by

$$\dot{x}_i = \frac{x_{i+1} - x_{i-1}}{\tau_i + \tau_{i-1}} \tag{14.73}$$

$$\ddot{x}_i = \frac{4}{\tau_i^2 + \tau_{i-1}^2} \left( \frac{\tau_{i-1}}{\tau_i + \tau_{i-1}} x_{i+1} + \frac{\tau_i}{\tau_i + \tau_{i-1}} x_{i-1} - x_i \right). \tag{14.74}$$

These equations indicate that the velocity and acceleration at point $i$ depend on $x_i$, and two adjacent points $x_{i-1}$, and $x_{i+1}$, as well as on the floating times $\tau_i$, and $\tau_{i-1}$. Therefore, two extra points, $x_{-1}$ and $x_{s+2}$, before the initial point and after the final point are needed to define velocity and acceleration at $x_0$ and $x_{s+1}$. These extra points and their corresponding floating times are shown in Figure 14.5.

The rest conditions at the beginning and at the end of motion require

$$x_{-1} = x_1 \qquad x_{s+2} = x_s \tag{14.75}$$

$$\tau_0 = \tau_{-1} \qquad \tau_{s+1} = \tau_s. \tag{14.76}$$

Using Equation (14.74), the equation of motion, $f_i = m\ddot{x}_i$, at the initial point is

$$f_0 = m\ddot{x}_0$$

$$= \frac{4m}{2\tau_0^2}(x_1 - x_0). \tag{14.77}$$

The minimum value of the first floating time $\tau_0$ is found by setting $f_0 = F$.

$$\tau_0 = \sqrt{\frac{2m(x_1 - x_0)}{F}} \tag{14.78}$$

It is the minimum value of the first floating time because, if $\tau_0$ is less than the value given by (14.78), then $f_0$ will be greater than $F$ and breaks the constraint (14.69). On the other hand, if $\tau_0$ is greater than the value given by (14.78), then $f_0$ will be less than $F$ and the input is not saturated yet. The same conditions exist at the final point where the equation of motion is

$$f_{s+1} = m\ddot{x}_{s+1}$$

$$= \frac{4m}{2\tau_s^2}(x_s - x_{s-1}). \tag{14.79}$$

The minimum value of the final floating-time, $\tau_s$, is achieved by setting $f_{s+1} = -F$.

$$\tau_s = \sqrt{\frac{2m(x_s - x_{s-1})}{-F}} \tag{14.80}$$

To find the minimum value of $\tau_1$, we develop the equation of motion at $x_1$

$$f_1 = \frac{4m}{\tau_1^2 + \tau_0^2}\left(\frac{\tau_0}{\tau_1 + \tau_0}x_2 + \frac{\tau_1}{\tau_1 + \tau_0}x_0 - x_1\right) \tag{14.81}$$

which is an equation with two unknowns $f_1$ and $\tau_1$. We are able to find $\tau_1$ numerically by adjusting $\tau_1$ to provide $f_1 = F$. Applying this procedure we are able to find the minimum floating times $\tau_{i+1}$ by applying the maximum force constraint $f_i = F$, and solving the equation of motion for $\tau_{i+1}$ numerically. When $\tau_i$ is known and the maximum force is applied to find the next floating-time, $\tau_{i+1}$, we are in the *forward path of the floating time algorithm*. In the last step of the forward path, $\tau_{s-1}$ is found at $x_{s-1}$. At this step, all the floating-times $\tau_i$, $(i = 0, 1, 2, ..., s)$ are known, while $f_i = F$ for $i = 0, 1, 2, ..., s - 1$, and $f_i = -F$ for $i = s$. Then, the force $f_s$ is

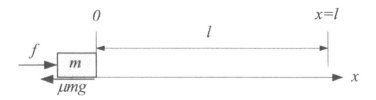

FIGURE 14.6. A rectilinear motion of a rigid mass $m$ under the influence of a control force $f(t)$ and a friction force $\mu mg$.

the only variable that is not calculated during the forward path by using the equation of motion at point $x_s$. It is actually dictated by the equation of motion at point $x_s$, because $\tau_{s-1}$ is known from the forward path procedure, and $\tau_s$ is known from Equation (14.80) to satisfy the final point condition. So, the value of $f_s$ can be found from the equation of motion at $i = s$ and substituting $\tau_s$, $\tau_{s-1}$, $x_{s-1}$, $x_s$, and $x_{s+1}$.

$$ f_s = \frac{4m}{\tau_s^2 + \tau_{s-1}^2} \left( \frac{\tau_{s-1}}{\tau_s + \tau_{s-1}} x_{s+1} + \frac{\tau_s}{\tau_s + \tau_{s-1}} x_{s-1} - x_s \right) \qquad (14.82) $$

Now, if $f_s$ does not break the constraint $|f(t)| \leq F$, the problem is solved and the minimum time motion is determined. The input signals, $f_i$, ($i = 0, 1, 2, ..., s + 1$), $i \neq s$, are always saturated, and also none of the floating-times $\tau_i$ can be reduced any more. However, it is expected that $f_s$ breaks the constraint $|f(t)| \leq F$ because accelerating in $s - 1$ steps with $f = F$ produces a large amount of kinetic energy and a huge deceleration is needed to stop the mass $m$ in the final step.

Now we reverse the procedure, and start a *backward path*. According to (14.82), $f_s$ can be adjusted to satisfy the constraint $f_s = -F$ by tuning $\tau_{s-1}$. Now $f_{n-1}$ must be checked for the constraint $|f(t)| \leq F$. This is because $\tau_{s-2}$ is already found in the forward path, and $\tau_{s-1}$ in the backward path. Hence, the value of $f_{n-1}$ is dictated by the equation of motion at point $x_{s-1}$. If $f_{s-1}$ does not break the constraint $|f(t)| \leq F$, the problem is solved and the time-optimal motion is achieved. Otherwise, the backward path must be continued to a point where the force constraint is satisfied. The position $x_k$ in the backward path, where $|f_k| \leq F$, is called switching point because $f_j = F$ for $j < k$, $0 \leq j < k$ and $f_j = -F$ for $j > k$, $k < j \leq s + 1$. ∎

**Example 318 ★** *Moving a mass on a rough surface.*

*Consider a rectilinear motion of a rigid mass $m$ under the influence of a variable force $f(t)$ and a friction force $\mu mg$, as shown in Figure 14.6. The force is bounded by $|f(t)| \leq F$, where $\pm F$ is the limit of available force. It is necessary to find a function $f(t)$ that moves $m$, from the initial conditions $x(0) = 0$, $v(0) = 0$ to the final conditions $x(t_f) = l > 0$, $v(t_f) = 0$*

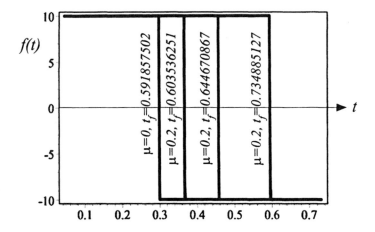

FIGURE 14.7. Time history of the optimal input $f(t)$ for different friction coefficients $\mu$.

in minimum total time $t = t_f$. The motion is described by the following equation of motion and boundary conditions:

$$f = m\ddot{x} - \mu m g \tag{14.83}$$

$$\begin{array}{ll} x(0) = 0 & v(0) = 0 \\ x(t_f) = l & v(t_f) = 0. \end{array} \tag{14.84}$$

Using the theory of optimal control, we know that a time optimal control solution for $\mu = 0$ is a piecewise constant function where the only discontinuity is at the switching point $t = \tau = t_f/2$ and

$$f(t) = \begin{cases} F & \text{if } t < \tau \\ -F & \text{if } t > \tau. \end{cases} \tag{14.85}$$

Therefore, the time optimal control solution for moving a mass $m$ from $x(0) = x_0 = 0$ to $x(t_f) = x_f = l$ on a smooth straight line is a bang-bang control with only one switching time. The input force $f(t)$ is on its maximum, $f = F$, before the switching point $x = (x_f - x_0)/2$ at $\tau = t_f/2$, and $f = -F$ after that. Any asymmetric characteristics, such as friction, will make the problem asymmetric by moving the switching point.

In applying the floating-time algorithm, we assume that a particle of unit mass, $m = 1\,\text{kg}$, slides under Coulomb friction on a rough horizontal surface. The magnitude of the friction force is $\mu m g$, where $\mu$ is the friction coefficient and $g = 9.81\,\text{m/s}^2$. We apply the floating-time algorithm using the following numerical values:

$$F = 10\,\text{N} \qquad l = 10\,\text{m} \qquad s + 1 = 200 \tag{14.86}$$

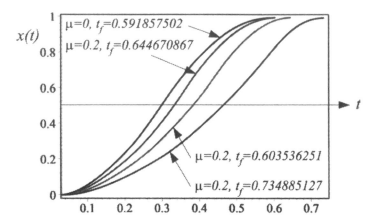

FIGURE 14.8. Time history of the optimal motion $x(t)$ for different friction coefficients $\mu$.

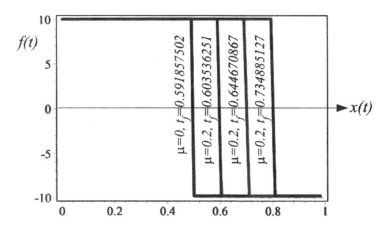

FIGURE 14.9. Position history of the optimal force $f(t)$ for different friction coefficients $\mu$.

*Figures 14.7, 14.8, and 14.9 show the results for some different values of μ. Figure 14.7 illustrates the time history of the optimal input force for different values of μ. Each curve is indicated by the value of μ and the corresponding minimum time of motion $t_f$. Time history of the optimal motions $x(t)$ are shown in Figure 14.8, while the time history of the optimal inputs $f(t)$ are shown in Figure 14.9. The switching times and positions are shown in Figures 14.7 and 14.9, respectively.*

*If μ = 0 then switching occurs at the midpoint of the motion $x(\tau) = l/2$ and halfway through the time $\tau = t_f/2$. Increasing μ delays both the switching times and the switching positions. The total time of motion also increases by increasing μ.*

**Example 319** ★ *First and second derivatives in central difference method.*
*Using a Taylor series, we expand x at points $x_{i-1}$ and $x_{i+1}$ as an extrapolation of point $x_i$*

$$x_{i+1} = x_i + \dot{x}_i\tau_i + \frac{1}{2}\ddot{x}_i\tau_i^2 + \cdots \tag{14.87}$$

$$x_{i-1} = x_i - \dot{x}_i\tau_{i-1} + \frac{1}{2}\ddot{x}_i\tau_{i-1}^2 - \cdots . \tag{14.88}$$

*Accepting the first two terms and calculating $x_{i+1} - x_{i-1}$ provides*

$$\dot{x}_i = \frac{x_{i+1} - x_{i-1}}{\tau_i + \tau_{i-1}}.$$

*Now, accepting the first three terms of the Taylor series and calculating $x_{i+1} + x_{i-1}$ provides*

$$\ddot{x}_i = \frac{4m}{\tau_i^2 + \tau_{i-1}^2}\left(\frac{\tau_{i-1}}{\tau_i + \tau_{i-1}}x_{i+1} + \frac{\tau_i}{\tau_i + \tau_{i-1}}x_{i-1} - x_i\right). \tag{14.89}$$

**Example 320** ★ *Convergence.*
*The floating time algorithm presents an iterative method hence, convergence criteria must be identified. In addition, a condition must be defined to terminate the iteration. In the forward path, we calculate the floating-time $\tau_i$ by adjusting it to a value that provides $f_i = F$. The floating-time $\tau_i$ converges to the minimum possible value, as long as $\partial\ddot{x}_i/\partial\tau_i < 0$ and $\partial\ddot{x}_i/\partial\tau_{i-1} > 0$. Figure 14.10 illustrates the behavior of $\ddot{x}_i$ as a function of $\tau_i$ and $\tau_{i-1}$. Using the Equation (14.74), the required conditions are fulfilled within a basin of convergence,*

$$Z_1 x_{s+1} + Z_2 x_s + Z_3 x_{s-1} < 0 \tag{14.90}$$

$$Z_4 x_{s+1} + Z_5 x_s + Z_6 x_{s-1} > 0 \tag{14.91}$$

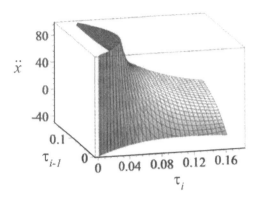

FIGURE 14.10. Behavior of $\ddot{x}_i$ as a function of $\tau_i$ and $\tau_{i-1}$.

*where,*

$$Z_1 = \frac{8\left(6\tau_i^4\tau_{i-1} + 8\tau_i^3\tau_{i-1}^2 + 6\tau_i^2\tau_{i-1}^3\right)}{\left(\tau_i^2 + \tau_{i-1}^2\right)^3 \left(\tau_i + \tau_{i-1}\right)^3} \tag{14.92}$$

$$Z_2 = \frac{8\left(\tau_{i-1}^5 - 3\tau_i^5 - 8\tau_i^3\tau_{i-1}^2 - 9\tau_i^4\tau_{i-1} + 3\tau_i\tau_{i-1}^4\right)}{\left(\tau_i^2 + \tau_{i-1}^2\right)^3 \left(\tau_i + \tau_{i-1}\right)^3} \tag{14.93}$$

$$Z_3 = \frac{8\left(-\tau_{i-1}^5 + 3\tau_i^5 + 3\tau_i^4\tau_{i-1} - 3\tau_i\tau_{i-1}^4 - 6\tau_i^2\tau_{i-1}^3\right)}{\left(\tau_i^2 + \tau_{i-1}^2\right)^3 \left(\tau_i + \tau_{i-1}\right)^3} \tag{14.94}$$

$$Z_4 = \frac{8\left(3\tau_{i-1}^5 - \tau_i^5 - 6\tau_i^3\tau_{i-1}^2 - 3\tau_i^4\tau_{i-1} + 3\tau_i\tau_{i-1}^4\right)}{\left(\tau_i^2 + \tau_{i-1}^2\right)^3 \left(\tau_i + \tau_{i-1}\right)^3} \tag{14.95}$$

$$Z_5 = \frac{8\left(-3\tau_{i-1}^5 + \tau_i^5 + 3\tau_i^4\tau_{i-1} - 9\tau_i\tau_{i-1}^4 - 8\tau_i^2\tau_{i-1}^3\right)}{\left(\tau_i^2 + \tau_{i-1}^2\right)^3 \left(\tau_i + \tau_{i-1}\right)^3} \tag{14.96}$$

$$Z_6 = \frac{8\left(6\tau_i^3\tau_{i-1}^2 + 8\tau_i^2\tau_{i-1}^3 + 6\tau_i\tau_{i-1}^4\right)}{\left(\tau_i^2 + \tau_{i-1}^2\right)^3 \left(\tau_i + \tau_{i-1}\right)^3}. \tag{14.97}$$

*The convergence conditions guarantee that $\ddot{x}_i$ decreases with an increase in $\tau_i$, and increases with an increase in $\tau_{i-1}$. Therefore, if either $\tau_i$ or $\tau_{i-1}$ is fixed, we are able to find the other floating time by setting $f_i = F$. Convergence conditions for backward path are changed to*

$$Z_1 x_{s+1} + Z_2 x_s + Z_3 x_{s-1} > 0 \tag{14.98}$$

$$Z_4 x_{s+1} + Z_5 x_s + Z_6 x_{s-1} < 0. \tag{14.99}$$

*A termination criterion may be defined by*

$$\|f_i| - F| \le \epsilon. \tag{14.100}$$

*where ε is a user-specified number. The termination criterion provides a good method to make sure that the maximum deviation is within certain bounds.*

**Example 321 ★** *Analytic calculating of floating times.*
*The rest condition at the beginning of the motion of an m on a straight line requires*

$$x_{-1} = x_1 \tag{14.101}$$

$$\tau_0 = \tau_{-1}. \tag{14.102}$$

*The first floating time $\tau_0$ is found by setting $f_0 = F$ and developing the equation of motion $f_i = m\ddot{x}_i$ at point $x_0$.*

$$\tau_0 = \sqrt{\frac{2m(x_1 - x_0)}{F}} \tag{14.103}$$

*Now the equation of motion at point $x_1$ is*

$$f_1 = \frac{4m}{\tau_1^2 + \tau_0^2}\left(\frac{\tau_0}{\tau_1 + \tau_0}x_2 + \frac{\tau_1}{\tau_1 + \tau_0}x_0 + x_1\right). \tag{14.104}$$

*Substituting $\tau_0$ from (14.103) into (14.104) and applying $f_1 = F$ provides the following equation*

$$F = \frac{4mF}{2mx_1 - 2mx_0 + F\tau_1^2}\left(\frac{\tau_0 x_2 + \tau_1 x_0}{\tau_1 + \sqrt{\frac{2m(x_1-x_0)}{F}}} + x_1\right) \tag{14.105}$$

*that must be solved for $\tau_1$. Then substituting $\tau_1$ from (14.105) into the equation of motion at $x_2$, and setting $f_2 = F$ leads to a new equation to find $\tau_2$. This procedure can similarly be applied to the other steps. However, calculating the floating times in closed form is not straightforward and getting more complicated step by step, hence, a numerical solution is needed. The equations for calculating $\tau_i$ are nonlinear and therefore have multiple solutions. Each positive solution must be examined for the constraint $f_i = F$. Negative solutions are not acceptable.*

**Example 322 ★** *Brachistochrone and path planning.*
*The floating-time method is sometimes applicable for path planning problems. As an illustrative example, we considered the well-known brachistochrone problem. As Johann Bernoulli says: "A material particle moves without friction along a curve. This curve connects point A with point B (point A is placed above point B). No forces affect it, except the gravitational attraction. The time of travel from A to B must be the smallest. This brings up the question: what is the form of this curve?"*

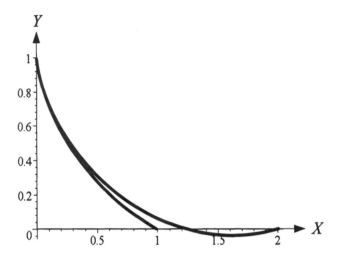

FIGURE 14.11. Time optimal path for a falling unit mass from $A(0,1)$ to two different destinations.

*The classical solution of the brachistochrone problem is a cycloid and its parametric equation is*

$$x = r(\beta - \sin\beta) \tag{14.106}$$
$$y = r(1 - \cos\beta). \tag{14.107}$$

*where $r$ is the radius of the corresponding cycloid and $\beta$ is the angle of rotation of $r$. When $\beta = 0$ the particle is at the beginning point $A(0,0)$. The particle is at the second point $B$ when $\beta = \beta_B$. The value of $\beta_B$ can be obtained from*

$$x_B = r(\beta_B - \sin\beta_B) \tag{14.108}$$
$$y_B = r(1 - \cos\beta_B). \tag{14.109}$$

*The total time of the motion is*

$$t_f = \beta_B\sqrt{\frac{r}{g}} \tag{14.110}$$

*In a path-planning problem, except for the boundaries, the path of motion is not known. Hence, the position of $x_i$ in Equations (14.73) and (14.74) are not given. Knowing the initial and final positions, we fix the $x_i$ coordinates while keeping the $y_i$ coordinates free. We will obtain the optimal path of motion by applying the known input force and searching for the optimum $y_i$ that minimizes the floating times.*

*Consider the points $B_1(1,0)$ and $B_2(2,0)$ as two different destinations of motion for a unit mass falling from point $A(0,1)$. Figure 14.11 illustrates* *the optimal path of motion for the two destinations, obtained by the floating-time method for s = 100. The total time of motion is $t_{f_1} = 0.61084$ s, and $t_{f_2} = 0.8057$ s respectively. In this calculation, the gravitational acceleration is assumed $g = 10$ m/ s$^2$ in $-Y$ direction.*

*An analytic solution shows that $\beta_{B_1} = 1.934563$ rad and $\beta_{B_2} = 2.554295$ rad. The corresponding total times are $t_{f_1} = 0.6176$ s, and $t_{f_2} = 0.8077$ s respectively. By increasing s, the calculated minimum time would be closer to the analytical results, and the evaluated path would be closer to a cycloid.*

*A more interesting and more realistic problem of brachistochrone can be brachistochrone with friction and brachistochrone with linear drag. Although there are analytical solutions for these two cases, no analytical solution has been developed for brachistochrone with nonlinear (say second degree) drag. Applying the floating-time algorithm for this kind of problem can be an interesting challenge.*

## 14.3   ★ Time-Optimal Control for Robots

Robots are multiple DOF dynamical systems. In case of a robot with $n$ DOF, the control force $\mathbf{f}$ and the output position $\mathbf{x}$ are vectors.

$$\mathbf{f} = \begin{bmatrix} f_1 & f_2 & \cdots & f_n \end{bmatrix}^T$$
$$\mathbf{x} = \begin{bmatrix} x_1 & x_2 & \cdots & x_n \end{bmatrix}^T$$

The constraint on the input force vector can be shown by

$$|\mathbf{f}_i| \leq \mathbf{F} \tag{14.111}$$

where the elements of the limit vector $\mathbf{F} \in \mathbb{R}^n$ may be different. The floating time algorithm is applied similar to the algorithm 14.2, however, at each step all the elements of the force vector $\mathbf{f}$ must be examined for their constraints. To attain the time optimal control, at least one element of the input vector $\mathbf{f}$ must be saturated at each step, while all the other elements are within their limits.

**Algorithm 14.2.** Floating time technique for the $n$ DOF systems.

1. *Divide the preplanned path of motion $\mathbf{x}(t)$ into $s + 1$ intervals and specify all coordinate vectors $\mathbf{x}_i$, $(i = 0, 1, 2, 3, ..., s + 1)$.*

2. *Develop the equations of motion at $\mathbf{x}_0$ and calculate $\tau_0$ for which only one component of the force vector $\mathbf{f}_0$ is saturated on its higher limit, while all the other components are within their limits.*

$$\begin{aligned} f_{0_k} &= F_k &, \quad k \in \{0, 1, 2, \cdots, n\} \\ f_{0_r} &\leq F_r &, \quad r = 0, 1, 2, \cdots, n &, \quad r \neq k \end{aligned}$$

3. *Develop the equations of motion at $\mathbf{x}_{s+1}$ and calculate $\tau_s$ for which only one component of the force vector $\mathbf{f}_{s+1}$ is saturated on its higher limit, while all the other components are within their limits.*

$$f_{s+1_k} = -F_k \quad , \quad k \in \{0, 1, 2, \cdots, n\}$$
$$f_{s+1_r} \leq F_r \quad , \quad r = 0, 1, 2, \cdots, n \quad , \quad r \neq k$$

4. *For $i$ from 1 to $s-1$, calculate $\tau_i$ such that only one component of the force vector $\mathbf{f}_i$ is saturated on its higher limit, while all the other components are within their limits.*

$$f_{i_k} = -F_k \quad , \quad k \in \{0, 1, 2, \cdots, n\}$$
$$f_{i_r} \leq F_r \quad , \quad r = 0, 1, 2, \cdots, n \quad , \quad r \neq k$$

5. *If $|\mathbf{f}_s| \leq \mathbf{F}$, then stop, otherwise set $j = s$.*

6. *Calculate $\tau_{j-1}$ such that only one component of the force vector $\mathbf{f}_j$ is saturated on its lower limit, while all the other components are within their limits.*

7. *If $|\mathbf{f}_{j-1}| \leq F$, then stop, otherwise set $j = j - 1$ and return to step 6.*

**Example 323 ★** *2R manipulator on a straight line.*
*Consider a 2R planar manipulator that its endpoint moves rest-to-rest from point $(1, 1.5)$ to point $(-1, 1.5)$ on a straight line $Y = 1.5$. Figure 14.12 illustrates a 2R planar manipulator with rigid arms. The manipulator has two rotary joints, whose angular positions are defined by the coordinates $\theta$ and $\varphi$. The joint axes are both parallel to the Z-axis of the global coordinate frame, and the robot moves in the XY-plane. Gravity acts in the $-Y$ direction and the lengths of the arms are $l_1$ and $l_2$.*
*We express the equations of motion for 2R robotic manipulators in the following form:*

$$P = A\ddot{\theta} + B\ddot{\varphi} + C\dot{\theta}\dot{\varphi} + D\dot{\varphi}^2 + M \quad (14.112)$$
$$Q = E\ddot{\theta} + F\ddot{\varphi} + G\dot{\theta}^2 + N \quad (14.113)$$

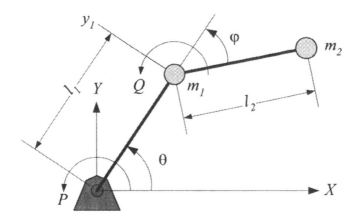

FIGURE 14.12. A 2R planar manipulator with rigid arms.

*where P and Q are the actuator torques and*

$$
\begin{aligned}
A &= A(\varphi) = m_1 l_1^2 + m_2 \left(l_1^2 + l_2^2 + 2l_1 l_2 \cos\varphi\right) \\
B &= B(\varphi) = m_2 \left(l_2^2 + l_1 l_2 \cos\varphi\right) \\
C &= C(\varphi) = -2m_2 l_1 l_2 \sin\varphi \\
D &= D(\varphi) = m_2 l_1 l_2 \sin\varphi \\
E &= E(\varphi) = B \\
F &= m_2 l_2^2 \\
G &= G(\varphi) = -D \\
M &= M(\theta, \varphi) = (m_1 + m_2) g l_1 \cos\theta + m_2 g l_2 \cos(\theta + \varphi) \\
N &= N(\theta, \varphi) = m_2 g l_2 \cos(\theta + \varphi).
\end{aligned}
\tag{14.114}
$$

*Following Equations (14.73) and (14.74), we define two functions to discretize the velocity and acceleration.*

$$
v(\dot{x}_i) = \frac{x_{i+1} - x_{i-1}}{\tau_i + \tau_{i-1}}
\tag{14.115}
$$

$$
a(\ddot{x}_i) = \frac{4}{\tau_i^2 + \tau_{i-1}^2} \left( \frac{\tau_{i-1}}{\tau_i + \tau_{i-1}} x_{i+1} + \frac{\tau_i}{\tau_i + \tau_{i-1}} x_{i-1} - x_i \right).
\tag{14.116}
$$

*Then, the equations of motion at each instant may be written as*

$$
P_i(t) = A_i a(\ddot{\theta}) + B_i a(\ddot{\varphi}) + C_i v(\dot{\theta}) v(\dot{\varphi}) + D_i v^2(\dot{\varphi}) + M_i
\tag{14.117}
$$

$$
Q_i(t) = E_i a(\ddot{\theta}) + F_i a(\ddot{\varphi}) + G_i v^2(\dot{\theta}) + N_i
\tag{14.118}
$$

*where $P_i$ and $Q_i$ are the required actuator torques at instant $i$. Actuators are assumed to be bounded by*

$$
|P_i(t)| \le P_M \qquad |Q_i(t)| \le Q_M.
\tag{14.119}
$$

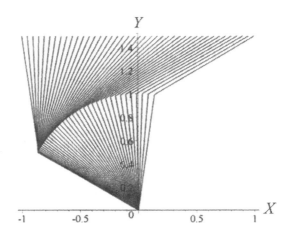

FIGURE 14.13. A 2R planar manipulator, moving from point $(1, 1.5)$ to point $(-1, 1.5)$ on a straight line $Y = 1.5$.

*To move the manipulator time optimally along a known trajectory, motors must exert torques with a known time history. Using the floating-time method, the motion starts at point $i = 0$ and ends at point $i = s + 1$. Introducing two extra points at $i = -1$ and $i = n + 2$, and applying the rest boundary conditions, we find*

$$\tau_0 = \tau_1 \quad \theta_0 = \theta_2 \quad \varphi_0 = \varphi_2$$
$$\tau_s = \tau_{s+1} \quad \theta_s = \theta_{s+2} \quad \varphi_s = \varphi_{s+2}. \tag{14.120}$$

*All inputs of the manipulator at instant $i$ are controlled by the common floating-times $\tau_i$ and $\tau_{i-1}$. In the forward path, when one of the inputs saturates at instant $i$, while the others are less than their limits, the minimum $\tau_i$ is achieved. Any reduction in $\tau_i$ increases the saturated input and breaks one of the constraints (14.119). The same is true in the backward path when we search for $\tau_{i-1}$.*

*Consider the following numerical values and the path of motion illustrated in Figure 14.13.*

$$m_1 = m_2 = 1\,\text{kg} \quad l_1 = l_2 = 1\,\text{m} \quad P_M = Q_M = 100\,\text{N m} \tag{14.121}$$

*To apply the floating time algorithm, the path of motion in Cartesian space must first be transformed into joint space using inverse kinematics. Then, the path of motion in joint space must be discretized to an arbitrary interval, say 200, and the algorithm 14.2 should be applied.*

*Figure 14.14 depicts the actuators' torque for minimum time motion after applying the floating time algorithm. In this maneuver, there exists one switching point, where the grounded actuator switches from maximum to minimum. The ungrounded actuator never saturates, but as expected, one*

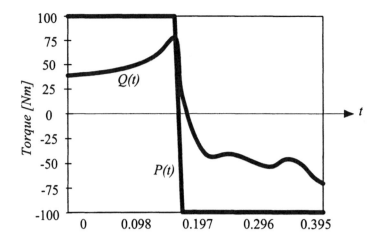

FIGURE 14.14. Time optimal control inputs for a 2R manipulator moving on line $Y = 1.5$, $-1 < X < 1$.

of the inputs is always saturated. Calculating the floating times allows us to calculate the kinematics information of motion in joint coordinate space. Time histories of the joint coordinates can be utilized to determine the kinematics of the end-effector in Cartesian space.

**Example 324 ★** *Multiple switching points.*

The 2R manipulator shown in Figure 14.12 is made to follow the path illustrated in Figure 14.15. The floating time algorithm is run for the following data:

$$
\begin{array}{llll}
m_1 = m_2 = 1\,\text{kg} & l_1 = l_2 = 1\,\text{m} & P_M = Q_M = 100\,\text{N\,m} \\
X(0) = 1.9\,\text{m} & X(t_f) = 0.5\,\text{m} & Y = 0
\end{array}
\tag{14.122}
$$

which leads to the solution shown in Figure 14.16. As shown in the Figure, there are three switching points for this motion. It is seen that the optimal motion starts while the grounded actuator is saturated and the ungrounded actuator applies a positive torque within its limits. At the first switching point, the ungrounded actuator reaches its negative limit. The grounded actuator shows a change from positive to negative until it reaches its negative limit when the second switching occurs. Between the second and third switching points, the grounded actuator is saturated. Finally, when the ungrounded actuator touches its negative limit for the second time, the third switching occurs.

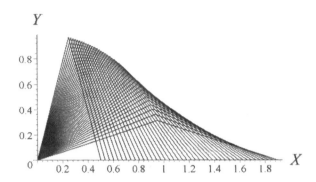

FIGURE 14.15. Illustration of motion of a 2R planar manipulator on line $y = 0$.

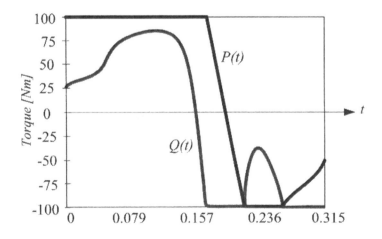

FIGURE 14.16. Time optimal control inputs for a 2R manipulator moving on line $Y = 0$, $0.5 < X < 1.9$.

# 14.4  Summary

Practically, every actuator can provide only a bounded output. When an actuator is working on its limit, we call it saturated. Time optimal control of an $n$ DOF robot has a simple solution: At every instant of time, at least one actuator must be saturated while the others are within their limits. Floating-time is an applied method to find the saturated actuator, the switching points, and the output of the non-saturated actuators. Switching points are the points that the saturated actuator switches with another one.

The floating-time method is based on discrete equations of motion, utilizing variable time increments. Then, following a recursive algorithm, it calculates the required output for the robot's actuators to follow a given path of motion.

# Exercises

1. Notation and symbols.

   Describe their meaning.

   a- $\tau_0$    b- $\tau_i$    c- $f_i$    d- $x_i$    e- $\ddot{x}_i$

   f- $\dot{x}_i$    g- $\tau_{-1}$    h- $\tau_s$    i- $x_{s+1}$    j- $f_s$

2. ★ Time optimal control of a 2 DOF system.

   Consider a dynamical system

   $$\begin{aligned} \dot{x}_1 &= -3x_1 + 2x_2 + 5Q \\ \dot{x}_2 &= 2x_1 - 3x_2 \end{aligned}$$

   that must start from an arbitrary initial condition and finish at $x_1 = x_2 = 0$, with a bounded control input $|Q| \leq 1$.

   Show that the functions

   $$\begin{aligned} f_1 &= -3x_1 + 2x_2 + 5Q \\ f_2 &= -2x_1 - 3x_2 \\ f_3 &= 1 \end{aligned}$$

   along with the Hamiltonian function $H$

   $$H = -1 + p_1 \left(-3x_1 + 2x_2 + 5Q\right) + p_2 \left(2x_1 - 3x_2\right)$$

   and the co-state variables $p_1$ and $p_2$ can solve the problem.

3. ★ Nonlinear objective function.

   Consider a one-dimensional control problem

   $$\dot{x} = -x + Q$$

   where $Q$ is the control command. The variable $x = x(t)$ must satisfy the boundary conditions

   $$\begin{aligned} x(0) &= a \\ x(t_f) &= b \end{aligned}$$

   and minimize the objective function $J$.

   $$J = \frac{1}{2} \int_0^{t_f} Q^2 dt$$

Show that the functions

$$
\begin{aligned}
f_1 &= \frac{1}{2}Q^2 \\
f_2 &= -x + Q \\
f_3 &= 0
\end{aligned}
$$

along with the Hamiltonian function $H$

$$
H = \frac{1}{2}p_0 Q^2 + p_1 (-x + Q)
$$

and the co-state variables $p_0$ and $p_1$ can solve the problem.

4. ★ Time optimal control to origin.

Consider a dynamical system

$$
\begin{aligned}
\dot{x}_1 &= x_2 \\
\dot{x}_2 &= -x_1 + Q
\end{aligned}
$$

that must start from an arbitrary initial condition and finish at the origin of the phase plane, $x_1 = x_2 = 0$, with a bounded control $|Q| \leq 1$. Find the control command to do this motion in minimum time.

5. ★ A linear dynamical system.

Consider a linear dynamical system

$$
\dot{x} = [A]\,x + bQ
$$

where

$$
x = \begin{bmatrix} x_1 \\ x_2 \end{bmatrix} \qquad [A] = \begin{bmatrix} 0 & 1 \\ 0 & 0 \end{bmatrix} \qquad b = \begin{bmatrix} 0 \\ 1 \end{bmatrix}
$$

subject to a bounded constraint on the control command

$$
Q \leq 1.
$$

Find the time optimal control command $Q$ to move from the system from $x_0$ to $x_1$.

(a)

$$
x_0 = \begin{bmatrix} -1 \\ -1 \end{bmatrix} \qquad x_1 = \begin{bmatrix} 1 \\ 1 \end{bmatrix}
$$

(b)

$$
x_0 = \begin{bmatrix} -1 \\ -1 \end{bmatrix} \qquad x_1 = \begin{bmatrix} 3 \\ 1 \end{bmatrix}
$$

6. ★ Constraint minimization.

Find the local minima and maxima of

$$f(x) = x_1^2 + x_2^2 + x_3^2$$

subject to the constraints

(a)
$$g_1 = x_1 + x_2 + x_3 - 3 = 0$$

(b)
$$\begin{aligned} g_1 &= x_1^2 + x_2^2 + x_3^2 - 5 = 0 \\ g_2 &= x_1^2 + x_2^2 + x_3^2 - 2x_1 - 3 = 0. \end{aligned}$$

7. ★ A control command with different limits.

Consider a rectilinear motion of a point mass $m = 1\,\text{kg}$ under the influence of a control force $f(t)$ on a smooth surface. The force is bounded to $F_1 \leq f(t) \leq F_2$. The mass is supposed to move from the initial conditions $x(0) = 0$, $v(0) = 0$ to the final conditions $x(t_f) = 10\,\text{m}$, $v(t_f) = 0$ in minimum total time $t = t_f$. Use the floating time algorithm to find the required control command $f(t) = m\ddot{x}$ and the switching time for

(a)
$$F_1 = 10\,\text{N} \qquad F_2 = 10\,\text{N}$$

(b)
$$F_1 = 8\,\text{N} \qquad F_2 = 10\,\text{N}$$

(c)
$$F_1 = 10\,\text{N} \qquad F_2 = 8\,\text{N}.$$

8. ★ A control command with different limits.

Find the optimal control command $|f(t)| \leq 20\,\text{N}$ to move the mass $m = 2\,\text{kg}$ from rest at point $P_1$ to $P_2$, and return to stop at point $P_3$, as shown in Figure 14.17. The value of $\mu$ is

(a)
$$\mu = 0$$

(b)
$$\mu = 0.2.$$

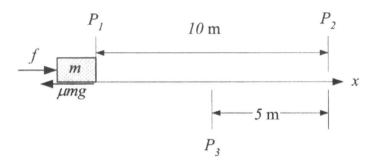

FIGURE 14.17. A rectilinear motion of a mass $m$ from rest at point $P_1$ to $P_2$, and a return to stop at point $P_3$.

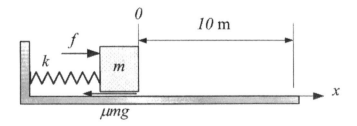

FIGURE 14.18. A rectilinear motion of a mass $m$ on a rough surface and attached to a wall with a spring.

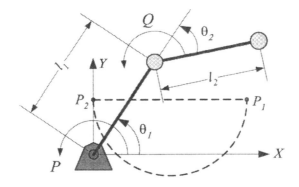

FIGURE 14.19. A 2R manipulator moves between two points on a line and a semi-circle.

9. ★ Motion of a mass under friction and spring forces.

Find the optimal control command $|f(t)| \le 100\,\mathrm{N}$ to move the mass $m = 1\,\mathrm{kg}$ rest-to-rest from $x(0) = 0$ to $x(t_f) = 10\,\mathrm{m}$. The mass is moving on a rough surface with coefficient $\mu$ and is attached to a wall by a linear spring with stiffness $k$, as shown in Figure 14.18. The value of $\mu$ and $k$ are

(a)
$$\mu = 0.1 \qquad k = 2\,\mathrm{N/m}$$

(b)
$$\mu = 0.5 \qquad k = 5\,\mathrm{N/m}.$$

10. ★ Convergence conditions.

Verify Equations (14.92) to (14.97) for the convergence condition of the floating-time algorithm.

11. ★ 2R manipulator moving on a line and a circle.

Calculate the actuators' torque for the 2R manipulator, shown in Figure 14.19, such that the end-point moves time optimally from $P_1(1.5\,\mathrm{m},\ 0.5\,\mathrm{m})$ to $P_2(0,\ 0.5\,\mathrm{m})$. The manipulator has the following characteristics:

$$
\begin{aligned}
m_1 &= m_2 = 1\,\mathrm{kg} \\
l_1 &= l_2 = 1\,\mathrm{m} \\
|P(t)| &\le 100\,\mathrm{N\,m} \\
|Q(t)| &\le 80\,\mathrm{N\,m}
\end{aligned}
$$

The path of motion is

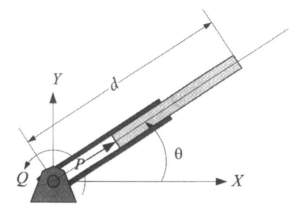

FIGURE 14.20. A polar manipulator, controlled by a torque $Q$ and a force $P$.

(a) a straight line

(b) a semi-circle with a center at $(0.75\,\text{m},\ 0.5\,\text{m})$.

12. ★ Time optimal control for a polar manipulator.

Figure 14.20 illustrates a polar manipulator that is controlled by a torque $Q$ and a force $P$. The base actuator rotates the manipulator and a force $P$ slides the second link on the first link. Find the optimal controls to move the endpoint from $P_1(1.5\,\text{m},\ 1\,\text{m})$ to $P_2(-1,\ 0.5\,\text{m})$ for the following data:

$$m_1 = 5\,\text{kg} \qquad\qquad m_2 = 3\,\text{kg}$$
$$|Q(t)| \le 100\,\text{N}\,\text{m} \qquad |P(t)| \le 80\,\text{N}\,\text{m}$$

13. ★ Control of an articulated manipulator.

Find the time optimal control of an articulated manipulator, shown in Figure 5.25, to move from $P_1 = (1.1, 0.8, 0.5)$ to $P_2 = (-1, 1, 0.35)$ on a straight line. The geometric parameters of the manipulator are given below. Assume the links are made of uniform bars.

$$d_1 = 1\,\text{m} \qquad\qquad d_2 = 0$$
$$l_2 = 1\,\text{m} \qquad\qquad\qquad\qquad\qquad l_3 = 1\,\text{m}$$
$$m_1 = 25\,\text{kg} \qquad m_2 = 12\,\text{kg} \qquad m_3 = 8\,\text{kg}$$
$$|Q_1(t)| \le 180\,\text{N}\,\text{m} \quad |Q_2(t)| \le 100\,\text{N}\,\text{m} \quad |Q_3(t)| \le 50\,\text{N}\,\text{m}$$

# 15

# Control Techniques

## 15.1 Open and Closed-Loop Control

A robot is a mechanism with an actuator at each joint $i$ to apply a force or torque to derive the link $(i)$. The robot is instrumented with position, velocity, and possibly acceleration sensors to measure the joint variables' kinematics. The measured values are usually relative. They are kinematics information of the frame $B_i$, attached to the link $(i)$, relative to the frame $B_{i-1}$ or $B_0$.

To cause each joint of the robot to follow a desired motion, we must provide the required torque command. Assume that the path of joint variables $q_d = q(t)$ are given as functions of time. Then, the required torques that cause the robot to follow the desired motion are calculated by the equations of motion and are equal to

$$\mathbf{Q}_c = \mathbf{D}(\mathbf{q}_d)\,\ddot{\mathbf{q}}_d + \mathbf{H}(\mathbf{q}_d, \dot{\mathbf{q}}_d) + \mathbf{G}(\mathbf{q}_d) \qquad (15.1)$$

where the subscripts $d$ and $c$ stand for *desired* and *controlled* respectively.

In an ideal world, the variables can be measured exactly and the robot can perfectly work based on the equations of motion (15.1). Then, the actuators' *control command* $\mathbf{Q}_c$ can cause the desired path $\mathbf{q}_d$ to happen. This is an *open-loop control algorithm*, that the control commands are calculated based on the equations of motion and a known desired path of motion. Then, the control commands are fed to the system to generate the desired path. Therefore, in an open-loop control algorithm, we expect the robot to follow the designed path, however, there is no mechanism to compensate any possible error.

Now assume that we are watching the robot during its motion by measuring the joints' kinematics. At any instant there can be a difference between the actual joint variables and the desired values. The difference is called *error* and is measured by

$$\mathbf{e} \;=\; \mathbf{q} - \mathbf{q}_d \qquad (15.2)$$
$$\dot{\mathbf{e}} \;=\; \dot{\mathbf{q}} - \dot{\mathbf{q}}_d. \qquad (15.3)$$

Let's define a control law and calculate a new control command vector by

$$\mathbf{Q} = \mathbf{Q}_c + \mathbf{k}_D \dot{\mathbf{e}} + \mathbf{k}_P \mathbf{e} \qquad (15.4)$$

where $\mathbf{k}_P$ and $\mathbf{k}_D$ are constant control gains. The control law compares the actual joint variables $(\mathbf{q}, \dot{\mathbf{q}})$ with the desired values $(\mathbf{q}_d, \dot{\mathbf{q}}_d)$, and generates

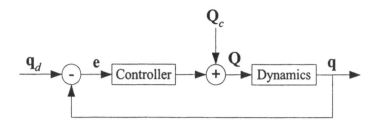

FIGURE 15.1. Illustration of feedback control algorithm.

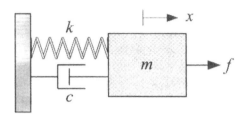

FIGURE 15.2. A linear mass-spring-damper oscillator.

a command proportionally. Applying the new control command changes the dynamic equations of the robot to produce the actual joint variables $\mathbf{q}$.

$$\mathbf{Q}_c + \mathbf{k}_D\dot{\mathbf{e}} + \mathbf{k}_P\mathbf{e} = \mathbf{D}(\mathbf{q})\,\ddot{\mathbf{q}} + \mathbf{H}(\mathbf{q}, \dot{\mathbf{q}}) + \mathbf{G}(\mathbf{q}) \qquad (15.5)$$

Figure 15.1 illustrates the idea of this control method in a *block diagram*. This is a *closed-loop control algorithm*, in which the control commands are calculated based on the difference between actual and desired variables. Reading the actual variables and comparing with the desired values is called *feedback*, and because of that, the closed-loop control algorithm is also called a *feedback control algorithm*.

The controller supplies a signal proportional to the error. This signal is added to the predicted command $\mathbf{Q}_c$ to compensate the error.

The principle of feedback control can be expressed as: *Increase the control command when the actual variable is smaller than the desired value and decrease the control command when the actual variable is larger than the desired value.*

**Example 325** *Mass-spring-damper oscillator.*

*Consider a linear oscillator made by a mass-spring-damper system shown in Figure 15.2. The equation of motion for the oscillator under the effect of an external force f is*

$$m\ddot{x} + c\dot{x} + kx = f \qquad (15.6)$$

*where, f is the control command, m is the mass of the oscillating, c is the viscous damping, and k is the stiffness of the spring. The required force to*

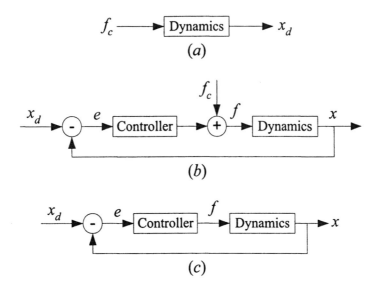

FIGURE 15.3. Open-loop and closed-loop control algorithms for a linear oscillator.

achieve a desired displacement $x_d = x(t)$ is calculated from the equation of motion.

$$f_c = m\ddot{x}_d + c\dot{x}_d + kx_d \qquad (15.7)$$

The open-loop control algorithm is shown in Figure 15.3(a). However, we may use the difference between the desired and actual outputs $e = x - x_d$, and define a control law

$$f = f_c + k_D\dot{e} + k_P e \qquad (15.8)$$
$$k_D > 0 \quad , \quad k_P > 0$$

to compensate a possible error. The new control law uses a feedback command as shown in Figure 15.3(b).

It is also possible to define a new control law only based on the error signal such as

$$f = -k_D\dot{e} - k_P e. \qquad (15.9)$$

Employing this law, we can define a more compact feedback control algorithm and change the equation of motion to

$$m\ddot{x} + (c + k_D)\dot{x} + (k + k_P)x = k_D\dot{x}_d + k_P x_d. \qquad (15.10)$$

The equation of the system can be summarized in a block diagram as shown in Figure 15.3(c).

A general scheme of a feedback control system may be explained so that a signal from the output feeds back to be compared to the input. This feedback *signal closes a loop and makes it reasonable to use the words **feedback** and **close-loop**. The principle of a closed loop control is to detect any error between the actual output and the desired input. As long as the error signal is not zero, the controller keeps changing the control command so that the error signal converges to zero.*

**Example 326** *Stability of a controlled system.*

*Consider a linear mass-spring-damper oscillator as shown in Figure 15.2 with the equation of motion given by*

$$m\ddot{x} + c\dot{x} + kx = f. \tag{15.11}$$

*We define a control law based on the actual output*

$$f = -k_D\dot{x} - k_Px \tag{15.12}$$
$$k_D > 0 \quad , \quad k_P > 0$$

*and transform the equation of motion to*

$$m\ddot{x} + (c + k_D)\dot{x} + (k + k_P)x = 0. \tag{15.13}$$

*By comparison with the open-loop equation (15.11), the closed loop equation shows that the oscillator acts as a free vibrating system under the action of new stiffness $k + k_P$ and damping $c + k_D$. Hence, the control law has changed the apparent stiffness and damping of the actual system. This example introduces the most basic application of control theory to improve the characteristics of a system and run the system to behave in a desired manner.*

*A control system must be stable when the desired output of the system changes, and also be able to eliminate the effect of a disturbance. Stability of a control system is defined as: The output must remain bounded for a given input or a bounded disturbance function.*

*To investigate the stability of the system, we must solve the closed loop differential equation (15.13). The equation is linear and therefore, it has an exponential solution.*

$$x = e^{\lambda t} \tag{15.14}$$

*Substituting the solution into the equation (15.13) provides the characteristic equation*

$$m\lambda^2 + (c + k_D)\lambda + (k + k_P) = 0 \tag{15.15}$$

*with two solutions*

$$\lambda_{1,2} = -\frac{c + k_D}{2m} \pm \frac{\sqrt{(c + k_D)^2 - 4m(k + k_P)}}{2m}. \tag{15.16}$$

The nature of the solution (15.14) depends on $\lambda_1$ and $\lambda_2$, and therefore on $k_D$ and $k_P$. The roots of the characteristic equation are complex

$$\lambda = -a \pm bi \tag{15.17}$$

$$a = \frac{c + k_D}{2m} \tag{15.18}$$

$$b = \frac{\sqrt{4m(k + k_P) - (c + k_D)^2}}{2m} \tag{15.19}$$

provided the gains are such that

$$(c + k_D)^2 < 4m(k + k_P). \tag{15.20}$$

In this case, the solution of the equation of motion is

$$x = Ce^{-\xi\omega_n t} \sin\left(\omega_n\sqrt{1 - \xi^2}t + \varphi\right) \tag{15.21}$$

where,

$$\omega_n = \sqrt{\frac{k + k_P}{m}} = \sqrt{a^2 + b^2} \tag{15.22}$$

$$\xi = \frac{c + k_D}{2\sqrt{m(k + k_P)}} = \frac{a}{\sqrt{a^2 + b^2}}. \tag{15.23}$$

The parameter $\omega_n$ is called **natural frequency**, and $\xi$ is the **damping ratio** of the system. The damping ratio controls the behavior of the system according to the following categories:

1. If $\xi = 0$, then the characteristic values are purely imaginary, $\lambda_{1,2} = \pm bi = \pm i\frac{1}{2m}\sqrt{4m(k + k_P)}$. In this case, the system has no damping, and therefore, it oscillates with a constant amplitude around the equilibrium, $x = 0$, forever.

2. If $0 < \xi < 1$, then the system is **under-damped** and it oscillates around the equilibrium with a decaying amplitude. The system is asymptotically stable in this case.

3. If $\xi = 1$, then the system is **critically-damped**. A critically damped oscillator has the fastest return to the equilibrium in an unoscillatory manner.

4. If $\xi > 1$, then the system is **over-damped** and it slowly returns to the equilibrium in an unoscillatory manner. The characteristic values are real and the solution of an over-damped oscillator is

$$x = Ae^{\lambda_1 t} + Be^{\lambda_2 t} \tag{15.24}$$

$$(\lambda_{1,2}) \in \mathbb{R}.$$

5. *If $\xi < 0$, then the system is unstable because the solution is*

$$x = Ae^{\lambda_1 t} + Be^{\lambda_2 t} \qquad (15.25)$$
$$\mathrm{Re}(\lambda_{1,2}) > 0$$

*and shows a motion with an increasing amplitude.*

**Example 327** *Solution of a characteristic equation.*
*Consider a system with the following characteristic equation:*

$$\lambda^2 + 6\lambda + 10 = 0. \qquad (15.26)$$

*Solutions of this equation are*

$$\lambda_{1,2} = -3 \pm i \qquad (15.27)$$

*showing a stable system because* $\mathrm{Re}\,(\lambda_{1,2}) = -3 < 0$ .
*Characteristic equations are linear polynomials. Hence, it is possible to use numerical methods, such as Newton-Raphson, to find the solution and determine the stability of the system.*

**Example 328** *Complex roots.*
*In case the characteristic equation has complex roots*

$$\lambda_{1,2} = a \pm bi \qquad (15.28)$$

*we may employ the Euler formula*

$$e^{i\theta} = \cos\theta + i\sin\theta \qquad (15.29)$$

*and show that the solution can be written in the form*

$$x = C_1 e^{at}(\cos bt + i\sin bt) + C_2 e^{at}(\cos bt - i\sin bt)$$
$$= e^{at}(A\cos bt + B\sin bt) \qquad (15.30)$$

*where, $C_1$ and $C_2$ are complex, and $A$ and $B$ are real numbers according to*

$$A = C_1 + C_2 \qquad (15.31)$$
$$B = (C_1 - C_2)\,i. \qquad (15.32)$$

**Example 329** ★ *Robot Control Algorithms.*
*Robots are nonlinear dynamical systems, and there is no general method for designing a nonlinear controller to be suitable for every robot in every mission. However, there are a variety of alternative and complementary methods, each best applicable to particular class of robots in a particular mission. The most important control methods are as follows:*
*    **Feedback Linearization** or **Computed Torque Control Technique.***
*In feedback linearization technique, we define a control law to obtain a linear*

*differential equation for error command, and then use the linear control design techniques. The feedback linearization technique can be applied to robots successfully, however, it does not guarantee robustness according to parameter uncertainty or disturbances.*

*This technique is a model-based control method, because the control law is designed based on a nominal model of the robot.*

**Linear Control Technique.** *The simplest technique for controlling robots is to design a linear controller based on the linearization of the equations of motion about an operating point. The linearization technique locally determines the stability of the robot. Proportional, integral, and derivative, or any combination of them, are the most practical linear control techniques.*

**Adaptive Control Technique.** *Adaptive control is a technique for controlling uncertain or time-varying robots. Adaptive control technique is more effective for low DOF robots.*

**Robust Control Technique, Adaptive** *In the robust control method, the controller is designed based on the nominal model plus some uncertainty. Uncertainty can be in a lot of parameters, such as the load carrying by the end-effector. For example, we develop a control technique to be effective for loads in a range $1 - 10$ kg.*

**Gain-Scheduling Control Technique.** *Gain-scheduling is a technique that tries to apply the linear control techniques to the nonlinear dynamics of robots. In gain-scheduling, we select a number of control points to cover the range of robot operation. Then at each control point, we make a linear time-varying approximation to the robot dynamics and design a linear controller. The parameters of the controller are then interpolated or scheduled between control points.*

## 15.2  Computed Torque Control

Dynamics of a robot can be described in the form

$$\mathbf{Q} = \mathbf{D(q)}\,\ddot{\mathbf{q}} + \mathbf{H(q, \dot{q})} + \mathbf{G(q)} \tag{15.33}$$

where $\mathbf{q}$ is the joint variable vector, and $\mathbf{Q(q, \dot{q}}, t)$ is the torques applied at joints. Assume a desired path in joint space is given by a twice differentiable function $\mathbf{q} = \mathbf{q}_d(t) \in C^2$. Hence, the desired time history of joints' position, velocity, and acceleration are known. We can control the robot to follow the desired path, by introducing a *computed torque control law* as below

$$\mathbf{Q} = \mathbf{D(q)}\,(\ddot{\mathbf{q}}_d - \mathbf{k}_D\dot{\mathbf{e}} - \mathbf{k}_P\mathbf{e}) + \mathbf{H(q, \dot{q})} + \mathbf{G(q)} \tag{15.34}$$

where $\mathbf{e}$ is the error vector

$$\mathbf{e} = \mathbf{q} - \mathbf{q}_d \tag{15.35}$$

and $\mathbf{k}_D$ and $\mathbf{k}_P$ are constant gain diagonal matrices. The control law is stable and applied as long as all the eigenvalues of the following matrix have negative real part.

$$[A] = \begin{bmatrix} 0 & \mathbf{I} \\ -\mathbf{k}_P & -\mathbf{k}_D \end{bmatrix} \tag{15.36}$$

**Proof.** The required $\mathbf{Q}_c$ to track $\mathbf{q}_d(t)$ can directly be found by substituting the path function into the equations of motion.

$$\mathbf{Q}_c = \mathbf{D}(\mathbf{q}_d)\,\ddot{\mathbf{q}}_d + \mathbf{H}(\mathbf{q}_d, \dot{\mathbf{q}}_d) + \mathbf{G}(\mathbf{q}_d) \tag{15.37}$$

The calculated torques are called *control inputs*, and the control is based on the *open-loop control law*. In an open-loop control, we have the equations of motion for a robot and we need the required torques to move the robot on a given path. Open-loop control is a blind control method, since the current state of the robot is not used for calculating the inputs.

Due to non-modeled parameters and also errors in adjustment, there is always a difference between the desired and actual paths. To make the robot's actual path converge to the desired path, we must introduce a feedback control. Let us use the feedback signal of the actual path and apply the computed torque control law (15.34) to the robot. Substituting the control law in the equations of motion (15.33), gives us

$$\ddot{\mathbf{e}} + \mathbf{k}_D\dot{\mathbf{e}} + \mathbf{k}_P\mathbf{e} = 0. \tag{15.38}$$

This is a linear differential equation for the error variable between the actual and desired outputs. If the $n \times n$ gain matrices $\mathbf{k}_D$ and $\mathbf{k}_P$ are assumed to be diagonal, then we may rewrite the error equation in a matrix form

$$\frac{d}{dt}\begin{bmatrix} \mathbf{e} \\ \dot{\mathbf{e}} \end{bmatrix} = \begin{bmatrix} 0 & \mathbf{I} \\ -\mathbf{k}_P & -\mathbf{k}_D \end{bmatrix}\begin{bmatrix} \mathbf{e} \\ \dot{\mathbf{e}} \end{bmatrix}$$

$$= [A]\begin{bmatrix} \mathbf{e} \\ \dot{\mathbf{e}} \end{bmatrix}. \tag{15.39}$$

The linear differential equation (15.39) is asymptotically stable when all the eigenvalues of $[A]$ have negative real part. The matrix $\mathbf{k}_P$ has the role of natural frequency, and $\mathbf{k}_D$ acts as damping.

$$\mathbf{k}_P = \begin{bmatrix} \omega_1^2 & 0 & 0 & 0 \\ 0 & \omega_2^2 & 0 & 0 \\ 0 & 0 & \cdots & 0 \\ 0 & 0 & 0 & \omega_n^2 \end{bmatrix} \tag{15.40}$$

$$\mathbf{k}_D = \begin{bmatrix} 2\xi_1\omega_1 & 0 & 0 & 0 \\ 0 & 2\xi_2\omega_2 & 0 & 0 \\ 0 & 0 & \cdots & 0 \\ 0 & 0 & 0 & 2\xi_n\omega_n \end{bmatrix} \tag{15.41}$$

Since $\mathbf{k}_D$ and $\mathbf{k}_P$ are diagonal, we can adjust the gain matrices $\mathbf{k}_D$ and $\mathbf{k}_P$ to control the response speed of the robot at each joint independently. A simple choice for the matrices is to set $\xi_i = 0, i = 1, 2, \cdots, n$, and make each joint response equal to the response of a critically damped linear second order system with natural frequency $\omega_i$.

The computed torque control law (15.34) has two components as shown below.

$$\mathbf{Q} = \underbrace{\mathbf{D}(\mathbf{q})\ddot{\mathbf{q}}_d + \mathbf{H}(\mathbf{q}, \dot{\mathbf{q}}) + \mathbf{G}(\mathbf{q})}_{\mathbf{Q}_{ff}} + \underbrace{\mathbf{D}(\mathbf{q})(-\mathbf{k}_D\dot{\mathbf{e}} - \mathbf{k}_P\mathbf{e})}_{\mathbf{Q}_{fb}} \qquad (15.42)$$

The first term, $\mathbf{Q}_{ff}$, is the *feedforward* command, which is the required torques based on open-loop control law. When there is no error, the control input $\mathbf{Q}_{ff}$ makes the robot follow the desired path $\mathbf{q}_d$. The second term, $\mathbf{Q}_{fb}$, is the *feedback* command, which is the correction torques to reduce the errors in the path of the robot.

Computed torque control is also called *feedback linearization*, which is an applied technique for robots' nonlinear control design. To apply the feedback linearization technique, we develop a control law to eliminate all nonlinearities and reduce the problem to the linear second-order equation of error signal (15.38) ∎

**Example 330** *Computed force control for an oscillator.*
*Figure 15.2 depicts a linear mass-spring-damper oscillator under the action of a control force. The equation of motion for the oscillator is*

$$m\ddot{x} + c\dot{x} + kx = f. \qquad (15.43)$$

*Applying a computed force control law*

$$f = m(\ddot{x}_d - k_D\dot{e} - k_P e) + c\dot{x} + kx \qquad (15.44)$$
$$e = x - x_d \qquad (15.45)$$

*reduces the error differential equation to*

$$\ddot{e} + k_D\dot{e} + k_P e = 0. \qquad (15.46)$$

*The solution of the error equation is*

$$e = Ae^{\lambda_1 t} + Be^{\lambda_2 t} \qquad (15.47)$$
$$\lambda_{1,2} = -k_D \pm \sqrt{k_D^2 - 4k_P} \qquad (15.48)$$

*where A and B are functions of initial conditions, and $\lambda_{1,2}$ are solutions of the characteristic equation*

$$m\lambda^2 + k_D\lambda + k_P = 0. \qquad (15.49)$$

*The solution (15.47) is stable and $e \to 0$ exponentially as $t \to \infty$ if $k_D > 0$.*

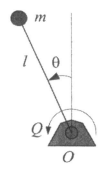

FIGURE 15.4. A controlled inverted pendulum.

**Example 331** *Inverted pendulum.*

Consider an inverted pendulum shown in Figure 15.4. Its equation of motion is

$$ml^2\ddot{\theta} - mgl\sin\theta = Q. \qquad (15.50)$$

To control the pendulum and bring it from an initial angle $\theta = \theta_0$ to the vertical-up position, we may employ a feedback control law as

$$Q = -k_D\dot{\theta} - k_P\theta - mgl\sin\theta. \qquad (15.51)$$

The parameters $k_D$ and $k_P$ are positive gains and are assumed constants. The control law (15.51) transforms the dynamics of the system to

$$ml^2\ddot{\theta} + k_D\dot{\theta} + k_P\theta = 0 \qquad (15.52)$$

showing that the system behaves as a stable mass-spring-damper.

In case the desired position of the pendulum is at a nonzero angle, $\theta = \theta_d$, we may employ a feedback control law based on the error $e = \theta - \theta_d$ as below,

$$Q = ml^2\ddot{\theta}_d - k_D\dot{e} - k_P e - mgl\sin\theta. \qquad (15.53)$$

Substituting this control law in the equation of motion (15.50) shows that the dynamic of the controlled system is governed by

$$ml^2\ddot{e} + k_D\dot{e} + k_P e = 0. \qquad (15.54)$$

**Example 332** *Control of a 2R planar manipulator.*

A 2R planar manipulator is shown in Figure 15.5 with dynamic equations given below.

$$\begin{bmatrix} Q_1 \\ Q_2 \end{bmatrix} = \begin{bmatrix} D_{11} & D_{12} \\ D_{21} & D_{22} \end{bmatrix} \begin{bmatrix} \ddot{\theta}_1 \\ \ddot{\theta}_2 \end{bmatrix}$$
$$+ \begin{bmatrix} C_{11} & C_{12} \\ C_{21} & C_{22} \end{bmatrix} \begin{bmatrix} \dot{\theta}_1 \\ \dot{\theta}_2 \end{bmatrix} + \begin{bmatrix} G_1 \\ G_2 \end{bmatrix} \qquad (15.55)$$

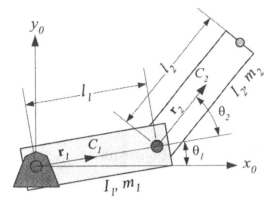

FIGURE 15.5. A 2R planar manipulator with massive links.

*where*

$$D_{11} = m_1 r_1^2 + I_1 + m_2 \left(l_1^2 + l_1 r_2 \cos\theta_2 + r_2^2\right) + I_2 \qquad (15.56)$$

$$D_{21} = D_{12} = m_2 l_1 r_2 \cos\theta_2 + m_2 r_2^2 + I_2 \qquad (15.57)$$

$$D_{22} = m^2 r_2^2 + I_2 \qquad (15.58)$$

$$C_{11} = -m_2 l_1 r_2 \dot\theta_2 \sin\theta_2 \qquad (15.59)$$

$$C_{21} = -m_2 l_1 r_2 (\dot\theta_1 + \dot\theta_2) \sin\theta_2 \qquad (15.60)$$

$$C_{12} = m_2 l_1 r_2 \dot\theta_1 \sin\theta_2 \qquad (15.61)$$

$$C_{22} = 0 \qquad (15.62)$$

$$G_1 = m_1 g r_1 \cos\theta_1 + m_2 g \left(l_1 \cos\theta_1 + r_2 \cos(\theta_1 + \theta_2)\right) \qquad (15.63)$$

$$G_2 = m_2 g r_2 \cos(\theta_1 + \theta_2). \qquad (15.64)$$

*Let's write the equations of motion in the following form:*

$$\mathbf{D(q)}\ddot{\mathbf{q}} + \mathbf{C(q,\dot q)}\dot{\mathbf{q}} + \mathbf{G(q)} = \mathbf{Q} \qquad (15.65)$$

*and multiply both sides by* $\mathbf{D}^{-1}$ *to transform the equations of motion to*

$$\ddot{\mathbf{q}} + \mathbf{D}^{-1}\mathbf{C}\dot{\mathbf{q}} + \mathbf{D}^{-1}\mathbf{G} = \mathbf{D}^{-1}\mathbf{Q}. \qquad (15.66)$$

*To control the manipulator to follow a desired path* $\mathbf{q} = \mathbf{q}_d(t)$, *we apply the following control law:*

$$\mathbf{Q} = \mathbf{D(q)}\,\mathbf{U} + \mathbf{C(q,\dot q)}\dot{\mathbf{q}} + \mathbf{G(q)} \qquad (15.67)$$

*where*

$$\mathbf{U} = \ddot{\mathbf{q}}_d - 2k\dot{\mathbf{e}} - k^2\mathbf{e} \qquad (15.68)$$

$$\mathbf{e} = \mathbf{q} - \mathbf{q}_d. \qquad (15.69)$$

*The vector* **U** *is the controller input,* **e** *is the position error, and k is a positive constant gain number. Substituting the control law into the equation of motion shows that the error vector satisfies a linear second-order ordinary differential equation*

$$\ddot{\mathbf{e}} + 2k\dot{\mathbf{e}} + k^2\mathbf{e} = 0 \qquad (15.70)$$

*and therefore, exponentially converges to zero.*

# 15.3    Linear Control Technique

Linearization of a robot's equations of motion about an operating point while applying a linear control algorithm is a practical robot control. This technique works well in a vicinity of the operating point. Hence, it is only a locally stable method. The linear control techniques are proportional, integral, derivative, and any combination of them.

The idea is to linearize the nonlinear equations of motion about some reference operating points to make a linear system, design a controller for the linear system, and then, apply the control to the robot. This technique will always result in a stable controller in some neighborhood of the operating point. However, the stable neighborhood may be quite small and hard to be determined.

A proportional-integral-derivative (PID) control algorithm employs a position error, derivative error, and integral error to develop a control law. Hence, a PID control law has the following general form for the input command:

$$Q = k_P e + k_I \int_0^t e\, dt + k_D \dot{e} \qquad (15.71)$$

where $e = q - q_d$ is the error signal, and $k_P$, $k_I$, and $k_D$ are positive constant gains associated to the proportional, integral, and derivative controllers.

The control command $Q$ is thus a sum of three terms: the P-term ,which is proportional to error $e$, the I-term, which is proportional to the integral of the error, and the D-term, which is proportional to the derivative of the error.

## 15.3.1    Proportional Control

In the case of proportional control, the PID control law (15.71) reduces to

$$Q = k_P e + Q_d. \qquad (15.72)$$

The variable $Q_d$ is the desired control command, which is called a *bias* or *reset factor*. When the error signal is zero, the control command is equal to the desired value. The proportional control has a drawback that results in a constant error at steady state condition.

## 15.3.2  Integral Control

The main function of an integral control is to eliminate the steady state error and make the system follow the set point at steady state conditions. The integral controller leads to an increasing control command for a positive error, and a decreasing control command for a negative error. An integral controller is usually used with a proportional controller. The control law for a PI controller is

$$Q = k_P e + k_I \int_0^t e \, dt. \tag{15.73}$$

## 15.3.3  Derivative Control

The purpose of derivative control is to improve the closed-loop stability of a system. A derivative controller has a predicting action by extrapolating the error using a tangent to the error curve. A derivative controller is usually used with a proportional controller. The PD control law is

$$Q = k_P e + k_D \dot{e}. \tag{15.74}$$

**Proof.** Any linear system behaves linearly if it is sufficiently near a reference operating point. Consider a nonlinear system

$$\dot{\mathbf{q}} = \mathbf{f}(\mathbf{q}, \mathbf{Q}) \tag{15.75}$$

where $\mathbf{q}_d$ is a solution generated by a specific input $\mathbf{Q}_c$

$$\dot{\mathbf{q}}_d = \mathbf{f}(\mathbf{q}_d, \mathbf{Q}_c). \tag{15.76}$$

Assume $\delta\mathbf{q}$ is a small change from the reference point $\mathbf{q}_d$ because of a small change $\delta\mathbf{Q}$ from $\mathbf{Q}$.

$$\begin{align}
\mathbf{q} &= \mathbf{q}_d + \delta\mathbf{q} \tag{15.77}\\
\mathbf{Q} &= \mathbf{Q}_c + \delta\mathbf{Q} \tag{15.78}
\end{align}$$

If the changes $\delta\mathbf{q}$ and $\delta\mathbf{Q}$ are assumed small for all times, then the equation (15.75) can be approximated by its Taylor expansion and $\mathbf{q}$ be the solution of

$$\dot{\mathbf{q}} = \frac{\partial \mathbf{f}}{\partial \mathbf{q}_d} \mathbf{q} + \frac{\partial \mathbf{f}}{\partial \mathbf{Q}_d} \mathbf{Q}. \tag{15.79}$$

The partial derivative matrices $\left[\frac{\partial \mathbf{f}}{\partial \mathbf{q}_d}\right]$ and $\left[\frac{\partial \mathbf{f}}{\partial \mathbf{Q}_d}\right]$ are evaluated at the reference point $(\mathbf{q}_d, \mathbf{Q}_d)$. ∎

**Example 333** *Linear control for a pendulum.*
*Figure 11.12 illustrates a controlled pendulum as a one-arm manipulator. The equation of motion for the arm is*

$$Q = I\ddot{\theta} + c\dot{\theta} + mgl \sin\theta \tag{15.80}$$

*where I is the arm's moment of inertia about the pivot joint and m is the mass of the arm. The joint has a viscous damping c and kinematic length, the distance between the pivot and CM, is l. Introducing a new set of variables*

$$\theta = x_1 \tag{15.81}$$
$$\dot{\theta} = x_2 \tag{15.82}$$

*converts the equation of motion to*

$$\dot{x}_1 = x_2 \tag{15.83}$$
$$\dot{x}_2 = \frac{Q - c x_2 - mgl \sin x_1}{I}. \tag{15.84}$$

*The linearized form of these equations is*

$$\begin{bmatrix} \dot{x}_1 \\ \dot{x}_2 \end{bmatrix} = \begin{bmatrix} 0 & 1 \\ -mgl/I & -c \end{bmatrix} \begin{bmatrix} x_1 \\ x_2 \end{bmatrix} + \begin{bmatrix} 0 & 0 \\ 0 & 1/I \end{bmatrix} \begin{bmatrix} 0 \\ Q \end{bmatrix}. \tag{15.85}$$

*Assume that the reference point is*

$$\mathbf{x}_d = \begin{bmatrix} x_1 \\ x_2 \end{bmatrix} = \begin{bmatrix} \pi/2 \\ 0 \end{bmatrix} \tag{15.86}$$
$$Q_c = mgl. \tag{15.87}$$

*The coefficient matrices in Equation (15.85) must then be evaluated at the reference point. We use a set of sample data*

$$\begin{aligned} m &= 1\,\text{kg} \\ l &= 0.35\,\text{m} \\ I &= 0.07\,\text{kg.\,m}^2 \\ c &= 0.01\,\text{N s/\,m} \end{aligned} \tag{15.88}$$

*and find*

$$\frac{\partial \mathbf{f}}{\partial \mathbf{q}_d} = \begin{bmatrix} 0 & 1 \\ -49.05 & -0.01 \end{bmatrix} \tag{15.89}$$

$$\frac{\partial \mathbf{f}}{\partial \mathbf{Q}_c} = \begin{bmatrix} 0 & 0 \\ 0 & 14.286 \end{bmatrix}. \tag{15.90}$$

*Now we have a linear system and we may apply any control law that applies to linear systems. For instance, a PID control law*

$$\mathbf{Q} = \mathbf{Q}_c - k_D \dot{\mathbf{e}} - k_P \mathbf{e} + k_I \int_0^t \mathbf{e}\, dt \tag{15.91}$$

*where*

$$\mathbf{e} = \mathbf{q} - \mathbf{q}_d = \begin{bmatrix} x_1 - \pi/2 \\ x_2 \end{bmatrix} \tag{15.92}$$

*can control the arm around the reference point.*

**Example 334** *PD control.*
*Let us define a PD control law as*

$$\mathbf{Q} = -k_D\dot{\mathbf{e}} - k_P\mathbf{e} \qquad (15.93)$$

$$\mathbf{e} = \mathbf{q} - \mathbf{q}_d \qquad (15.94)$$

*Applying the PD control to a robot with dynamic equations as*

$$\begin{aligned}\mathbf{Q} &= \mathbf{D}(\mathbf{q})\,\ddot{\mathbf{q}} + \mathbf{H}(\mathbf{q},\dot{\mathbf{q}}) + \mathbf{G}(\mathbf{q}) \\ &= \mathbf{D}(\mathbf{q})\,\ddot{\mathbf{q}} + \mathbf{C}(\mathbf{q},\dot{\mathbf{q}})\dot{\mathbf{q}} + \mathbf{G}(\mathbf{q}) \end{aligned} \qquad (15.95)$$

*will produce the following control equation:*

$$\mathbf{D}(\mathbf{q})\,\ddot{\mathbf{q}} + \mathbf{C}(\mathbf{q},\dot{\mathbf{q}})\dot{\mathbf{q}} + \mathbf{G}(\mathbf{q}) + k_D\,(\dot{\mathbf{q}} - \dot{\mathbf{q}}_d) - k_P\,(\mathbf{q} - \mathbf{q}_d) = 0. \qquad (15.96)$$

*This control is ideal when* $\mathbf{q}_d$ *is a constant vector associated with a specific configuration of a robot, and therefore* $\dot{\mathbf{q}}_d = 0$. *In this case the PD controller can make the configuration* $\mathbf{q}_d$ *globally stable.*

*In case of a path given by* $\mathbf{q} = \mathbf{q}_d(t)$, *we define a modified PD controller in the form*

$$\mathbf{Q} = \mathbf{D}(\mathbf{q})\ddot{\mathbf{q}}_d + \mathbf{C}(\mathbf{q},\dot{\mathbf{q}})\dot{\mathbf{q}}_d + \mathbf{G}(\mathbf{q}) - k_D\dot{\mathbf{e}} - k_P\mathbf{e} \qquad (15.97)$$

*and reduce the closed-loop equation to*

$$\mathbf{D}(\mathbf{q})\ddot{\mathbf{e}} + (\mathbf{C}(\mathbf{q},\dot{\mathbf{q}}) + k_D)\,\dot{\mathbf{e}} + k_P\mathbf{e} = 0. \qquad (15.98)$$

*The linearization of this equation about a control point* $\mathbf{q} = \mathbf{q}_d = cte$ *provides a stable dynamics for the error signal*

$$\mathbf{D}(\mathbf{q}_d)\ddot{\mathbf{e}} + k_D\dot{\mathbf{e}} + k_P\mathbf{e} = 0. \qquad (15.99)$$

## 15.4   Sensing and Control

Position, velocity, acceleration, and force sensors are the most common sensors used in robotics. Consider the inverted pendulum shown in Figure 15.6 as a one DOF manipulator with the following equation of motion:

$$ml^2\ddot{\theta} - c\dot{\theta} - mgl\sin\theta = Q. \qquad (15.100)$$

From an open-loop control viewpoint, we need to provide a moment $Q_c(t)$ to force the manipulator to follow a desired path of motion $\theta_d(t)$ where

$$Q_c = ml^2\ddot{\theta}_d - c\dot{\theta}_d - mgl\sin\theta_d. \qquad (15.101)$$

In robotics, we usually calculate $Q_c$ from the dynamics equation and dictate it to the actuator.

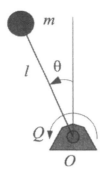

FIGURE 15.6. An inverted pendulum.

The manipulator will respond to the applied moment and will move. The equation of motion (15.100) is a model of the actual manipulator. In other words, we want the manipulator to work based on this equation. However, there are so many unmodeled phenomena that we cannot include them in our equation of motion, or we cannot model them. Some are temperature, air pressure, exact gravitational acceleration, or even the physical parameters such as $m$ and $l$ that we think have good accuracy. So, applying a control command $Q_c$ will move the manipulator and provide a real value for $\theta$, $\dot{\theta}$, and $\ddot{\theta}$, which are not necessarily equal to $\theta_d$, $\dot{\theta}_d$, and $\ddot{\theta}_d$. Sensing is now important because we need to measure the actual angle $\theta$, angular velocity $\dot{\theta}$, and angular acceleration $\ddot{\theta}$ to compare them with $\theta_d$, $\dot{\theta}_d$, and $\ddot{\theta}_d$ and make sure that the manipulator is following the desired path. This is the reason why the feedback control systems and the error signal $e = \theta - \theta_d$ were introduced.

Robots are supposed to do a job in an environment, so they can interact with the environment. Therefore, a robot needs two types of sensors: 1-sensing the robot's internal parameters, which are called *proprioceptors*, and 2-sensing the robot's environmental parameters, which are called *exteroceptors*. The most important interior parameters are position, velocity, acceleration, force, torque, and inertia.

### 15.4.1  *Position Sensors.*

**Rotary encoders**. In robotics, almost all kinds of actuators provide a rotary motion. Then we may provide a rotation motion for a revolute joint, or a translation motion for a prismatic joint, by using gears. So, it is ideally possible to sense the relative position of the links connected by the joint based on the angular position of the actuator. The error in this position sensing is due to non-rigidity and backlash. The most common position sensor is a *rotary encoder* that can be *optical*, *magnetic*, or *electrical*. As an example, when the encoder shaft rotates, a disk counting a pattern of fine lines interrupts a light beam. A photodetector converts the light pulses into a countable binary waveform. The shaft angle is then determined by counting the number of pulses.

**Resolvers**. We may design an electronic device to provide a mathematical function of the joint variable. The mathematical function might be sine, cosine, exponential, or any combination of mathematical functions. The joint variable is then calculated indirectly by resolving the mathematical functions. Sine and cosine functions are more common.

**Potentiometers**. Using an electrical bridge, the potentiometers can provide an electric voltage proportional to the joint position.

**LVDT and RVDT**. LVDT/RVDT or a Linear/Rotary Variable Differential Transformer operates with two transformers sharing the same magnetic core. When the core moves, the output of one transformer increases while the other's output decreases. The difference of the current is a measure of the core position.

### 15.4.2  Speed Sensors.

**Tachometers**. Generally speaking, a tachometer is a name for any velocity sensor. Tachometers usually provide an analog signal proportional to the angular velocity of a shaft. There are a vast amount of different designs for tachometers, using different physical characteristics such as magnetic field.

**Rotary encoders**. Any rotary sensor can be equipped with a time measuring system and become an angular velocity sensor. The encoder counts the light pulses of a rotating disk and the angular velocity is then determined by time between pulses.

**Differentiating devices**. Any kind of position sensor can be equipped with a digital differentiating device to become a speed sensor. The digital or numerical differentiating needs a simple processor. Numerical differentiating is generally an erroneous process.

**Integrating devices**. The output signal of an accelerometer can be numerically integrated to provide a velocity signal. The digital or numerical integrating also needs a simple processor. Numerical differentiating is generally a smooth and reliable process.

### 15.4.3  Acceleration Sensors.

Acceleration sensors work based on Newton's second law of motion. They sense the force that causes an acceleration of a known mass. There are many types of accelerometers. Stress-strain gage, capacitive, inductive, piezoelectric, and micro-accelerometers are the most common. In any of these types, force causes a proportional displacement in an elastic material, such as deflection in a micro-cantilever beam, and the displacement is proportional to the acceleration.

Applications of accelerometers include measurement of acceleration, angular acceleration, velocity, position, angular velocity, frequency, impulse, force, tilt, and orientation.

**Force and Torque Sensors.** Any concept and method that we use in sensing acceleration may also be used in force and torque sensing. We equip the wrists of a robot with at least three force sensors to measure the contact forces and moments with the environment. The wrist's force sensors are important especially when the robot's job is involved with touching unknown surfaces and objects.

**Proximity Sensors.** Proximity sensors are utilized to detect the existence of an object, field, or special material before interacting with it. Inductive, capacitive, Hall effect, sonic, ultrasonic, and optical are the most common proximity sensors.

The inductive sensors can sense the existence of a metallic object due to a change in inductance. The capacitive sensors can sense the existence of gas, liquid, or metals that cause a change in capacitance. Hall effective sensors work based on the interaction between the voltage in a semiconductor material and magnetic fields. These sensors can detect the existence of magnetic fields and materials. Sonic, ultrasonic, and optical sensors work based on the reflection or modification in an emitted signal by objects.

## 15.5  Summary

In an open-loop control algorithm, we calculate the robot's required torque commands $\mathbf{Q}_c$ for a given joint path $\mathbf{q}_d = \mathbf{q}(t)$ based on the equations of motion

$$\mathbf{Q}_c = \mathbf{D}(\mathbf{q}_d)\,\ddot{\mathbf{q}}_d + \mathbf{H}(\mathbf{q}_d, \dot{\mathbf{q}}_d) + \mathbf{G}(\mathbf{q}_d). \qquad (15.102)$$

However, there can be a difference between the actual joint variables and the desired values. The difference is called error $\mathbf{e}$

$$\mathbf{e} \;=\; \mathbf{q} - \mathbf{q}_d \qquad (15.103)$$
$$\dot{\mathbf{e}} \;=\; \dot{\mathbf{q}} - \dot{\mathbf{q}}_d. \qquad (15.104)$$

By measuring the error command, we may define a control law and calculate a new control command vector

$$\mathbf{Q} = \mathbf{Q}_c + \mathbf{k}_D\dot{\mathbf{e}} + \mathbf{k}_P\mathbf{e} \qquad (15.105)$$

to compensate for the error. The parameters $\mathbf{k}_P$ and $\mathbf{k}_D$ are constant gain diagonal matrices.

The control law compares the actual joint variables $(\mathbf{q}, \dot{\mathbf{q}})$ with the desired values $(\mathbf{q}_d, \dot{\mathbf{q}}_d)$, and generates a command proportionally. Applying the new control command changes the dynamic equations of the robot to

$$\mathbf{Q}_c + \mathbf{k}_D\dot{\mathbf{e}} + \mathbf{k}_P\mathbf{e} = \mathbf{D}(\mathbf{q})\,\ddot{\mathbf{q}} + \mathbf{H}(\mathbf{q}, \dot{\mathbf{q}}) + \mathbf{G}(\mathbf{q}). \qquad (15.106)$$

This is a closed-loop control algorithm, in which the control commands are calculated based on the difference between actual and desired variables.

Computed torque control

$$\mathbf{Q} = \mathbf{D}(\mathbf{q}) \left( \ddot{\mathbf{q}}_d - \mathbf{k}_D \dot{\mathbf{e}} - \mathbf{k}_P \mathbf{e} \right) + \mathbf{H}(\mathbf{q}, \dot{\mathbf{q}}) + \mathbf{G}(\mathbf{q}) \tag{15.107}$$

is an applied closed-loop control law in robotics to make a robot follow a desired path.

# Exercises

1. Response of second-order systems.

   Solve the characteristic equations and determine the response of the following second-order systems at $x(1)$, if they start from $x(0) = 1$, $\dot{x}(0) = 0$.

   (a)
   $$\ddot{x} + 2\dot{x} + 5x = 0$$

   (b)
   $$\ddot{x} + 2\dot{x} + x = 0$$

   (c)
   $$\ddot{x} + 4\dot{x} + x = 0$$

2. Modified PD control.

   Apply a modified PD control law
   $$\begin{aligned} f &= -k_P e - k_d \dot{x} \\ e &= x - x_d \end{aligned}$$

   to a second-order linear system
   $$m\ddot{x} + c\dot{x} + kx = f$$

   and reduce the system to a second-order equation in an error signal.
   $$m\ddot{e} + (c + k_D)\dot{e} + (k + k_P)e = kx_D$$

   Then, calculate the steady state error for a step input
   $$x = x_d = cte.$$

3. Modified PID control.

   Apply a modified PD control law
   $$\begin{aligned} f &= -k_P e - k_d \dot{x} - k_I \int_0^t e \, dt \\ e &= x - x_d \end{aligned}$$

   to a second-order linear system
   $$m\ddot{x} + c\dot{x} + kx = f$$

and reduce the system to a third-order equation in an error signal.

$$m\ddot{e} + (c + k_D)\ddot{e} + (k + k_P)\dot{e} + k_I e = 0$$

Then, find the PID gains such that the characteristic equation of the system simplifies to

$$\left(\lambda^2 + 2\xi\omega_n\lambda + \omega_n^2\right)(\lambda + \beta) = 0.$$

4. Linearization.

Linearize the given equations and determine the stability of the linearized set of equations.

$$\begin{aligned}
\dot{x}_1 &= x_2^2 + x_1 \cos x_2 \\
\dot{x}_2 &= x_2 + (1 + x_1 + x_2)x_1 + x_1 \sin x_2
\end{aligned}$$

5. Expand the control equations for a 2R planar manipulator using the following control law:

$$\mathbf{Q} = \mathbf{D}(\mathbf{q})\left(\ddot{\mathbf{q}}_d - k_D\dot{\mathbf{e}} - k_P\mathbf{e}\right) + \mathbf{H}(\mathbf{q}, \dot{\mathbf{q}}) + \mathbf{G}(\mathbf{q})$$

6. One-link manipulator control.

A one-link manipulator is shown in Figure 15.6.

(a) Derive the equation of motion.

(b) Determine a rest-to-rest joint path between $\theta(0) = 45\,\mathrm{deg}$ and $\theta(0) = -45\,\mathrm{deg}$.

(c) Solve the time optimal control of the manipulator and determine the torque $Q_c(t)$ for

$$\begin{aligned}
m &= 1\,\mathrm{kg} \\
l &= 1\,\mathrm{m} \\
|Q| &\leq 120\,\mathrm{N\,m}.
\end{aligned}$$

(d) Now assume the mass is $m = 1.01\,\mathrm{kg}$ and solve the equation of motion numerically by feeding the calculated torques $Q_c(t)$. Determine the position and velocity errors at the end of the motion.

(e) Design a computed torque control law to compensate the error during the motion.

7. ★ Mass-spring control.

Solve Exercise 14.9 and calculate the optimal control input. Increase the stiffness %10, and design a computed torque control law to eliminate error during the motion.

8. ★ 2R manipulator control.

   (a) Solve Exercise 13.16 and calculate the optimal control inputs.

   (b) Increase the masses by 10%, and solve the dynamic equations numerically.

   (c) Determine the position and velocity error in Cartesian and joint spaces by applying the calculated optimal inputs.

   (d) Design a computed torque control law to eliminate error during the motion.

9. ★ PR planar manipulator control.

   (a) Solve Exercise 14.12 and calculate the optimal control inputs.

   (b) Increase the gravitational acceleration by 10%, and solve the dynamic equations numerically.

   (c) Determine the position and velocity error in Cartesian and joint spaces by applying the calculated optimal inputs.

   (d) Design a computed torque control law to eliminate error during the motion.

10. Sensing and measurement.

    Consider the one DOF manipulator in Figure (15.100). To control the manipulator, we need to sense the actual angle $\theta$, angular velocity $\dot{\theta}$, and angular acceleration $\ddot{\theta}$ and compare them with $\theta_d$, $\dot{\theta}_d$, and $\ddot{\theta}_d$ to make sure that the manipulator is following the desired path. Can we measure the actual moment $Q$, that the actuator is providing, and compare with the predicted value $Q_c$ instead? Does making $Q$ equal to $Q_c$ guarantee that the manipulator does what it is supposed to do?

# References

## Chapter 1

Asimov, I., 1950, *I, Robot*, Doubleday & Company, Inc., New York.

Aspragathos, N. A., and Dimitros, J. K., 1998, A comparative study of three methods for robot kinematics, *IEEE Transaction on Systems, Man and Cybernetic-PART B: CYBERNETICS, 28(2), 115-145.*

Chernousko, F. L., Bolotnik, N. N., and Gradetsky, V. G., 1994, *Manipulation Robots: Dynamics, Control, and Optimization*, CRC Press, Boca Raton, Florida.

Denavit, J., and Hartenberg, R. S., 1955, A kinematic notation for lower-pair mechanisms based on matrices, *Journal of Applied Mechanics, 22(2), 215-221.*

Dugas, R., 1995, *A History of Mechanics* (English translation), Switzerland, Editions du Griffon, Central Book Co., New York.

Erdman, A. G., 1993, *Modern Kinematics: Developed in the Last Forty Years*, John Wiley & Sons, New York.

Hunt, K. H., 1978, *Kinematic Geometry of Mechanisms*, Oxford University Press, London.

Milne, E. A., 1948, *Vectorial Mechanics*, Methuen & Co. LTD., London.

Niku, S. B., 2001, *Introduction to Robotics: Analysis, Systems, Applications*, Prentice Hall, New Jersey.

Rosheim, M. E., 1994, *Robot Evolution: The Development of Anthrobotics*, John Wiley & Sons, New York.

Tsai, L. W., 1999, *Robot Analysis*, John Wiley & Sons, New York.

Veit, S., 1992, Whatever happened to ... personal robots?, *The Computer Shopper, 12(11), 794-795.*

## Chapter 2

Buss, S. R., 2003, *3-D Computer Graphics: A Mathematical Introduction with OpenGL*, Cambridge University Press, New York.

Cheng, H., and Gupta, K. C., 1989, A historical note on finite rotations, *Journal of Applied Mechanics, 56, 139-145.*

Coe, C. J., 1934, Displacement of a rigid body, *American Mathematical Monthly, 41(4), 242-253.*

Denavit, J., and Hartenberg, R. S., 1955, A kinematic notation for lower-pair mechanisms based on matrices, *Journal of Applied Mechanics, 22(2), 215-221.*

Hunt, K. H., 1978, *Kinematic Geometry of Mechanisms*, Oxford University Press, London.

Mason, M. T., 2001, *Mechanics of Robotic Manipulation*, MIT Press, Cambridge, MA.

Murray, R. M., Li, Z., and Sastry, S. S. S., 1994, *A Mathematical Introduction to Robotic Manipulation*, CRC Press, Boca Raton, Florida.

Nikravesh, P., 1988, *Computer-Aided Analysis of Mechanical Systems*, Prentice Hall, New Jersey.

Niku, S. B., 2001, *Introduction to Robotics: Analysis, Systems, Applications*, Prentice Hall, New Jersey.

Paul, B., 1963, On the composition of finite rotations, *American Mathematical Monthly*, *70(8), 859-862.*

Paul, R. P., 1981, *Robot Manipulators: Mathematics, Programming, and Control*, MIT Press, Cambridge, Massachusetts.

Rimrott, F. P. J., 1989, *Introductory Attitude Dynamics*, Springer-Verlag, New York.

Rosenberg, R., M. 1977, *Analytical Dynamics of Discrete Systems*, Plenum Publishing Co., New York.

Schaub, H., and Junkins, J. L., 2003, *Analytical Mechanics of Space Systems*, AIAA Educational Series, American Institute of Aeronautics and Astronautics, Inc., Reston, Virginia.

Suh, C. H., and Radcliff, C. W., 1978, *Kinematics and Mechanisms Design*, John Wiley & Sons, New York.

Spong, M. W., Hutchinson, S., and Vidyasagar, M., 2006, *Robot Modeling and Control*, John Wiley & Sons, New York.

Tsai, L. W., 1999, *Robot Analysis*, John Wiley & Sons, New York.

# Chapter 3

Buss, S. R., 2003, *3-D Computer Graphics: A Mathematical Introduction with OpenGL*, Cambridge University Press, New York.

Denavit, J., and Hartenberg, R. S., 1955, A kinematic notation for lower-pair mechanisms based on matrices, *Journal of Applied Mechanics, 22(2), 215-221.*

Hunt, K. H., 1978, *Kinematic Geometry of Mechanisms*, Oxford University Press, London U.K.

Mason, M. T., 2001, *Mechanics of Robotic Manipulation*, MIT Press, Cambridge, Massachusetts.

Murray, R. M., Li, Z., and Sastry, S. S. S., 1994, *A Mathematical Introduction to Robotic Manipulation*, CRC Press, Boca Raton, Florida.

Nikravesh, P., 1988, *Computer-Aided Analysis of Mechanical Systems*, Prentice Hall, New Jersey.

Paul, B., 1963, On the composition of finite rotations, *American Mathematical Monthly*, *70(8), 859-862.*

Paul, R. P., 1981, *Robot Manipulators: Mathematics, Programming, and Control*, MIT Press, Cambridge, Massachusetts.

Rimrott, F. P. J., 1989, *Introductory Attitude Dynamics*, Springer-Verlag, New York.

Rosenberg, R., M. 1977, *Analytical Dynamics of Discrete Systems*, Plenum Publishing Co., New York.

Schaub, H., and Junkins, J. L., 2003, *Analytical Mechanics of Space Systems*, AIAA Educational Series, American Institute of Aeronautics and Astronautics, Inc., Reston, Virginia.

Spong, M. W., Hutchinson, S., and Vidyasagar, M., 2006, *Robot Modeling and Control*, John Wiley & Sons, New York.

Suh, C. H., and Radcliff, C. W., 1978, *Kinematics and Mechanisms Design*, John Wiley & Sons, New York.

Tsai, L. W., 1999, *Robot Analysis*, John Wiley & Sons, New York.

Wittenburg, J., and Lilov, L., 2003, Decomposition of a finite rotation into three rotations about given axes, *Multibody System Dynamics, 9, 353-375.*

# Chapter 4

Ball, R. S., 1900, *A Treatise on the Theory of Screws*, Cambridge University Press, USA.

Bottema, O., and Roth, B., 1979, *Theoretical Kinematics*, North-Holland Publication, Amsterdam, The Netherlands.

Chernousko, F. L., Bolotnik, N. N., and Gradetsky, V. G., 1994, Manipulation Robots: Dynamics, Control, and Optimization, CRC press, Boca Raton, Florida.

Davidson, J. K., and Hunt, K. H., 2004, *Robots and Screw Theory: Applications of Kinematics and Statics to Robotics*, Oxford University Press, New York.

Denavit, J., and Hartenberg, R. S., 1955, A kinematic notation for lower-pair mechanisms based on matrices, *Journal of Applied Mechanics, 22(2), 215-221.*

Hunt, K. H., 1978, *Kinematic Geometry of Mechanisms*, Oxford University Press, London.

Mason, M. T., 2001, *Mechanics of Robotic Manipulation*, MIT Press, Cambridge, Massachusetts.

Murray, R. M., Li, Z., and Sastry, S. S. S., 1994, *A Mathematical Introduction to Robotic Manipulation*, CRC Press, Boca Raton, Florida.

Niku, S. B., 2001, *Introduction to Robotics: Analysis, Systems, Applications*, Prentice Hall, New Jersey.

Plücker, J., 1866, Fundamental views regarding mechanics, *Philosophical Transactions, 156, 361-380.*

Selig, J. M., 2005, *Geometric Fundamentals of Robotics*, 2nd ed., Springer, New York.

Schaub, H., and Junkins, J. L., 2003, *Analytical Mechanics of Space Systems*, AIAA Educational Series, American Institute of Aeronautics and Astronautics, Inc., Reston, Virginia.

Schilling, R. J., 1990, *Fundamentals of Robotics: Analysis and Control*, Prentice Hall, New Jersey.

Suh, C. H., and Radcliff, C. W., 1978, *Kinematics and Mechanisms Design*, John Wiley & Sons, New York.

Spong, M. W., Hutchinson, S., and Vidyasagar, M., 2006, *Robot Modeling and Control*, John Wiley & Sons, New York.

## Chapter 5

Asada, H., and Slotine, J. J. E., 1986, *Robot Analysis and Control*, John Wiley & Son, New York.

Ball, R. S., 1900, *A Treatise on the Theory of Screws*, Cambridge University Press, USA.

Bernhardt, R., and Albright, S. L., 2001, *Robot Calibration*, Springer, New York.

Bottema, O., and Roth, B., 1979, *Theoretical Kinematics*, North-Holland Publication, Amsterdam, The Netherlands.

Davidson, J. K., and Hunt, K. H., 2004, *Robots and Screw Theory: Applications of Kinematics and Statics to Robotics*, Oxford University Press, New York.

Denavit, J., and Hartenberg, R. S., 1955, A kinematic notation for lower-pair mechanisms based on matrices, *Journal of Applied Mechanics, 22(2), 215-221.*

Hunt, K. H., 1978, *Kinematic Geometry of Mechanisms*, Oxford University Press, London.

Mason, M. T., 2001, *Mechanics of Robotic Manipulation*, MIT Press, Cambridge, Massachusetts.

Paul, R. P., 1981, *Robot Manipulators: Mathematics, Programming, and Control*, MIT Press, Cambridge, Massachusetts.

Schilling, R. J., 1990, *Fundamentals of Robotics: Analysis and Control*, Prentice Hall, New Jersey.

Schroer, K., Albright, S. L., and Grethlein, M., 1997, Complete, minimal and model-continuous kinematic models for robot calibration, *Rob. Comp.-Integr. Manufact., 13(1), 73-85.*

Spong, M. W., Hutchinson, S., and Vidyasagar, M., 2006, *Robot Modeling and Control*, John Wiley & Sons, New York.

Suh, C. H., and Radcliff, C. W., 1978, *Kinematics and Mechanisms Design*, John Wiley & Sons, New York.

Tsai, L. W., 1999, *Robot Analysis*, John Wiley & Sons, New York.

Wang, K., and Lien, T., 1988, Structure, design & kinematics of robot manipulators, *Robotica, 6, 299-306.*

Zhuang, H., Roth, Z. S., and Hamano, F., 1992, A complete, minimal and model-continuous kinematic model for robot manipulators, *IEEE Trans. Rob. Automation, 8(4), 451-463.*

## Chapter 6

Asada, H., and Slotine, J. J. E., 1986, *Robot Analysis and Control*, John Wiley & Sons, New York.

Paul, R. P., 1981, *Robot Manipulators: Mathematics, Programming, and Control*, MIT Press, Cambridge, Massachusetts.

Spong, M. W., Hutchinson, S., and Vidyasagar, M., 2006, *Robot Modeling and Control*, John Wiley & Sons, New York.

Tsai, L. W., 1999, *Robot Analysis*, John Wiley & Sons, New York.

Wang, K., and Lien, T., 1988, Structure, design and kinematics of robot manipulators, *Robotica*, *6, 299-306.*

# Chapter 7

Bottema, O., and Roth, B., 1979, *Theoretical Kinematics*, North-Holland Publications, Amsterdam, The Netherlands.

Geradin, M., and Cardonna, A. , *Kinematics and Dynamics of Rigid and Flexible Mechanisms Using Finite Elements and Quaternion Algebra*, Computational Mechanics, 1987.

Hunt, K. H., 1978, *Kinematic Geometry of Mechanisms*, Oxford University Press, London.

Mason, M. T., 2001, *Mechanics of Robotic Manipulation*, MIT Press, Cambridge, Massachusetts.

Schaub, H., and Junkins, J. L., 2003, *Analytical Mechanics of Space Systems*, AIAA Educational Series, American Institute of Aeronautics and Astronautics, Inc., Reston, Virginia.

Spong, M. W., Hutchinson, S., and Vidyasagar, M., 2006, *Robot Modeling and Control*, John Wiley & Sons, New York.

Suh, C. H., and Radcliff, C. W., 1978, *Kinematics and Mechanisms Design*, John Wiley & Sons, New York.

Tsai, L. W., 1999, *Robot Analysis*, John Wiley & Sons, New York.

# Chapter 8

Hunt, K. H., 1978, *Kinematic Geometry of Mechanisms*, Oxford University Press, London.

Kane, T. R., Likins, P. W., and Levinson, D. A., 1983, *Spacecraft Dynamics*, McGraw-Hill, New York.

Kane, T. R., and Levinson, D. A., 1980, *Dynamics: Theory and Applications*, McGraw-Hill, New York.

Mason, M. T., 2001, *Mechanics of Robotic Manipulation*, MIT Press, Cambridge, Massachusetts.

Rimrott, F. P. J., 1989, *Introductory Attitude Dynamics*, Springer-Verlag, New York.

Schilling, R. J., 1990, *Fundamentals of Robotics: Analysis and Control*, Prentice-Hall, New Jersey.

Spong, M. W., Hutchinson, S., and Vidyasagar, M., 2006, *Robot Modeling and Control*, John Wiley & Sons, New York.

Suh, C. H., and Radcliff, C. W., 1978, *Kinematics and Mechanisms Design*, John Wiley & Sons, New York.

Talman, R., 2000, *Geometric Mechanics*, John Wiley & Sons, New York.

Tsai, L. W., 1999, *Robot Analysis*, John Wiley & Sons, New York.

## Chapter 9

Carnahan, B., Luther, H. A., and Wilkes, J. O., 1969, *Applied Numerical Methods*, John Wiley & Sons, New York.

Eich-Soellner, E., and Führer, C., 1998, *Numerical Methods in Multibody Dynamics*, B.G. Teubner Stuttgart.

Gerald, C. F., and Wheatley, P. O., 1999, *Applied Numerical Analysis*, 6th ed., Addison Wesley, New York.

Nikravesh, P., 1988, *Computer-Aided Analysis of Mechanical Systems*, Prentice Hall, New Jersey.

## Chapter 10

Mason, M. T., 2001, *Mechanics of Robotic Manipulation*, MIT Press, Cambridge, Massachusetts.

Nikravesh, P., 1988, *Computer-Aided Analysis of Mechanical Systems*, Prentice Hall, New Jersey.

Rimrott, F. P. J., 1989, *Introductory Attitude Dynamics*, Springer-Verlag, New York.

Spong, M. W., Hutchinson, S., and Vidyasagar, M., 2006, *Robot Modeling and Control*, John Wiley & Sons, New York.

Suh, C. H., and Radcliff, C. W., 1978, *Kinematics and Mechanisms Design*, John Wiley & Sons, New York.

Tsai, L. W., 1999, *Robot Analysis*, John Wiley & Sons, New York.

## Chapter 11

Goldstein, H., Poole, C., and Safko, J., 2002, *Classical Mechanics*, 3rd ed., Addison Wesley, New York.

MacMillan, W. D., 1936, *Dynamics of Rigid Bodies*, McGraw-Hill, New York.

Meirovitch, L., 1970, *Methods of Analytical Dynamics*, McGraw-Hill, New York.

Nikravesh, P., 1988, *Computer-Aided Analysis of Mechanical Systems*, Prentice Hall, New Jersey.

Rimrott, F. P. J., 1989, *Introductory Attitude Dynamics*, Springer-Verlag, New York.

Rosenberg, R. M., 1977, *Analytical Dynamics of Discrete Systems*, Plenum Publishing Co., New York.

Schaub, H., and Junkins, J. L., 2003, *Analytical Mechanics of Space Systems*, AIAA Educational Series, American Institute of Aeronautics and Astronautics, Inc., Reston, Virginia.

Thomson, W. T., 1961, *Introduction to Space Dynamics*, John Wiley & Sons, New York.

Tsai, L. W., 1999, *Robot Analysis*, John Wiley & Sons, New York.

Wittacker, E. T., 1947, *A Treatise on the Analytical Dynamics of Particles and Rigid Bodies*, 4th ed., Cambridge University Press, New York.

# Chapter 12

Brady, M., Hollerbach, J. M., Johnson, T. L., Lozano-Prez, T., and Mason, M. T., 1983, *Robot Motion: Planning and Control*, MIT Press, Cambridge, Massachusetts.

Murray, R. M., Li, Z., and Sastry, S. S. S., 1994, *A Mathematical Introduction to Robotic Manipulation*, CRC Press, Boca Raton, Florida.

Nikravesh, P., 1988, *Computer-Aided Analysis of Mechanical Systems*, Prentice Hall, New Jersey.

Niku, S. B., 2001, *Introduction to Robotics: Analysis, Systems, Applications*, Prentice Hall, New Jersey.

Paul, R. P., 1981, *Robot Manipulators: Mathematics, Programming, and Control*, MIT Press, Cambridge, MA.

Spong, M. W., Hutchinson, S., and Vidyasagar, M., 2006, *Robot Modeling and Control*, John Wiley & Sons, New York.

Suh, C. H., and Radcliff, C. W., 1978, *Kinematics and Mechanisms Design*, John Wiley & Sons, New York.

Tsai, L. W., 1999, *Robot Analysis*, John Wiley & Sons, New York.

# Chapter 13

Asada, H., and Slotine, J. J. E., 1986, *Robot Analysis and Control*, John Wiley & Sons, New York.

Murray, R. M., Li, Z., and Sastry, S. S. S., 1994, *A Mathematical Introduction to Robotic Manipulation*, CRC Press, Boca Raton, Florida.

Niku, S. B., 2001, *Introduction to Robotics: Analysis, Systems, Applications*, Prentice Hall, New Jersey.

Spong, M. W., Hutchinson, S., and Vidyasagar, M., 2006, *Robot Modeling and Control*, John Wiley & Sons, New York.

# Chapter 14

Ailon, A., and Langholz, G., 1985, On the existence of time optimal control of mechanical manipulators, *Journal of Optimization Theory and Applications*, *46(1)*, 1-21.

Bahrami M., and Nakhaie Jazar G., 1991, Optimal control of robotic manipulators: optimization algorithm, *Amirkabir Journal*, *5(18)*, *(in Persian)*.

Bahrami M., and Nakhaie Jazar G., 1992, Optimal control of robotic manipulators: application, *Amirkabir Journal*, *5(19)*, *(in Persian)*.

Bassein, R., 1989, An Optimization Problem, *American Mathematical Monthly*, *96(8)*, 721-725.

Bobrow, J. E., Dobowsky, S., and Gibson, J. S., 1985, Time optimal control of robotic manipulators along specified paths, *The International Journal of Robotics Research, 4(3), 495-499.*

Courant, R., and Robbins, H., 1941, *What is Mathematics?*, Oxford University Press, London.

Fotouhi, C. R., and Szyszkowski W., 1998, An algorithm for time optimal control problems, *Transaction of the ASME, Journal of Dynamic Systems, Measurements, and Control, 120, 414-418.*

Fu, K. S., Gonzales, R. C., and Lee, C. S. G., 1987, *Robotics, Control, Sensing, Vision and Intelligence*, McGraw-Hill, New York.

Gamkrelidze, R. V., 1958, *The theory of time optimal processes in linear systems*, Izvestiya Akademii Nauk SSSR, 22, 449-474.

Garg, D. P., 1990, The new time optimal motion control of robotic manipulators, *The Franklin Institute, 327(5), 785-804.*

Hart P. E., Nilson N. J., and Raphael B., 1968, A formal basis for heuristic determination of minimum cost path, *IEEE Transaction Man, System & Cybernetics, 4, 100-107.*

Kahn, M. E., and Roth B., 1971, The near minimum time control of open loop articulated kinematic chain, *Transaction of the ASME, Journal of Dynamic Systems, Measurements, and Control, 93, 164-172.*

Kim B. K., and Shin K. G., 1985, Suboptimal control of industrial manipulators with weighted minimum time-fuel criterion, *IEEE Transactions on Automatic Control, 30(1), 1-10.*

Krotov, V. F., 1996, *Global Methods in Optimal Control Theory*, Marcel Decker, Inc.

Kuo, B. C., and Golnaraghi, F., 2003, *Automatic Control Systems*, John Wiley & Sons, New York.

Lee, E. B., and Markus, L., 1961, Optimal Control for Nonlinear Processes, *Archive for Rational Mechanics and Analysis, 8, 36-58.*

Lee, H. W. J., Teo, K. L., Rehbock, V., and Jennings, L. S., 1997, Control parameterization enhancing technique for time optimal control problems, *Dynamic Systems and Applications, 6, 243-262.*

Lewis, F. L., and Syrmos, V. L., 1995, *Optimal Control*, John Wiley & Sons, New York.

Meier E. B., and Bryson A. E., 1990, Efficient algorithm for time optimal control of a two-link manipulator, *Journal of Guidance, Control and Dynamics, 13(5), 859-866.*

Mita T., Hyon, S. H., and Nam, T. K., 2001, Analytical time optimal control solution for a two link planar acrobat with initial angular momentum, *IEEE Transactions on Robotics and Automation, 17(3), 361-366.*

Nakamura, Y., 1991, *Advanced Robotics: Redundancy and Optimization*, Addison Wesley, New York.

Nakhaie Jazar G., and Naghshinehpour A., 2005, Time optimal control algorithm for multi-body dynamical systems, *IMechE Part K: Journal of Multi-Body Dynamics, 219(3), 225-236.*

Pinch E. R., 1993, *Optimal Control and the Calculus of Variations*, Oxford University Press, New York.

Pontryagin, L. S., Boltyanskii, V. G., Gamkrelidze, R. V., and Mishchenko, E. F., 1962, *The Mathematical Theory of Optimal Processes*, John Wiley & Sons, New York.

Roxin, E., 1962, The existence of optimal controls, *Michigan Mathematical Journal, 9, 109-119.*

Shin K. G., and McKay N. D., 1985, Minimum time control of a robotic manipulator with geometric path constraints, *IEEE Transaction Automatic Control, 30(6), 531-541.*

Shin K. G., and McKay N. D., 1986, Selection of near minimum time geometric paths for robotic manipulators, *IEEE Transaction Automatic Control, 31(6), 501-511.*

Skowronski, J. M., 1986, *Control Dynamics of Robotic Manipulator*, Academic Press Inc., U.S.

Slotine, J. J. E., and Yang, H. S., 1989, Improving the efficiency of time optimal path following algorithms, *IEEE Trans. Robotics Automation, 5(1), 118-124.*

Spong, M. W., Thorp, J. S., and Kleinwaks, J., M., 1986, The control of robot manipulators with bounded input, *IEEE Journal of Automatic Control, 31(6), 483-490.*

Sundar, S., and Shiller, Z., 1996, A generalized sufficient condition for time optimal control, *Transaction of the ASME, Journal of Dynamic Systems, Measurements, and Control, 118(2), 393-396.*

Takegaki, M., and Arimoto, S., 1981, A new feedback method for dynamic control of manipulators, *Transaction of the ASME, Journal of Dynamic Systems, Measurements, and Control, 102, 119-125.*

Vincent T. L., and Grantham W. J., 1997, *Nonlinear and Optimal Control Systems*, John Wiley & Sons, New York.

# Chapter 15

Åström, K. J., and Hägglund, T., *PID Controllers*, 2nd ed., 1995, Instrument Society of America, Research Triangle Park, North Carolina.

Fu, K. S., Gonzales, R. C., and Lee, C. S. G., 1987, *Robotics, Control, Sensing, Vision and Intelligence*, McGraw-Hill, New York.

Kuo, B. C., and Golnaraghi, F., 2003, *Automatic Control Systems*, John Wiley & Sons, New York.

Lewis, F. L., and Syrmos, V. L., 1995, *Optimal Control*, John Wiley & Sons, New York.

Paul, R. P., 1981, *Robot Manipulators: Mathematics, Programming, and Control*, MIT Press, Cambridge, Massachusetts.

Spong, M. W., Hutchinson, S., and Vidyasagar, M., 2006, *Robot Modeling and Control*, John Wiley & Sons, New York.

Takegaki, M., and Arimoto, S., 1981, A new feedback method for dynamic control of manipulators, *Transaction of the ASME, Journal of Dynamic Systems, Measurements, and Control, 102, 119-125.*

Vincent T. L., and Grantham W. J., 1997, *Nonlinear and Optimal Control Systems*, John Wiley & Sons, New York.

# Appendix A

# Global Frame Triple Rotation

In this appendix, the 12 combinations of triple rotation about global fixed axes are presented.

$1\text{-}Q_{X,\gamma}Q_{Y,\beta}Q_{Z,\alpha}$

$$= \begin{bmatrix} c\alpha c\beta & -c\beta s\alpha & s\beta \\ c\gamma s\alpha + c\alpha s\beta s\gamma & c\alpha c\gamma - s\alpha s\beta s\gamma & -c\beta s\gamma \\ s\alpha s\gamma - c\alpha c\gamma s\beta & c\alpha s\gamma + c\gamma s\alpha s\beta & c\beta c\gamma \end{bmatrix} \qquad (A.1)$$

$2\text{-}Q_{Y,\gamma}Q_{Z,\beta}Q_{X,\alpha}$

$$= \begin{bmatrix} c\beta c\gamma & s\alpha s\gamma - c\alpha c\gamma s\beta & c\alpha s\gamma + c\gamma s\alpha s\beta \\ s\beta & c\alpha c\beta & -c\beta s\alpha \\ -c\beta s\gamma & c\gamma s\alpha + c\alpha s\beta s\gamma & c\alpha c\gamma - s\alpha s\beta s\gamma \end{bmatrix} \qquad (A.2)$$

$3\text{-}Q_{Z,\gamma}Q_{X,\beta}Q_{Y,\alpha}$

$$= \begin{bmatrix} c\alpha c\gamma - s\alpha s\beta s\gamma & -c\beta s\gamma & c\gamma s\alpha + c\alpha s\beta s\gamma \\ c\alpha s\gamma + c\gamma s\alpha s\beta & c\beta c\gamma & s\alpha s\gamma - c\alpha c\gamma s\beta \\ -c\beta s\alpha & s\beta & c\alpha c\beta \end{bmatrix} \qquad (A.3)$$

$4\text{-}Q_{Z,\gamma}Q_{Y,\beta}Q_{X,\alpha}$

$$= \begin{bmatrix} c\beta c\gamma & -c\alpha s\gamma + c\gamma s\alpha s\beta & s\alpha s\gamma + c\alpha c\gamma s\beta \\ c\beta s\gamma & c\alpha c\gamma + s\alpha s\beta s\gamma & -c\gamma s\alpha + c\alpha s\beta s\gamma \\ -s\beta & c\beta s\alpha & c\alpha c\beta \end{bmatrix} \qquad (A.4)$$

$5\text{-}Q_{Y,\gamma}Q_{X,\beta}Q_{Z,\alpha}$

$$= \begin{bmatrix} c\alpha c\gamma + s\alpha s\beta s\gamma & -c\gamma s\alpha + c\alpha s\beta s\gamma & c\beta s\gamma \\ c\beta s\alpha & c\alpha c\beta & -s\beta \\ -c\alpha s\gamma + c\gamma s\alpha s\beta & s\alpha s\gamma + c\alpha c\gamma s\beta & c\beta c\gamma \end{bmatrix} \qquad (A.5)$$

$6\text{-}Q_{X,\gamma}Q_{Z,\beta}Q_{Y,\alpha}$

$$= \begin{bmatrix} c\alpha c\beta & -s\beta & c\beta s\alpha \\ s\alpha s\gamma + c\alpha c\gamma s\beta & c\beta c\gamma & -c\alpha s\gamma + c\gamma s\alpha s\beta \\ -c\gamma s\alpha + c\alpha s\beta s\gamma & c\beta s\gamma & c\alpha c\gamma + s\alpha s\beta s\gamma \end{bmatrix} \qquad (A.6)$$

$7\text{-}Q_{X,\gamma}Q_{Y,\beta}Q_{X,\alpha}$

$$= \begin{bmatrix} c\beta & s\alpha s\beta & c\alpha s\beta \\ s\beta s\gamma & c\alpha c\gamma - c\beta s\alpha s\gamma & -c\gamma s\alpha - c\alpha c\beta s\gamma \\ -c\gamma s\beta & c\alpha s\gamma + c\beta c\gamma s\alpha & -s\alpha s\gamma + c\alpha c\beta c\gamma \end{bmatrix} \qquad (A.7)$$

$8\text{-}Q_{Y,\gamma}Q_{Z,\beta}Q_{Y,\alpha}$

$$= \begin{bmatrix} -s\alpha s\gamma + c\alpha c\beta c\gamma & -c\gamma s\beta & c\alpha s\gamma + c\beta c\gamma s\alpha \\ c\alpha s\beta & c\beta & s\alpha s\beta \\ -c\gamma s\alpha - c\alpha c\beta s\gamma & s\beta s\gamma & c\alpha c\gamma - c\beta s\alpha s\gamma \end{bmatrix} \qquad (A.8)$$

$9\text{-}Q_{Z,\gamma}Q_{X,\beta}Q_{Z,\alpha}$

$$= \begin{bmatrix} c\alpha c\gamma - c\beta s\alpha s\gamma & -c\gamma s\alpha - c\alpha c\beta s\gamma & s\beta s\gamma \\ c\alpha s\gamma + c\beta c\gamma s\alpha & -s\alpha s\gamma + c\alpha c\beta c\gamma & -c\gamma s\beta \\ s\alpha s\beta & c\alpha s\beta & c\beta \end{bmatrix} \qquad (A.9)$$

$10\text{-}Q_{X,\gamma}Q_{Z,\beta}Q_{X,\alpha}$

$$= \begin{bmatrix} c\beta & -c\alpha s\beta & s\alpha s\beta \\ c\gamma s\beta & -s\alpha s\gamma + c\alpha c\beta c\gamma & -c\alpha s\gamma - c\beta c\gamma s\alpha \\ s\beta s\gamma & c\gamma s\alpha + c\alpha c\beta s\gamma & c\alpha c\gamma - c\beta s\alpha s\gamma \end{bmatrix} \qquad (A.10)$$

$11\text{-}Q_{Y,\gamma}Q_{X,\beta}Q_{Y,\alpha}$

$$= \begin{bmatrix} c\alpha c\gamma - c\beta s\alpha s\gamma & s\beta s\gamma & c\gamma s\alpha + c\alpha c\beta s\gamma \\ s\alpha s\beta & c\beta & -c\alpha s\beta \\ -c\alpha s\gamma - c\beta c\gamma s\alpha & c\gamma s\beta & -s\alpha s\gamma + c\alpha c\beta c\gamma \end{bmatrix} \qquad (A.11)$$

$12\text{-}Q_{Z,\gamma}Q_{Y,\beta}Q_{Z,\alpha}$

$$= \begin{bmatrix} -s\alpha s\gamma + c\alpha c\beta c\gamma & -c\alpha s\gamma - c\beta c\gamma s\alpha & c\gamma s\beta \\ c\gamma s\alpha + c\alpha c\beta s\gamma & c\alpha c\gamma - c\beta s\alpha s\gamma & s\beta s\gamma \\ -c\alpha s\beta & s\alpha s\beta & c\beta \end{bmatrix} \qquad (A.12)$$

# Appendix B

# Local Frame Triple Rotation

In this appendix, the 12 combinations of triple rotation about local axes are presented.

1-$A_{x,\psi}A_{y,\theta}A_{z,\varphi}$

$$= \begin{bmatrix} c\theta c\varphi & c\theta s\varphi & -s\theta \\ -c\psi s\varphi + c\varphi s\theta s\psi & c\varphi c\psi + s\theta s\varphi s\psi & c\theta s\psi \\ s\varphi s\psi + c\varphi s\theta c\psi & -c\varphi s\psi + s\theta c\psi s\varphi & c\theta c\psi \end{bmatrix} \quad \text{(B.1)}$$

2-$A_{y,\psi}A_{z,\theta}A_{x,\varphi}$

$$= \begin{bmatrix} c\theta c\psi & s\varphi s\psi + c\varphi s\theta c\psi & -c\varphi s\psi + s\theta c\psi s\varphi \\ -s\theta & c\theta c\varphi & c\theta s\varphi \\ c\theta s\psi & -c\psi s\varphi + c\varphi s\theta s\psi & c\varphi c\psi + s\theta s\varphi s\psi \end{bmatrix} \quad \text{(B.2)}$$

3-$A_{z,\psi}A_{x,\theta}A_{y,\varphi}$

$$= \begin{bmatrix} c\varphi c\psi + s\theta s\varphi s\psi & c\theta s\psi & -c\psi s\varphi + c\varphi s\theta s\psi \\ -c\varphi s\psi + s\theta c\psi s\varphi & c\theta c\psi & s\varphi s\psi + c\varphi s\theta c\psi \\ c\theta s\varphi & -s\theta & c\theta c\varphi \end{bmatrix} \quad \text{(B.3)}$$

4-$A_{z,\psi}A_{y,\theta}A_{x,\varphi}$

$$= \begin{bmatrix} c\theta c\psi & c\varphi s\psi + s\theta c\psi s\varphi & s\varphi s\psi - c\varphi s\theta c\psi \\ -c\theta s\psi & c\varphi c\psi - s\theta s\varphi s\psi & c\psi s\varphi + c\varphi s\theta s\psi \\ s\theta & -c\theta s\varphi & c\theta c\varphi \end{bmatrix} \quad \text{(B.4)}$$

5-$A_{y,\psi}A_{x,\theta}A_{z,\varphi}$

$$= \begin{bmatrix} c\varphi c\psi - s\theta s\varphi s\psi & c\psi s\varphi + c\varphi s\theta s\psi & -c\theta s\psi \\ -c\theta s\varphi & c\theta c\varphi & s\theta \\ c\varphi s\psi + s\theta c\psi s\varphi & s\varphi s\psi - c\varphi s\theta c\psi & c\theta c\psi \end{bmatrix} \quad \text{(B.5)}$$

6-$A_{x,\psi}A_{z,\theta}A_{y,\varphi}$

$$= \begin{bmatrix} c\theta c\varphi & s\theta & -c\theta s\varphi \\ s\varphi s\psi - c\varphi s\theta c\psi & c\theta c\psi & c\varphi s\psi + s\theta c\psi s\varphi \\ c\psi s\varphi + c\varphi s\theta s\psi & -c\theta s\psi & c\varphi c\psi - s\theta s\varphi s\psi \end{bmatrix} \quad \text{(B.6)}$$

7-$A_{x,\psi}A_{y,\theta}A_{x,\varphi}$

$$= \begin{bmatrix} c\theta & s\theta s\varphi & -c\varphi s\theta \\ s\theta s\psi & c\varphi c\psi - c\theta s\varphi s\psi & c\psi s\varphi + c\theta c\varphi s\psi \\ s\theta c\psi & -c\varphi s\psi - c\theta c\psi s\varphi & -s\varphi s\psi + c\theta c\varphi c\psi \end{bmatrix} \quad \text{(B.7)}$$

$8\text{-}A_{y,\psi}A_{z,\theta}A_{y,\varphi}$

$$= \begin{bmatrix} -s\varphi s\psi + c\theta c\varphi c\psi & s\theta c\psi & -c\varphi s\psi - c\theta c\psi s\varphi \\ -c\varphi s\theta & c\theta & s\theta s\varphi \\ c\psi s\varphi + c\theta c\varphi s\psi & s\theta s\psi & c\varphi c\psi - c\theta s\varphi s\psi \end{bmatrix} \quad (B.8)$$

$9\text{-}A_{z,\psi}A_{x,\theta}A_{z,\varphi}$

$$= \begin{bmatrix} c\varphi c\psi - c\theta s\varphi s\psi & c\psi s\varphi + c\theta c\varphi s\psi & s\theta s\psi \\ -c\varphi s\psi - c\theta c\psi s\varphi & -s\varphi s\psi + c\theta c\varphi c\psi & s\theta c\psi \\ s\theta s\varphi & -c\varphi s\theta & c\theta \end{bmatrix} \quad (B.9)$$

$10\text{-}A_{x,\psi}A_{z,\theta}A_{x,\varphi}$

$$= \begin{bmatrix} c\theta & c\varphi s\theta & s\theta s\varphi \\ -s\theta c\psi & -s\varphi s\psi + c\theta c\varphi c\psi & c\varphi s\psi + c\theta c\psi s\varphi \\ s\theta s\psi & -c\psi s\varphi - c\theta c\varphi s\psi & c\varphi c\psi - c\theta s\varphi s\psi \end{bmatrix} \quad (B.10)$$

$11\text{-}A_{y,\psi}A_{x,\theta}A_{y,\varphi}$

$$= \begin{bmatrix} c\varphi c\psi - c\theta s\varphi s\psi & s\theta s\psi & -c\psi s\varphi - c\theta c\varphi s\psi \\ s\theta s\varphi & c\theta & c\varphi s\theta \\ c\varphi s\psi + c\theta c\psi s\varphi & -s\theta c\psi & -s\varphi s\psi + c\theta c\varphi c\psi \end{bmatrix} \quad (B.11)$$

$12\text{-}A_{z,\psi}A_{y,\theta}A_{z,\varphi}$

$$= \begin{bmatrix} -s\varphi s\psi + c\theta c\varphi c\psi & c\varphi s\psi + c\theta c\psi s\varphi & -s\theta c\psi \\ -c\psi s\varphi - c\theta c\varphi s\psi & c\varphi c\psi - c\theta s\varphi s\psi & s\theta s\psi \\ c\varphi s\theta & s\theta s\varphi & c\theta \end{bmatrix} \quad (B.12)$$

# Appendix C

# Principal Central Screws Triple Combination

In this appendix, the six combinations of triple principal central screws are presented.

$1\text{-}\check{s}(h_X,\gamma,\hat{I})\,\check{s}(h_Y,\beta,\hat{J})\,\check{s}(h_Z,\alpha,\hat{K})$

$$
=\begin{bmatrix}
c\alpha c\beta & -c\beta s\alpha & s\beta & \gamma p_X + \alpha p_Z s\beta \\
c\gamma s\alpha + c\alpha s\beta s\gamma & c\alpha c\gamma - s\alpha s\beta s\gamma & -c\beta s\gamma & \beta p_Y c\gamma - \alpha p_Z c\beta s\gamma \\
s\alpha s\gamma - c\alpha c\gamma s\beta & c\alpha s\gamma + c\gamma s\alpha s\beta & c\beta c\gamma & \beta p_Y s\gamma + \alpha p_Z c\beta c\gamma \\
0 & 0 & 0 & 1
\end{bmatrix}
$$

(C.1)

$2\text{-}\check{s}(h_Y,\beta,\hat{J})\,\check{s}(h_Z,\alpha,\hat{K})\,\check{s}(h_X,\gamma,\hat{I})$

$$
=\begin{bmatrix}
c\alpha c\beta & s\beta s\gamma - c\beta c\gamma s\alpha & c\gamma s\beta + c\beta s\alpha s\gamma & \alpha p_Z s\beta + \gamma p_X c\alpha c\beta \\
s\alpha & c\alpha c\gamma & -c\alpha s\gamma & \beta p_Y + \gamma p_X s\alpha \\
-c\alpha s\beta & c\beta s\gamma + c\gamma s\alpha s\beta & c\beta c\gamma - s\alpha s\beta s\gamma & \alpha p_Z c\beta - \gamma p_X c\alpha s\beta \\
0 & 0 & 0 & 1
\end{bmatrix}
$$

(C.2)

$3\text{-}\check{s}(h_Z,\alpha,\hat{K})\,\check{s}(h_X,\gamma,\hat{I})\,\check{s}(h_Y,\beta,\hat{J})$

$$
=\begin{bmatrix}
c\alpha c\beta - s\alpha s\beta s\gamma & -c\gamma s\alpha & c\alpha s\beta + c\beta s\alpha s\gamma & \gamma p_X c\alpha - \beta p_Y c\gamma s\alpha \\
c\beta s\alpha + c\alpha s\beta s\gamma & c\alpha c\gamma & s\alpha s\beta - c\alpha c\beta s\gamma & \gamma p_X s\alpha + \beta p_Y c\alpha c\gamma \\
-c\gamma s\beta & s\gamma & c\beta c\gamma & \alpha p_Z + \beta p_Y s\gamma \\
0 & 0 & 0 & 1
\end{bmatrix}
$$

(C.3)

$4\text{-}\check{s}(h_Z,\alpha,\hat{K})\,\check{s}(h_Y,\beta,\hat{J})\,\check{s}(h_X,\gamma,\hat{I})$

$$
=\begin{bmatrix}
c\alpha c\beta & c\alpha s\beta s\gamma - c\gamma s\alpha & s\alpha s\gamma + c\alpha c\gamma s\beta & \gamma p_X c\alpha c\beta - \beta p_Y s\alpha \\
c\beta s\alpha & c\alpha c\gamma + s\alpha s\beta s\gamma & c\gamma s\alpha s\beta - c\alpha s\gamma & \beta p_Y c\alpha + \gamma p_X c\beta s\alpha \\
-s\beta & c\beta s\gamma & c\beta c\gamma & \alpha p_Z - \gamma p_X s\beta \\
0 & 0 & 0 & 1
\end{bmatrix}
$$

(C.4)

$5\text{-}\check{s}(h_Y,\beta,\hat{J})\,\check{s}(h_X,\gamma,\hat{I})\,\check{s}(h_Z,\alpha,\hat{K})$

$$
=\begin{bmatrix}
c\alpha c\beta + s\alpha s\beta s\gamma & c\alpha s\beta s\gamma - c\beta s\alpha & c\gamma s\beta & \gamma p_X c\beta + \alpha p_Z c\gamma s\beta \\
c\gamma s\alpha & c\alpha c\gamma & -s\gamma & \beta p_Y - \alpha p_Z s\gamma \\
c\beta s\alpha s\gamma - c\alpha s\beta & s\alpha s\beta + c\alpha c\beta s\gamma & c\beta c\gamma & \alpha p_Z c\beta c\gamma - \gamma p_X s\beta \\
0 & 0 & 0 & 1
\end{bmatrix}
$$

(C.5)

$6\text{-}\check{s}(h_X,\gamma,\hat{I})\,\check{s}(h_Z,\alpha,\hat{K})\,\check{s}(h_Y,\beta,\hat{J})$

$$= \begin{bmatrix} c\alpha c\beta & -s\alpha & c\alpha s\beta & \gamma p_X - \beta p_Y s\alpha \\ s\beta s\gamma + c\beta c\gamma s\alpha & c\alpha c\gamma & c\gamma s\alpha s\beta - c\beta s\gamma & \beta p_Y c\alpha c\gamma - \alpha p_Z s\gamma \\ c\beta s\alpha s\gamma - c\gamma s\beta & c\alpha s\gamma & c\beta c\gamma + s\alpha s\beta s\gamma & \alpha p_Z c\gamma + \beta p_Y c\alpha s\gamma \\ 0 & 0 & 0 & 1 \end{bmatrix}$$

$$(\text{C.6})$$

# Appendix D

# Trigonometric Formula

Definitions in terms of exponentials.

$$\cos z = \frac{e^{iz} + e^{-iz}}{2} \tag{D.1}$$

$$\sin z = \frac{e^{iz} - e^{-iz}}{2i} \tag{D.2}$$

$$\tan z = \frac{e^{iz} - e^{-iz}}{i\left(e^{iz} + e^{-iz}\right)} \tag{D.3}$$

$$e^{iz} = \cos z + i \sin z \tag{D.4}$$

$$e^{-iz} = \cos z - i \sin z \tag{D.5}$$

Angle sum and difference.

$$\sin(\alpha \pm \beta) = \sin\alpha\cos\beta \pm \cos\alpha\sin\beta \tag{D.6}$$

$$\cos(\alpha \pm \beta) = \cos\alpha\cos\beta \mp \sin\alpha\sin\beta \tag{D.7}$$

$$\tan(\alpha \pm \beta) = \frac{\tan\alpha \pm \tan\beta}{1 \mp \tan\alpha\tan\beta} \tag{D.8}$$

$$\cot(\alpha \pm \beta) = \frac{\cot\alpha\cot\beta \mp 1}{\cot\beta \pm \cot\alpha} \tag{D.9}$$

Symmetry.

$$\sin(-\alpha) = -\sin\alpha \tag{D.10}$$

$$\cos(-\alpha) = \cos\alpha \tag{D.11}$$

$$\tan(-\alpha) = -\tan\alpha \tag{D.12}$$

Multiple angle.

$$\sin(2\alpha) = 2\sin\alpha\cos\alpha = \frac{2\tan\alpha}{1 + \tan^2\alpha} \tag{D.13}$$

$$\cos(2\alpha) = 2\cos^2\alpha - 1 = 1 - 2\sin^2\alpha = \cos^2\alpha - \sin^2\alpha \tag{D.14}$$

$$\tan(2\alpha) = \frac{2\tan\alpha}{1 - \tan^2\alpha} \tag{D.15}$$

$$\cot(2\alpha) = \frac{\cot^2\alpha - 1}{2\cot\alpha} \tag{D.16}$$

$$\sin(3\alpha) = -4\sin^3\alpha + 3\sin\alpha \tag{D.17}$$

$$\cos(3\alpha) = 4\cos^3\alpha - 3\cos\alpha \tag{D.18}$$

$$\tan(3\alpha) = \frac{-\tan^3\alpha + 3\tan\alpha}{-3\tan^2\alpha + 1} \tag{D.19}$$

$$\sin(4\alpha) = -8\sin^3\alpha\cos\alpha + 4\sin\alpha\cos\alpha \tag{D.20}$$

$$\cos(4\alpha) = 8\cos^4\alpha - 8\cos^2\alpha + 1 \tag{D.21}$$

$$\tan(4\alpha) = \frac{-4\tan^3\alpha + 4\tan\alpha}{\tan^4\alpha - 6\tan^2\alpha + 1} \tag{D.22}$$

$$\sin(5\alpha) = 16\sin^5\alpha - 20\sin^3\alpha + 5\sin\alpha \tag{D.23}$$

$$\cos(5\alpha) = 16\cos^5\alpha - 20\cos^3\alpha + 5\cos\alpha \tag{D.24}$$

$$\sin(n\alpha) = 2\sin((n-1)\alpha)\cos\alpha - \sin((n-2)\alpha) \tag{D.25}$$

$$\cos(n\alpha) = 2\cos((n-1)\alpha)\cos\alpha - \cos((n-2)\alpha) \tag{D.26}$$

$$\tan(n\alpha) = \frac{\tan((n-1)\alpha) + \tan\alpha}{1 - \tan((n-1)\alpha)\tan\alpha} \tag{D.27}$$

Half angle.

$$\cos\left(\frac{\alpha}{2}\right) = \pm\sqrt{\frac{1+\cos\alpha}{2}} \tag{D.28}$$

$$\sin\left(\frac{\alpha}{2}\right) = \pm\sqrt{\frac{1-\cos\alpha}{2}} \tag{D.29}$$

$$\tan\left(\frac{\alpha}{2}\right) = \frac{1-\cos\alpha}{\sin\alpha} = \frac{\sin\alpha}{1+\cos\alpha} = \pm\sqrt{\frac{1-\cos\alpha}{1+\cos\alpha}} \tag{D.30}$$

$$\sin\alpha = \frac{2\tan\frac{\alpha}{2}}{1+\tan^2\frac{\alpha}{2}} \tag{D.31}$$

$$\cos\alpha = \frac{1-\tan^2\frac{\alpha}{2}}{1+\tan^2\frac{\alpha}{2}} \tag{D.32}$$

Powers of functions.

$$\sin^2\alpha = \frac{1}{2}\left(1-\cos(2\alpha)\right) \tag{D.33}$$

$$\sin\alpha\cos\alpha = \frac{1}{2}\sin(2\alpha) \tag{D.34}$$

$$\cos^2\alpha = \frac{1}{2}\left(1+\cos(2\alpha)\right) \tag{D.35}$$

$$\sin^3\alpha = \frac{1}{4}\left(3\sin(\alpha) - \sin(3\alpha)\right) \tag{D.36}$$

$$\sin^2 \alpha \cos \alpha = \frac{1}{4} \left( \cos \alpha - 3 \cos(3\alpha) \right) \tag{D.37}$$

$$\sin \alpha \cos^2 \alpha = \frac{1}{4} \left( \sin \alpha + \sin(3\alpha) \right) \tag{D.38}$$

$$\cos^3 \alpha = \frac{1}{4} \left( \cos(3\alpha) + 3 \cos \alpha \right) \tag{D.39}$$

$$\sin^4 \alpha = \frac{1}{8} \left( 3 - 4 \cos(2\alpha) + \cos(4\alpha) \right) \tag{D.40}$$

$$\sin^3 \alpha \cos \alpha = \frac{1}{8} \left( 2 \sin(2\alpha) - \sin(4\alpha) \right) \tag{D.41}$$

$$\sin^2 \alpha \cos^2 \alpha = \frac{1}{8} \left( 1 - \cos(4\alpha) \right) \tag{D.42}$$

$$\sin \alpha \cos^3 \alpha = \frac{1}{8} \left( 2 \sin(2\alpha) + \sin(4\alpha) \right) \tag{D.43}$$

$$\cos^4 \alpha = \frac{1}{8} \left( 3 + 4 \cos(2\alpha) + \cos(4\alpha) \right) \tag{D.44}$$

$$\sin^5 \alpha = \frac{1}{16} \left( 10 \sin \alpha - 5 \sin(3\alpha) + \sin(5\alpha) \right) \tag{D.45}$$

$$\sin^4 \alpha \cos \alpha = \frac{1}{16} \left( 2 \cos \alpha - 3 \cos(3\alpha) + \cos(5\alpha) \right) \tag{D.46}$$

$$\sin^3 \alpha \cos^2 \alpha = \frac{1}{16} \left( 2 \sin \alpha + \sin(3\alpha) - \sin(5\alpha) \right) \tag{D.47}$$

$$\sin^2 \alpha \cos^3 \alpha = \frac{1}{16} \left( 2 \cos \alpha - 3 \cos(3\alpha) - 5 \cos(5\alpha) \right) \tag{D.48}$$

$$\sin \alpha \cos^4 \alpha = \frac{1}{16} \left( 2 \sin \alpha + 3 \sin(3\alpha) + \sin(5\alpha) \right) \tag{D.49}$$

$$\cos^5 \alpha = \frac{1}{16} \left( 10 \cos \alpha + 5 \cos(3\alpha) + \cos(5\alpha) \right) \tag{D.50}$$

$$\tan^2 \alpha = \frac{1 - \cos(2\alpha)}{1 + \cos(2\alpha)} \tag{D.51}$$

Products of sin and cos.

$$\cos \alpha \cos \beta = \frac{1}{2} \cos(\alpha - \beta) + \frac{1}{2} \cos(\alpha + \beta) \tag{D.52}$$

$$\sin \alpha \sin \beta = \frac{1}{2} \cos(\alpha - \beta) - \frac{1}{2} \cos(\alpha + \beta) \tag{D.53}$$

$$\sin \alpha \cos \beta = \frac{1}{2} \sin(\alpha - \beta) + \frac{1}{2} \sin(\alpha + \beta) \tag{D.54}$$

$$\cos \alpha \sin \beta = \frac{1}{2} \sin(\alpha + \beta) - \frac{1}{2} \sin(\alpha - \beta) \tag{D.55}$$

$$\sin(\alpha + \beta) \sin(\alpha - \beta) = \cos^2 \beta - \cos^2 \alpha = \sin^2 \alpha - \sin^2 \beta \tag{D.56}$$

$$\cos(\alpha + \beta)\cos(\alpha - \beta) = \cos^2 \beta + \sin^2 \alpha \tag{D.57}$$

Sum of functions.

$$\sin \alpha \pm \sin \beta = 2 \sin \frac{\alpha \pm \beta}{2} \cos \frac{\alpha \pm \beta}{2} \tag{D.58}$$

$$\cos \alpha + \cos \beta = 2 \cos \frac{\alpha + \beta}{2} \cos \frac{\alpha - \beta}{2} \tag{D.59}$$

$$\cos \alpha - \cos \beta = -2 \sin \frac{\alpha + \beta}{2} \sin \frac{\alpha - \beta}{2} \tag{D.60}$$

$$\tan \alpha \pm \tan \beta = \frac{\sin(\alpha \pm \beta)}{\cos \alpha \cos \beta} \tag{D.61}$$

$$\cot \alpha \pm \cot \beta = \frac{\sin(\beta \pm \alpha)}{\sin \alpha \sin \beta} \tag{D.62}$$

$$\frac{\sin \alpha + \sin \beta}{\sin \alpha - \sin \beta} = \frac{\tan \frac{\alpha + \beta}{2}}{\tan \frac{\alpha - + \beta}{2}} \tag{D.63}$$

$$\frac{\sin \alpha + \sin \beta}{\cos \alpha - \cos \beta} = \cot \frac{-\alpha + \beta}{2} \tag{D.64}$$

$$\frac{\sin \alpha + \sin \beta}{\cos \alpha + \cos \beta} = \tan \frac{\alpha + \beta}{2} \tag{D.65}$$

$$\frac{\sin \alpha - \sin \beta}{\cos \alpha + \cos \beta} = \tan \frac{\alpha - \beta}{2} \tag{D.66}$$

Trigonometric relations.

$$\sin^2 \alpha - \sin^2 \beta = \sin(\alpha + \beta)\sin(\alpha - \beta) \tag{D.67}$$

$$\cos^2 \alpha - \cos^2 \beta = -\sin(\alpha + \beta)\sin(\alpha - \beta) \tag{D.68}$$

# Index

Printed in the United States of America.

Printed in the United States
151604LV00004B/4/P